Visual Walkthrough for
Fundamental Molecular Biology

The next few pages will introduce you to the many pedagogical resources, tools, and features that this text utilizes to convey the key concepts of molecular biology

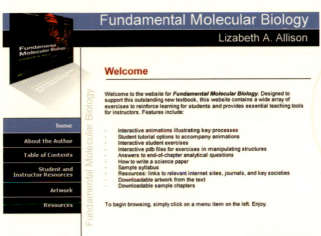

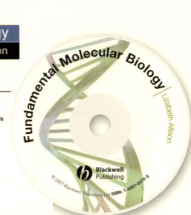

Multimedia Support

Accompanying *Fundamental Molecular Biology* are student and instructor media resources – an **Interactive Website** and **CD-ROM** – designed to reinforce learning for students and provide essential teaching tools for instructors

The **Interactive Website** at **http://www.blackwellpublishing.com/Allison** includes:

- Interactive animations illustrating key processes
- Student tutorial options to accompany animations
- Interactive student exercises
- Manipulating structures: interactive pdb files
- Answers to end-of-chapter analytical questions
- How to write a science paper
- Sample syllabus
- Resources: links to relevant internet sites, journals, and key societies
- Downloadable artwork from the text
- Downloadable sample chapters

And the **CD-ROM** includes:

- Sample interactive animation and tutorial
- List of all animations
- Sample syllabus
- Downloadable artwork from the text
- Link to the website

How to log on to **http://www.blackwellpublishing.com/Allison**

Below is your unique web-access token that will give you full access to the *Fundamental Molecular Biology* companion website. To access the site, first scratch off the silver foil coating below to reveal your unique web-access token. Next, visit **http://www.blackwellpublishing.com/Allison**, click **Student & Instructor Resources** from the left menu and then follow the onscreen instructions to complete your registration. Once completed, write down your personal login name and password because you will need them for future visits.

YOUR WEB-ACCESS TOKEN IS:

Her [Rosalind Franklin's] photographs are among the most beautiful X-ray photographs of any substance ever taken.

Obituary of Rosalind Franklin, *Nature* (1958), 182:154.

Outline

1.1 Introduction
1.2 Historical perspective
  Insights into heredity from round and wrinkled peas: Mendelian genetics
  Insights into the nature of hereditary material: the transforming principle is DNA
  Creativity in approach leads to the one gene–one enzyme hypothesis

The importance of technological advances: the Hershey–Chase experiment
A model for the structure of DNA: the DNA double helix
Chapter summary
Analytical questions
Suggestions for further reading

1.1 Introduction

For decades, DNA was largely an academic subject and not the source of dinner table conversation in the average household. In 1995 this changed when media coverage of the O.J. Simpson murder trial brought DNA fingerprinting to homes across the world. Two years later, the cloning of Dolly the sheep was headline news. Then, in 2001, scientists announced the rough draft of the human genome sequence. In commenting on this landmark achievement, former US President Clinton likened the "decoding of the book of life" to a medical version of the moon landing. Increasingly, DNA has captivated Hollywood and the general public, excited scientists and science fiction writers alike, inspired artists, and challenged society with emerging ethical issues (Fig. 1.1).

Chapter Openers

An engaging quote, outline of key topics, and brief conceptual introduction set the stage for what you will learn in each chapter

---

Chapter 1

The beginnings of molecular biology

Her [Rosalind Franklin's] photographs are among the most beautiful X-ray photographs of any substance ever taken.

Obituary of Rosalind Franklin, *Nature* (1958), 182:154

Outline

1.1 Introduction
1.2 Historical perspective
  Insights into heredity from round and wrinkled peas: Mendelian genetics
  Insights into the nature of hereditary material: the transforming principle is DNA
  Creativity in approach leads to the one gene–one enzyme hypothesis

The importance of technological advances: the Hershey–Chase experiment
A model for the structure of DNA: the DNA double helix
Chapter summary
Analytical questions
Suggestions for further reading

1.1 Introduction

For decades, DNA was largely an academic subject and not the source of dinner table conversation in the average household. In 1995 this changed when media coverage of the O.J. Simpson murder trial brought DNA fingerprinting to homes across the world. Two years later, the cloning of Dolly the sheep was headline news. Then, in 2001, scientists announced the rough draft of the human genome sequence. In commenting on this landmark achievement, former US President Clinton likened the "decoding of the book of life" to a medical version of the moon landing. Increasingly, DNA has captivated Hollywood and the general public, excited scientists and science fiction writers alike, inspired artists, and challenged society with emerging ethical issues (Fig. 1.1).

1.2 Historical perspective

The last 5–10 years mark the beginning of public awareness of molecular biology. However, the real starting point of this field occurred half a century ago when James D. Watson and Francis Crick suggested a structure for the salt of deoxyribonucleic acid (DNA). The history of the discovery of DNA – from

---

Chapter summary

Mutations result from changes in the nucleotide sequence of DNA or from trinucleotide repeat expansions, deletions, insertions, or rearrangements of DNA sequences in the genome. Mutations that alter a single nucleotide pair are called point mutations. Transition mutations replace one pyrimidine base with the other, or one purine base with the other. Transversion mutations replace a pyrimidine with a purine or vice versa. Such mismatches can become permanently incorporated in the DNA sequence during DNA replication. Whether nucleotide substitutions have a phenotypic effect depends on whether they alter a critical nucleotide in a gene regulatory region, the template for a functional RNA molecule, or the codons in a protein-coding gene.

Analytical questions

1 What would be the effect on reading frame and gene function if:
  (a) Two nucleotides were inserted into the middle of an mRNA?
  (b) Three nucleotides were inserted into the middle of an mRNA?
  (c) One nucleotide was inserted into one codon and one subtracted from the next?
  (d) A transition mutation occurs from G → A during DNA replication. What is the effect after a second round of DNA replication?
  (e) Exposure to an alkylating agent leads to the formation of O⁶-methylguanine. What is the effect after DNA replication?

Suggestions for further reading

Abraham, R.T., Tibbetts, R.S. (2005) Guiding ATM to broken DNA. *Science* 308:510–511.

Acharya, S., Foster, P.L., Brooks, P., Fishel, R. (2003) The coordinated functions of the *E. coli* MutS and MutL proteins in mismatch repair. *Molecular Cell* 12:233–246.

Banerjee, A., Yang, W., Karplus, M., Verdine, G.L. (2005) Structure of a repair enzyme interrogating undamaged DNA elucidates recognition of damaged DNA. *Nature* 434:612–618.

Costa, R.M.A., Chigança, V., da Silva-Galhardo, R., Carvalho, H., Menck, C.F.M. (2003) The eukaryotic nucleotide excision repair pathway. *Biochimie* 85:1083–1099.

Crick, F.H.C. (1974) The double helix: a personal view. *Nature* 248:766–769.

End-of-Chapter Study Aids

End-of-chapter summaries, analytical questions, and suggested readings reinforce the material you learned in the chapter and promote further study outside of the classroom

---

DNA repair and recombination   171

nonhomologous end-joining events that need fill-in of gaps or extension of the 3′ or 5′ overhangs. The rejoining of the broken ends is carried out by DNA ligase IV in association with XRCC4.

Chapter summary

Mutations result from changes in the nucleotide sequence of DNA or from trinucleotide repeat expansions, deletions, insertions, or rearrangements of DNA sequences in the genome. Mutations that alter a single nucleotide pair are called point mutations. Transition mutations replace one pyrimidine base with the other, or one purine base with the other. Transversion mutations replace a pyrimidine with a purine or vice versa. Such mismatches can become permanently incorporated in the DNA sequence during DNA replication. Whether nucleotide substitutions have a phenotypic effect depends on whether they alter a critical nucleotide in a gene regulatory region, the template for a functional RNA molecule, or the codons in a protein-coding gene.

A spontaneous mutation is one that occurs as a result of natural process in cells such as DNA replication errors that are not corrected by proofreading. Induced mutations occur as a result of interaction with DNA-damaging agents such as oxygen free radicals, ultraviolet or ionizing radiation, and various chemicals. There are three major classes of DNA damage: single base changes, structural distortion, and DNA backbone damage. Single base changes arise through the deamination, alkylation, and oxidation of bases. Bulky adducts that cause structural distortion of the DNA double helix, such as thymine dimers, can be induced by ultraviolet radiation or by chemical mutagenesis. DNA damage is also caused by intercalating agents and base analogs. Backbone damage includes the formation of abasic sites and double-stranded DNA breaks. There are a variety of complex cellular responses to different types of DNA damage: those that bypass the damage, those that directly reverse the damage, and those that remove that damaged section of DNA and replace it with undamage

172   Chapter 7

recognition and exonuclease processing. Strand invasion of the 3′ tail that is generated is initiated by Rad51. Strand exchange generates a joint molecule between damaged and undamaged duplex DNA and repair synthesis occurs. The interlinked molecules are then processed by branch migration and Holliday junction resolution by an enzyme complex called the resolvesome, followed by ligation of the repaired DNA strands. Nonhomologous end-joining rejoins double-strand breaks via direct ligation of the DNA ends without any sequence homology requirements. Following a double–strand break, the broken ends of DNA are recognized by a Ku heterodimer. The ends are processed by an endonuclease complex (Artemis/DNA-PK$_{cs}$) and repair polymerases. The rejoining of broken ends is carried out by DNA ligase.

Analytical questions

1 What would be the effect on reading frame and gene function if:
  (a) Two nucleotides were inserted into the middle of an mRNA?
  (b) Three nucleotides were inserted into the middle of an mRNA?
  (c) One nucleotide was inserted into one codon and one subtracted from the next?
  (d) A transition mutation occurs from G → A during DNA replication. What is the effect after a second round of DNA replication?
  (e) Exposure to an alkylating agent leads to the formation of O⁶-methylguanine. What is the effect after DNA replication?

2 A friend of yours with xeroderma pigmentosum seeks your advice about participating in a suntanning competition in Florida during Spring Break. Provide appropriate advice.

3 Draw a diagram of a Holliday junction during double-strand break repair. Starting with that diagram, illustrate branch migration and resolution. Is the resulting DNA duplex "repaired" to its original state?

4 You have isolated a novel protein factor you suspect is involved for efficient mismatch repair in mammalian cells. Design an experiment to test for repair activity *in vitro*. Show sample positive results.

Suggestions for further reading

Abraham, R.T., Tibbetts, R.S. (2005) Guiding ATM to broken DNA. *Science* 308:510–511.

Acharya, S., Foster, P.L., Brooks, P., Fishel, R. (2003) The coordinated functions of the *E. coli* MutS and MutL proteins in mismatch repair. *Molecular Cell* 12:233–246.

Banerjee, A., Yang, W., Karplus, M., Verdine, G.L. (2005) Structure of a repair enzyme interrogating undamaged DNA elucidates recognition of damaged DNA. *Nature* 434:612–618.

Costa, R.M.A., Chigança, V., da Silva-Galhardo, R., Carvalho, H., Menck, C.F.M. (2003) The eukaryotic nucleotide excision repair pathway. *Biochimie* 85:1083–1099.

Crick, F.H.C. (1974) The double helix: a personal view. *Nature* 248:766–769.

David, S.S. (2005) DNA search and rescue. *Nature* 434:569–570.

Dudáš, A., Chovanec, M. (2004) DNA double-strand break repair by homologous recombination. *Mutation Research* 566:131–167.

Dzantiev, L., Constantin, N., Genschel, J., Iyer, R.R., Burgers, P.M., Modrich, P. (2004) A defined human system that supports bidirectional mismatch-provoked excision. *Molecular Cell* 15:31–41.

Friedberg, E.C., Lehmann, A.R., Fuchs, R.P.P. (2005) Trading places: how do DNA polymerases switch during translesion DNA synthesis? *Molecular Cell* 18:499–505.

Genschel, J., Modrich, P. (2003) Mechanism of 5′-directed excision in human mismatch repair. *Molecular Cell* 12:1077–1086.

Heyer, W.-D., Ehmsen, K.T., Solinger, J.A. (2003) Holliday junction in the eukaryotic nucleus: resolution in sight? *Trends in Biochemical Sciences* 28:548–557.

TOOL BOX 9.1

Production of recombinant proteins

Many proteins which may be used for medical treatment or for research are normally expressed at very low concentrations. Through recombinant DNA technology, a large quantity of a protein of interest can be produced. This is called "overexpression" of the recombinant protein. Production of recombinant proteins involves the cloning of the cDNA encoding the desired protein into an "expression vector." The expression vector contains a promoter so that the cDNA can be expressed. Next, the recombinant expression vector is introduced into a bacterial or eukaryotic host cell to allow overexpression of the protein. There are also systems available for *in vitro* translation.

Overexpression of recombinant proteins in bacteria
A commonly used bacterial expression vector contains the *lac* promoter upstream of a multiple cloning site (see Section 10.5 for details). The lactose analog isopropylthiogalactoside (IPTG) stimulates the expression of a cloned cDNA encoding the protein of interest (Fig. 1A). Depending upon the particular application, cloned genes can be expressed to produce native proteins (in their natural state) or fusion proteins, where the foreign polypeptide is fused to a vector-encoded epitope tag (e.g. GST, His, or FLAG). Native proteins are preferred for therapeutic use because fusion proteins can be immunogenic in humans.

Figure 1 Comparison of the production of recombinant proteins in a bacterial host and by *in vitro* translation. (A) Production of recombinant proteins involves the cloning of the cDNA encoding the protein of interest into an expression vector. The expression vector typically contains a *lac* promoter so that the cDNA can be expressed in the *E. coli* host cell. Upon induction with IPTG the recombinant protein is expressed. (B) Protein can be translated *in vitro* in a rabbit reticulocyte lysate. The lysate is a cell-free extract made from red blood cell precursors. The lysate contains the cellular components necessary for protein synthesis (tRNA, ribosomes, initiation, elongation, and termination factors). In this system, the cDNA is cloned into an expression vector under control of a phage RNA polymerase promoter, such as SP6, T3, or T7. Upon addition of the appropriate phage polymerase to the cell lysate, transcription of an mRNA transcript occurs, followed by translation. If a radioactively labeled amino acid, such as ^{35}S-methionine, is included in the reaction mixture, then a labeled protein is produced.

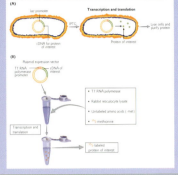

FOCUS BOX 15.5

Genetically modified crops: are you eating genetically engineered tomatoes?

Most people are familiar with the flavorless tomatoes sold in supermarkets. Their lack of flavor results from the practice of shipping green tomatoes and chemically ripening them later. Vine-ripened tomatoes cannot be shipped very far because they spoil too quickly. In the early 1990s, scientists used transgenic technology to try to counteract this problem. They genetically engineered tomatoes to alter an aspect of fruit ripening called "softening." The process of fruit softening is caused in part by the breakdown of pectins (compounds which give support to the cell walls of fruit). The tomatoes were engineered to have reduced levels of polygalacturonase, an enzyme that breaks down pectin. In addition, for selection purposes, the tomatoes carried resistance to the antibiotic kanamycin. Tomato plants were transformed with an antisense polygalacturonase gene sequence, which encodes an RNA transcript that is complementary to polygalacturonase mRNA. The two RNAs (sense and antisense) bind to one another so that they are degraded and translation of the polygalacturonase protein is prevented. The result was the "Flavr-Savr" tomato which spoiled less quickly after harvesting and thus could be left to ripen on the vines longer, developing more flavor and allowing later shipment to stores.

In late 1991, Calgene had a variety of Flavr-Savr tomato ready for marketing. They requested the opinion of the Food and Drug Administration (FDA), since this was the first genetically modified food to reach this point in product development. Public concerns about food safety also prompted Calgene to request a ruling from the FDA regarding the safety of antibiotic resistance genes in food. Calgene received FDA approval for its tomatoes in mid-1994 and started selling tomatoes in markets in the Chicago area. The tomatoes were clearly labeled as "genetically modified" and supplied with information pamphlets. Despite growing protests by activist groups, the tomatoes were well received. However, it was difficult to ship the delicate tomatoes without damage, the tomatoes did not grow well in Florida production fields the first year, and high development costs were followed by several years of low tomato prices. By March of 1997, the Flavr-Savr tomato had been taken off the market, and Calgene sold its interests in the tomato to Monsanto.

So, is that slice of tomato on your sandwich genetically modified? The answer is "No." Several companies are currently developing new varieties of GM tomatoes, but there are currently no GM tomatoes present in US markets either as whole tomatoes or in processed tomato foods.

not available for monocotyledonous plants, alternative techniques have been developed. Monocotyledons (monocots) are a class of flowering plants having an embryo with one cotyledon, such as daffodils, lilies, cereals, and grasses.

T-DNA-mediated gene delivery

Living plants and plants cells in culture can be transformed by transferred DNA (T-DNA) excised from the Ti (tumor-inducing) plasmid of *Agrobacterium tumefaciens*. *A. tumefaciens* is a Gram-negative soil bacterium that causes crown gall disease in many dicotyledonous plants. Thus, recombinant T-DNA is a highly efficient gene transfer vector for dicotyledonous plants. The T-DNA region normally carries genes encoding plant hormones and opine (modified amino acids) synthesis enzymes. Crown gall disease results from transformation of the plant genome with this part of the Ti plasmid in a process analogous to bacterial conjugation. This tumor-inducing portion of the plasmid can be replaced with foreign genes and is still transferred into the host genome.

The general strategy is to cut leaf disks (causing cell injury) and then incubate with *Agrobacterium* carrying recombinant disarmed Ti vectors (Fig. 15.16). The disks can then be transferred to shoot-inducing medium

DISEASE BOX 12.1

Cancer and epigenetics

CpG island hypermethylation and genome-wide hypomethylation are common epigenetic features of cancer cells. Too little methylation across the genome or too much methylation in the CpG islands can cause problems, the former by activating nearby oncogenes, and the latter by silencing tumor suppressor genes (see Section 17.2). Global hypomethylation may stimulate oncogene expression. However, since it takes places primarily in DNA repetitive sequences its main effect may be linked to chromosomal instability. Genome-wide demethylation appears to progress with age. This loss of methylation may, in part, explain the higher incidence of cancer among the elderly. Loss of methylation has also been linked to poor nutrition. For example, S-adenosylmethionine, a derivative of folic acid, is the primary methyl donor in the cell (Fig. 1). A lack of folic acid in the diet has been shown to predispose cells of an organism to cancer.

In addition to having aberrant DNA methylation patterns, cancer cells also have a histone H4 "cancer signature." This signature is found in many types of cancer, ranging from leukemia to cancer of the bone, cervix, prostate, and testis, and is characterized by a loss of monoacetylation at lysine 16 and trimethylation at lysine 20. The deacetylation occurs mainly in regions of repetitive DNA that also undergo hypomethylation in cancer cells. Because of this loss of acetylated histones in cancer cells, there is great interest in using histone deacetylase (HDAC) inhibitors alone, or in combination with DNA demethylating agents, to reactivate silenced tumor suppressor genes. A large number of clinical trials are underway to determine whether these inhibitors are safe and mediate the desired effect in treatment of various types of cancer.

Figure 1 Folic acid and related pathways for synthesis of *S*-adenosylmethionine (SAM). SAM is the methyl donor for the methylation of DNA and histones. DHFR, dihydrofolate reductase; MTHF, methyltetrahydrofolate; MTHFD, methylenetetrahydrofolate dehydrogenase; SAH, S-adenosylhomocysteine; THF, tetrahydrofolate.

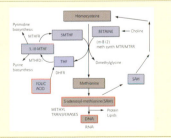

Imprinting occurs in mammals, but no other vertebrates looked at so far. There are around 80 different genes currently known to be imprinted in mammals. These genes tend to have important roles in development and the loss of imprinting is implicated in a number of genetic diseases and types of cancer in humans (Table 12.2, Disease box 12.3). Nearly all imprinted genes are organized in clusters in the genome.

Pedagogical Boxes

A variety of insightful boxes throughout the text explain additional concepts and topics in molecular biology:

- **Tools Boxes** explore key experimental methods and techniques in molecular biology

- **Focus Boxes** offer more detailed treatment of topics and delve into experimental strategies, historical background, and areas for further exploration

- **Disease Boxes** illustrate key principles of molecular biology by examining diseases that result from gene defects

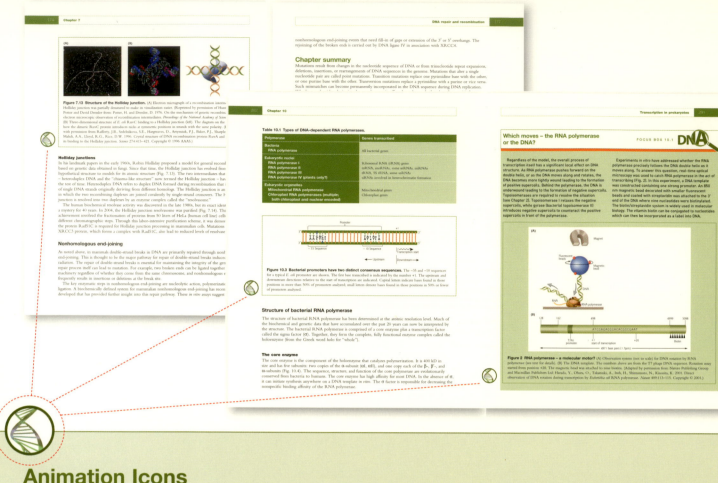

Animation Icons

Throughout the text, animation icons indicate key processes and concepts that are further illustrated and explained online

Key Terms and Glossary

Key terms used throughout the text are defined in the glossary at the end of the book, helping readers to navigate the complex language of molecular biology.

Fundamental Molecular Biology

Fundamental Molecular Biology

Lizabeth A. Allison
Department of Biology
College of William and Mary
Williamsburg
VA 23185, USA

Blackwell
Publishing

BLACKWELL PUBLISHING
350 Main Street, Malden, MA 02148-5020, USA
9600 Garsington Road, Oxford OX4 2DQ, UK
550 Swanston Street, Carlton, Victoria 3053, Australia

First published 2007 by Blackwell Publishing Ltd

1   2007

Library of Congress Cataloging-in-Publication Data
Allison, Lizabeth.
    Fundamental molecular biology / Lizabeth Allison.
        p.   ; cm.
    Includes bibliographical references and index.
    ISBN 13: 978-1-4051-0379-4 (hardback : alk. paper)
    ISBN 10: 1-4051-0379-5 (hardback : alk. paper)
    1. Molecular biology–Textbooks.   I. Title.
    [DNLM: 1. Molecular Biology.   QU 450 A438f 2007]
QH506.A45 2007
572.8–dc22   2006026641

A catalogue record for this title is available from the British Library.

Set in 11/13pt Bembo
by Graphicraft Limited, Hong Kong
Printed and bound in Singapore
by Markono Print Media Pte Ltd

The publisher's policy is to use permanent paper from mills that operate a sustainable forestry policy, and which has been manufactured from pulp processed using acid-free and elementary chlorine-free practices. Furthermore, the publisher ensures that the text paper and cover board used have met acceptable environmental accreditation standards.

For further information on
Blackwell Publishing, visit our website:
www.blackwellpublishing.com

Contents

Chapter summary
Analytical questions
Suggestions for further reading

Preface

The fast pace of modern molecular biology research is driven by intellectual curiosity and major challenges in medicine, agriculture, and industry. No discipline in biology has ever experienced the explosion in growth and popularity that molecular biology is now undergoing. There is intense public interest in the Human Genome Project and genetic engineering, due in part to fascination with how our own genes influence our lives. With this fast pace of discovery, it has been difficult to find a suitable, up-to-date textbook for a course in molecular biology. Other textbooks in the field fall into two categories: they are either too advanced, comprehensive, and overwhelmingly detailed, with enough material to fill an entire year or more of lectures, or they are too basic, superficial, and less experimental in their approach. It is possible to piece together literature for a molecular biology course by assigning readings from a variety of sources. However, some students are poorly prepared to learn material strictly from lectures and selected readings in texts and the primary literature that do not match exactly the content of the course. At the other end, instructors may find it difficult to decide what topics are the most important to include in a course and what to exclude when presented with an extensive array of choices. This textbook aims to fill this perceived gap in the market. The intent is to keep the text to a manageable size while covering the essentials of molecular biology. Selection of topics to include or omit reflects my view of molecular biology and it is possible that some particular favorite topic may not be covered to the desired extent. Students often complain when an instructor teaches "straight from the textbook," so adding favorite examples is encouraged to allow instructors to enrich their course by bringing to it their own enthusiasm and insight.

Approach

A central theme of the textbook is the continuum of biological understanding, starting with basic properties of genes and genomes, RNA and protein structure and function, and extending to the complex, hierarchical interactions fundamental to living organisms. A comprehensive picture of the many ways molecular biology is being applied to the analysis of complex systems is developed, including advances that reveal fundamental features of gene regulation during cell growth and differentiation, and in response to a changing nvironment, as well as developments that are more related to commercial and medical applications. Recent advances in technology, the process and thrill of discovery, and ethical considerations in molecular biology research are emphasized.

The text highlights the process of discovery – the observations, the questions, the experimental designs to test models, the results and conclusions – not just presenting the "facts." At the same time the language of molecular biology is emphasized, and a foundation is built that is based in fact. It is not feasible to examine every brick in the foundation and still have time to view the entire structure. However, as often as possible real examples of data are shown, e.g. actual results of an EMSA, Western blot, or RNA splicing assay. Experiments are selected either because they are classics in the field or because they illustrate a particular approach frequently used by molecular biologists to answer a diversity of questions.

Organization

The textbook is designed for a one-term course on molecular biology (or molecular genetics) for undergraduate students who are primarily majoring in biology or chemistry, with a large percentage of premedical students. First-year graduate students with a minimal background in molecular genetics/biology would also benefit from this course. Students would be expected to have completed, at a minimum, a two-term introductory biology course and to have completed at least 1 year of chemistry. Each chapter opens with a conceptual statement and historical perspective, followed by explanation and elaboration. Chapters end with a list of key references from the primary literature, and a series of analytical questions.

The book begins with a five chapter sequence that should be a review for most students, but with more detail than they would have encountered in an introductory biology or genetics course. Students of molecular biology need to have a solid grasp of these concepts so they may need to refresh their understanding of them. Depending on the curriculum at a particular institution, more or less time may need to be spent on these introductory chapters. Chapter 1 is a brief history of genetics and the beginnings of molecular biology. Chapter 2 discusses the structure and chemical properties of DNA. Chapter 3 discusses the organization of genomes and eukaryotic chromatin. Chapter 4 deals with the versatility of RNA structure and function, and Chapter 5 provides an overview of the flow of genetic information from DNA to RNA to protein and covers basic protein structure and function. The genetic code for amino acids is presented along with protein structure and function. This order reflects the view that to understand protein structure and function it is essential to first understand the flow of genetic information and where the primary sequence of amino acids derives from, and the consequences of alterations in the genetic code.

Chapters 6 and 7 cover DNA replication, telomere maintenance, and DNA repair and recombination. Although some instructors prefer to cover DNA replication later in a course, my view is that the information is essential early on, particularly to be able to understand many of the experimental strategies used in studying genes and their activities at the molecular level.

There is always debate on where to place methods and techniques – scattered throughout, in an appendix, or as specific chapters. I have mainly taken the latter approach, with the intent that this textbook will be a useful resource for students well beyond the course. Many undergraduate programs now include a research component and having a compilation of the standard techniques in molecular biology along with theoretical background and how they arose from basic research provides an essential aid. My approach in teaching is to cover some of the very basic methods in a series of "recombinant DNA technology" lectures, but to introduce others as needed to understand experiments discussed throughout the course. For example, Chapters 8 and 9, which cover recombinant DNA technology, molecular cloning, and tools for analyzing gene expression would certainly not be covered from start to finish. Covering method after method would become tedious. In addition, appreciation of the concepts behind techniques is much greater after students have acquired more experience in molecular biology.

Eukaryotic molecular biology is emphasized, although where details are better understood from bacteria, these are included. When fundamental processes such as DNA replication, repair, and recombination are discussed, the focus is on eukaryotes because the basic process is similar to that in prokaryotes, although the components of the machinery and the specific names of the players may differ. Prokaryotic transcription is given a separate chapter (Chapter 10), however, since some aspects of transcriptional regulation are fundamentally different than in eukaryotes, e.g. the concept of the operon, attenuation, and riboswitches. The basic transcription apparatus is introduced in Chapter 10 and how transcripts are initiated, elongated, and terminated is covered.

Chapter 11 covers the control of transcription in eukaryotes, introducing the regulatory elements, the general transcription factors, the interaction of DNA-binding proteins and DNA targets, the role of coactivators and corepressors, and regulated nuclear import of transcription factors. Chapter 12 covers the emerging field of epigenetics and monoallelic gene expression.

Chapter 13 introduces RNA processing and post-transcriptional gene regulation in eukaryotes, while Chapter 14 covers the mechanism of translation, with a focus on eukaryotic translation.

Chapters 15–17 cover some of the many applications of molecular biology. Chapter 15 introduces genetically modified organisms and their use in basic and applied research. Chapter 16 covers genome analysis, including DNA typing, genomics, and proteomics. Chapter 17 covers aspects of medical molecular biology including the molecular biology of cancer, gene therapy, and human behavior.

The course length is easily adjustable. The book is designed so that more or less time can be spent on particular topics according to an instructor's preference. The material in boxes can be treated as supplementary material if the course is too long for the needs of a particular class. On the other hand, there will be additional readings in these sections for students who want to go beyond the material in the main text to gain a deeper understanding of a particular topic.

Special features

Unique aspects of the book include a cohesive discussion of epigenetics and medical molecular biology, and the use of boxes to highlight molecular tools (Tool boxes), and to provide a more detailed treatment of material that will be of interest to the very keen student (Focus boxes). In addition, the textbook has a strong emphasis on biomedical research, which will appeal to the many premedical students who are likely to take the course prior to taking the MCATs. "Disease boxes" use diseases resulting from defects in a key gene to illustrate many principles of molecular biology. These examples place complex regulatory pathways, such as nucleotide excision repair, in a relevant context, making them more memorable for students.

- Book features:
  - Tool boxes explore key experimental methods and techniques in molecular biology
  - Focus boxes offer more detailed treatment of topics, delve into experimental strategies, and suggest areas for further exploration
  - Disease boxes illustrate key principles of molecular biology by examining diseases that result from key gene defects
  - Chapter-opening quotes, outlines, and introductions
  - End-of-chapter analytical questions
  - End-of-book glossary.
- Interactive website features:
  - Interactive animations (based on art from the book and identified in the book with a special icon)
  - Interactive student tutorials and pdb files
  - Interactive student exercises
  - Answers to end-of-chapter analytical questions
  - Additional student and instructor resources
  - Downloadable artwork from the text.
- CD-ROM features:
  - Downloadable artwork from the text
  - Sample interactive animation and tutorial
  - Sample syllabus
  - Link to website.

Acknowledgments

I am forever indebted to my undergraduate mentor L. Gerard Swartz, my master's thesis advisor, Gerald Shields, my PhD thesis advisor, Aimee Bakken, and my faculty mentors, Frank Sin and Larry Wiseman for their inspiration and belief in me throughout my education and career. My husband, Michael Levine and my son Andrew (born May 2003) deserve special thanks for their patience and encouragment ("good job Mommy!"). I thank my parents for nurturing my creativity and teaching me to follow my dreams. This book is dedicated to my mother Marjorie Allison (1929–1999) who remembered I was a molecular biologist as opposed to a microbiologist by thinking of the moles in her flower garden, and to my father Jack Allison (1918–2004) who sparked my interest in science while allowing me to wash beakers and flasks in his high school chemistry lab during the summer. At Blackwell Publishing, Nancy Whilton was my visionary cheerleader and Elizabeth Frank dealt very efficiently with all the nuts and bolts of the process. Rosie Hayden capably managed all the behind-the-scenes editorial work, Sarah Edwards heroically orchestrated the design and production aspects, and Jane Andrew skillfully handled the copy editing with an excellent eye for detail. In addition, Kieran Thomas designed a creative and easy-to-use website and Matt Payne spearheaded the marketing and publicity efforts. I also thank the members of the Allison lab who have put up with me sequestering myself in my office for many months at a time.

Finally, I acknowledge the contributions of my outside reviewers: Brian Ashburner (University of Toledo), Alice Cheung (University of Massachusetts, Amherst), Robert S. Dotson (Tulane University), Jutta Heller

(Loyola University), Daniel Herman (University of Wisconsin–Eau Claire), Jerry Honts (Drake University), Jason Kahn (University of Maryland), Chentao Lin (University of California, Los Angeles), Alison Liu (Rutgers, The State University of New Jersey), Hao Nguyen (California State University, Sacramento), Rekha C. Patel (University of South Carolina), Ravinder Singh (University of Colorado), and Scott A. Strobel (Yale University).

I appreciate greatly the time spent by these reviewers and thank them for their insightful and exceptionally helpful comments, most of which I hope I have addressed. Any remaining errors are mine and I welcome comments and suggestions for improvement.

Lizabeth A. Allison
Williamsburg, VA 2006

Chapter 1
The beginnings of molecular biology

Her [Rosalind Franklin's] photographs are among the most beautiful X-ray photographs of any substance ever taken.

Obituary of Rosalind Franklin, *Nature* (1958), 182:154.

Outline

1.1 Introduction

For decades, DNA was largely an academic subject and not the source of dinner table conversation in the average household. In 1995 this changed when media coverage of the O.J. Simpson murder trial brought DNA fingerprinting to homes across the world. Two years later, the cloning of Dolly the sheep was headline news. Then, in 2001, scientists announced the rough draft of the human genome sequence. In commenting on this landmark achievement, former US President Clinton likened the "decoding of the book of life" to a medical version of the moon landing. Increasingly, DNA has captivated Hollywood and the general public, excited scientists and science fiction writers alike, inspired artists, and challenged society with emerging ethical issues (Fig. 1.1).

1.2 Historical perspective

The last 5–10 years mark the beginning of public awareness of molecular biology. However, the real starting point of this field occurred half a century ago when James D. Watson and Francis Crick suggested a structure for the salt of deoxyribonucleic acid (DNA). The history of the discovery of DNA – from

Figure 1.1 DNA in art. (Susan Rankaitis © 2002, "DNA 2" from SPR Synthesis Project, 8′ × 16′, combined media. Courtesy of Robert Mann Gallery.)

its isolation as "nuclein" from soiled bandages, to proof that it is the universal hereditary material, to elucidation of the double helix structure in 1953 – is a riveting story (Fig. 1.2).

The details of this history are beyond the scope of this textbook. However, some highlights are presented to illustrate four important principles of scientific discovery:

1 Some great discoveries are not appreciated or communicated to a wide audience until years after the discoverers are dead and their discoveries are "rediscovered."

2 A combined approach of *in vivo* and *in vitro* studies has led to significant advances.

3 Major breakthroughs often follow technological advances.

4 Progress in science may result from competition, collaboration, and the tenacity and creativity of individual investigators.

Insights into heredity from round and wrinkled peas: Mendelian genetics

Heredity is the transmission of characteristics from parent to offspring by means of genes. In his 1893 book entitled *Germ-Plasm: a Theory of Heredity* August Weismann concludes that:

> **The more deeply . . . we penetrate into the phenomena of heredity, the more firmly are we convinced that something of the kind [germplasm or hereditary substance] does exist, for it is impossible to explain the observed phenomena by means of much simpler assumptions. We are thus reminded afresh that we have to deal not only with the infinitely great, but also with the infinitely small . . .**

The connection between DNA and heredity, however, was not demonstrated until the middle of the 20th century.

Much of the inspiration for the study of heredity originated from the research of Gregor Johann Mendel, an Augustinian monk working in Austria in the 1860s. Mendel bred different varieties of garden peas (*Pisum sativum*), such as those with round seeds and wrinkled seeds. He then compared the characteristics of parents and offspring. Results from his experiments led to the formulation of what Mendel described as "the law of combination of different characters" (Fig. 1.3).

I. HEREDITY AND GENES

384-322 B.C.

Aristotle proposes the theory of pangenesis (hereditary characteristics are carried and transmitted by gemmules from individual body cells)

1866

Gregor Mendel (the "Father of Modern Genetics") publishes his paper on inheritance of traits in peas

1884

E. Strasburger, Oscar Hertwig, R.A. von Kölliker and August Weismann independently identify the cell nucleus as the basis of inheritance

1889

Hugo DeVries hypothesizes the existence of "pangenes"

1893

August Weismann proposes his "germ plasm" theory and challenges the widely held idea that acquired characteristics can be inherited

1900

Hugo DeVries, Karl Correns, and Erich Von Tschermak independently rediscover and verify Mendel's Laws

II. CHROMOSOMES

1875

E. Strasburger describes what will later be called chromosomes

1882

Walter Flemming describes behavior of chromosomes during mitosis

1888

W. Waldeyer names chromosomes ("color bodies")

1902

Walter S. Sutton and Theodore Boveri observe that chromosomes in cells behave in ways parallel to Mendel's characters during meiosis, and propose a chromosomal basis for heredity

1905

Nettie M. Stevens and Edmund B. Wilson independently develop the idea of sex determination by chromosomes

III. DNA

1869

Friedrich (Fritz) Miescher isolates an acidic, phosphorus-rich substance he called "nuclein" from the nuclei of white blood cells in pus from soiled bandages

1928

Frederick Griffith demonstrates a heritable "transforming principle" that transmits the ability of bacteria to cause pneumonia in mice

1929

Phoebus Aaron Levene characterizes and names the compounds ribonucleic acid and deoxyribonucleic acid, and a "tetranucleotide" structure of DNA, in which the 4 bases of DNA are arranged one after another in a set of 4

1938

Rudolf Signer, Torbjorn Caspersson and Einer Hammarstein find molecular weights for DNA between 500,000 and 1,000,000 daltons, suggesting that DNA must be a polynucleotide

Proteins and DNA are studied by many scientists using X-ray crystallography.
The term "molecular biology" is coined by **Warren Weaver**

Figure 1.2 Three lines of research led to the discovery that DNA is the hereditary material. Selected landmarks in the study of heredity and the nature of genes, chromosome structure and function, and DNA structure from 384 BC to 1953.

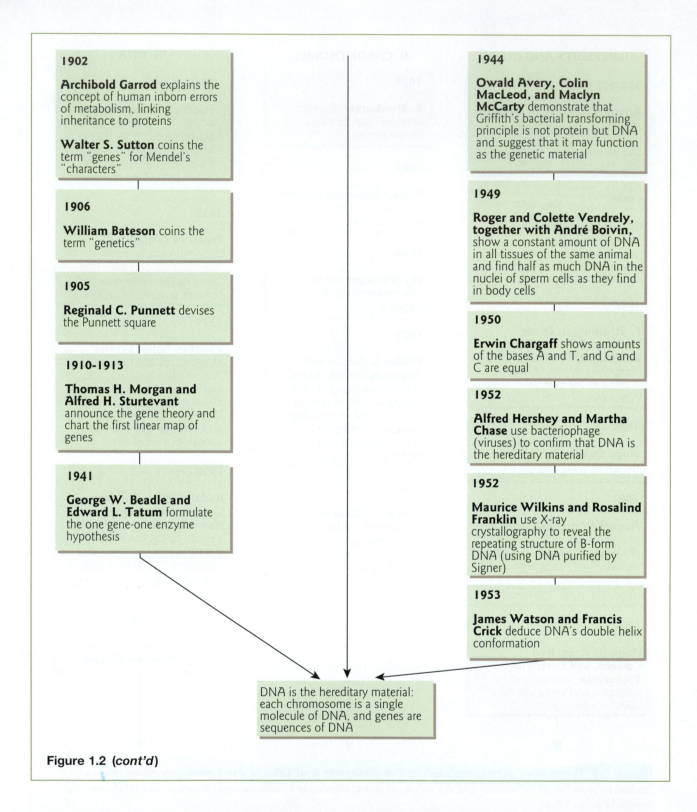

1902

Archibold Garrod explains the concept of human inborn errors of metabolism, linking inheritance to proteins

Walter S. Sutton coins the term "genes" for Mendel's "characters"

1906

William Bateson coins the term "genetics"

1905

Reginald C. Punnett devises the Punnett square

1910–1913

Thomas H. Morgan and Alfred H. Sturtevant announce the gene theory and chart the first linear map of genes

1941

George W. Beadle and Edward L. Tatum formulate the one gene-one enzyme hypothesis

1944

Owald Avery, Colin MacLeod, and Maclyn McCarty demonstrate that Griffith's bacterial transforming principle is not protein but DNA and suggest that it may function as the genetic material

1949

Roger and Colette Vendrely, together with André Boivin, show a constant amount of DNA in all tissues of the same animal and find half as much DNA in the nuclei of sperm cells as they find in body cells

1950

Erwin Chargaff shows amounts of the bases A and T, and G and C are equal

1952

Alfred Hershey and Martha Chase use bacteriophage (viruses) to confirm that DNA is the hereditary material

1952

Maurice Wilkins and Rosalind Franklin use X-ray crystallography to reveal the repeating structure of B-form DNA (using DNA purified by Signer)

1953

James Watson and Francis Crick deduce DNA's double helix conformation

DNA is the hereditary material: each chromosome is a single molecule of DNA, and genes are sequences of DNA

Figure 1.2 (cont'd)

Mendel's report was greeted by a disinterest that lasted 36 years. The significance of his work was finally recognized in 1900, upon independent rediscovery of his principles by Hugo DeVries (the Netherlands), Karl Correns (Germany), and Erich Von Tschermak (Austria). This marked the age of classic or Mendelian genetics. The basic principles of genetics – the law of segregation, the law of independent assortment, and the concept of dominant and recessive traits – are attributed to Gregor Mendel.

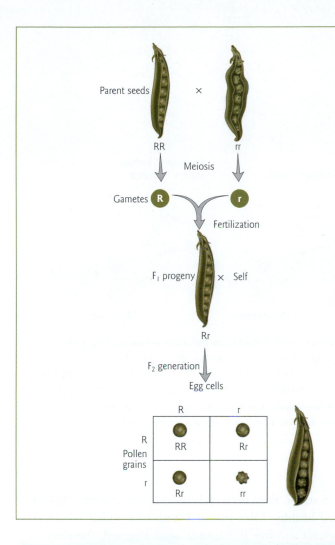

Figure 1.3 The law of combination of different characters. Mendel studied the inheritance of seed change. We know today that wrinkled seeds possess an abnormal form of starch (see Fig. 12.10). The diagram shows Mendel's genetic hypothesis to explain the 3 : 1 ratio of dominant : recessive phenotypes observed in the F_2 generation of a monohybrid cross. True-breeding (homozygous) round (R) seeds and true-breeding wrinkled (r) seeds were planted. Plants were cross-pollinated and allowed to grow and mature. The heterozygous F_1 plants (Rr) were allowed to self-pollinate and the F_2 generation was analyzed: $^3/_4$ round, $^1/_4$ wrinkled (3 : 1 ratio).

Insights into the nature of hereditary material: the transforming principle is DNA

From 1900 on, investigators continued to explore the nature of genes, the behavior of chromosomes during mitosis, and the chemical composition of DNA. But it took a combined approach of *in vivo* and *in vitro* studies to finally make the link between the hereditary material and DNA. A recurrent theme throughout this textbook will be this powerful experimental approach of combining studies performed within a living organism (*in vivo*) with studies performed in cells or tissues grown in culture, or in cell extracts or synthetic mixtures of cell components (*in vitro*).

In vivo experiments

In 1928, Frederick Griffith described a transforming principle that transmitted the ability of bacteria to cause pneumonia in mice (Fig. 1.4). In an elegant *in vivo* experiment, Griffith used pathogenic and nonpathogenic strains of *Streptococcus pneumoniae* to infect mice. Pathogenic strains form glistening "smooth" colonies (visible clumps of cells) when grown on nutrient agar in the laboratory, due to the polysaccharide coats they synthesize. They are designated "S" to distinguish them from nonpathogenic strains. The nonpathogenic strains lack the polysaccharide coat and form "rough" (R) colonies. These R strains do not cause pneumonia because without the protective coat, the bacteria are attacked by the immune system of the infected animal.

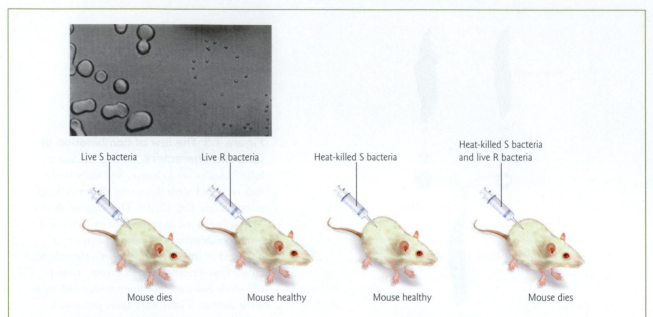

Figure 1.4 The transforming principle. Griffith's experiment with *Streptococcus pneumoniae*. Smooth (S)-type bacterial cells will kill mice as will heat-killed S-type cells injected along with live rough (R)-type cells. Insets above show colonies of *S. pneumoniae* growing on nutrient medium. The small colonies (right) are the nonpathogenic R type, and the large, glistening mucoid colonies (left) are the pathogenic S type. (Reproduced from Avery, O., MacLeod, C., McCarty, M. 1944. Studies on the chemical nature of the substance inducing transformation of *Pneumococcal* types. *Journal of Experimental Medicine* 79:137–158, by copyright permission of The Rockefeller University Press.)

When Griffith injected mice with live bacteria of an S strain, they invariably died of pneumonia. When he injected mice with live R bacteria, the mice remained healthy. Mice injected with heat-killed S bacteria also remained healthy. However, when heat-killed S bacteria and live R bacteria were injected, the mice died. Griffith called this the "transforming principle." He concluded there was transfer of some component of the pathogenic (S) bacteria which allowed the nonpathogenic (R) bacteria to make the polysaccharide coat and evade the mouse immune response.

Griffith's model of genetic transformation was met with almost universal skepticism, in part because he was not able to effectively communicate his ideas to the scientific community. Apparently, he was so shy that he had trouble even reading his papers in front of a small audience.

In vitro experiments

After nearly 16 years with no further advances in characterizing Griffith's transforming principle, an important breakthrough occurred. An *in vitro* assay was developed that provided the means by which the nature of the transforming factor in heat-killed S cells could be directly investigated without having to inject mice and wait for them to die. The assay involved selection of transformed cells from untransformed cells by their resistance to agglutination (clumping) by serum containing antibodies directed against R cells. Oswald Avery, Colin MacLeod, and Maclyn McCarty used this assay to show in 1944 that Griffith's transforming principle was DNA. They demonstrated that purified DNA was sufficient to cause transformation, and that the transforming factor could be destroyed by enzymes that degrade DNA (deoxyribonucleases) but not by protease or ribonuclease enzymes.

The discovery that the transforming principle was DNA came as a surprise. At the time, scientists had not yet learned that bacteria contained DNA or genes, though DNA was known to be a component of eukaryotic chromosomes. In addition, Phoebus Levene's tetranucleotide model for the structure of DNA was still widely accepted and it was thus thought that DNA was too simple a molecule to direct the development of plants and animals. Sadly, Griffith never did hear the explanation. He was killed by bombs falling on London during World War II, after refusing to leave the lab during an air raid.

Creativity in approach leads to the one gene–one enzyme hypothesis

Pioneering work was performed by George Beadle and Edward L. Tatum in 1941. They were the first to demonstrate a link between a gene and a step in a metabolic pathway catalyzed by an enzyme (protein). Their approach was novel. Instead of attempting to work out the chemistry of known genetic differences, they worked backwards and selected mutants of the pink bread mold, *Neurospora crassa*, in which known chemical reactions were blocked (Fig. 1.5).

Normally, *Neurospora* can be grown on a defined minimal medium consisting of sugar, some inorganic acids and salts, a nitrogen source, and niacin (vitamin B_3). Beadle and Tatum induced mutations in the mold by means of X-irradiation and isolated mutants that required specific supplementary compounds in order to grow (auxotrophs). In each case, a metabolic step leading to the synthesis of a specific compound had been blocked. There was a one-to-one correspondence between a genetic mutation and the lack of a specific enzyme required in a biochemical pathway. The first pathway elucidated by Beadle and Tatum was the conversion of the amino acid tryptophan via kynurenine and 3-hydroxyanthranilic acid to niacin. From their results, they formulated the "one gene–one enzyme hypothesis."

Beadle and Tatum's hypothesis was later revised to the "one gene–one polypeptide hypothesis." With some updates, to take into account functional RNA molecules, this hypothesis still holds true. In addition, the study of mutations continues to be a driving force in genetics and in modern molecular biology.

The importance of technological advances: the Hershey–Chase experiment

An important event in the history of the characterization of DNA was the emerging availability and utility of radioisotopes in basic science research in the early post-World War II years. Radioisotopes allowed Alfred Hershey and Martha Chase to carry out a classic experiment in 1952 showing that the genetic material of a virus that infects bacteria, bacteriophage T2 (literally "bacterium eater"), is DNA (Fig. 1.6).

The DNA of bacteriophage T2 (phage for short) was known to be contained within a protein coat, so Hershey and Chase designed an experiment to determine whether the protein or DNA carried the genetic information to make a new phage. First, they selectively labeled phage DNA with the radioactive isotope 32-phosphorus (^{32}P) and phage protein with 35-sulfur (^{35}S). DNA contains phosphorus but no sulfur; while protein is composed of some sulfur (in the amino acids methionine and cysteine) but no phosphorus. Next, they incubated bacteria (*Escherichia coli*) with the labeled phage. During infection, the phage attaches to the bacterium and injects its DNA. At this point, Hershey and Chase encountered a major problem. They were not able to tear the empty phage coat away from the bacterial cell wall after injection of its DNA. Without this step they could not complete their experiment. In an ingenious moment, they tried the recently invented kitchen blender and found they could separate the empty phage coats from the bacteria. Fred Waring, a popular band leader, financially backed development of the blender which bears his name. This step removed the ^{35}S since the phage protein did not enter the bacterial cell, but left the ^{32}P phage DNA inside the bacteria. After synthesis of phage components from the phage genetic material and their assembly, lysis of the bacteria occurred. Isolated progeny phage particles only contained ^{32}P, showing that all the information required to make new phage was contained within the injected DNA.

This finding demonstrated clearly that DNA is the genetic material in a system other than bacteria – further suggesting that DNA could be the universal hereditary material. The next challenge, however, was to explain how DNA could contain enough information to control the life of an organism. As noted above, DNA was thought to be a fairly simple (tetranucleotide) molecule at that time. How could it replicate, pass itself along cell after cell, and retain the message?

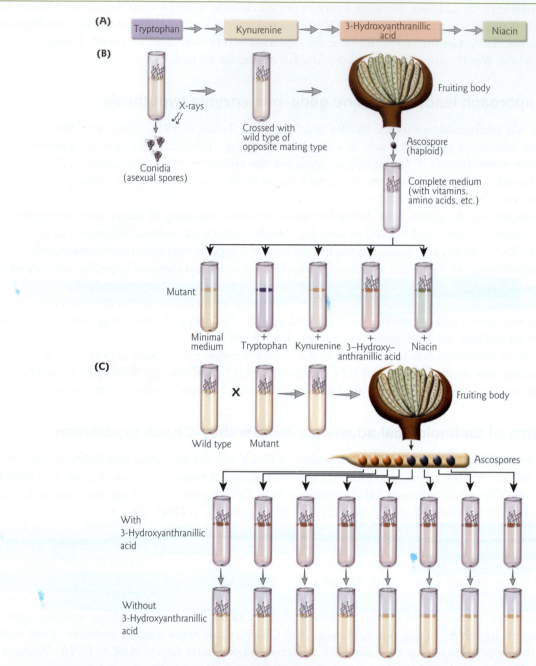

Figure 1.5 The one gene–one enzyme hypothesis. (A) Pathway of niacin synthesis in *Neurospora* determined by Beadle and Tatum. Arrows represent an enzyme-mediated step. The pathway is now known to involve seven steps. (B) Method for detecting mutants with enzyme deficiencies in the niacin synthesis pathway in *Neurospora*. Conidia are exposed to X-rays and crossed to the wild type. Haploid spores are then grown on complete medium, and cultures from this are grown on minimal medium. Failure to grow on minimal medium indicates a growth defect. The nature of the defect is investigated by growing the mutant strain on minimal medium supplemented with various intermediates in the pathway of niacin synthesis. In this example, the mutant can grow if given niacin, or alternatively 3-hydroxyanthranilic acid. It could not grow if given only kynurenine. Therefore, the mutation affects the pathway between kynurenine and 3-hydroxyanthranilic acid. Since the mutant cannot grow when given only tryptophan, Beadle and Tatum knew that tryptophan occurred in the pathway before the step with the deficient enzyme. (C) Method for confirming the genetic character of the mutation. The mutant strain is crossed to the wild type and placed on minimal medium without 3-hydroxyanthranilic acid. The observation that four haploid spores are unable to grow and four grow confirms that the defect is due to a genetic mutation, since this follows the principles of Mendelian genetics.

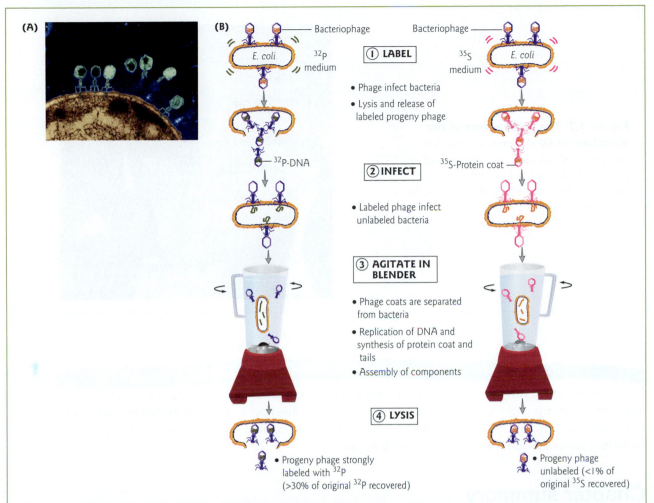

Figure 1.6 DNA is the genetic material of bacteriophage T2. (A) Bacteriophage T2 on *E. coli*. Color-enhanced transmission electron micrograph (TEM) of bacteriophages (green-blue) attacking a bacterium (brown). The bacteriophage uses its tail as a syringe to inject its own DNA (blue) into the bacterium. The bacteriophage then replicates within the bacterium. (Credit: Eye of Science / PhotoResearchers, Inc.) (B) The Hershey–Chase experiment using ^{35}S- and ^{32}P-labeled bacteriophage T2. The DNA label (^{32}P) enters the *E. coli* bacterium during infection. The protein label (^{35}S) does not.

A model for the structure of DNA: the DNA double helix

In 1953, James Watson and Francis Crick proposed the double helix as a model for the structure of DNA (Fig. 1.7). Their discovery was based, in part, on X-ray diffraction analysis performed by Rosalind Franklin in Maurice Wilkins' laboratory (Fig. 1.8) – a fascinating story of a competitive race to the finish. Other scientific findings at the time provided the necessary context for the proposed structure of DNA. These included: (i) the proposed α-helical secondary structure of proteins by Linus Pauling in 1951 (see Chapter 5); and (ii) the disproving of Levene's tetranucleotide hypothesis by Edwin Chargaff by showing that the four bases of DNA were not in a 1 : 1 : 1 : 1 ratio (see Section 2.2).

What is not always mentioned, in retrospect, is that the contemporary response of the scientific community to the proposed structure of DNA was lukewarm. It was only when the role of DNA in protein synthesis was elucidated that the biochemical community became seriously interested in the structure of DNA. The double helix marks the starting point for fundamental molecular biology. Molecular biology is the study of biological phenomena at the molecular level, in particular the study of the molecular structure

Figure 1.7 The discoverers of the structure of DNA. James Watson (left) and Francis Crick seen with their model of part of a DNA molecule in 1953. (Credit: A. Barrington Brown / Photo Researchers, Inc.).

of DNA and the information it encodes, and the biochemical basis of gene expression and its regulation. From this point we turn from the discovery of the hereditary material to its organization in the genome of different organisms. Subsequent chapters are devoted to its function in replication, transcription, and translation. Finally, the applications of recombinant DNA technology to many aspects of our daily lives, including medicine and forensic science, will be addressed.

Chapter summary

Heredity is the transmission of characteristics from parent to offspring by means of genes. Mendel described the law of combination of different characters, which defined the gene as the unvarying bearer of hereditary traits. Three lines of research on the nature of genes, the behavior of chromosomes during mitosis, and the chemical composition of DNA finally converged to show that DNA is the hereditary material: chromosomes are a single molecule of DNA and genes are sequences of DNA. Avery and his colleagues demonstrated that DNA was the genetic material when they showed that the transforming agent was DNA. Griffith had originally demonstrated the phenomenon of transformation of *Streptococcus* bacteria in mice. Hershey and Chase demonstrated that it was the DNA of bacteriophage T2 that entered the bacterial cell. By 1953, the evidence was strongly supportive of DNA as the genetic material. Several lines of evidence, including the X-ray diffraction photographs of Franklin, led Watson and Crick to suggest the double helical model of the structure of DNA.

Analytical questions

1 Sodium hydroxide degrades both proteins and nucleic acids, while phenol denatures proteins but not nucleic acids. In the transformation experiments performed by Griffith with *Streptococcus pnemoniae*, what result would be expected if an extract of S-strain bacteria was treated with phenol? What would be expected if it was treated with sodium hydroxide?

2 When Hershey and Chase infected bacterial cells that had been grown in the presence of radioactive phosphorus with bacteriophage T2, both the phage RNA and DNA must also have incorporated the labeled phosphorus and yet the experimental results were not affected. Why not?

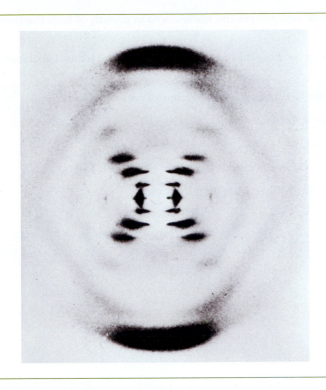

Figure 1.8 X-ray diffraction photograph of DNA. This image of the DNA double helix was obtained by Rosalind Franklin in 1953, the year in which Watson and Crick discovered DNA's structure, aided by Franklin's work. The image results from a beam of X-rays being scattered onto a photographic plate by the crystalline DNA. Various features about the structure of the DNA can be determined from the pattern of spots and bands. The cross of bands indicates the helical nature of DNA. (Credit: Omikron / Photo Researchers, Inc.)

3 Four independent methionine auxotrophs of *Neurospora* were studied to determine which related compounds might substitute for their methionine requirement. In the table below "+" indicates growth on minimal medium supplemented with the indicated compound, and "−" indicates no growth.

Mutant strain	Nothing added [minimal medium]	O-acetyl homoserine	Methionine	Homocysteine	Cystathione
Wild type	+	+	+	+	+
Mutant A	−	+	+	+	+
Mutant B	−	−	+	+	+
Mutant C	−	−	+	+	−
Mutant D	−	−	+	−	−

Based on the results, give the order of these four compounds in the biosynthetic pathway of methionine.

Suggestions for further reading

Avery, O., MacLeod, C., McCarty, M. (1944) Studies on the chemical nature of the substance inducing transformation of *Pneumococcal* types. *Journal of Experimental Medicine* 79:137–158.

Beadle, G.W., Tatum, E.L. (1941) Genetic control of biochemical reactions in *Neurospora*. *Proceedings of the National Academy of Sciences USA* 27:499–506.

Chargaff, E. (1951) Structure and function of nucleic acids as cell constituents. *Federation Proceedings* 10:654–659.

Fuller, W. (2003) Who said "helix"? *Nature* 424:876–878.

Griffith, F. (1928) Significance of pneumococcal types. *Journal of Hygiene* 27:113–159.

Hershey, A., Chase, M. (1952) Independent functions of viral protein and nucleic acid in growth of bacteriophage. *Journal of General Physiology* 36:39–56.

Judson, H.F. (1996) *The Eighth Day of Creation. Makers of the Revolution in Biology.* Cold Spring Harbor Laboratory Press, New York.

Lee, T.F. (1991) *The Human Genome Project. Cracking the Genetic Code of Life.* Plenum Press, New York.

Olby, R. (2003) Quiet debut for the double helix. *Nature* 421:402–405.

Watson, J.D. (1968) *The Double Helix.* Signet, New York.

Watson, J.D., Crick, F.H.C. (1953) A structure for deoxyribose nucleic acid. *Nature* 171:737–738.

Weismann, A. (1893) *The Germ-Plasm: a Theory of Heredity.* Walter Scott, London.

Chapter 2
The structure of DNA

DNA neither cares nor knows. DNA just is. And we dance to its music.

Richard Dawkins, *River Out of Eden: A Darwinian View of Life* (1995), p. 133.

Outline

2.1 Introduction

The DNA double helix is an icon for modern biology; a form represented from art galleries to corporate logos. The structure of DNA is certainly aesthetically pleasing, but is relatively meaningless without an understanding of how it relates to function. Likewise, function cannot truly be understood without knowledge of structure. One goal of this chapter will be to describe the structure of DNA, while keeping its function within living cells in mind.

DNA is often perceived as simply a very long sequence of base pairs organized into an invariant, ladder-like, double-helical structure. However, at the 1988 European Molecular Biology Organization (EMBO) Workshop on DNA curvature and bending, two sessions were scheduled to establish a common nomenclature for parameters used to describe the geometry of nucleic acid chains and helices. As a result, DNA can tip, propeller twist, buckle, roll, tilt, stagger, stretch, shear, rise, slide, and shift! Before considering the flexibility of nucleic acids, the more familiar primary structure will be reviewed.

2.2 Primary structure: the components of nucleic acids

While the acronyms DNA and RNA are considered standard nomenclature these days, their names underwent a long evolution, from nuclein to thymus nucleic acid to nucleic acid to deoxyribonucleic acid and ribonucleic acid. DNA was first called thymus nucleic acid because it was isolated from the thymus gland of cattle, while RNA was first identified as yeast nucleic acid. Nucleic acids are a long chain or polymer of repeating subunits, called nucleotides. Each nucleotide subunit is composed of three parts: a five-carbon sugar, a phosphate group, and a base. The chemical structures of these basic components are shown in Fig. 2.1.

Five-carbon sugars

Nucleotide subunits of RNA contain a pentose (five-carbon) sugar called ribose. Nucleotide subunits of DNA contain the sugar deoxyribose. The five carbon atoms in each pentose sugar are assigned numbers 1′ through 5′. Primes are used in the numbering of the ring positions in the sugars to differentiate them from the ring positions of the bases. Both sugars have an oxygen as a member of the five-member ring;

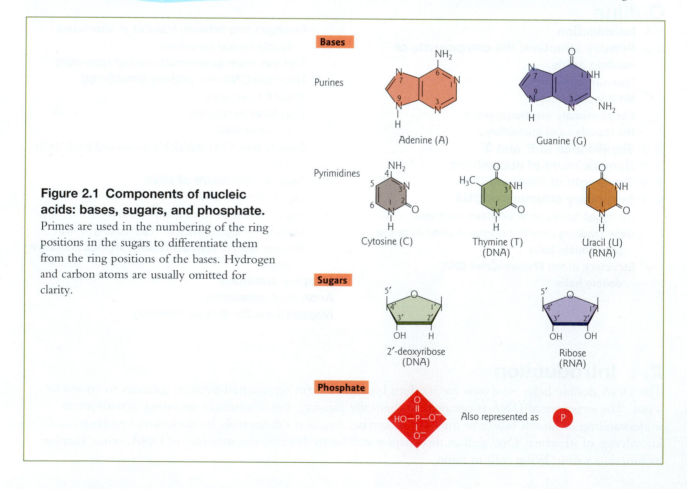

Figure 2.1 Components of nucleic acids: bases, sugars, and phosphate.
Primes are used in the numbering of the ring positions in the sugars to differentiate them from the ring positions of the bases. Hydrogen and carbon atoms are usually omitted for clarity.

Bases

Purines

Adenine (A)   Guanine (G)

Pyrimidines

Cytosine (C)   Thymine (T) (DNA)   Uracil (U) (RNA)

Sugars

2′-deoxyribose (DNA)   Ribose (RNA)

Phosphate

Also represented as   P

the 5′-carbon is outside the ring. The sugars differ only in the presence or absence ("deoxy") of an oxygen in the 2′ position. This minor distinction between RNA and DNA dramatically influences their function. The remarkable versatility of RNA is critically dependent on this 2′-hydroxyl group (see Chapter 4).

Nitrogenous bases

As their name suggests, the bases are nitrogen-containing molecules having the chemical properties of a base (a substance that accepts an H$^+$ ion or proton in solution). Two of the bases, adenine (A) and guanine (G) have a double carbon–nitrogen ring structure; these are called purines. The other three bases, thymine (T), cytosine (C), and uracil (U) have a single ring structure; these are called pyrimidines. Thymine is found in DNA only, while uridine is specific for RNA.

   A few years before Watson and Crick proposed the three-dimensional structure of DNA, Erwin Chargaff developed a paper chromatography method to analyze the amount of each base present in DNA. Chargaff observed certain regular relationships among the molar concentrations of the different bases (Fig. 2.2). These relationships are now called Chargaff's rules. He showed that the number of A residues in all DNA samples was equal to the number of T residues; i.e. [A] = [T], where the molar concentration of the base is denoted by the symbol for the base enclosed in square brackets. Accordingly, the number of G residues equals that of C: [G] = [C]. Finally, the amount of the purine bases equals that of the pyrimidine bases: [A] + [G] = [T] + [C]. Chargaff also found that the base composition of DNA, defined as the "percent G + C," differs among species but is constant in all cells of an organism within a species. The G + C content can vary from 22 to 73%, depending on the organism.

The phosphate functional group

The phosphate functional group (PO$_4$) gives DNA and RNA the property of an acid (a substance that releases an H$^+$ ion or proton in solution) at physiological pH, hence the name "nucleic acid." The linking bonds that are formed from phosphates are esters that have the additional property of being stable, yet are

SOURCE	ADENINE TO GUANINE	THYMINE TO CYTOSINE	ADENINE TO THYMINE	GUANINE TO CYTOSINE	PURINES TO PYRI- MIDINES	AMINO GROUPS TO ENOLIC HYDROXYLS
Ox	1.29	1.43	1.04	1.00	1.1	1.4
Man	1.56	1.75	1.00	1.00	1.0	1.3
Hen	1.45	1.29	1.06	0.91	0.99	1.5
Salmon	1.43	1.43	1.02	1.02	1.02	1.4
Wheat	1.22	1.18[1]	1.00	0.97[1]	0.99	1.4
Yeast	1.67	1.92	1.03	1.20	1.0	1.3
Hemophilus influenzae, type C	1.74	1.54	1.07	0.91	1.0	1.5
B. coli K-12	1.05	0.95	1.09	0.99	1.0	1.6
Avian tubercle bacillus	0.4	0.4	1.09	1.08	1.1	1.7
Serratia marcescens	0.7	0.7	0.95	0.86	0.9	1.6
Hydrogen organism Bacillus Schatz	0.7	0.6	1.12	0.89	1.0	1.7

[1] In these computations the sum of cytosine and methylcytosine was used. If cytosine alone is considered, the thymine to cytosine ratio is 1.62 and that of guanine to cytosine 1.33.

Figure 2.2 Chargaff's data showing the base composition of DNA. (Reproduced from Chargaff, E. 1951. Structure and function of nucleic acids as cell constituents. *Federation Proceedings* 10:654–659 by copyright permission of Blackwell Publishing Ltd.)

easily broken by enzymatic hydrolysis. When a nucleotide is removed from a DNA or RNA chain, the nucleotide is not destroyed in the process. Further, after the phosphodiester bond is formed (see below), one oxygen atom of the phosphate group is still negatively ionized. The negatively charged phosphates are extremely insoluble in lipids, which ensures the retention of nucleic acids within the cell or nuclear membrane.

Nucleosides and nucleotides

A DNA or RNA chain is formed in a series of three steps (Fig. 2.3). In the first reaction, each base is chemically linked to one molecule of sugar at the 1'-carbon of the sugar, forming a compound called a nucleoside. When a phosphate group is also attached to the 5'-carbon of the same sugar, the nucleoside

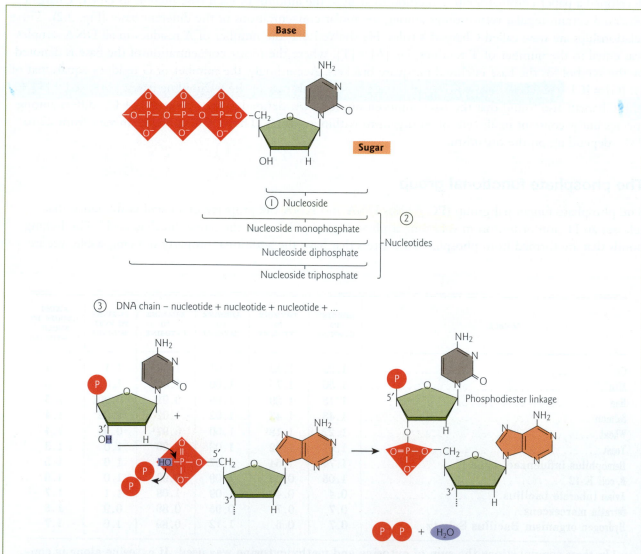

Figure 2.3 Formation of nucleic acid chains. DNA and RNA chains are formed through a series of three steps: (1) a base attached to a sugar is a nucleoside, (2) a nucleoside with one or more phosphates attached is a nucleotide, and (3) nucleotides are linked by 5' to 3' phosphodiester bonds between adjacent nucleotides to form a DNA or RNA chain. Bond formation occurs by a condensation reaction involving the removal of H_2O and pyrophosphate. The structure of a nucleoside and three nucleotides with differing numbers of phosphates is shown for DNA.

becomes a nucleotide. Finally, nucleotides are joined (polymerized) by condensation reactions to form a chain. The hydroxyl group on the 3′-carbon of a sugar of one nucleotide forms an ester bond to the phosphate of another nucleotide, eliminating a molecule of water. This chemical bond linking the sugar components of adjacent nucleotides is called a phosphodiester bond, or 5′ → 3′ phosphodiester bond, indicating the polarity of the strand.

2.3 Significance of 5′ and 3′

The ends of a DNA or RNA chain are distinct and have different chemical properties. The two ends are designated by the symbols 5′ and 3′. The symbol 5′ refers to the carbon in the sugar to which a phosphate (PO$_4$) functional group is attached. The symbol 3′ refers to the carbon in the sugar ring to which a hydroxyl (OH) functional group is attached. The asymmetry of the ends of a DNA strand implies that each strand has a polarity determined by which end bears the 5′-phosphate and which end bears the 3′-hydroxyl group.

This 5′ → 3′ directionality of a nucleic acid strand is an extremely important property of the molecule. Understanding this directionality (polarity) is critical for understanding aspects of replication and transcription, for reading a DNA sequence, and for carrying out experiments in the lab. By convention, a DNA sequence is written with the 5′ end to the left, and the 3′ end to the right. This is also the direction of synthesis (see Section 6.3).

2.4 Nomenclature of nucleotides

The rather complicated nomenclature of the nucleoside and nucleotide derivatives of the DNA and RNA bases is summarized in Table 2.1. Nucleotides may contain one phosphate unit (monophosphate), two such units (diphosphate), or three (triphosphate). When incorporated into a nucleic acid chain, a nucleotide contains one each of the three components. When free in the cell pool, nucleotides usually occur as triphosphates. The triphosphate form serves as the precursor building block for a DNA or RNA chain during synthesis. Because the nomenclature is rather complicated, scientists use shorthand for the names of the nucleotides. For example, deoxycytidine triphosphate (DNA) and cytidine triphosphate (RNA) are abbreviated to dCTP and CTP, respectively. The letters A, G, C, T, and U stand for the bases but, in practice, they are commonly used to represent the whole nucleotides containing these bases.

Many of these terms are not needed in this book; however, they are included because they are likely to be encountered in further reading or by students involved in research. Knowing the more complex names for deoxynucleoside triphosphates (dNTPs) and nucleoside triphosphates (NTPs) is important for understanding

Table 2.1 Nucleic acid nomenclature.

Base	Nucleoside		Nucleotide*	
	DNA	RNA	DNA	RNA
Adenine (A)	Deoxyadenosine	Adenosine	Deoxyadenosine 5′-triphosphate (dATP)	Adenosine 5′-triphosphate (ATP)
Guanine (G)	Deoxyguanosine	Guanosine	Deoxyguanosine 5′-triphosphate (dGTP)	Guanosine 5′-triphosphate (GTP)
Cytosine (C)	Deoxycytidine	Cytidine	Deoxycytidine 5′-triphosphate (dCTP)	Cytidine 5′-triphosphate (CTP)
Thymine (T)	Deoxythymidine	–	Deoxythymidine 5′-triphosphate (dTTP)	–
Uracil (U)	–	Uridine	–	Uridine 5′-triphosphate (UTP)
Generic (N)	Deoxynucleoside	Nucleoside	Deoxynucleoside 5′-triphosphate (dNTP)	Nucleoside 5′-triphosphate (NTP)

* Nucleotides may contain one phosphate unit (monophosphate), two such units (diphosphate), or three (triphosphate). The triphosphate form shown in the table serves as the precursor building block for nucleic acid synthesis.

scientific literature, and for ordering the correct compound from a catalog when planning an experiment. Note also that ATP and GTP, the energy currency of cells, are both nucleoside triphosphates.

2.5 The length of RNA and DNA

Cellular RNAs range in length from less than one hundred to many thousands of nucleotides; the number of nucleotides (nt) or "bases" is used as a measure of length. Cellular DNA molecules can be as long as several hundred million nucleotides. The number of base pairs (bp) is used as a measure of length of a double-stranded DNA. In practice, the unit of length used for DNA is the kilobase pair (kb or kbp), corresponding to 1000 base pairs, or the megabase pair (Mb or Mbp) corresponding to 1,000,000 base pairs. In the laboratory, researchers often make use of short chains of single-stranded DNA (usually less than 50 bases) called oligonucleotides.

2.6 Secondary structure of DNA

The primary structures of DNA and RNA are generally quite similar, however their conformations are quite different. RNA commonly exists as a single polynucleotide chain, or strand. In contrast, as Watson and Crick deduced, the biologically active structure of DNA is more complex than a single string of nucleotides linked by phosphodiester bonds. DNA generally exists as two interwound strands. This structural difference is critical to the different functions of the two types of nucleic acids. The secondary and tertiary structure of RNA will be discussed along with the versatility of its function in Chapter 4. The remainder of this chapter will focus on DNA. Various chemical forces drive the formation of the DNA double helix. These include hydrogen bonds between the bases and base stacking by hydrophobic interactions.

Hydrogen bonds form between the bases

Thermodynamically stable hydrogen bonds form between the nitrogenous bases on opposite strands of the interwound DNA chains (Fig. 2.4). Hydrogen bonds are very weak bonds that involve the sharing of a hydrogen between two electronegative atoms, such as oxygen and nitrogen. The hydrogen bonds provide one type of force holding the strands together. Although individually very weak, hydrogen bonds give structural stability to a molecule with large numbers of them. The hydrogen bonding between bases is referred to as "Watson–Crick" or "complementary" base pairing. It occurs in such a way that adenine (A) normally pairs with thymine (T) by two hydrogen bonds, and guanine (C) pairs with cytosine (C) by three hydrogen bonds. Watson–Crick base pairing allows the 1′-carbons on the two strands to be exactly the same distance apart (1.08 nm). This results in the regular, symmetric framework of the DNA double helix. The Watson–Crick structure of paired complementary strands explains Chargaff's rules – the relationships among the molar concentrations of the different bases.

   Why aren't there other stable base pairs present in DNA, such as G with A, or C with T? Some of these are ruled out by the difficulty of making two or more hydrogen bonds. But others, such as G with T, are not excluded for that reason (Fig. 2.5). The hydrogen bonding between G and T produces a pair with a similar overall shape to Watson–Crick base pairs (see Fig. 2.4). Likewise, GU base pairing is stable, and is of importance in RNA structure and RNA–protein interactions (see Section 4.2). However, if G-C to G-T changes were to occur in DNA, the sequence of bases in the DNA could change drastically with each cell division. In fact, there are proofreading mechanisms and DNA repair mechanisms that recognize non-Watson–Crick base pairs and correct the majority of mistakes (see Section 6.6 and 7.6).

Base stacking provides chemical stability to the DNA double helix

The molecular processes of cellular life generally take place in a watery solution, and intracellular components are largely molecules that are easily dissolved in water. The nitrogenous bases are an exception as they are

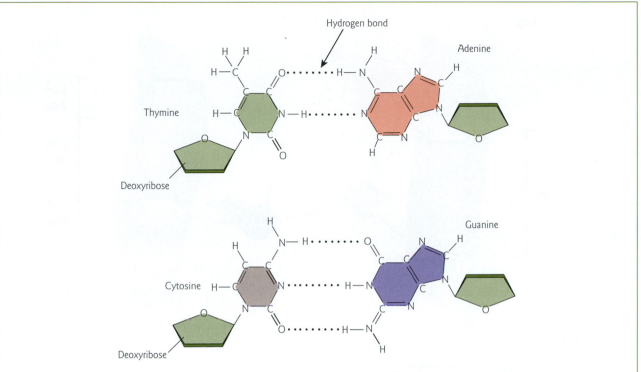

Figure 2.4 The two common Watson–Crick base pairs of DNA. Adenine (A) is joined to thymine (T) by two hydrogen bonds, while guanine (G) is joined to cytosine (C) by three hydrogen bonds.

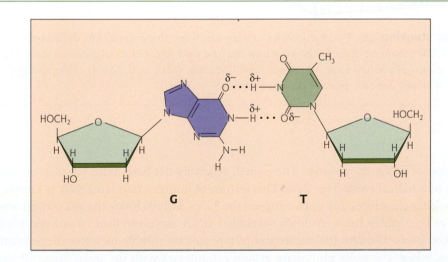

Figure 2.5 Pairing of G with T allows two hydrogen bonds. The overall shape of a GT base pair is similar to that of the Watson–Crick pairs shown in Fig. 2.4. A closely related base pair, that of G with U (uracil), is used to specify amino acids for the synthesis of proteins. As shown in Fig. 2.1, U is like T but without the CH_3 group.

nonpolar and thus hydrophobic ("water hating"). On their own they are practically insoluble in the aqueous environment of cells. The asymmetric distribution of charge across a polar water molecule thus has important consequences for the structure of DNA. Once the bases are attached to a sugar and a phosphate to form a nucleotide, they become soluble in water, but even so their insolubility still places strong constraints on the

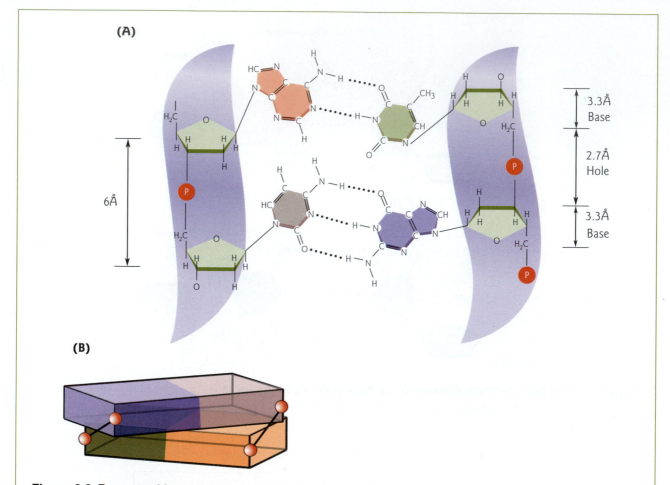

(A)

6Å

3.3Å
Base

2.7Å
Hole

3.3Å
Base

(B)

Figure 2.6 Base stacking. (A) Two nucleotides in schematic form, showing the key dimensions and the "hole" between the bases. (B) Schematic diagram showing how the base pairs (colored rectangles) can stack onto each other without a gap by means of a helical twist. (Redrawn from Calladine, C.R., et al. (2004) *Understanding DNA. The Molecule and How it Works*, 3rd edn. Elsevier Academic Press, New York.)

overall conformation of DNA in solution. The paired, relatively flat bases tend to stack on top of one another by means of a helical twist (Fig. 2.6). This feature of double-stranded DNA is known as "base stacking." This stacking eliminates any gaps between the bases and excludes the maximum amount of water from the interior of the double helix. A double-stranded DNA molecule thus has a hydrophobic core composed of stacked bases. Because the sugars and phosphates are soluble in water they orient towards the outside of the helix where the polar phosphate groups can interact with the polar environment (Fig. 2.7). The solubility of bases in water provides a driving force for DNA to form a double helix, and the energy of base stacking provides much of the chemical stability of the double helix. In general, the stacking of the bases often contributes as much as, or more than, half of the free energy of the total base pair.

Structure of the Watson–Crick DNA double helix

Alternating deoxyribose sugars and phosphate groups form the backbone of DNA. The bases are attached to the sugars and are located between the backbones of the DNA strands, lying perpendicular to the long axis of the strands. As the backbones of the two strands wind around each other, they form a double helix

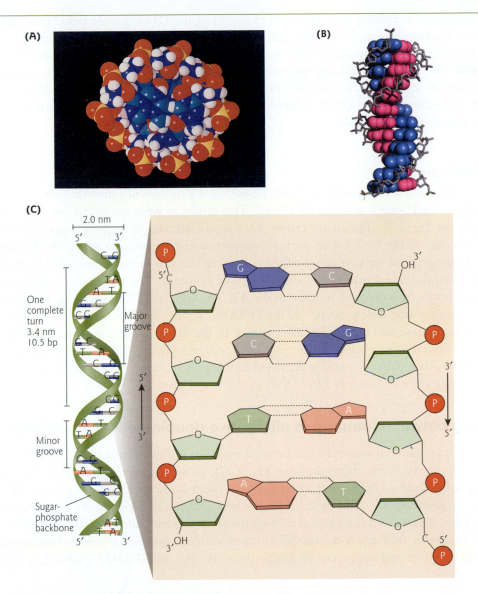

Figure 2.7 Various representations of the DNA double helix. (A) The helix axis is most apparent from a view directly down the axis. The sugar–phosphate backbone is on the outside of the helix where the polar phosphate groups (red and yellow) can interact with the polar environment. The nitrogen-containing bases (blue) are inside, stacking perpendicular to the helix axis (Reprinted with permission of the University of California, Lawrence Livermore National Laboratory, and the Department of Energy. Image courtesy of Nelson L. Max.) (B) This picture shows how DNA bases are stacked together along the double helix. The atoms of the sugar–phosphate backbone are shown in gray, and the ultraviolet-absorbing portions of DNA – the bases – are shown in blue or magenta, depending on which side of the DNA strand they are located. (Image courtesy of Bern Kohler, The Ohio State University.) (C) The ribbon-like strands represent the sugar–phosphate backbones, and the horizontal rungs are the nitrogenous base pairs, of which there are 10.5 per complete turn. The major and minor grooves are apparent. The inset highlights the antiparallel nature of the two strands of the helix. The DNA sequence shown reads 5′-GCTA-3′.

(Fig. 2.7). The ladder terminology that is often used to describe DNA structure implies that there are open spaces between successive "rungs" of base pairs. A better analogy would be a stack of coins, since the bases are stacked right on top of each other. The genetic code is the sequence of bases, which varies from one DNA molecule to another.

As noted above, there is polarity in each strand ($5' \rightarrow 3'$) of the DNA double helix: one end of a DNA strand will have a $5'$-phosphate and the other end will have a $3'$-hydroxyl group. Watson and Crick found that hydrogen bonding could only occur if the polarity of the two strands ran in opposite directions. Thus, the two strands of the DNA double helix are antiparallel ($5' \rightarrow 3'$ and $3' \rightarrow 5'$). This structural feature has important implications for the mechanism of DNA replication (see Section 6.5). The DNA double helix is also referred to as double-stranded DNA (dsDNA) or duplex DNA to distinguish it from the single-stranded DNA (ssDNA) found in some viruses.

Major and minor grooves

The two bonds that attach a base pair to its deoxyribose sugar rings are not directly opposite. Therefore, the sugar–phosphate backbone is not equally spaced. This results in what are called the major and minor grooves of DNA (see Fig. 2.7). The major groove has a significant role in sequence-specific DNA–protein interactions. The edges of the base-paired purines and pyrimidines are solvent accessible. In particular, the solvent-exposed nitrogen and oxygen atoms of the bases that line the major grooves of DNA can make hydrogen bonds with the side chains of the amino acids of a protein. The pattern of these hydrogen-bonding groups (whether donors or acceptors) is different for AT, TA, GC, and CG base pairs. Thus, the major groove carries a message (the base sequence of the DNA) in a form that can be read by DNA-binding proteins. The minor groove of DNA is less informative. In the minor groove, the hydrogen bonding patterns are the same regardless of which way the base pair is flipped, and there is only one difference in the pattern between AT and GC base pairs. As will be discussed in Chapter 11, most transcription factors (proteins involved in regulating gene expression) bind DNA in the major groove.

Distinguishing between features of alternative double-helical structures

Oligonucleotides of any desired sequence can be synthesized in sufficient quantity and pure enough to be studied by single-crystal X-ray diffraction (see Fig. 9.17). However, even with highly pure material, X-ray crystallography is still a challenging technique. It is often likened to gardening by scientists – some researchers have a "green thumb" and others do not when it comes to growing crystals successfully. Structures have been determined for a number of synthetic oligonucleotides (usually 4–24 self-complementary bases that fold into a double helix) under various conditions. Studies have revealed the basic structure of three fundamental types of double helix: B-, A-, and Z-DNA (Table 2.2, Fig. 2.8).

Table 2.2 Alternative forms of the DNA double helix.*

	A-DNA	B-DNA	Z-DNA
Orientation	Right-handed	Right-handed	Left-handed
Major groove	Deep and narrow	Moderate depth, wide	Very shallow, virtually nonexistent, sometimes called a "single groove"
Minor groove	Shallow and broad (superficial)	Moderate depth, narrow	Very deep and narrow
Bases/turn	11	10.5	12
Conditions	Low humidity (75%), high salt	High humidity (95%), low salt	High $MgCl_2$ (> 3 M), NaCl, or ethanol. In the presence of methylated cytosine: high humidity and low salt

* Other forms of DNA have been crystallized, including B′, C, C′, C″, D, E, and T. All of these are right-handed structures, and occur under unique conditions. For example, C-DNA forms in the presence of lithium salts and low humidity.

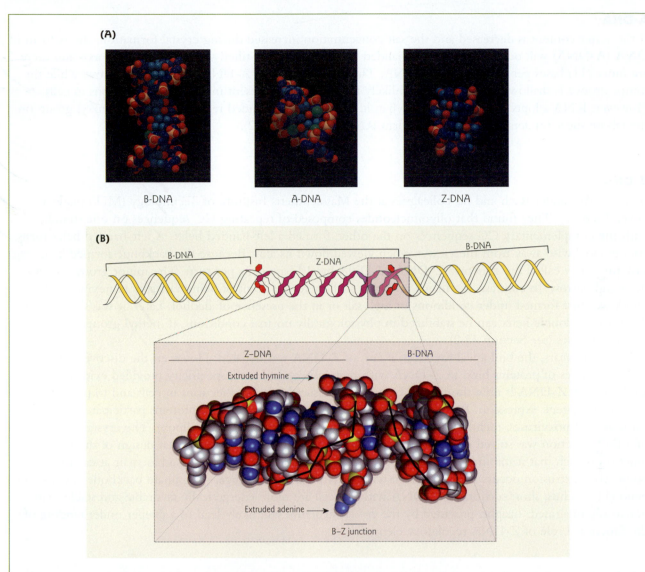

Figure 2.8 Alternative double-helical structures of DNA. (A) Three types of DNA double helix are displayed in these space-filling models: B-DNA (Watson–Crick DNA), A-DNA, and Z-DNA. (Reproduced from Dickerson, R.E. 1983. The DNA helix and how it is read. *Scientific American* 249:94–111, with permission of the University of California, Lawrence Livermore National Laboratory, and the Department of Energy. Images courtesy of Nelson L. Max. (B) The structure of the B–Z junction. A region of Z-DNA is connected to B-DNA through a junction in which one base pair is flipped out, or extruded, from the DNA helix. (Reprinted by permission from Nature Publishing Group and Macmillan Publishers Ltd: Sinden, R.R. 2005. DNA twists and flips *Nature* 437:1097–1098. Copyright © 2005.)

B-DNA (Watson–Crick DNA)

B-DNA is a right-handed helix; it turns in a clockwise manner when viewed down its axis. The bases are stacked almost exactly perpendicular to the main axis with 10.5 bases per turn. The major groove is wide and of moderate depth, while the minor groove is of moderate depth but is much narrower. B-DNA occurs under conditions of high humidity (95%) and relatively low salt. Since the inside of a cell is mostly water with relatively low salt concentration, it follows that the predominant form *in vivo* is B-DNA.

A-DNA

If the water content is decreased and the salt concentration increased during crystal formation, the A form of DNA (A-DNA) will occur. In this right-handed helix the bases are tilted with respect to the axis and there are more (11) bases per turn than in B-DNA. The major groove of A-DNA is deep and narrow, while the minor groove is shallow and broad. It is unlikely that A-DNA is present in any lengthy sections in cells. However, RNA adopts an A-form helix when it forms double-stranded regions. The 2′-hydroxyl group on the ribose sugar hinders formation of B-form RNA (see Section 4.2).

Z-DNA

In 1979, Alexander Rich and his colleagues at the Massachusetts Institute of Technology (MIT) made a novel discovery. They found that oligonucleotides composed of repeating GC sequences on one strand, with the complementary CG sequences on the other, formed a left-handed helix. A left-handed helix turns counterclockwise away from the viewer when viewed down its axis. Because the backbone formed a zig-zag structure, they called the structure Z-DNA. Z-DNA has 12 base pairs per turn. The minor groove is very deep and narrow. In contrast, the major groove is shallow to the point of being virtually nonexistent. Z-DNA was first formed under conditions of high salt or in the presence of alcohol. Later, it was shown that this form of double helix can be stabilized in physiologically normal conditions, if methyl groups are added to the cytosines (see Section 12.2).

For many years, the biological function, if any, of Z-DNA was debated. However, the discovery that certain families of proteins bind to Z-DNA with high affinity and great specificity provided evidence for a role *in vivo*. Z-DNA is now thought to be present transiently in short sections in cells and to play a role in regulating gene expression. Recent data show that some Z-DNA-binding proteins participate in the pathology of poxviruses, including vaccinia virus and variola (the agent of smallpox). The crystal structure of a B–Z junction was solved in 2005 (see Fig. 2.8). What came as a surprise is that design of the B–Z junction is such that it minimizes structural distortion of the helix, by extruding a base pair at each junction point. Base extrusion occurs in conjunction with a sharp turn in the sugar–phosphate backbone and a slight bend (11°). Thus, short sections of Z-DNA within a cell are more energetically favorable and stable than previously imagined. Insights provided by the crystal structure will likely lead to a deeper understanding of the functional role of Z-DNA regulatory elements.

DNA can undergo reversible strand separation

The replica of each strand of DNA has the base sequence of its complementary strand, and from one strand, the other can be made. This important characteristic of the molecule allows for the fidelity of DNA replication, transcription (making an RNA copy of the DNA), and translation (decoding the RNA message to make a protein). During DNA replication and transcription, the strands of the helix must separate transiently and reversibly.

The same feature that allows DNA to fulfill these biological roles also makes it possible to manipulate DNA *in vitro*. The unwinding and separation of DNA strands, referred to as denaturation or "melting," can be induced in the laboratory. The hydrogen bonds can be broken and the DNA strands separated by heating the DNA molecule, whereas the phosphodiester bonds remain intact (Fig. 2.9). A point is reached in which the thermal agitation overcomes the hydrogen bonds, hydrophobic interactions, and other forces that stabilize the double helix, and the molecule "melts." This strand separation of DNA changes its absorption of ultraviolet (UV) light in the 260 nm range. Native double-stranded DNA absorbs less light at 260 nm by about 40% than does the equivalent amount of single-stranded DNA. Thus, as DNA denatures, its absorption of UV light increases, a phenomenon known as "hyperchromicity." In contrast, base stacking in duplex DNA quenches the capacity of the bases to absorb UV light. The temperature at which half the bases in a double-stranded DNA sample have denatured is denoted the melting temperature (T_m) (Fig. 2.10). Near the denaturation

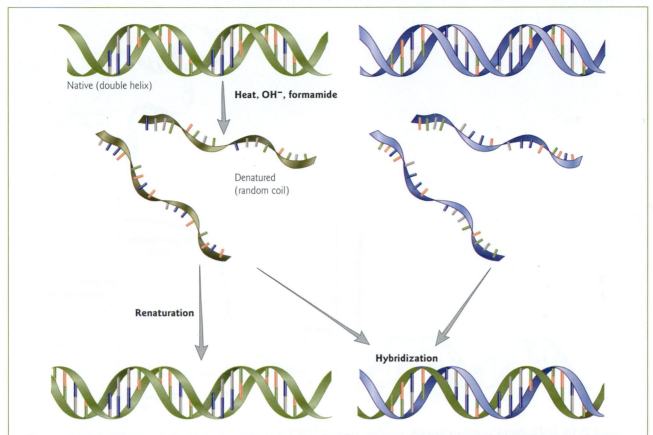

Figure 2.9 The denaturation, renaturation, and hybridization of double-stranded DNA molecules.
DNA is denatured to separate the two strands. The denatured DNA molecules are allowed to renature (anneal) by incubation just below the melting temperature. Alternatively, denatured complementary DNA from two different sources can be hybridized.

temperature, a small increase in temperature causes an abrupt loss of the multiple, weak interactions holding the two strands together, so that denaturation occurs rapidly along the entire length of the DNA.

The G + C content of a DNA molecule has a significant effect on its T_m. Since a GC base pair has three hydrogen bonds to every two in an AT base pair, the higher the GC content in a given molecule of DNA, the higher the temperature required to denature the DNA (Fig. 2.11). More importantly, the stacking interactions of GC bases pairs with neighboring base pairs are more favorable energetically than interactions of AT base pairs with their adjacent base pairs.

In addition to heat, other methods can be used to denature DNA. Lowering the salt concentration of a DNA solution promotes denaturation by removing the cations that shield the negative charges on the two strands from each other. At low ionic strength, the mutually repulsive forces of these negative charges from the backbone phosphoryl groups are enough to denature the DNA, even at a relatively low temperature. In addition, high pH or organic solvents such as formamide disrupt the hydrogen bonding between DNA strands and promote denaturation.

When heated solutions of denatured DNA are slowly cooled, single strands often meet their complementary strands and form a new double helix. This is called "renaturation" or "annealing." The capacity to renature denatured DNA molecules permits hybridization – the complementary base pairing of strands from two different sources (see Fig. 2.9). These important principles will be returned to in Chapter 8 when we discuss the use of denaturation, renaturation, and hybridization as tools for molecular biology research.

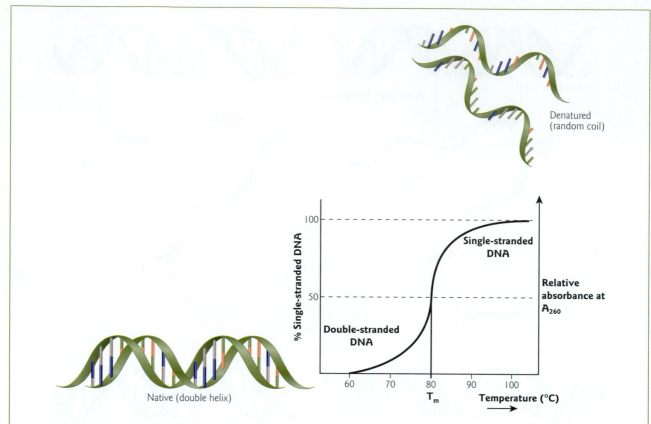

Figure 2.10 DNA denaturation curve. Double-stranded DNA is heated and its melting is measured by the increase in absorbance at 260 nm. The point at which 50% of the DNA is single-stranded is the melting temperature, or T_m. In this example, the T_m is about 80°C.

2.7 Unusual DNA secondary structures

Originally, DNA was thought of as a static, linear string of genetic information. Since the mid–1960s, there has been a rapid expansion in awareness of its heterogeneity and flexibility in form. Through dynamic changes in secondary and tertiary structures, the DNA molecule can regulate expression of its linear sequence information. Unusual secondary structures, such as slipped structures, cruciforms, and triple helix DNA are generally sequence-specific (Fig. 2.12). Some of these may be dependent on DNA supercoiling (see below). Supercoiling provides the necessary driving energy for their formation, due to the release of torsional strain.

Slipped structures

Slipped structures have been postulated to occur at tandem repeats. A tandem repeat (sometimes called a direct repeat) in DNA is two or more adjacent, approximate copies of a pattern of nucleotides, arranged in a head to tail fashion. For example, the sequence 5′-TACGTACGTACGTACG-3′ contains four tandem repeats of "TACG" (Fig. 2.12A).

Slipped structures are found upstream of regulatory sequences (e.g. for gene transcription) *in vitro*. They were characterized by using enzymes that cut phosphodiester bonds in single-stranded DNA but not in double-stranded DNA. It is possible that they have importance for DNA–protein interactions. In addition, there are a number of hereditary neurological diseases caused by the expansion of simple triplet repeat sequences in either coding or noncoding regions (Disease box 2.1; see also Section 7.2). The triplet repeats

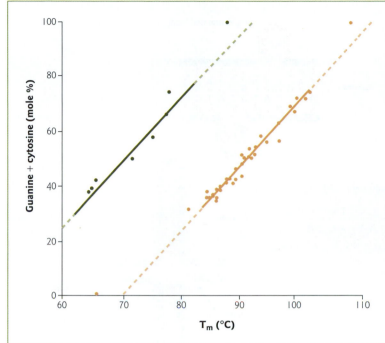

Figure 2.11 Dependence of DNA denaturation on G + C content and on salt concentration. DNA from many different sources was dissolved in solutions of low (green line) and high (orange line) concentrations of salt at pH 7.0. The points represent the temperature at which the DNA samples melted, graphed against their G + C content. (Adapted from Marmur, J. and Doty, P. 1962. Determination of the base composition of deoxyribonucleic acid from its thermal denaturation temperature. *Journal of Molecular Biology* 5:109–118, Copyright © 1962, with permission from Elsevier.)

cause DNA to assume unusual DNA secondary structures, which, in turn influence replication and transcription by blocking replication forks and promoting repair. The formation of DNA slipped structures within the long repeating CTG, GGC, and GAA strands, compared with their complementary strands, plays an important role in their expansion.

Cruciform structures

The formation of short bubbles of unpaired single-stranded DNA from negative supercoiling (see below) can be stabilized by cruciform structures. Cruciform structures are paired stem-loop formations (see Fig. 2.12B).

Friedreich's ataxia and triple helix DNA   DISEASE BOX 2.1

Friedreich's ataxia is a rare inherited neurological disease characterized by the progressive loss of voluntary muscular coordination (ataxia) and heart enlargement. Named after the German doctor, Nikolaus Friedreich, who first described the disease in 1863, Friedreich's ataxia is generally diagnosed in childhood and affects both males and females. The disorder is caused by a 5′-GAA-3′ trinucleotide repeat expansion in the first intron of the Friedreich's ataxia gene, which is located on chromosome 9. A normal individual has 8–30 copies of this trinucleotide repeat, while Friedreich's ataxia patients have as many as 1000. The larger the number of repeat copies, the earlier the onset of the disease and the quicker the decline of the patient.

This expanded GAA tract is well known to adopt the triplex conformation. The formation of the triple helix is involved in the inhibition of transcription of the Friedrich's ataxia gene and a corresponding reduction in the amount of the frataxin protein. Frataxin is found in the mitochondria of humans. While the precise role of human frataxin remains to be determined, the protein appears to be involved in regulating the export and/or import of iron into mitochondria.

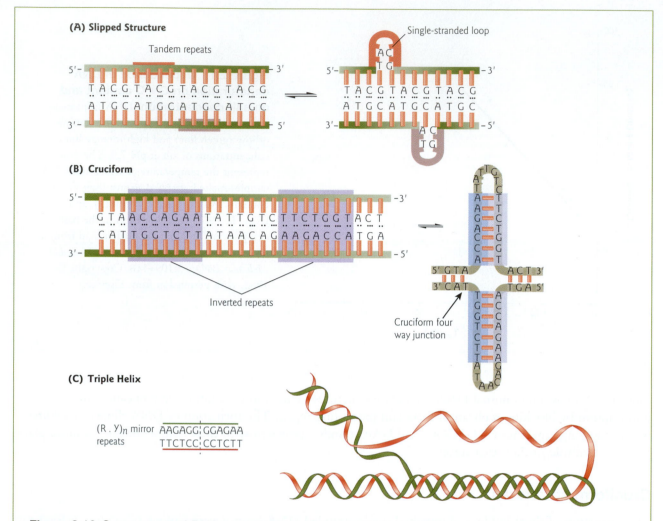

(A) Slipped Structure

Tandem repeats

Single-stranded loop

(B) Cruciform

Inverted repeats

Cruciform four way junction

(C) Triple Helix

(R · Y)ₙ mirror repeats

AAGAGG GGAGAA
TTCTCC CCTCTT

Figure 2.12 Some unusual DNA secondary structures. (A) Slipped structure with compensating loops in alternate strands at a tandem repeat. (B) Cruciform with paired stem-loop formation at an inverted repeat. (C) Triple helix DNA at a purine–pyrimidine (R · Y) stretch containing a mirror repeat symmetry.

They have been characterized *in vitro* for many inverted repeats in plasmids (small circular DNA) and bacteriophages. Inverted repeats are base sequences of identical composition on the complementary strands. They read exactly the same from 5′ → 3′ on each strand (in other words, the sequence reads the same from left to right as from right to left). Sometimes inverted repeats are referred to as "palindromes" because of their similarity to a word or phase that reads identically when spelled backward, such as "rotor" or "nurses run." Cruciform structures have been visualized by electron microscopy. The DNA becomes rearranged so each repeat pairs with the complementary sequence on its own strand of DNA, instead of with the complement on the other strand. Experimental evidence has led to the hypothesis that cruciform structures can act as regulatory elements in DNA replication and gene expression in various prokaryotic and eukaryotic systems. Confirmation of a functional role *in vivo* awaits further investigation.

Triple helix DNA

In a triple helix, a third strand of DNA joins the first two to form triplex DNA. Triple helix DNA occurs at purine–pyrimidine stretches in DNA and is favored by sequences containing a mirror repeat symmetry (see Fig. 2.12C). In this type of repeat, the sequences on each side of the axis of symmetry are mirror images of each other. The participating third strand may originate from within a single purine–pyrimidine tract (intramolecular triplex) or from a separate sequence (intermolecular triplex).

The purine strand of the Watson–Crick duplex associates with the third strand through Hoogsteen hydrogen bonds in the major groove. The original duplex structure is maintained in a B-like conformation. Hoogsteen AT and GC base pairs (named after their discoverer Karst Hoogsteen) have altered patterns of hydrogen bonding compared with Watson–Crick base pairs (Fig. 2.13; compare with Fig. 2.4). In the Hoogsteen AT pair, the adenine base is rotated through 180° about the bond to the sugar, and the Hoogsteen GC pair only forms two hydrogen bonds, compared with three in the Watson–Crick GC pair. Hoogsteen GC base pairs are not stable at the neutral pH of cells (pH 7–8). One of the nitrogens on the cytosine must have a hydrogen added to it for this type of base pair to form, and this protonation requires a lower pH (pH 4–5). Hoogsteen base pairs have gained importance recently because they are occasionally found in complexes of DNA with anticancer drugs and they show up in triple helices associated with genetic disease (Disease box 2.1).

2.8 Tertiary structure of DNA

Many naturally occurring DNA molecules are circular, with no free 5′ or 3′ end. Due to the polarity of the strands of the DNA double helix, the 5′ end of one strand can only join its own 3′ end to covalently close a circle. Thus, circular, double-stranded DNA is essentially two circles of single-stranded DNA twisted around

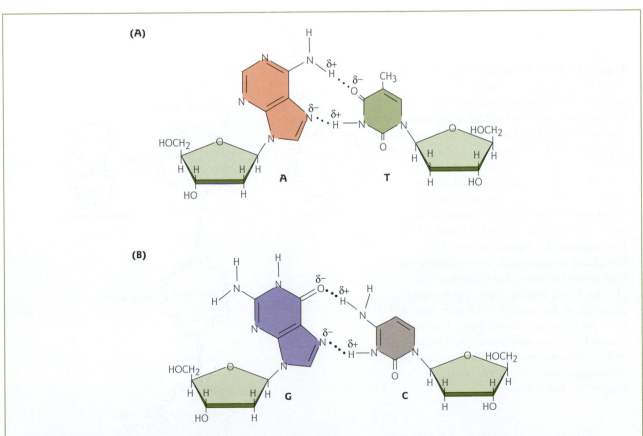

Figure 2.13 Hoogsteen base pairs. (A) AT, and (B) GC pairs. Symbols + and − are partial charges. Hydrogen bonds are represented by dotted lines. For simplicity, not all hydrogens are shown.

each other. Such circular DNA molecules often become overwound or underwound, with respect to the number of complete turns of the DNA double helix. This DNA can then become supercoiled (under torsional stress). Supercoils are a twisted, three-dimensional structure which is more favorable energetically. Supercoiling of DNA can be visualized by electron microscopy and occurs both *in vivo* and *in vitro*.

Supercoiling of DNA

Consider a double-stranded linear DNA molecule of 10 complete turns (or twists, T = 10) with 10.5 bp/turn (Fig. 2.14). If the ends of the DNA molecule are sealed together, the result is an energetically relaxed circle that lies flat. Since each chain is seen to cross the other 10 times, this relaxed circle has a linking number (L) of 10. But, if the double helix is underwound by one full turn to the left and then the ends are sealed together, the result is a strained circle with 11.67 bp/turn, where L = 9 and T = 9. Once the two ends become covalently linked, the linking number cannot change. Generally, changes in the average number of base pairs per turn of the double helix will be counteracted by the formation of an appropriate number of supercoils in the opposite direction. In this example, one negative (left-handed) supercoil is introduced spontaneously, re-establishing the total number of original "turns" of the helix (T = 10, L = 9). Overtwisting of the double helix usually leads to positive (right-handed) supercoiling. For example, if the double helix is overwound by one full turn to the right and then the ends are sealed together, the result is a

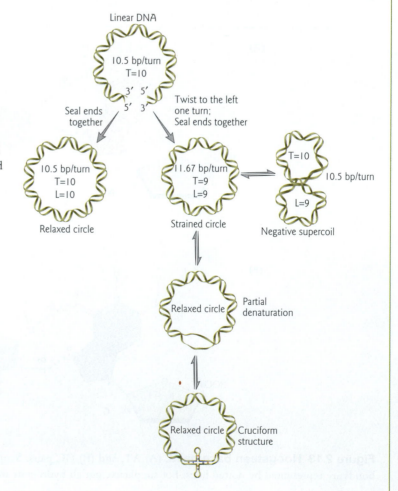

Figure 2.14 DNA supercoiling. A linear DNA molecule of 10 complete turns (or twists, T = 10) is assumed to have 10.5 bp/turn, as is usual for B-DNA in solution. The ends of the DNA molecule can be sealed together to make a relaxed circle. This relaxed circle has a linking number of L = 10. But, if the double helix is underwound by one full turn to the left and then the ends are sealed together, the result is a strained circle with 11.67 bp/turn, where L = 9 and T = 9. One negative (left-handed) supercoil is introduced spontaneously, re-establishing the total number of original turns of the helix (T = 10, L = 9, 10.5 bp/turn). Upon partial denaturation, supercoiled DNA may convert to its unsupercoiled relaxed form. The single-stranded area may then convert to a more stable cruciform structure if there are inverted homologous sequences in the denatured regions (see Fig. 2.12).

strained circle with 9.5 bp/turn, where L = 11, T = 11. The introduction of one positive (right-handed) supercoil restores the total number of original turns of the helix (T = 10, L = 11).

The supercoiled state is inherently less stable than relaxed DNA. The stress present within supercoiled DNA molecules sometimes leads to localized denaturation, in which the complementary strands come apart in a short section. This has important implications for cellular processes such as replication and transcription.

Topoisomerases relax supercoiled DNA

Forms of DNA that have the same sequence yet differ in their linkage number are referred to as topological isomers (topoisomers). Topoisomers can be visualized by their differing mobilities when separated by gel electrophoresis (Fig. 2.15) (see Tool box 8.6 for methods). Topoisomerases are highly conserved enzymes that convert (isomerize) one topoisomer of DNA to another. They do so by changing the linking number (L). DNA topoisomerases fall into two major categories, type I and type II. The two types can be further subdivided into four subfamilies: IA, IB, IIA, and IIB. So far, at least five different topoisomerases have been reported to be present in higher eukaryotes, including humans (Table 2.3). The first topoisomerase was

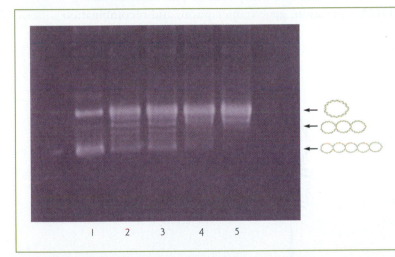

Figure 2.15 Relaxation of supercoiled plasmid DNA by topoisomerase I.
Lane 1: Relaxed and supercoiled topoisomers of pUC18 plasmid DNA. Lanes 2–6: pUC18 plasmid DNA after treatment with 2, 4, 6, or 8 units of topoisomerase I, respectively, for 45 minutes at 37°C. DNA samples were separated by agarose gel electrophoresis and visualized by staining with ethidium bromide (see Tool box 8.6 for methods). The speed with which the DNA molecules migrate increases as the number of superhelical turns increases.

Table 2.3 Human DNA topoisomerases.

DNA topoisomerase	Type	DNA cleavage	Structural role	Function
I	IB	ssb	Relax both negatively and positively supercoiled DNA	Replication Transcription Recombination
IIIα	IA	ssb	Relax only negatively supercoiled DNA	Recombination Transcription of ribosomal RNA genes
IIIβ	IA	ssb	Relax only negatively supercoiled DNA	Recombination
IIα	IIA	dsb	Relax both positively and negatively supercoiled DNA Facilitate unknotting or decatenation of entangled DNA	Chromosome condensation Chromosome segregation Replication
IIβ	IIA	dsb	Relax both positively and negatively supercoiled DNA Facilitate unknotting or decatenation of entangled DNA	Not well defined

dsb, double-stranded break in DNA; ssb, single-stranded break in DNA.

discovered by Jim Wang, while he was looking for the enzyme that relieves the supercoils that form during DNA replication (see Section 6.6).

Type I topoisomerases are proficient at relaxing supercoiled DNA. They do not require the energy of ATP. Type IA can only relax negative supercoils, while type IB can relax both negative and positive supercoils. They act by forming a transient single-stranded break in the DNA (cleavage of a phosphodiester bond between adjacent nucleotides) and, while winding the broken ends, pass the other strand through the break. Topoisomerases do not create free ends, but instead become themselves covalently attached to one of the two broken ends of the DNA (which one depends on the specific enzyme) (Fig. 2.16). The nick is then sealed, after relaxation increases the linking number by one.

Type II topoisomerases are usually ATP-dependent. They form transient double-stranded breaks in the double helix and pass another double helix through the temporary gap or DNA-linked "protein gate." They are proficient in relaxing both negatively and positively supercoiled DNA and, in addition, can unknot or decatenate entangled DNA molecules (catenation is the interlocking of circles of DNA like links in a chain). Prokaryotic topoisomerase II (sometimes called gyrase) has the special property of being able to introduce negative supercoils.

Both type I and type II topoisomerases play important roles in many cellular processes, including chromosome condensation and segregation, DNA replication, gene transcription, and recombination (Disease box 2.2).

Figure 2.16 Mechanism of action of a type I topoisomerase. The enzyme binds to a circular DNA molecule with one negative supercoil (see Fig. 2.14) and unwinds the double helix. It nicks one strand and prevents free rotation of the helix by remaining bound to each broken end. The 5′ broken end is covalently attached to the amino acid tyrosine (see inset), and the 3′ end is noncovalently bound to another region of the enzyme. The enzyme passes the unbroken strand of DNA through the break and ligates the cut ends, thereby increasing the linking number of the DNA by one. The enzyme falls away and the strands renature, leaving a relaxed circle. Inset: in the strand breakage reaction by the topoisomerase, the oxygen of the tyrosine hydroxyl group in the active site of the enzyme attacks a DNA phosphorus, forming a covalent phosphotyrosine link between the DNA and the enzyme, and breaking a DNA phosphodiester bond at the same time. Rejoining of the DNA strand occurs by the reverse. The oxygen of the free DNA 3′-OH group attacks the phosphorus of the phosphotyrosine link, breaking the covalent bond between the protein and DNA, and reforming the phosphodiester bond between adjacent nucleotides in the DNA chain.

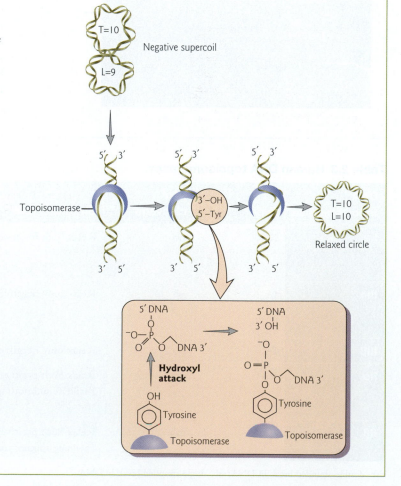

Topoisomerase-targeted anticancer drugs

Topoisomerases are the targets for a number of important anticancer drugs. Topoisomerase II-targeted drugs are used in approximately half of all chemotherapy regimens. Topoisomerase I is the target for a number of promising agents, including camptothecin, a drug derived from the bark of *Camptotheca acuminata*, a Chinese yew tree. Camptothecin was first isolated and found to kill certain types of cancer cells in 1966 by Dr Monroe E. Wall and Dr Mansukh C. Wani. It was not until 1985 that another group of researchers determined that camptothecin is a topoisomerase I poison. Now, camptothecin analogs are being used for the treatment of ovarian and colon cancer.

Topoisomerase-targeted anticancer drugs act in one of two ways, either as an inhibitor of at least one step in the catalytic cycle, or as poisons. Breaks created by topoisomerases are normally transient in nature (see Fig. 2.16). When the anticancer drug acts as a poison, the broken DNA is trapped as a stable intermediate bound to topoisomerase. There is a high likelihood of this being converted to a permanent break in the genome.

These drugs preferentially target cancer cells because cancer cells are rapidly growing and usually contain higher levels of topoisomerases than slower growing cells. Due to their mode of action, the higher the cellular level of topisomerases, the more lethal these drugs become. In addition, cancer cells often have impaired DNA repair pathways, so they are more susceptible to the effects of DNA-damaging agents. Like all chemotherapeutic agents, drugs targeted to topoisomerases also affect normal fast growing cells, such as white blood cells, the lining of the gastointestinal tract, and hair follicles. This is why cancer patients undergoing chemotherapy often experience nausea, diarrhea, and hair loss, and are susceptible to infections.

What is the significance of supercoiling *in vivo*?

Virtually all DNA within both prokaryotic and eukaryotic cells exists in the negative supercoiled state. If these domains are unrestrained (not supercoiled around DNA-binding proteins), there is an equilibrium between tension and unwinding of the helix. If the supercoils are restrained with proteins, they are stabilized by the energy of interaction between the proteins and the DNA. Experimental evidence suggests that DNA supercoiling plays an important role in many genetic processes, such as replication, transcription, and recombination.

Negative supercoiling puts energy into DNA. Underwinding makes it easier to pull the two strands of the double helix apart. Therefore, negative supercoiling makes it easier to open replication origins and gene promoters. The potential energy in the supercoils also promotes formation of unusual DNA secondary structures, like cruciforms (see Figs 2.12 and 2.14). In addition, it is possible that a B-DNA → Z-DNA transition is triggered by increased negative supercoiling. This is because switching a portion of the DNA from a right-handed to left-handed helix releases the strain imposed by the negative supercoils, since the twist (base pairs per turn) in a portion of the DNA has been reversed. Positive supercoiling occurs ahead of replication forks and transcription complexes. Positive supercoiling makes it much harder to open the double helix and therefore blocks essential DNA processes.

Supercoiled DNA is a well-characterized feature of the circular genome of some small viruses, for example, bacteriophage PM2. The genome of this bacteriophage can exist as a relaxed circle (covalent closure) or a supercoiled circle, which has a twisted appearance. This is the native form *in vivo* (Fig. 2.17). In bacteriophages it has been shown that relaxed circular DNA correlates with reduced activity in replication and transcription, whereas negative supercoiling leads to increased activity in replication and transcription. Similarly, bacteria have large, circular genomes that can form independent DNA loop domains. A protein complex holds each domain in place to form "subcircles" with an average size of ~40 kb. These domains may form supercoiled structures of importance for replication and transcription. Supercoiling also greatly facilitates chromosome condensation in bacteria.

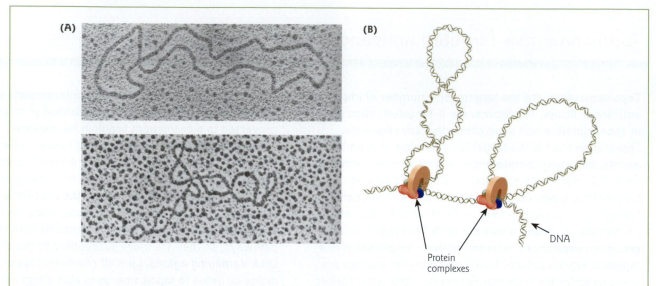

Figure 2.17 Supercoiling occurs in nature. (A) The DNA of the bacteriophage PM2 in two topological forms: relaxed circle (upper panel) and supercoiled (lower panel). The latter is the native form. (Reproduced with permission from Wang, J.C. 1982. DNA topoisomerases. *Scientific American* 247:97.) (B) Schematic representation of DNA loop domains (subcircles) in bacterial (circular genome) or eukaryotic (linear) genomic DNA.

While chromosomes in eukaryotes are not usually circular, supercoils are made possible when sections of linear DNA are embedded in a lattice of proteins associated with the chromatin. This association can create anchored ends that form independent loop domains, as described above for bacteria. The significance of supercoiling in eukaryotes is not as readily apparent as in bacteriophages or bacteria. There are some examples of increased transcription with negative supercoiling, but it is difficult to study conclusively because eukaryotes have such large, complex genomes. In addition, because replication and transcription processes themselves generate DNA supercoiling (see Fig. 6.7 and Focus box 10.1), figuring out the mechanism by which DNA supercoiling modulates genetic processes poses a challenge.

Chapter summary

DNA and RNA are chain-like molecules composed of subunits called nucleotides joined by phosphodiester bonds. Each nucleotide subunit is composed of three parts: a five-carbon sugar, a phosphate group, and a nitrogenous base. Natural RNAs comes in sizes ranging from less than one hundred to many thousands of nucleotides, while DNA can be as long as several kilobases to thousands of megabases. The 5′-PO_4 and 3′-OH ends of a DNA or RNA chain are distinct and have different chemical properties.

DNA has a double-helical structure with sugar–phosphate backbones on the outside and base pairs on the inside. The predominant form in cells is B-DNA, a right-handed helix with 10.5 bases per turn. The double helix is stabilized by hydrogen bonds between base pairs and base stacking by hydrophobic interactions. The bases pair in a specific way: adenine (A) with thymine (T) and guanine (G) with cytosine (C). The G + C content of a natural DNA can vary from 22 to 73%, and this can have a strong effect on the physical properties of DNA, particularly its melting temperature. The melting temperature (T_m) of a DNA is the temperature at which the two strands are 50% denatured. Separated DNA strands can be induced to renature, or anneal. Complementary strands from different sources can form a double helix in a process called hybridization.

Repetitive nucleotide sequences in the DNA double helix sometimes adopt unusual secondary structures, such as slipped structures, cruciforms, and triple helix DNA. These secondary structures can affect DNA replication and transcription. Some of these structures are dependent on DNA supercoiling. Almost all DNA within both prokaryotic and eukaryotic cells exists in the negative supercoiled state. Negative supercoiling makes it easier to separate the DNA strands during replication and transcription. Forms of DNA that have the same sequence yet differ in their linking number are referred to as topoisomers. Topoisomerases are enzymes that relax supercoiled DNA and convert one topoisomer of DNA to another by changing the linking number. Topoisomerases play important roles in DNA replication and gene transcription.

Analytical questions

1 The DNA duplexes below are denatured and then allowed to reanneal. Which of the two molecules would have the highest T_m? Which of the two is least likely to reform the original structure? Why?

  (a) 5′-ATATCATATGATATGTA-3′
  3′-TATAGTATACTATACAT-5′

  (b) 5′-CGGTACTCGTGCAGGT-3′
  3′-GCCATGAGCACGTCCA-5′

2 You are studying a protein that you suspect has DNA topoisomerase activity. Describe how you would test the protein for this activity *in vitro*. Show sample results.

3 When the base composition of DNA from a grasshopper was determined, 29% of the bases were found to be adenine.

  (a) What is the percentage of cytosine?

  (b) What is the entire base composition of the DNA?

  (c) What is the [G] + [C] content?

Suggestions for further reading

Bacolla, A., Wells, R.D. (2004) Non-B DNA conformations, genomic rearrangements, and human disease. *Journal of Biological Chemistry* 279:47411–47414.

Blackburn, G.M., Gait, M.J. (1990) *Nucleic Acids in Chemistry and Biology*. IRL Press, Oxford.

Calladine, C.R., Drew, H.R., Luisi, B.F., Travers, A.A. (2004) *Understanding DNA. The Molecule and How it Works*, 3rd edn. Elsevier Academic Press, New York.

Chargaff, E. (1951) Structure and function of nucleic acids as cell constituents. *Federation Proceedings* 10:654–659.

Cortés, F., Pastor, N., Mateos, S., Domínguez, I. (2003) Roles of DNA topoisomerases in chromosome segregation and mitosis. *Mutation Research* 543:59–66.

Dickerson, R.E. (1983) The DNA helix and how it is read. *Scientific American* 249:94–111.

Dickerson, R.E., Bansal, M., Calladine, C.R. et al. (1989) Definitions and nomenclature of nucleic acid structure and parameters. *EMBO Journal* 8:1–4.

Ha, S.C., Lowenhaupt, K., Rich, A., Kim, Y.G., Kim, K.K. (2005) Crystal structure of a junction between B-DNA and Z-DNA reveals two extruded bases. *Nature* 437:1183–1186.

Herbert, A.G., Spitzner, J.R., Lowenhaupt, K., Rich, A. (1993) Z-DNA binding protein from chicken blood nuclei. *Proceedings of the National Academy of Sciences USA* 90:3339–3342.

Kim, Y.G., Lowenhaupt, K., Oh, D.B., Kim, K.K., Rich, A. (2004) Evidence that vaccinia virulence factor E3L binds Z-DNA *in vivo*: implications for development of a therapy for poxvirus infection. *Proceedings of the National Academy of Sciences USA* 101:1514–1518.

Kool, E.T. (2001) Hydrogen bonding, base stacking, and steric effects in DNA replication. *Annual Review of Biophysical and Biomolecular Structure* 30:1–22.

Larsen, A.K., Escargueil, A.E., Skladanowski, A. (2003) Catalytic topoisomerase II inhibitors in cancer therapy. *Pharmacology and Therapeutics* 99:167–181.

Marmur, J., Doty, P. (1962) Determination of the base composition of deoxyribonucleic acid from its thermal denaturation temperature. *Journal of Molecular Biology* 5:109–118.

Oberlies, N.H., Kroll, D.J. (2004) Camptothecin and taxol: historic achievements in natural products research. *Journal of Natural Products* 67:129–135.

Oussatcheva, E.A., Pavlicek, J., Sankey, O.F., Sinden, R.R., Lyubchenko, Y.L., Potaman, V.N. (2004) Influence of global DNA topology on cruciform formation in supercoiled DNA. *Journal of Molecular Biology* 338:735–743.

Sinden, R.R. (2005) DNA twists and flips. *Nature* 437:1097–1098.

Wang, J.C. (2002) Cellular roles of DNA topoisomerases: a molecular perspective. *Nature Reviews: Molecular and Cellular Biology* 3:430–440.

Wittig, B. Wölfl, S., Dorbic, T., Vahrson, W., Rich, A. (1992) Transcription of human c-myc in permeabilized nuclei is associated with formation of Z-DNA in three discrete regions of the gene. *EMBO Journal* 11:4653–4663.

Chapter 3

Genome organization: from nucleotides to chromatin

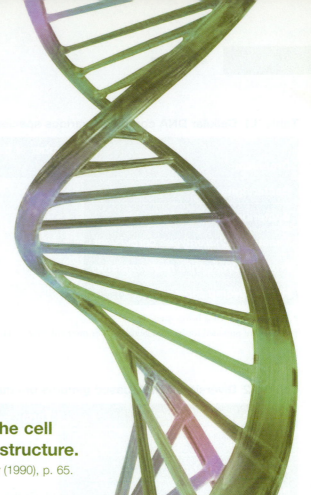

The way in which eukaryotic DNA is packaged in the cell nucleus is one of the wonders of macromolecular structure.

G. Michael Blackburn, *Nucleic Acids in Chemistry and Biology* (1990), p. 65.

Outline

3.1 Introduction

In this chapter we will compare and contrast the higher order organization of genomes that will be encountered frequently throughout the rest of this textbook. As such, this chapter should serve as a reference point for review as needed. The emphasis is on the eukaryotic genome, but bacterial genomes, plasmid DNA, mammalian DNA viruses and bacteriophages, organelle genomes, and RNA-based genomes are also described. The diversity of mechanisms for packaging very long molecules of DNA into very small cellular spaces is truly remarkable (Tables 3.1 and 3.2).

Table 3.1 Cellular DNA content of various species.

Organism	Number of base pairs	DNA length (mm)*	Size of cellular space (mm)	Number of chromosomes†
Bacteriophage λ	4.85×10^4	0.017	< 0.0001	1
Bacterium (*Escherichia coli*)	4.7×10^6	1.4	0.001	1
Yeast (*Saccharomyces cerevisiae*)	1.25×10^7	4.6	0.005	16
Fruit fly (*Drosophila melanogaster*)	1.65×10^8	56.0	0.010	4
Human (*Homo sapiens*)	3×10^9	999.0	0.010	23

* The length given is before packing.

† Values are provided for haploid genomes. For most eukaryotes, diploid somatic cells would have twice this number of chromosomes.

Table 3.2 Diversity of DNA-based genome organization.

Genome	Form	Size (kb)
Eukaryotes	ds linear	10^4 to 10^6
Bacteria	ds circular	10^3
Plasmids	ds circular (some ds linear)	2–15
Mammalian DNA viruses	ss linear, ds linear, ds circular	3–280
Bacteriophage	ss circular, ds linear	~50
Chloroplast DNA	ds circular (or ds linear?)	120–160
Mitochondrial DNA	ds circular (some ds linear)	Animals: 16.5 Plants: 100–2500

ds, double-stranded; ss, single-stranded.

3.2 Eukaryotic genome

Eukaryotic cells must fit approximately up to 2 m of unpacked DNA into the spherical nucleus, which is a less than 10 micron diameter space (1 μm = 10^{-6} m) (Fig. 3.1). This represents a packaging ratio of approximately 1000- to 10,000-fold. This wondrous feat is accomplished by the packing of linear DNA molecules into chromatin (DNA with its associated proteins). DNA is first coiled around a histone complex called a nucleosome. Runs of nucleosomes are formed into a zig-zagging string of chromatin which is then folded into loop domains, and finally the metaphase chromosome. The compaction achieved at each level in this hierarchy is shown in Fig. 3.2.

In eukaryotes, the term genome is often used interchangeably with the terms genomic DNA, chromosomal DNA, or nuclear DNA (to distinguish it from organelle or plasmid DNA). The compaction of the genome into chromatin forms the substrate relevant to the vital processes of DNA replication, recombination, transcription, repair, and chromosome segregation.

Chromatin structure: historical perspective

In 1928, Albrecht Kossel isolated small basic proteins from the nuclei of goose erythrocytes (red blood cell precursors). He named these proteins histones. It was not until the 1970s that it was determined by a

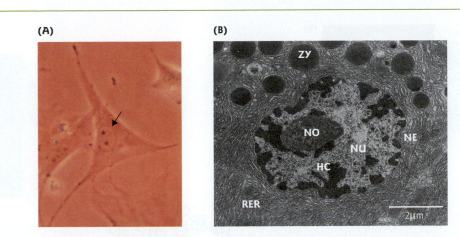

(A) **(B)**

Figure 3.1 The eukaryotic genome is located in the cell nucleus. (A) A typical animal cell viewed by light microscopy. The diameter of the nucleus (arrowed) is approximately 10–15 nm. The two dark spheres within the nucleus are the nucleoli. (B) A pancreatic acinar cell viewed by electron microscopy. The cell nucleus (NU) is bounded by a typical double-membrane envelope (NE). Electron dense heterochromatin (HC) is mainly located at the nuclear periphery whereas a moderately dense, oval-shaped nucleolus (NO) is more centrally positioned. Abundant rough endoplasmic reticulum (RER) and electron dense zymogen granules (ZY), consisting of a mixture of nascent digestive enzymes can also be seen in the cytoplasm. (Photograph courtesy of Julia P. Galkiewicz, College of William and Mary.)

combination of electron microscopic and biochemical studies by the laboratories of Roger Kornberg and Pierre Chambon that the fundamental packing unit of chromatin is the nucleosome. The first clear insights into chromatin structure came about by serendipity. Nuclease experiments performed for other reasons revealed that the DNA in chromatin degrades into a series of discrete fragment sizes separated by 180 bp (Fig. 3.3). This fragment size is now known to represent the DNA associated with a single nucleosome (mononucleosome).

Histones

Two types of histones exist, the highly conserved core histones (molecular weight 11,000–16,000 daltons or Da) and the much more variable linker histones (slightly larger, > 20,000 Da) (Fig. 3.4). For example, cow histone H4 differs from pea H4 in only two places, where the amino acids valine and lysine are exchanged for isoleucine and arginine, respectively. The core histones are present in the nucleosome as an octamer composed of a dimer of histones H2A and H2B at each end and a tetramer of histones H3 and H4 in the center, around which 146 bp of genomic DNA is wound (Fig. 3.5). The linker histone H1 (or alternative forms such as histone H5 and H1°) occurs between core octamers, where the DNA enters and exits the nucleosome.

All four core histones contain an extended histone-fold domain at the carboxyl (C) terminal end of the protein through which histone–histone interactions and histone–DNA interactions occur, and the charged tails at the amino (N) terminal end which contain the bulk of the lysine residues. These charged tails are the sites of many post-translational modifications of the histone proteins, including acetylation and methylation. The importance of these modifications will be discussed in detail in Section 11.6.

Nucleosomes

The term "nucleosome" specifically refers to the core octamer of histones plus the linker histone and approximately 180 bp of DNA. The core particle is comprised of 146 bp DNA plus the core histone

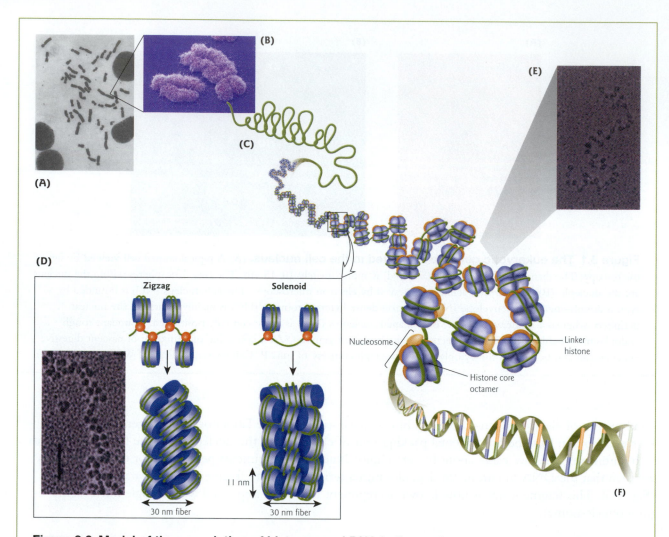

Figure 3.2 Model of the association of histones and DNA in the nucleosome. The way in which the chromatin fiber becomes packaged into a more condensed structure, ultimately forming a mitotic chromosome, is illustrated. (A) Light micrograph of human chromosomes stained with the dye Giemsa. The large ovals are intact (unlysed) white blood cells (Photograph provided by the author). (B) Colored scanning electron micrograph (SEM) of human X (right) and Y (left) sex chromosomes. Each chromatid is 700 nm in diameter. (Credit: Andrew Syred / Photo Researchers, Inc.) (C) DNA loop domains form 200 nm diameter chromatin fiber. (D) The panel on the left shows an electron micrograph of a 30 nm fiber. (Reproduced from Thoma, F., Koller T.H., Klug, A. 1979. Involvement of histone H1 in the organization of the nucleosome and of the salt-dependent superstructures of chromatin. *Journal of Cell Biology* 83:403–427, by copyright permission of The Rockefeller University Press.) The panel on the right shows a schematic of the fiber interpreted as a zig-zag or solenoid model. (Adapted from Khorasanizadeh, S. 2004. The nucleosome: from genomic organization to genomic regulation. *Cell* 116:259–272, Copyright © 2004, with permission from Elsevier.) (E) A 10 nm fiber showing the beads-on-a-string structure. (Reproduced from Thoma et al. 1979. *Journal of Cell Biology* 83:403–427, by copyright permission of The Rockefeller University Press.) (F) The DNA double helix, 2 nm diameter.

octamer. The core histone octamer acts like a spool; the negatively charged DNA wraps nearly twice around the positively charged histones in 1.67 left-handed superhelical turns (Fig. 3.5). A recent model suggests that the conformation of nucleosomal DNA differs from naked DNA. The double helix is tightly curved around the nucleosome core and is "stretched" on average by 1–2 bp, resulting in a difference of

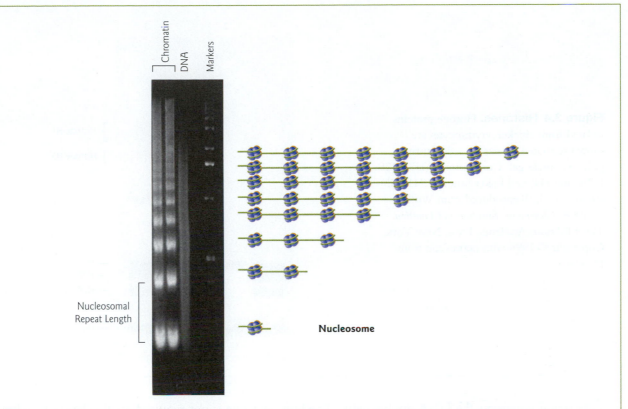

Chromatin
DNA
Markers

Nucleosomal Repeat Length

Nucleosome

Figure 3.3 Micrococcal nuclease cleavage of chromatin reveals nucleosome repeats. Chromatin and naked DNA were treated with a low concentration of the enzyme micrococcal nuclease, followed by removal of proteins and separation of the DNA fragments on an agarose gel (see Tool box 8.6). The visible ladder of bands in the chromatin samples are separated from each other by a single nucleosomal repeat length of DNA (multiples of ~180 bp). A ladder is observed because the micrococcal nuclease-treated chromatin is only partially digested. More extensive digestion would result in all linker DNA being cleaved and the formation of nucleosomal core particles and a single ~147 bp fragment. The cleavage of naked DNA generates a wide distribution of fragment sizes (visible as a smear), because the entire length of the DNA is unprotected and accessible to the enzyme. (Reproduced from Wolffe, A. 1998. *Chromatin. Structure and Function.* Third Edition. Academic Press, New York, Copyright © 1998 with permission from Elsevier.)

10.17 bp/turn for nucleosomes, compared with 10.5 bp/turn for naked DNA. Histones can be removed from DNA by high salt concentration, so the major interactions between DNA and the core histones appear to be electrostatic in nature.

Beads-on-a-string: the 10 nm fiber

Nearly 30 years ago, Pierre Chambon's laboratory published striking electron microscope images of the eukaryotic genome. These 1975 images clearly showed the existence of uniformly sized particles in a repeating pattern, with the appearance of beads on a string. Such beads had been hinted at earlier in lower resolution images. In 1939, Amram Scheinfeld in his book entitled *You and Heredity* writes that "at certain times they [the chromosomes] may stretch out into filaments ever so much longer, and then we find that what they consist of apparently are many gelatinous beads closely strung together. These beads either are, in themselves, or contain the 'genes'. . . ."

The beads-on-a-string appearance of chromatin can be visualized by electron microscopy as a 10–11 nm fiber after low salt extraction (Fig. 3.2). The beads represent DNA wrapped around the histone core octamer

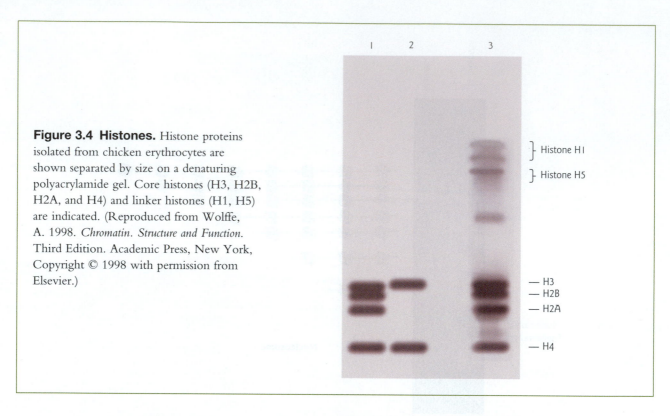

Figure 3.4 Histones. Histone proteins isolated from chicken erythrocytes are shown separated by size on a denaturing polyacrylamide gel. Core histones (H3, H2B, H2A, and H4) and linker histones (H1, H5) are indicated. (Reproduced from Wolffe, A. 1998. *Chromatin. Structure and Function.* Third Edition. Academic Press, New York, Copyright © 1998 with permission from Elsevier.)

and the string represents the DNA double helix. The linker histone is not required for this level of packing. Since isolation occurs under nonphysiological conditions, the question remains as to whether the 10–11 nm fiber is present *in vivo* or is a consequence of the extraction procedure.

The 30 nm fiber

Chromatin released from nuclei by nuclease digestion has the canonical beads-on-a-string structure in lower salt concentrations (≤ 10 mM). On the addition of salt (i.e. increasing the ionic strength), or when observed *in situ*, nucleosomal arrays form a compact fiber of approximately 30 nm in diameter. Because of the difficulty in maintaining the integrity of this fragile chromatin structure during isolation, the structure remains poorly understood. Models range from a relatively ordered helical solenoid to a zig-zag aggregation of nucleosomes. The classic solenoid model was accepted for many years and involves six consecutive nucleosomes arranged in a turn of a helix. However, solenoids are not seen at physiological salt concentrations (i.e. ~150 mM monovalent cation). Recently, techniques for the preservation of chromatin for electron microscopic studies, biochemical analyses, and X-ray crystallography have improved. Studies now suggest that, at least in transcriptionally active cells, nucleosomes do not form a solenoid. Instead, they adopt a zig-zag ribbon structure that twists or supercoils (see Fig. 3.2).

Loop domains

The 30 nm fiber is, in turn, further compacted into structures not yet fully understood. Looped domains form that contain 50–100 kb of DNA (the length of loops is approximately 0.25 µm) (see Fig. 3.2). These loops may be created by attaching DNA to proteins associated with an underlying nuclear scaffold. In the interphase of the cell cycle, the packing ratio is 1000-fold. The chromatin fiber in a typical human chromosome is long enough to pass many times around the nucleus, even when condensed into loops. Experimental evidence suggests that individual interphase chromosomes are compacted into distinct regions or "territories" in the nucleus.

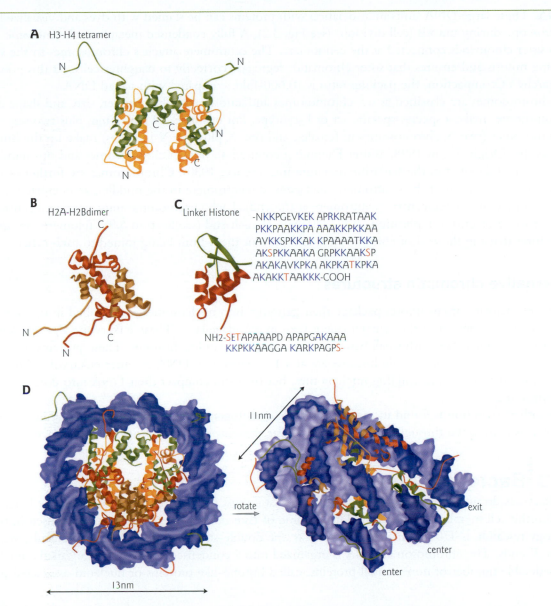

A H3-H4 tetramer

B H2A-H2Bdimer

C Linker Histone

-NKKPGEVKEK APRKRATAAK
PKKPAAKKPA AAAKKPKKAA
AVKKSPKKAK KPAAAATKKA
AKSPKKAAKA GRPKKAAKSP
AKAKAVKPKA AKPKATKPKA
AKAKKTAAKKK-COOH

NH2-SETAPAAAPD APAPGAKAAA
KKPKKAAGGA KARKPAGPS-

D

11nm

13nm

rotate

enter

center

exit

Figure 3.5 Atomic structure of the core and linker histones. (A) A tetramer of histones H3 (green) and H4 (orange). (B) A dimer of histones H2A (red) and H2B (brown). (C) The linker histone has a conserved wing helix fold; the variant H5 is shown. The N- and C-terminal tails of linker histones are disordered and consist of numerous lysines and serines; the amino acid sequence corresponding to a human H1 is shown. (D) The atomic structure of the nucleosome core particle, shown from two angles. Each strand of DNA is shown in a different shade of blue. The DNA makes 1.7 turns round the histone octamer to form an overall particle with a disk-like structure. Histones are colored as in (A) and (B). (Reproduced from Khorasanizadeh, S. 2004. The nucleosome: from genomic organization to genomic regulation. *Cell* 116:259–272, Copyright © 2004, with permission from Elsevier.)

Metaphase chromosomes

Further condensation requires a number of ATP-hydrolyzing enzymes, including topoisomerase II and the condensin complex. Condensin is a large protein complex composed of five subunits, and is one of the most abundant structural components of metaphase chromosomes. Each chromosome contains a single DNA molecule, i.e. a double–stranded DNA double helix. The typical human chromosome contains ~100 Mb of

DNA. These large DNA units in association with proteins can be stained with dyes and visualized in the light microscope during mitosis (cell division) (see Fig. 3.2). A fully condensed metaphase chromosome consists of two sister chromatids connected at the centromere. The centromere attaches chromosomes to the spindle during mitosis and ensures that sister chromatids segregate correctly to daughter cells. At this point in the hierarchy of compaction, the packing ratio is 10,000-fold, compared with naked DNA.

Chromosomes are classified as sex chromosomes and autosomes. The number, size, and shape of the chromosomes make a species-specific set or karyotype. For example, 44 autosomes plus two sex chromosomes (two X chromosomes in females, and one X plus one Y in males) make up the human karyotype. Originally, in 1898, Walter Flemming counted 48 human chromosomes and this number was still erroneously cited in the scientific literature into the late 1930s. Chromosomes are further classified based on the location of the centromere: metacentric (centromere in the middle), acrocentric (centromere toward one end), or telocentric (centromere at the end). Each chromosome must contain a centromere, one or more origins of replication, and a telomere at each end (see Section 6.9). Telomeres are specialized structures that cap the end of chromosomes and prevent them from being joined to each other.

Alternative chromatin structures

The vast majority of eukaryotes package their genomes into nucleosomes as described in the preceding sections. An exception to this general rule is seen in dinoflagellates. These eukaryotic algae package the majority of their DNA with small basic proteins completely unlike histones. These proteins only represent 10% of the mass of DNA, while histones are at a 1 : 1 ratio with DNA in other eukaryotes. Dinoflagellates do not form structures resembling nucleosomes, but they still compact their DNA into distinct chromosomes.

Another exception is found in mammals during gametogenesis. The majority of DNA in spermatozoa (sperm) is compacted through interaction with basic proteins known as protamines, in place of histones.

3.3 Bacterial genome

Prokaryotes do not have a nucleus. However, they still must fit DNA that is 1000 times the length of the cell within the cell membrane (Fig. 3.6). The genome of *Escherichia coli*, a bacterium widely used in molecular biology research, is 4700 kb in size and exists as one double-stranded circular DNA molecule, with no free 5′ or 3′ ends. The chromosomal DNA is organized into a condensed ovoid structure called a nucleoid. A considerable number of nonessential proteins, called histone-like proteins or nucleoid-associated proteins, are

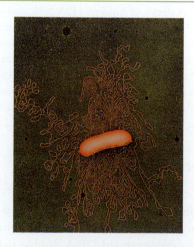

Figure 3.6 The bacterial genome. False-color transmission electron micrograph (TEM) of a lysed bacterial cell (*E. coli*). The DNA is visible as the gold colored fibrous mass lying around the bacterium. Magnification: ×15,700. (Credit: G. Murti / Photo Researchers, Inc.)

thought to be involved in DNA compaction and genome organization. These include HU (heat–unstable protein), IHF (integration host factor), HNS (heat-stable nucleoid structuring), and SMC (structural maintenance of chromosomes). HU and HNS are particularly abundant. Further condensation packs the bacterial genome into supercoiled domains of 20–100 kb. Approximately 50% of DNA supercoiling is unrestrained. These domains are dynamic and unlikely to have sequence-specific domain boundaries. Negative superhelicity is maintained by the action of topoisomerases, in particular by the ability of gyrase to remove the positive supercoils generated during replication and transcription (see Section 2.8).

3.4 Plasmids

Plasmids are small, double-stranded circular or linear DNA molecules carried by bacteria, some fungi, and some higher plants. They are extrachromosomal (meaning separate from the host cell chromosome), independent, and self-replicating. At least one copy of a plasmid is passed on to each daughter cell during cell division. Their relationship with their host cell could be considered as either parasitic or symbiotic. They range in size from 2 to 100 kb (Fig. 3.7). The majority of plasmids are circular; however, a variety of linear plasmids have been isolated. A notable example is the linear plasmid pC1K1 carried by *Claviceps purpurea*, a fungus found on rye. The fungus contains poisonous alkaloids that cause ergotism – hallucinations and sometimes death – in humans who eat the infected grain and was a likely contributor to the Salem Witch Trials.

The focus in this textbook will be on the circular plasmids from bacteria. Plasmids are important for our study for two main reasons: they are carriers of resistance to antibiotics, and they provide convenient vehicles for recombinant DNA technology (see Section 8.4).

3.5 Bacteriophages and mammalian DNA viruses

The motivation for most of the early studies on viruses centered on their pathogenicity, but they have also proved extremely useful systems for analysis of fundamental principles of molecular biology. For example, DNA viruses provide a cloned set of genes, organized in a physiologically meaningful array on a single DNA molecule. Before the advent of gene cloning technologies, viruses provided a readily available source of pure

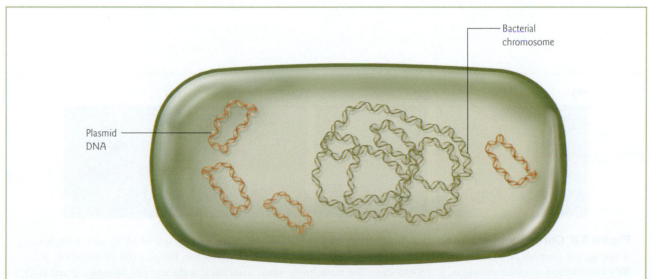

Figure 3.7 Schematic representation of a bacterium containing plasmid DNA. Plasmids are small, circular molecules of DNA that are extrachromosomal and self-replicating within the host bacterium.

DNA for studies of genomic expression, function, and replication. Bacteriophages and mammalian DNA viruses have DNA genomes that occur in a myriad of forms, ranging from double-stranded to single-stranded DNA and linear to circular forms (Table 3.2).

Bacteriophages

The chromosome of bacterial viruses usually consists of a single DNA molecule, largely devoid of associated proteins. For example, bacteriophage lambda (λ) has a double-stranded linear genome. Upon infection of a host bacterium, the DNA closes to form a circle. This phage is used widely as a tool for molecular biology research (see Chapter 8). In contrast, another phage commonly used by molecular biologists, M13, has a single-stranded circular genome. All bacteriophages have the ability to package an exceedingly long DNA molecule into a relatively small volume. A case in point is bacteriophage λ that uses the enzyme terminase to package 17 μm of DNA into a preformed protein "head" (capsid) which is less that 0.1 μm on any side. After packaging of the genome, a preformed tail is attached to the viral head.

Mammalian DNA viruses

Mammalian DNA viruses infect mammalian cells and make use of the host cell machinery for their replication. For this reason, the papovaviruses (for *papilloma*, *polyoma*, and *vacuolating*), in particular, have been one of the most important model systems for understanding molecular and genetic characteristics of eukaryotes. Their genomes come in a diversity of forms. For example, human papilloma virus (HPV), a causative agent of cervical and other cancers (see Chapter 16), has a double-stranded circular genome. Likewise, simian virus 40 (SV40) from rhesus monkey, also has a double-stranded circular genome. In contrast, adenovirus, a vector used for human gene therapy (see Section 17.3), has a double-stranded linear genome.

Little is known about how mammalian DNA viruses package their genome into the viral capsid (the protein shell encoded by viral genes). Some viruses encode their own basic proteins, while others usurp the host cell machinery. For example, papovavirus uses the host cell histones, H2A, H2B, H3, and H4 to package its genome. Histone H1 is absent from the nucleosome-like particles. Electron micrographs of SV40 show that the covalently closed, circular, double-stranded DNA is organized in a chromatin-like structure called a minichromosome (Fig. 3.8).

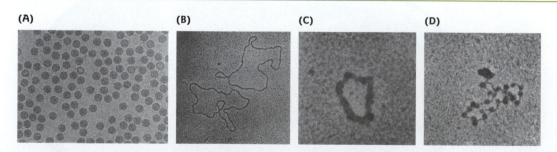

(A)   **(B)**   **(C)**   **(D)**

Figure 3.8 Chromatin formation in simian virus 40 (SV40). (A) Electron micrograph of SV40 viral particles. (Photograph courtesy of Norm Olson and Timothy Baker, University of California, San Diego.) (B) SV40 DNA. (C) SV40 condensed minichromosome. (D) SV40 extended minichromosome associated with host cell histones. (Parts B-D reproduced with permission from Singer, M. and Berg, P. 1997. *Exploring Genetic Mechanisms*, University Science Books, Sausalito, CA. Copyright © 1997 by University Science Books.)

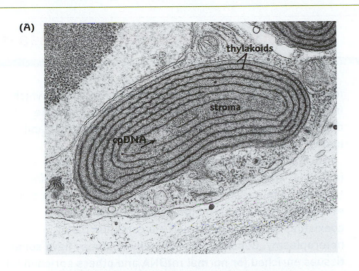

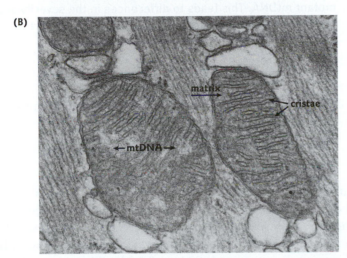

Figure 3.9 Organelle DNA.
(A) Chloroplast in the freshwater red alga *Compsopogon*. Red algal chloroplasts are similar to green algae and land plants in that they are bounded by a typical double membrane envelope. The small, somewhat electron translucent region in the middle of the chloroplast stroma is one of many chloroplast DNA (cpDNA) sites. Thylakoids are typically unstacked and often reveal small attached granules knowns as phycobilisomes, the site of the red and/or blue accessory pigments. (B) Two mitochondria of mouse (*Peromyscus*) heart tissue. Two possible mitochondrial DNA (mtDNA) regions are evident in the matrix, as well as the double membrane envelope and shelf-like cristae. (Photographs courtesy of Joe Scott, College of William and Mary.)

3.6 Organelle genomes: chloroplasts and mitochondria

Both mitochondria and chloroplasts contain their own genetic information (Fig. 3.9). The genomes are usually, but not always, circular. In circular form, the mitochondrial and chloroplast genomes look remarkably similar to bacterial genomes. This similarity, along with other observations, led to the "endosymbiont hypothesis" – the idea that both mitochondria and chloroplasts are derived from primitive organisms that were free-living and much like bacterial organisms. Organelle genomes are inherited independently of the nuclear genome and they exhibit a uniparental mode of inheritance, with traits being passed to offspring only from their mother. The organelles are only contributed from the maternal gamete (e.g. egg cell), and not from the paternal gamete (e.g. sperm cell or pollen grain).

Chloroplast DNA (cpDNA)

Chloroplasts are found in higher plants, some protozoans, and algae. The cpDNA encodes enzymes involved in photosynthesis. The most standard depiction of cpDNA is as a circular, double-stranded DNA molecule, ranging in size from 120 to 160 kb, with 20–40 copies per organelle. However, this is a subject of debate. Recent studies suggest that, in fact, most cpDNA is linear and only a minor amount is in a circular form.

DISEASE BOX 3.1

Mitochondrial DNA and disease

In 1988, 25 years after the discovery that mitochondria have their own genes, researchers made a link between certain human diseases and mtDNA mutations. Most mtDNA defects lead to degenerative disorders, especially of the brain and muscles, but because of the essential function of mitochondria in cellular ATP production, the effects can be widespread.

One of the first diseases to be linked to a small inherited mutation in a mitochondrial gene was a form of young adult blindness (Leber's hereditary optic neuropathy, LHON). The most common defects associated with LHON occur in genes coding for protein components of complex I of the electron transport chain. mtDNA mutations such as deletions or duplications that affect many genes at once have also been identified. One example is Kearns–Sayre syndrome, which involves paralysis of eye muscles, progressive muscle degeneration, heart disease, hearing loss, diabetes, and kidney failure.

Normally, all of the mtDNA within the cells of an individual are identical – a condition called homoplasmy. However, a mutation occurring in one copy of mtDNA can eventually result in both mutant and normal mtDNA coexisting within the same cell – a condition called heteroplasmy. Consequently, an individual may have some tissues enriched for normal mtDNA and others enriched for mutant mtDNA. This leads to differences in the severity and the kind of symptoms that may be displayed for a particular disease.

Whatever the form, cpDNA is free of the associated proteins characteristic of eukaryotic DNA. Compared with nuclear DNA of the same organism, it has a different buoyant density and base composition.

Mitochondrial DNA (mtDNA)

Mitochondria are found in plants, animals, fungi, and aerobic protists. The mtDNA encodes essential enzymes involved in ATP production (Disease box 3.1). mtDNA is usually a circular, double-stranded DNA molecule that is not packaged with histones. There are a few exceptions where mtDNA is linear, generally in lower eukaryotes such as yeast and some other fungi. mtDNA differs greatly in size among organisms. In animals, it is typically 16–18 kb, while in plants it ranges in size from 100 kb to 2.5 Mb. There are multiple copies of mtDNA per organelle, with anywhere from several to as many as 30 copies in *Euglena* protozoans.

3.7 RNA-based genomes

RNA serves as the genome for a number of infectious agents, including eukaryotic RNA viruses, retroviruses, viroids, and other subviral pathogens. RNA-based genomes have rates of mutation that are 1000-fold higher than DNA-based genomes. This is due mainly to the lack of exonuclease proofreading activity displayed by RNA polymerases. The assumption has been that high mutation rates are advantageous because this allows RNA viruses to alter their proteins rapidly so that they can evade recognition by host defense systems. However, there is no direct experimental proof showing a positive correlation between mutation and adaptation rates. Instead, the mutation rate may be explained by a fitness trade-off between replication rate and replication fidelity. Proofreading activity would increase replication fidelity and decrease deleterious mutations, but it would come at a cost. Pausing to proofread would slow down the rate of replication by RNA polymerase. Whatever the case, the overall success of this strategy is apparent – several well-known viral diseases are due to RNA viruses (Table 3.3).

Eukaryotic RNA viruses

Eukaryotic RNA viruses are a very diverse group. They infect many different hosts, including plants and animals. Medically, they are an extremely important group, with many significant human or veterinary

Table 3.3 Major types of RNA viruses.

Type of virus	Genome	Mode of replication	Virus family	Some pathogenic members
Eukaryotic RNA viruses	RNA	RNA → RNA	Togaviridae	Rubella (German measles)
			Picornaviridae	Rhinovirus (common cold)
				Foot and mouth disease
				Polio
			Coronaviridae	Coronaviruses: common cold, severe acute respiratory syndrome (SARS)
			Rhabdoviridae	Rabies
			Paramyxoviridae	Measles
				Mumps
			Filoviridae	Ebola
				Marburg
			Reoviridae	Rotavirus
			Orthomyxoviridae	Influenza
Retroviruses	RNA	RNA → DNA → RNA	Lentiviridae	HIV-1 (AIDS)

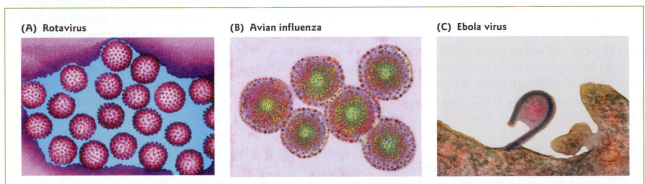

(A) Rotavirus   **(B) Avian influenza**   **(C) Ebola virus**

Figure 3.10 Examples of the diversity of eukaryotic RNA viruses. (A) Colored transmission electron micrograph (TEM) of a cluster of rotaviruses. Rotaviruses get their name from their wheel-like appearance, with a rounded core and radiating spikes of the outer protein shell. They do not have an envelope. These viruses are associated with gastroenteritis and diarrhea in humans and other mammals. (Credit: Dr. Linda Stannard, UCT / Photo Researchers, Inc.) (B) Micrograph showing avian influenza virus, a member of the Orthomyxoviridae family. The virus has an enveloped capsid. The flu virus causes an infectious and contagious respiratory disease. (Credit: James Cavallini / Photo Researchers, Inc.) (C) Ebola virus. Colored TEM of the release of an Ebola virus (hook, blue) from a host cell (red). This enveloped filovirus, which causes Ebola fever, removes part of the host cell membrane (pink, center) as it leaves, ensuring that the host's defenses do not recognize it as foreign. Ebola virus causes a fever, severe hemorrhaging, and central nervous system damage. (Credit: LSHTM / Photo Researchers, Inc.)

pathogens (Table 3.3). They come in a variety of sizes and shapes, with enveloped or nonenveloped capsids (external protein coat) (Fig. 3.10). The envelope is a layer of lipid and protein surrounding the capsid. Typical RNA viruses replicate without forming DNA intermediates, a feature that distinguishes them from the retroviruses, which also have an RNA genome. There are three main categories of RNA virus based on the type of RNA genome: plus-strand viruses, minus-strand viruses, and double-stranded RNA. The terms plus and minus refer to the coding strand and the noncoding strand, respectively. Plus-strand viruses make protein directly because their genomic RNA also serves as the mRNA for about a dozen genes. Plus-strand viruses are exemplified by the picornaviruses, which include pathogens that cause polio and the common cold in humans, and foot and mouth disease in livestock. Minus-strand viruses must first make the complementary plus-strand before using the RNA in protein synthesis. The minus-strand RNA viruses are

DISEASE BOX 3.2

Avian flu

Even the most optimistic scenarios for how the next worldwide flu epidemic (pandemic) might proceed are grim. Predictions are that 20% of the world's population will become ill, with close to 30 million people needing hospitalization, a quarter of whom will die. Pandemics result when a virus to which most people have no immunity, usually an avian strain, acquires the ability to transmit readily from animal to person, and then from person to person. This can happen by the virus mutating so that it can be passed between people, or it could exchange genes with a common human flu strain. The genes for two proteins that make up the viral outer coat – hemagglutinin (H) and neuraminidase (N) – are constantly mutating and come in many different varieties. Each flu virus is named after the types of these two proteins. In 1918 the extremely pathogenic H1N1 flu virus left as many as 40 million dead. Despite better standards of health care, in 1968 the relatively mild H3N2 virus still killed some 750,000 people worldwide. Worldwide attention is now focused on the H5N1 avian influenza virus, although there are other avian influenza strains, including H7N7 and H9N2, that have occasionally infected humans in recent years.

H5N1 and the threat of an influenza pandemic

H5N1 has not yet gained the ability to transmit readily from human to human. Because human-to-human transmission is still rare, the World Health Organization (WHO) categorizes the current outbreak as a "phase 3" pandemic threat, meaning the beginning of the "pandemic alert period," with "phase 6" marking the actual pandemic. The first documented instance of bird-to-human infection with the H5N1 flu virus was in 1997 in Hong Kong. Hong Kong reacted by destroying its entire poultry population of 1.5 million birds within 3 days. Even with this rapid response, the outbreak killed six of 18 infected people. Since 1997, the H5N1 virus has continued to evolve and has acquired the ability to kill its natural host, wild waterfowl, and has spread across South East Asia and to Europe and Africa. In addition, it has expanded its host range to include chickens, ducks, tigers, leopards, domestic cats, and pigs. Since late 2003, H5N1 has led to the deaths of more than 50 people in Vietnam, Thailand, Cambodia, and Turkey. Ducks can be symptomless carriers of H5N1, making them particularly dangerous, and migratory birds may carry the virus to other countries.

Vaccination and the use of antiviral drugs are two of the most important responses to a potential flu pandemic. Because the H5N1 virus is constantly changing, vaccines must be updated regularly to remain effective. About 40 countries have published plans to deal with a pandemic. At least 18 countries have ordered stockpiles of an anti-influenza drug (oseltamivir), but the percentage of the population that would be covered varies widely by country. It may take years for Roche, the drug's only supplier, to fill all the orders.

the most widespread, including the agents of well-known diseases such as rabies, mumps, measles, influenza (Disease box 3.2), severe acute respiratory syndrome (SARS), and the more exotic Ebola and Marburg filoviruses, which have caused epidemics of fatal hemorrhagic fever in Africa. dsRNA viruses are relatively uncommon. The reoviruses are the best-known dsRNA viruses. Rotavirus is a member of this family, which causes infant diarrhea. Its genome contains a dozen or so separate dsRNA molecules, each coding for a single viral protein.

Retroviruses

Retroviruses are also called "RNA tumor viruses" because many members play a role in cancer (see Section 17.2). Retroviruses have single-stranded RNA genomes that replicate through a DNA intermediate by reverse transcription. Upon infecting an animal cell, the retrovirus converts the single-stranded RNA (ssRNA) into a double-stranded DNA copy. The retrovirus DNA is then inserted into the host cell DNA. Once integrated, the retrovirus DNA remains permanently inserted in the host genome. Consequently, retroviruses currently are impossible to get rid of completely after infection and integration. They include many well-known animal pathogens, and one prominent human pathogen, human immunodeficiency virus

1 (HIV-1) (Table 3.3). Retroviruses are important vectors for gene therapy and will be discussed in more detail in Chapter 16.

Viroids

Viroids are "subviral" pathogens that cause infectious disease in higher plants. The surprise discovery was that the viroid RNA is itself the infectious agent. Because of their small size and unique properties, viroids are now among some of the best-studied RNA molecules (see Section 4.7). The viroid genome consists of a single, very small, circular molecule of RNA, ranging in size from 250 to 400 nt. Unlike viral RNAs, viroid RNAs do not encode any proteins and they are not protected by a protein coat. There are approximately 30 known viroid species and hundreds of variants that cause disease in more than two dozen crop plants, including the avocado sunblotch viroid and coconut cadang cadang viroid.

Other subviral pathogens

Other subviral pathogens include satellite RNAs and virusoids. Viroids replicate autonomously by using host-encoded RNA polymerase. In contrast, satellite RNAs multiply only in the presence of a helper virus that provides the appropriate RNA-dependent RNA polymerase. Some of the larger satellite RNAs may encode a protein. Satellite RNAs are found in plants (e.g. satellite tobacco necrosis virus) and animals. A well known human satellite RNA is hepatitis delta virus (HDV). HDV is a small single-stranded RNA satellite of hepatitis B virus.

A virusoid is an RNA molecule that does not encode any proteins and depends on a helper virus for replication and capsid formation. Virusoids occur in association with viruses causing plant diseases such as velvet tobacco mottle and subterranean clover mottle. They are sometimes regarded as a subtype of satellite RNA. The virusoid genome resembles a viroid and consists of circular, single-stranded RNA with self-cleaving activity (see Section 4.7).

Chapter summary

The genomes of most organisms are made of DNA; certain viruses and subviral pathogens have RNA genomes. Eukaryotic DNA combines with basic protein molecules called histones to form structures known as nucleosomes. Each nucleosome contains four pairs of core histones (H2A, H2B, H3, and H4) in a wedge-shaped disk, around which is wrapped 146 bp of DNA. The linker histone H1 is bound to DNA between the core histone octamers, where the DNA enters and exits the nucleosome. The first order of chromatin folding is represented by a string of nucleosomes. This 10 nm nucleosome fiber is further folded into a 30 nm fiber in a zig-zag ribbon structure, which is then folded into loop domains, and finally the metaphase chromosome. Each chromosome is composed of one linear, double-stranded DNA molecule.

Bacterial chromosomal DNA exists as one double-stranded, circular DNA molecule organized into a condensed structure called a nucleoid. Plasmids are self-replicating small, double-stranded, circular or linear DNA molecules carried by bacteria, some fungi, and some higher plants. Plasmids are important tools for recombinant DNA technology. Bacteriophages and mammalian DNA viruses have DNA genomes that occur in a variety of forms, ranging from double-stranded to single-stranded DNA and linear to circular forms. Viruses either package their genomes with their own basic proteins, or use host cell histones.

Both mitochondria and chloroplasts contain their own genetic information. The small, double-stranded DNA genomes are usually, but not always, circular and there are multiple copies per organelle. Organelle genomes are maternally inherited.

RNA serves as the genome for a number of important infectious agents, including eukaryotic RNA viruses, retroviruses, viroids, and other subviral pathogens. Eukaryotic viruses have either single-stranded or double-stranded RNA genomes and replicate without forming DNA intermediates. This feature distinguishes them from the retroviruses. Retroviruses have a single-stranded RNA genome that is replicated through a DNA intermediate. The genomes of viroids and other subviral pathogens such as satellite RNAs and virusoids are composed of single-stranded RNA. Viroids and virusoids are plant pathogens that do not encode any proteins. Satellite RNAs are found in both plants and animals. Some of the larger satellite RNAs may encode a protein.

Analytical questions

1 Brief digestion of eukaryotic chromatin with micrococcal nuclease gives DNA fragments ~200 bp long. You repeat the experiment, but incubate the samples for a longer period of time while you are in class. This longer digestion yields 146 bp fragments. Why?

2 Do the 10 and 30 nm eukaryotic chromatin fibers exist *in vivo*? Discuss electron microscopic and biochemical evidence in support of your answer.

3 You are asked to characterize the genome of a newly isolated virus, and to determine whether it is composed of DNA or RNA. After using nucleases to completely degrade the sample to its constituent nucleotides, you determine the approximate relative proportions of nucleotides. The results of your assay are as follows:

0% dGTP	15% GTP	0% dCTP	33% CTP	0% dATP	22% ATP	0% dTTP	30% UTP

What can you conclude about the composition of the viral genome?

Suggestions for further reading

Bendich, A.J. (2004) Circular chloroplast chromosomes: The grand illusion. *Plant Cell* 16:1661–1666.

Blackburn, G.M., Gait, M.J., eds (1990) *Nucleic Acids in Chemistry and Biology*. IRL Press, Oxford.

Check, E. (2005) Avian flu. Is this our best shot? *Nature* 435:404–406.

Cook, P.R. (2001) *Principles of Nuclear Structure and Function*. Wiley-Liss, Inc., New York.

Flores, R., Hernández, C., Martínez de Alba, A.E., Daròs, J.A., Di Serio, F. (2005) Viroids and viroid–host interactions. *Annual Reviews in Phytopathology* 43:4.1–4.23.

Furió, V., Moya, A., Sanjuán, R. (2005) The cost of replication fidelity in an RNA virus. *Proceedings of the National Academy of Sciences USA* 102:10233–10237.

Grunstein, M. (1992) Histones as regulators of genes. *Scientific American* 267:40–47.

Khorasanizadeh, S. (2004) The nucleosome: from genomic organization to genomic replication. *Cell* 116:259–272.

McFarland, R., Taylor, R.W., Turnbull, D.M. (2002) The neurology of mitochondrial DNA disease. *Lancet Neurology* 1:343–351.

Normile, D. (2005) New focus. Vietnam battles bird flu . . . and critics. *Science* 309:368–373.

Oldenberg, D.J., Bendich, A.J. (2004) Most chloroplast DNA of maize seedlings in linear molecules with defined ends and branched forms. *Journal of Molecular Biology* 335:953–970.

Oudet, P., Chambon, P. (2004) Seeing is believing. *Cell* S116:S79–S80.

Richmond, T.J., Davey, C.A. (2003) The structure of DNA in the nucleosome core. *Nature* 423:145–150.

Schalch, T., Duda, S., Sargent, D.F., Richmond, T.J. (2005) X-ray structure of a tetranucleosome and its implications for the chromatin fibre. *Nature* 436:138–141.

Scheinfeld, A. (1939) *You and Heredity*. Frederick A. Stokes Co., New York.

Sherratt, D.J. (2003) Bacterial chromosome dynamics. *Science* 301:780–785.

Singer, M., Berg, M. (1997) *Exploring Genetic Mechanisms*. University Science Books: Sausalito, CA.

Swedlow, J.R., Hirano, T. (2003) The making of the mitotic chromosome: modern insights into classical questions. *Molecular Cell* 11:557–569.

Wallace, D.C. (1997) Mitochondrial DNA in aging and disease. *Scientific American* 277:22–29.

Wargo, M.J., Rizzo, P.J. (2001) Exception to eukaryotic rules. *Science* 294:2477.

Wolffe, A. (1998) *Chromatin. Structure and Function*, 3rd edn. Academic Press, New York.

Yang, Q., Catalano, C.E. (2004) A minimal kinetic model for a viral DNA packaging machine. *Biochemistry* 43:290–299.

Chapter 4
The versatility of RNA

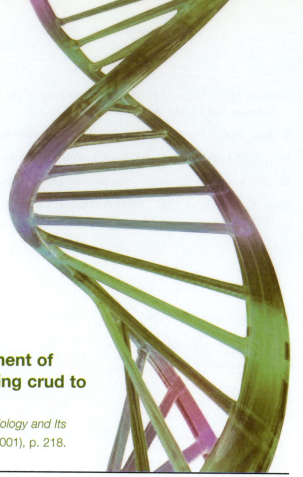

The final stage in the exaltation of the RNA component of RNase P occurred in 1983 – converting contaminating crud to catalytic component after a decade.

Harrison Echols, *Operators and Promoters: The Story of Molecular Biology and Its Creators* (2001), p. 218.

Outline

4.1 Introduction

Initial studies on RNA structure were pursued side by side with that of DNA. What became increasingly apparent is that RNA has a much greater structural and functional versatility compared with DNA. The growing database of RNA structures has led to characterization of numerous RNA secondary and tertiary structural motifs. RNA is now viewed as a modular structure built from a combination of these building blocks and tertiary linkers. RNA chains fold into unique three-dimensional structures which act similarly to globular proteins. The folding patterns provide the basis for their chemical reactivity and specific interactions with other molecules, including proteins, nucleic acids, and small ligands. RNA is involved in a wide range of essential cellular processes from DNA replication to protein synthesis.

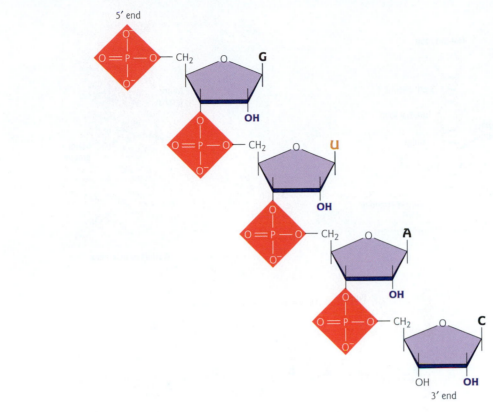

Figure 4.1 Components of RNA. The figure shows the structure of the backbone of RNA, composed of alternating phosphates and ribose sugars. The features of RNA that distinguish it from DNA are highlighted. The ribose has a hydroxyl group at the 2′ position and RNA contains the base uracil in place of thymine.

4.2 Secondary structure of RNA

As introduced in Chapter 2, RNA is a chain-like molecule composed of subunits called nucleotides joined by phosphodiester bonds (Fig. 4.1). Each nucleotide subunit is composed of three parts: a ribose sugar, a phosphate group, and a nitrogenous base. The common bases found in RNA are adenine (A), guanine (G), cytosine (C), and uracil (U). Single-stranded RNA folds into a variety of secondary structural motifs that are stabilized by both Watson–Crick and unconventional base pairing.

Secondary structure motifs in RNA

Secondary structures of RNA can be predicted with good accuracy by computer analysis, based on thermodynamic data for the free energies of various conformations, comparative sequence analysis, and solved crystal structures. Some of the common secondary structures that form the building blocks of RNA architecture are shown in Fig. 4.2. These include bulges, base-paired helices or "stems," single-stranded hairpin or internal loops, and junctions. RNA structure was once envisioned as a collection of relatively rigid stems comprised of Watson–Crick bases pairs and the single-stranded loops defined by these stems. The first structure of transfer RNA showed otherwise. In fact, as we shall see RNA adopts structures that Harry Noller described in a 2005 *Science* review article as "breathtakingly intricate and graceful."

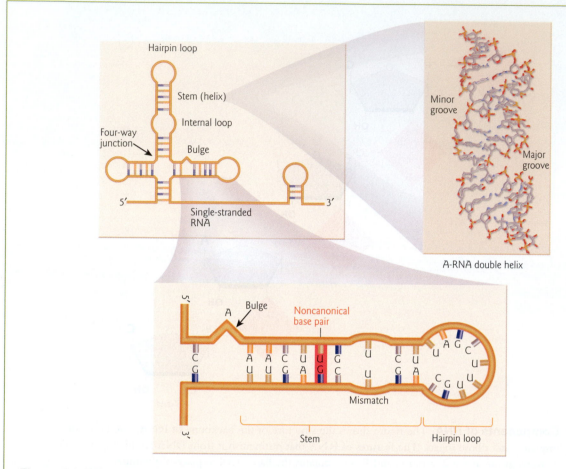

Figure 4.2 RNA secondary structure. Schematic representation of the structural motifs in a typical secondary structure of RNA. Motifs include base-paired stems with noncanonical base pairs (lower inset), hairpin loops, internal loops, bulges, and junctions. Base-paired stems form an A-type helix (top inset).

Base-paired RNA adopts an A-type double helix

In DNA the double helix forms from two separate DNA strands. In RNA, helix formation occurs by hydrogen bonding between base pairs and base stacking hydrophobic interactions within one single-stranded chain of nucleotides. X-ray crystallography studies have shown that base-paired RNA primarily adopts a right-handed A-type double helix with 11 bp per turn (Fig. 4.2). The 2′-hydroxyl group of the ribose sugar in RNA hinders formation of a B-type helix – the predominant form in double-stranded DNA – but can be accommodated within an A-type helix. Regular A-type RNA helices with Watson–Crick base pairs have a deep, narrow major groove that is not well suited for specific interactions with ligands. On the other hand, although the minor groove does not display sequence specificity, it includes the ribose 2′-OH groups which are good hydrogen bond acceptors, and it is shallow and broad, making it accessible to ligands. Because of these structural features, it is common for RNA to be recognized by RNA-binding proteins in the minor groove.

RNA helices often contain noncanonical base pairs

In addition to conventional Watson–Crick base pairs, RNA double helices often contain noncanonical (non-Watson–Crick) base pairs. There are more than 20 different types of noncanonical base pairs, involving two

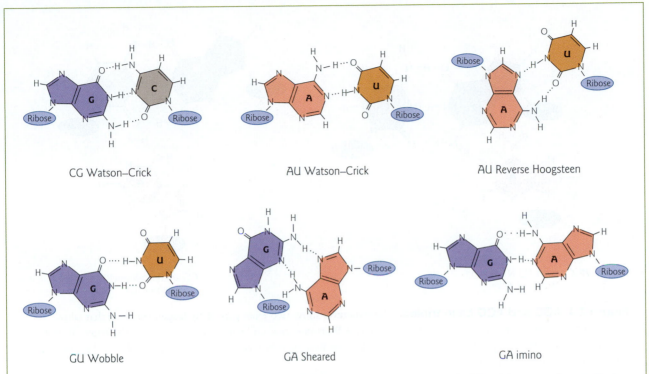

CG Watson–Crick   AU Watson–Crick   AU Reverse Hoogsteen

GU Wobble   GA Sheared   GA imino

Figure 4.3 Base pairs found in RNA double helices. Hydrogen bonding (dashed lines) between the standard Watson–Crick base pairs (CG, AU) is compared with hydrogen bonds that form between noncanonical pairs. Shown are the structures of four commonly found pairs: AU reverse Hoogsteen, GU wobble, GA sheared, and GA imino.

or more hydrogen bonds, that have been encountered in RNA structures. The most common are the GU wobble, the sheared GA pair, the reverse Hoogsteen pair, and the GA imino pair (Fig. 4.3). Because the GU pair only has two hydrogen bonds (compared with three for a GC pair), this requires a sideways shift of one base relative to its position in the regular Watson–Crick geometry. Weaker interactions from the reduction in hydrogen bonding may be countered by the improved base stacking that results from each sideways base displacement. In addition, RNA structures frequently involve unconventional base pairing such as base triples (Fig. 4.4). These base triples typically involve one of the standard base pairs, most commonly either a Watson–Crick or a reverse Hoogsteen pair. The third base can interact in a variety of unconventional ways. Noncanonical base pairs and base triples are important mediators of RNA self-assembly and of RNA–protein and RNA–ligand interactions. For example, noncanonical base pairs widen the major groove and make it more accessible to ligands.

4.3 Tertiary structure of RNA

RNA chains fold into unique three-dimensional structures that act similarly to globular proteins. Indeed, Francis Crick wrote in his 1966 paper in the *Cold Spring Harbor Symposium on Quantitative Biology* "tRNA looks like Nature's attempt to make RNA do the job of a protein." These remarks were made by Crick 2 years after the "cloverleaf" secondary structure of the transfer RNA (tRNA) for alanine in yeast was published by R.W. Holley and colleagues (Fig. 4.5). The actual shape of the functional tRNA in the cell is not an open cloverleaf. X-ray crystallography studies 10 years later showed that tRNA twists into an L-shaped three-dimensional structure. Many basic principles of RNA structure were learned from detailed analysis of both the secondary and tertiary structures of tRNAs. Obtaining crystal structures of larger RNA

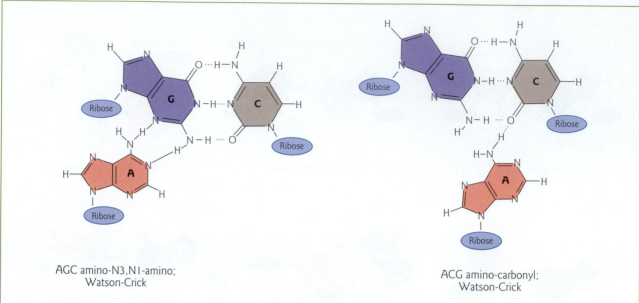

AGC amino-N3,N1-amino;
Watson-Crick

ACG amino-carbonyl;
Watson-Crick

Figure 4.4 AGC and ACG base triples. The structures show two examples of hydrogen bonding that allow unusual triple base pairing. In both examples, a standard Watson–Crick GC pair forms the core of the triple. In the example on the left, the third base A is joined to G by two hydrogen bonds, while in the base triple on the right, A is joined to C by only one hydrogen bond.

molecules has proved to be a challenge. It was not until over 20 years later that structures were solved for larger RNAs, such as the 160 nt P4–P6 domain of a group I ribozyme (see below) and the ribosome subunits that are comprised of over 4500 nt of ribosomal RNA (rRNA) and more than 50 proteins. The structure of the ribosome will be discussed in detail in Chapter 14.

tRNA structure: important insights into RNA structural motifs

tRNA is transcribed as a molecule about twice as long as its final form. The pre–tRNA transcript is then processed by various nucleases at both the 5′ and 3′ end (see Section 4.6). After processing, the average tRNA is about 76 nt long, and all of the different tRNAs of a cell fold into the same general shape. One of the important insights into RNA structure came from the observation that the processed tRNA is further altered by the modification of bases.

Modified bases

In general, tRNAs contain more than 50 modified bases. Modifications range from simple methylation to complete restructuring of the purine ring (Fig. 4.6). Inosine (I) was the first modified nucleoside in tRNA to be identified. Nucleoside modifications are not unique to tRNA; for example, extensive base modification occurs during maturation of the ribosomal RNAs (see Section 13.9). The first modified nucleoside to be identified in any RNA was the ubiquitous pseudouridine (ψ). Pseudouridine was discovered over 20 years earlier than inosine, but its role in tRNA function was not characterized until much later.

tRNA loops each have a separate function

Certain structural elements, of course, are unique to the function of tRNA. For example, every tRNA so far examined has the sequence ACC on the 3′ end to which the amino acid is attached (see Fig. 4.5).

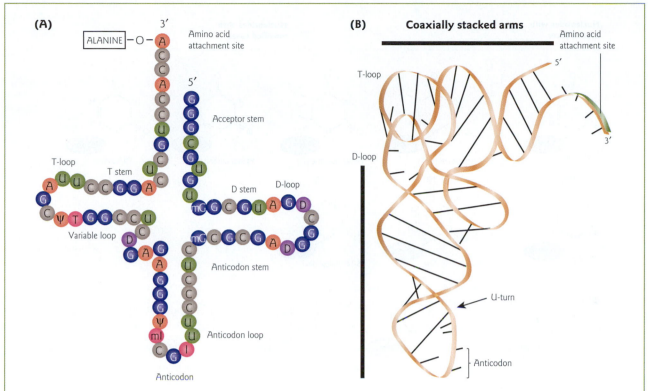

Figure 4.5 Secondary and tertiary structure of tRNA. (A) "Cloverleaf" secondary structure of alanine tRNA from yeast. The key structural features are labeled; note the modified bases in the loops. (B) L-shaped three-dimensional structure of tRNA showing the "arms" formed by coaxial stacking of the acceptor stem with the T stem, and the D stem with the anticodon stem. The arrow points to the U turn motif in the anticodon loop, which causes an abrupt reversal in the direction of the RNA chain.

However, a general principle gleaned from studies of tRNAs is the importance of loop motifs. Each of the three tRNA loops that form the "cloverleaf" secondary structure seems to serve a specific purpose. These functions will be described in detail in Chapter 14 in the context of the mechanism of translation. In brief, the t-loop (or t-ψ-C loop) is involved in recognition by the ribosomes, the D loop (or dihydrouridine loop) is associated with recognition by the aminoacyl tRNA synthetases, and the anticodon loop base pairs with the codon in mRNA. The anticodon loop in all tRNA is bounded by uracil on the 5′ side and a modified purine on the 3′ side. This purine is always modified, but the modification varies widely. Another commonly observed motif in many RNAs is the "U turn." In the anticodon loop of tRNA, hydrogen bonding of the N3 position of uridine with the phosphate group of a nucleotide three positions downstream causes an abrupt reversal or "U turn" in the direction of the RNA chain (see Fig. 4.5).

Coaxial stacking of stems

Another important principle learned from the study of tRNA structure is that base–paired stems often are involved in long-range interactions with other stems by coaxial stacking. In tRNA, the 7 bp acceptor stem stacks on the 5 bp T stem to form one continuous A-type helical arm of 12 bp (see Fig. 4.5). The other two helices, the D stem and anticodon stem, also stack, although imperfectly, to form a second helical arm. The two coaxially stacked arms are what form the familiar L shape of tRNA. Coaxial stacking is a common feature of RNA. It is widespread in ribosomal RNA where continuous coaxial stacking of as many as 70 bp is found, and underpins the formation of pseudoknots (see below).

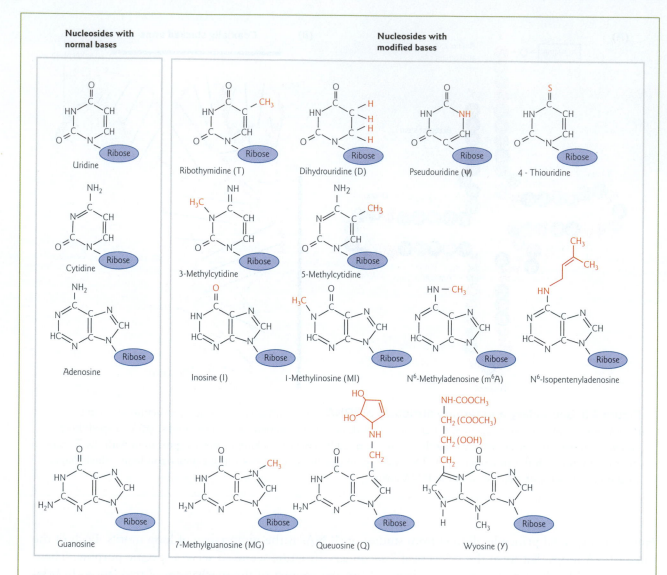

Figure 4.6 Structure of modified bases found in tRNA. The structures of nucleosides with normal bases and with modified bases are compared. Base modifications are highlighted in red.

Common tertiary structure motifs in RNA

Large RNAs are composed of a number of structural domains that assemble and fold independently. RNA folding uses the two principal devices that were first seen in the double-helical structures of DNA and RNA: hydrogen bonding and base stacking. Preformed secondary structural domains of RNA interact to form the tertiary structure. Bases in loops and bulges that are supposedly unpaired are often involved in a variety of long-range interactions, forming noncanonical base pairs. The three-dimensional structure is maintained through these interactions between distant nucleotides and interactions between 2′-OH groups. These long-range interactions are less stable than standard Watson–Crick base pairs and can be broken under mild denaturing conditions. RNA is negatively charged, which makes tertiary structure formation a process that requires charge neutralization, either through binding of basic proteins, or binding of monovalent and/or divalent metal ions. There are a number of highly conserved, complex RNA folding motifs. Common motifs include the pseudoknot, the A-minor motif, tetraloops, ribose zippers, and kink-turns. The examples provided below highlight how these different motifs interact with each other in a modular fashion to form intricate folding patterns.

Pseudoknot motif

A pseudoknot motif forms when a single-stranded loop base pairs with a complementary sequence outside this loop and folds into a three-dimensional structure by coaxial stacking (Fig. 4.7). The first experimental evidence for pseudoknot formation came from studies of a plant RNA virus in 1987. Now, many other pseudoknots have been identified in a wide variety of RNAs. For example, the 5′ half of human telomerase RNA consists of the RNA template for telomere synthesis and a highly conserved pseudoknot that is required for telomerase activity. The function of telomerase is described in detail in Section 6.9. Solution structure of the pseudoknot from human telomerase RNA was determined by using nuclear magnetic resonance (NMR) spectroscopy (see Section 9.10 for method). An intricate network of tertiary interactions was shown to form a triple helix structure, stabilized by base triples. Mutations in the pseudoknot region are involved in the human genetic disease, dyskeratosis congenita (see Disease box 6.3). Although vertebrates, ciliates, and yeast have widely divergent telomerase RNAs, they all contain a pseudoknot motif.

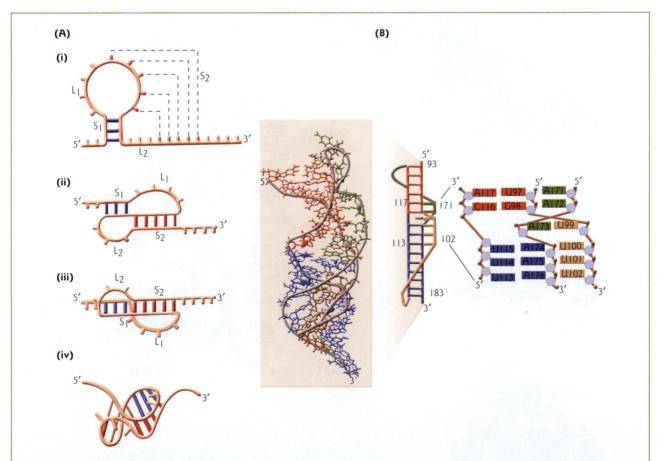

Figure 4.7 RNA pseudoknot motif. (A) Schematic presentation of a pseudoknot found in the tRNA-like structure of the turnip yellow mosaic virus. S_1 and S_2 represent double helical stem regions. L_1 and L_2 indicate single-stranded loops. (i) Conventional secondary structure. (ii) Formation of stem S_1, simultaneously with S_2. (iii) Coaxial stacking of S_1 and S_2, forming a quasicontinuous double helix. (iv) Schematic three-dimensional representation. (Redrawn from Pleij, C.W.A., Rietveld, K., Bosch, L. 1985. A new principle of RNA folding based on pseudoknotting. *Nucleic Acids Research* 13:1717–1730.) (B) Solution structure of the human telomerase RNA pseudoknot. The phosphate backbone is identified by a gray ribbon. Inset: schematic representation of the pseudoknot junction and tertiary structure, showing details of the extended triple helix surrounding the helical junction. Such multiple base triple interactions between loop 1 and stem 2 are unique among pseudoknot structures determined to date. Nucleotides are colored by structural element: stem 1 (red), stem 2 (blue), loop 1 (orange), and loop 2 (green). (Reproduced from Theimer, C.A., Blois, C.A., Feigon, J. 2005. Structure of the human telomerase RNA pseudoknot reveals conserved tertiary interactions essential for function. *Molecular Cell* 17:671–682, Copyright © 2005, with permission from Elsevier.)

A-minor motif

The A-minor motif is one of the most abundant long-range interactions in large RNA molecules. This motif was first observed in the hammerhead ribozyme (see Section 4.7) and the P4–P6 domain of the group I ribozyme, and is found extensively in ribosomal RNAs. In this folding pattern, single-stranded adenosines make tertiary contacts with the minor grooves of RNA double helices by hydrogen bonding and van der Waals contacts (Fig. 4.8). The minor groove interactions have been likened to a "lock and key" because of the precise way in which the adenosines fit into the groove. The motif is stabilized by both base–base interactions and nucleoside–nucleoside interactions. Critical contacts are made within the riboses as well as the bases.

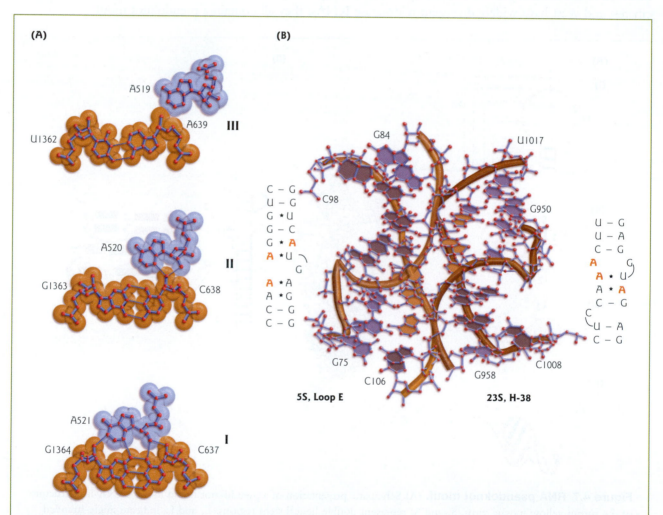

Figure 4.8 The A-minor motif. (A) Examples of the three most important kinds of A-minor motifs from the 23S ribosomal RNA (rRNA) of the archaeon *Haloarcula marismortui*, showing the precise lock-and-key minor groove interactions. Types I and II are adenine (A)-specific. Type III interactions involving other base types are seen, but A is preferred. (B) The interaction between helix 38 of 23S rRNA and 5S rRNA in *H. marismortui*. The only direct contact between these two molecules includes six A-minor interactions, involving three As in 23S rRNA and three As in 5S rRNA. Secondary structure diagrams are provided for the interacting sequences, with the As indicated in orange. (Reproduced from Nissen, P., Ippolito, J.A., Ban, N., Moore, P.B., Steitz, T.A. 2001. RNA tertiary interactions in the large ribosomal subunit: the A-minor motif. *Proceedings of the National Academy of Sciences USA* 98:4899–4903. Copyright © 2001 National Academy of Sciences, USA.)

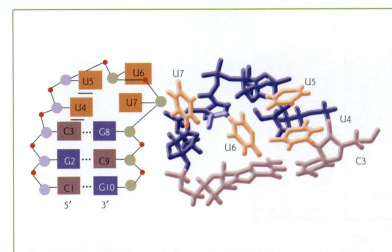

Figure 4.9 Tetraloop motif. A stem loop with the tetraloop sequence UUUU is shown. Base-stacking interactions promote and stabilize the tetraloop structure. The red circles between the riboses (blue and green circles) represent phosphate groups of the RNA backbone. Dashed lines denote Watson–Crick pairings and a thick line represents base-stacking interactions. (Reproduced from Koplin, J., Mu, Y., Richter, C., Schwalbe, H., Stock, G. 2005. Structure and dynamics of an RNA tetraloop: a joint molecular dynamics and NMR study. Structure 13:1255–1267, Copyright © 2005, with permission from Elsevier.)

Tetraloop motif

The stability of a stem-loop stucture is often enhanced by the special properties of the loop. For example, a stem loop with the "tetraloop" sequence UUUU is particularly stable due to special base–stacking interactions in the loop (Fig. 4.9). Tetraloops often include "G turns" in which a stabilizing hydrogen bond to the backbone phosphate is made from the 1–nitrogen position of a guanine base. Tetraloops are a prominent feature within the P4-P6 group I intron domain (Fig. 4.10).

Ribose zipper motif

Helix–helix interactions are often formed by "ribose zippers" involving hydrogen bonding between the 2′-OH of a ribose in one helix and the 2-oxygen of a pyrimidine base (or the 3-nitrogen of a purine base) of the other helix between their respective minor groove surfaces. Two ribose zippers are found in the P4-P6 group I intron domain (Fig. 4.10). One ribose zipper mediates the interaction between an A–rich bulge and the P4 stem. Another ribose zipper mediates a long-range interaction involving a tetraloop motif.

Kink-turn motif

Another type of motif first found in ribosomal RNA is the kink-turn or "K turn." Kink–turns are asymmetric internal loops embedded in RNA double helices. The most striking feature is the sharp bend (the "kink") in the phosphodiester backbone of the three-nucleotide bulge associated with this structure. In a kink-turn from the large ribosomal RNA of the extreme halophilic (salt-tolerant) archaean *Haloarcula marismortui*, each asymmetric loop is flanked by CG base pairs on one side and sheared GA base pairs on the other. Further illustrating how various structural motifs work together to define RNA shape, an A-minor interaction brings together these two helical stems (Fig. 4.11).

4.4 Kinetics of RNA folding

The structural flexibility of the RNA backbone and the propensity of nucleotides to base pair with short stretches of complementary regions can lead to difficulty in defining a single native structure, since there are many possible structures that a particular RNA chain can adopt. Misfolding, for example due to incorrect base pairing, is a problem for both secondary and tertiary structures. This "RNA folding problem" is not just a problem for the molecular biologist trying to determine the significance of predicted RNA secondary structures for function. Since only a single or a few possible structures lead to function, RNA itself must

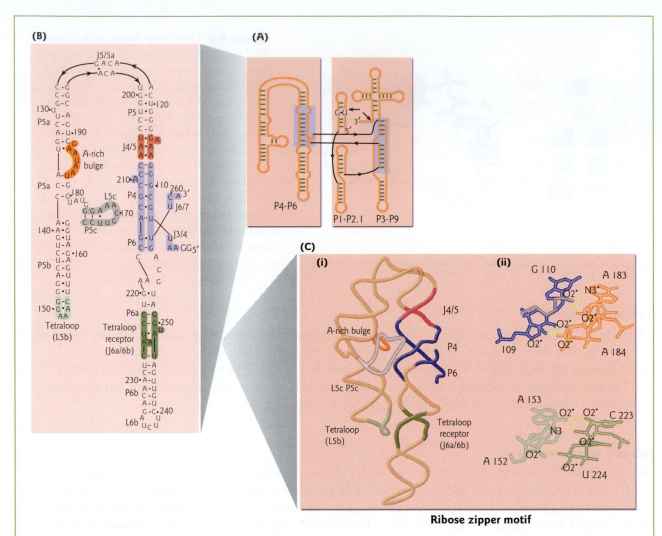

Ribose zipper motif

Figure 4.10 Ribose zipper motif. (A) The secondary structure of the *Tetrahymena thermophila* ribozyme. The phylogenetically conserved catalytic core of the ribozyme is shaded in blue. Arrows indicate the 5′- and 3′-splice sites of this self-splicing group I intron. (B) The secondary structure of the P4–P6 domain is shown in more detail. Helical regions are numbered sequentially through the sequence; J, joining region; P, paired region. Nucleotides are highlighted as follows: blue and red, part of the conserved core; orange, the A-rich bulge; light green, the GAAA tetraloop; dark green, the conserved 11 nt tetraloop receptor; gray, P5c. (Reprinted with permission from Cate, J.H., Gooding, A.R., Podell, E., et al. (1996) Crystal structure of a group I ribozyme domain: principles of RNA packing. *Science* 273:1678–1685. Copyright © 1996 AAAS.) (C) Structure of the P4–P6 group I intron domain and its two ribose zippers. (i) One ribose zipper mediates the interaction between the A-rich bulge (orange) and the P4 stem (blue). The other ribose zipper mediates the interaction between the tetraloop (light green) and the tetraloop receptor (dark green). (ii) In the ribose zippers, there are two residues on each side (109–110, 184–183 and 152–153, 223–224) in which riboses interact by hydrogen bonding (yellow broken line) between the 2′-hydroxyl groups (O2′) of the two chain segments in an antiparallel orientation. The 2′-hydroxyl groups of the 3′ end residues also form minor groove hydrogen bonds to either the N3 atom of a purine (G110, A152) or the O_2 atom of a pyrimidine (C109, C223) of the 5′ end residues on the opposite chain segment. (Reproduced from Tamura, M., Holbrook, S.R. 2002. Sequence and structural conservation in RNA ribose zippers. *Journal of Molecular Biology* 320:455–474, Copyright © 2002, with permission from Elsevier.)

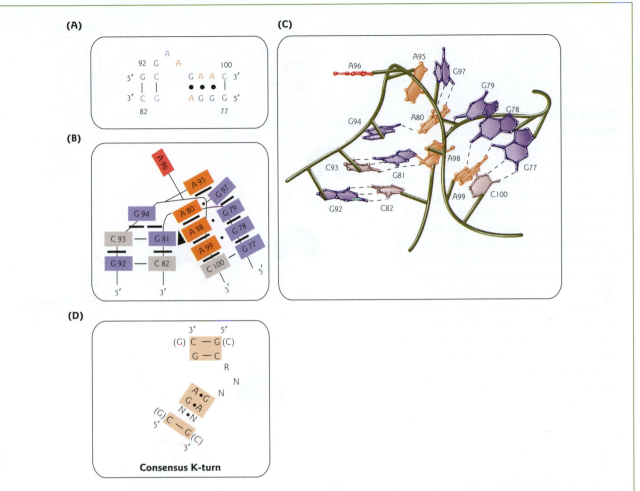

Figure 4.11 Kink-turn motif. (A) Secondary structure of a kink-turn motif in 23S rRNA of the archaeon *Haloarcula marismortui*. (B) Schematic representation of the relative base-stacking and base-pairing interactions. A black triangle represents an A-minor interaction. (C) Three-dimensional representation of the kink-turn. Hydrogen bonds are indicated by dashed lines. (D) Consensus secondary structure diagram derived from the eight K turns found in ribosomal RNA. (Adapted by permission from Nature Publishing Group and Macmillan Publishers Ltd: Klein, D.J., Schmeing, T.M., Moore, P.B., Steitz, T.A. 2001. The kink-turn: a new RNA secondary structure motif. *EMBO Journal* 20:4214–4221. Copyright © 2001.)

avoid the problem of misfolding into alternative, nonfunctional structures *in vivo*. Specific RNA-binding proteins form tight complexes with their target RNAs and act as chaperones to aid in RNA folding (Fig. 4.12). One example is the family of heterogenous nuclear ribonucleoprotein (hnRNP) proteins. This group of more than 20 different proteins assists in preventing misfolding and aggregation of pre-mRNA. Another example is that of rRNA which folds correctly only by assembly with ribosomal proteins. The ribosomal proteins stage the order of folding of rRNA during ribosome assembly to avoid losing improperly folded RNA in kinetic folding traps.

Kinetic folding profiles were first established for tRNAs. The secondary structure for these small RNA molecules forms first within 10^{-4} to 10^{-5} seconds, followed by the tertiary structure in 10^{-2} to 10^{-1} seconds. The folding of the *Tetrahymena thermophila* ribozyme (see Section 4.6) was recently analyzed using a hydroxyl radical footprinting assay (Fig. 4.13). The researchers generated hydroxyl radicals by radiolysis of water with a synchrotron X-ray beam. The short-lived hydroxyl radicals were able to break the ribozyme RNA

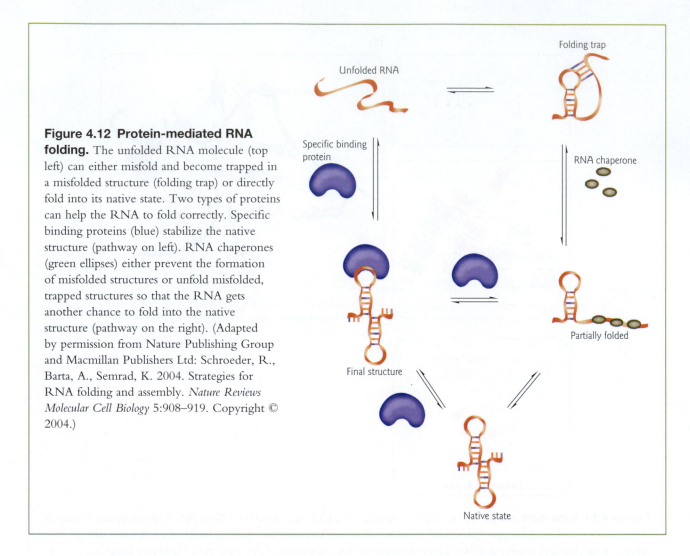

Figure 4.12 Protein-mediated RNA folding. The unfolded RNA molecule (top left) can either misfold and become trapped in a misfolded structure (folding trap) or directly fold into its native state. Two types of proteins can help the RNA to fold correctly. Specific binding proteins (blue) stabilize the native structure (pathway on left). RNA chaperones (green ellipses) either prevent the formation of misfolded structures or unfold misfolded, trapped structures so that the RNA gets another chance to fold into the native structure (pathway on the right). (Adapted by permission from Nature Publishing Group and Macmillan Publishers Ltd: Schroeder, R., Barta, A., Semrad, K. 2004. Strategies for RNA folding and assembly. *Nature Reviews Molecular Cell Biology* 5:908–919. Copyright © 2004.)

backbone only in places where it was accessible. As soon as the RNA formed a three-dimensional structure, the backbone region that was located inside the structure became inaccessible and was protected from cleavage. RNA folding could then be monitored by the appearance of the protected regions with time. The most stable domain was shown to form within several seconds, but the catalytic center of this large ribozyme required several minutes to complete folding. A portion of the catalytic center is susceptible to misfolding and the formation of an alternative helix. The resolution of this helix into the correct helix is a slow step.

4.5 RNA is involved in a wide range of cellular processes

Five major types of RNA serve unique roles in mediating the flow of genetic information (Fig. 4.14). Ribosomal RNA (rRNA) is an essential component of the ribosome. Messenger RNA (mRNA) is a copy of the genomic DNA sequence that encodes a gene product and binds to ribosomes in the cytoplasm. Transfer RNAs (tRNAs) are "charged" with an amino acid. They deliver to the ribosome the appropriate amino acid via interaction of the tRNA anticodon with the mRNA codon. Small nuclear RNA (snRNA) has a role in pre-mRNA splicing, a process which prepares the mRNA for translation, and small nucleolar RNA (snoRNA) has a role in rRNA processing. The role of RNA in RNA processing and translation is discussed in detail in Chapters 13 and 14, respectively.

This general way of thinking about the pathway of gene expression from DNA to functional product via an RNA intermediate overemphasizes proteins as the ultimate goal. What came as a surprise early on was

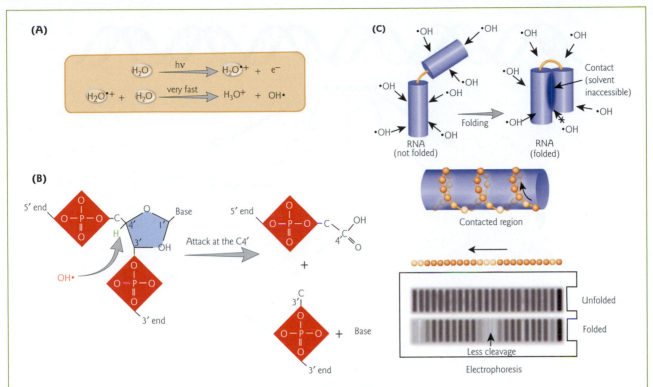

Figure 4.13 Hydroxyl radical footprinting of an RNA structure. (A) Production of hydroxyl radicals (OH·) by ionizing radiation. (B) Cleavage of the RNA backbone after hydrogen removal at the C4′ atom by the electrophilic, highly reactive hydroxyl radical. (C) A large RNA can fold into a tertiary structure under the appropriate solution conditions. The tertiary contacts within the folded RNA molecule result in local reductions in solvent accessibility (the shaded interface). The hydroxyl radicals cannot react with the protected backbone sugar, and hence there is a reduced cleavage of protected nucleotides. The circles indicate individual nucleotides. Their shading reflects the observed intensity of the electrophoretic bands that would be observed for this hypothetical structure. These regions of reduced intensity are termed "footprints." Only one strand of a helix is shown for clarity. (Redrawn from Brenowitz, M., Chance, M.R., Dhavan, G., Takamoto, K. 2002. Probing the structural dynamics of nucleic acids by quantitative time-resolved and equilibrium hydroxyl radical "footprinting." *Current Opinion in Structural Biology* 12:648–653.)

the discovery of the tremendous variety and versatility of functional RNA products (Table 4.1). RNA is involved in a wide range of cellular processes along the pathway of gene expression, including DNA replication, RNA processing, mRNA turnover, protein synthesis, and protein targeting. One of the most important findings in molecular biology in the last 25 years was the discovery that RNA molecules can catalyze chemical reactions in living cells. This led to the hypothesis that the prebiological world was an "RNA world," populated by RNAs that performed both the informational function of DNA and the catalytic function of proteins (Focus box 4.1).

Contributing to the versatility of RNA function is the ability of RNA to form complementary base pairs with other RNA molecules and with single-stranded DNA. The ability of RNA to make specific base pairs is key to understanding its role in everything from post-transcriptional gene silencing to translation. RNA–protein interactions are also of central importance. Most of the RNA in a eukaryotic cell is associated with protein as part of RNA–protein complexes termed ribonucleoprotein (RNP) particles. In addition, most, if not all, RNA-based catalytic reactions are thought to take place in conjunction with proteins. In other chapters, some of these important RNP complexes, such as the ribosome, are discussed in detail. Functional outcomes of RNA–nucleic acid and RNA–protein interactions are categorized below. Specific examples are highlighted in Table 4.1.

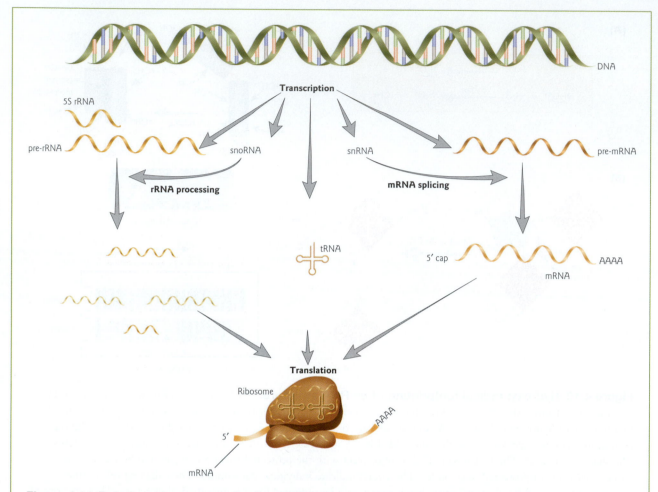

Figure 4.14 Relationships among the five major types of RNA during gene expression. Overview of the role of ribosomal RNA (rRNA), messenger RNA (mRNA), transfer RNA (tRNA), small nuclear RNA (snRNA), and small nucleolar RNA (snoRNA) in RNA processing and protein synthesis.

1 RNA can serve as a "scaffold." An RNA molecule may act as a scaffold or framework upon which proteins can be assembled in an orderly fashion, as is the case in the signal recognition particle (SRP). Proteins recognize the primary nucleotide sequence of RNA and/or secondary and tertiary structural motifs.

2 RNA–protein interactions can influence the catalytic activity of proteins. In some catalytic RNPs, the protein functions as the enzyme, but the RNA is required to target or bind the enzyme to the substrate. An example of this is telomerase, where the RNA serves as the template for the addition of deoxynucleoside triphosphates (dNTPs) by the reverse transcriptase protein. In contrast, in other catalytic RNPs, such as ribonuclease (RNase) P and the ribosome, the RNA is catalytic, not the protein.

3 RNA can be catalytic. RNA molecules termed "ribozymes" can catalyze a number of the chemical reactions that take place in living cells (see Sections 4.6 and 4.7 below).

4 Small RNAs can directly control gene expression. Examples of how RNA plays a role in gene regulation will be discussed in detail in later chapters. These include differential RNA folding and riboswitches (Section 10.7), and RNA interference and microRNAs (Section 13.10).

5 RNA can be the hereditary material. Many viruses have RNA genomes and are either self-replicating or replicate through a DNA intermediate (see Section 3.7).

Table 4.1 RNPs are involved in a wide range of cellular processes.

RNP	Point in pathway of gene expression	Function of RNP	Composition of RNP	Role of RNA component	Cross reference
Telomerase	DNA replication	Adds telomeric repeats to the ends of chromosomes during DNA replication	Telomerase RNA + protein (reverse transcriptase)	Template for reverse transcriptase	Chapter 6
RNase MRP (ribonuclease mitochondrial RNA processing)	DNA replication and RNA processing	Cleaves RNA primer in mtDNA replication; role in processing 5.8S ribosomal RNA in the nucleolus	7-2 RNA + proteins	Catalytic RNP	Chapter 6
Spliceosome	RNA processing	Removal of introns from nuclear pre-mRNA	snRNAs + ~200 proteins	Strong evidence that U6 and U2 snRNA catalyze splicing	Chapter 13
RNase P	RNA processing	Generates 5′ end of mature tRNAs	*E. coli*: M1 RNA + C5 protein Human: H1 RNA + ~10 proteins	*E. coli*: catalytic RNA Human: catalytic RNP	Chapter 4
Ribosome	Translation	Protein synthesis machinery	Four rRNAs + > 50 ribosomal proteins	23S rRNA catalyzes peptide bond formation	Chapter 14
Signal recognition particle (SRP)	Protein targeting	Mediates protein targeting to the endoplasmic reticulum	7S RNA + six proteins	RNA serves as a scaffold for organized binding of proteins	Chapter 14

4.6 Historical perspective: the discovery of RNA catalysis

Thousands of different chemical reactions are required to carry out essential processes in living cells. These reactions may take place spontaneously, but they rarely occur at a rate fast enough to support life. Catalysis is necessary for these biochemical reactions to proceed at a useful rate. In the presence of a catalyst, reactions can be accelerated by a factor of a billion or even a trillion under physiological conditions, in a highly specific, regulated manner. For a very long time it was assumed that biological catalysis depended exclusively on protein enzymes. Then, in a landmark discovery at the beginning of the 1980s, two labs demonstrated independently that RNA can also possess catalytic activity.

Thomas Cech and co-workers published a report in 1982 that generated great excitement within the scientific community. In their paper they demonstrated that the single intron of the large ribosomal RNA of *Tetrahymena thermophila* has self-splicing activity *in vitro*. A year later, Sidney Altman and co-workers showed that the RNA component of RNase P from *Escherichia coli* is able to carry out processing of pre-tRNA in the absence of its protein subunit *in vitro*. The discovery of self-splicing RNA was completely unexpected. Needless to say, many control experiments had to be performed to convince all skeptics that RNA itself could possess catalytic activity. In 1989 Cech and Altman were awarded the Nobel Prize in chemistry for this revolutionary discovery. The following sections provide a brief synopsis of the experiments leading to this breakthrough and highlight some of the current research in the field.

Tetrahymena group I intron ribozyme

In 1979, Thomas Cech was studying transcription of ribosomal RNA genes from the ciliated protozoan *Tetrahymena thermophila*. After using the "R looping" technique of electron microscopy to hybridize 17S and 26S rRNA with ribosomal DNA, he saw the expected R loops caused by the rRNA hybridizing to the

Molecular biologists who speculate on the origins of life on earth are faced with a classic "chicken and egg problem" – which came first, proteins or nucleic acids? In the modern world, the replication of DNA and RNA is dependent on protein enzymes, and the synthesis of protein enzymes is dependent on DNA and RNA. The term "RNA world" was introduced by Walter Gilbert in 1986 to describe a hypothetical stage in the evolution of life some 4 billion years ago when RNA both carried the genetic information and catalyzed its own replication. The origin and prebiotic chemistry of this RNA world, of course, remains open to speculation.

According to the RNA world hypothesis, "life" first existed in the form of replicating RNA molecules (Fig. 1). In this ancient world neither protein nor DNA existed yet. Evidence in support of this hypothesis is that proteins cannot replicate themselves, except via mechanisms that involve an RNA intermediate. In contrast, RNA has all the structural prerequisites necessary for self-replication. RNA genomes are widespread among viruses and their replication in infected cells proceeds via complementary RNA chains. Compared with DNA or protein, RNA is clearly the most

self-sufficient molecule. RNA molecules are capable of doing basically all that proteins can do. They can self-fold into specific three-dimensional structures, recognize other macromolecules and small ligands with precision, and perform catalysis of covalent reactions. Ribozymes can catalyze a diversity of reactions including polymerizing nucleotides, ligating DNA, cleaving DNA phosphodiester bonds, and synthesizing peptides. The later discovery that the ribosome – the catalyst still responsible for synthesizing nearly all proteins in cells – is, in fact, a ribozyme provides strong evidence for an RNA world.

At some point, requirements for enzymes with a greater repertoire of functional groups, more stable tertiary structure, and superior catalytic powers are thought to have favored the transition from RNA-based catalysis to protein-based catalysis that is present in the current DNA/RNA/protein world (Fig. 1). The original RNA world, if it ever existed on earth, is long gone. But a modern RNA world exists that has been vastly underestimated. Each year, more and more new species of noncoding RNAs with important roles in cells are being discovered (see Section 13.10).

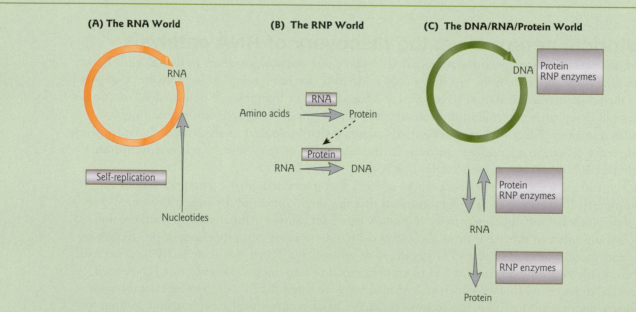

Figure 1 The RNA world and the transition to the present DNA/RNA/protein world. (A) In the RNA world, RNA functioned as both a carrier of information and an enzyme. It catalyzed its own replication. (B) During the transitional period, RNA catalyzed the synthesis of proteins, and these proteins catalyzed the transition from RNA to DNA. (C) Today, proteins and RNPs catalyze the replication of DNA. They also catalyze the transcription of DNA into RNA, and the reverse transcription of RNA into DNA. The translation of mRNA into proteins is mediated by the ribosome, a large ribozyme.

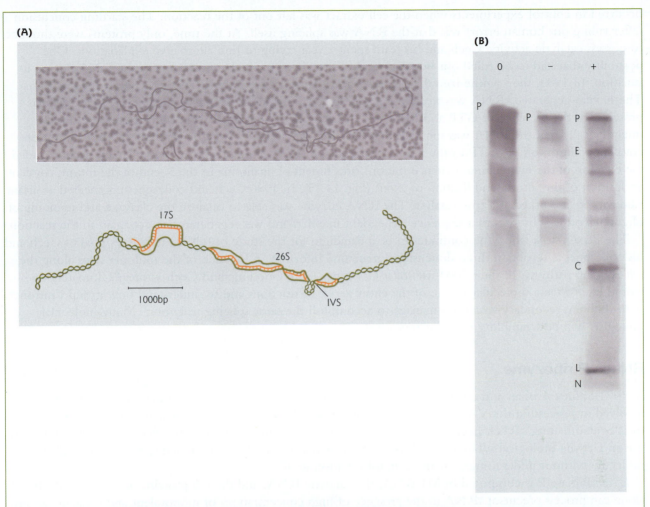

Figure 4.15 Self-splicing of *Tetrahymena* preribosomal RNA (pre-rRNA). (A) When 17S and 26 rRNA were hybridized with rDNA, the two expected R loops were seen by electron microscopy. Each R loop consists of an RNA–DNA hybrid that displaces one strand of the duplex DNA. A small loop structure interrupted the R loop between 26S rRNA and DNA. This looped out stretch of DNA was an intervening sequence (IVS) or intron, which is spliced out in the final RNA product. In the schematic shown, the green line is the single-stranded DNA and the orange line is RNA. (Redrawn from Echols, H. 2001. *Operators and Promoters. The Story of Molecular Biology and its Creators*. University of California Press, Berkeley, CA.) (B) Radiolabeled *Tetrahymena* pre-26S rRNA was transcribed *in vitro* with SP6 RNA polymerase. The pre-rRNA was then tested for splicing of the intron under various conditions. 0, no further incubation; −, incubation at 30°C for 75 minutes in splicing buffer with GTP omitted; +, incubation under the same conditions with GTP. Samples were separated by polyacrylamide gel electrophoresis and visualized by autoradiography. P, precursor RNA containing intron; E, ligated exons; C, spliced circular intron RNA; L, spliced linear intron RNA; N, spliced nicked circular intron RNA. The experiment shows that GTP is required for intron splicing. (Reproduced from Price, J.V., Kieft, G.L., Kent, J.R., Sievers, E.L., Cech, T.R. 1985. Sequence requirements for self-splicing of the *Tetrahymena thermophila* pre-ribosomal RNA. *Nucleic Acids Research* 13:1871–1889, by permission of Oxford University Press.)

complementary DNA. He also saw a small loop structure that interrupted the R loop between 26S rRNA and DNA. This looped out stretch of DNA within the RNA–DNA hybrid was an intervening sequence or "intron," which is spliced out in the final RNA product (Fig. 4.15). To follow up on this observation, Cech and colleagues attempted to develop an *in vitro* assay in which they could fractionate cell extracts and determine the proteins required for splicing. Completely unexpectedly, splicing of the rRNA intron occurred in control experiments when the cell extract was left out of the reaction. The startling conclusion (after ruling out human error) was that the RNA was splicing itself. At the time, only proteins were thought to possess catalytic activity. Cech and his team spent a year trying to find alternative explanations. One possibility that had to be ruled out was that residual proteins remained associated with the RNA during its isolation. In 1982, they synthesized the precursor rRNA from a recombinant rDNA gene cloned in *E. coli*. The *in vitro*-generated rRNA was made using pure RNA polymerase in the absence of any other cellular proteins. In the presence of GTP and Mg^{2+} the "naked" rRNA still underwent splicing, demonstrating unequivocally that the RNA was splicing itself. Self-splicing activities were determined by the amount of covalent addition of ^{32}P-GTP to the 5′ end of the intron RNA. Reactions that were characterized included the excision of the intervening sequence (intron), attachment of guanosine to the 5′ end of the intron, covalent cyclization of the intron, and ligation of exons (Fig. 14.15). In 1986 Cech and colleagues engineered a variant ribozyme that worked as a true catalyst. The RNA enzyme was able to catalyze the cleavage and rejoining of oligonucleotide substrates in a sequence-dependent manner, and was regenerated to act again in the reaction.

The *Tetrahymena* ribozyme continues to be a paradigm for the study of RNA catalysis. A goal of Cech and his colleagues is to obtain three-dimensional structural information for each of the multiple steps along the self-splicing pathway. In their 2004 *Molecular Cell* paper Guo, Gooding, and Cech wrote: "Ultimately one would like to see a molecular movie of the entire series of reactions and to understand how group I introns with different secondary structures manage to accomplish the same splicing reactions." Many molecular biologists will be scrambling for front row seats!

RNase P ribozyme

In 1971 Sidney Altman and co-workers began trying to purify and characterize RNase P, the enzyme involved in processing the 5′-leader sequence of precursor tRNA in *E. coli*. After many attempts to remove the "contaminating" RNA from the preparation, 12 years later they demonstrated that the RNA component was in fact the biological catalyst. The RNase P RNA is a true RNA catalyst, acting on another RNA molecule without undergoing a chemical transformation itself.

E. coli RNase P is composed of M1 RNA, the catalytic RNA, and the C5 protein. *In vitro*, the M1 RNA alone can process precursor tRNA in the presence of high concentrations of monovalent and divalent cations. *In vivo*, the C5 protein is required to enhance the catalytic efficiency of M1 RNA and increase its substrate versatility. In contrast, in human cells, H1 RNA associates with at least 10 distinct protein subunits to form RNase P. The proposed tertiary structure of H1 RNA conforms to the catalytic core configuration of *E. coli* M1 RNA. However, H1 RNA shows no catalytic activity *in vitro*, unless associated with protein subunits Rpp21 and Rpp29 (Fig. 4.16). Eukaryotic RNase P is assembled in the nucleolus and shares some subunits with RNase MRP (mitochondrial RNA processing), including Rpp29 (see Table 4.1 and Section 6.7). Thus, while bacterial RNase P is an RNA enzyme, its eukaryotic counterpart acts as a catalytic ribonucleoprotein.

4.7 Ribozymes catalyze a variety of chemical reactions

RNA molecules with catalytic activity are called RNA enzymes or "ribozymes." Naturally occurring ribozymes are often autocatalytic, which leads to their own modification. This characteristic contradicts the classic definition of an enzyme, which is "a substance that increases the rate, or velocity, of a chemical reaction without itself being changed in the overall process." However, catalytic RNAs have been discovered that are true enzymes. For example, the 23S rRNA in the ribosome catalyzes peptide bond formation without being modified in the process (see Section 14.5).

Mode of ribozyme action

Ribozymes catalyze reactions essentially in the same ways that proteins do. They form substrate-binding sites and lower the activation energy of a reaction, thus allowing the reaction to proceed much faster.

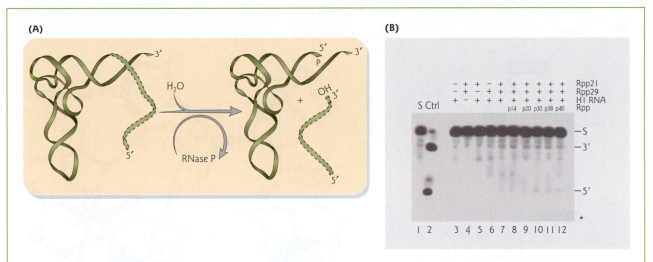

Figure 4.16 Maturation of tRNA catalyzed by RNase P. (A) The 5′-leader sequence (dashed ribbon) of tRNA is removed in a processing reaction catalyzed by RNase P. (B) Reconstitution of endonuclease activity of human RNase P. The indicated recombinant RNase P-associated protein subunits (Rpp) and H1 RNA were incubated with radiolabeled precursor tRNA^{Tyr} in a cleavage reaction. Cleavage products, tRNA (3′) and 5′-leader sequence (5′), were separated by polyacrylamide gel electrophoresis and visualized by autoradiography. Uncleaved substrate (S) and a control assay with purified human RNase P (Ctrl) are shown in lanes 1 and 2, respectively. The experiment shows that protein subunits of RNase P are required for its catalytic activity; H1 RNA alone cannot remove the 5′ leader sequence of tRNA. (Reprinted from Mann, H., Ben-Asouli, Y., Schein, A., Moussa, S., Jarrous, N. 2003. Eukaryotic RNaseP: role of RNA and protein subunits of a primordial catalytic ribonucleoprotein in RNA-based catalysis. *Molecular Cell* 12:925–935. Copyright © 2003, with permission from Elsevier.)

Many ribozymes are metalloenzymes. Binding of divalent cations (e.g. Mg^{2+}) in the active site is critical for their folding into an active state. Interestingly, even though RNA enzymes and protein enzymes are not evolutionarily related, the active site of a self-splicing group I intron has the same orientation of two metal ions as found in a protein-based DNA polymerase (Fig. 4.17). This observation points to the importance of the two-metal-ion mechanism of catalysis in reactions involving phosphate transfer. However, ribozymes are not limited to using metal ions as functional groups in catalysis. Some ribozymes may use general acid–base chemistry, in which nucleotide bases, sugar hydroxyl groups, and even the phosphate backbone directly contribute to catalysis by donating or accepting protons during the chemical step of the reaction.

Naturally occurring ribozymes are classified into two different groups, the large and small ribozymes, based on differences in size and reaction mechanism.

Large ribozymes

The RNA component of RNase P, and members of the group I and group II intron family, belong to the group of large ribozymes. Group I and group II ribozymes are self-splicing introns that are discussed in detail in a later chapter on RNA processing (see Section 13.3). They vary in size from a few hundred nucleotides up to about 3000 nucleotides, and are further distinguished from the small ribozymes by all cleaving RNA to generate 3′-OH termini, as opposed to a product with a 2′,3′-cyclic phosphate and a product with a 5′-OH terminus (Table 4.2). Additional large ribozymes are the RNA components of the spliceosome, which also have enzymatic properties (see Section 13.5), and the ribosomal RNAs, characterized by their ability to catalyze peptide bond formation (see Section 14.5).

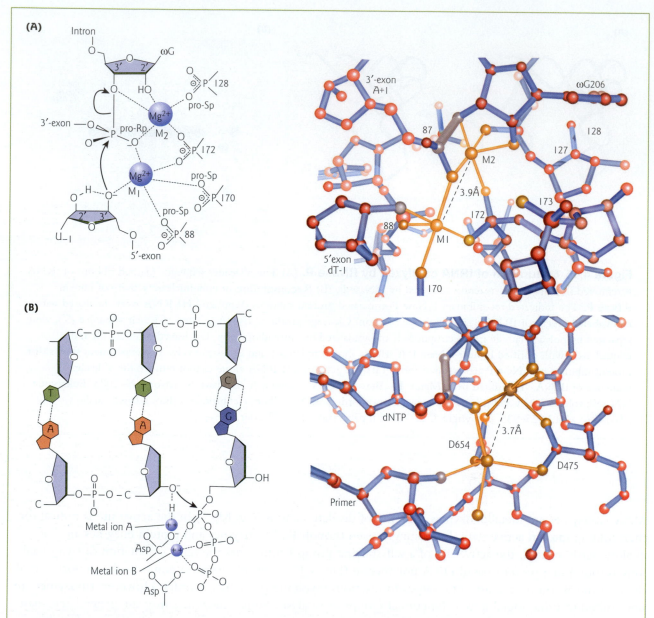

Figure 4.17 Similarity between group I intron and protein-based DNA polymerase active sites. The active sites of a self-splicing group I intron and bacteriophage T7 DNA polymerase are compared. The 5′-exon is analogous to the primer oligonucleotide strand, the 3′-exon to the incoming deoxynucleotide triphosphate (dNTPs), and the ωG (the last nucleotide of the intron) to the pyrophosphate leaving group. Both sites contain two metal ions, M_1 (Metal ion A) and M_2 (Metal ion B), and coordinate those metals in a similar manner. In DNA polymerase, the two metals are held in place by interaction with two highly conserved aspartate residues. The active site Mg^{2+} ions are shown as large blue spheres, the predicted inner and outer sphere ligands are shown as small orange spheres, and the metal-to-metal distance is labeled. Orange lines indicate inner sphere coordinations. (A) Two-metal active site coordination within the group I intron active site. The splicing reaction involving attack on the phosphodiester bond between the exon and intron, with loss of ωG, is shown with curved arrows. (B) Two-metal active site coordination within the T7 DNA polymerase. M_1 (Metal ion A) interacts with the triphosphates of incoming dNTPs to neutralize their negative charge. After catalysis, the pyrophosphate product is stabilized through similar interactions with M_2 (Metal ion B). (Structures reprinted with permission from Stahley, M.R., Strobel, S.A. 2005. Structural evidence for a two-metal-ion mechanism of group I intron splicing. *Science* 309:1587–1590. Copyright © 2005 AAAS.)

Table 4.2 Types of naturally occurring ribozymes.

Ribozyme	Source	Function	Reaction products
Small ribozymes			
Hammerhead	Plant viroids and newt satellite RNAs	Replication	5′-OH; 2′,3′-cyclic phosphate
Hairpin	Plant satellite RNAs	Replication	5′-OH; 2′,3′-cyclic phosphate
HDV	Hepatitis delta virus (human)	Replication	5′-OH; 2′,3′-cyclic phosphate
VS	*Neurospora crassa* mitochondria	Replication	5′-OH; 2′,3′-cyclic phosphate
Riboswitch ribozyme	*Bacillus subtilis*	*glmS* mRNA self-degradation	5′-OH; 2′,3′-cyclic phosphate
Large ribozymes			
RNase P	Eukaryotes, prokaryotes	tRNA processing	5′-phosphate and 3′-OH
Group I introns	Eukaryotes (nucleus, organelles), prokaryotes, bacteriophages	Splicing	Intron with 5′-guanosine and 3′-OH, 5′/3′-ligated exons
Group II introns	Eukaryotes (organelles), prokaryotes	Splicing	Intron with 2′–5′-lariat and 3′-OH; 5′/3′-ligated exons
Spliceosome	Eukaryotes (nucleus)	Pre-mRNA splicing	Intron with 2′–5′-lariat and 3′-OH; 5′/3′-ligated exons
Ribosome	Eukaryotes, prokaryotes	Translation	Peptide bond

Small ribozymes

The group of small ribozymes includes the hammerhead and hairpin motif, the hepatitis delta virus (HDV) RNA, the Varkud satellite (VS) RNA, and the *glmS* riboswitch ribozyme (Table 4.2). These five different ribozymes range in size from about 40 nt up to 154 nt. The hammerhead, so called for its three helices in a T shape, is the most frequently found catalytic motif in plant pathogenic RNAs, such as viroids (Fig. 4.18) (see Section 3.7). The hairpin ribozyme has only been found in some virusoids. The HDV RNA is a viroid-like satellite virus of the human hepatitis B virus (HBV) that when present causes an exceptionally strong type of hepatitis in infected patients. The VS ribozyme is part of a larger RNA that is transcribed from a plasmid found in the mitochondria of some strains of *Neurospora crassa*, a filamentous fungus. The *glmS* riboswitch ribozyme is involved in regulating bacterial gene expression (see Section 10.7 for details).

With the exception of the riboswitch ribozyme, which is involved in gene regulation, the catalytic motifs in the small ribozymes are all involved in their self-replication. Replication of the circular RNAs occurs via a "rolling circle" mechanism (see Section 6.8). This leads to the formation of long linear transcripts consisting of monomers joined in tandem. These are self-cleaved into monomers by the catalytic motifs. The self-cleavage of phosphodiester bonds occurs by "in line" nucleophilic substitution (Fig. 4.18). The internal 2′-OH group of the ribose next to the phosphodiester bond to be cleaved attacks the phosphate, leading to an inversion of the configuration around the phosphorus. The incoming group is "in line" with the hydroxyl group in the transition state leaving the reaction center. The reaction yields a product with a 2′,3′-cyclic phosphate and a product with a 5′-OH terminus. This catalytic property suggests that viroids and other subviral pathogens may have an ancient evolutionary origin independent of viruses, dating back to the RNA world (see Focus box 4.1).

Since their discovery, small ribozymes have received much attention for their potential as tools to combat viral diseases. For example, ribozymes are being tested for their ability to inhibit the replication of human immunodeficiency virus type 1 (HIV-1), the causative agent of acquired immune deficiency syndrome (AIDS) (see Section 17.3).

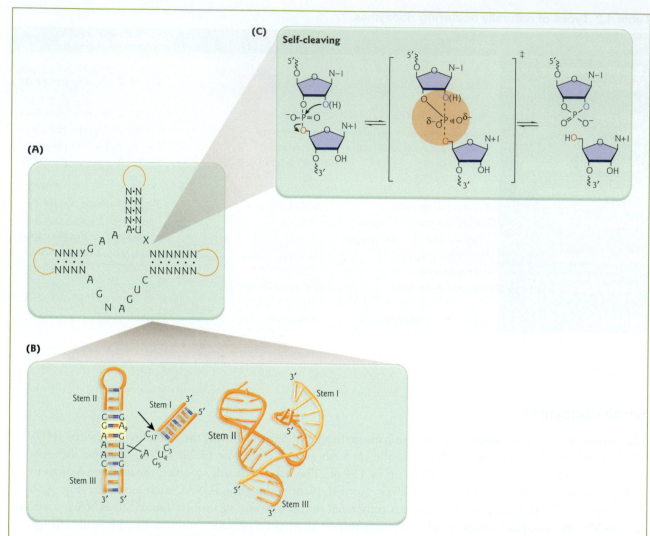

Figure 4.18 Hammerhead ribozyme. The secondary structure of the hammerhead ribozyme consensus sequence is represented according to the original scheme (A), and according to recent X-ray crystallography data (B). The tertiary structure is also depicted. The arrow shows the site of self-cleavage. Y = C, U; X = A, C, U. In the new model, stems I, II, and III are base-paired helices oriented in a Y shape around a core of conserved nucleotides. (C) The self-cleavage reaction proceeds by "in-line" (S_N2 type) nucleophilic substitution. The 2′-hydroxyl is the attacking nucleophile (blue) and the bridging 5′-oxygen (red) is the leaving group. There is an inversion of the stereochemical configuration of the nonbridging oxygen atoms that are bound to the phosphorus which is undergoing attack, leading to an intermediate or transition state, in which five electronegative oxygens form transient bonds with phosphorus (yellow shading). N − 1 and N + 1 are the nucleotide bases on the 5′ and 3′ sides of the reactive phosphodiester bond, respectively. The symbol ‡ indicates the transition state, and (H) represents hydrogens for which it is not clear whether, or how closely, they are associated with the oxygens. (Redrawn from Fedor, M.J. and Williamson, J.R. 2005. The catalytic diversity of RNAs. *Nature Reviews Molecular Cell Biology* 6:399–412.)

Chapter summary

RNA is a chain-like molecule composed of subunits called nucleotides joined by phosphodiester bonds. Some of the common secondary structures that form the building blocks of RNA structure are bulges, base-paired A-type double helices (stems), single-stranded hairpin or internal loops, junctions, and turns. Base-paired stems often contain noncanonical base pairs, such as GU pairs or base triples. In addition,

RNA often contains a variety of modified nucleosides, such as inosine or pseudouridine. RNA chains fold into unique three-dimensional structures that act similarly to globular proteins. Important insights in RNA folding motifs have come from X-ray crystallographic studies of the structure of tRNA, group I introns, and rRNA. Preformed secondary structural domains of RNA fold to form a tertiary structure stabilized by many long-range interactions including coaxial stacking of helices, and formation of pseudoknots, A-minor motifs, tetraloops, ribose zippers, and kink-turn motifs. The structural flexibility of the RNA backbone and the tendency of nucleotides to base pair with complementary regions can lead to misfolding of RNA. Specific RNA-binding proteins form tight complexes with their target RNAs *in vivo* and act as chaperones to aid in proper RNA folding.

In addition to the five major types of RNA – rRNA, mRNA, tRNA, snRNA, and snoRNA – there is a tremendous diversity of functional RNA products. RNA is involved in a wide range of cellular processes along the pathway of gene expression from DNA replication to protein synthesis. Contributing to this versatility is the ability of RNA to form complementary base pairs with other RNAs and with single-stranded DNA, and to interact with proteins as part of RNPs.

A landmark discovery in the late 1970s to early 1980s was that RNA can be catalytic. RNA molecules termed ribozymes catalyze a number of chemical reactions that take place in a living cell, ranging from cleavage of phosphodiester bonds to peptide bond formation. The first ribozymes discovered were a self-splicing intron in *Tetrahymena thermophila* rRNA and the RNA component of RNase P in *E. coli*. Many other ribozymes have been characterized since that time, including other self-splicing introns, components of the spliceosome, the rRNAs, and small ribozymes such as the hammerhead ribozyme which plays a role in self-replication.

Analytical questions

1 Make up an RNA sequence that will form a hairpin with a 9 bp stem and a 7 bp loop. Draw both the primary structure and the secondary structure.

2 What addition(s) would you need to make to the primary sequence in Question 1 to allow pseudoknot formation?

3 You suspect that a tetraloop is critical for the folding of a ribozyme into its active form. Describe an experiment to demonstrate whether the RNA folds into a similar tertiary structure when the tetraloop is deleted.

4 You have discovered a small RNA involved in the removal of a novel type of intron from another RNA transcript. Design an experiment to determine whether the small RNA functions as a catalytic RNA or RNP. Show sample positive results.

Suggestions for further reading

Brenowitz, M., Chance, M.R., Dhavan, G., Takamoto, K. (2002) Probing the structural dynamics of nucleic acids by quantitative time-resolved and equilibrium hydroxyl radical "footprinting." *Current Opinion in Structural Biology* 12:648–653.

Cate, J.H., Gooding, A.R., Podell, E., et al. (1996) Crystal structure of a group I ribozyme domain: principles of RNA packing. *Science* 273:1678–1685.

Correll, C.C., Swinger, K. (2003) Common and distinctive features of GNRA tetraloops based on the GUAA tetraloop structure at 1.4 Å resolution. *RNA* 9:355–363.

Crick, F.H. (1966) The genetic code – yesterday, today, and tomorrow. *Cold Spring Harbor Symposium on Quantitative Biology* 31:1–9.

Doublie, S., Tabor, S., Long, A.M., Richardson, C.C., Ellenberger, T. (1998) Crystal structure of a bacteriophage T7 DNA replication complex at 2.2 Å resolution. *Nature* 391:251–258.

Doudna, J.A., Lorsch, R.A. (2005) Ribozyme catalysis: not different, just worse. *Nature Structural and Molecular Biology* 12:395–402.

Echols, H. (2001) *Operators and Promoters. The Story of Molecular Biology and its Creators*. University of California Press, Berkeley, CA.

Fedor, M.J., Williamson, J.R. (2005) The catalytic diversity of RNAs. *Nature Reviews Molecular Cell Biology* 6:399–412.

Gesteland, R.F., Cech, T.R., Atkins, J.F. (1999) *The RNA World*, 2nd edn. Cold Spring Harbor Laboratory Press, Cold Spring Harbor, NY.

Guo, F., Gooding, A.R., Cech, T.R. (2004) Structure of the *Tetrahymena* ribozyme: base triple sandwich and metal ion at the active site. *Molecular Cell* 16:351–362.

Klein, D.J., Schmeing, T.M., Moore, P.B., Steitz, T.A. (2001) The kink-turn: a new RNA secondary structure motif. *EMBO Journal* 20:4212–4221.

Kruger, K., Grabowski, P.J., Zaug, A.J., Sands, J., Gottschling, D.E., Cech, T.R. (1982) Self-splicing RNA: autoexcision and autocyclization of the ribosomal RNA intervening sequence of *Tetrahymena*. *Cell* 31:147–157.

Mann, H., Ben-Asouli, Y., Schein, A., Moussa, S., Jarrous, N. (2003) Eukaryotic RNaseP: role of RNA and protein subunits of a primordial catalytic ribonucleoprotein in RNA-based catalysis. *Molecular Cell* 12:925–935.

Moore, P.B., Steitz, T.A. (2003) The structural basis of large ribosomal subunit function. *Annual Review of Biochemistry* 72:813–850.

Nagaswamy, U., Voss, N., Zhang, Z., Fox, G.E. (2000) Database of non-canonical base pairs found in known RNA structures. *Nucleic Acids Research* 28:375–376.

Noller, H.F. (2005) RNA structure: Reading the ribosome. *Science* 309:1508–1514.

Orgel, L.E. (2004) Prebiotic chemistry and the origin of the RNA world. *Critical Reviews in Biochemistry and Molecular Biology* 39:99–123.

Pleij, C.W.A. (1990) Pseudoknots: a new motif in the RNA game. *Trends in Biochemical Sciences* 15:143–147.

Price, J.V., Kieft, G.L., Kent, J.R., Sievers, E.L., Cech, T.R. (1985) Sequence requirements for self-splicing of the *Tetrahymena thermophila* pre-ribosomal RNA. *Nucleic Acids Research* 13:1871–1889.

Schroeder, R., Barta, A., Semrad, K. (2004) Strategies for RNA folding and assembly. *Nature Reviews Molecular Cell Biology* 5:908–919.

Sclavi, B., Sullivan, M., Chance, M.R., Brenowitz, M., Woodson, S.A. (1998) RNA folding at millisecond intervals by synchrotron hydroxyl radical footprinting. *Science* 279:1940–1943.

Stahley, M.R., Strobel, S.A. (2005) Structural evidence for a two-metal-ion mechanism of group I intron splicing. *Science* 309:1587–1590.

Tamura, M., Holbrook, S.R. (2002) Sequence and structural conservation in RNA ribose zippers. *Journal of Molecular Biology* 320:455–474.

Theimer, C.A., Blois, C.A., Feigon, J. (2005) Structure of the human telomerase RNA pseudoknot reveals conserved tertiary interactions essential for function. *Molecular Cell* 17:671–682.

Spirin, A.S. (2002) Omnipotent RNA. *FEBS Letters* 530:4–8.

Waas, W.F., de Crécy-Lagard, V., Schimmel, P. (2005) Discovery of a gene family critical to wyosine base formation in a subset of phenylalanine-specific transfer RNAs. *Journal of Biological Chemistry* 280:37616–37622.

Zaug, A.J., Cech, T.R. (1986) The intervening sequence RNA of *Tetrahymena* is an enzyme. *Science* 231:470–475.

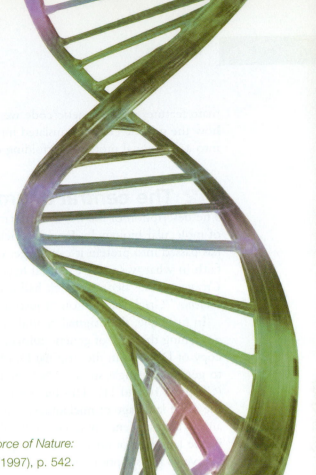

Chapter 5

From gene to protein

Life . . . is a relationship between molecules.

Linus Pauling, as quoted in T. Hager, *Force of Nature: The Life of Linus Pauling* (1997), p. 542.

Outline

5.1 Introduction

In the late 1930s, geneticists knew that genes were the basic unit of inheritance. They knew a great deal about how genes function, but they did not know the composition of a gene. Descriptions were vague and ranged from "a molecule of living stuff made up of many atoms held together" to the "ultimate unit of life." We now know that a gene is a specific stretch of nucleotides in DNA (or in some viruses, RNA) that contains information for making a particular RNA molecule that in most cases is used to make a particular protein. Each cell contains thousands of different genes and makes thousands of different RNAs and proteins. Gene expression is the process by which the information in DNA is converted to RNA and then to protein. The main features of the genetic code were established nearly 50 years ago. This chapter gives an overview of how the genetic code is translated into a sequence of amino acids, and how this chain of amino acids folds into a functional protein. Misfolding of proteins is associated with a number of diseases.

5.2 The central dogma

The description of the flow of information involving the genetic material was termed the "central dogma" of molecular biology by Francis Crick in 1957. The central dogma was stated by Crick as "once information has passed into protein it cannot get out again." Crick's choice of the word dogma was not a call for blind faith in what was really a central hypothesis. According to Horace Judson in his book *The Eighth Day of Creation*, it was because Crick had it in his mind that "a dogma was an idea for which there was *no reasonable evidence.*" Crick told Judson "I just didn't *know* what dogma *meant* . . . Dogma was just a catch phase."

In principle, the original central dogma still holds true with some modern updates (Fig. 5.1). Each process governing the flow of genetic information has been given a specific name. The process of making an exact copy of DNA from the original DNA is called "replication" (Chapter 6). The process of DNA being copied to generate a single-strand RNA identical in sequence to one strand of duplex DNA is called "transcription" (Chapters 10 and 11). This term is used because the information is rewritten (transcribed), but in basically the same language of nucleotides. The process of the RNA nucleotide sequence being converted into the amino acid sequence of a protein is called "translation" (Chapter 14). This term denotes that the information in the language of nucleotides is copied (translated) into another language – the language of amino acids. "Reverse transcription" is the process of a single-stranded DNA copy being generated from single-stranded RNA. Reverse transcriptase activity is a function of the enzyme telomerase (Chapter 6) and is present in retroviruses for the replication of their genome (Chapter 3). Finally, RNA may be copied directly into RNA. As discussed in Chapter 3, some viruses have an RNA genome and duplicate their genome without any DNA intermediate.

5.3 The genetic code

This section addresses the basic question of how the genetic code is translated into a specific sequence of amino acids. Full appreciation of the process of translation requires an understanding of the intricate machinery involved in deciphering the genetic code. The mechanism of translation is described in detail in Chapter 14. Before reaching the key step where a peptide bond forms between two amino acids, many events must take place. Ribosomes must be assembled in the nucleolus of the cell from a diversity of different gene products. tRNAs must be "charged" with their appropriate amino acid, and all the players must join together in the cytoplasm. In addition to the ribosome, other key players in the process of protein synthesis are the messenger RNA (mRNA) carrying the genetic information in the form of the genetic (triplet) code and transfer RNA (tRNA).

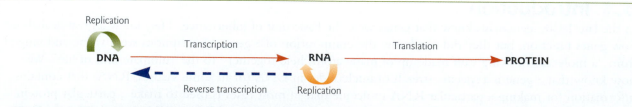

Figure 5.1 The central dogma of molecular biology. Solid arrows show transfers of information that occur in all cells; dashed arrows show transfers that can occur in special cases.

Translating the genetic code

A DNA sequence is read in triplets using the antisense (non-coding) strand as a template that directs the synthesis of RNA via complementary base pairing. The other non-template strand of DNA is called the sense (coding) strand and is the strand of DNA that bears the same sequence as the RNA (except for possessing T instead of U) (Fig. 5.2). An open reading frame (ORF) in the mRNA indicates the presence of a start codon followed by codons for a series of amino acids and ending with a termination codon. There is typically a transcribed but untranslated region of the mRNA (UTR). The start codon codes for the amino acid methionine, which is generally cleaved during or after translation to result in the N-terminus of the completed polypeptide.

The genetic code, deciphered almost 50 years ago, provides the fundamental clues for decoding of genetic information into polypeptides by way of translation. In the historical presentation of the genetic code, each "codon box" is composed of four three-letter codes, 64 in all (Table 5.1). Sixty-one codons are recognized by tRNAs for the incorporation of the 20 common amino acids (Fig. 5.3). Three codons signal termination of protein synthesis, or code for selenocysteine and pyrrolysine, the 21st and 22nd amino acids, respectively. The genetic code is said to be "degenerate" because it requires that tRNAs specific for a particular amino acid respond to multiple coding triplets that differ only in the third letter. For example, leucine is coded for by four different codons, while methionine has only one codon. This observation gave rise to the "wobble hypothesis" proposed by Francis Crick (Table 5.2). The hypothesis states that the pairing between codon and anticodon at the first two codon positions always follows the usual rule for complementary base pairing, but that exceptional "wobbles" (non-Watson–Crick base pairing) can occur at the third position.

Initially, the genetic code was thought to be universal. Now, it is known that in certain organisms and organelles the meaning of select codons has been changed; for example, the specification of serine by CUG in *Candida albicans*, or of tryptophan by UGA in mitochondria (Table 5.3).

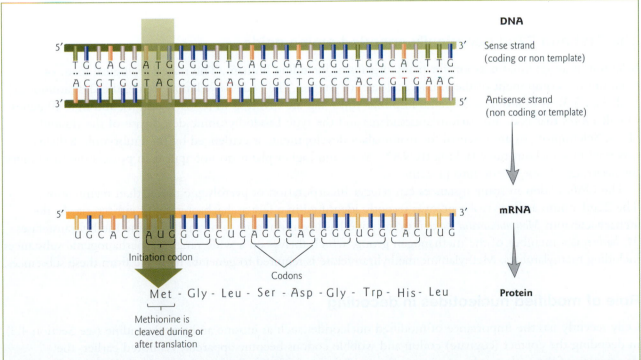

Figure 5.2 Translation of the genetic code. The diagram depicts the relationship between the DNA sense and antisense strands and the mRNA. The triplet codons of the mRNA are translated into a sequence of amino acids to make a protein.

Table 5.1 The genetic code.

UUU Phe	UCU Ser	UAU Tyr	UGU Cys
UUC Phe	UCC Ser	UAC Tyr	UGC Cys
UUA Leu	UCA Ser	UAA **Stop**	UGA **Stop** $\rightarrow$ Sel
UUG Leu	UCG Ser	UAG **Stop** $\rightarrow$ Pyr	UGG Trp
CUU Leu	CCU Pro	CAU His	CGU Arg
CUC Leu	CCC Pro	CAC His	CGC Arg
CUA Leu	CCA Pro	CAA Gln	CGA Arg
CUG Leu	CCG Pro	CAG Gln	CGG Arg
AUU Ile	ACU Thr	AAU Asn	AGU Ser
AUC Ile	ACC Thr	AAC Asn	AGC Ser
AUA Ile	ACA Thr	AAA Lys	AGA Arg
AUG Met **(Start)**	ACG Thr	AAG Lys	AGG Arg
GUU Val	GCU Ala	GAU Asp	GGU Gly
GUC Val	GCC Ala	GAC Asp	GGC Gly
GUA Val	GCA Ala	GAA Glu	GGA Gly
GUG Val	GCG Ala	GAG Glu	GGG Gly

The 21st and 22nd genetically encoded amino acids

Selenium is an essential nutrient for many organisms including humans. The major biological form of selenium is a component of the amino acid selenocysteine. The code for selenocysteine, the 21st amino acid (Fig. 5.3), is found in > 15 genes in prokaryotes that are involved in redox reactions, and in > 40 genes in eukaryotes that code for various antioxidants and the type I iodothyronine deiodinase of the thyroid gland. Selenoproteins are essential for mammalian development, as evidenced by the embryonic lethality observed in knockout mice lacking tRNA^{Sel}. Yeast and higher plants do not appear to possess the machinery for inserting selenocysteine into proteins.

The UAG codon in some instances can trigger incorporation of pyrrolysine rather than termination. The 22nd amino acid, pyrrolysine, was recently identified in a few archaebacteria and eubacteria. In the archaebacterium *Methanocarcina barkeri*, pyrrolysine has been found in some methylamine methyltransferases. *M. barkeri* is a member of the methanogen group, which thrives on a wide range of methanogenic substances including methylamines. Methylamine methyltransferase is required to generate methane from these substances.

Role of modified nucleotides in decoding

Only recently has the importance of modified nucleotides such as inosine and pseudouridine (see Section 4.3) in decoding the correct (cognate) codon and wobble codons become apparent. As noted earlier, the degeneracy of the code results in some amino acids being coded for by as many as six codons (e.g. leucine, serine), whereas others are coded for by as few as one (e.g. methionine, tryptophan). When bases in the anticodon are modified, further pairing patterns become possible in addition to those predicted by the regular and wobble pairing involving A, C, U, and G (see Table 5.2). Some modifications selectively restrict

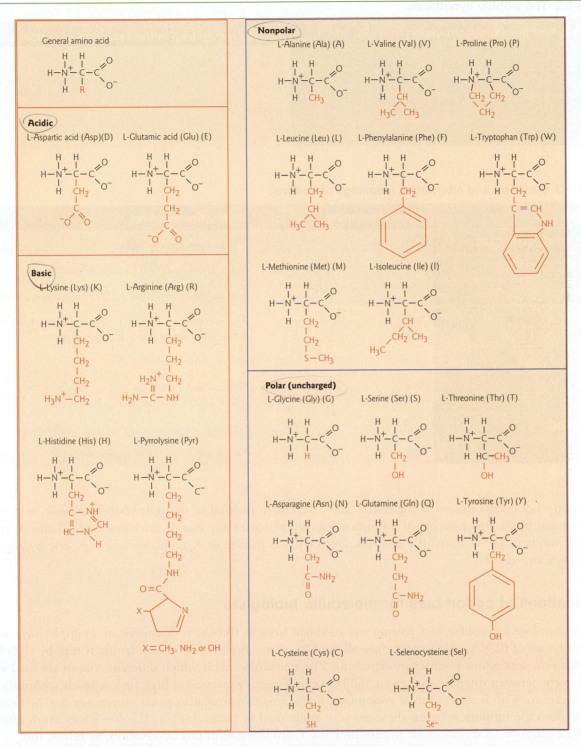

Figure 5.3 The 22 genetically encoded amino acids found in proteins. At physiological pH, the amino acids usually exist as ions. Note the groupings of the various R groups (shown in red). The standard three-letter abbreviations and one-letter symbols are indicated in brackets.

Table 5.2 The wobble hypothesis.

Base in first position of anticodon	Base(s) recognized in third position of codon
U	A or G
C	G only
A	U only
G	C or U

Table 5.3 Common and alternative meanings of codons.

Codon	General meaning	Alternative meaning
CUU, CUC, CUA, CUG	Leu	Thr in yeast mitochondria
CUG	Leu	Ser in *Candida albicans* (yeast that causes thrush)
AUA	Ile	Met in yeast mitochondria, *Drosophila*, and vertebrates
UGA	Stop	Trp in mycoplasma, and mitochrondria of higher plants Sel in some eukaryotic and bacterial genes
UAA	Stop	Gln in ciliated protozoa
UAG	Stop	Gln in ciliated protozoa Pyr in some archaebacteria and eubacteria
AGA/AGG	Arg	Stop in mitochondria of yeast and vertebrates
CGG	Arg	Trp in mitochondria of higher plants

anticodon–codon interactions, while others allow tRNAs to respond to multiple codons. Inosine, which is often present in the first position of the anticodon, can pair with any one of three bases, U, C, and A. In contrast, modification of uracil (U) to 2-thiouracil restricts pairing to A alone, because only one hydrogen bond can form with G.

Implications of codon bias for molecular biologists

The bottom line for wobble base pairing and modified bases in tRNAs is that there are multiple ways to construct a set of tRNAs able to recognize all the 61 codons. A particular codon family is read by tRNAs with different anticodons in different organisms. The frequencies with which different codons are used vary significantly between different organisms and between proteins expressed at high or low levels within the same organism. This is referred to as codon bias. For example, mammalian genes commonly use AGG and AGA codons for arginine, whereas these are very rarely used in *Escherichia coli*. *E. coli* is a bacterium often used for expression of recombinant human proteins. Correlating with this observation, in *E. coli*, the tRNA^{Arg} that reads the infrequently used AGG and AGA codons for arginine is present only at very low levels. The expression of functional proteins in heterologous hosts (i.e. hosts of a different species), is a cornerstone of molecular biology research. Codon bias can have a major impact on the efficiency of expression of proteins if they contain codons that are rarely used in the desired host. Notably, *Tetrahymena*, the ciliate that played an important role in the discovery of telomerase (see Section 6.9), possesses tRNAs that read the canonical stop codons UAA and UAG as glutamine (Gln), making these genes impossible to express heterologously without some type of redesign strategy of the gene or host.

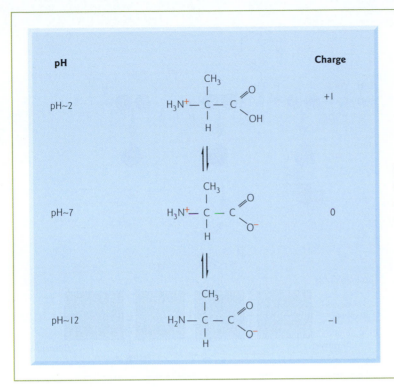

pH | | Charge
pH~2 | | +1
pH~7 | | 0
pH~12 | | −1

Figure 5.4 The acid–base properties of amino acids. The three major forms of alanine occuring in titrations between pH 2 and pH 12 are shown.

5.4 Protein structure

Proteins, like nucleic acids, are chain-like polymers of small subunits. In the case of DNA and RNA, the links of the chain are nucleotides. In proteins, the links of the chain are the amino acids specified by the genetic code. Whereas DNA is composed of only four different nucleotides, proteins have a repertoire of 20 common amino acids and, in some special cases, two additional amino acids. Some proteins contain an abundance of one amino acid, while others may lack one or two types of amino acid. Each amino acid has an amino group (NH_3^+), and a carboxyl group (COO^-) attached to a central carbon called the α-carbon (Fig. 5.3). At pH 7 the amino and carboxyl groups are charged but over a pH range from 1 to 14 these groups exhibit binding and dissociation of a proton (Fig. 5.4). The weak acid–base behavior of amino acids provides the basis for many techniques for identification of different amino acids and protein separations (see Chapter 9). The remaining groups attached to the α-carbon are a hydrogen atom (H) and the side chain or "R group." The only difference between any two amino acids is in their different side chains. Each side chain has distinct properties, including charge, hydrophobicity, and polarity. It is the arrangement of amino acids, with their distinct side chains, that gives each protein its characteristic structure and function.

Primary structure

Amino acids joined together by peptide bonds forms the primary structure of a protein (Fig. 5.5). The amino group of one molecule reacts with the carboxyl group of the other in a condensation reaction resulting in the elimination of water and the formation of a dipeptide. A short sequence of amino acids is called a peptide, with the term polypeptide applied to longer chains of amino acids, usually of known sequence and length. When joined in a series of peptide bonds, amino acids are called "residues" to distinguish between the free form and the form found in proteins. The peptide bond has a partial double bond character (Fig. 5.6). Free rotation occurs only between the α-carbon and the peptide unit. The peptide chain is thus flexible, but more rigid than it would be if there were free rotation about all of the bonds.

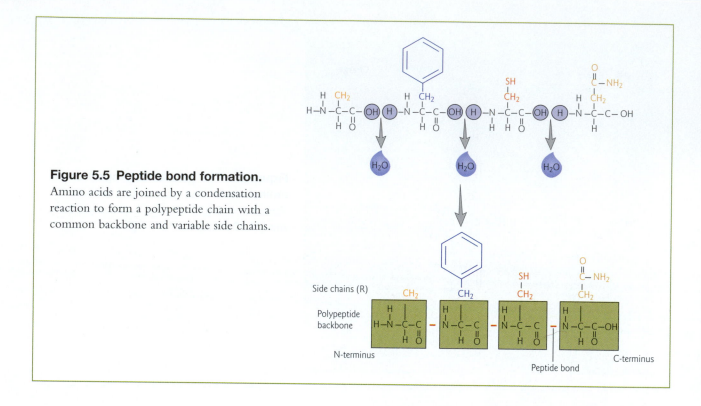

Figure 5.5 Peptide bond formation.
Amino acids are joined by a condensation reaction to form a polypeptide chain with a common backbone and variable side chains.

Figure 5.6 Rigidity of the peptide bond.
(A) The peptide bond acts as a partial double bond as a result of resonance. (B) *Trans-* and *cis*-configurations are possible about the rigid peptide bond.

Protein primary structure is divided into two main components, the polypeptide backbone that has the same composition in all proteins, and the variable side chain groups (Fig. 5.5). A polypeptide chain has polarity and by convention is depicted with the free amino group at its left end. This is termed the amino terminus, or N-terminus. The polypeptide chain also has a free carboxyl group at its right end which is the carboxyl terminus, or C-terminus. Like the corresponding DNA and mRNA sequences, amino acids sequences of proteins are read from left to right. The individual amino acids have three-letter abbreviations, but for the presentation of long protein sequences a single-letter code is used for each amino acid residue. Both single– and three-letter abbreviations are shown alongside the R groups in Fig. 5.3. In some cases, the single-letter code is easy to remember; e.g. "A" stands for alanine and "L" for leucine. In other cases, the code is more cryptic; e.g. "Q" stands for glutamine and "W" for tryptophan.

Secondary structure

Interactions of amino acids with their neighbors gives a protein its secondary structure. These interactions are primarily stabilized by hydrogen bonds, but also depend on disulfide bridges, van der Waals interactions, hydrophobic contacts, hydrogen bonds between nonbackbone groups, and electrostatic interactions. For example, two R groups having the same charge, either positive or negative, will repel one another. Thus, like charges tend to cause extension, rather than folding, of the chain. The three basic elements of protein secondary structure are the α-helix, the β-pleated sheet, and the unstructured turns that connect these elements. All other structures represent variations on one of these basic themes.

α-Helix

The right-handed α-helix is the most common structural motif found in proteins (Fig. 5.7). Approximately 30% of all residues in globular proteins are found in α-helices. The structure was derived from theoretical models by Linus Pauling and Robert Corey. Publication of the crystallographic structure of myoglobin in 1960 confirmed that α-helices do, in fact, occur in proteins and were largely as predicted. These α-helices are stabilized by hydrogen bonding among near neighbor amino acids with each residue being hydrogen bonded to two other residues. The structure has a pitch of 5.4 Å, which is the repeat distance, a diameter of 2.3 Å, and contains 3.6 amino acids per turn, forming a tight helix. Most amino acids can contribute to the α-helical structure. However, because of its cyclic chemical structure, proline cannot participate as a donor in the hydrogen bonding that stabilizes an α-helix (see Fig. 5.3). Thus, proline is referred to as a "helix-breaking residue."

β-Pleated sheet

Another common secondary structure found in proteins is the β-pleated sheet or β-strand (Fig. 5.7). The Greek letter "β" is an historical designation, indicating that the β-pleated sheet was the second type of secondary structure predicted from the model-building studies of Pauling and Corey. The structure involves extended amino acid chains in a protein that interact by hydrogen bonding. The chains are packed side by side to create a pleated, accordion-like appearance with a repeat distance of 7.0 Å. Two segments of a polypeptide chain (or two individual polypeptide chains) can form two different types of β-structures. If both segments are aligned in the N-terminal to C-terminal direction, or in the C-terminal to N-terminal direction, the β-structure is said to be parallel. If one segment is N-terminal to C-terminal and the other is C-terminal to N-terminal, the β-structure is termed antiparallel.

Turns

Connecting the α-helices and β-pleated sheets elements in protein are "turns." Turns are relatively short loops of amino acids that do not exhibit a defined secondary structure themselves, but are essential for the overall folding of a protein. Other disordered or irregular structures in proteins are normally confined to the N- and C-terminals or more rarely to loop regions within a protein or linker region connecting one or more domains.

Tertiary structure

The folded three-dimensional shape of a polypeptide is its tertiary structure (Fig. 5.7). This spatial arrangement of amino acid residues that are widely separated in the primary sequence is stabilized by covalent and noncovalent bonds. Most interactions are noncovalent. The principal covalent bonds within and between polypeptides are disulfide (S–S) bonds or "bridges" between cysteines (Fig. 5.8). Disulfide bonds are only broken at high temperature, at acidic pH, or in the presence of reducing agents. The

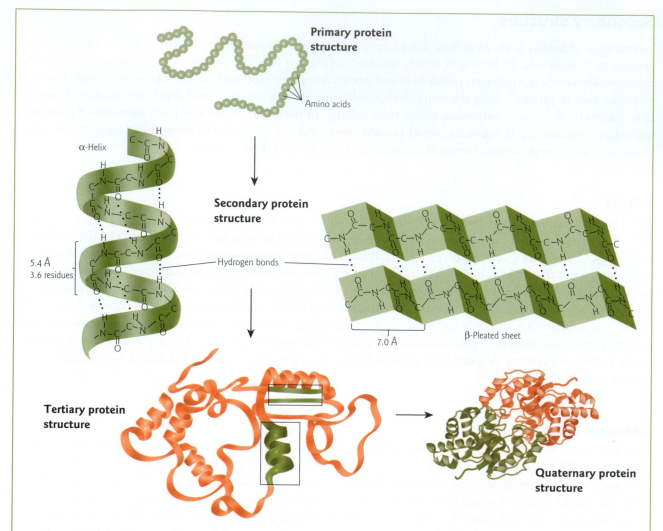

Figure 5.7 Four levels of protein structure. The primary protein structure is the sequence of a chain of amino acids. Secondary structures such as the α-helix and the β-pleated sheet are stabilized by hydrogen bonding between nearby amino acids in the chain. The secondary structure folds into a three-dimensional tertiary structure through noncovalent and covalent interactions. The quaternary protein structure is a protein consisting of more than one amino acid chain.

noncovalent bonds are primarily hydrophobic and hydrogen bonds. Predictably, hydrophobic amino acids cluster together in the interior of a polypeptide, or at the interface between polypeptides, so they can avoid contact with water. Hydrophobic interactions play a major role in tertiary and quaternary structures of proteins. A striking example is green fluorescent protein (GFP), which folds into a cylinder that protects the fluorophore from exposure to solvent (see Fig. 9.5). The three main categories of tertiary structure are illustrated by globular, fibrous, and membrane proteins.

Globular proteins

The overall shape of most proteins is roughly spherical. Proteins that adopt this form are called globular proteins. Figure 5.9A illustrates how the enzyme lysozyme folds up into a globular tertiary structure, forming the active site within a deep pocket between folded regions. The structure of lysozyme was

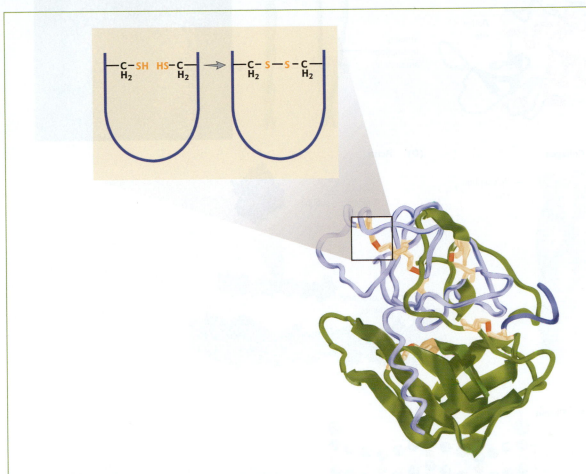

Figure 5.8 Disulfide bonds in tertiary folding. The backbone structure of α-chymotrypsin, an enzyme involved in digesting proteins in the small intestine, is shown (Protein Data Bank, PDB: 5CHA). Its structure contains five disulfide bonds (red bars). Cysteines are shown in light orange. The inset shows two cysteine side chains on the opposite side of a loop domain. The two thiol groups can undergo a reaction, involving the loss of two hydrogens and the formation of a covalent disulfide bond between them. Chymotrypsin is activated by cleavage of the inactive precusor chymotrypsinogen, which is secreted by the pancreas. The three segments of polypeptide chain (green, light blue, and dark blue) produced by proteolytic processing remain linked by disulfide bridges.

reported in 1974; it was the first enzyme ever to have its structure solved by X-ray diffraction. Lysozyme is a widespread enzyme found in animal secretions such as tears and in egg white. It catalyzes the breaking of gycosidic bonds between certain residues in components of bacterial cell walls, resulting in lysis of the bacteria. Because of its catalytic properties, lysozyme is often used by molecular biologists to lyse bacteria in the first step of protein or nucleic acid purification protocols.

Fibrous proteins

Fibrous proteins have properties distinct from globular proteins. A common feature of most fibrous proteins is their long filamentous or "rod-like" structure. They include a number of major designs (Fig. 5.9B–E). A triple helical arrangement of polypeptide chains is exemplified by the collagen family of proteins, which are a major structural component of skin, tendons, ligaments, teeth, and bone. The α-keratins, which are structural components of mammalian hooves, nails, and hair, adopt a structure composed of "coiled coils" of

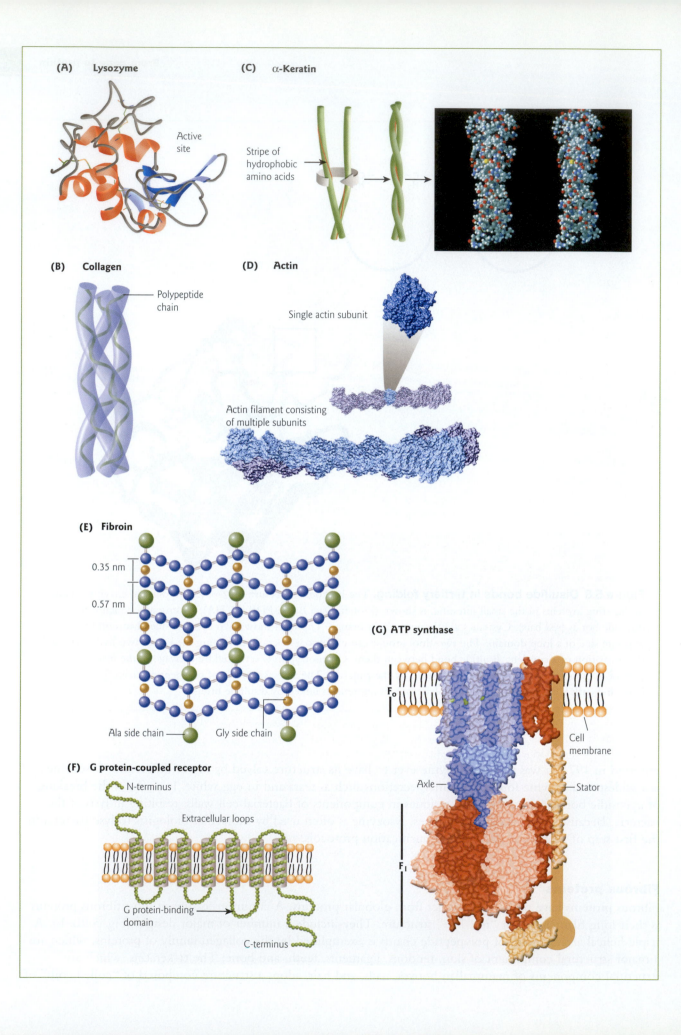

(A) Lysozyme

Active site

(C) α-Keratin

Stripe of hydrophobic amino acids

(B) Collagen

Polypeptide chain

(D) Actin

Single actin subunit

Actin filament consisting of multiple subunits

(E) Fibroin

0.35 nm

0.57 nm

Ala side chain

Gly side chain

(G) ATP synthase

F_O

Cell membrane

Axle

Stator

F_1

(F) G protein-coupled receptor

N-terminus

Extracellular loops

G protein-binding domain

C-terminus

α-helices. A variation on this helical theme is illustrated by the actin filament. This component of the cell cytoskeleton consists of two filaments of polymerized actin monomers, twisted around each other into a helix. In contrast, silk fibroin, a collection of proteins made by spiders or silkworms, is composed of structures made from extended antiparallel β-pleated sheets.

Membrane proteins

The second large class of proteins distinct from globular proteins are the membrane proteins. The primary sequence of these proteins folds into characteristic transmembrane helical structures. The major differences with soluble proteins lie in the relative distribution of hydrophobic amino acid residues. The seven transmembrane helix structure is a common motif in membrane proteins, exemplified by G protein-coupled receptors (Fig. 5.9F). Another example is the cystic fibrosis transmembrane conductance regulator (CFTR) which has two membrane-spanning domains that form a chloride ion channel (see Fig. 17.20). CFTR is defective in patients with cystic fibrosis. One of the most remarkable examples of proteins that are embedded in the cell membrane is the "molecular motor" ATP synthase (Fig. 5.9G). This enzyme is composed of two rotary motors, F_o and F_1. F_o stands for "factor oligomycin" referring to the binding of the antibiotic oligomycin to this motor, while F_1 stands for "factor one." The two motors are connected by a stator (stationary part), so that when F_o turns, F_1 turns as well. The end result of this elegant design is that F_1 generates ATP as it turns, using the free energy of the electrochemical gradient of protons.

Quaternary structure

A functional protein can be composed of one or more polypeptides, forming a quaternary structure (see Fig. 5.7). The stabilizing interactions for quaternary structure are the same as those responsible for tertiary structure, namely disulfide bonds, hydrophobic interactions, charge-pair interactions, and hydrogen bonds, with the exception that they occur between one or more polypeptide chains. The term subunit is generally used to refer to individual polypeptide chains in a complex protein. Quaternary structure can be based on proteins with identical subunits or nonidentical subunits. The presence of this higher order structure allows

Figure 5.9 (*opposite*) Examples of the structures of some globular, fibrous, and transmembrane proteins. (A) A ribbon model depicts how the α-helices (coiled ribbons) and β-pleated sheets (flat arrows) present in lysozyme interact to form a globular shape. The location of the enzyme active site is shown (Protein Data Bank, PDB: 1HEW). (B) Collagen is a fibrous protein composed of three polypeptide chains wound around each other in a helical arrangement. Collagen has a repetitive primary sequence in which every third residue is glycine, and there are also repeating proline residues (Protein Data Bank, PDB: 1BKV). (C) α-Keratin is composed of a coiled coil of two α-helices, stabilized by hydrophobic interactions. Hydrophobic residues are located at positions 1 and 4 in a repeating unit of seven residues that occurs in a "stripe" that twists about each helix. In the molecular graphic, colored spheres represent atoms: carbon is grey, hydrogen white, nitrogen blue, oxygen red, and sulfur yellow. (Credit: National Institutes of Health / Photo Researchers, Inc.) (D) An actin filament is composed of two strands of polymerized actin subunits (Protein Data Bank, PDB: 1NWK, 1RDW). (E) Fibroin consists of layers of antiparallel β-pleated sheets rich in alanine (green) and glycine (orange) residues. The spacing between strands alternates between 0.35 and 0.57 nm. The small side chains interdigitate and allow close packing of each layered sheet, as shown in this side view (Protein Data Bank, PDB: 2SLK). (F) G protein-coupled receptor. Members of this family of cell surface receptors share a seven-helical transmembrane structure. The G protein-binding domain is located in an intracellular loop. (G) ATP synthase. The "electric motor" (F_o) is embedded in the cell membrane and is powered by the flow of protons (hydrogen ions) across the membrane. As the protons flow through the motor, they turn a circular rotor (shown in blue). This rotor is connected by an axle and stator (orange) to the intracellular "chemical motor" (F_1) which generates ATP as it turns (Protein Data Bank, PDB: 1C17, 1E79, 2A7U, 112P).

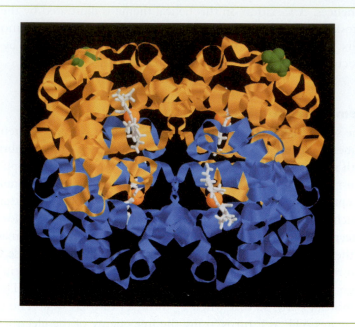

Figure 5.10 The quaternary structure of hemoglobin. Computer graphic of the hemoglobin molecule. It consists of four globin polypeptide chains (α-globin in blue, β-globin in yellow), each carrying a heme group (white) with a central iron atom, which binds to oxygen. The green structure is the amino acid glutamic acid at residue 6 on the β-chain. In sickle cell anemia this is replaced by valine due to a mutation (see Fig. 8.14). (Credit: Kenneth Eward / BioGrafx / Photo Researchers, Inc.).

greater versatility of function. For example, catalytic or binding sites are often formed at the interface between subunits. A classic example of the functional result of quaternary structure is hemoglobin, a tetramer containing two different subunits that join to form a binding site for a heme group (Fig. 5.10). The protein contains two α- and two β-subunits in its tetrameric state, and is usually written as $\alpha_2\beta_2$. Similarly, antibodies contain two heavy and two light chains, with the antigen-binding site formed by the interaction of the two chains (see Fig. 12.23).

Size and complexity of proteins

There is tremendous variation in the size and complexity of proteins. The molecular weight of proteins and the number of subunits (polypeptide chains) shows much diversity. Dalton units (1 Da is equivalent to 1 atomic mass unit) are used frequently in the protein literature to describe the molecular weight. More accurately, this is the absolute molecular weight representing the mass in grams of 1 mole of protein. Molecular weight or relative molecular mass (M_r) is the mass of a molecule relative to 1/12th the mass of the carbon (^{12}C) isotope (which is 12 atomic mass units), and, by definition, is a dimensionless quantity and does not possess any units. However, for most purposes, the term molecular weight is used loosely in protein molecular biology. Typical polypeptide chains have molecular weights ranging from 20 to 70 kDa (20,000–70,000 Da). The smallest polypeptides that form folded proteins have molecular weights of about 11 kDa. The average molecular weight of an amino acid is 110, which means that the typical polypeptide chain contains in the range of 181 to 636 amino acids.

Proteins contain multiple functional domains

Proteins larger than about 20 kDa are often formed from two or more domains that are generally associated with specific functions. A single domain is usually formed from a continuous amino acid sequence and not portions of sequence scattered throughout the polypeptide. This will be a recurrent theme in subsequent chapters. For example, when the binding of transcription factors to DNA is discussed in detail in Section 11.5, we will see that domains can contain common structural–functional motifs. Examples include a finger-shaped motif called a zinc finger that is involved in DNA binding, and the leucine zipper family of DNA-binding proteins that have two subunits that come together to form a dimer through the use of a coiled-coil region.

Prediction of protein structure

The three-dimensional structures of proteins are determined by their amino acid sequences, but the prediction of these structures remains a challenge for molecular biologists. Structural models seek to find the lowest free-energy structure for a specified amino acid sequence using computer algorithms. This is difficult, however, because of the vast size of the conformational space involved and because of a lack of accurate calculations of the free energies of protein conformations in solvent. Some progress has been made recently in predicting structures for small protein domains (< 85 residues) at < 1.5 Å resolution, and predictions of certain structural elements, such as the α-helix, are becoming more reliable. In addition, the increasingly large number of structures determined by X-ray crystallography or nuclear magnetic resonance (NMR) (see Section 9.10 for methods) has helped to define families of amino acid sequences that share related tertiary structures. By comparing the sequences of proteins of unknown structure with those that have been determined, it is often possible to make structural predictions based on the identified similarity.

The Protein Structure Initiative was launched 5 years ago by the National Institutes of Health with the ultimate goal of obtaining the three-dimensional structures of 10,000 proteins in a decade. Among the technological advances aiding this project is the development of robotic systems for setting up protein crystals for measurement by X-ray diffraction. Results so far suggest that proteins come in a relatively limited variety of shapes.

5.5 Protein function

Proteins have an amazing diversity of biological functions. Many of these functions will become of central importance in understanding material presented in subsequent chapters. Proteins provide the structures that give cells integrity and shape, such as components of the cytoskeleton and the architecture of the nucleus. Others serve as hormones to carry signals from one cell to another, or to transport oxygen around the bodies of multicellular organisms. Of particular significance for molecular biologists are the suites of proteins that mediate the activities of genes at all points in the flow of genetic information from replication to transcription to translation. One vital role of proteins is to serve as enzymes that catalyze the hundreds of chemical reactions necessary for life.

Enzymes are biological catalysts

In Chapter 4, we saw that some RNA molecules and ribonucleoprotein (RNP) particles can act as catalysts. Most enzymes in the cell, however, are globular proteins. Enzymes lower the activation energies of the chemical groups that participate in a reaction, and thereby speed up the reaction (Fig. 5.11). The substrate forms a tight complex with the enzyme by binding to a region of the enzyme called the active site. The active site is often a cleft or pocket in the enzyme (see, for example, Fig. 5.9A), with some side chains of amino acids contributing to the binding of the substrate and others to the catalysis of the reaction. After the enzyme–substrate complex forms, the substrate itself usually undergoes a small change in shape to hold it in a reactive configuration that facilitates catalysis. The activated enzyme–substrate complex then engages in one or a series of chemical transformations, which result in conversion of the substrate to the product. The product then dissociates from the enzyme, allowing the enzyme to participate in another cycle of substrate binding and catalysis. Most enzymes act through an induced-fit mechanism: the enzyme changes shape upon binding the substrate, and the active site has a shape that is complementary to that of the substrate only after the substrate is bound (Fig. 5.11). Details of the mode of action of specific enzymes are presented where relevant at a number of points throughout this textbook.

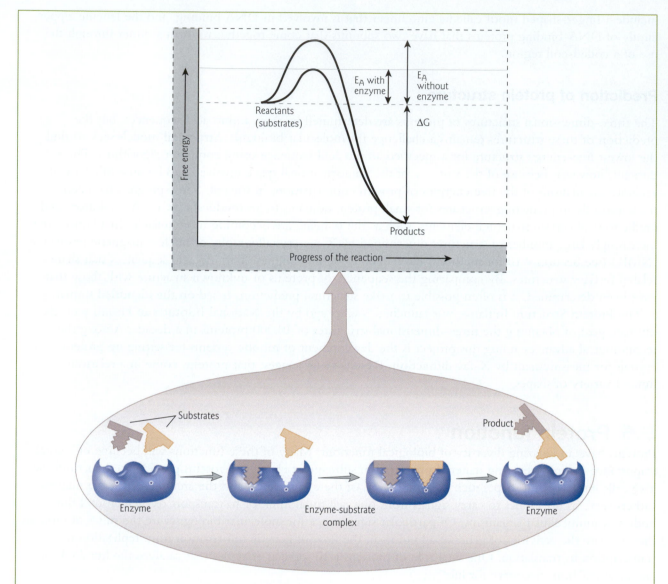

Figure 5.11 Enzymes lower activation energies. The activation energy (E_A) with an enzyme is lower than the E_A of an uncatalyzed reaction and thus speeds up the rate of the reaction. The change in free energy (ΔG) remains the same because the equilibrium position remains unaltered. In the induced-fit model, the enzyme changes shape upon binding substrates. The active site has a shape complementary to the substrates only after the substrates are bound.

The functional activity of enzymes and other proteins can be regulated at several different levels, including at the level of gene transcription and protein synthesis (Fig. 5.12). Of importance for our discussion of protein structure and function here are the roles of post-translational modifications and allosteric regulation in controlling protein activity.

Regulation of protein activity by post-translational modifications

After translation, proteins are joined covalently and noncovalently with other molecules. Complexes that form between lipids and proteins are called lipoproteins, those proteins with a carbohydrate moiety attached are called glycoproteins, while complexes with metal ions are termed metalloproteins, and so on. In

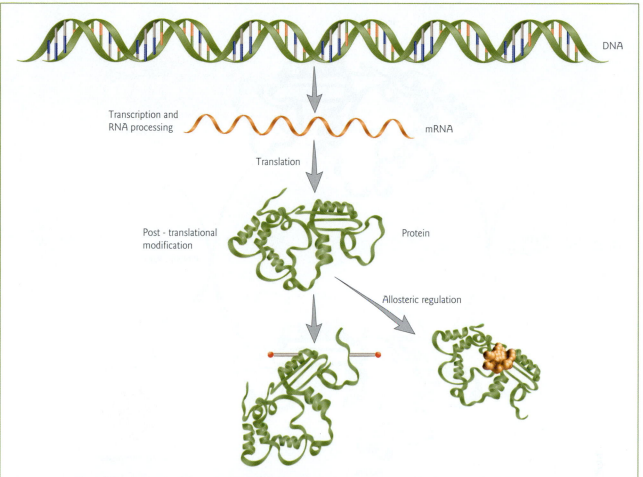

DNA

Transcription and RNA processing

mRNA

Translation

Post - translational modification

Protein

Allosteric regulation

Figure 5.12 Levels of regulation of protein activity. Protein activity can be regulated at the level of transcription, RNA processing, and translation, or by post-translational modifications such as phosphorylation (red symbol) or allosteric effectors (brown symbol).

addition, amino acids are often modified after their incorporation into polypeptides. These post-translational modifications can have both structural and regulatory functions. Important modifications include methylation, acetylation, ubiquitinylation, and sumoylation. Details of these types of post-translational modifications and their impact on gene expression are discussed in Chapter 11 (see Fig. 11.19).

The most common regulatory reaction in molecular biology is the reversible phosphorylation of amino acid side chains (Fig. 5.13). Many steps in gene expression and cell signaling pathways involve post-translational modification of proteins by phosphorylation. This will be a theme highlighted at numerous points in the remainder of this textbook. Kinases catalyze the addition of phosphate groups, whereas enzymes called phosphatases remove phosphates. Kinases tend to be very specific, acting on a very few substrates. In contrast, phosphatases tend to be nonspecific. Many kinases self-regulate through autophosphorylation; many also can initiate reactions that are part of other negative feedback systems. Two protein kinase groups have been widely studied in eukaryotes, those that phosphorylate tyrosine side chains, and those that phosphorylate serine or threonine side chains. Adding phosphate to a protein can cause it to change its shape, for example by masking or unmasking a catalytic domain; or the phosphorylated side chain itself can be part of a binding motif recognized by other proteins allowing proteins to dock and facilitating multiprotein complexes to form, or conversely to promote dissociation of a complex.

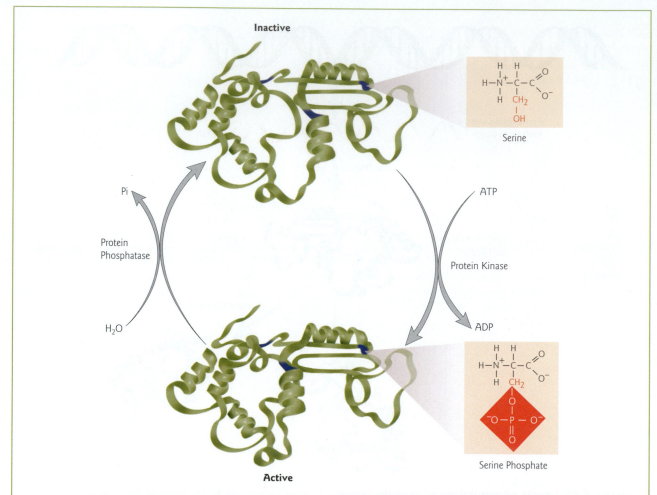

Figure 5.13 Protein phosphorylation. Reversible phosphorylation and dephosphorylation is a common cellular mechanism for regulating protein activity. In this example, the target protein is active when phosphorylated (e.g. at the amino acid serine) and inactive when dephosphorylated; the opposite pattern occurs in some proteins.

Allosteric regulation of protein activity

The binding of a ligand to a protein at one site on a protein can cause a substantial change in the conformation of that protein. Such ligand-induced conformational changes are known as allosteric regulation (allostery means "other shape"). As a result of the shape change, an active site, or another binding site, elsewhere on the protein is altered in a way that increases or decreases its activity (Fig. 5.14). Examples of proteins controlled in this way range from metabolic enzymes to transcriptional regulatory proteins. The ligand (the allosteric effector) is often a small molecule – a sugar or an amino acid. For example, the Lac transcriptional repressor is regulated by the small molecule effector allolactose (see Section 10.5). Allosteric regulation of a given protein can also be mediated by the binding of another protein, and a very similar effect can, in some cases, be triggered by enzymatic modification of a single amino acid residue within the regulated protein, such as by phosphorylation.

Cyclin-dependent kinase activation

A classic example of post-translational regulation of protein activity occurs in the family of kinases known as cyclin-dependent kinases (CDKs). CDK activity is regulated by both phosphorylation and allosteric

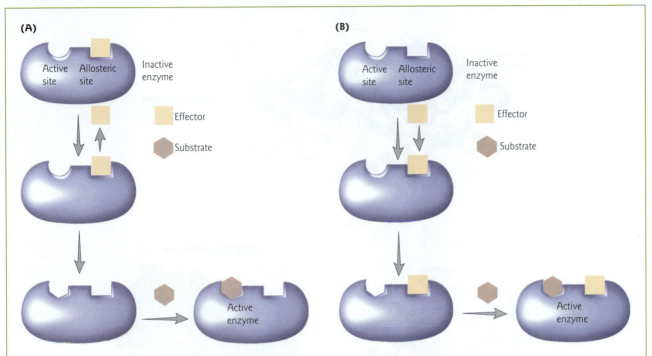

Figure 5.14 Allosteric regulation of enzyme activity. (A) Negative control. Binding of the effector molecule to the allosteric site on the enzyme locks the enzyme in an inactive shape that cannot bind substrate. (B) Positive control. Binding of an effector causes a change in shape of the enzyme to an active form that can bind substrate.

modification induced by the interaction between the enzyme and a regulatory protein called cyclin. The key role of cyclin–CDK complexes in regulating progression through the cell cycle is discussed further in Chapters 6 and 17 (see Figs 6.10 and 17.6).

CDK is composed of two major domains, a small N-terminal domain and a much larger C-terminal domain (Fig. 5.15). Two elements of CDK structure are critical for its regulation: (i) the PSTAIRE α-helix (named for the sequence of conserved residues Pro-Ser-Thr-Ala-Ile-Arg-Glu) in the N-terminal domain; and (ii) a flexible loop, called the T loop, which contains a phosphorylation site at the threonine residue in position 160 of the amino acid chain (Thr160). A molecule of ATP binds in the active site cleft between the two domains. It is this ATP that donates the phosphate group to a polypeptide substrate. In the absence of cyclin, CDK is inactive. In the inactive conformation, the T loop is located at the entrance to the active site, thereby blocking polypeptide substrates from gaining access to the ATP molecule. In addition, a glutamate residue in the PSTAIRE helix that is critical for catalysis is held at a distance from the active site. Binding of cyclin to CDK induces a conformational change that moves the T loop away from the entrance of the active site, allowing access of the polypeptide substrate, and exposing the phosphorylation site in the T loop. The shape change also moves the PSTAIRE helix into the active site, allowing the critical glutamate residue to take part in catalysis. However, this first allosteric change only partially activates the enzyme. Phosphorylation of Thr160 in the T loop is mediated by another kinase called CDK–activating kinase (CAK). Once added, the phosphate group on the threonine interacts with three arginine residues, each from a different region around the catalytic cleft. These interactions reorganize and stabilize the catalytic cleft in the conformation favorable for full activity. In the fully activated state, the cyclin–CDK complex phosphorylates target proteins at specific serine or threonine residues.

Macromolecular assemblages

Cells are much more than collections of individual macromolecules. Single proteins can catalyze biochemical reactions, but higher cellular functions depend on carefully orchestrated protein interaction networks.

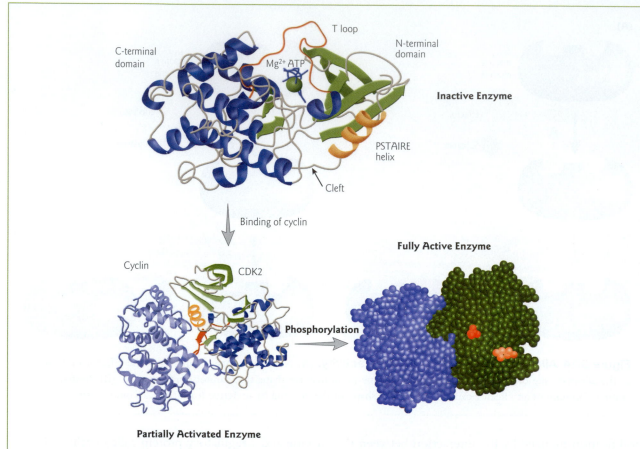

Figure 5.15 Activation of CDK by cyclin binding and phosphorylation. The monomeric cyclin–dependent kinase (CDK) structure is inactive. The enzyme has a bilobed structure. The C-terminal domain (or lobe) is shown in dark blue and the N-terminal domain (or lobe) in green. The position of the PSTAIRE α-helix (orange) holds a critical residue out of the catalytic center, where ATP (dark blue) and an Mg^{2+} ion (green) are located, and the T loop (red) blocks access of the polypeptide substrate. Repositioning of the PSTAIRE helix occurs upon binding of cyclin (light blue), and the T loop is removed from the opening of the catalytic center. This complex is partially active. Upon phosphorylation of the T loop at Thr160, the cyclin–CDK complex becomes fully active. In the space-filling representation of the complex of cyclin (light blue) and CDK (green), phosphorylated Thr160 (red) together with ATP (orange) are shown (Protein Data Bank, PDB: 1FIN, 1JST).

Expression of the genetic information in eukaryotes relies on the sequential action of large and dynamic macromolecular assemblages or "molecular machines." One protein can recruit another to particular locations or substrates and in that way can control what that protein acts on. There are many examples of cooperative binding of proteins in the pathway of gene expression. Such macromolecular assemblages will be discussed in detail in subsequent chapters, in relation to their role in mediating DNA replication, repair, recombination, transcription, chromatin remodeling, RNA processing, and translation.

5.6 Protein folding and misfolding

In some cases, protein folding is initiated before the completion of protein synthesis on ribosomes. Other proteins undergo the major part of their folding after release from the ribosome in either the cytoplasm or in specific compartments such as mitochondria or the endoplasmic reticulum. Most proteins require other proteins called "molecular chaperones" to fold properly *in vivo*. Incorrectly folded proteins are generally targeted for degradation. The accumulation of misfolded proteins is associated with a number of human diseases.

Molecular chaperones

Molecular chaperones increase the efficiency of the overall process of protein folding by reducing the probability of competing reactions, such as aggregation. ATP is required for most of the molecular chaperones to function with full efficiency. Molecular chaperones include heat shock proteins, such as Hsp40, Hsp70, and Hsp90, which promote protein folding and aid in the destruction of misfolded proteins (Fig. 5.16). The designation of these as heat shock proteins reflects the fact that their concentrations are substantially increased during cellular stress. There are several classes of folding catalysts that accelerate potentially slow steps in the folding process. The most important are peptidylprolyl isomerases, which increase the rate of *cis–trans* isomerization of peptide bonds involving proline residues, and protein disulfide isomerases, which enhance the rate of formation and reorganization of disulfide bonds.

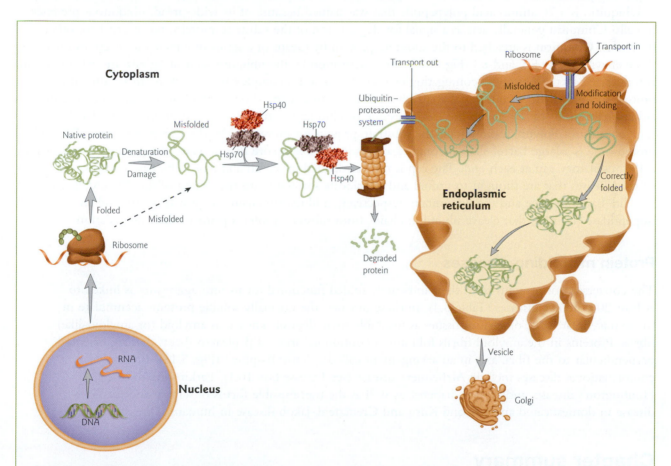

Figure 5.16 Regulation of protein folding. By associating with exposed hydrophobic domains, molecular chaperones Hsp70 and Hsp40 promote the folding of newly synthesized proteins. Alternatively, they can interact with misfolded proteins, promoting their degradation by the ubiquitin–proteasome system (see Fig. 5.17). Many newly synthesized proteins are translocated to the endoplasmic reticulum where they fold with the help of molecular chaperones. Correctly folded proteins are transported to the Golgi complex and then secreted from the cell. Misfolded proteins are detected by a quality control mechanism and targeted for degradation by proteasomes in the cytoplasm. (Adapted from Goldberg, A.L. 2003. Protein degradation and protection against misfolded or damaged proteins. *Nature* 426:895–899; and Dobson, C.M. 2003. Protein folding and misfolding. *Nature* 426:884–890.)

Successful targeting to the cellular protein degradation machinery requires that a protein remains soluble; chaperones help to maintain this solubility. For example, some secreted proteins are translocated into the endoplasmic reticulum (ER) where folding takes place before secretion through the Golgi apparatus to the extracellular environment. The ER contains a wide range of molecular chaperones and folding catalysts. Importantly, the proteins are subjected to a "quality control" check before being exported from the ER (Fig. 5.16). Incorrectly folded proteins are detected by the "unfolded protein response" and sent along another pathway in which they are ubiquitinylated and then degraded in the cytoplasm by proteasomes.

Ubiquitin-mediated protein degradation

Cells have several intracellular proteolytic pathways for degrading normal proteins whose concentration must be rapidly decreased, misfolded or denatured proteins, and foreign proteins taken up by a cell. One major intracellular pathway involves degradation by enzymes within lysosomes – membrane-bound organelles whose interior is acidic. Distinct from the lysosomal pathway are cytoplasmic and nuclear mechanisms for degrading proteins. The best characterized pathway is the ubiquitin-mediated protein degradation pathway.

Ubiquitin is a 76 amino acid polypeptide that was named because of its widespread, ubiquitous presence in cells. Ubiquitin generally acts as a signal for degradation of the substrate protein, but it can have other functions. Ubiquitin is attached to the substrate protein by means of a series of enzymatic reactions involving three enzymes, E1, E2, and E3 (Fig. 5.17). After activation by the ubiquitin-activating enzyme E1, ubiquitin is transferred to the ubiquitin-conjugating enzyme E2. The E2 complex then binds to an E3 ubiquitin protein ligase to form a complex that attaches ubiquitin to a lysine (Lys) residue on the substrate protein. Additional rounds of transfer to the Lys48 residue of ubiquitin result in a chain of ubiquitin molecules (polyubiquitin). Polyubiquitin targets the substrate protein for proteolysis by the 26S proteasome. The 26S proteosome is a large barrel-shaped complex consisting of a 20S core that contains the protease activities plus a 19S regulatory cap at each end. The cap is further divided into a lid and base. The base consists of a ring of six ATPases that are thought to unfold and translocate substrate into the lumen of the 20S core particle. The lid–base interface and the lid contain, respectively, a ubiquitin chain receptor (Rpn10) and an isopeptidase (Rpn11) that cleave ubiquitin chains from substrate proteins prior to their degradation.

Protein misfolding diseases

The conversion of proteins from their intricately folded functional forms into aggregates is linked to at least 20 different diseases (Table 5.4). In these diseases, the normally soluble proteins accumulate in the extracellular space of various tissues as insoluble toxic deposits known as amyloid (or amyloid-like) fibrils. Proteins in the amyloid fibrils fold into a continuous array of β-pleated sheets that are oriented perpendicular to the fibril axis in an arrangement called a "cross β-spine" (Fig. 5.18). Protein conformational diseases include Alzheimer's disease (see Disease box 16.1), Parkinson's disease, Huntington's disease, and type II diabetes, as well as the transmissible forms of scrapie and mad cow disease in domesticated animals, and Kuru and Creutzfeld–Jakob disease in humans (Disease box 5.1).

Chapter summary

A gene is transcribed into mRNA, which carries the genetic information to the ribosomes. The mRNA sequence is translated into the amino acid sequence of a protein. Messenger RNAs are read in the $5' \rightarrow 3'$ direction, the same direction in which they are transcribed. Proteins are made in the amino $\rightarrow$ carboxyl direction, which means that the amino acid at the amino terminal is added first.

The genetic code is a set of three-base codons in mRNA that instruct the ribosome to incorporate specific amino acids into a polypeptide. Each base is part of only one codon; i.e. the triplet code is nonoverlapping. There are 64 codons in all, including three stop signals. Under special circumstances the stop codons encode selenocysteine and pyrrolysine, the 21st and 22nd amino acids, respectively. The remainder code for the

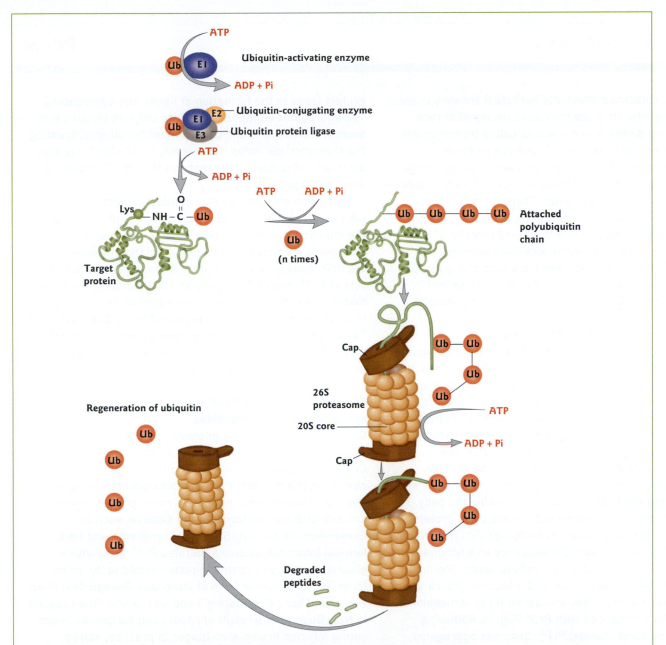

Figure 5.17 Ubiquitin-mediated protein degradation. Ubiquitin (Ub) is attached to a protein by a series of enzyme-mediated reactions, and the ubiquitin-conjugated protein is then targeted to the 26S proteasome. Ubiquitin is released and the target protein is degraded.

20 common amino acids. This means that the code is highly degenerate. Some tRNAs bind the same amino acid but recognize different mRNA codons. In addition, the third base of a codon is allowed to form a non-Watson–Crick base pair with the anticodon, such as the GU wobble pair, or the modified base inosine with U, C, and A. The frequencies with which different codons are used vary significantly between different organisms and between proteins within the same organism. This is referred to as codon bias.

The genetic code is not strictly universal. In certain eukaryotic nuclei and mitochondria and in mycoplasma, codons that cause termination in the standard genetic code can code for amino acids such as

Prions

Prions (proteinaceous infectious particles) are very unusual infectious agents. They are the causative agent of rare, brain-wasting diseases in mammals called transmissible spongiform encephalopathies. These progressive neurodegenerative diseases are characterized by sponge-like holes in the brain, dementia, and loss of muscle control of voluntary movements (ataxia). There is no cure yet, and no reliable diagnostic test until it is too late. Once symptoms appear, death results in 6–12 months. Prion diseases occur in both humans and other animals. The prototype disease was discovered in sheep and goats, and was called "scrapie" because of the observation that affected animals rubbed against fences to stay upright.

The "prion only" hypothesis of infection

When investigators began trying to isolate the infectious agent for transmissible spongiform encephalopathies, they noted a number of unusual characteristics: a lack of an immune response characteristic of infectious disease, a long incubation time (up to 40 years), and resistance of the infectious agent to radiation, which destroys living microorganisms such as viruses and bacteria. The conclusion of many avenues of research was that the infectious agent is not a living organism but a protein with the surprising ability to replicate itself within the body. The infectious protein was designated scrapie prion protein (PrP^{Sc}). The next surprising discovery was that the prion PrP^{Sc} has the same amino acid sequence as a normal host protein (PrP^{C}) encoded by a cellular gene. The only difference between the normal and infectious proteins lies in their structure. Prp^{Sc} has an altered three-dimensional folding pattern compared with Prp^{C} (Fig. 1). Following this conformational change, PrP^{Sc} becomes aggregated, insoluble in nondenaturing detergents, resistant to proteases and heat, and can survive standard sterilization techniques. Most importantly, PrP^{Sc} is infectious and can convert endogenous PrP^{C} to the PrP^{Sc} form in a host. These characteristics of prion proteins make them dangerous and particularly difficult to work with.

The normal PrP^{C} protein is a 28 kDa cell surface glycoprotein expressed in neurons; however, its function is unknown. The "prion only" or "protein only" model for infection is as follows. Prions propagate through a chain reaction in which a host protein PrP^{C} is post-translationally misfolded to form new prions (Figs 1 and 2). Conversion of the normal cellular prion protein into an abnormally folded protein leads to the formation of fibrils and aggregates. The aggregates clump together into amyloid plaques that surround brain cells and causes them to collapse, creating the characteristic holes in the brain. The Nobel Prize was awarded to Stanley B. Prusiner in 1997 for this model of prion pathogenesis.

The "prion only" hypothesis of infection was greeted with skepticism and some researchers maintained that the infectious agent must be a virus. Experiments with transgenic animals have strengthened the link between the PrP^{C} gene product and prion disease. The essential role of PrP^{C} in prion disease was established by the finding that PrP^{C} knockout mice were resistant to disease and incapable of propagating prions (see Fig. 15.4 for methods). The strongest evidence to date comes from experiments demonstrating that *in vitro*-generated PrP^{Sc} can cause a prion disease in wild-type hamsters.

Sporadic, inherited, and infectious transmissable spongiform encephalopathies

Due to their unique mode of operation, prions can be sporadic, inherited, or infectious. A sporadic form called Creutzfeldt–Jakob disease (CJD) affects one in a million people. In this disease, PrP^{C} misfolds spontaneously and then by "autoinfection" generates more prions. Inherited autosomal dominant forms of the disease, such as Gerstmann–Sträussler–Scheinker syndrome and fatal familial insomnia, involve a mutated PrP^{C} gene with a greater tendency to spontaneously misfold to the prion form. The first human form of infectious disease described was called kuru ("trembling") and was at one time rampant in New Guinea, as a result of ritual cannibalism. Although eating infected brains is no longer in practice, eating beef products from a cow with bovine spongiform encephalopathy ("mad cow disease") is linked to an outbreak of novel variant of CJD (vCJD) in humans. Muscle meat alone appears safe, but muscle meat contaminated with brain or spinal tissue from infected cows can be deadly. Since 1995, more than 150 people have contracted the human version of mad cow disease in Europe. One case of mad cow disease has been reported in the USA so far, and the infected animal was destroyed. Chronic wasting disease is a similar disease in the USA in elk and deer. There is no evidence yet for transmission of chronic wasting disease to humans. Prions have also been found in yeast and other fungi. Yeast prions are not functionally or

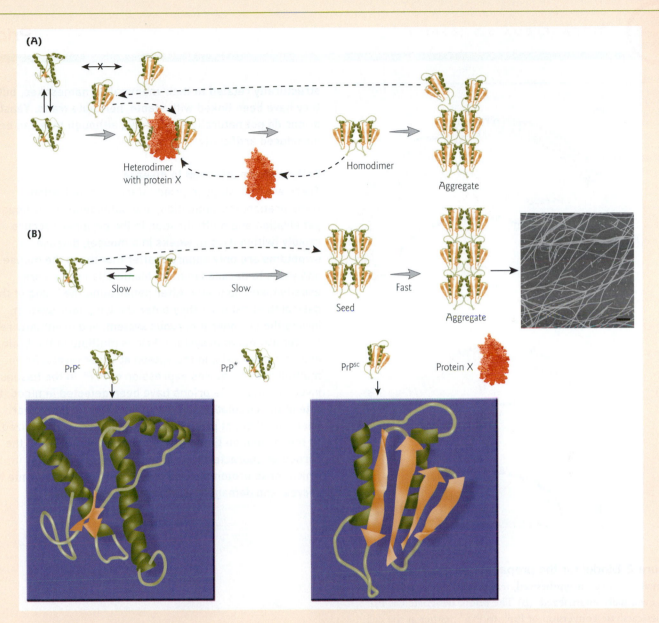

Figure 1 Model for the conformational conversion of PrP^C into the prion PrP^{Sc}. (A) The refolding model. A high energy barrier prevents spontaneous conversion of PrP^C to PrP^{Sc}, but PrP^C is in equilibrium with an intermediate form PrP★. PrP★ can bind to exogenously introduced PrP^{Sc} in association with protein X, a hypothetical factor. The heterodimer is then converted to a PrP^{Sc} homodimer and forms large aggregates. The autoinfection process is maintained by recycling of protein X and breaking off of PrP^{Sc} monomers and oligomers from the aggregates to form new heterodimers. (B) The seeding model. In this model PrP^C and exogenously introduced Prp^{Sc} are in reversible thermodynamic equilibrium. When several PrP^{Sc} monomers form an aggregate or "seed," PrP^C misfolds into a conformation with a greater number of β-pleated sheets (flat arrows) and fewer α-helices (coiled ribbons). The misfolded protein is then rapidly recruited into the PrP^{Sc} aggregate. (Redrawn from: Zou, W.Q. and Gambetti, P. 2005. From microbes to prions: the final proof of the prion hypothesis. *Cell* 121:155–157.) (Inset) Network of amyloid fibrils of aggregrated PrP^{Sc} as visualized by atomic force microscopy (see Fig. 9.17 for methods). Scale bar, 500 nm. (Reprinted from Jones, E.M., Surewicz, W.K. 2005. Fibril conformation as the basis of species- and strain-dependent seeding specificity of mammalian prion amyloids. *Cell* 121:63–72, Copyright © 2005, with permission from Elsevier.)

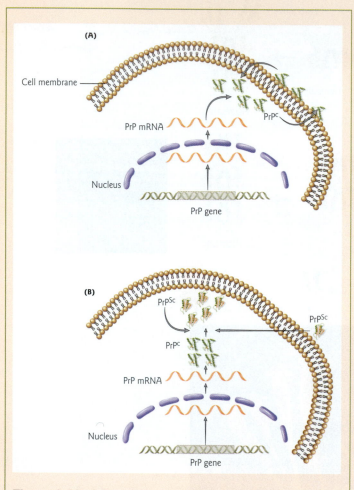

Figure 2 Model for the propagation of prions. (A) In a normal cell, PrP^C is synthesized, transported to the cell surface, and eventually internalized. (B) The prion, designated as PrP^{Sc}, causes catalytic conversion of PrP^C to PrP^{Sc}, either at the cell surface or after internalization. (Redrawn from Weissmann, C. 2004. The state of the prion *Nature Reviews Microbiology* 2:861–871.)

structurally related to their mammalian namesakes, but they have been linked with stable, heritable traits. Yeast prions do not naturally infect cells, although they can be introduced artificially.

Pathway from infection to disease
There are four steps in prion infection: penetration, translocation, multiplication, and pathogenesis. Although penetration and multiplication in the periphery occurs rapidly (within days to weeks in a mouse), disease symptoms are only apparent after months in the mouse and years to decades in man. Mammalian prions are usually taken up orally. After penetrating the lining of the gastrointestinal tract they enter the lymphatic system, invade the peripheral nervous system, and eventually reach the central nervous system. Prions multiple in the brain and, in some hosts, in the spleen at lower levels. Prion multiplication requires expression of PrP^C in the tissues involved, and while prions have been detected in blood, the role of the circulation in spreading infection is unclear. There are distinct prion strains, originally characterized by the incubation time until onset of symptoms, and the particular characteristics of the symptoms. The way in which these proteins elude the immune system, invade the nerves, and damage the brain remains unclear.

tryptophan and glutamine. In several mitochondrial genomes and in the nuclei of at least one yeast, the sense of a codon is changed from one amino acid to another.

Proteins are polymers of amino acids linked through peptide bonds. The sequence of amino acids in a polypeptide is the primary structure. It is the arrangement of amino acids, with their distinct side chains, that gives each protein its characteristic structure and function. The three-dimensional structures of proteins are determined by their amino acid sequences, but prediction of these structures by scientists remains a challenge. Interactions of amino acids with their neighbors gives rise to the secondary structure elements

Table 5.4 Some human protein misfolding diseases.

Disease	Misfolded protein	Nature and location of lesions
Alzheimer's disease	Amyloid β-protein Tau	Extracellular plaques and tangles in neuronal cytoplasm
Parkinson's disease	α-Synuclein	Neuronal cytoplasm
Huntington's disease	Polyglutamine expansion in huntingtin	Neuronal nuclei and cytoplasm
Type II diabetes	Islet amyloid polypeptide (amylin)	Aggregates in pancreas
Creuktzfeldt–Jakob disease	Prion protein (PrpSc)	Extracellular plaques and oligomers, inside and outside of neurons

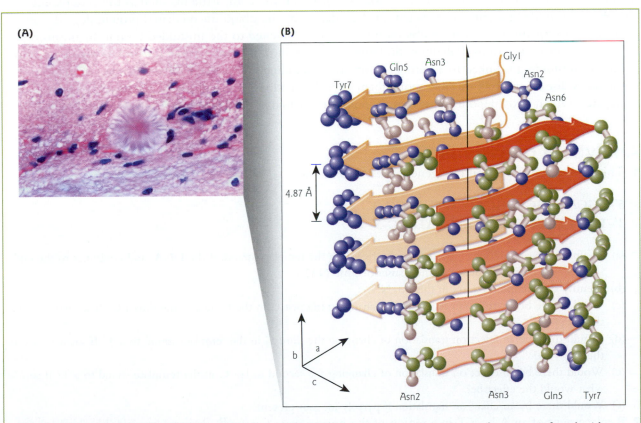

Figure 5.18 Amyloid-like fibrils. (A) Light micrograph of a section through the brain of a mouse infected with new variant Creutzfeldt–Jakob disease. Nuclei (black dots) of neuronal cells are seen. At the center is a circular amyloid plaque, a deposit of misfolded protein. (Credit: James King-Holmes / Photo Researchers, Inc.) (B) Atomic structure of the cross β-spine of amyloid-like fibrils. The structure is a pair of antiparallel β-sheets with a dry interface containing no water in between. Each sheet is formed from parallel segments stacked in register. The backbone of each β-strand is shown as a flat arrow, with side chains protruding. The two sheets are tightly bonded by interdigitating side chains that act like the teeth of a zipper. (Reprinted by permission from Nature Publishing Group and Macmillan Publishers, Ltd: Nelson, R., Sawaya, M.R., Balbirnie, M., Madsen, A., Riekel, C., Grothe, R., Eisenberg, D. 2005. Structure of the cross-β spine of amyloid-like fibrils. *Nature* 435:773–778. Copyright © 2005.)

including the α-helix and β-pleated sheet. The overall three-dimensional shape of a polypeptide is its tertiary structure, and interaction with other polypeptides (subunits) forms the quaternary structure. The tertiary and quaternary structures are stabilized by both noncovalent and covalent interactions, such as the disulfide bonds that form between cysteine residues. The overall shape of most proteins is globular. Other distinct classes of tertiary structure are fibrous and membrane proteins. Proteins contain multiple domains that are associated with specific functions.

Proteins have tremendous diversity of structure and function, including serving as enzymes that catalyze the chemical reactions essential for life. Enzymes speed up the rate of reactions by lowering the activation energies of the chemical groups that participate in a reaction. Most enzymes act through an induced-fit mechanism in which the enzyme changes shape upon binding the substrate.

Post-translational modification, such as the reversible phosphorylation of specific tyrosine, serine, or threonine residues, is an important mechanism for regulating the activity of enzymes and other proteins. Allosteric modification is another important regulatory mechanism. For example, the binding of an allosteric effector molecule to an enzyme can either activate or inactivate the enzyme. Activation of cyclin-dependent kinases requires both binding of the regulatory protein cyclin and modification by phosphorylation.

Most proteins require molecular chaperones to fold properly within the cell, either while still associated with the ribosome, or after release in the cytoplasm or in specific compartments such as the endoplasmic reticulum. Misfolded proteins are targeted for degradation by the ubiquitin-mediated protein degradation pathway. In this pathway, a chain of ubiquitin molecules is attached to the misfolded protein by means of a series of enzymatic reactions. Polyubiquitin targets the protein to the proteasome for proteolytic degradation.

Protein misfolding is linked to a number of diseases, including neurodegenerative disorders such as Alzheimer's disease and transmissible spongiform encephalopathies (prion diseases). Misfolded proteins aggregate to form toxic deposits known as amyloid (or amyloid-like) fibrils.

Analytical questions

1 The sequence of a portion of a gene is presented below:

5′-AGCAATGCATGCATCGTTATGG-3′
3′-TCGTTACGTACGTAGCAATACC-5′

(a) Assuming that transcription starts with the first T in the template strand of the DNA, and continues to the end, what would be the sequence of the transcribed mRNA?
(b) Identify the initiation codon in this mRNA.
(c) Would there be an effect on translation of changing the first C in the template strand to a G? If so, what would the effect be?
(d) Would there be an effect on translation of changing the third G in the template strand to a T? If so, what would the effect be?
(e) Would there be an effect on translation of changing the second to last C in the template strand to a T? If so, what would the effect be?
(f) Would there be an effect on translation if pyrrolysine was present?

2 Replacement of an A by a T in a region of the human gene for the β-chain of hemoglobin is associated with sickle cell anemia:

Normal: **5′-ATGGTGCACCTGACTCCTGAGGAGAAGTCT-3′**
Sickle cell: **5′-ATGGTGCACCTGACTCCTGTGGAGAAGTCT-3′**

(a) What is the nucleotide sequence of the normal and sickle cell hemoglobin mRNA?
(b) What is the amino acid sequence in this part of the β-polypeptide chain, and what is the amino acid replacement that results in sickle cell hemoglobin?
(c) Why might this amino acid substitution make a difference in protein structure?

3 You determine the structure of a protein of unknown function. The protein adopts a filamentous coiled coil of two α-helices. Is the protein likely to be an enzyme? Explain.

4 Provide an explanation for why you should politely decline a serving of brains at a party hosted by cannibals.

Suggestions for further reading

Agris, P.F. (2004) Decoding the genome: a modified view. *Nucleic Acids Research* 32:223–238.

Aguzzi, A., Polymenidou, M. (2004) Mammalian prion biology: one century of evolving concepts. *Cell* 116:313–327.

Atkins, J.F., Gesteland, R.F. (2000) The twenty-first amino acid. *Nature* 407:463–465.

Atkins, J.F., Gesteland, R.F. (2002) The twenty-second amino acid. *Science* 296:1409–1411.

Bradley, P., Misura, K.M.S., Baker, D. (2005) Toward high-resolution *de novo* structure prediction for small proteins. *Science* 309:1868–1871.

Castilla, J., Saá, P., Hetz, C., Soto, C. (2005) *In vitro* generation of infectious scrapie prion. *Cell* 121:195–206.

Dickerson, R.E. (2005) *Present at the Flood. How Structural Molecular Biology came About.* Sinauer Associates, Sunderland, MA.

Dobson, C.M. (2003) Protein folding and misfolding. *Nature* 426:884–890.

Glatzel, M., Giger, O., Seeger, H., Aguzzi, A. (2004) Variant Creutzfeldt–Jakob disease: between lymphoid organs and brain. *Trends in Microbiology* 12:51–53.

Goldberg, A.L. (2003) Protein degradation and protection against misfolded or damaged proteins. *Nature* 426:895–899.

Gustafsson, C., Govindarajan, S., Minshull, J. (2004) Codon bias and heterologous protein expression. *Trends in Biotechnology* 22:346–353.

Hayden, M.R., Tyagi, S.C., Kerklo, M.M., Nicholls, M.R. (2005) Type 2 diabetes mellitus as a conformational disease. *Journal of the Pancreas* 6:287–302.

Jones, E.M., Surewicz, W.K. (2005) Fibril conformation as the basis of species- and strain-dependent seeding specificity of mammalian prion amyloids. *Cell* 121:63–72.

Judson, H.F. (1996) *The Eighth Day of Creation. Makers of the Revolution in Biology.* Cold Spring Harbor Laboratory Press, Cold Spring Harbor, NY.

Lee, T.F. (1991) *The Human Genome Project. Cracking the Genetic Code of Life.* Plenum Press, New York.

Ma, J., Lindquist, S. (2002) Conversion of PrP to a self-perpetuating PrP^{Sc}-like conformation in the cytosol. *Science* 298:1785–1788.

Nelson, R., Sawaya, M.R., Balbirnie, M., Madsen, A., Riekel, C., Grothe, R., Eisenberg, D. (2005) Structure of the cross-β spine of amyloid-like fibrils. *Nature* 435:773–778.

Selkoe, D.J. (2003) Folding proteins in fatal ways. *Nature* 426:900–904.

Tuite, M.F. (2004) The strain of being a prion. *Nature* 428:265–267.

Weissmann, C. (2004) The state of the prion. *Nature Reviews Microbiology* 2:861–871.

Weissmann, C. (2005) Birth of a prion: spontaneous generation revisited. *Cell* 122:165–168.

Whitford, D. (2005) *Proteins. Structure and Function.* John Wiley & Sons, Chichester, UK.

Chapter 6
DNA replication and telomere maintenance

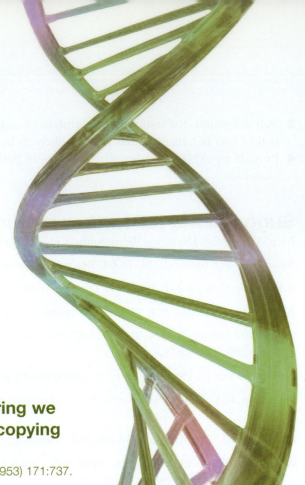

It has not escaped our notice that the specific pairing we have postulated immediately suggests a possible copying mechanism for the genetic material.

James D. Watson and Francis Crick, *Nature* (1953) 171:737.

Outline

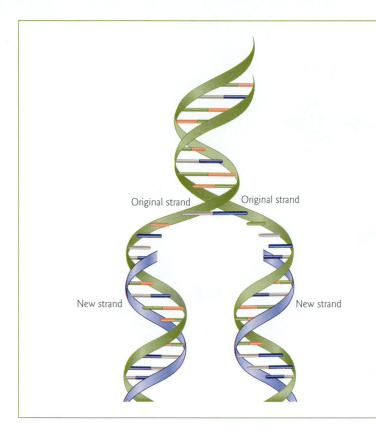

Original strand   Original strand

New strand   New strand

Figure 6.1 Semiconservative DNA replication. Semiconservative DNA replication gives two daughter duplex DNAs, each of which contains one original strand and one new strand.

6.1 Introduction

The regulation of DNA replication is fundamental to understanding the continuity of life. As cells multiply and give rise to new cells, the genome must be accurately duplicated so that information is passed to each new generation with minimal error. In essence DNA replication simply involves the melting apart of the two strands of the double helix followed by the polymerization of new complementary strands on the resulting single-stranded templates (Fig. 6.1). During the past several decades a much more complex view of DNA replication has emerged. For example, around 30 to 40 proteins are involved in the process of DNA replication in eukaryotes. One of the first requirements is for the replication machinery to gain access to nucleosome-bound DNA. In addition, eukaryotic cells need to coordinate their proliferation and differentiation during development. Thus, replication involves decisions of when, where, and how to initiate DNA replication to ensure that only one complete and accurate copy of the genome is made before a cell divides. Replication at telomeres (chromosomal ends) poses a special problem for linear chromosomes. The molecular machinery that regulates telomere maintenance is discussed in Section 6.9. The focus of this chapter is on eukaryotic DNA replication, but comparisons are made with bacterial and viral DNA replication where appropriate.

6.2 Historical perspective

Three possible modes of replication could be hypothesized based on Watson and Crick's model for the structure of the DNA double helix: semiconservative, conservative, and dispersive. In semiconservative replication each new DNA molecule is comprised of one original (template or parental) strand and one new (daughter) strand. In conservative replication, one daughter molecule would consist of the original parent and the other daughter would be totally new DNA. In dispersive replication, some parts of the original helix are conserved and some parts are not. Daughter molecules would consist of part template and part newly synthesized DNA.

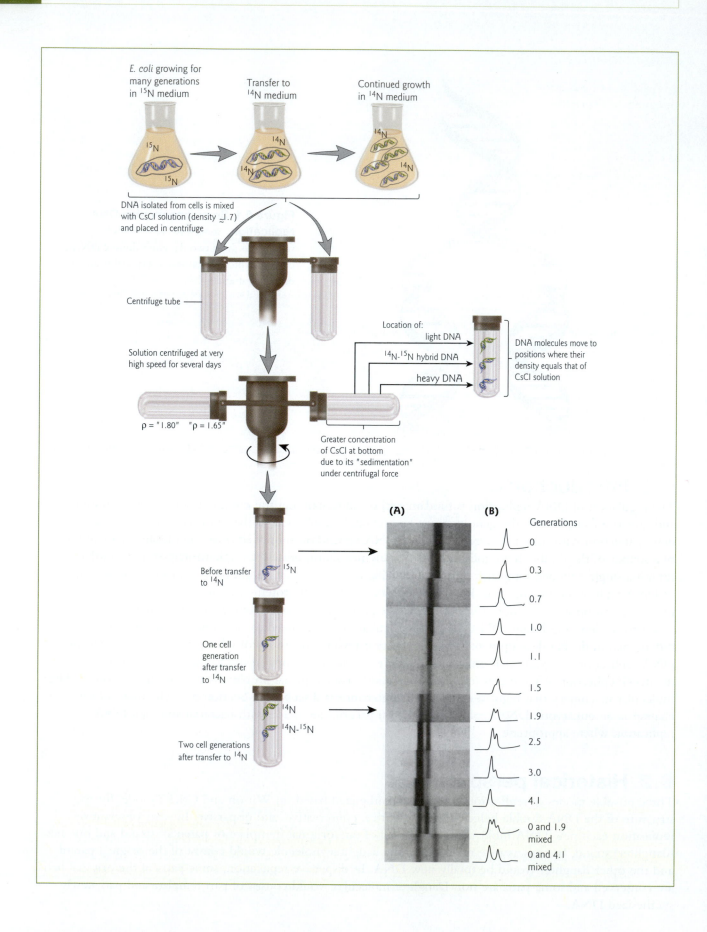

E. coli growing for many generations in ^{15}N medium

Transfer to ^{14}N medium

Continued growth in ^{14}N medium

DNA isolated from cells is mixed with CsCl solution (density ≃1.7) and placed in centrifuge

Centrifuge tube

Solution centrifuged at very high speed for several days

$\rho = "1.80"$ $"\rho = 1.65"$

Greater concentration of CsCl at bottom due to its "sedimentation" under centrifugal force

Location of:
light DNA
^{14}N-^{15}N hybrid DNA
heavy DNA

DNA molecules move to positions where their density equals that of CsCl solution

(A)   **(B)**   Generations

Before transfer to ^{14}N ^{15}N

0
0.3
0.7
1.0
1.1
1.5
1.9

One cell generation after transfer to ^{14}N

2.5
3.0
4.1

Two cell generations after transfer to ^{14}N ^{14}N ^{14}N-^{15}N

0 and 1.9 mixed

0 and 4.1 mixed

Insight into the mode of DNA replication: the Meselson–Stahl experiment

The mode of replication was determined in 1958 by Matthew Meselson and Franklin W. Stahl. They designed an experiment to distinguish between semiconservative, conservative, and dispersive replication. First, they needed a way to tell original DNA from newly synthesized DNA. To this end, they grew *Escherichia coli* in medium containing ^{15}N, a heavy isotope of nitrogen. ^{15}N contains one more neutron than the naturally occurring ^{14}N. Unlike radioisotopes, ^{15}N is stable and is not radioactive. After growing several generations of bacteria in the ^{15}N medium, the DNA of *E. coli* became denser because the nitrogenous bases had incorporated the heavy isotope.

The density of the strands was determined using a technique known as density-gradient centrifugation. A solution of cesium chloride (CsCl) – a heavy metal salt – containing the DNA samples is spun in an ultracentrifuge at high speed for several hours. Eventually, an equilibrium between centrifugal force and diffusion occurs, such that a gradient forms with a high concentration of CsCl at the bottom of the tube and a low concentration at the top. DNA forms a band in the tube at the point where its density is the same as that of the CsCl. The bands are detected by observing the tubes with ultraviolet light at a wavelength of 260 nm, in which DNA absorbs strongly.

After many generations, Meselson and Stahl transferred the bacteria with heavy (^{15}N) DNA to a medium containing only ^{14}N. What they found was that DNA replicated in the ^{14}N medium was intermediate in density between light (^{14}N) and heavy (^{15}N). In the next generation, only DNA of intermediate and light density was present. The results shown in Fig. 6.2 are consistent only with semiconservative replication. If replication had been conservative, there would have been two bands at the first generation of replication – an original ^{15}N (heavy) double helix and a new ^{14}N (light) double helix. Additionally, throughout the experiment, the original DNA would have continued to show up as a ^{15}N (heavy) band. If the method of replication had been dispersive, the result would have been various multiple-banded patterns, depending on the degree of dispersiveness.

Insight into the mode of DNA replication: visualization of replicating bacterial DNA

The semiconservative method of replication was visually verified by J. Cairns in 1963 using the technique of autoradiography. This technique makes use of the fact that radioactive emissions expose photographic film. The visible silver grains on the film can then be counted to provide an estimate of the quantity of radioactive material present. Cairns grew *E. coli* in a medium containing the base thymine labeled with tritium, a radioactive isotope of hydrogen (^3H). The DNA was then extracted from the bacteria and autoradiographs were made. By analysis of DNA at different time points during replication (it takes approximately 42 minutes to replicate the entire genome), Cairns showed that replication of the circular

Figure 6.2 (*opposite*) The Meselson and Stahl experiment to determine the mode of DNA replication. Meselson and Stahl shifted ^{15}N-labeled *E. coli* cells to a ^{14}N medium for several generations, and then subjected the bacterial DNA to CsCl gradient ultracentrifugation. The bands after centrifugation come about from semiconservative replication of ^{15}N DNA (blue) replicating in a ^{14}N medium (green). (Inset) (A) Photographs of the centrifuge tubes under ultraviolet illumination. The dark bands correspond to heavy ^{15}N-labeled DNA (right) and light ^{14}N-labeled DNA (left). A band of intermediate density was also observed between these two and is the predominant band observed at 1.0 and 1.1 generations. This band corresponds to double-stranded DNA molecules in which one strand is labeled with ^{15}N, and the other with ^{14}N. After 1.9 generations, there were approximately equal quantities of the $^{15}N/^{14}N$ band and the ^{14}N band. After three or four generations, there was a progressive depletion of the $^{15}N/^{14}N$ band and a corresponding increase in the ^{14}N band, as expected for semiconservative replication. (B) Densitometer tracings of the bands in panel (A), which can be used to quantify the amount of DNA in each band. (Reprinted by permission of Matthew Meselson from: Meselson, M., Stahl, F. 1958. The replication of DNA in *Escherichia coli*. *Proceedings of the National Academy of Sciences USA* 44:671–682.)

genome was bidirectional. The two strands in the double helix separate at an origin of replication, exposing bases to form a cytologically visible replication "eye" or "bubble" that contains two replication forks. The two replication forks proceed in opposite directions around the circle (Fig. 6.3). During replication, the chromosome looks like the Greek letter theta (θ) by electron microscopy. Replication intermediates are thus termed "theta structures." Cairns' findings have subsequently been verified by both autoradiographic and genetic analysis.

6.3 DNA synthesis occurs from 5′ to 3′

Before exploring the complexity of eukaryotic DNA replication, some basic principles common to most DNA replication pathways will be described. We now know that during semiconservative replication, the new strand of DNA is synthesized from 5′ to 3′. Nucleotides are added one at a time to the 3′ hydroxyl end of the DNA chain, forming new phosphodiester bonds. Deoxynucleoside 5′ triphosphates (dNTPs) are the building blocks. The terminal two phosphates are lost in the reaction, making the reaction essentially irreversible (Fig. 6.4). The choice of nucleotide to add to the chain is determined by complementary base pairing with the template strand. Details of the exact mechanism for how this process occurs varies for cells, organelle genomes, plasmids, and viruses. The mode of replication depends, in part, on whether the genome is circular or linear. The most common mode of replication is semidiscontinuous DNA replication. Other mechanisms include continuous DNA replication and rolling circle replication.

6.4 DNA polymerases are the enzymes that catalyze DNA synthesis

Enzymes that polymerize nucleotides into a growing strand of DNA are called DNA polymerases. Over the past few years, the number of known DNA polymerases in both prokaryotes and eukaryotes has grown tremendously. Bacteria have five different DNA polymerases (Focus box 6.1), whereas mammalian cells are now known to contain at least 14 distinct DNA polymerases (Table 6.1). In eukaryotes, three different DNA polymerases are involved in chromosomal DNA replication: DNA polymerase α, DNA polymerase δ, and DNA polymerase ε. DNA polymerase γ is used strictly for mitochondrial DNA (mtDNA) replication. These four enzymes are referred to as the replicative polymerases, to distinguish them from the remaining polymerases that are involved in repair processes. The repair polymerases will be discussed in detail in Chapter 7.

All the known DNA polymerases can only add nucleotides in the 5′ → 3′ direction. In other words, a DNA polymerase can catalyze the formation of a phosphodiester bond between the first 5′-phosphate group of a new dNTP and the 3′-hydroxyl group of the last nucleotide in the newly synthesized strand (Fig. 6.4). But the DNA polymerases cannot act in the opposite orientation to create a phosphodiester bond with the 5′-phosphate of a nucleotide already in the DNA and the 3′-hydroxyl of a new dNTP.

Another feature of DNA polymerases is that they cannot initiate DNA synthesis de novo (Latin for "from the beginning"). With the exception of DNA polymerase α (the polymerase involved in primer synthesis), they all require a "primer." The primer is usually a short RNA chain which must be synthesized on the DNA template before DNA polymerase can start elongation of a new DNA chain (primers are discussed in detail below). DNA polymerases recognize and bind to the free 3′-hydroxyl group at the end of the primer. Once primed, polymerases can extend pre-existing chains rapidly and with high fidelity. Bacterial and mammalian DNA polymerases can add ~500 and ~50 nt/second, respectively.

6.5 Semidiscontinuous DNA replication

The major form of replication that occurs in nuclear DNA (eukaryotes), some viruses (e.g. the papovavirus SV40), and bacteria is called semidiscontinuous DNA replication. Fundamental features are conserved from *E. coli* to humans. The differences are in the details; that is, in the specific enzymes and other proteins that are involved in the process (Table 6.2).

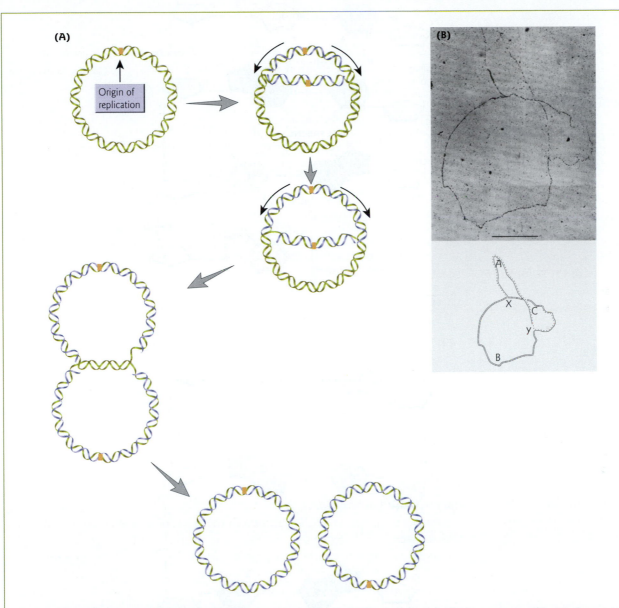

Figure 6.3 Bacterial DNA replication. (A) Bidirectional replication of the *E. coli* chromosome. The black arrows indicate the advancing replication forks. The intermediate figures are called theta (θ) structures. (B) Autoradiograph of *E. coli* DNA during replication. The DNA was allowed to replicate for one whole generation and a portion of a second in the presence of radioactive nucleotides. The lower explanatory diagram has labels for the three loops, A, B, and C, created by the existence of two replication forks, X and Y, in the DNA. Forks are created when the circle opens for replication. The length of the chromosome is about 1300 μm. The part of the DNA that has replicated only once has one labeled strand (solid line) and one unlabeled strand (dashed line) (loops A and C). The DNA that has replicated twice has one part that is doubly labeled (loop B, two solid lines) and one part with only one labeled strand (loop A, solid and dashed lines). (Reprinted from Cairns, J. 1963. The chromosomes of *E. coli*. *Cold Spring Harbor Symposium in Quantitative Biology* 28:43–46. Copyright ©1963, with permission from Cold Spring Harbor Laboratory Press and John Cairns.)

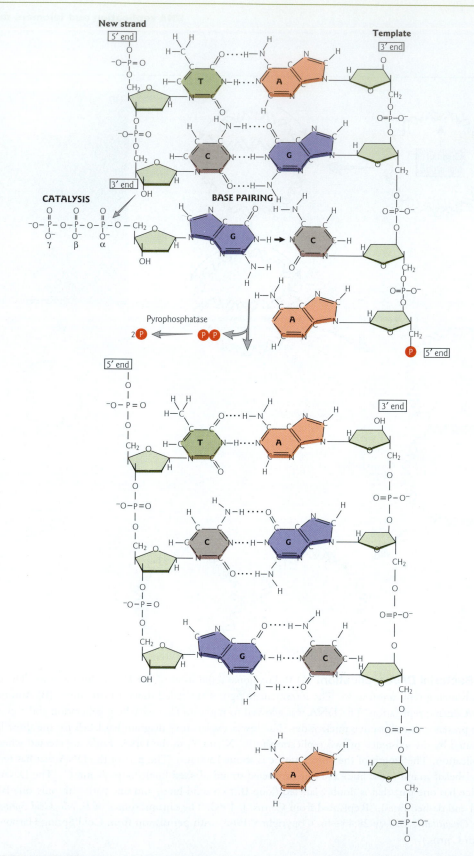

Figure 6.4 DNA synthesis occurs from 5′ to 3′. The template strand directs which of the four dNTPs is added. The dNTP that base pairs (black arrow) with the template strand is highly favored for addition to the growing strand. DNA synthesis is initiated by the nucleophilic attack of the α–phosphate of the incoming dNTP (gray arrow). This results in the extension of the 3′ end of the new strand by one nucleotide and the release of one molecule of pyrophosphate. Pyrophosphatase rapidly hydrolyzes the pyrophosphate into two phosphate molecules.

Bacterial DNA polymerases

There are five major DNA polymerases in the bacterium *E. coli*: DNA polymerases I, II, III, IV, and V. These five polymerases collectively carry out the same general suite of functions as the 14 eukaryotic DNA polymerases.

DNA polymerase I

DNA polymerase I was the very first polymerase from any organism to be purified and characterized. It is used extensively in molecular biology research because of its availability and unique properties. It is the most abundant polymerase in *E. coli*, but it turns out that it is not the enzyme responsible for most of replication. Instead, it plays a role in primer removal and gap filling between Okazaki fragments, and in the nucleotide excision repair pathway (see Section 7.6). DNA polymerase I has two subunits. One subunit called the Klenow fragment has $5' \rightarrow 3'$ polymerase activity. The other subunit has both $3' \rightarrow 5'$ and $5' \rightarrow 3'$ exonuclease activity. The two subunits together are called the holoenzyme. The holoenzyme has the unique ability to start replication at a nick (broken phosphodiester bond) in the DNA sugar–phosphate backbone. This property is exploited in the laboratory to make radioisotope-labeled DNA by a technique called "nick translation." The Klenow fragment is also widely used in molecular biology research.

For example, it is used for labeling DNA by a technique called "random priming" (see Tool box 8.5).

DNA polymerase III

It came as a surprise to researchers when they realized that DNA polymerase I, the most abundant polymerase, is not the main replicative polymerase in *E. coli*. In fact, a much less abundant enzyme, DNA polymerase III, catalyzes genome replication. The holoenzyme contains 10 different polypeptide subunits. The α-subunit has the replicase activity and the ε-subunit has the proofreading activity ($3' \rightarrow 5'$ exonuclease). DNA polymerase III also plays a role in nucleotide excision repair pathways (see Section 7.6).

DNA polymerases II, IV, and V

DNA polymerase II is involved in DNA repair mechanisms (see Section 7.4). Both DNA polymerases IV and V mediate translesion DNA synthesis. DNA polymerase IV, also called DinB, is encoded by the *dinB* gene. DNA polymerase V, also known as the UmuD$_2$C complex, is encoded by the *umuDC* operon. Both polymerases can bypass DNA damage that has blocked replication by DNA polymerase III. These polymerases may play a role in adaptive mutagenesis, since they are prone to making mistakes.

Leading strand synthesis is continuous

Since DNA polymerase can only add nucleotides in the $5' \rightarrow 3'$ direction, the antiparallel structure of the two strands of the DNA double helix poses a problem for replication. Both strands of DNA end up being synthesized from 5′ to 3′ but the process involves different mechanisms (Fig. 6.5). Once primed, continuous replication is possible on the $3' \rightarrow 5'$ template strand. The strand in which DNA replication is continuous is called the "leading strand." Leading strand synthesis occurs in the same direction as movement of the replication fork. The DNA polymerase on the leading strand template has what is called high processivity: once it attaches, it does not release until it meets a replication fork moving in the opposite direction, or until the entire strand is replicated.

Lagging strand synthesis is discontinuous

A discontinuous form of replication takes place on the complementary or "lagging strand." On this strand the DNA is copied in short segments (1000–2000 nt in prokaryotes and 100–200 nt in eukaryotes) moving in the opposite direction to the replication fork. These short segments were first described in 1969 by Reiji and Tuneko Okazaki, and are thus called "Okazaki fragments." Lagging strand replication requires the repetition of four steps: primer synthesis, elongation, primer removal with gap filling, and joining of the Okazaki fragments. Despite these extra steps, synthesis of both new strands occurs concurrently. Nucleotides

Table 6.1 The eukaryotic DNA polymerases.*

Name	Function
High fidelity replicases Pol α (alpha)	Priming DNA synthesis during replication and repair
Pol δ (delta)	DNA replication of leading (and lagging?) strand during replication and repair (BER, DSBR, MMR, NER)
Pol ε (epsilon)	DNA replication of lagging strand during replication and repair (BER, DSBR, NER)
Pol γ (gamma)	Mitochondrial DNA replication and repair
High fidelity repair Pol β (beta)	BER, DSBR
Pol η (eta)	Translesion DNA synthesis (relatively accurate replication past thymine–thymine dimers)
Error-prone repair Pol ζ (zeta)	Translesion DNA synthesis (thymine dimer bypass)
Pol θ (theta)	Repair of DNA interstrand cross-links
Pol ι (iota)	Translesion DNA synthesis (required during meiosis)
Pol κ (kappa)	Translesion DNA synthesis (deletion and base substitution), DSBR (nonhomologous end joining)
Pol λ (lambda)	Translesion DNA synthesis
Pol μ (mu)	DSBR (nonhomologous end joining)
Pol ν (nu)	DNA cross-link repair?
Rev1	Abasic site synthesis (deoxycytidyl transferase activity inserts C across from a nucleotide lacking a base)

* Terminal deoxynucleotidyl transferase (TdT) is sometimes included in the list of DNA polymerases. This enzyme is a lymphoid, cell-specific, template-independent polymerase that adds nucleotides nearly randomly to coding ends during V(D)J recombination (see Fig. 12.25). BER, base excision repair; DSBR, double-stranded break repair; MMR, mismatch repair; NER, nucleotide excision repair.

Table 6.2 The players in DNA replication.

Function	E. coli	Human (SV40 model)
Helicase	DnaB	Mcm2-7 (T antigen)
Loading helicase/primase	DnaC	Mcm2-7 (T antigen)
Single strand maintenance	SSB	RPA
Priming	DnaG (primase)	Pol α/primase
Sliding clamp	β	PCNA
Clamp loading (ATPase)	γδ complex	RFC
Strand elongation	Pol III	Pol δ/pol ε
RNA primer removal	Pol I	FEN-1, RNase H1
Ligation of Okazaki fragments	Ligase	Ligase I

PCNA, proliferating cell nuclear antigen; Pol, DNA polymerase; RFC, replication factor C; RPA, replication protein A; SSB, single-stranded DNA-binding protein.

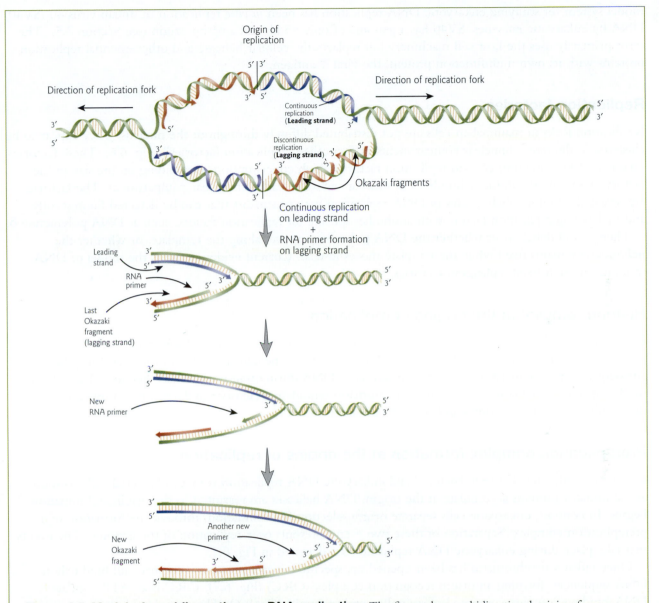

Figure 6.5 Model of semidiscontinuous DNA replication. The figure shows a bidirectional origin of replication with two replication forks proceeding in opposite directions. Continuous replication from 5′ to 3′ occurs on the leading strand in the direction of the moving replication fork. For simplicity only one replication fork is depicted. RNA primer formation and elongation create Okazaki fragments during discontinuous DNA replication on the lagging strand.

are added to the leading and lagging strands at the same time and rate, by two DNA polymerases, one for each strand.

6.6 Nuclear DNA replication in eukaryotic cells

Cellular mechanisms for DNA replication were first studied in bacteria since their genomes are smaller and therefore easier to manipulate. In 1984, a cell-free system finally allowed scientists to make progress with studying replication in eukaryotic cells. The focus of the rest of this section will be on eukaryotic cells. The model system for studying eukaryotic DNA replication has been *in vitro* replication of simian virus 40 (SV40) DNA by eukaryotic enzymes. SV40 has a genome of only ~5 kb and a 65 bp origin (see Section 3.5). The virus primarily uses the host cell machinery but replaces the cellular helicase and other essential replication proteins with its own multifunction protein, the viral T antigen.

Replication factories

Replication forks in mammalian cells are not distributed diffusely throughout the nucleus. They appear to be clustered in discrete subnuclear compartments or foci called "replication factories" (Fig. 6.6). These factories (diameter 100–1000 nm) contain replication factors at high concentration. Depending on the size of the factory, between 40 and many hundreds of forks are active in each subnuclear compartment. The factories are revealed by pulse-labeling sites of DNA replication with precursors that can be detected fluorescently, and by labeling replication factors with antibodies specific for replication factors, such as DNA polymerase δ.

There is still debate over whether the DNA polymerase moves along the template, or whether the polymerase remains fixed while the template moves instead. Current evidence favors the model of DNA spooling through fixed replication factories.

Histone removal at the origins of replication

During DNA replication, the cellular machinery needs to gain access to the DNA that is packaged into chromatin in the nucleus. How this is achieved has yet to be determined. But the events may be analogous to how transcription factors gain access to DNA during transcription (see Section 11.6). Histone modifications (particularly acetylation) and chromatin remodeling factors may loosen the chromatin to allow disassembly of the nucleosomes and access to the template DNA.

Prereplication complex formation at the origins of replication

One major difference between bacterial and eukaryotic DNA replication is that in bacterial cells, as soon as the initiator proteins accumulate at the origin, DNA helicases are recruited to the origin and initiation begins. In contrast, eukaryotic cells separate origin selection from initiation, through the formation of a prereplication complex. Separation of these two events prevents over-replication of the genome. The events that take place during eukaryotic DNA replication are depicted in Fig. 6.7.

Once eukaryotic chromatin has been opened up, specific initiator proteins recognize and bind origin DNA sequences, forming an origin recognition complex (ORC) (Fig. 6.7). ORC is an ATP-regulated, DNA-binding complex composed of six polypeptide subunits (Orc1–6). The complex binds to the origin of replication and then recruits Cdc6 (cell division cycle 6) (see Focus box 6.2) and Mcm (minichromosome maintenance) proteins, other components of the prereplication complex that are essential for initiation of DNA replication. The SV40 T antigen (Tag) functions as a viral ORC comparable to the cellular ORC. Accessory proteins also play a role in initiation, such as single-stranded DNA-binding protein (SSB) in *E. coli* and replication protein A (RPA) in mammals.

Origins of replication

An origin of replication is a site on chromosomal DNA where a bidirectional pair of replication forks initiate. In eukaryotes, bidirectional replication does not start at the ends of the linear chromosomes, but at many internal sites spaced approximately every 50 kb (see Fig. 6.6). For example, the mouse has 25,000 origins (about 150 kb each). Humans have 10,000–100,000 origins. Origin DNA sequences usually have many adenine–thymine (AT) base pairs and are said to be "AT rich." This makes sense because less energy is required to melt the two hydrogen bonds joining A with T, compared with the three hydrogen bonds joining guanine–cytosine (GC) base pairs. In addition, the stacking interactions of GC base pairs with

(A)

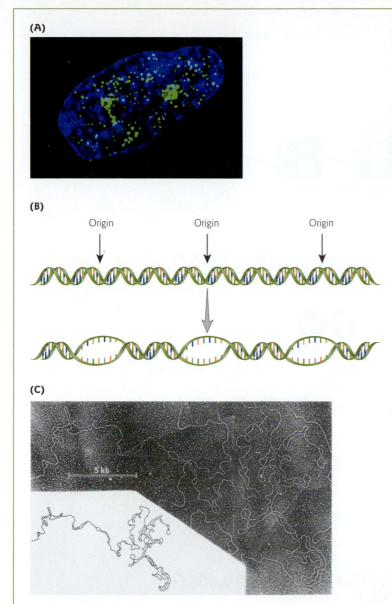

(B)

Origin   Origin   Origin

(C)

5 kb

Figure 6.6 Eukaryotic DNA replication occurs in replication factories from multiple origins. (A) BrdU (5-bromo-2-deoxyuridine) labeling of cells shows multiple "replication factories" which appear as discrete nuclear subcompartments or foci. S phase cells were labeled for 6 hours with BrdU. BrdU is an analog of thymidine. In newly synthesized DNA thymidine is (partly) replaced by BrdU, which is added to the cells in high concentration. The BrdU can be made visible by making use of a fluorescently labeled antibody against the BrdU (indirect immunofluorescence). Each factory may contain 40 to several hundred replication forks. (Reprinted from Barbie, D.A., Kudlow, B.A., Frock, R. et al. 2004. Nuclear reorganization of mammalian DNA synthesis prior to cell cycle exit. *Molecular and Cellular Biology* 24:595–607. Copyright © 2004, with permission from the American Society for Microbiology. Photograph courtesy of Brian K. Kennedy, University of Washington). (B) Diagram showing the formation of replication bubbles (eyes) in eukaryotic DNA because of multiple sites of origin of DNA replication. (C) Electron micrograph (and explanatory line drawing) of replicating *Drosophila* DNA showing these replication bubbles. (Reprinted by permission of David S. Hogness from: Kriegstein, H.J., Hogness, D.S. 1974. Mechanism of DNA replication in *Drosophila* chromosomes: Structure of replication forks and evidence for bidirectionality. *Proceedings of the National Academy of Sciences USA* 71:135–139.)

The naming of genes involved in DNA replication

FOCUS BOX 6.2

Many of the genes involved in eukaryotic DNA replication were first characterized for their role in cell cycle control in the budding yeast, S*accharomyces cerevisiae*. Their names reflect this history. The search for mutations that affected the yeast cell cycle began in the late 1960s. Because cell division is essential for life, mutations that affect the cell cycle were isolated as conditional, temperature-sensitive mutants in which the gene product can function at the permissive temperature (usually room temperature, 20–23°C) but not at the restrictive temperature (35–37°C). After shifting to the higher temperature, researchers look for yeast that accumulate at a particular point in the cell cycle. Uniform cell cycle arrest suggests that a gene product is required at that particular point in the cell cycle.

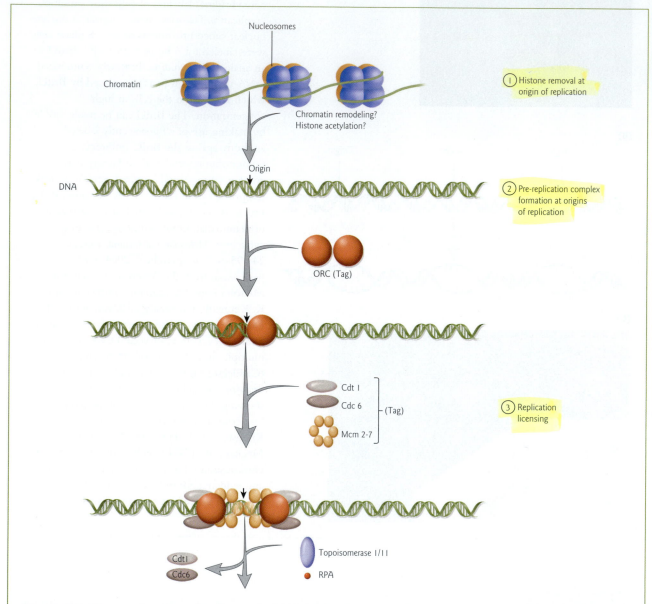

Figure 6.7 The mechanism of eukaryotic DNA replication. The process of DNA replication in eukaryotic cells involves 12 major steps:

1 Histone removal at the origin of replication allows access of the replication machinery to the template DNA.

2 Prereplication complex (pre-RC) formation at the origins of replication. The assembly of the pre-RC is an ordered process that is initiated by the association of the origin recognition complex (ORC) with the origin. Detailed analysis of simian virus 40 (SV40) DNA replication has led to a model for DNA replication of eukaryotic chromosomal DNA. The viral T antigen (Tag) replaces some of the cellular proteins.

3 Replication "licensing." Once bound, ORC recruits at least two additional proteins, Cdc6 and Cdt1. ORC and these two proteins function together to recruit the Mcm2-7 helicase complex to complete the formation of the pre-RC. Cell cycle regulation of pre-RC complex formation and activation ensures that DNA only replicates once per cell cycle (see Figure 6.10 for details).

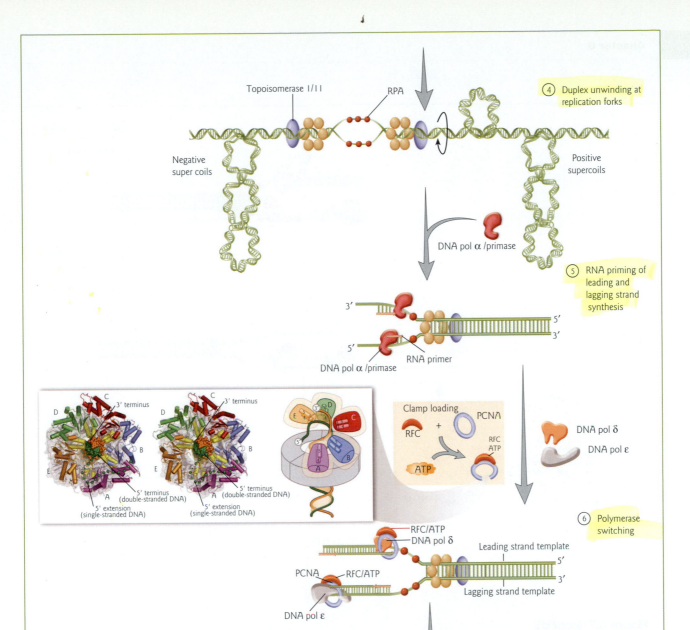

Figure 6.7 (*cont'd*)

4 Duplex unwinding at replication forks, and relaxing of positive supercoils. Cdc6 and Cdt1 are released from the complex, and other replication factors are recruited (e.g. replication protein A, RPA). The helicase activity of Mcm2-7 unwinds the DNA duplex. Topoisomerase I and/or topoisomerase II resolve positive supercoils ahead of the replication fork.

5 RNA priming of leading and lagging strand synthesis. For simplicity, only one replication fork is depicted from this point on. Once present at the origin, DNA pol α/primase synthesizes an RNA primer and briefly extends it.

6 Polymerase switching. The resulting primer–template junction is recognized by the sliding clamp loader (RFC), which assembles a sliding clamp (PCNA) at these sites (see inset). Either DNA pol δ or pol ε recognizes this primer and begins leading or lagging strand synthesis, respectively. (Inset) Clamp loading. (Left) Model for primed DNA interacting with the RFC–PCNA complex. Stereoview of the DNA–RFC–PCNA model. A potential exit path for the 5′ end of the template strand is indicated by green spheres. (Right) Schematic representation of the DNA–RFC–PCNA model. Alignment of RFC-A with the minor groove of the double helix positions the 5′ terminus of the template strand near the opening between RFC-E and RFC-A. (Protein Data Bank, PDB:1SXJ. Adapted by permission from Nature Publishing Group and Macmillan Publishers Ltd: Bowman, G.D., O'Donnell, M., Kurlyan, J. 2004. Structural analysis of a eukaryotic sliding DNA clamp–clamp loader complex. *Nature* 429:724–730. Copyright © 2004.)

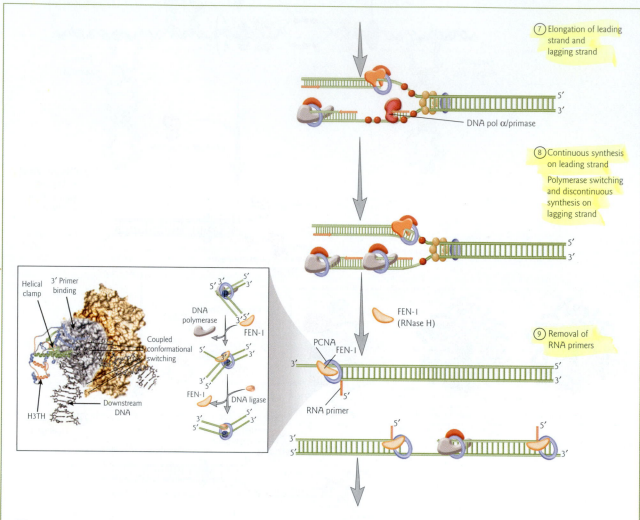

Figure 6.7 (cont'd)

7 Elongation of leading strand and lagging strand.

8 Continuous synthesis on the leading strand; polymerase switching on the lagging strand.

9 Removal of RNA primers. RNA primers are degraded by the endonuclease activity of FEN-1 (see inset). (Inset) PCNA coordinated rotary handoff mechanism. (Left) Composite structure of FEN-1–DNA and FEN-1–PCNA complexes. Downstream DNA (5′ end) passes through the central cavity of PCNA with the upstream duplex (3′ end) kinked 90° by FEN-1. DNA polymerase and DNA ligase I may occupy or sequentially bind the two additional binding sites on PCNA (blue). H3TH, helix–three turn–helix motif. (Right) PCNA (blue) may coordinate the sequential activities of DNA polymerase (gray), FEN-1 (orange), and DNA ligase I (pink) during DNA replication. Each enzyme can potentially recognize a kinked DNA intermediate (green lines). Rotation about the phosphate bond (P, blue sphere) in the kink may facilitate the sequential handoff of DNA intermediates on PCNA. (Protein Data Bank, PDB:1RWZ. Adapted from Chapados, B.R., Hosfield, D.J., Han, S., Qiu, J., Yelent, B., Shen, B., Tainer, J.A. 2004. Structural basis for FEN-1 substrate specificity and PCNA-mediated activation in DNA replication and repair. *Cell* 116:39–50. Copyright © 2004, with permission from Elsevier.)

neighboring base pairs are more favorable energetically than interaction of AT base pairs with their adjacent base pairs. Structural features of DNA such as negative supercoiling (easily unwound sequences) are also important for origins. In *E. coli*, the initiator protein (DnaA) can only bind to negatively supercoiled origin DNA.

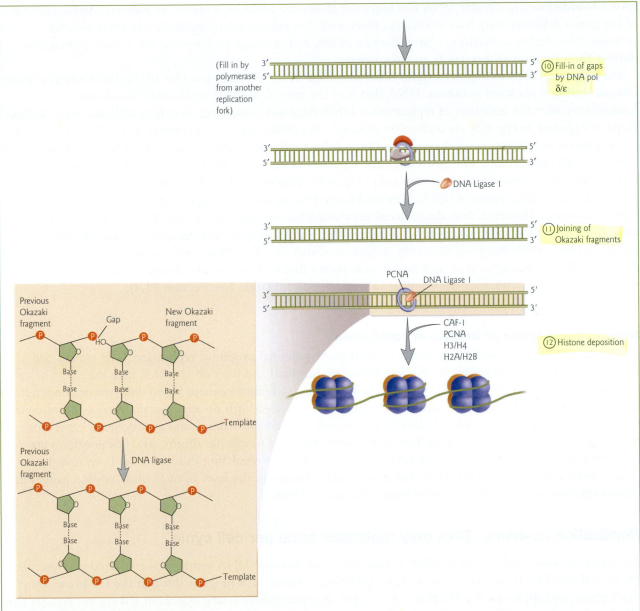

Figure 6.7 (cont'd)

10 Fill-in of gaps left by primer removal is mediated by either DNA pol δ or pol ε.

11 Joining of Okazaki fragments. (Inset) The formation of a phosphodiester bond between adjacent Okazaki fragments by the action of DNA ligase I in association with PCNA.

12 Histone deposition. Nucleosomes are reassembled on nascent DNA via interaction with chromatin assembly factor-1 (CAF-1) and PCNA (see Fig. 6.15 for details).

Mammalian origin sequences lack an easily identifiable consensus sequence. A consensus sequence (or canonical sequence) is the "ideal" form of a DNA sequence found in slightly different forms in different organisms, but which is believed to have the same function. The consensus sequence gives, for each position, the nucleotide most often found. In part because of the lack of a consensus sequence, the nature of mammalian replication origins has been a subject of much debate for years. In contrast, in the budding yeast *Saccharomyces cerevisiae*, there is a consensus sequence called an autonomous replicating sequence (ARS). Similarly, most bacteria have a single, well-defined origin. For example, when the origin sequence of *E. coli*

(oriC) is added to any circular DNA this sequence allows the DNA to replicate in bacteria. Archaebacteria of the genus *Sulfolobus* may have as many as three well-defined origins of replication in their circular genome. The Archaea constitute a third domain of life, and although prokaryotic, are as phylogenetically distinct from bacteria as they are from eukaryotes.

One approach used to identify origins takes advantage of the unusual structure of the DNA intermediates formed during replication initiation. DNA that is in the process of being replicated is not linear. Immediately after the initiation of replication, a DNA fragment containing an origin will take on a "bubble" shape as depicted in Fig. 6.8. As replication proceeds, the bubble shape will convert to a "Y" shape. DNA in the process of replication can be separated from fully replicated or unreplicated DNA by two-dimensional agarose gel electrophoresis. The first dimension separates DNA fragments by size and shape, while the second dimension separates them by size only (Fig. 6.8). This method has been used for mapping origins of replication in many types of DNA, from multicopy yeast plasmids to single-copy chromosome regions in mammalian cells. However, two-dimensional gel electrophoresis does not answer the question of whether replication initiates at specific sites or anywhere within a given origin. A technique called replication initiation point (RIP) mapping allows the detection of start sites for DNA synthesis at the nucleotide level (Fig. 6.9). RIP mapping in yeast and human cells shows that there is a single, defined start point at which replication initiates, a situation very similar to transcription initiation (see Section 11.4).

Selective activation of origins of replication

Metazoan (multicellular animal) genomes contain many potential origins of replication. The overall rate of replication is determined largely by the number of origins used and the rate at which they initiate. The rate of elongation of different DNA chains varies little. During early development when embryos are undergoing rapid cleavages (cell divisions), origin sites are uniformly activated with no apparent preference for sequence. Later in development at the mid-blastula transition, cell division slows down and zygotic gene expression begins. At this stage, initiation of replication becomes restricted to specific origin sites. Some origins are selectively activated while others are suppressed. The parameters regulating this transition from nonspecific to site-specific initiation are not clear, but may include changes in the level of nucleotide pools, changes in chromatin structure, and the ratio of initiator proteins to DNA.

Replication licensing: DNA only replicates once per cell cycle

All organisms must replicate their DNA before every cell division. DNA synthesis is restricted to a specific phase of the cell cycle known as the S phase. Following mitosis, cells progress through the G1 phase, into the S phase, and then into the G2 phase (Fig. 6.10). A replication licensing system in eukaryotes ensures that DNA only replicates once per cell cycle. This is mediated through tight regulation of the formation and activation of prereplication complexes by the levels of cyclin-dependent kinases (CDKs).

Cyclin-dependent kinases are key activators of the cell cycle transitions. Their kinase activity phosphorylates selected serines and threonines on specific proteins, thereby activating these proteins to carry out their function (see Fig. 5.15). For catalysis, CDKs must each associate with a regulatory subunit called a cyclin. Cyclins were first discovered in rapidly dividing sea urchin and surf clam embryos as proteins that accumulate gradually during interphase and are abruptly destroyed during mitosis. CDK activity tracks the rise and fall of cyclins. CDKs are activated during late G1, where they eventually induce cells to progress through the cell cycle. They are inactivated during late mitosis (Fig. 6.10).

Mcm2-7 is the licensing protein complex

During the G1 cell cycle phase, ORC binds to each origin first and then recruits two other proteins called Cdc6 and Cdt1 (Fig. 6.10). Cdt1 was originally isolated from the fission yeast, *Saccharomyces pombe*, as a gene whose expression in the cell cycle is regulated in a Cdc10-dependent manner. *In vitro* assembly reactions

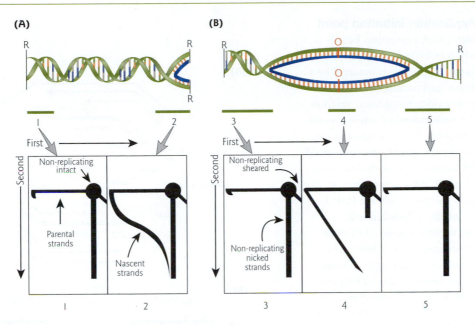

Figure 6.8 Mapping eukaryotic DNA replication origins. DNA is isolated from growing cells. This DNA is then digested with one or two restriction enzymes so that the region of interest is fragmented into pieces of usable size (2–10 kb). Samples are subject to two-dimensional agarose gel electrophoresis, blotted to a membrane, and hybridized with probes derived from the left and right ends of the fragment of interest (see Tool box 8.7 for more detailed methods). (A) Distribution in the final two-dimensional gel of the signals produced by hybridization probes (1 and 2) located at the left and right ends of a hypothetical restriction fragment (defined by restriction sites, R) that does not contain a replication origin. The fragment is assumed to be replicated from right to left. The diagram shows a replication intermediate (RI) in which the replication fork is near the right end of the fragment. The complete set of RIs produced during replication would contain members with replication forks at all positions from the right end to the left end. The large spot represents excess nonreplicating restriction fragments. The diagonal smear below and to the right of the spot of intact fragments represents nonreplicating restriction fragments that were broken by shearing during DNA preparation. The continuous smear extending downward from the position of the intact nonreplicating restriction fragments represents nonreplicating strands with a single (or occasionally more than one) nick. The parental strands of the RIs, which are constant in size, form a horizontal line extending backward from the spot of nonreplicating restriction fragments. The nascent strands, which vary in size depending on the extent of replication, form an arc. This arc extends from very small strands released by the smallest RIs to nearly full length strands released from the largest RIs. In the example shown, only probe 2 can detect all the strands in the nascent strand arc. Probe 1 detects only the largest nascent strands, indicating that this fragment is replicated primarily from right to left. (B) When a restriction fragment contains a replication origin near its center, both of the probes near the ends of the fragment (probes 3 and 5) detect only long nascent strands, but an internal probe located near the origin (probe 4) detects a complete nascent strand arc. (Redrawn from Huberman, J.A. 1997. Mapping replication origins, pause sites, and termini by neutral/alkaline two-dimensional gel electrophoresis. *Methods: a Companion to Methods in Enzymology* 13:247–257.)

in yeast cell extracts have provided insight into the molecular mechanisms of licensing complex assembly (Fig. 6.11). In association with Cdc6 and Cdt1, ORC functions as an ATP-dependent "molecular machine" that loads the licensing protein complex, Mcm2-7. Mcm2-7 is a hexameric (six subunits, numbered 2–7) complex with helicase activity (see Fig. 6.7). When ORC is unable to hydrolyze ATP, assembly of the Mcm2-7 complex on the origin DNA is inhibited (Fig. 6.11). Once the Mcm2-7 complex is loaded, it becomes tightly associated with origin DNA, and the ORC–Cdc6–Cdt1 complex is not required to maintain its binding to DNA. In the SV40 model of DNA replication, origin recognition and helicase activity are carried out by the viral T antigen (Tag).

Figure 6.9 Replication initiation point (RIP) mapping. (A) A replication bubble is depicted. Newly synthesized DNA (leading strand and Okazaki fragments) is initiated by a small RNA primer (red rectangles) and used as the template in a primer extension reaction (green rectangles with extending arrows outside the replication bubble). Primer extension stops at DNA–RNA junctions on the nascent strand because the DNA polymerase included in the reaction cannot use RNA as a template. Extension stops at the points labeled RIP 1, RIP 2, etc. The origin of bidirectional replication is the transition point (TP) from discontinuous to continuous synthesis. The smallest fragments, RIP 1 and RIP 1′, mark the transition point between the leading and lagging strand. (B) Primer extension products (replication intermediates) are fractionated on sequencing gels adjacent to corresponding sequencing lanes (see Chapter 8 for DNA sequencing method). Reactions for the top and bottom strand are shown. The distance between individual RIPs can vary from several nucleotides to more than 100 nt. RIP mapping shows that leading strand synthesis starts at a unique site, in both small and large origins of replication. (Adapted from Bielinsky, A.K., Gerbi, S.A. 2001. Where it all starts: eukaryotic origins of DNA replication. *Journal of Cell Science* 114:643–651. Copyright © 2001, with permission from The Company of Biologists Ltd.)

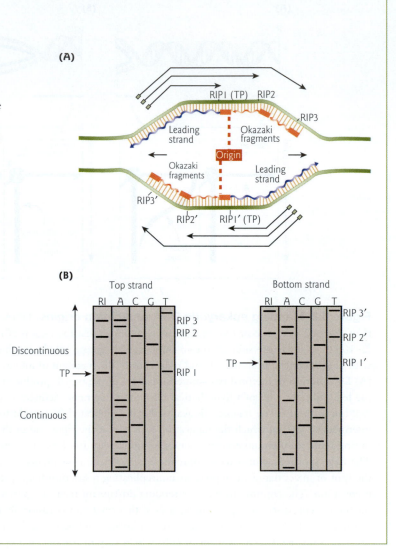

Only licensed origins containing Mcm2-7 can initiate a pair of replication forks. The helicase activity of the complex unwinds DNA ahead of each replication fork (see below). Once the forks are initiated, Mcm2-7 is displaced from the origin and moves with the replication fork.

Regulation of the replication licensing system by CDKs

ORC, Cdc6, Cdt1, and Mcm2-7 can each be independently downregulated as a consequence of CDK activity. This means that no further Mcm2-7 can be loaded onto origins in the S phase, G2, and early mitosis, when CDK activity is high. Removal of the licensing proteins from the origins during the S phase blocks the formation of prereplication complexes and ensures that origins "fire" (initiate a bidirectional pair of forks) only once per cell cycle.

The mode of downregulation differs for each protein and may vary between yeast and vertebrates. As an illustration, the same CDK/cyclin that initiates mitosis in mammalian cells also at the same time binds to the large subunit of ORC (Orc1) and inhibits assembly of the licensing complex at the origins. Experimental evidence also suggests that CDK activity phosphorylates Mcm2-7. This modification causes it to associate with RanGTP and exportin 1/CRM1 – a nuclear export mediator (see Section 11.9). The formation of this

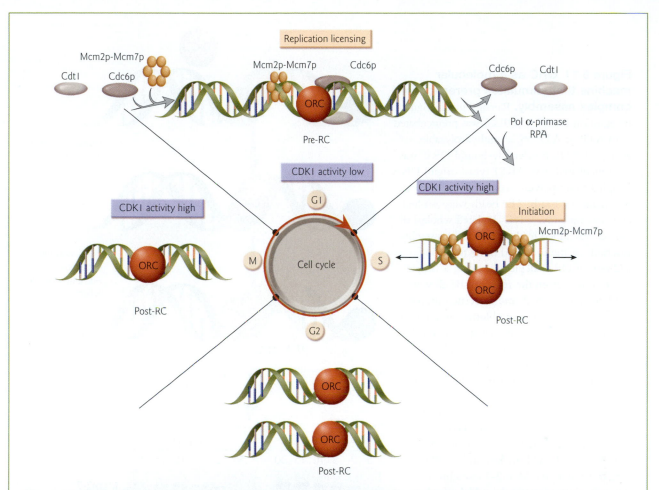

Figure 6.10 Replication licensing events lead to origin activation only once during the cell cycle.
DNA synthesis is restricted to a small period known as the synthetic (S) phase. The S phase is preceded by one gap (G) phase and succeeded by another. A cell can grow by passing through the G_1, S, and G_2 phases before it divides at mitosis (M). Each phase is under control of a specific cyclin–dependent kinase (CDK)/cyclin complex. The activity of the replication licensing system and the Cdks occur at different stages. The diagram depicts events leading to origin activation in the budding yeast. Origin recognition complex (ORC) binds to an ARS element in yeast. The stepwise assembly of the prereplication complex (pre-RC) occurs during the G_1 phase when Cdk activity is low. When Cdk activity rises at the G_1/S transition, the pre-RC is disassembled. The post-RC remains stable until the end of mitosis, and, due to high Cdk activity, the pre-RC cannot reassociate during this time but must await the next G_1 phase. (Adapted from Bielinsky, A.K., Gerbi, S.A. 2001. Where it all starts: eukaryotic origins of DNA replication. *Journal of Cell Science* 114:643–651. Copyright © 2001, with permission from The Company of Biologists Ltd.)

complex sequesters Mcm2-7 in the nucleus (it is not exported) and prevents it from binding to origins of replication. Finally, in vertebrate cells, the activity of Cdt1 is controlled through a specific inhibitor protein called geminin, which itself is regulated by CDK–dependent degradation.

Duplex unwinding at replication forks

DNA helicases are enzymes that use the energy of ATP to melt (separate the two strands of the double helix) the DNA duplex. They progressively catalyze the transition from double-stranded to single-stranded DNA in the direction of the moving replication fork. SV40 T antigen (or the Mcm2-7 helicase) is bound to the leading strand template and moves in the $3' \rightarrow 5'$ direction (see Fig. 6.7). In contrast, the *E. coli* helicase

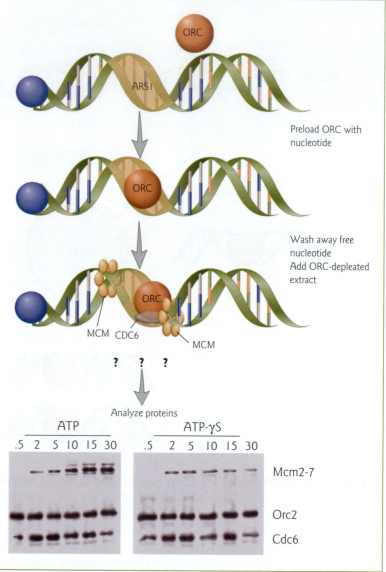

Figure 6.11 ORC as a molecular machine that stimulates prereplication complex assembly. Recombinant origin recognition complex (ORC) was preincubated with ATP or ATP-γS, a nonhydrolyzable analog of ATP. Nucleotide-bound ORC was then incubated with ARS1 (yeast origin) DNA coupled to streptavidin–coated (see Tool box 8.4) magnetic beads. The beads were washed and, subsequently, ORC-depleted whole-cell extract was added to the beads in an assembly reaction. At the indicated times following addition of ORC-depleted extract, a sample was removed from the reaction. Beads were collected and washed, and associated proteins were analyzed by immunoblotting (see Fig. 9.9 for methods), with antibodies specific for the following components of the prereplication complex: Mcm2-7, Orc2, and Cdc6. The data show that ATP hydrolysis by ORC stimulates prereplication complex assembly, and that prereplication complex assembly is inhibited by ATP-γS-bound ORC. (Adapted from Bowers, J.L., Randell, J.C.W., Chen, S., Bell, S.P. 2004. ATP hydrolysis by ORC catalyzes reiterative Mcm2-7 assembly at a defined origin of replication. *Molecular Cell* 16:967–978. Copyright © 2004, with permission from Elsevier.)

(DnaB) translocates in a 5′ → 3′ direction while unwinding DNA and therefore is bound to the lagging strand template during DNA replication.

Movement of the replication fork machinery along the DNA molecule results in the generation of positive supercoiling ahead of the fork, while the already replicated parental strands in its wake become negatively supercoiled (see Fig. 6.7). The resulting accumulation of torsional strain could lead to inhibition of fork movement if not relieved by a DNA topoisomerase. Either type I or type II topoisomerases are capable of removing (relaxing) the positive supercoils ahead of the fork. However, the progeny DNA molecules that are formed remain multiply intertwined because of failure to remove all of the links between the parental strands during DNA synthesis. Topoisomerase II is required to resolve this tangled structure into two separate progeny genomes.

RNA priming of leading strand and lagging strand DNA synthesis

In most DNA replication systems the process of starting new DNA chains is distinct from the process of elongation of established chains. In bacteria, eukaryotic nuclear DNA, and some viruses, synthesis of an

RNA primer is required to start leading strand synthesis and for each Okazaki fragment to be synthesized on the lagging strand. Mitochondrial DNA replication starts with a preformed RNA that base pairs with the template strand. In bacteriophages, linear plasmids, and some viruses such as adenovirus, a priming nucleotide is provided by a protein that binds DNA. A deoxynucleoside monophosphate (dNMP) covalently attaches to a specific serine, threonine, or tyrosine residue. In DNA repair and parvovirus replication, a DNA primer is used. The DNA is nicked to provide a free 3'-OH for the polymerase.

The RNA primer is created *de novo*. In *E. coli*, an enzyme called primase (the product of the *dnaG* gene) catalyzes the priming reaction. It is an RNA polymerase that is only used for this specific purpose. In eukaryotes, the RNA primer is synthesized by DNA polymerase α and its associated primase activity (see Fig. 6.7). The eukaryotic enzyme exists as a complex consisting of a subunit with DNA polymerase activity, a subunit necessary for assembly, and two small proteins that together provide the primase activity. The complex is usually referred to as pol α/primase. The pol α/primase enzyme binds to the initiation complex at the origin and synthesizes a short strand consisting of approximately 10 bases of RNA, followed by 20–30 bases of DNA (called iDNA, for initiator DNA).

Polymerase switching

Multiple dynamic protein interactions are involved in DNA replication. A key feature of the replication process is the ordered hand-off or "trading places" of DNA from one protein to another, or from complex to complex. A striking example of such "trading places" occurs after primer synthesis is complete – the pol α/primase complex is replaced by the DNA polymerase that will extend the chain. This hand-off of the DNA template from one DNA polymerase to another is called polymerase switching. On the leading strand, the switch is to DNA polymerase δ. Recent data suggest that DNA polymerase ϵ elongates the lagging strand (see Fig. 6.7).

Elongation of leading strands and lagging strands

DNA replication in eukaryotes involves the highly regulated and coordinated action of at least three distinct DNA polymerases. Once DNA polymerase δ is recruited to the leading strand, synthesis is continuous. However, lagging strand synthesis requires repeated cycles of polymerase switching from DNA polymerase α to DNA polymerase ϵ, each time a new Okazaki fragment is initiated (see Fig. 6.7). To some extent the choice of polymerase is regulated by expression, activity, and localization of the DNA polymerases. Recent work suggests the polymerase switching is also regulated by competitive protein–protein interactions involving DNA polymerases α, δ, and ϵ, proliferating cell nuclear antigen (PCNA), replication factor C (RFC), and RPA. The central figure in this process is PCNA.

PCNA: a sliding clamp with many protein partners

The name proliferating cell nuclear antigen (PCNA) reflects the protein's original characterization as an abundant component in the nucleus of dividing cells (Disease box 6.1). PCNA has the ability to interact with multiple partners. In addition to its roles for DNA replication, PCNA is also involved in DNA repair, translesion DNA synthesis, DNA methylation, chromatin remodeling, and cell cycle regulation (Fig. 6.12).

In DNA replication, PCNA acts as a "sliding clamp" to increase DNA polymerase processivity. Three identical PCNA monomers are joined in a head-to-tail arrangement to form a ring-shaped trimer. The central hole is large enough to encircle the double helix of DNA. In the presence of ATP, RFC (the "clamp loader") opens the PCNA trimer, passes DNA into the ring, and then reseals it. RFC consists of five subunits. Its ATPase domains extend in a spiral arrangement above the central channel of PCNA. Structural analysis suggests that the clamp loader complex locks onto primed DNA in a screwcap-like arrangement, with the RFC spiral matching the minor grooves of the DNA double helix (see Fig. 6.7).

DISEASE BOX 6.1

Systemic lupus erythematosus and PCNA

Individuals sometimes form antibodies against their own proteins (called autoimmune antibodies), so that their sera react with different parts of cells. The targets for these antibodies can be revealed by allowing an individual's serum to react with cells, and then indirectly immunolabeling any bound antibodies with fluorescently tagged secondary antibodies (see Tool box 9.4). As a result, various autoimmune antibodies reacting against different cellular components have been described.

Some individuals with systemic lupus erythematosus (SLE) possess autoantibodies directed against PCNA (3% of cases). Why this correlation exists is not clear. Nevertheless, the presence of such antibodies is useful in the diagnosis of this autoimmune disease and the antibodies are widely used by molecular biologists. It is not clear exactly what causes SLE but it can be triggered by medications, hormonal factors, infections, exposure to chemicals, and sunlight. In some cases, individuals inherit a genetic predisposition to developing SLE. SLE is most common in women between ages 15 and 40 and affects approximately one in 1000 people. Symptoms range from mild to severe. They include swollen glands, joint pain, fatigue, skin rashes, light sensitivity, migraines, fever, and tissue damage to the kidneys, heart, lungs, blood cells, and digestive system. There is no known cure for SLE but symptoms can be managed with lifestyle changes and medication.

Without PCNA, the considerable torque generated from the production of double-helical DNA would cause the polymerase to lose its place at the replication fork. PCNA allows the polymerase to relax and regain its hold. It keeps the polymerase from falling off the DNA template, so that many thousands of nucleotides are polymerized before the enzyme dissociates. The efficient movement of the replication fork also relies on rapid placement of PCNA at newly primed sites on the lagging DNA strand by RFC. This allows Okazaki fragment synthesis to keep pace with the continous DNA synthesis on the leading strand.

Proofreading

Despite being classified as high-fidelity enzymes, the replicative polymerases are not perfect. They generate errors spontaneously when copying DNA, with mutation rates ranging from 10^{-4} to 10^{-5} per base pair. This means that during each round of replication, there is one mistake for every 10,000 to 100,000 bp. Many replicative polymerases have an associated proofreading exonuclease that excises 90–99% of misincorporated nucleotides, reducing the spontaneous polymerase error rates to within the range of 10^{-7} to 10^{-8}. For example, DNA polymerase δ has a subunit with $3' \rightarrow 5'$ exonuclease activity (Fig. 6.13). The structure of DNA polymerase, determined by X-ray crystallography, has been likened to a hand holding the DNA. The polymerase activity is within the fingers and thumb, and the exonuclease domain is at the base of the palm. Incorporation of an incorrect base at the $3'$ end causes a melting of the end of the duplex. As a result, the polymerase pauses and excises the mispaired base, then elongation resumes. DNA polymerase α (involved in primer synthesis) does not have $3' \rightarrow 5'$ exonuclease activity.

Nucleotide selectivity largely depends on the geometry of Watson–Crick base pairs

For a long time after Watson and Crick noted that complementary base pairs form specific hydrogen bonds, these were thought to be the major contributors to the fidelity of DNA replication. Base–base hydrogen bonding does contribute to fidelity; however, selection of the correct nucleotide is now thought to largely depend on the shape and size of the Watson–Crick base pairs. The shape and size of AT and GC base pairs are remarkably similar to each other (see Fig. 2.4), but differ from mismatched base pairs. The abnormal geometry of mismatched base pairs results in steric hindrance at the active site that inhibits efficient catalysis.

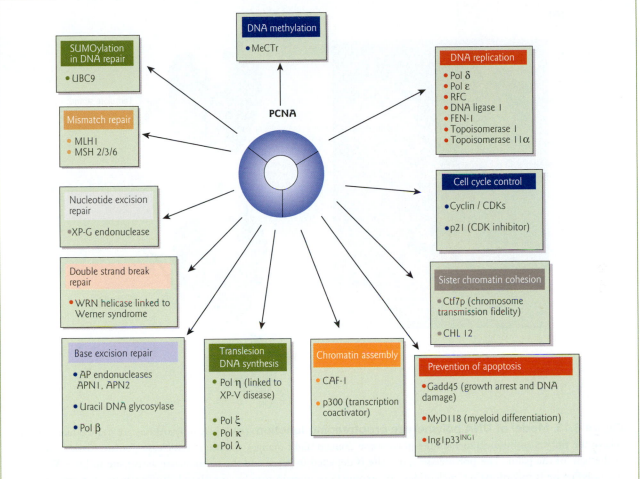

Figure 6.12 PCNA-interacting proteins. Proliferating cell nuclear antigen (PCNA) is a sliding clamp with many protein partners. In addition to its roles for DNA replication, PCNA is also involved in cell cycle control, sister chromatin cohesion during cell division, prevention of apoptosis, chromatin assembly, translesion DNA synthesis, various DNA repair pathways, and DNA methylation.

Maturation of nascent DNA strands

Maturation of newly synthesized DNA involves several different steps: RNA primer removal, gap fill-in, and joining of Okazaki fragments on the lagging strand (see Fig. 6.7).

RNA primer removal

Two different pathways have been proposed for RNA primer removal. In one model, ribonuclease (RNase) H1 nicks the RNA primer leaving one nucleotide upstream of the RNA–DNA junction. The primer is then degraded by the $5' \rightarrow 3'$ exonuclease activity of FEN-1 (flap endonuclease 1). FEN-1 is a structure-specific $5'$ nuclease (with both exo- and endonuclease activities) that acts in association with PCNA. An exonuclease is an enzyme that removes dNMPs from the end of a nucleotide chain by breaking the terminal phosphodiester bond. An endonuclease is an enzyme that cleaves the phosphodiester bond joining adjacent nucleotides at an internal site in the DNA chain.

In a second model, DNA polymerase ε causes strand displacement of the downstream Okazaki fragment. This is followed by the endonuclease activity of FEN-1 which removes the entire RNA-containing $5'$ flap

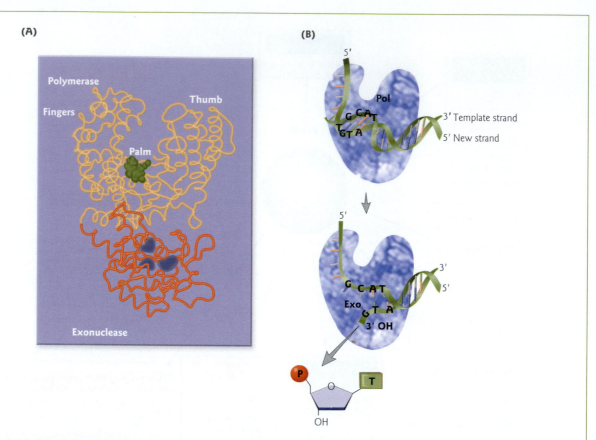

Figure 6.13 Model of DNA polymerase proofreading function. (A) DNA polymerases are shaped like a hand where the fingers and thumb include the polymerase domain (light orange) and the exonuclease domain (dark orange) is at the base of the palm. The polymerase active site is depicted in green and the exonuclease active site in blue. When the polymerase is incorporating nucleotides, the exonuclease domain may be in a closed conformation that prevents binding of single-stranded DNA, but changes to a more open conformation for proofreading. (Adapted from Kunkel, T.A. and Bebenek, K. 2000. DNA replication fidelity. *Annual Review of Biochemistry* 69:497–529. Copyright © 2000, with permission from Annual Reviews.) (B) As long as the correct nucleotides are added to the 3′ end of the new strand, the 3′ end remains in the polymerase active site. Incorporation of a mismatch causes a melting of the newly formed double-stranded DNA. The polymerase pauses, and the 3′ end of the new strand is transferred to the exonuclease domain, where the mismatched base is excised and released as a deoxynucleoside monophosphate (dNMP).

(see Fig. 6.7). In this model, RNase H1 is not required. Recent studies suggest that another enzyme, Dna2, which has both helicase and endonclease activity, may be required in some cases.

Gap fill-in and joining of the Okazaki fragments

The remaining gap left by primer removal on the lagging strand is filled in by DNA polymerase ε, resulting in a nicked double-stranded DNA. Ligation then occurs by the action of DNA ligase I, which joins the Okazaki fragments by catalyzing the formation of new phosphodiester bonds (see Fig. 6.7). X-ray crystallographic analysis shows a unique feature of mammalian ligases – a DNA-binding domain that allows ligase to encircle its DNA substrate, stabilize that DNA in a distorted structure, and position the catalytic core on the nick (Fig. 6.14). The DNA immediately upstream of the nick adopts an A-form helix with an expanded minor groove, whereas the downstream DNA is in the normal B form. The DNA-binding

(A)

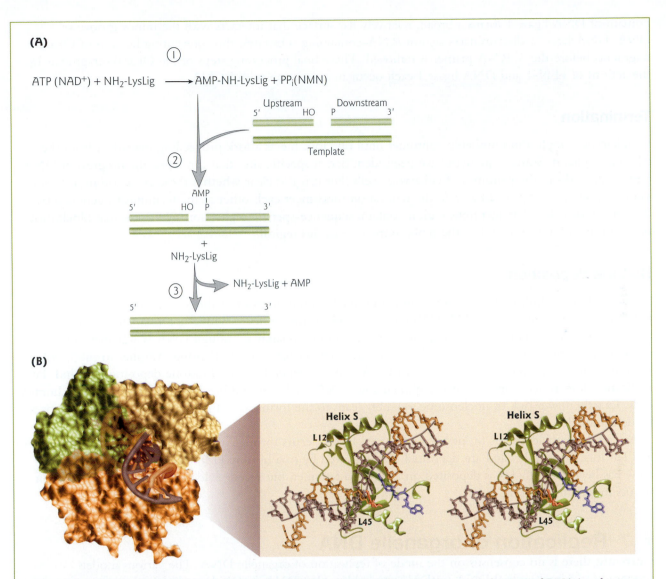

(B)

Figure 6.14 Human DNA ligase I encircles DNA during ligation. (A) Enzymatic ligation of DNA involves three steps. (1) An enzyme–AMP complex is formed by the attack of lysine (lys) on the α-phosphate of ATP (or NAD$^+$) releasing inorganic pyrophosphate (PP$_i$) or nicotinamide mononucleotide (NMN). (2) The 5′ phosphate (5′ P) of the nicked DNA strand (downstream) attacks the lys–AMP intermediate to form an AppDNA intermediate (pyrophosphate linkage, 5′ P to the 5′ phosphate of AMP). (3) The 3′-OH terminated end of the nicked strand (upstream) attacks the 5′ P of AppDNA, covalently joining the DNA strands and releasing AMP. (B) Molecular surface of the ligase I–DNA complex. Three domains of ligase I surround the AppDNA reaction intermediate. The catalytic core is composed of the adenylation domain (green) and the OB fold (yellow). The DNA-binding domain is shown in orange. The adenylation domain is semitransparent to highlight the AMP cofactor held within the active site. (Inset) Stereo view of the OB fold domain (yellow) as it distorts the DNA duplex, resulting in an A- to B-form transition of DNA structure across the nick (red to blue nucleotides). The DNA helical axis (gray line) shifts by more than 5 Å at the nick site. (Protein Data Bank, PDB:1X9N. Adapted by permission from Nature Publishing Group and Macmillan Publishers Ltd: Pascal, J.M., O'Brien, P.J., Tomkinson, A.E., Ellenberger, T. 2004. Human DNA ligase I completely encircles and partially unwinds nicked DNA. *Nature* 432:473–478. Copyright © 2004.)

domain of DNA ligase 1 forms a broad, relatively flat surface that interacts with the minor groove of DNA. DNA ligase 1 discriminates against RNA-containing substrates, thus preventing ligation of Okazaki fragments before the 5′ RNA primer is removed. These final processing steps of the Okazaki fragments by the actions of FEN-1 and DNA ligase I each occur in association with PCNA.

Termination

In eukaryotes, replication probably continues until one fork meets a fork proceeding towards it from the adjacent replicon. Some sequences have been identified at specific sites that can arrest the progress of DNA replication forks in the genomes of eukaryotic cells, but it is not clear whether these are a common feature associated with all origins. In *E. coli*, the replication forks meet each other at the terminus to generate two daughter molecules. The terminus region contains sequence-specific replication arrest sites that block fork progression and limit the end of the replication cycle to this region.

Histone deposition

As the replication fork moves through chromatin, nucleosomes in front of the fork are disassembled into H3-H4 tetramers (or dimers) and H2A-H2B dimers. Nucleosomes re-form within approximately 250 bp behind the replication fork. Thus, histone deposition occurs almost as soon as enough DNA is available to form nucleosomes (approximately 180 bp). Chromatin assembly factor 1 (CAF-1) brings histones to the DNA replication fork via direct interaction with PCNA. CAF-1-dependent rapid histone deposition behind the replication fork is necessary to prevent spontaneous DNA double-strand breaks and S phase arrest in human cells. Exactly how a lack of nucleosome assembly leads to the formation of double-strand breaks remains to be determined.

Assembly of nucleosomes on newly synthesized DNA occurs through a stepwise mechanism. Histones H3 and H4 form a complex and are deposited first, followed by two histone H2A-H2B dimers. The general view has been that H3-H4 is deposited on DNA as a tetramer, but recent evidence suggest that deposition occurs in dimer form (Fig. 6.15).

6.7 Replication of organelle DNA

Currently, there is no consensus on the mode of replication of organelle DNA. The various models proposed for mitochondrial DNA (mtDNA) and chloroplast DNA (cpDNA) replication remain controversial.

Models for mtDNA replication

How mtDNA replicates remains a subject of debate. What is known for certain is that DNA polymerase γ is used exclusively for mtDNA replication and proofreading. Two models have been proposed for the mode of mtDNA replication. One, called the strand displacement model, invokes continuous DNA replication. The other, called the strand coupled model, proposes semidiscontinuous DNA replication.

Strand displacement model

The strand displacement model (also called the strand asynchronous model) for mammalian mtDNA replication is the most widely accepted, longest standing model (Fig. 6.16A). In this model, replication is unidirectional around the circle and there is one replication fork for each strand. One strand is called the light strand (L) and the other the heavy strand (H). The designation of the strands as H and L arose from the observation that the two strands have different buoyant densities in denaturing CsCl density gradients (see Fig. 6.2 for methods) due to a strand bias in base composition.

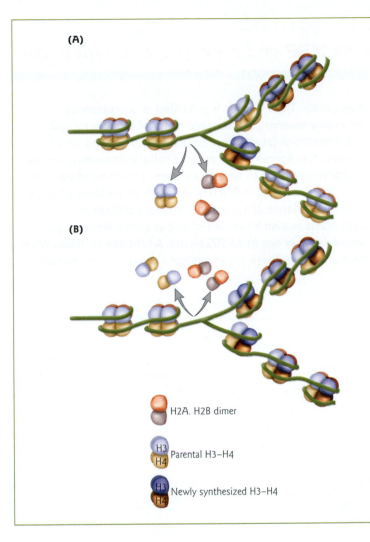

(A)

(B)

H2A. H2B dimer

H3
H4 Parental H3–H4

H3
H4 Newly synthesized H3–H4

Figure 6.15 Possible mechanisms for nucleosome assembly after DNA replication. (A) The tetrameric model. Parental H3–H4 tetramers are dissociated from nucleosomes and deposited randomly onto one of the daughter DNA strands. Newly synthesized H3–H4 tetramers are deposited on the daughter strands in a CAF-1-dependent manner. In this model, parental and newly synthesized tetramers are deposited into distinct nucleosomes. (B) The dimeric model: Parental H3–H4 tetramers are dissociated into dimers and are paired with newly synthesized H3–H4 dimers on each daughter DNA strand by the action of a CAF-1-containing complex. In this model, H3–H4 dimers from parental nucleosomes are segregated evenly onto daughter DNA strands. (Adapted from Tagami, H., Ray-Gallet, D., Almouzni, G., Nakatani, Y. 2004. Histone H3.1 and H3.3 complexes mediate nucleosome assembly pathways dependent or independent of DNA synthesis. *Cell* 116:51–61. Copyright © 2004, with permission from Elsevier.)

There is one priming event per template strand – thus, there are two origins (O_H and O_L). Replication begins in a region called the displacement or "D" loop – a region of 500–600 nt. A preformed RNA primer is required at both origins. The RNA primer is thought to be a processed RNA transcript synthesized at another location in the mtDNA by the mitochondrial RNA polymerase. The H strand is used as a template first to make a new L strand, starting at O_H. After approximately two-thirds of the mtDNA has been copied by DNA polymerase, the replication fork passes the major origin of L strand synthesis. Only after the displaced H strand passes the origin on the L strand, exposing this site in single-stranded form, does synthesis of a new H strand start from O_L. Synthesis is continuous around the circle on both strands. The RNA primers are cleaved by the multifunction endoribonuclease RNase MRP (Disease box 6.2).

Strand coupled model

In recent years, another model for mammalian mtDNA replication has been proposed called the strand coupled model (Fig. 6.16B). Analysis of mtDNA replication intermediates by two-dimensional agarose gel electrophoresis (see Fig. 6.8) suggest that lagging strand replication (L strand) initiates at multiple sites, probably involving discontinuous synthesis of short Okazaki fragments, and requiring multiple primers. Thus, in this model the coupled leading (H strand) and lagging strand synthesis represents a semidiscontinous, bidirectional mode of DNA replication.

RNase MRP and cartilage-hair hypoplasia

The endoribonuclease RNase MRP (mitochondrial RNA processing) has at least two functions, namely cleavage of RNA primers in mtDNA replication and nucleolar processing of pre-ribosomal RNA. Needless to say, this caused some controversy and confusion in the literature, while researchers sorted out its dual function in two completely separate organelles.

RNase MRP is a ribonucleoprotein complex. Recently, it was shown that mutations in the RNA component of RNase MRP cause a rare form of dwarfism called cartilage-hair

hypoplasia. The disorder is inherited in an autosomal recessive manner. The disease has multiple phenotypic manifestations (pleoitropy) including short limbs, short stature, fine sparse hair, impaired cellular immunity, anemia, and predisposition to several cancers. This form of dwarfism was first described in Old Order Amish in the United States with an incidence of 1.5 in 1000 births. Cartilage-hair hypoplasia is also found in Finland at a high frequency, approximately one in 23,000 births. A function of RNase MRP that affects multiple organ systems is likely to be disrupted.

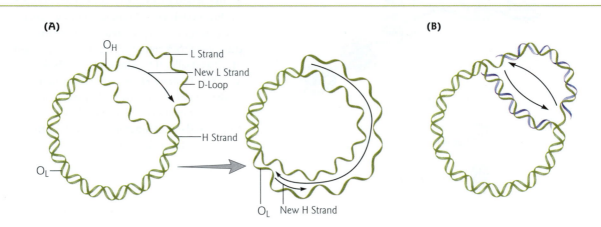

Figure 6.16 Models for mitochondrial DNA replication. (A) The strand displacement model is the widely accepted model. Replication is unidirectional from a heavy (H) strand origin (O_H) and a light (L) strand origin (O_L); both strands can be replicated continuously. (B) The strand coupled model was proposed more recently. In this model, replication is bidirectional and occurs in a semidiscontinuous mode, involving the synthesis of Okazaki fragments on the lagging strand.

Replication of cpDNA

How cpDNA replicates also remains a subject of debate – in particular, because there is a continuing debate over whether the majority of cpDNA is circular or linear. For most organisms studied, two different replication origins, named oriA and oriB, have been proposed, supporting a strand displacement model as in mtDNA. However, another model proposes initiation at two sites forming D loops, merging of the D loops to form a theta (θ) replication intermediate, and then conversion to a rolling circle mechanism (see Section 6.8). A more recent model proposes a mode of recombination–dependent replication.

6.8 Rolling circle replication

Some circular DNA molecules replicate by rolling circle replication – a process that does not include a theta (θ) shaped intermediate (Fig. 6.17, compare with Fig. 6.3). Rolling circle replication occurs in the

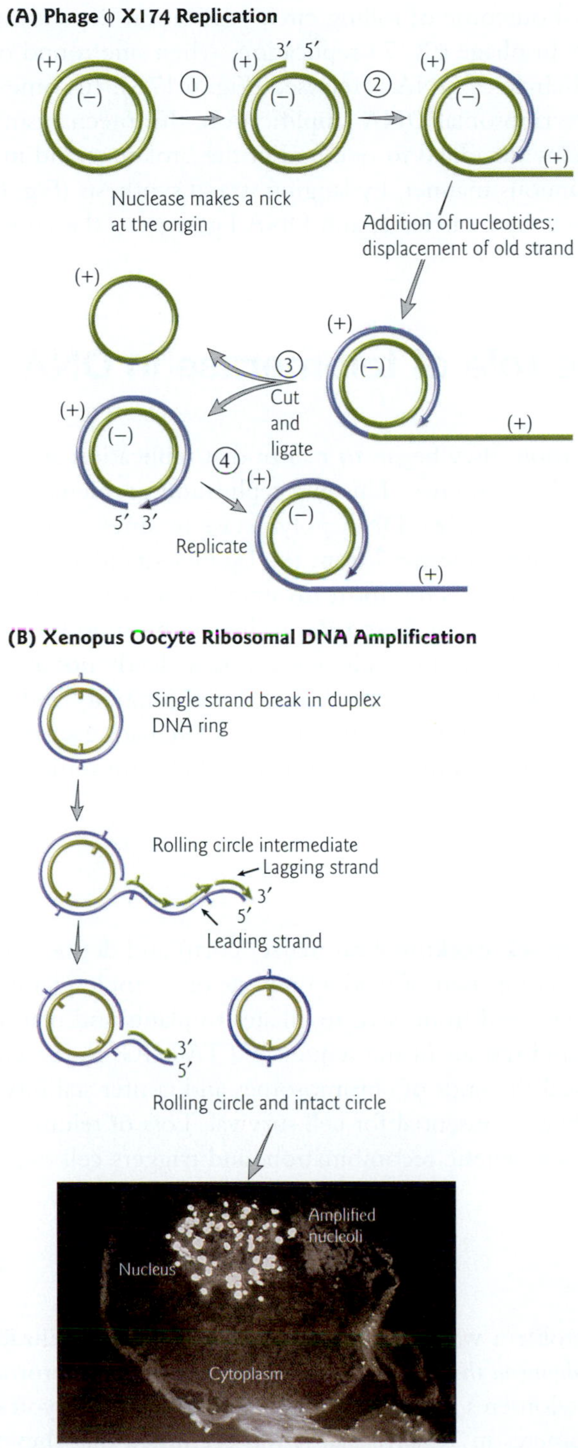

(A) Phage φ X174 Replication

(1) Nuclease makes a nick at the origin

(2) Addition of nucleotides; displacement of old strand

(3) Cut and ligate

(4) Replicate

(B) Xenopus Oocyte Ribosomal DNA Amplification

Single strand break in duplex DNA ring

Rolling circle intermediate
— Lagging strand

Leading strand

Rolling circle and intact circle

Amplified nucleoli

Nucleus

Cytoplasm

Figure 6.17 Two examples of the rolling circle model of DNA replication. (A) Phage φX174 replication involves four main steps. (1) An endonuclease creates a nick in the positive (+) strand of the double-stranded replicative form of the phage. (2) The free 3′ end created by the nick serves as the primer for the addition of nucleotides to the positive strand (green), as the 5′ end of the positive strand is displaced. The negative (−) strand serves as the template. Further replication occurs as the positive strand approaches unit length. (3) The unit length of single-stranded (+) DNA that has been displaced is cleaved off by an endonuclease and the ends are ligated to form a circle (blue). (4) Replication continues, producing another new positive strand, using the negative strand as a template. The process repeats over and over to yield many copies of the circular phage genome. (B) *Xenopus* oocyte ribosomal DNA (rDNA) amplification by a rolling circle mechanism. As the circle rolls to the left in the diagram, the leading strand (blue) elongates continuously. The lagging strand (green) elongates discontinuously, using the unrolled leading strand as a template and RNA primers for each Okazaki fragment. The double-stranded DNA thus produced grows to many rDNA repeat units in length before one rDNA repeat's worth is cleaved off and ligated to form a circle. Very large numbers of extrachromosomal rDNA circles are needed to produce sufficient ribosomal RNA components for the massive stock of ribosomes found in a mature frog oocyte. The photograph shows nucleoli that have formed around each amplified rDNA circle in a section from a *Xenopus* oocyte. The amplified nucleoli were visualized by immunostaining with a nucleolus-specific antibody. (Photograph courtesy of Aimee Hayes Bakken, University of Washington.)

multiplication of many bacterial and eukaryotic viral DNAs, of bacterial F (fertility) factors during mating, and of the DNA in certain cases of gene amplification, and possibly as part of chloroplast DNA replication, as noted in Section 6.7.

In rolling circle replication, a phosphodiester bond is broken in one of the strands of a circular DNA. This creates a free 3′-hydroxyl end and a free 5′-phosphate end. Synthesis of a new circular strand occurs by the addition of nucleotides to the 3′ end using the complementary intact strand as a template. The other end is displaced as a 5′-phosphate tail. The final outcome of rolling circle replication depends upon the type of circular DNA which has been replicated. In phage φX174 replication, when one round of replication is complete, a full-length, single-stranded circle of DNA is released (Fig. 6.17A). In some cases, such as phage λ replication and *Xenopus* oocyte ribosomal DNA amplification, this mechanism is used to replicate double-stranded DNA. As nucleotides are added to one end of the broken strand in a continuous fashion, the tail is replicated in a discontinuous manner, by lagging strand synthesis (Fig. 6.17B). The tail is cleaved from the new double-stranded circle by a nuclease, and DNA ligase joins the ends to form a circle.

6.9 Telomere maintenance: the role of telomerase in DNA replication, aging, and cancer

As researchers worked out the details of DNA replication, they began to realize that replication at chromosome ends poses a special problem for linear chromosomes. This end replication problem was defined by James Watson and A. Olovnikov in the early 1970s (Fig. 6.18). DNA polymerase requires a short RNA primer and proceeds only 5′ to 3′. When the final primer is removed from the lagging strand at the end of a chromosome, this 8–12 nt region is left unreplicated. There is no upstream strand onto which DNA polymerase ε (or δ) can build to fill the gap. Strict application of these rules to linear chromosomes predicts that chromosomes would get shorter with each round of replication. However, this is clearly not always the case, since progressive shortening of telomeres would eventually lead to chromosome instability and cell or organismal death. The discovery of the solution to the end replication problem is a fascinating story that has unfolded with new twists each year. The story begins with telomeres, the functional chromosome elements that protect the ends of eukaryotic linear chromosomes.

Telomeres

Telomeres were identified in 1938 by Barbara McClintock working with maize (corn) and defined by H.M. Muller working with *Drosophila*. Telomeres are comprised of tandem repeats of a simple guanine (G) rich sequence. The sequence is evolutionarily conserved from yeast to ciliates to plants and mammals. For example, human telomeres contain several thousand repeats of the sequence TTAGGG. In the ciliate *Tetrahymena*, the sequence is TTGGGG. Telomeres seal the ends of chromosomes and confer stability by keeping the chromosomes from ligating together. They are essential for cell survival. Loss of telomeres leads to end-to-end chromosome fusions, facilitates increased genetic recombination, and triggers cell death through apoptosis.

Solution to the end replication problem

The elegant solution to the incomplete replication problem was reported by Carol Greider and Elizabeth Blackburn in 1985. They studied this puzzle in *Tetrahymena thermophila*, a pond-dwelling ciliate protozoan (single-celled eukaryote) because it has over 40,000 telomeres. By comparison, most other eukaryotes have less than 100 telomeres. Since there are so many telomeres in *Tetrahymena*, it was presumed that they would also be an abundant source of the machinery required for solving the end replication problem.

   Greider and Blackburn discovered an enzyme activity they called telomerase terminal transferase which catalyzed *de novo* addition of telomeric repeats to the ends of chromosomes (Fig. 6.19). Later they shortened the name to "telomerase." Telomerase is now known to be a ribonucleoprotein (RNP) complex – the enzyme has an RNA subunit and a protein subunit that are both essential for activity. The first telomerase component purified was the telomerase RNA or telomerase RNA component (TERC) from *Tetrahymena* in 1989. That same year, telomerase activity was documented in human cells, and interest in this RNP escalated because of the potential medical relevance. Ten years later, the protein component of telomerase

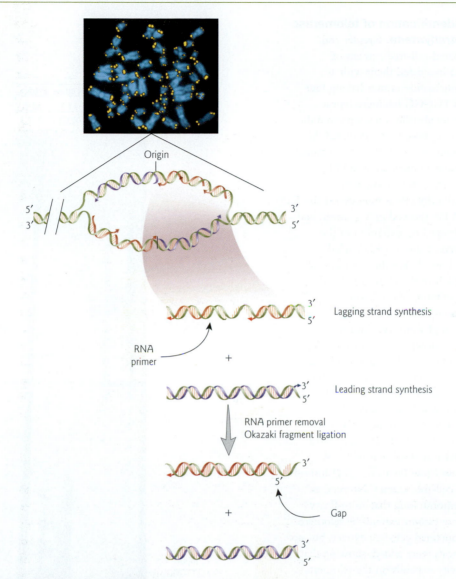

Figure 6.18 DNA end replication problem. A replication fork moving from an interior position on the chromosome is shown moving towards the end of the chromosome. Leading strand synthesis can copy the template strand all the way to the last nucleotide. When the final primer is removed from the lagging strand, however, an 8–12 nt region is left uncopied leaving a 5′-terminal gap. Following subsequent rounds of replication, if only the semiconservative DNA replication machinery operates, this gap will result in progressively shorter chromosomes. (Inset) Fluorescence *in situ* hybridization (FISH) analysis of individual telomeres from human cells. (Ning, Y., Xu, J., Li, Y., Chavez, L., Riethman, H.C., Lansdorp, P.M., Weng, N. 2003. Telomere length and the expression of natural telomeric genes in human fibroblasts. *Human Molecular Genetics* 12:1329–1336. Copyright © 2003. Reprinted with permission of the Oxford University Press, and Yi Ning et al.)

was purified and characterized as a reverse transcriptase – a polymerase that synthesizes DNA using an RNA template. The protein component is called telomerase reverse transcriptase (TERT).

Maintenance of telomeres by telomerase

At the completion of lagging strand synthesis and primer removal, there is a shortened 5′ strand (often called the C-rich strand or C strand), and a 12–16 base overhang on the 3′ strand (G-rich strand or G strand).

Figure 6.19 Identification of telomerase activity in *Tetrahymena*. Greider and Blackburn prepared cell-free extracts of *Tetrahymena* and incubated them with a synthetic oligonucleotide primer having four repeats of the TTGGGG telomere repeat sequence. After incubation, they separated the products by electrophoresis and detected them by autoradiography. Lanes 1–4 each contained a different ^{32}P-labeled nucleotide (dATP, cCTP, dGTP, and dTTP) along with unlabeled ("cold") dNTPs as indicated. Lane 1, with labeled dATP, showed only a smear, and lanes 2 and 4 showed no extension of the synthetic telomere. Lane 3, with labeled dGTP, clearly showed periodic extension of the telomere. Each of the clusters of bands represents that addition of one more TTGGGG sequence (with some variation in the degree of completion). Additional experiments (not shown) demonstrated that at a higher concentration, dTTP also could be incorporated into telomeres. Lanes 5–8 show the results of an experiment with one labeled and only one unlabeled nucleotide. This experiment verified that telomerase activity requires both dGTP and dTTP. Lanes 9–12 contained the Klenow fragment of *E. coli* DNA polymerase I (see Focus box 6.1) instead of *Tetrahymena* cell-free extract. No repeats were added, demonstrating that an ordinary DNA polymerase cannot extend the telomere. Lanes 13–16 contained cell-free extract, but no primer. No repeats were added, showing that telomerase activity depends on the telomere-like primer. (Reprinted from Greider, C.W. and Blackburn, E.H. 1985. Identification of a specific telomere terminal transferase activity in *Tetrahymena* extracts. *Cell* 43:405–413. Copyright © 1985, with permission from Elsevier.)

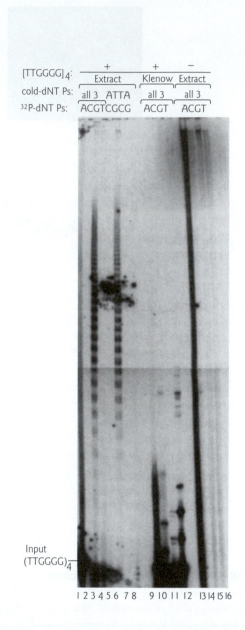

Researchers found that there is a region complementary to the C strand telomere repeats in the telomerase RNA (Fig. 6.20). The RNA provides the template for telomere repeat synthesis. A pseudoknot structure in the RNA (see Fig. 4.7) is important for the processivity of repeat addition.

Counterintuitively, telomerase does not extend the short 5′ (lagging) strand. Instead, telomerase causes elongation of the 3′ template for the lagging strand (5′ → 3′). Extension of the telomere terminus results in the addition of one telomeric repeat. Repositioning allows another round of copying onto the template for the lagging strand. Repeated translocation and elongation steps result in chromosome ends with an array of tandem repeats. After elongation of the 3′ G strand, synthesis of the shortened 5′ C strand is presumably

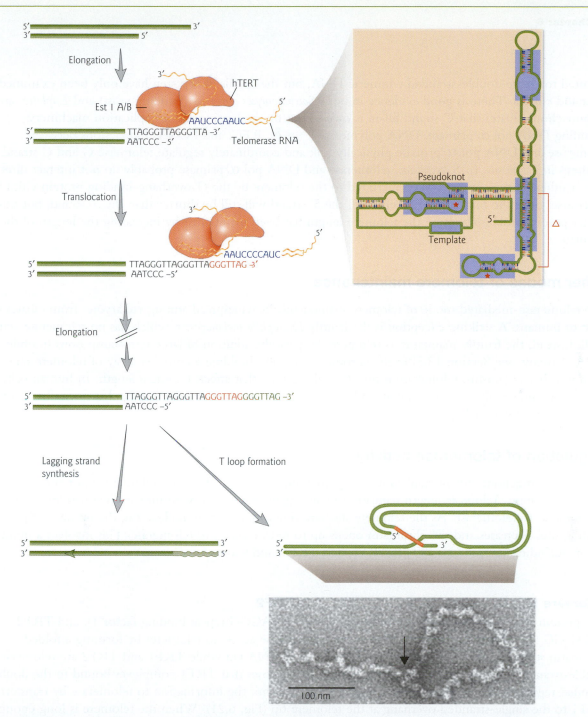

Figure 6.20 Synthesis of telomeric DNA by human telomerase. The 3′ nucleotides of the G-rich overhang at the end of the chromosome (shown ending as TTA-3′) base pair with the complementary sequence in the telomerase RNA. The 3′ end is extended by addition of dGTP, dTTP, and dATP using the RNA as a template. The extended DNA end becomes available for another round of elongation by telomerase and/or DNA pol α/primase, which uses it as a template for lagging strand synthesis of the C-rich telomere strand. Alternatively, the 3′ overhang folds into a T loop structure. (Upper inset) Conserved structural motifs in vertebrate telomerase RNA (blue boxes). The positions of mutations in the telomerase RNA gene linked with dyskeratosis congenita are indicated by red symbols (see Disease box 6.3). (Adapted from Smogorzewska, A. and de Lange, T. 2004. Regulation of telomerase by telomeric proteins. *Annual Review of Biochemistry* 73:177–208. Copyright © 2004, with permission from Annual Reviews.) (Lower inset) Electron micrograph showing a telomeric T loop in chromatin from chicken erythrocytes. The arrow denotes a loop-tail junction. (Reproduced from Nikitina, T. and Woodcock, C.L. 2004. Closed chromatin loops at the ends of chromosomes. *The Journal of Cell Biology* 166:161–165, by copyright permission of The Rockefeller University Press, and by author permission.)

required to create double-stranded telomeric DNA, but the details of this step have only been examined in yeast and ciliates. There is good evidence in both *Saccharomyces cerevisiae* (budding yeast) and *Euplotes crassus* (hypotrichous ciliate) that C strand fill-in is carried out by the lagging strand replication machinery, including DNA pol α/primase, DNA polymerase ε (or δ), RFC, and PCNA. Interactions between telomerase and DNA pol α/primase physically link and coordinately regulate telomeric G and C strand synthesis in *E. crassus*. In *S. cerevisiae*, telomerase and DNA pol α/primase probably do not interact directly – one or other enzyme appears to be recruited to the telomere by the G-overhang-binding protein Cdc13p. Of course, once the final primer is removed, the 5′ strand will still be shorter than the 3′ strand, but this no longer poses a problem because the telomere length has been maintained by increasing the length of the template.

Other modes of telomere maintenance

The telomerase-mediated mode of telomere maintenance is widespread among eukaryotes from ciliates to yeast to humans. A striking exception is the fruitfly *Drosophila melanogaster*, which has no telomerase activity at all. Instead, the fruitfly maintains its telomeres by periodic addition of large retrotransposons (mobile DNA elements, see Section 12.5) to the chromosomal ends, building a complex array of telomere repeats. Even in other organisms, telomerase is not the only activity that affects telomere length. In human cells and in fungi, telomeres also can be maintained by a recombination-based mechanism, and they can be shortened by the action of exonucleases.

Regulation of telomerase activity

One aspect of genome maintenance involves protecting telomeres. However, telomeres must also not become too long. Telomere length regulation in all organisms from yeast to humans involves the accessibility of telomeres to telomerase. As the telomere shortens due to incomplete replication, the number of protein-binding sites decreases and the chromatin opens up to restore access to telomerase. This involves a number of factors including the proteins POT1, TRF1, and TRF2, and t-loop formation at the telomeres.

Telomere length control by POT1, TRF1, and TRF2

The proteins POT1 (protection of telomeres), TRF1 (TTAGGG repeat binding factor 1), and TRF2 (TTAGGG repeat binding factor 2) may prevent telomerase access to telomeres by forming a folded chromatin structure. POT1 binds the 3′ single-stranded DNA tail while TRF1 and TRF2 are telomeric double-stranded DNA-binding proteins. One model proposes that TRF1 complexes bound to the double-stranded region of the telomere sense the length and transmit the information to telomerase by transferring POT1 to the single-stranded overhang at the telomere tip (Fig. 6.21). When the telomere is long enough, the levels of POT1 on the overhang are high, and telomerase is inhibited. When the telomere is too short, little or no POT1 is transferred to the end and telomerase is no longer inhibited, allowing it to add DNA back to the telomere. In other words, TRF1 and TRF2 can "count" the number of G-rich repeats, and when the telomere becomes overly long, they inhibit further telomerase activity.

t-loop formation

In a wide range of eukaryotes, including mammals, birds (chickens), protozoans (*Trypanosoma brucei* and *Oxytricha fallax*), and the garden pea (*Pisum sativum*), telomeric DNA has been shown to form a unique t-loop structure. In this structure, the 3′ single-stranded DNA tail invades the double-stranded telomeric DNA to form the loop in which the 3′ overhang is base paired to the C strand sequence (see Fig. 6.20). The loop in association with TRF1 and TRF2 may aid in preventing telomerase access, since telomerase requires an unpaired 3′ end for function.

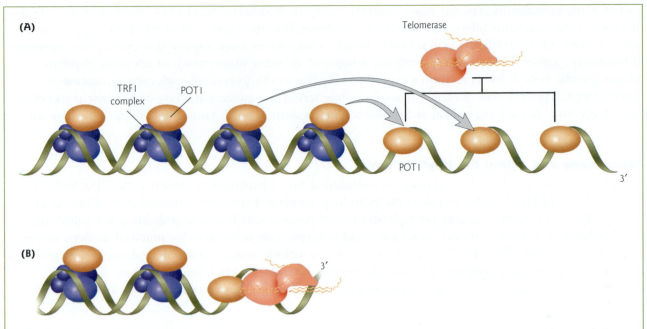

Figure 6.21 Model for telomere length control. POT1 binds to the TRF1 complex on the double-stranded portion of telomeres. TRF1 complexes sense the length of the telomere and transmit this information to telomerase via POT1, by transferring POT1 to the 3′ overhang at the telomere tip. (A) When the telomere is long enough, POT1 levels are high at the 3′ end and the action of telomerase is blocked. (B) When the telomere is too short, little or no POT1 is present at the 3′ end and telomerase is no longer inhibited. (Adapted by permission from Nature Publishing Group and Macmillan Publishers Ltd: Lundblad, V. 2003. Telomeres: taking the measure. *Nature* 423:926. Copyright © 2003.)

Telomerase, aging, and cancer

In most unicellular organisms, telomerase has a "housekeeping" function, meaning that its core components are always expressed. In contrast, most human somatic cells do not express enough telomerase to maintain a constant telomere length during cycles of chromosomal replication (Table 6.3). High levels of telomerase activity are restricted to ovaries, testes, some proliferating epithelial cells, and lymphocytes. Adult stem cells (undifferentiated cells that can undergo unlimited division and can give rise to one or several different cell types) have weak telomerase activity. In human fibroblasts (connective tissue cells), the level of telomerase

Table 6.3 Telomerase activity.

Cell type	Telomere length	Telomerase activity
Single-celled eukaryotes, e.g. *Tetrahymena*	Maintained	+
Human germline cells, e.g. sperm, oocytes (eggs)	Maintained	+
Human somatic cells:		
Not rapidly dividing, e.g. fibroblast cells	Progressively shorten (50–200 nt/division)	−*
Rapidly dividing, e.g. epidermis, bone marrow, gastrointestinal mucosa	Maintained	+
Human malignant cancer cells (advanced tumors)	Maintained	+

* There may be low levels of transient, periodic expression in fibroblasts.

activity is undetectable by standard assays. Recently, using more sensitive techniques, researchers have shown that there is, in fact, some telomerase activity in fibroblasts. But, the expression of telomerase appears to be transient, and fails to stabilize overall telomere length. These observations suggest that continuous expression of telomerase, rather than periodic expression, is required for stable maintenance of telomere length in human somatic cells. It has been shown experimentally that in the absence of a telomere maintenance system, fungi, trypanosomes, flies, and mosquitoes lose terminal sequences at a rate of 3–5 base pairs per end per cell division. In contrast, human and mouse telomeres shorten much faster (50–150 bp/end/division).

Telomerase and aging: the Hayflick limit

The observations described above provide an explanation for a phenomenon known as the "Hayflick limit." In 1962, Leonard Hayflick discovered, contrary to long-standing dogma, that cultured normal human and animal cells have a limited capacity for replication – the point at which cells stop dividing was called the Hayflick limit. This distinction between mortal and immortal cells is the basis for much of modern cancer research (see Section 17.2). The limit to the number of doublings somatic cells can undergo is now proposed to be a consequence of progressive telomere shortening. After the Hayflick limit, telomere shortening triggers an irreversible state of cellular aging or senescence, a state characterized by continued cell viability without further cell division.

Telomere shortening: a molecular clock for aging?

Based on the correlation between cellular senescence and telomerase activity, scientists proposed that telomere shortening is a "molecular clock" that triggers aging. This proposal captured the imagination of the general public. Telomerase reactivation was hailed as a "fountain of youth" by the media. However, the flip side is that cellular senescence may be an important mechanism to protect multicellular organisms from cancer. One of the hallmark features of cancer cells is that they become immortalized and can grow uncontrollably (see Fig. 17.1). In most human cancer cells, telomerase has been reactivated. Telomerase may thus be a more attractive target for anticancer therapy rather than anti-aging therapy.

Direct evidence for a relationship between telomere shortening and aging

From the observations discussed above, a model has been suggested that telomeres function as molecular counting devices that register the number of cell divisions. They then trigger proliferative arrest when telomeres shorten to specific lengths. A number of lines of evidence provide support for this model of a causal relation between telomere shortening and aging. These include experiments in human cells in culture and transgenic mice. However, relationships among telomere length, telomerase expression, and cellular lifespan turn out to be much more complex than first thought. In cells without telomerase activity, shortening of telomeres cannot always be used as a measure of the number of cell divisions. For example, there are reports of instances in which short telomere length does not correlate with entry into cellular senescence.

Effect of experimental activation of telomerase on normal human somatic cells   In 1998, an experiment was carried out to test the effect of experimental activation of telomerase on normal human somatic cells. The limiting component in somatic cells is the reverse transcriptase (hTERT). Transcriptional repression of the hTERT gene leads to a loss of telomerase activity. In contrast, expression of telomerase RNA is virtually ubiquitous.

Similar experiments were carried out in two telomerase-negative normal human cell types: retinal pigment epithelial cells and foreskin fibroblasts. In control cells in which a plasmid DNA vector alone ("empty vector") was introduced (see Section 9.2 for methods), the cells had a normal lifespan, no detectable telomerase activity, and their telomeres were shortened with a loss of 100 nt/division. After about 20 divisions they underwent senescence (when several kilobase pairs were lost from the chromosome ends) and

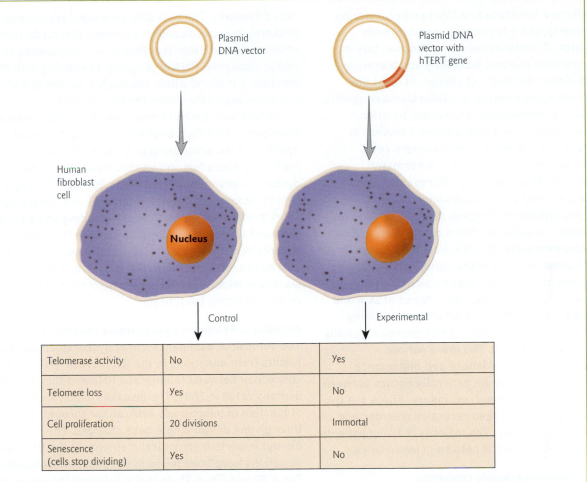

Figure 6.22 Telomerase activity increases the lifespan. Schematic diagram of an experiment demonstrating a link between telomerase activity and cellular immortality. Either plasmid DNA vector alone or plasmid DNA vector containing the hTERT gene was introduced into human fibroblast cells that did not have telomerase activity (see Section 9.2 for methods). When telomerase activity was restored, cells continued to divide and the telomeres were maintained. (Bodnar, A.G., Ouellette, M., Frolkis, M. et al. 1998. Extension of life-span by introduction of telomerase into normal human cells. *Science* 279:349–352.)

entered a nondividing state (Fig. 6.22). In the experimental cells, a plasmid DNA vector containing the gene for hTERT was introduced. In these cells, telomerase activity was restored. The result was that the cells had a greatly extended lifespan, apparently limitless. After over 20 doublings past the normal lifespan they were still phenotypically youthful, meaning they had a normal karyotype (e.g. no end-to-end fusions) and the telomeres remained long.

Insights from telomerase-deficient mice   It is possible to engineer mice in which both alleles of a gene are deleted (see Fig. 15.4 for method). Such mice are called "knockout" mice. A mouse knockout for telomerase RNA was constructed. When cells from mice lacking the telomerase RNA component were grown in culture, they were still proliferating after 20 divisions. Thus, it was initially thought that mice and humans were fundamentally different with regard to their requirement for telomerase. However, after 300 divisions, the cells showed severe growth defects and progressive telomere shortening. The explanation is that mice start with very long telomeres, over three times as long as human telomeres (10–60 kb). So, when the

Dyskeratosis congenita: loss of telomerase activity

Changes in telomere structure and function occur during aging. In the elderly, short telomeres correlate with diminished health. Diminished telomere function late in life may even promote genome instability and therefore contribute to a higher incidence of cancer. At least one human premature aging syndrome, dyskeratosis congenita, is associated with compromised telomerase function.

Dyskeratosis congenita is a rare inherited disease in which patients have problems in tissues where cells multiply rapidly and where telomerase is normally expressed (Table 1). There are only 180 reported cases worldwide. Patients may suffer from abnormal skin pigmentation, nail dystrophy (ridging, destruction, and loss of nails), premature graying, cirrhosis of the liver, and gut disorders. In approximately 70% of cases, the disease is associated with abnormalities of the bone marrow leading eventually to anemia and an increased risk of bleeding or infection. There is also an increased incidence of skin and gastrointestinal cancer. Symptoms often appear during childhood. The skin pigmentation and nail changes typically appear first, usually by age 10. The more serious complications of bone marrow failure and malignancy develop around age 20–30. Death generally occurs between 16 and 50 years, due to bone marrow failure. There are two forms of this disease, an X-linked recessive disorder and an autosomal dominant disorder. The majority (90%) of patients are male and show the X-linked pattern of inheritance.

X-linked recessive dyskeratosis congenita

The X-linked recessive form of dyskeratosis congenita results from mutations in the gene coding for a protein called dyskerin. The dyskerin gene was identified in 1998. Dyskerin is a pseudouridine synthase that binds to many small nucleolar RNAs (snoRNAs) and is proposed to play a role in ribosomal RNA processing. In keeping with this function, it is found within the nucleolus, the site of ribosome assembly within the cell nucleus.

Patients with dyskerin mutations have five-fold less telomerase RNA than unaffected siblings, implicating dyskerin in the processing or stability of telomerase RNA. Further analysis has shown that telomerase RNA has a dyskerin-binding motif (see Fig. 6.20). Correlating with a loss of telomerase activity, telomeres are abnormally short in many patient cell types, including white blood cells and fibroblasts. Most mutations in the gene result in substitution of a single amino acid – they alter but do not eliminate protein function. For example, mutations that inactivate telomerase have no impact on dyskerin's other role in ribosome biogenesis.

Autosomal dominant dyskeratosis congenita

The autosomal dominant form of dyskeratosis congenita results from mutations in the telomerase RNA gene. The connection between telomerase RNA and this disorder was demonstrated in 2001. The disease results from partial loss of function of telomerase RNA, either through deletions or through one or two single base changes. The mutations disrupt important structural elements in the RNA, e.g. mutations have been found in the pseudoknot domain (see Fig. 6.20 and Fig. 4.7). As in the X-linked recessive form of the disease, patients have abnormally short telomeres in dividing cells types where telomerase is normally expressed.

Table 1 Symptoms of dyskeratosis congenita.

Organ system	Defect in dyskeratosis congenita
General	Reduced telomerase activity, abnormally short telomeres, age-dependent increase in chromosomal rearrangements
Hair	Hair loss (including eyelashes and eyebrows), premature graying
Mouth	Precancerous oral lesions (leukoplakia), tooth loss, and cavities
Skin	Abnormal pigmentation, skin cancer
Finger and toenails	Nail dystrophy
Lung	Fibrosis
Liver	Cirrhosis
Intestine	Gut disorders, cancer
Testes	Hypogonadism (defects in sperm formation)
Bone marrow	Poor wound healing, frequent infections, failure to produce blood cells

researchers performed long-term culture of the cells, to take into account the longer telomeres, they did see an effect. After 450 divisions, cell growth stopped.

Similarly, in knockout mice lacking the telomerase RNA, the early generations of mice appeared completely normal. However, by the sixth generation, when the telomeres had shortened sufficiently to have an effect, changes associated with premature aging and cellular senescence began to be observed. There were defects in spermatogenesis, impaired proliferation of hematopoietic cells (blood cell precursors), impaired wound healing, premature graying, hair loss, and changes in the lining of the gut. Patients with the genetic disorder dyskeratosis congenita have symptoms that are very similar to the phenotype of late generation telomerase-deficient mice (Disease box 6.3).

Gene therapy for liver cirrhosis   In humans, chronic alcoholism leads to cirrhosis – heavy scarring of the liver. Excessive telomere shortening has been shown in patients with liver cirrhosis. When possible treatment is by liver transplant. In 2000, experiments were performed to attempt to inhibit liver cirrhosis in mice by telomerase gene delivery (Fig. 6.23). Knockout mice that were telomerase-deficient were used in this experiment. First, liver cirrhosis was experimentally induced by toxin-mediated liver injury (treatment with the solvent carbon tetrachloride). This resulted in high cellular turnover and telomere shortening. Upon injection of an adenovirus vector carrying the telomerase RNA gene, there was restored telomerase activity, a reduction in scarring of the liver, and improved liver function. The double-edged sword, of course, is the possibility of tumor formation, so these gene therapy strategies have not yet progressed to human trials.

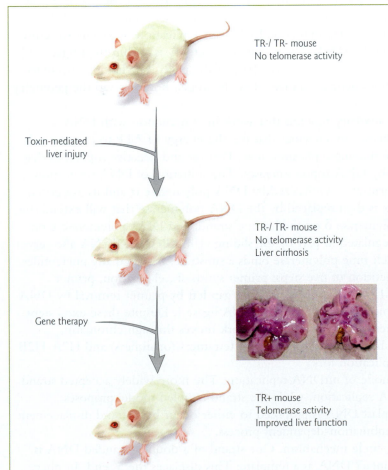

TR-/ TR- mouse
No telomerase activity

Toxin-mediated liver injury

TR-/ TR- mouse
No telomerase activity
Liver cirrhosis

Gene therapy

TR+ mouse
Telomerase activity
Improved liver function

Figure 6.23 Telomerase gene therapy. Schematic diagram of an experiment showing that the reactivation of telomerase by introducing the telomerase RNA gene (see Fig. 15.4 for methods) can improve liver function in a mouse with liver cirrhosis. (Inset) Photograph of a liver from a mouse with telomerase activity (TR+/TR+) (left) compared with a liver from a mouse lacking telomerase activity (TR−/TR−) (right). Smaller regenerative nodules (purplish-red) in the TR−/TR− mouse are apparent. (Reprinted with permission from Rudolph, K.L., Chang, S., Millard, M., Schreiber-Agus, N., DePinho, R.A. 2000. Inhibition of experimental liver cirrhosis in mice by telomerase gene delivery. *Science* 287:1253–1258. Copyright © 2000 AAAS.)

Chapter summary

Double-stranded DNA replicates in a semiconservative manner. When the parental strands separate, each serves as the template for making a new complementary strand. The most common mode of DNA replication is semidiscontinuous. Once primed, continuous replication is possible on the leading strand in the direction of the movement of the replicating fork. The other strand is replicated discontinuously, forming Okazaki fragments in the opposite direction. This allows both strands to be replicated in a $5' \rightarrow 3'$ direction, but the process involves different mechanisms.

DNA replication in eukaryotic cells occurs in discrete subnuclear compartments or replication factories. Histones are removed at the origin of replication to allow access of the replication machinery to the DNA. An origin of replication is a site on chromosomal DNA where a bidirectional pair of replication forks initiate. In *E. coli* there is a single, well-defined origin. In eukaryotes, bidirectional replication starts at many AT-rich internal sites spaced approximately every 50 kb. During the development of multicellular animals, there is selective activation or suppression of origins of replication. In yeast there is a consensus sequence called an autonomous replicating sequence (ARS), but mammalian origins lack a well-defined consensus sequence.

Once chromatin has been opened up in eukaryotic cells, specific initiator proteins recognize and bind origin DNA sequences forming an origin recognition complex (ORC). The ORC recruits other replication factors and, in an ATP-dependent process, loads the Mcm2-7 complex on the origin DNA. Only "licensed" origins containing Mcm2-7 can initiate a pair of replication forks. This replication licensing system is regulated by levels of cyclin-dependent kinases and ensures that DNA only replicates once during the cell cycle.

DNA synthesis is catalyzed by an enzyme called DNA polymerase that adds any of the four dNTP precursors into a nascent strand of DNA. DNA polymerase III is the main replicative polymerase in *E. coli*. In eukaryotes, three different DNA polymerases are involved in chromosomal DNA replication: DNA polymerase α, δ, and ε. DNA polymerase γ is used for mitochondrial DNA replication only. The structure of DNA polymerase resembles a hand holding the DNA. The polymerase activity is within the fingers and thumb, and the proofreading $3' \rightarrow 5'$ exonuclease activity is at the base of the palm. Nucleotide selectivity and recognition of incorrectly added nucleotides depends on base–base hydrogen bonding and the geometry of Watson–Crick base pairs.

The replication machinery includes many auxiliary proteins that work in conjunction with DNA polymerase to mediate replication. DNA helicases are enzymes that use the energy of ATP to progressively melt the DNA duplex in the direction of the moving replication fork. Positive and negative supercoils that are generated during replication are relieved by DNA topoisomerases. The initiation of DNA replication requires a primer. In eukaryotes, the RNA primer is synthesized by DNA polymerase α and its associated primase activity. The pol α/primase complex is then replaced by the DNA polymerase that will extend the chain. The polymerase switch is to DNA polymerase δ on the leading strand, and DNA polymerase ε on the lagging strand. Polymerase switching is regulated in part by the sliding clamp PCNA. PCNA also serves to increase DNA polymerase processivity: each time polymerase binds a substrate, it adds many nucleotides.

Lagging strand replication requires the repetition of five steps: primer synthesis, elongation, primer removal by the exonuclease activity of FEN-1, filling of the remaining gap left by primer removal by DNA polymerase, and joining of the Okazaki fragments by the action of DNA ligase I. Despite these extra steps, synthesis of both new strands occurs concurrently. As the replication fork moves through chromatin, nucleosomes in front of the fork are disassembled into histone H3-H4 tetramers (or dimers) and H2A-H2B dimers. Nucleosomes re-form behind the replication fork.

Two models have been proposed for the mode of mtDNA replication. The more widely accepted strand displacement model invokes continuous DNA replication, while the strand coupled model proposes semidiscontinuous DNA replication. Chloroplast DNA is proposed to either occur by a strand displacement model, a rolling circle mechanism, or a recombination-dependent process.

Some circular DNAs replicate by a rolling circle mechanism. One strand of a double-stranded DNA is nicked and the $3'$ end is extended using the intact DNA as a template. This displaces the $5'$ end. In phage ϕX174 replication, when one round of replication is complete, a full length, single-stranded circle of DNA is released. In phage λ replication and *Xenopus* oocyte ribosomal DNA amplification, the displaced strand serves as the template for discontinous, lagging strand synthesis.

At the completion of lagging strand synthesis and primer removal on a linear chromosome, there is a shortened 5′ strand and a 12–16 base overhang on the 3′ strand. By extending the 3′ ends of chromosomes (telomeres), the enzyme telomerase eliminates the progressive loss of chromosome ends that conventional synthesis by the replication fork machinery would cause. Telomerase is a ribonucleoprotein complex composed of a reverse transcriptase and an RNA that provides the template for telomere repeat synthesis. When telomeres become too long, the proteins POT1, TRF1, and TRF2 and a folded chromatin structure called a t-loop inhibit further telomerase activity by preventing access to telomeres. Most human somatic cells do not express enough telomerase to maintain a constant telomere length during cycles of DNA replication. Experimental evidence suggests a relationship between telomere shortening and aging. In most cancer cells, which are characterized by uncontrolled growth, telomerase has been reactivated.

Analytical questions

1 Diagram the results that Meselson and Stahl would have obtained if DNA replication were conservative.

2 The diagram below shows a replication fork in nuclear DNA.

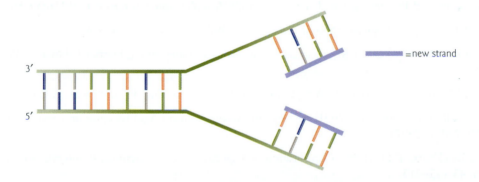

(a) Label the "leading strand" and "lagging strand" and indicate to which strand of DNA telomerase adds repeats.

(b) Show on the drawing what happens next on each strand as more of the duplex DNA unwinds at the replication fork. Use arrows to show the direction of synthesis for each strand. You do not need to show all the protein components of the replication machinery.

3 You are studying a protein that you suspect functions to recruit other components of the licensing protein complex for eukaryotic DNA replication. Describe how you would assay the protein for this activity and show sample positive results.

4 You are studying a mammalian DNA virus with a 200 kb double-stranded genome. Based on the size, you suspect that the genome has more than one origin of replication. Propose experiments to test your hypothesis and map the origins.

5 Cells you have been culturing usually undergo senescence after about 20 divisions. But, some cells have become immortal. You suspect that telomerase has been activated. Propose an experiment to assay for telomerase activity and show sample positive results.

Suggestions for further reading

Barbie, D.A., Kudlow, B.A., Frock, R. et al. (2004) Nuclear reorganization of mammalian DNA synthesis prior to cell cycle exit. *Molecular and Cellular Biology* 24:595–607.

Bendich, A.J. (2004) Circular chloroplast chromosomes: the grand illusion. *Plant Cell* 16:1661–1666.

Bielinsky, A.K., Gerbi, S.A. (2001) Where it all starts: eukaryotic origins of DNA replication. *Journal of Cell Science* 114:643–651.

Blow, J.J., Hodgson, B. (2002) Replication licensing – defining the proliferative state? *Trends in Cell Biology* 12:72–78.

Bodnar, A.G., Ouellette, M., Frolkis, M. et al. (1998) Extension of life-span by introduction of telomerase into normal human cells. *Science* 279:349–352.

Bogenhagen, D.F., Clayton, D.A. (2003) The mitochondrial DNA replication bubble has not burst. *Trends in Biochemical Sciences* 28:357–405.

Bowers, J.L., Randell, J.C.W., Chen, S., Bell, S.P. (2004) ATP hydrolysis by ORC catalyzes reiterative Mcm2-7 assembly at a defined origin of replication. *Molecular Cell* 16:967–978.

Bowman, G.D., O'Donnell, M., Kurlyan, J. (2004) Structural analysis of a eukaryotic sliding DNA clamp–clamp loader complex. *Nature* 429:724–730.

Cairns, J. (1963) The chromosomes of *E. coli. Cold Spring Harbor Symposium in Quantitative Biology* 28:43–46.

Chan, S.R., Blackburn, E.H. (2004) Telomeres and telomerase. *Philosophical Transactions of the Royal Society of London B* 359:109–121.

Chen, J.L., Greider, C.W. (2004) Telomerase RNA structure and function: implications for dyskeratosis congenita. *Trends in Biochemical Sciences* 29:183–192.

Chapados, B.R., Hosfield, D.J., Han, S., Qiu, J., Yelent, B., Shen, B., Tainer, J.A. (2004) Structural basis for FEN-1 substrate specificity and PCNA-mediated activation in DNA replication and repair. *Cell* 116:39–50.

Cook, P.R. (2001) *Principles of Nuclear Structure and Function.* Wiley Liss, Inc., New York.

DePamphilis, M.L., ed. (1996) *DNA Replication in Eukaryotic Cells.* Cold Spring Harbor Laboratory Press, Cold Spring Harbor, NY.

DePamphilis, M.L. (2003) Eukaryotic DNA replication origins: Reconciling disparate data. *Cell* 114:274–275.

Dhar, S.K., Delmolino, L., Dutta, A. (2001) Architecture of the human origin recognition complex. *Journal of Biological Chemistry* 276:29067–29071.

Greider, C.W., Blackburn, E.H. (1985) Identification of a specific telomere terminal transferase activity in *Tetrahymena* extracts. *Cell* 43:405–413.

Henneke, G., Friedrich-Heineken, E., Hübscher, U. (2003) Flap endonuclease 1: A novel tumor suppressor protein. *Trends in Biochemical Sciences* 28:384–390.

Hingorani, M.M., O'Donnell, M. (2000) Sliding clamps: A (tail)ored fit. *Current Biology* 10:R25–R29.

Huberman, J.A. (1997) Mapping replication origins, pause sites, and termini by neutral/alkaline two-dimensional gel electrophoresis. *Methods: a Companion to Methods in Enzymology* 13:247–257.

Hübscher, U., Nasheuer, H.-P., Syväoja, J.E. (2000) Eukaryotic DNA polymerases, a growing family. *Trends in Biochemical Sciences* 25:143–147.

Hunt, T. (2004) The discovery of cyclin. *Cell* S116:S63–S64.

Kaguni, L.S. (2004) DNA polymerase γ, the mitochondrial replicase. *Annual Reviews of Biochemistry* 73:293–320.

Kriegstein, H.J., Hogness, D.S. (1974) Mechanism of DNA replication in *Drosophila* chromosomes: Structure of replication forks and evidence for bidirectionality. *Proceedings of the National Academy of Sciences USA* 71:135–139.

Kunkel, T.A, Bebenek, K. (2000) DNA replication fidelity. *Annual Review of Biochemistry* 69:497–529.

Lai, C.K., Miller, M.C., Collins, K. (2003) Roles for RNA in telomerase nucleotide and repeat addition processivity. *Molecular Cell* 11:1673–1683.

Li, C.J., Vassilev, A., DePamphilis, M.L. (2004) Role for Cdk1 (Cdc2)/cyclin A in preventing the mammalian origin recognition complex's largest subunit (Orc1) from binding to chromatin during mitosis. *Molecular and Cellular Biology* 24:5875–5886.

Li, X., Rosenfeld, M.G. (2004) Origins of licensing control. *Nature* 427:687–688.

Lundgren, M., Andersson, A., Chen, L., Nilsson, P., Bernander, R. (2004) Three replication origins in *Sulfolobus* species: synchronous initiation of chromosome replication and asynchronous termination. *Proceedings of the National Academy of Sciences USA* 101:7046–7051.

Maga, G., Hübscher, U. (2003) Proliferating cell nuclear antigen (PCNA): A dancer with many partners. *Journal of Cell Science* 116:3051–3060.

Masutomi, K., Yu, E.Y., Khurts, S. et al. (2003) Telomerase maintains telomere structure in normal human cells. *Cell* 114:241–253.

Meselson, M., Stahl, F. (1958) The replication of DNA in *Escherichia coli*. *Proceedings of the National Academy of Sciences USA* 44:671–682.

Mühlbauer, S.K., Lössl, A., Tzekova, L., Zou, Z., Koop, H.U. (2002) Functional analysis of plastid DNA replication origins in tobacco by targeted inactivation. *Plant Journal* 32:175–184.

Niida, H., Matsumoto, T., Satoh, H., Shiwa, M., Tokutake, Y., Furuichi, Y., Shinkai, Y. (1998) Severe growth defect in mouse cells lacking the telomerase RNA component. *Nature Genetics* 19:203–206.

Nikitina, T., Woodcock, C.L. (2004) Closed chromatin loops at the ends of chromosomes. *Journal of Cell Biology* 166:161–165.

Ning, Y., Xu, J., Li, Y., Chavez, L., Riethman, H.C., Lansdorp, P.M., Weng, N. (2003) Telomere length and the expression of natural telomeric genes in human fibroblasts. *Human Molecular Genetics* 12:1329–1336.

Pascal, J.M., O'Brien, P.J., Tomkinson, A.E., Ellenberger, T. (2004) Human DNA ligase I completely encircles and partially unwinds nicked DNA. *Nature* 432:473–478.

Ridanpää, M., van Eenennaam, H., Pelin, K. et al. (2001) Mutations in the RNA component of RNase MRP cause a pleiotropic human disease, cartilage-hair hypoplasia. *Cell* 104:195–203.

Robinson, N.P., Dionne, I., Lundgren, M., Marsh, V.L., Bernander, R., Bell, S.D. (2004) Identification of two origins of replication in the single chromosome of the archaeon *Sulfolobus solfataricus*. *Cell* 116:25–38.

Rochaix, J.-D., Bird, A., Bakken, A. (1974) Ribosomal RNA gene amplification by rolling circles. *Journal of Molecular Biology* 87:473–587.

Rudolph, K.L., Chang, S., Millard, M., Schreiber-Abus, DePinho, R.A. (2000) Inhibition of experimental liver cirrhosis in mice by telomerase gene delivery. *Science* 287:1253–1258.

Smogorzewska, A., deLange, T. (2004) Regulation of telomerase by telomeric proteins. *Annual Review of Biochemistry* 73:177–208.

Tagami, H., Ray-Gallet, D., Almouzni, G., Nakatani, Y. (2004) Histone H3.1 and H3.3 complexes mediate nucleosome assembly pathways dependent or independent of DNA synthesis. *Cell* 116:51–61.

Yamaguchi, R., Newport, J. (2003) A role for RanGTP and Crm1 in blocking re-replication. *Cell* 113:115–125.

Yasukawa, T., Yang, M.Y., Jacobs, H.T., Holt, I.J. (2005) A bidirectional origin of replication maps to the major noncoding region of human mitochondrial DNA. *Molecular Cell* 18:651–662.

Ye, X., Franco, A.A., Santos, H., Nelson, D.M., Kaufman, P.D., Adams, P.D. (2003) Defective S phase chromatin assembly causes DNA damage, activation of the S phase checkpoint, and S phase arrest. *Molecular Cell* 11:341–351.

Chapter 7
DNA repair and recombination

We totally missed the possible role of enzymes in DNA repair. . . . I later came to realize that DNA is so precious that probably many distinct repair mechanisms would exist. Nowadays one could hardly discuss mutation without considering repair at the same time.

Francis Crick, *Nature* (1974) 248:766.

Outline

7.1 Introduction

As a cell multiplies and divides, in most cases the genome is accurately copied and information is passed on to the next generation with minimal error. When DNA polymerases do make mistakes, their proofreading activity generally, but not always, corrects the error. However, forms of DNA damage unrelated to the process of replication occur relatively commonly. Such damage is induced by exposure to a number of different types of agents, for example, oxygen free radicals, ultraviolet or ionizing radiation, and various chemicals. DNA damage poses a continuous threat to genomic integrity. To cope with this problem cells have evolved a range of DNA repair enzymes and repair polymerases as complex as the DNA replication apparatus itself, which indicates their importance for the survival of a cell.

In Chapter 6, we saw that DNA replication involves multiple biochemical steps mediated by a complex suite of proteins. Like DNA replication, DNA repair and recombination are also performed by multiprotein assemblies. Repair and recombination processes share many common features and are intimately intertwined with each other and with DNA replication. For example, DNA replication is required for synthesizing new stretches of DNA during various types of repair and recombination, repair of all types occurs in tandem with replication, and recombination not only promotes genetic crossing-over during meiosis but is also a major DNA repair mechanism. This chapter begins with a discussion of types of mutations resulting from errors in DNA replication and DNA damage, and their phenotypic consequences. Following that is a discussion of the general types of DNA damage and the cellular responses to DNA damage.

7.2 Types of mutations and their phenotypic consequences

High-fidelity DNA replication is beneficial for maintaining genetic information over many generations; unrepaired DNA damage may lead to mutations that promote disease or cell death. On the other hand, low-fidelity DNA replication is beneficial for the evolution of species, and for generating diversity leading to increased survival when organisms are subjected to changing environments. Mutations result from changes in the nucleotide sequence of DNA or from deletions, insertions, or rearrangements of DNA sequences in the genome. A spontaneous mutation is one that occurs as a result of natural processes in cells, for example DNA replication errors. These can be distinguished from induced mutations; those that occur as a result of interaction of DNA with an outside agent or mutagen that causes DNA damage. Moreover, some sites on chromosomes are "hotspots" where mutations arise at a higher frequency than other regions of the DNA. Mutations are of fundamental importance in molecular biology for several reasons:

1 As noted above, mutations are important as the major source of genetic variation that drives evolutionary change.
2 Mutations may have deleterious or (rarely) advantageous consequences to an organism or its descendants. Mutations in germ cells can lead to heritable genetic disorders, while mutations in somatic cells may lead to acquired diseases such as cancer or neurodegenerative disorders.
3 Mutant organisms are important tools for molecular biologists in characterizing the genes involved in cellular processes.

At the molecular level, the simplest type of mutation is a nucleotide substitution, in which a nucleotide pair in a DNA duplex is replaced with a different nucleotide pair. Mutations that alter a single nucleotide pair are called point mutations. Different types of nucleotide substitutions and their phenotypic consequences are discussed in the following sections. Other kinds of mutations cause more drastic changes in DNA, such as expansions of trinucleotide repeats (see below), extensive insertions and deletions, and major chromosomal rearrangements. Such changes can be caused by the insertion of a transposable DNA element (see Section 12.5) or by errors in cellular recombination processes. Major chromosomal rearrangements associated with cancer are discussed in Chapter 17 (Section 17.2, Table 17.3).

Transitions and transversions can lead to silent, missense, or nonsense mutations

Transition mutations replace one pyrimidine base with another, or one purine base with another. In contrast, transversion mutations replace a pyrimidine with a purine or vice versa (Table 7.1). For example, in an A → G substitution, an A is replaced with a G in one of the DNA strands. This substitution temporarily creates a mismatched GT base pair. During the next round of replication, this transition mutation becomes

Table 7.1 Types of nucleotide substitutions.

Nucleotide substitution	Mutation
Transition mutation Pyrimidine → pyrimidine	T → C or C → T
Purine → purine	A → G or G → A
Transversion mutation Pyrimidine → purine	T → A, T → G, C → A, or C → G
Purine → pyrimidine	A → T, A → C, G → T, or G → C

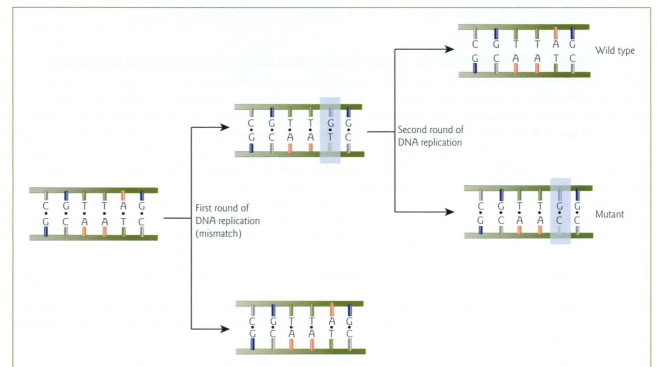

Figure 7.1 A point mutation can be permanently incorporated by DNA replication. A point mutation may be introduced by incorporation of an incorrect nucleotide in the first round of replication creating a mismatch. If the error is not repaired, in the second round of replication, the nucleotide substitution becomes permanently incorporated in the DNA sequence.

permanently incorporated in the DNA sequence. The mismatch is resolved as a GC base pair in one daughter molecule and an AT base pair in the other daughter molecule. In this example, the GC base pair is the mutant and the AT base pair is the wild type (nonmutant) (Fig. 7.1). Spontaneous nucleotide substitutions are often biased in favor of transitions. In the human genome, the ratio of transitions to transversions is approximately 2 : 1. Whether nucleotide substitutions have a phenotypic effect depends on whether they alter a critical nucleotide in a gene regulatory region, in the template for a functional RNA molecule, or whether they are silent, missense, or nonsense mutations in a protein–coding gene (Fig. 7.2).

Silent mutations

Nucleotide substitutions in a protein-coding gene may or may not change the amino acid in the encoded protein. Mutations that change the nucleotide sequence without changing the amino acid sequence are called

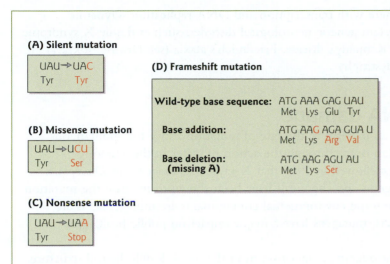

(A) Silent mutation

UAU → UAC
Tyr   Tyr

(B) Missense mutation

UAU → UCU
Tyr   Ser

(C) Nonsense mutation

UAU → UAA
Tyr   Stop

(D) Frameshift mutation

Wild-type base sequence:	ATG AAA GAG UAU
	Met Lys Glu Tyr
Base addition:	ATG AAG AGA GUA U
	Met Lys Arg Val
Base deletion: (missing A)	ATG AAG AGU AU
	Met Lys Ser

Figure 7.2 Examples of types of point mutations in protein-coding sequences. (A) Silent mutation: the altered codon codes for the same amino acid. (B) Missense mutation: the altered codon codes for a different amino acid and the protein is often nonfunctional. (C) Nonsense mutation: the new codon is a termination codon. Protein synthesis stops and the protein is nonfunctional. (D) Frameshift mutation: the addition or deletion of one or more base pairs results in a shift in the reading frame of the resulting mRNA, and leads to production of a nonfunctional protein. Red letters indicate the affected codons and corresponding amino acids.

synonymous mutations or silent mutations (Fig. 7.2A). Mutational changes in nucleotides that are outside of coding regions can also be silent. However, some noncoding sequences do have essential functions in gene regulation (see Section 13.10) and, in this case, mutations in these sequences would have phenotypic effects.

Missense mutations

Nucleotide substitutions in protein-coding regions that do result in changed amino acids are called missense mutations or nonsynonymous mutations (Fig. 7.2B). A change in the amino acid sequence of a protein may alter the biological properties of the protein. A classic example of a phenotypic effect of a single amino acid change is the change responsible for the human hereditary disease sickle cell anemia (see Fig. 8.14). The molecular basis of the sickle cell anemia mutation is an AT → TA transversion causing the normal glutamic acid codon in the β-globin chain of hemoglobin to be replaced with valine.

Nonsense mutations

A nucleotide substitution that creates a new stop codon is called a nonsense mutation (Fig. 7.2C). Because nonsense mutations cause premature chain termination during protein synthesis, the remaining polypeptide fragment is nearly always nonfunctional.

Insertions or deletions can cause frameshift mutations

Insertions or deletions of nucleotides can also occur in DNA, but at a rate considerably lower than that of nucleotide substitution. A classic example of a phenotypic effect of a small deletion is the change responsible for the human hereditary disease cystic fibrosis. The deletion of three base pairs in the nucleotide sequence of the cystic fibrosis transmembrane conductance regulator (CFTR) gene results in the loss of the codon for phenylalanine (see Fig. 17.20).

If the length of an insertion or deletion is not an exact multiple of three nucleotides, the mutation shifts the phase in which the ribosome reads the triplet codons and, consequently, alters all of the amino acids downstream from the site of the mutation. Such mutations are called frameshift mutations because they "shift" the reading frame of the codons in the mRNA (Fig. 7.2D).

Expansion of trinucleotide repeats leads to genetic instability

Some regions of DNA have unusual genetic instability because of the presence of trinucleotide repeats. As discussed in Chapter 2, trinucleotide repeat expansions can adopt triple helix conformations and assume unusual DNA secondary structures that interfere with transcription and DNA replication. Dynamic expansion of trinucleotide repeats leads to certain genetic neurological disorders such as fragile X syndrome (see Disease box 12.2), Huntington's disease, Kennedy's disease, Friedreich's ataxia (see Disease box 2.1), spinocerebellar ataxia type 1, and myotonic dystrophy.

7.3 General classes of DNA damage

A mutagen is any chemical agent that causes an increase in the rate of mutation above the spontaneous background. Spontaneous damage to DNA can occur through the action of water in the aqueous environment of the cell. Hermann Muller first showed in 1927 that X-rays are mutagenic in *Drosophila*. Since that time, a large number of physical agents and chemicals have been shown to increase the mutation rate by causing damage to the DNA. Because some environmental contaminants are mutagenic, as are numerous chemicals found in tobacco products, mutagens have a major impact on public health (see Section 17.2, Fig. 17.14).

Damage to DNA consists of any change introducing a deviation from the usual double-helical structure. There are three major classes of DNA damage: single base changes, structural distortion, and DNA backbone damage.

Single base changes

A single base change or "conversion" affects the DNA sequence, but has only a minor effect on overall structure. For example, replacement of the amino group of cytosine with oxygen converts cytosine to uracil, a base that should only be present in an RNA chain (Fig. 7.3A). This type of conversion process is called deamination. Deamination is the most frequent and important kind of hydrolytic damage, and can occur spontaneously from the action of water, or be induced by a chemical mutagen. When a UG base pair replaces a CG base pair this causes only a minor structural distortion in the DNA double helix. This type of damage is not likely to completely block the process of replication or transcription, but may lead to the production of a mutant RNA or protein product. Vertebrate DNA frequently contains 5-methylcytosine in place of cytosine. Methylated cytosines are hotspots for spontaneous mutations in vertebrate DNA because deamination of 5-methylcytosine generates thymine. This results in the change of a GC base pair into an AT when damaged DNA is replicated. Some of the consequences of changes in patterns of methylated DNA are discussed in Chapter 12.

DNA is also subject to damage by alkylation, oxidation, and radiation. Alkylating agents such as nitrosamines lead to the formation of O^6-methylguanine (Fig. 7.3B). This modified base often mispairs with thymine, resulting in the change of a GC base pair into an AT base pair when the damaged DNA is replicated. Potent oxidizing agents are generated by ionizing radiation and by chemical agents that generate free radicals. These reactive oxygen species (for example O_2^-, H_2O_2, and $OH[\cdot]$) can generate 8-oxoguanine (oxoG), a damaged guanine base containing an extra oxygen atom (Fig. 7.3C). OxoG is highly mutagenic because it can form a Hoogsteen base pair (see Fig. 2.13) with adenine. This gives rise to a GC → TA transversion, which is one of the most common mutations found in human cancers.

Structural distortion

Ultraviolet (UV) light has a detrimental effect on cells due to selective absorption of the UV rays. Radiation with a wavelength of about 260 nm is strongly absorbed by the bases. The most frequent UV-induced lesions of DNA are the induction of pyrimidine dimers between two neighboring thymine bases by UV irradiation (Fig. 7.3D). These are also termed cyclobutane–pyrimidine dimers (CPD), because a cyclobutane ring is generated by links between carbon atoms 5 and 6 of adjacent thymines. Because covalent bonds form between thymines on the same strand, this disrupts the complementary base pairs that form the double helix. Thymine dimers thus distort the structure of the duplex DNA. Structural distortion may impede

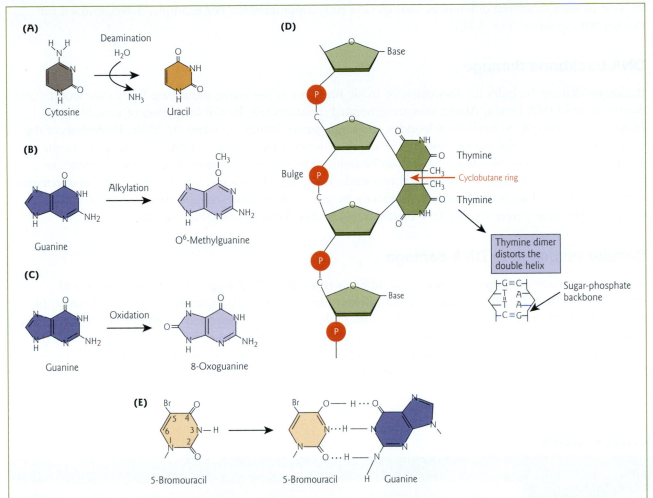

Figure 7.3 Types of DNA damage. (A) Single base change by deamination of cytosine to uracil. (B) Alkylation of the oxygen on carbon atom 6 of guanine generates O^6-methylguanine. (C) Oxidation of guanine generates 8-oxoguanine. (D) UV radiation induces the formation of a cyclobutane ring between adjacent thymines, forming a thymine dimer. This leads to structural distortion of the duplex DNA. (E) 5-Bromouracil, a base analog of thymine, can mispair with guanine.

transcription and replication by blocking the movement of polymerases. Consequently, the induction of pyrimidine dimers is a more severe defect than a single base change. UV irradiation can also induce dimers between cytosine and thymine, called pyrimidine (6,4)-pyrimidone photoproducts. Other bulky adducts can be induced by chemical mutagenesis; e.g. by exposure to large polycyclic hydrocarbons or alkylating agents (see Section 17.2, Fig. 17.14).

DNA damage is also caused by intercalating agents and base analogs. Intercalating agents such as ethidium bromide (see Tool box 8.6) contain several polycyclic rings. These flat rings insert between the DNA bases, binding and stacking with the DNA bases, just as the bases bind or stack with each other in the double helix. Due to the resulting distortion of the double helix, intercalating agents can cause insertion or deletion of one or more base pairs during DNA replication. Base analogs are compounds such as 5-bromouracil, an analog of thymine, that substitute for normal bases. They are similar enough to the normal bases to be taken up by cells, converted into deoxynucleotides, and incorporated into DNA during replication. However, because of their structural differences, analogs base pair inappropriately. For example, 5-bromouracil can mispair with guanine (Fig. 7.3E).

DNA backbone damage

Backbone damage includes the formation of abasic sites (loss of the nitrogenous base from a nucleotide) and double-strand DNA breaks. Abasic sites are generated spontaneously by the formation of unstable base adducts. For example, in purine nucleotides, the sugar–purine bonds are relatively labile. Hydrolysis of the N-glycosyl linkage in the purine base by the action of water leaves a hydroxyl (-OH) in its place in the depurinated DNA. Double-strand breaks can be induced by ionizing radiation (e.g. X-rays, radioactive materials) and a wide range of chemical compounds. Ionizing radiation can attack (ionize) the deoxyribose sugar in the DNA backbone directly or indirectly by generating reactive oxygen species. Double-strand breaks are the most severe type of DNA damage, since they disrupt both DNA strands.

Cellular response to DNA damage

There are a variety of complex responses to different types of DNA damage in both prokaryotes and eukaryotes (Table 7.2). The responses fall into three main categories: (i) those that bypass the damage; (ii) those that directly reverse the damage; and (iii) those that remove the damaged section of DNA and replace it in with undamaged DNA, by excision or recombinational repair systems. The focus of the remaining sections in this chapter is on the major repair pathways present in mammalian cells, in particular human cells. Comparisons are made where appropriate with the repair pathways present in *Escherichia coli*, in particular where details are better understood.

Table 7.2 DNA repair systems.

Type	Damage	Repair proteins
Damage bypass		
Translesion DNA synthesis	Pyrimidine dimer or apurinic site	DNA polymerases IV and V in *E. coli* Pol ζ, pol η, pol ι, pol κ, and pol λ in humans
Damage reversal		
Photoreactivation	Pyrimidine dimers	DNA photolyase★
Removal of methyl groups	O^6-methylguanine	Methyltransferase
Damage removal		
Base excision repair	Damaged base	DNA glycosylases
Mismatch repair	Replication errors	MutS, MutL, and MutH in *E. coli* MutSα, MutLα, and EXO1 in humans
Nucleotide excision repair	Pyrimidine dimer Bulky adduct on base	UvrA, UvrB, UvrC, and UvrD in *E. coli* XPA, XPB, XPC, XPD, ERCC1/XPF, and XPG in humans
Double-strand break repair	Double-strand breaks	RecA and RecBCD in *E. coli* MRN complex, Rad51, Rad 52, BRCA1, BRCA2, XRCC3, etc. in humans for homologous recombination Ku proteins, Artemis/DNA-PK$_{CS}$, XRCC4 in humans for nonhomologous end-joining

★ Not present in placental mammals, including humans.

7.4 Lesion bypass

The normal replication machinery uses high-fidelity DNA polymerases. These high-fidelity polymerases accurately copy nondamaged template DNA, but are unable to bypass DNA lesions that cause structural distortion of the DNA helix. Specialized low-fidelity, "error-prone" DNA polymerases (see Table 6.1) transiently replace the replicative polymerases and copy past damaged DNA in a process called translesion synthesis (TLS) (Fig. 7.4). In the simplest of models, the replicative polymerase, being unable to bypass a lesion in the DNA either "falls off" the DNA or simply translocates downstream of the lesion to continue replication. This allows for proliferating cell nuclear antigen (PCNA) mediated loading of another DNA polymerase capable of replicating the lesion. In keeping with the "factory" model for DNA replication, TLS polymerases and their auxiliary factors are stored in these subnuclear compartments for rapid recruitment. Eventually, the replicative polymerase regains control of the template until such time as replication is completed or the polymerase again encounters a replication-blocking DNA lesion.

The error-prone DNA polymerases are able to copy damaged DNA templates permissively and efficiently. However, because they are error-prone they may generate mutations. Typical error rates range from 10^{-1} to 10^{-3} per base pair (on undamaged DNA). Most lesions completely alter the pairing properties of the pre-existing bases. In this case, the polymerases insert incorrect nucleotides opposite the lesions, giving rise to nucleotide substitution mutants. Alternatively, when the polymerases skip past lesions, they insert a correct nucleotide opposite bases downstream from the lesion, generating a frameshift mutation. An exception to the general tendency of repair polymerases to be error-prone is DNA polymerase eta (η). DNA polymerase η performs TLS past a thymine–thymine (TT) dimer by inserting two adenine residues. This results in the lesion being bypassed in an error-free manner. The presence of DNA polymerase η thus protects cells from UV damage. In contrast, DNA polymerase iota (ι) achieves TLS by a highly error-prone bypass of a TT dimer.

Translesion synthesis enables a cell to survive what might be a fatal block to replication, but with the risk of a higher mutation rate. For this reason, in *E. coli*, the translesion polymerases (DNA polymerases IV and V; see Focus box 6.1) are not present under normal circumstances. Their synthesis is only induced in response to DNA damage, as part of a pathway known as the "SOS response."

7.5 Direct reversal of DNA damage

In most organisms, UV radiation damage to DNA (pyrimidine dimers) can be directly repaired by a process called photoreactivation or "light repair." However, placental mammals, including humans, do not have a photoreactivation pathway. During photoreactivation, the enzyme DNA photolyase uses energy from near-UV to blue light to break the covalent bonds holding the two adjacent pyrimidines together (Fig. 7.5).

Another example of direct reversal of DNA damage is the repair of the methylated base O^6-methylguanine. Methyltransferases present in organisms ranging from *E. coli* to humans catalyze removal of the methyl group (Fig. 7.6). A sulfhydryl group of a cysteine residue of the methyltransferase accepts the methyl group from guanine. This process is very costly to the cell because once the methyltransferase accepts the methyl group from guanine, the enzyme cannot be used again.

7.6 Repair of single base changes and structural distortions by removal of DNA damage

Repair systems that remove damaged DNA include repair of single base changes, structural distortions, and double-stranded DNA damage (see Table 7.2). A key feature of each of these pathways is that multiple dynamic protein interactions are involved in the repair process. There is an ordered hand-off of DNA from one protein or complex to another. The DNA repair proteins are modular, composed of multiple structural domains with distinct biochemical functions. They also have multiple binding sites of modest affinity that mediate interaction with other repair proteins. The presence of multiple binding sites facilitates "trading places" by direct competition for binding sites or allosteric structural rearrangements.

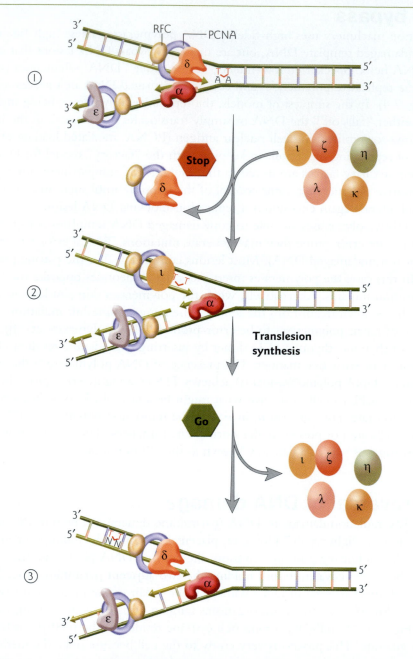

Figure 7.4 Model for translesion DNA synthesis. For simplicity, the replication machinery is depicted as DNA polymerase (pol) δ on the leading strand, DNA pol ε on the lagging strand, the clamp-loader RFC, and the sliding clamp PCNA. (1) The replication machinery is shown arrested (stop sign) at a thymine dimer on the template for leading strand synthesis. Multiple specialized error-prone DNA repair polymerases are stored in a subnuclear compartment near the arrested replication fork (DNA pol ι, λ, ζ, η, and κ). (2) When the replication fork stalls at the lesion, DNA pol δ is replaced by one of the specialized polymerases (DNA pol ι in this example). (3) When translesion synthesis has bypassed the damage and extended to a suitable position downstream, another polymerase switch occurs. DNA pol ι is released and replaced by DNA pol δ. High-fidelity DNA replication continues (go sign).

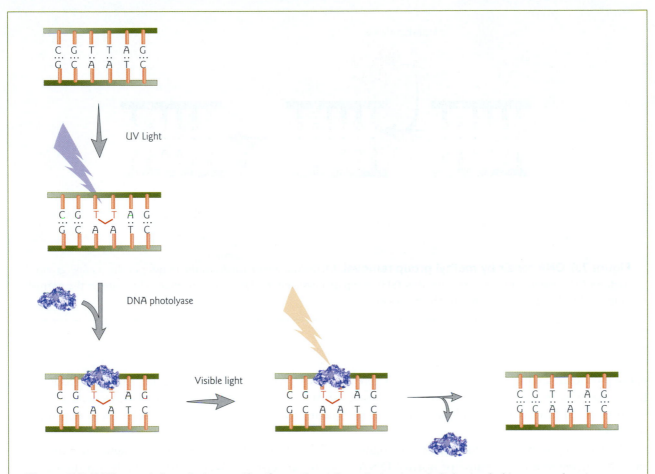

Figure 7.5 DNA repair by photoreactivation. Ultraviolet radiation (purple lightning bolt) causes a thymine–thymine dimer to form (red). The DNA photolyase enzyme (blue) binds to this region of the DNA. The enzyme absorbs near UV to visible light (peach lightning bolt) and breaks the rings formed by the pyrimidine dimers to restore the two thymine residues. The enzyme dissociates from the repaired DNA.

The first step in all the pathways is that the repair machinery must gain access to the DNA. A model for chromatin alterations during mammalian DNA repair proposes the following. Upon sensing DNA damage in the DNA, chromatin is loosened by the combined action of acetylation of histones and chromatin remodeling (see Section 11.6) to allow removal of the damage by the repair machinery. The resulting gap is filled in by DNA polymerase. PCNA then recruits chromatin assembly factors, such as chromatin assembly factor 1 (CAF-1), which together restore chromatin on the new DNA, in the same manner as after DNA replication (see Fig. 6.15).

There are two main pathways for repair of single base changes: base excision repair and mismatch repair. Base excision repair involves the correction of single base changes that are due to conversion. Mismatched base pairs that result from DNA polymerase errors during replication are corrected by mismatch repair. The nucleotide excision repair pathway is used for repair of structural distortion, for example bulges from thymine dimers induced by UV irradiation. The repair of double-strand breaks is discussed in Section 7.7.

Base excision repair

Base excision repair is initiated by a group of enzymes called DNA glycosylases. These enzymes cleave the *N*-glycosidic bond connecting the deoxyribose sugar to the damaged base. The damaged base is then

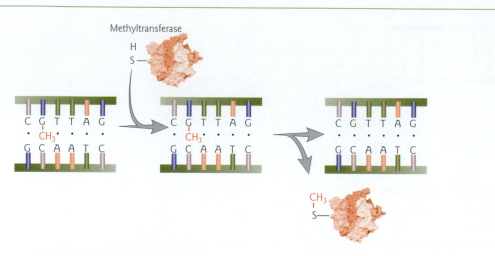

Figure 7.6 DNA repair by methyl group removal. Methyltransferase catalyzes the transfer of the methyl group (red) on O^6-methylguanine to the sulfhydryl (SH) group of a cysteine residue on the enzyme. This restores the normal G in the DNA sequence and inactivates the enzyme.

excised to form an abasic site in the DNA. There are DNA glycosylases that recognize oxidized/reduced bases, methylated bases, deaminated bases, or base mismatches. For example, uracil DNA glycosylase removes uracil from UA or UG base pairs, and the human oxoG repair enzyme, hOGG1, catalyzes excision of oxoG.

The first step in base excision repair is recognition of the lesion. How DNA repair proteins find the rare sites of damage in a vast expanse of normal DNA is poorly understood. Recognition of a lesion such as oxoG, which differs by only two atoms from its normal counterpart guanine (G), is particularly remarkable. Recent X-ray structural analysis of a repair complex suggests that the damaged base goes through a series of "gates" or checkpoints within the enzyme. hOGG1 first binds nonspecifically to DNA. If the enzyme encounters an oxoG base paired with cytosine (C), the oxoG is extruded from the double helix into a "G-specific pocket" (Fig. 7.7). Subsequently, the oxoG is inserted deeply into a lesion recognition pocket (oxoG pocket) on the enzyme. Amino acids lining this pocket contact oxoG directly, providing a basis for specific recognition. When the enzyme encounters a normal GC base pair in DNA, the G is transiently extruded into the G-specific pocket in the enzyme, but is restricted from access to the oxoG pocket, and is returned to the double helix.

Depending on the initial events in base removal in mammalian cells, the repair patch may be a single nucleotide (short patch) or 2–10 nt (long patch) (Fig. 7.8). Both short patch and long patch repairs are characterized by "hand-off" of the DNA from endonucleases, which excise the damaged base, to DNA polymerase and ligase for repair synthesis. In short patch repair, the enzymes glycosylase-associated β-lyase and APE1 (apurinic/apyrimidinic endonuclease) make nicks 3′ and 5′ to the abasic site in the DNA, respectively. DNA polymerase β replaces the missing nucleotide, and the DNA ligase 3–XRCC1 complex seals the gaps in the sugar–phosphate backbone. The name XRCC1 derives from the use of rodent (Chinese hamster) mutants for the cloning of human repair genes. The human genes are isolated by their ability to correct the mutation in rodent complementation groups. The correcting human genes are designated as XRCC for "X-ray repair cross complementing rodent repair deficiency."

In long patch repair, APE1 makes an incision 5′ to the abasic site (Fig. 7.8). Then, DNA polymerase δ or ε, and PCNA, displace the strand 3′ to the nick to produce a flap of 2–10 nt. The flap is cut at the junction of the single to double strand transition by FEN-1 (flap endonuclease-1). A patch of the same size is then synthesized by DNA polymerase δ or ε with the aid of PCNA and ligated by DNA ligase I.

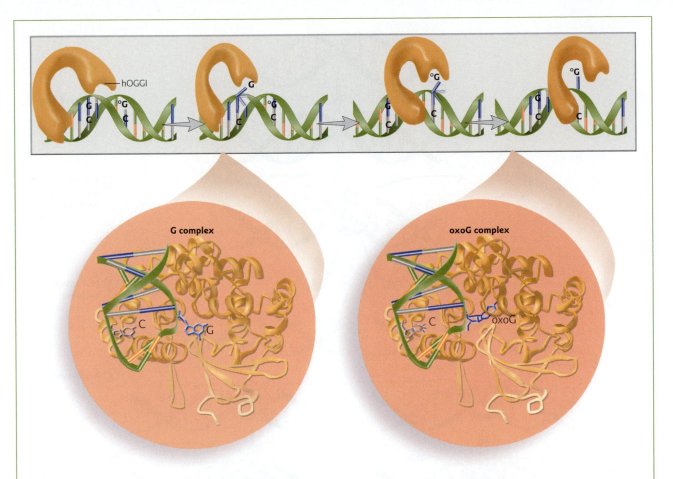

Figure 7.7 Model for DNA damage recognition by 8-oxoguanine DNA glycosylase 1 (hOGG1). The enzyme hOGG1 first binds nonspecifically to DNA. If the enzyme encounters a normal GC base pair, contacts between the enzyme and the C base result in transient extrusion of the G into a G-specific pocket in the enzyme, followed by return to the DNA double helix. When hOGG1 encounters an oxoGC base pair, the oxoG is first extruded into the G-specific pocket, and then inserted into the oxoG-specific lesion recognition pocket where it is excised from the DNA. Comparison of the overall structures of complexes with G-containing DNA (left) or oxoG-containing DNA (right). Both protein (gold) and DNA (green) are represented as backbone ribbon traces. The C (light blue) and oxoG or G (dark blue) are rendered in ball-and-stick representations. The oxoG is bound in the lesion recognition pocket, whereas the G is bound at the alternative, extrahelical G-specific pocket. (Adapted by permission from Nature Publishing Group and Macmillan Publishers Ltd: David, S.S. 2005. Structural biology: DNA search and rescue. *Nature* 434:569–570; and Banerjee, A., Yang, W., Karplus, M., Verdine, G.L. 2005. Structure of a repair enzyme interrogating undamaged DNA elucidates recognition of damaged DNA. *Nature* 434:612–618. Copyright © 2005.)

Mismatch repair

Basic features of the mismatch repair pathway are conserved from *E. coli* to humans, but only MutS and MutL homologs appear to have been conserved throughout evolution. Hereditary deficiency in mismatch repair causes an increased rate of gene mutations and susceptibility to certain types of cancer, including hereditary nonpolyposis colon cancer (Disease box 7.1). Mismatch repair depends on a number of activities in human cells, including MutSα (MSH2–MSH6 heterodimer) or MutSβ (MSH2–MSH3), MutLα (MLH1–PMS2 heterodimer), the exonuclease EXO1, DNA polymerase δ, the replication clamp PCNA, the clamp-loader RFC (replication factor C), the single-strand DNA-binding protein RPA (replication protein A) (see Chapter 6) and the nonhistone chromosomal protein HMGB1 (high-mobility group B protein 1).

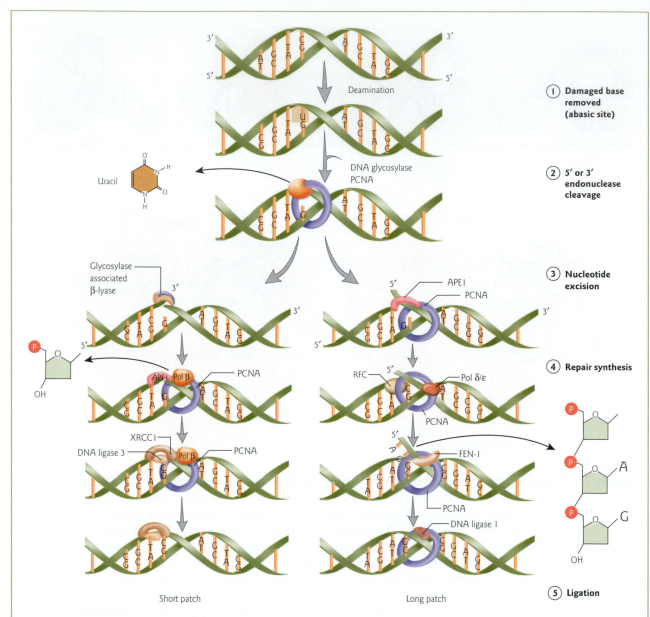

Figure 7.8 Base excision repair pathway in mammalian cells. The diagram shows the five major steps involved in repair of the deamination of cytosine to uracil. (1) The uracil is removed by uracil DNA glycosylase to create an abasic site. (2) Either a 3′ nick by glycosylase-associated β-lyase, or a 5′ nick by APE1 in association with the sliding clamp PCNA, cleaves the phosphodiester bond adjacent to the abasic site. (3) In a short patch repair, the single damaged nucleotide is excised by APE1. In a long patch repair, 2–10 nt are excised by FEN-1. (4) Repair synthesis is mediated by DNA polymerase β for short patch repairs, and by DNA polymerase δ or ε for long patch repairs. (5) The gap in the DNA backbone is ligated by DNA ligase 3–XRCC1 for short patch repairs, and DNA ligase 1 for long patch repairs.

Reconstitution of mismatch repair in an *in vitro* system has shown that while all these factors are necessary for efficient repair, MutSα, EXO1, and DNA polymerase δ are indispensible (Fig. 7.9). MutSα appears to mediate most mismatch recognition events in mammalian cells, whereas MutSβ plays a limited role in the repair of base–base mismatches, but repairs insertion/deletion mispairs more efficiently than MutSα.

Hereditary nonpolyposis colorectal cancer: a defect in mismatch repair

DISEASE BOX 7.1

Colorectal cancer is the second leading cause of cancer death in the United States. In about 80% of affected individuals the cancer develops sporadically, while the remaining 20% have an inherited susceptibility to the disease. Hereditary nonpolyposis colorectal cancer (HPNCC or Lynch syndrome) is the most common hereditary form of colorectal cancer, accounting for 20–25% of such cancers and 3–5% of all colorectal cancers. It affects about one in 200 individuals. The phenotype is characterized by few polyps in the colon and early onset (before age 45) of multiple tumors in the transverse and ascending portions of the colon. HNPCC is an autosomal dominant condition resulting from mutations in one of several DNA mismatch repair genes. Mutations in either the *MSH2* (*MutSα*) or *MLH1* (*MutLα*) genes account for more than 90% of mutations in HNPCC families, although other defects may also occur in other repair genes. Individuals who inherit these mutations have an approximate 80% lifetime risk of developing colorectal cancer, and an increased risk of other forms of cancer, such as endometrial cancer.

The progression to HNPCC is a multistep process:

1 **Germline mutation in one allele of the mismatch repair genes**. Individuals with a germline mutation of a mismatch repair gene are heterozygous. An individual inherits one inactivated allele, which establishes a predisposition for the development of colorectal cancer.

2 **Somatic loss of the wild-type allele**. There is spontaneous somatic loss or inactivation of the wild-type (normal) allele, due to the probability of mutation being increased in dividing cells.

3 **Defective mismatch repair mechanism**. Two inactivated alleles in dividing cells of the colon or another tissue lead to a completely defective mismatch repair pathway.

4 **Accumulation of mistakes in DNA replication**. Because of the defect in mismatch repair, there is an accumulation of mistakes in DNA replication, and thus an increased spontaneous mutation rate. This is an early event in tumor progress, and accelerates the accumulation of mutations in tumor suppressor genes and oncogenes, which leads to deregulated cell growth (see Section 17.2).

5 **Microsatellite instability**. Defects in mismatch repair genes also result in an accumulation of mutations in the microsatellite regions of DNA. Microsatellites are tandemly repeated DNA sequences that are located throughout the genome (see Section 16.2). Mutations in mismatch repair genes cause alterations in the number of repeats in these sequences of DNA. Variability in the number of repeats between cancerous and normal tissue is called microsatellite instability. Studies have shown that in more than 90% of tumors with microsatellite instability, this results in the mutation of a gene involved in growth regulation, the type II transforming growth factor (TGF)-β receptor. Mutations in the microsatellite region of this gene leave the receptor unable to bind TGF-β, a growth inhibitor of normal epithelial cells of the colon. Microsatellite instability also leads to mutations in the BAX gene, an important promoter of apoptosis (programmed cell death).

At-risk individuals can be identified through family history and genetic testing. Currently at-risk individuals are monitored by colonoscopy and in some cases prophylactic surgery is recommended. In the near future, chemopreventive agents will likely play a greater role in colorectal cancer prevention.

The first step in mismatch repair is recognition of the error. The method of strand discrimination (i.e. which strand is the daughter strand with the mistake) in mammalian cells is unknown. One model proposes that the 5′ end of an Okazaki fragment and the polymerase machinery associated with the 3′ end of the nascent strand may provide markers for strand discrimination in postreplication mismatch repair (Fig. 7.10). In *E. coli* the newly synthesized strand with the mistake is identified by the absence of methyl groups on GATC sequences (see Section 8.2).

Once the mismatch is recognized, repair does not just involve the simple removal of the mispaired nucleotide. Instead, a large region of DNA including the mismatch is excised. Exactly how this occurs is

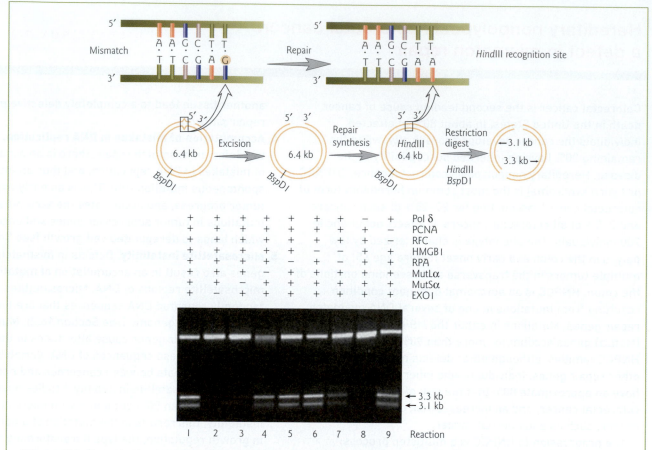

Figure 7.9 Reconstitution of human mismatch repair in an *in vitro* system. The *in vitro* system was tested with a GT heteroduplex circular DNA substrate containing a single-strand break 128 bp 5′ to the GT mismatch. Repair of the mismatch was scored by assaying for restoration of a *Hin*dIII restriction enzyme recognition site (see Section 8.3). If the circular DNA is repaired, then restriction digests with *Hin*dIII and *Bsp*DI yield two repair products of 3.1 and 3.3 kb. If the mismatch is not repaired, then *Hin*dIII cannot cut the DNA substrate, and the single cut with *Bsp*DI yields a 6.4 kb fragment. Repair assays were performed in reactions containing the GT heteroduplex, and purified DNA polymerase δ, PCNA, RFC, HMGB1, RPA, MutLα, MutSα, and EXO1 as indicated. DNA fragments were analyzed by agarose gel electrophoresis (see Tool box 8.6). The data show that MutSα, EXO1, and DNA pol δ are essential for repair. (Reprinted from: Zhang, Y., Yuan, F., Presnell, S.R. et al. 2005. Reconstitution of 5′-directed human mismatch repair in a purified system. *Cell* 122:693–705. Copyright © 2005, with permission from Elsevier.)

still under study. The "molecular switch model" proposes that ATP-bound MutSα forms a sliding clamp (analogous to PCNA) on mismatched DNA. The rate-limiting step in the pathway would thus be exchange of bound ADP for ATP on MutSα at the mismatch site. In conjunction with MutLα, MutSα is then proposed to diffuse either 5′ or 3′ for several thousand nucleotides along the DNA backbone. Energy from ATP hydrolysis is not required for movement of the complex. Instead, ATP hydrolysis occurs when MutSα finally dissociates from the DNA. Another model suggests that MutSα stays at the mismatch during repair. The development of the *in vitro* system that reconstitutes mismatch repair should help in evaluating these differing models.

Whether mobile or stationary, the MutSα–MutLα complex somehow triggers activation of the repair machinery. A recurrent theme again emerges – "hand-off" of damaged DNA from a complex with nuclease activity to a complex with polymerase activity (Fig. 7.10). A single-strand break by the exonuclease EXO1

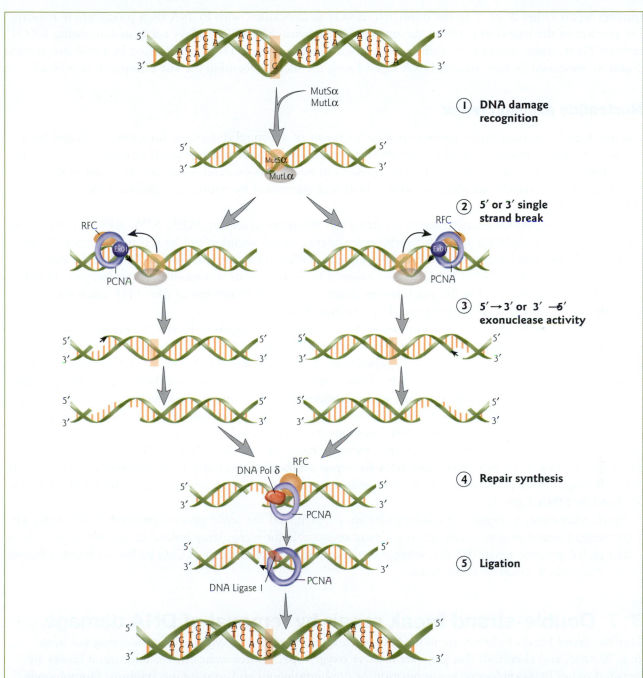

Figure 7.10 Mismatch repair pathway in mammalian cells. The five major steps in the pathway are depicted. (1) DNA damage is recognized by the MutSα–MutLα complex. Movement of the complex away from the mismatch may signal excision system activation at the strand break. (2) A 5′ or 3′ single-strand break is generated by EXO1 in association with PCNA and RFC. (3) Progressive exonuclease activity of EXO1 removes the mismatch. (4) 5′ → 3′ repair synthesis is mediated by DNA polymerase δ and associated factors. (5) Ligation of the remaining gap in the DNA backbone is catalyzed by DNA ligase 1.

initiates repair either 3′ or 5′ to the mismatch. EXO1 in association with PCNA then progressively removes the portion of the strand in between the nick and the mismatch. HMGB1 plays a role in stimulating EXO1 activity. Next, repair synthesis to replace the excised strand with new DNA is mediated by DNA polymerase δ and its associated factors. Finally, DNA ligase I seals the nick by forming the last phosphodiester bond.

Nucleotide excision repair

The nucleotide excision repair pathway is used for repair of structural distortion, for example, bulges from thymine–thymine dimers induced by UV irradiation. Given the human desire to achieve the perfect suntan, this repair pathway is of particular relevance. Defects in nucleotide excision repair are the cause of the hereditary disease xeroderma pigmentosum, which is characterized by unusually high sensitivity to UV light (Disease box 7.2).

In humans, nucleotide excision repair is carried out by six repair factors: RPA, XPA, XPC, TFIIH, XPG, and XPF/ERCC1. ERCC1 stands for *excision repair cross complementing rodent repair deficiency*. Three of the six factors also have other essential roles in cells. RPA plays an essential role in DNA replication (see Section 6.6), while XPF/ERCC1 is also essential for homologous recombination (see below). TFIIH is a multiprotein complex that also plays an important role in gene transcription (the TFII stands for *transcription factor for RNA polymerase II*, see Section 11.4).

In the first step of nucleotide excision repair, DNA damage is recognized by the cooperative binding of RPA, XPA, XPC, and the TFIIH complex which assemble in random order (Fig. 7.11). The TFIIH complex is composed of a number of polypeptides including XPB and XPD. The 5′ → 3′ helicase activity of XPB and XPD unwinds the DNA double helix. Next, an incision is made on the 3′ side of the damage, by an endonuclease (XPG), approximately six nucleotides from the damage. Another incision is made by a second endonuclease (XPF-ERCC1) on the 5′ side of the damage, approximately 20 nt from the damage. After the helicase activity of XPB and XPD unwinds the duplex DNA, the damage-containing DNA sequence (24–32 nt) is released. Subsequently, there is "hand-off" of the DNA to DNA polymerase ε or DNA polymerase δ, PCNA, RFC, and RPA for repair synthesis. The choice of polymerase depends on cell type, although DNA polymerase ε may be most commonly used. The remaining gap in the DNA backbone is closed by DNA ligase I.

Nucleotide excision repair and transcription are coupled, and the repair process proceeds most rapidly via a transcribed strand of genes. The repair pathway responsible for recognizing lesions in the whole genome is called global genome repair (GGR), while the transcription-coupled repair (TCR) pathway identifies lesions in the transcribed strand of active genes.

7.7 Double-strand break repair by removal of DNA damage

Double-strand breaks in DNA are induced by such agents as reactive oxygen species, ionizing radiation (e.g. X-rays), and chemicals that generate reactive oxygen species (free radicals). Double-strand breaks are repaired either by homologous recombination or nonhomologous end-joining mechanisms. Homologous recombination repairs double-strand breaks by retrieving genetic information from an undamaged homologous chromosome. In cases where the two chromosomes are not exactly homologous, gene conversion may take place (see Section 12.7). In contrast, nonhomologous end-joining rejoins double-strand breaks via direct ligation of the DNA ends without any requirement for sequence homology.

Double-strand break repair mechanisms have been conserved through evolution and operate in both prokaryotes and eukaryotes. While homologous recombination plays a major role in double-strand break repair in prokaryotes and single cell eukaryotes, it plays a more minor, although important, role in multicellular eukaryotes. In mammalian cells, double-strand breaks in DNA are primarily repaired through nonhomologous end-joining. This repair pathway functions throughout the cell cycle. In contrast, the main function of homologous recombination is to repair double-strand breaks at the replication fork. Homologous recombination takes place in the late S–G2 phase of the cell cycle.

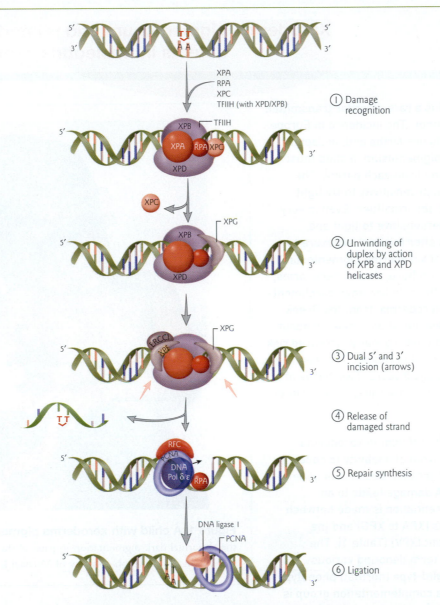

Figure 7.11 Mammalian nucleotide excision repair pathway. The diagram shows the six major steps to repair a thymine dimer. (1) The DNA damage is recognized by the cooperative binding of RPA, XPA, XPC, and TFIIH. (2) Unwinding of the duplex DNA is promoted by the action of XPB and XPD helicases which are subunits of the TFIIH complex. (3) XPC is released and the endonuclease XPG joins the repair complex. XPG makes a 3′ incision, and a 5′ incision is made by the endonuclease XPF–ERCC1. (4) The damaged strand is released. (5) Repair synthesis is mediated by DNA polymerase δ or ε in association with PCNA, RFC, and RPA. (6) Ligation of the remaining gap in the DNA backbone is mediated by DNA ligase 1.

Homologous recombination

Homologous recombination plays many essential roles in eukaryotic organisms. This process maintains the integrity of the genome by mediating proper chromosome segration during meiosis. Meiotic recombination often gives rise to crossing-over between genes on the two homologous parental chromosomes, thus ensuring variation in the sets of genes passed on to the next generation. In addition, homologous recombination is

DISEASE BOX 7.2

Xeroderma pigmentosum and related disorders: defects in nucleotide excision repair

Xeroderma pigmentosum (XP) is a rare disorder transmitted in an autosomal recessive manner. The incidence in Europe and North America is 1/250,000 live births and, in Japan, 1/40,000. To have xeroderma pigmentosum, a child must receive two defective genes, one from each parent. The symptoms include unusually high sensitivity to UV light (photosensitivity) and pigment abnormalities. Even in very small children, the skin is hypersensitive to light and children get serious sunburns after minimal exposure to sunlight. The first indications of xeroderma pigmentosum are freckling in sun-exposed skin (face, throat, neck, arms, hands) (Fig. 1). The surface of the skin becomes parchment-like and dry (hence the name xeroderma, from the Greek for "dry skin"). Neurological degeneration is seen in about 14–40% of patients. Individuals have a greatly increased risk (> 1000-fold) of sunlight-induced skin cancer. The mean age for developing malignancies is age 8 years. Two-thirds of patients die before reaching adulthood because of skin cancer.

XP complementation groups

The unusually high sensitivity to UV light in xeroderma pigmentosum patients results from an inability to cope properly with UV-induced DNA lesions. Because of gene defects in repair proteins, DNA damage leads to an increased mutation rate. Differentiation is made between seven complementation groups (XPA to XPG) and the xeroderma pigmentosum variant (XPV) (Table 1). The complementation group is the term denoting various mutations that do not form a wild-type (normal) phenotype after crossing. In this case, the complementation group is defined as when fibroblast (skin) cells of two different patients with the same defect are fused *in vitro* and the DNA damage is maintained. The capacity to carry out nucleotide excision repair after UV irradiation is determined by the uptake of radiolabeled thymidine into the DNA. Nucleotide excision repair activity is reduced in patients with xeroderma pigmentosum, so less radiolabeled thymidine is incorporated into the DNA. If the two patients have different gene defects, the cells correct each other reciprocally and the DNA damage is repaired.

The gene coding for DNA polymerase η is defective in individuals having the xeroderma pigmentosum variant (XPV) defect. XPV is unique in that it is the only one of the eight XP complementation groups that is not deficient in

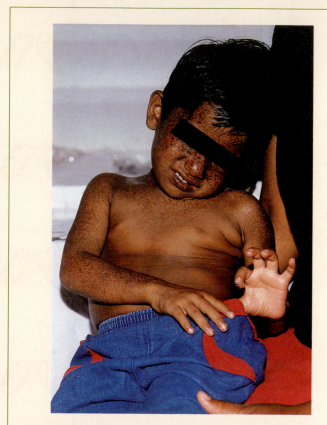

Figure 1 A child with xeroderma pigmentosum. Note the abnormal dark pigmentation in parts of the body exposed to sunlight. (Photograph courtesy of Michael J. Levine, M.D.)

nucleotide excision repair of DNA lesions. DNA polymerase η can bypass thymine–thymine dimers in a relatively accurate fashion by inserting two adenines opposite the lesion. In the absence of a functional DNA polymerase η, UV-induced thymine–thymine dimers are presumably bypassed by a different polymerase such as the error-prone DNA polymerase ζ. The reduced accuracy of translesion DNA synthesis over thymine–thymine dimers leads to an increased mutation frequency that contributes to the XPV disorder.

Treating xeroderma pigmentosum

Early diagnosis and prevention is the basis of treatment for xeroderma pigmentosum. If protected from sunlight at an

Table 1 Genes defective in xeroderma pigmentosum (XP).

Complementation group	Relative frequency	Symptoms	Function
XPA	High (most common)	Very severe	DNA damage recognition (GGR and TCR)
XPB	Rare	Severe	Helicase
XPC	Third most common	Severe	DNA damage recognition (GGR only)
XPD	Intermediate	Variable	Helicase
XPE	Rare	Mild	DNA damage recognition (GGR only)
XPF	Rare/intermediate	Mild	Endonuclease
XPG	Rare	Variable	Endonuclease
XPV	Second most common	Severe	Polymerase η (translesion repair)

GGR, global genome repair; TCR, transcription-coupled repair.

early age, individuals can remain completely free of skin lesions. A strict light-protective lifestyle must be adopted and the application of broadspectrum UV-protective sunscreen is essential. There are reports of successful application of a topical DNA repair enzyme. The medication, containing recombinant T4 endonuclease V encapsulated in liposomes (artificial lipid spheres), is applied once daily locally to the skin. The liposomes deliver the endonuclease to the skin cells by fusion with the cell membrane and endocytosis. The endonuclease can then repair UV-induced pyrimidine dimers in the damaged DNA. The therapy reportedly reduces the skin cancer rate of xeroderma pigmentosum patients by 30% and the rate of precancerous lesions by as much as 68%. In the future, gene therapy may be possible. The introduction of an intact repair gene into skin cells would allow the cells to synthesize a functional component of the nucleotide excision repair pathway.

Two other DNA repair deficiency syndromes

Xeroderma pigmentosum is distinguished from two other DNA repair deficiency syndromes, trichothiodystrophy and Cockayne syndrome, by differential diagnoses. The existence of two other nucleotide excision repair deficiency syndromes further highlights the essential nature of this repair pathway.

Trichothiodystrophy

Trichothiodystrophy (TTD) is an autosomal recessive disorder that is noncancerous, but individuals have an exaggerated sensitivity to light. The symptoms include brittle hair and nails and dry scaly skin. Patients tend to be short in stature, intellectually impaired, and have impaired sexual development. Twenty percent of patients are photosensitive. Trichothiodystrophy is divided into three complementation groups. Certain defects in the XP genes, *XPB* and *XPD*, result in TTD. There appears to be at least one other gene, *TTD-A*, which causes this disease, but it has not yet been characterized. Recently, patients with combined features of XP and TTD have been reported to have a defect in XPD.

Cockayne syndrome

The clinical features of Cockayne syndrome (CS) have little in common with xeroderma pigmentosum. This disease is characterized by photosensitivity, cataracts, and deafness, but no pigmentation abnormalities and no increased risk of skin cancer. Patients suffer from severe mental and physical retardation. They have skeletal deformation and short stature, resulting in a wizened appearance. They usually die at age 20 years. CS is divided into two complementation groups, CS-A and CS-B. CS-A and CS-B are both components of complexes associated with RNA polymerase II and there is a defect in transcription-coupled repair. A combined phenotype of XP and CS is found in several XPG patients and in rare XPB and XPD patients. These patients have the skin features of XP and neurological abnormalities of CS.

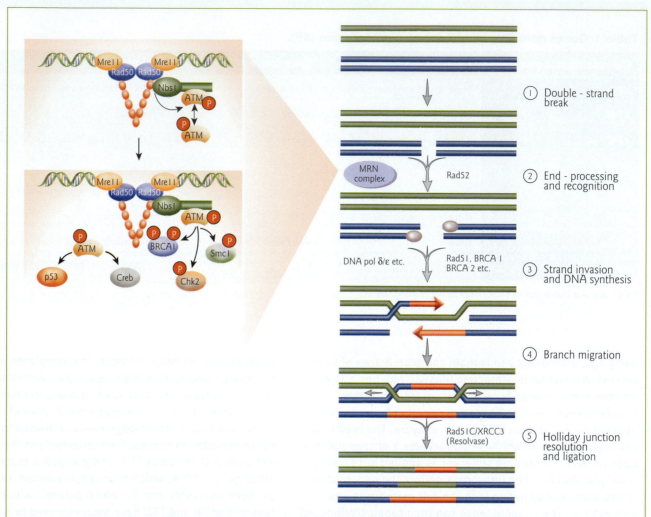

Figure 7.12 Model for mammalian DNA double-strand break repair by homologous recombination.
In this model, green and blue double-stranded DNAs (dsDNAs) represent homologous sequences. (1) A double-strand break (DSB) is induced by ionizing radiation. (2) The MRN (Mre11–Rad50–Nbs1) complex is rapidly recruited to the DSB site. The 3′,5′-exonuclease activity of Mre11 generates 3′ single-strand DNA tails that are recognized by Rad52. (3) The 3′ tails invade homologous intact sequences. Strand exchange generates a hybrid molecule between damaged and undamaged duplex DNAs. Sequence information that is missing at the DSB site is restored by DNA synthesis (newly synthesized DNA is shown in red). (4) The interlinked molecules are then processed by branch migration (indicated by right and left arrows). (5) Finally, Holliday junction resolution and ligation occur (see Fig. 7.13). (Inset) ATM activation at DSBs. MRN complexes form a bridge between free DNA ends via the coiled-coil arms of Rad50 dimers. Inactive ATM dimers are recruited to the DSBs through interaction with the carboxyl terminus of Nbs1, and by a less stable interaction with Rad50. Activating signals are delivered to ATM dimers, possibly through a conformational change in Nbs1. ATM undergoes phosphorylation accompanied by its conversion from a dimer to a monomer. Activated ATM monomers either remain near the DSB, where they phosphorylate proteins involved in DNA repair, or diffuse away from the DSB sites to phosphorylate nuclear substrates, such as p53 and Creb that are involved in cell cycle control. (Reprinted with permission from Abraham, R.T., Tibbetts, R.S. 2005. Guiding ATM to broken DNA. *Science* 308:510–511. Illustration: Katharine Sutliff. Copyright © 2005 AAAS.)

Hereditary breast cancer syndromes: mutations in *BRCA1* and *BRCA2*

DISEASE BOX 7.3

Breast cancer is a disease affecting one in eight women in the USA. Most commonly, cancers arise as somatic mutations with a number of different genes involved. About 5–10% of all cases are linked to a single gene mutation that increases the susceptibility to develop breast cancer. These women are said to have a genetic predisposition to breast cancer. Approximately 80–90% of hereditary breast cancers are caused by mutations in the *BRCA1* and *BRCA2* genes. In addition, some hereditary breast cancer families with mutations in these genes also have a history of ovarian cancer. Both *BRCA1* and *BRCA2* function as tumor suppressor genes that play roles in the repair of damaged DNA. However, the exact roles of these proteins in DNA repair are not yet understood. BRCA1 was recently shown to bind to and enhance the activity of topoisomerase II, an enzyme that helps untangle DNA and segregate chromosomes when cells are replicating their DNA (see Section 6.6). Both *BRCA1* and *BRCA2* are inherited in an autosomal dominant fashion. The lifetime risks for breast and ovarian cancers for carriers of *BRCA1* and *BRCA2* range from 50 to 87% and 15 to 44%, respectively. Genetic testing for mutations in these genes is available. Options for women at high risk for breast cancer include surveillance, chemoprevention, and prophylactic surgery.

central to transposition, mating-type switching in yeast, and antigen-switching in trypanosomes. The role of homologous recombination in these processes is discussed in detail in Section 12.7.

Of relevance to this chapter, homologous recombination is also an important mechanism for the repair of double-strand breaks in DNA. Double-strand breaks are a particularly lethal form of DNA damage. For example, hereditary deficiencies in double-strand break repair are linked to an increased susceptibility to breast cancer (Disease box 7.3). To coordinate cell cycle progression with repair, the cellular response to such damage must be rapid and finely orchestrated. A serine–threonine kinase in the nucleus called ATM (ataxia telangiectasia mutated) is a key signal transducer. Exposure of cells to ionizing radiation or other double-strand break–inducing agents triggers an increase in ATM kinase activity. ATM is recruited to the break site and phosphorylates some of the proteins involved in DNA repair (e.g. BRCA1, see below) and cell cycle control (Fig. 7.12). One important target of ATM is the tumor suppressor protein p53 (see Fig. 17.8). Humans that lack ATM suffer from a syndrome called ataxia telangiectasia, characterized by extreme sensitivity to radiation, increased susceptibility to developing cancer, immunodeficiency, premature aging, and neurodegenerative disorders.

The cellular response to double-strand breaks results in the localization of break sites to repair foci which, along with recruiting ATM, contain most of the repair protein Rad52 in the cell. These foci may encompass more than one DNA lesion. The Mre11–Rad50–Nbs1 (MRN) complex is recruited to the double-strand break site and initiates repair. The DNA is first processed by the $3' \rightarrow 5'$ exonuclease activity of Mre11 in the MRN complex to form single-stranded tails (Fig. 7.12). The MRN complex forms a bridge between DNA ends via the coiled-coil domains of Rad50 dimers. The single-stranded DNA tails are then recognized by Rad52. Strand invasion of the $3'$ tails with intact homologous sequences is initiated by Rad51. The proteins Rad54, Rad55, Rad57, BRCA1, and BRCA2 are also involved in homologous recombination, but their precise roles have yet to be determined (see Disease box 7.3). Strand exchange generates a joint molecule between damaged and undamaged duplex DNA. Sequence information is restored by DNA synthesis using the DNA replication machinery. The interlinked molecules are then processed by branch migration and Holliday junction resolution (see below), followed by ligation of the repaired DNA strands.

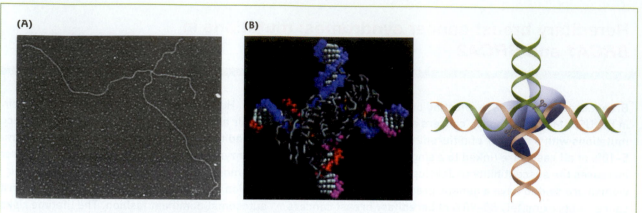

Figure 7.13 Structure of the Holliday junction. (A) Electron micrograph of a recombination intermediate. The Holliday junction was partially denatured to make its visualization easier. (Reprinted by permission of Huntington Potter and David Dressler from: Potter, H. and Dressler, D. 1976. On the mechanism of genetic recombination: electron microscopic observation of recombination intermediates. *Proceedings of the National Academy of Sciences USA.*) (B) Three-dimensional structure of *E. coli* RuvC binding to a Holliday junction (left). The diagram on the right shows how the dimeric RuvC protein introduces nicks at symmetric positions in strands with the same polarity. (Reprinted with permission from Rafferty, J.B., Sedelnikova, S.E., Hargreaves, D., Artymiuk, P.J., Baker, P.J., Sharples, G.J., Mahdi, A.A., Lloyd, R.G., Rice, D.W. 1996. Crystal structure of DNA recombination protein RuvA and a model for its binding to the Holliday junction. *Science* 274:415–421. Copyright © 1996 AAAS.)

Holliday junctions

In his landmark papers in the early 1960s, Robin Holliday proposed a model for general recombination based on genetic data obtained in fungi. Since that time, the Holliday junction has evolved from a hypothetical structure to models for its atomic structure (Fig. 7.13). The two intermediates that he proposed – heteroduplex DNA and the "chiasma-like structure" now termed the Holliday junction – have survived the test of time. Heteroduplex DNA refers to duplex DNA formed during recombination that is composed of single DNA strands originally deriving from different homologs. The Holliday junction is an intermediate in which the two recombining duplexes are joined covalently by single-strand crossovers. The Holliday junction is resolved into two duplexes by an enzyme complex called the "resolvasome."

The human biochemical resolvase activity was discovered in the late 1980s, but its exact identity remained a mystery for 40 years. In 2004, the Holliday junction resolvasome was purified (Fig. 7.14). This impressive achievement involved the fractionation of proteins from 50 liters of HeLa (human cell line) cells through six different chromatographic steps. Through this labor-intensive purification scheme, it was demonstrated that the protein Rad51C is required for Holliday junction processing in mammalian cells. Mutations in the related XRCC3 protein, which forms a complex with Rad51C, also lead to reduced levels of resolvase activity.

Nonhomologous end-joining

As noted above, in mammals double-strand breaks in DNA are primarily repaired through nonhomologous end-joining. This is thought to be the major pathway for repair of double-strand breaks induced by ionizing radiation. The repair of double-strand breaks is essential for maintaining the integrity of the genome, but the repair process itself can lead to mutation. For example, two broken ends can be ligated together by the repair machinery regardless of whether they come from the same chromosome, and nonhomologous end-joining frequently results in insertions or deletions at the break site.

The key enzymatic steps in nonhomologous end-joining are nucleolytic action, polymerization, and ligation. A biochemically defined system for mammalian nonhomologous end-joining has recently been

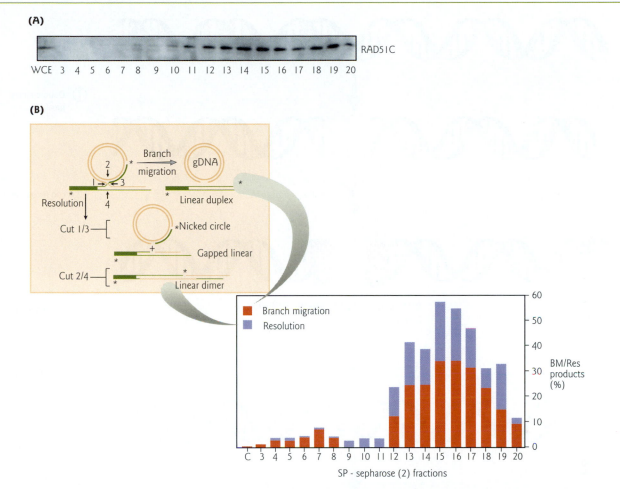

Figure 7.14 Rad51C is required for Holliday junction processing in mammalian cells. The human Holliday junction resolvasome was purified through a series of six different chromatographic steps. (A) Rad51C is a component of the resolvasome. Fractions from the final column (SP-sepharose) in the purification scheme were analyzed by Western blotting (see Section 9.6) with a Rad51C-specific monoclonal antibody. WCE, whole cell extract. (B) Each column fraction was assayed for resolvase activity. Recombination intermediates were made by strand exchange between a 3′-^{32}P end-labeled linear duplex DNA and a gapped circular plasmid (gDNA). ^{32}P end labels are indicated by asterisks. Branch migration dissociates the structure into a ^{32}P-labeled linear duplex and unlabeled gDNA products, and thus was measured by the increase in ^{32}P-labeled linear duplex DNA. Resolution occurs in one of two possible orientations. Cleavage in strands 1/3 produces two labeled products (^{32}P-labeled nicked circular and ^{32}P-labeled gapped linear DNA). Cleavage in strands 2/4 produces ^{32}P-labeled linear dimers. Reaction products were visualized by agarose gel electrophoresis and autoradiography. Product formation was quantified by phosphorimaging (see Tool box 8.4). The percentage of branch migration (BM) products compared with resolution (Res) products was plotted for each fraction. Fractions shown by Western blot to contain Rad51C had the greatest branch migration and resolution activities. (Reprinted with permission from Liu, Y., Masson, J.Y., Shah, R., O'Regan, P., West, S.C. 2004. RAD51C is required for Holliday junction processing in mammalian cells. *Science* 303:243–246. Copyright © 1996 AAAS.)

developed that has provided further insight into this repair pathway. These *in vitro* assays suggest that there is flexibility in the order of the three key enzymatic steps. For example, ligation on one strand can precede nucleolytic or polymerization action on the other strand. One possible scenario is presented in Fig. 7.15. Following a double-strand break, the broken ends of DNA are recognized by two heterodimers of the Ku70

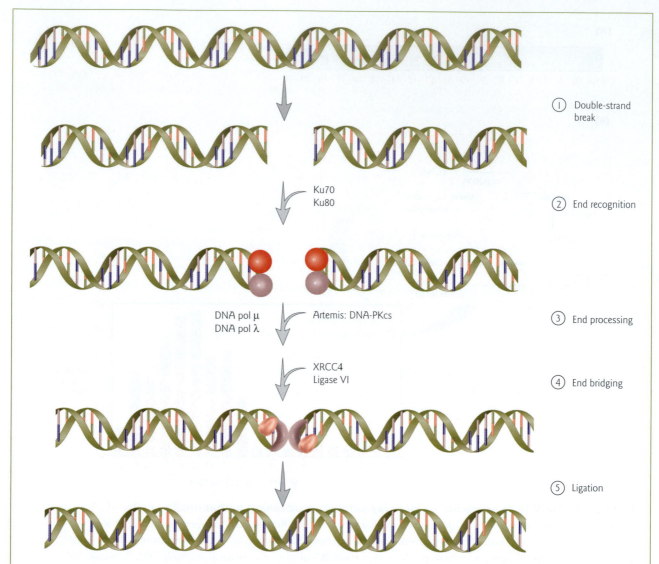

① Double-strand break

② End recognition

Ku70
Ku80

③ End processing

DNA pol μ   Artemis: DNA-PKcs
DNA pol λ

XRCC4
Ligase VI

④ End bridging

⑤ Ligation

Figure 7.15 Model for mammalian DNA double-strand break repair by nonhomologous end-joining.
(1) A double-strand break is induced in DNA by ionizing radiation. (2) The broken ends are recognized by heterodimers of Ku70/Ku80. (2) The endonuclease Artemis is activated by the DNA-dependent protein kinase catalytic subunit (DNA-PK$_{CS}$) and trims excess or damaged DNA at the break site. (3) DNA polymerase (pol) μ or DNA pol λ fill-in gaps and extend 3′ or 5′ overhangs. (4) The ligase complex (XRCC4–DNA ligase IV) is recruited to the damaged site and forms a bridge, bringing the broken ends of the DNA together. (5) The broken ends are ligated by the XRCC4–DNA ligase IV complex.

and Ku80 proteins. The heterodimers form a scaffold that holds the broken ends in close proximity, allowing other enzymes to act. The Ku heterodimer recruits the nuclease (Artemis/DNA-PK$_{CS}$), the polymerases (DNA polymerases μ and λ), and the ligase complex (XRCC4–DNA ligase IV) to the damaged site. The endonuclease Artemis is activated after it is phosphorylated by the DNA-dependent protein kinase catalytic subunit (DNA-PK$_{CS}$). The activated Artemis/DNA-PK$_{CS}$ complex then trims excess or damaged DNA at the break site. DNA polymerases μ and λ, or the enzyme TdT (see Table 6.1) are required for any nonhomologous end-joining events that need fill-in of gaps or extension of the 3′ or 5′ overhangs. The rejoining of the broken ends is carried out by DNA ligase IV in association with XRCC4.

Chapter summary

Mutations result from changes in the nucleotide sequence of DNA or from trinucleotide repeat expansions, deletions, insertions, or rearrangements of DNA sequences in the genome. Mutations that alter a single nucleotide pair are called point mutations. Transition mutations replace one pyrimidine base with the other, or one purine base with the other. Transversion mutations replace a pyrimidine with a purine or vice versa. Such mismatches can become permanently incorporated in the DNA sequence during DNA replication. Whether nucleotide substitutions have a phenotypic effect depends on whether they alter a critical nucleotide in a gene regulatory region, the template for a functional RNA molecule, or the codons in a protein-coding gene.

A spontaneous mutation is one that occurs as a result of natural processes in cells such as DNA replication errors that are not corrected by proofreading. Induced mutations occur as a result of interaction with DNA-damaging agents such as oxygen free radicals, ultraviolet or ionizing radiation, and various chemicals. There are three major classes of DNA damage: single base changes, structural distortion, and DNA backbone damage. Single base changes arise through the deamination, alkylation, and oxidation of bases. Bulky adducts that cause structural distortion of the DNA double helix, such as thymine dimers, can be induced by ultraviolet radiation or by chemical mutagenesis. DNA damage is also caused by intercalating agents and base analogs. Backbone damage includes the formation of abasic sites and double-stranded DNA breaks. There are a variety of complex cellular responses to different types of DNA damage: those that bypass the damage, those that directly reverse the damage, and those that remove that damaged section of DNA and replace it with undamaged DNA, by excision or recombinational repair systems.

In damage bypass, specialized DNA polymerases transiently replace the replicative polymerases and copy past the damaged DNA in a process called translesion synthesis. Most of the repair polymerases are error-prone, but one of them, mammalian DNA polymerase η, performs translesion synthesis past thymine dimers in an error-free manner.

In most organisms, except for placental mammals, pyrimidine dimers can be directly repaired by a process called photoreactivation in which the enzyme DNA photolyase reverses the damage. Methyltransferases can repair the methylated base O^6-methylguanine.

Repair systems that removed damaged DNA involve multiple dynamic protein interactions and the ordered hand-off of the damaged DNA from one protein complex to another. Base excision repair involves the correction of single base changes that are due to conversion. In mammals, the damaged base is recognized and excised by a DNA glycosylase. The abasic site is then repaired by endonucleolytic excision of the single damaged nucleotide or 2–10 adjacent nucleotides, repair synthesis by DNA polymerase, and ligation. Mismatched base pairs that result from DNA polymerase errors during replication are corrected by mismatch repair. In the mammalian cell pathway, the mismatch is recognized by the MutSα complex, and a large region of DNA including the mismatch is excised by exonuclease activity. Repair synthesis replaces the excised strand and DNA ligase seals the nick by forming the last phosphodiester bond. The nucleotide excision repair pathway is used for repair of bulky adducts, such as the thymine dimers induced by UV irradiation. Nucleotide excision repair in humans is carried out by six repair factors (RPA, XPA, XPC, TFIIH, XPG, and XPF/ERCC1), which include proteins that recognize the damage, 3′ and 5′ endonucleases, and helicase activity. After the damaged strand is removed by the repair factors, repair synthesis by DNA polymerase replaces the strand and the remaining gaps are closed by DNA ligase.

Double-strand breaks in DNA are repaired by either homologous recombination or nonhomologous end-joining. Homologous recombination plays a major role in repair in prokaryotes and single cell eukaryotes. In mammalian cells, double-strand breaks are primarily repaired through nonhomologous end-joining.

Homologous recombination repairs double-strand breaks by retrieving genetic information from an undamaged homologous chromosome. Damage is recognized by the MRN complex that mediates end recognition and exonuclease processing. Strand invasion of the 3′ tail that is generated is initiated by Rad51. Strand exchange generates a joint molecule between damaged and undamaged duplex DNA and repair synthesis occurs. The interlinked molecules are then processed by branch migration and Holliday junction resolution by an enzyme complex called the resolvasome, followed by ligation of the repaired DNA strands.

Nonhomologous end-joining rejoins double-strand breaks via direct ligation of the DNA ends without any sequence homology requirements. Following a double-strand break, the broken ends of DNA are recognized by a Ku heterodimer. The ends are processed by an endonuclease complex (Artemis/DNA-PK$_{CS}$) and repair polymerases. The rejoining of broken ends is carried out by DNA ligase.

Analytical questions

1 What would be the effect on reading frame and gene function if:
  (a) Two nucleotides were inserted into the middle of an mRNA?
  (b) Three nucleotides were inserted into the middle of an mRNA?
  (c) One nucleotide was inserted into one codon and one subtracted from the next?
  (d) A transition mutation occurs from G → A during DNA replication. What is the effect after a second round of DNA replication?
  (e) Exposure to an alkylating agent leads to the formation of O^6-methylguanine. What is the effect after DNA replication?

2 A friend of yours with xeroderma pigmentosum seeks your advice about participating in a suntanning competition in Florida during Spring Break. Provide appropriate advice.

3 Draw a diagram of a Holliday junction during double-strand break repair. Starting with that diagram, illustrate branch migration and resolution. Is the resulting DNA duplex "repaired" to its original state?

4 You have isolated a novel protein factor you suspect is essential for efficient mismatch repair in mammalian cells. Design an experiment to test for repair activity *in vitro*. Show sample positive results.

Suggestions for further reading

Abraham, R.T., Tibbetts, R.S. (2005) Guiding ATM to broken DNA. *Science* 308:510–511.

Acharya, S., Foster, P.L., Brooks, P., Fishel, R. (2003) The coordinated functions of the *E. coli* MutS and MutL proteins in mismatch repair. *Molecular Cell* 12:233–246.

Banerjee, A., Yang, W., Karplus, M., Verdine, G.L. (2005) Structure of a repair enzyme interrogating undamaged DNA elucidates recognition of damaged DNA. *Nature* 434:612–618.

Costa, R.M.A., Chiganças, V., da Silva-Galhardo, R., Carvalho, H., Menck, C.F.M. (2003) The eukaryotic nucleotide excision repair pathway. *Biochimie* 85:1083–1099.

Crick, F.H.C. (1974) The double helix: a personal view. *Nature* 248:766–769.

David, S.S. (2005) DNA search and rescue. *Nature* 434:569–570.

Dudáš, A., Chovanec, M. (2004) DNA double-strand break repair by homologous recombination. *Mutation Research* 566:131–167.

Dzantiev, L., Constantin, N., Genschel, J., Iyer, R.R., Burgers, P.M., Modrich, P. (2004) A defined human system that supports bidirectional mismatch-provoked excision. *Molecular Cell* 15:31–41.

Friedberg, E.C., Lehmann, A.R., Fuchs, R.P.P. (2005) Trading places: how do DNA polymerases switch during translesion DNA synthesis? *Molecular Cell* 18:499–505.

Genschel, J., Modrich, P. (2003) Mechanism of 5′-directed excision in human mismatch repair. *Molecular Cell* 12:1077–1086.

Heyer, W.D., Ehmsen, K.T., Solinger, J.A. (2003) Holliday junction in the eukaryotic nucleus: resolution in sight? *Trends in Biochemical Sciences* 28:548–557.

Holliday, R. (1974) Molecular aspects of genetic exchange and gene conversion. *Genetics* 78:273–287.

Lehmann, A.R. (2003) DNA repair-deficient diseases, xeroderma pigmentosum, Cockayne syndrome and trichothiodystrophy. *Biochimie* 85:1101–1111.

Lisby, M., Mortensen, U.H., Rothstein, R. (2003) Colocalization of multiple DNA double-strand breaks at a single Rad52 repair centre. *Nature Cell Biology* 5:572–577.

Liu, Y., Masson, J.Y., Shah, R., O'Regan, P., West, S.C. (2004) RAD51C is required for Holliday junction processing in mammalian cells. *Science* 303:243–246.

Liu, Y., West, S.C. (2004) Happy hollidays: 40th anniversary of the Holliday junction. *Nature Reviews Molecular Cell Biology* 5:937–946.

Ma, Y., Lu, H., Tippin, B. et al. (2004) A biochemically defined system for mammalian nonhomologous DNA end joining. *Molecular Cell* 16:701–713.

Maga, G., Hubscher, U. (2003) Proliferating cell nuclear antigen (PCNA): a dancer with many partners. *Journal of Cell Science* 116:3051–3060.

Norgauer, J., Idzko, M., Panther, E., Hellstern, O., Herouy, Y. (2003) Xeroderma pigmentosum. *European Journal of Dermatology* 13:4–9.

O'Driscoll, M., Jeggo, P.A. (2005) The role of double-strand break repair – insights from human genetics. *Nature Review Genetics* 7:45–54.

Pasternak, J.J. (1999) *An Introduction to Human Molecular Genetics. Mechanisms of Inherited Diseases.* Fitzgerald Science Press, Bethesda, MD.

Potter, H., Dressler, D. (1979) DNA recombination: *in vivo* and *in vitro* studies. *Cold Spring Harbor Symposium for Quantitative Biology* 43:969–985.

Sancar, A., Lindsey-Boltz, L.A., Ünsal-Kaçmaz, K., Linn, S. (2004) Molecular mechanisms of mammalian DNA repair and the DNA damage checkpoints. *Annual Review of Biochemistry* 73:39–85.

Stauffer, M.E., Chazin, W.J. (2004) Structural mechanisms of DNA replication, repair, and recombination. *Journal of Biological Chemistry* 279:30915–30918.

Tippin, B., Pham, P., Goodman, M.F. (2004) Error-prone replication for better or worse. *Trends in Microbiology* 12:288–295.

Yu, H.A., Lin, K.M., Ota, D.M., Lynch, H.T. (2003) Hereditary nonpolyposis colorectal cancer: preventive management. *Cancer Treatment Reviews* 29:461–470.

Zhang, Y., Yuan, F., Presnell, S.R. et al. (2005) Reconstitution of 5′-directed human mismatch repair in a purified system. *Cell* 122:693–705.

Chapter 8

Recombinant DNA technology and molecular cloning

Sometimes a good idea comes to you when you are not looking for it. Through an improbable combination of coincidences, naiveté and lucky mistakes, such a revelation came to me one Friday night in April, 1983, as I gripped the steering wheel of my car and snaked along a moonlit mountain road into northern California's redwood country. That was how I stumbled across a process that could make unlimited numbers of copies of genes, a process now known as the polymerase chain reaction (PCR).

Kary B. Mullis, *Scientific American* (1990) 262:36.

Outline

8.1 Introduction

The cornerstone of most molecular biology technologies is the gene. To facilitate the study of genes, they can be isolated and amplified. One method of isolation and amplification of a gene of interest is to clone the gene by inserting it into another DNA molecule that serves as a vehicle or vector that can be replicated in living cells. When these two DNAs of different origin are combined, the result is a recombinant DNA molecule. Although genetic processes such as crossing-over technically produce recombinant DNA, the term is generally reserved for DNA molecules produced by joining segments derived from different biological sources. The recombinant DNA molecule is placed in a host cell, either prokaryotic or eukaryotic. The host cell then replicates (producing a clone), and the vector with its foreign piece of DNA also replicates. The foreign DNA thus becomes amplified in number, and following its amplification can be purified for further analysis.

8.2 Historical perspective

In the early 1960s, before the advent of gene cloning, studies of genes often relied on indirect or fortuitous discoveries, such as the ability of bacteriophages to incorporate bacterial genes into their genomes. For example, a strain of phage phi 80 with the *lac* operator incorporated into its genome was used to demonstrate that the Lac repressor binds specifically to this DNA sequence (see Fig. 10.12). The synthesis of many disparate experimental observations into recombinant DNA technology occurred between 1972 and 1975, through the efforts of several research groups working primarily on bacteriophage lambda (λ).

Insights from bacteriophage lambda (λ) cohesive sites

In 1962, Allan Campbell noted that the linear genome of bacteriophage λ forms a circle upon entering the host bacterial cell, and a recombination (breaking and rejoining) event inserts the phage DNA into the host chromosome. Reversal of the recombination event leads to normal excision of the phage DNA. Rare excision events at different places can result in the incorporation of nearby bacterial DNA sequences (Fig. 8.1). Further analysis revealed that phage λ had short regions of single-stranded DNA whose base sequences were complementary to each other at each end of its linear genome. These single-stranded regions were called "cohesive" (*cos*) sites. Complementary base pairing of the *cos* sites allowed the linear genome to become a circle within the host bacterium. The idea of joining DNA segments by "cohesive sites" became the guiding principle for the development of genetic engineering. With the molecular characterization of restriction and modification systems in bacteria, it soon became apparent that the ideal engineering tools for making cohesive sites on specific DNA pieces were already available in the form of restriction endonucleases.

Insights from bacterial restriction and modification systems

Early on, Salvador Luria and other phage workers were intrigued by a phenomenon termed "restriction and modification." Phages grown in one bacterial host often failed to grow in different bacterial strains ("restriction"). However, some rare progeny phages were able to escape this restriction. Once produced in the restrictive host they had become "modified" in some way so that they now grew normally in this host. The entire cycle could be repeated, indicating that the modification was not an irreversible change. For example, phage λ grown on the C strain of *Escherichia coli* (λ·C) were restricted in the K-12 strain (the standard strain for most molecular work) (Fig. 8.2). However, the rare phage λ that managed to grow in the K-12 strain now had "K" modification (λ·K). These phages grew normally on both C and K-12; however, after growth on C, the phage λ with "C" modification (λ·C) was again restricted in K-12. Thus, the K-12 strain was able to mark its own resident DNA for preservation, but could eliminate invading DNA from another distantly related strain. In 1962, the molecular basis of restriction and modification was defined by Werner Arber and co-workers.

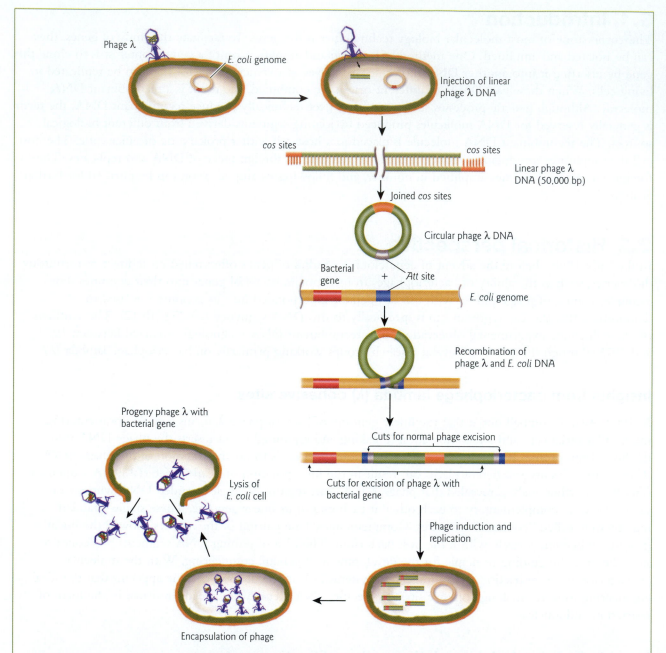

Figure 8.1 Bacteriophage lambda (λ) cohesive sites. Following the injection of a linear phage λ DNA into *E. coli* host cells, the phage λ genome circularizes by joining of the cohesive (*cos*) sites. In the lysogenic mode of replication, phage DNA is incorporated into the host genome by recombination at attachment (*Att*) sites on the phage and bacterial chromosome, and replicated as part of the host DNA. Under certain conditions, such as when the host encounters mutagenic chemicals or UV radiation, reversal of this recombination event leads to excision of the phage DNA. Rare excision events at different places allow phage λ to pick up bacterial genes. In the lytic mode of the phage life cycle, phage λ progeny with bacterial genes incorporated in their genomes are released from the lysed *E. coli*.

Restriction system

After demonstrating that phage λ DNA was degraded in a restricting host bacterium, Arber and co-workers hypothesized that the restrictive agent was a nuclease with the ability to distinguish whether DNA was resident or foreign. Six years later, such a nuclease was biochemically characterized in *E. coli* K-12 by Matt

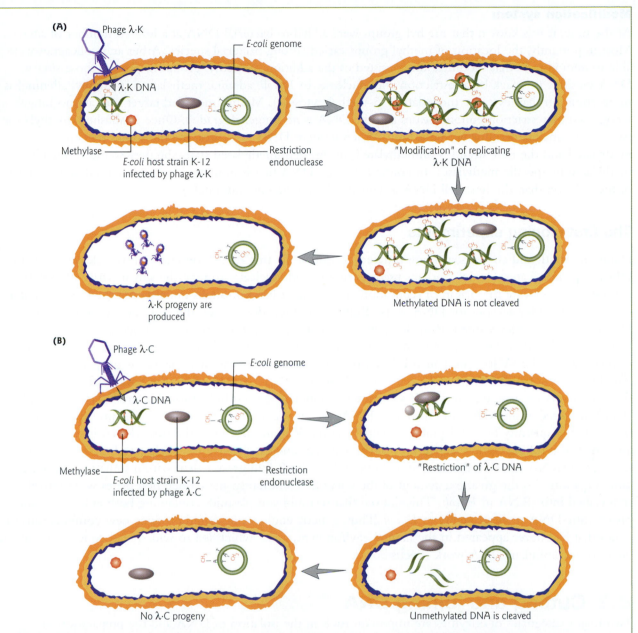

Figure 8.2 Restriction and modification systems in bacteria. Restriction endonucleases and their corresponding methylases function in bacteria to protect against bacteriophage infection. (A) Modification. When *E. coli* host strain K-12 is infected by phage λ·K, the phage DNA is not recognized as foreign because it has the same methylation pattern as the *E. coli* host genome. When the phage DNA replicates, the newly replicated DNA is modified by a specific methylase to maintain the pattern. Methylated DNA is not cleaved by restriction endonucleases, so progeny phage λ·K are produced. (B) When *E. coli* host strain K-12 is infected by phage λ·C, the phage DNA is recognized as foreign, because it does not have the same methylation pattern as the host genome. The phage DNA is cleaved by a specific restriction endonuclease, and no progeny phage λ·C are produced.

Meselson and Bob Yuan. The purified enzyme cleaved λ·C-modified DNA into about five pieces but did not attack λ·K-modified DNA (Fig. 8.2). Restriction endonucleases (also referred to simply as restriction enzymes) thus received their name because they restrict or prevent viral infection by degrading the invading nucleic acid.

Modification system

At the time, it was known that methyl groups were added to bacterial DNA at a limited number of sites. Most importantly, the location of methyl groups varied among bacterial species. Arber and colleagues were able to demonstrate that modification consisted of the addition of methyl groups to protect those sites in DNA sensitive to attack by a restriction endonuclease. In *E. coli*, adenine methylation (6-methyl adenine) is more common than cytosine methylation (5-methyl cytosine). Methyl-modified target sites are no longer recognized by restriction endonucleases and the DNA is no longer degraded. Once established, methylation patterns are maintained during replication. When resident DNA replicates, the old strand remains methylated and the new strand is unmethylated. In this hemimethylated state, the new strand is quickly methylated by specific methylases. In contrast, foreign DNA that is unmethylated or has a different pattern of methylation than the host cell DNA is degraded by restriction endonucleases.

The first cloning experiments

Hamilton Smith and co-workers demonstrated unequivocally that restriction endoncleases cleave a specific DNA sequence. Later, Daniel Nathans used restriction endonucleases to map the simian virus 40 (SV40) genome and to locate the origin of replication. These major breakthroughs underscored the great potential of restriction endonucleases for DNA work. Building on their discoveries, the cloning experiments of Herbert Boyer, Stanley Cohen, Paul Berg, and their colleagues in the early 1970s ushered in the era of recombinant DNA technology. One of the first recombinant DNA molecules to be engineered was a hybrid of phage λ and the SV40 mammalian DNA virus genome. In 1974 the first eukaryotic gene was cloned. Amplified ribosomal RNA (rRNA) genes or "ribosomal DNA" (rDNA) from the South African clawed frog *Xenopus laevis* were digested with a restriction endonuclease and linked to a bacterial plasmid. Amplified rDNA was used as the source of eukaryotic DNA since it was well characterized at the time and could be isolated in quantity by CsCl-gradient centrifugation. Within oocytes of the frog, rDNA is selectively amplified by a rolling circle mechanism from an extrachromosomal nucleolar circle (see Fig. 6.17). The number of rRNA genes in the oocyte is about 100- to 1000-fold greater than within somatic cells of the same organism. To the great excitement of the scientific community, the cloned frog genes were actively transcribed into rRNA in *E. coli*. This showed that recombinant plasmids containing both eukaryotic and prokaryotic DNA replicate stably in *E. coli*. Thus, genetic engineering could produce new combinations of genes that had never appeared in the natural environment, a feat which led to widespread concern about the safety of recombinant DNA work (Focus box 8.1).

8.3 Cutting and joining DNA

Two major categories of enzymes are important tools in the isolation of DNA and the preparation of recombinant DNA: restriction endonucleases and DNA ligases. Restriction endonucleases recognize a specific, rather short, nucleotide sequence on a double-stranded DNA molecule, called a restriction site, and cleave the DNA at this recognition site or elsewhere, depending on the type of enzyme. DNA ligase joins two pieces of DNA by forming phosphodiester bonds.

Major classes of restriction endonucleases

There are three major classes of restriction endonucleases. Their grouping is based on the types of sequences recognized, the nature of the cut made in the DNA, and the enzyme structure. Type I and III restriction endonucleases are not useful for gene cloning because they cleave DNA at sites other than the recognition sites and thus cause random cleavage patterns. In contrast, type II endonucleases are widely used for mapping and reconstructing DNA *in vitro* because they recognize specific sites and cleave just at these sites (Table 8.1). In addition, the type II endonuclease and methylase activities are usually separate, single subunit enzymes. Although the two enzymes recognize the same target sequence, they can be purified separately from each

Fear of recombinant DNA molecules

In the wake of the first cloning experiments, there was immediate concern from both scientists and the general public about the possible dangers of recombinant DNA work. Concerns primarily focused on the ethics of "tampering with nature" and the potential for the escape of genetically engineered pathogenic bacteria from a controlled laboratory environment. One fear was that *E. coli* carrying cloned tumor virus DNA could be transferred to humans and trigger a global cancer epidemic. Not everyone shared these fears. James Watson wrote in his chapter in the book *Genetics and Society* (1993):

> I was tempted then to put together a book called the Whole Risk Catalogue. It would contain risks for old people and young people and so on. It would be a very popular book in our semi-paranoid society. Under "D" I would put dynamite, dogs, doctors, dieldrin [an insecticide] and DNA. I must confess to being more frightened of dogs. But everyone has their own things to worry about.

In 1975 a landmark meeting was held at the Asilomar Conference Center near San Francisco. The meeting was attended by over 100 molecular biologists. Recommendations arising from this meeting formed the basis for official guidelines developed by the National Institutes of Health (NIH) regarding containment. As time passed, there were no disasters that occurred as a result of recombinant DNA technology, and it was concluded by most scientists that under these guidelines the technology itself did not pose any risk to human health or the environment. Containment works very well and engineered bacteria and vectors do very poorly under natural conditions.

Currently, activities involving the handling of recombinant DNA molecules and organisms must be conducted in accordance with the *NIH Guidelines for Research Involving Recombinant DNA Molecules*. Four levels of risk are recognized, from minimal to high, for which four levels of containment (physical and biological barriers to the escape of dangerous organisms) are outlined. The highest risk level is for experiments dealing with highly infectious agents and toxins that are likely to cause serious or lethal human disease for which preventive or therapeutic interventions are not usually available. Precautions include negative-pressure air locks in laboratories and experiments done in laminar-flow hoods, with filtered or incinerated exhaust air. The bacteria used routinely in molecular biology, such as nonpathogenic strains of *E. coli*, are "Risk group I" agents, which are not associated with disease in healthy adult humans. Standard vectors for recombinant DNA are genetically designed to decrease, by many orders of magnitude, the probability of dissemination of recombinant DNA outside the laboratory.

Today, fears focus not so much on the technology *per se*, but on the application of recombinant DNA technology to agriculture, medicine, and bioterrorism. For example, there is concern about the safety of genetically engineered foods in the marketplace, the spread of herbicide-resistant genes from transgenic crop plants to weeds, the use of gene therapy for eugenics (artificial human selection), and the construction of recombinant DNA "designer weapons." The latter refers to engineering infectious microbes to be even more virulent, antibiotic-resistant, and environmentally stable. On December 13, 2002, new federal regulations were published to implement the US Public Health and Security and Bioterroism Preparedness and Response Act of 2002 (http://www.fda.gov/oc/bioterrorism/bioact.html). The regulations apply to the possession, use, and transfer of select agents that are considered potential bioterrorist agents, such as *Yersinia pesti* (plague), *Bacillus anthracis* (anthrax), and variola virus (smallpox).

other. Some type II restriction endonucleases do not conform to this narrow definition, making it necessary to define further subdivisions. The discussion here will focus on the "orthodox" type II restriction endonucleases that are commonly used in molecular biology research.

Restriction endonclease nomenclature

Restriction endonucleases are named for the organism in which they were discovered, using a system of letters and numbers. For example, *Hind*III (pronounced "hindee-three") was discovered in *Haemophilus*

Table 8.1 Major classes of restriction endonucleases.

Class	Abundance	Recognition site	Composition	Use in recombinant DNA research
Type I	Less common than type II	Cut both strands at a nonspecific location > 1000 bp away from recognition site	Three-subunit complex: individual recognition, endonuclease, and methylase activities	Not useful
Type II	Most common	Cut both strands at a specific, usually palindromic, recognition site (4–8 bp)	Endonuclease and methylase are separate, single-subunit enzymes	Very useful
Type III	Rare	Cleavage of one strand only, 24–26 bp downstream of the 3′ recognition site	Endonuclease and methylase are separate two-subunit complexes with one subunit in common	Not useful

influenza (strain d). The *Hin* comes from the first letter of the genus name and the first two letters of the species name; d is for the strain type; and III is for the third enzyme of that type. *Sma*I is from *Serratia marcescens* and is pronounced "smah-one," *Eco*RI (pronounced "echo-r-one") was discovered in *Escherichia coli* (strain R), and *Bam*HI is from *Bacillus amyloliquefaciens* (strain H). Over 3000 type II restriction endonucleases have been isolated and characterized to date. Approximately 240 are available commercially for use by molecular biologists.

Recognition sequences for type II restriction endonucleases

Each orthodox type II restriction endonuclease is composed of two identical polypeptide subunits that join together to form a homodimer. These homodimers recognize short symmetric DNA sequences of 4–8 bp. Six base pair cutters are the most commonly used in molecular biology research. Usually, the sequence read in the 5′ → 3′ direction on one strand is the same as the sequence read in the 5′ → 3′ direction on the complementary strand. Sequences that read the same in both directions are called palindromes (from the Greek word *palindromos* for "run back"). Figure 8.3 shows some common restriction endonucleases and their recognition sequences. Some enzymes, such as *Eco*R1, generate a staggered cut, in which the single-stranded complementary tails are called "sticky" or cohesive ends because they can hydrogen bond to the single-stranded complementary tails of other DNA fragments. If DNA molecules from different sources share the same palindromic recognition sites, both will contain complementary sticky ends (single-stranded tails) when digested with the same restriction endonuclease. Other type II enzymes, such as *Sma*I, cut both strands of the DNA at the same position and generate blunt ends with no unpaired nucleotides when they cleave the DNA.

Restriction endonucleases exhibit a much greater degree of sequence specificity in the enzymatic reaction than is exhibited in the binding of regulatory proteins, such as the Lac repressor to DNA (see Section 10.6). For example, a single base pair change in a critical operator sequence usually reduces the affinity of the Lac repressor by 10- to 100-fold, whereas a single base pair change in the recognition site of a restriction endonuclease essentially eliminates all enzymatic activity.

Like other DNA-binding proteins, the first contact of a restriction endonuclease with DNA is nonspecific (Fig. 8.4). Nonspecific binding usually does not involve interactions with the bases but only with the DNA sugar–phosphate backbone. The restriction endonuclease is loosely bound and its catalytic center is kept at a safe distance from the phosphodiester backbone. Nonspecific binding is a prerequisite for efficient target site location. For example, *Bam*HI moves along the DNA in a linear fashion by a process called "sliding." Sliding involves helical movement due to tracking along a groove of the DNA over short distances (< 30–50 bp). This reduces the volume of space through which the protein needs to search to one dimension. However, the "random walk" nature of linear diffusion gives equal probabilities for forward and reverse steps, so if the distances between the nonspecific binding site and the recognition site are large (> 30–50 bp), the protein

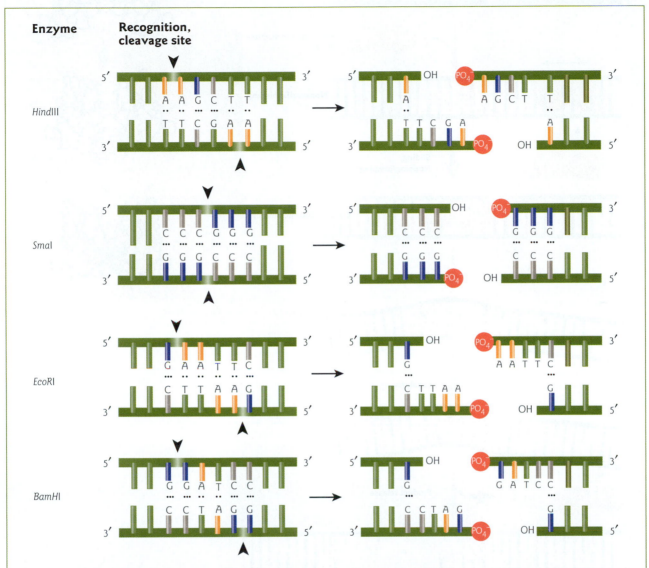

Figure 8.3 Cleavage patterns of some common restriction endonucleases. The recognition and cleavage sites, and cleavage patterns of *Hind*III, *Sma*I, *Eco*RI, and *Bam*HI are shown. Restriction endonucleases catalyze the hydrolysis of phosphodiester bonds in palindromic DNA sequences to produce double-strand breaks, resulting in the formation of $5'-PO_4^-$ and $3'-OH$ termini with "sticky" ends (*Hind*III, *Eco*RI, and *Bam*HI) or "blunt" ends (*Sma*I).

would return repeatedly to its start point. The main mode of translocation over long distances is thus by "hopping" or "jumping." In this process, the protein moves between binding sites through three-dimensional space, by dissociating from its initial site before reassociating elsewhere in the same DNA chain. Because of relatively small diffusion constants of proteins, most rebinding events will be short range "hops" back to or near the initial binding site. In the example of *Bam*HI, once the target restriction site is located, the recognition process triggers large conformational changes of the enzyme and the DNA (called coupling), which leads to the activation of the catalytic center (Fig. 8.4). In addition to indirect interaction with the DNA backbone, specific binding is characterized by direct interaction of the enzyme with the nitrogenous bases.

All structures of orthodox type II restriction endonucleases characterized by X-ray crystallography so far show a common structural core composed of four conserved β-strands and one α-helix (Focus box 8.2). In the presence of the essential cofactor Mg^{2+}, the enzyme cleaves the DNA on both strands at the same time within or in close proximity to the recognition sequence (restriction site). The enzyme cuts the DNA

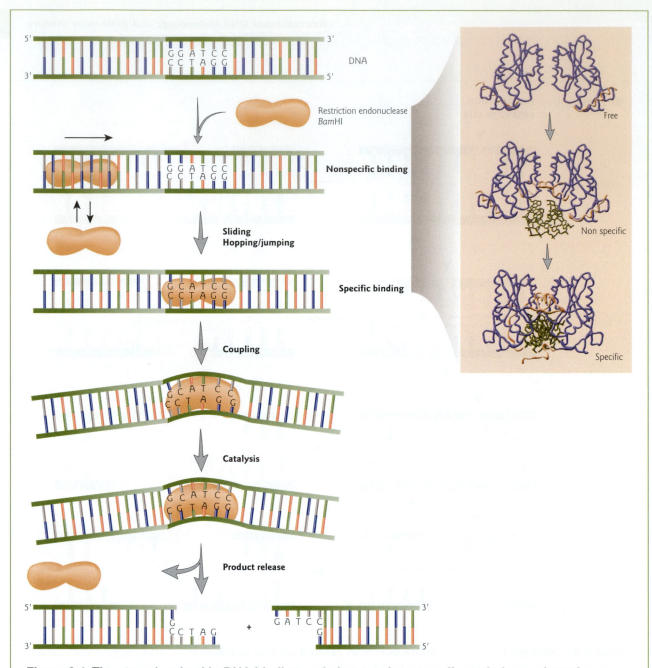

Figure 8.4 The steps involved in DNA binding and cleavage by a type II restriction endonuclease.
Type II restriction endonucleases, like *Bam*HI, bind DNA as dimers. The first contact with DNA is nonspecific. The target site is then located by a combination of linear diffusion or "sliding" of the enzyme along the DNA over short distances, and hopping/jumping over longer distances. Once the target restriction site is located, the recognition process (coupling) triggers large conformational changes of the enzyme and the DNA, which leads to activation of the catalytic center. Catalysis results in product release. (Pingoud, A., Jeltsch, A. 2001. Structure and function of type II restriction endonucleases. *Nucleic Acids Research* 29:3705–3727; and Gowers, D.M., Wilson, G.G., Halford, S.E. 2005. Measurement of the contributions of 1D and 3D pathways to the translocation of a protein along DNA. *Proceedings of the National Academy of Sciences USA* 102:15883–15888.) (Inset) Structures of free, nonspecific, and specific DNA-bound forms of *Bam*HI. The two dimers are shown in brown, the DNA backbone is in green and the bases in gray. *Bam*HI becomes progressively more closed around the DNA as it goes from the nonspecific to specific DNA binding mode. (Protein Data Bank, PDB:1ESG. Adapted from Viadiu, H., Aggarwal, A.K. 2000. Structure of *Bam*HI bound to nonspecific DNA: a model for DNA sliding. *Molecular Cell* 5:889–895. Copyright © 2000, with permission from Elsevier.)

duplex by breaking the covalent, phosphodiester bond between the phosphate of one nucleotide and the sugar of an adjacent nucleotide, to give free 5′-phosphate and 3′-OH ends. Type II restriction endonucleases do not require ATP hydrolysis for their nucleolytic activity. Although there are a number of models for how this nucleophilic attack on the phosphodiester bond occurs (Focus box 8.2), the exact mechanism by which restriction endonucleases achieve DNA cleavage has not yet been proven experimentally for any type II restriction endonuclease.

DNA ligase

The study of DNA replication and repair processes led to the discovery of the DNA-joining enzyme called DNA ligase. DNA ligases catalyze formation of a phosphodiester bond between the 5′-phosphate of a nucleotide on one fragment of DNA and the 3′-hydroxyl of another (see Fig. 6.14). This joining of linear DNA fragments together with covalent bonds is called ligation. Unlike the type II restriction endonucleases, DNA ligase requires ATP as a cofactor.

Because it can join two pieces of DNA, DNA ligase became a key enzyme in genetic engineering. If restriction-digested fragments of DNA are placed together under appropriate conditions, the DNA fragments from two sources can anneal to form recombinant molecules by hydrogen bonding between the complementary base pairs of the sticky ends. However, the two strands are not covalently bonded by phosphodiester bonds. DNA ligase is required to seal the gaps, covalently bonding the two strands and regenerating a circular molecule. The DNA ligase most widely used in the lab is derived from the bacteriophage T4. T4 DNA ligase will also ligate fragments with blunt ends, but the reaction is less efficient and higher concentrations of the enzyme are usually required *in vitro*. To increase the efficiency of the reaction, researchers often use the enyzme terminal deoxynucleotidyl transferase to modify the blunt ends. For example, if a single-stranded poly(dA) tail is added to DNA fragments from one source, and a single-stranded poly(dT) tail is added to DNA from another source, the complementary tails can hydrogen bond (Fig. 8.5). Recombinant DNA molecules can then be created by ligation.

8.4 Molecular cloning

The basic procedure of molecular cloning involves a series of steps. First, the DNA fragments to be cloned are generated by using restriction endonucleases, as described in Section 8.3. Second, the fragments produced by digestion with restriction enzymes are ligated to other DNA molecules that serve as vectors. Vectors can replicate autonomously (independent of host genome replication) in host cells and facilitate the manipulation of the newly created recombinant DNA molecule. Third, the recombinant DNA molecule is transferred to a host cell. Within this cell, the recombinant DNA molecule replicates, producing dozens of identical copies known as clones. As the host cells replicate, the recombinant DNA is passed on to all progeny cells, creating a population of identical cells, all carrying the cloned sequence. Finally, the cloned DNA segments can be recovered from the host cell, purified, and analyzed in various ways.

Vector DNA

Cloning vectors are carrier DNA molecules. Four important features of all cloning vectors are that they: (i) can independently replicate themselves and the foreign DNA segments they carry; (ii) contain a number of unique restriction endonuclease cleavage sites that are present only once in the vector; (iii) carry a selectable marker (usually in the form of antibiotic resistance genes or genes for enzymes missing in the host cell) to distinguish host cells that carry vectors from host cells that do not contain a vector; and (iv) are relatively easy to recover from the host cell. There are many possible choices of vector depending on the purpose of cloning. The greatest variety of cloning vectors has been developed for use in the bacterial host *E. coli*. Thus, the first practical skill generally required by a molecular biologist is the ability to grow pure cultures of bacteria.

*Eco*RI: kinking and cutting DNA

*Eco*RI functions as a homodimer of two identical 31,000 molecular weight subunits and catalyzes the cleavage of a double-stranded sequence d(GAATTC). The interaction of the restriction endonuclease *Eco*RI with DNA illustrates how subtle features of its shape and surface characteristics allow it to interact with complementary surfaces on the DNA.

The crystal structure of *Eco*RI complexed with a 12 bp DNA duplex was determined in 1986. One dimer contains a

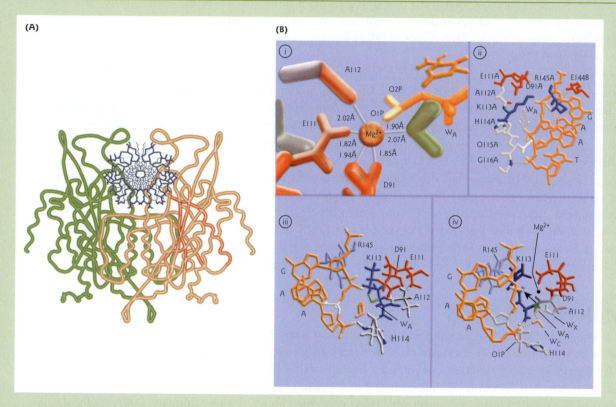

Figure 1 Structure of *Eco*RI. (A) Crystal structure of the two subunits (green and light orange) of *Eco*RI bound to DNA (blue). In one subunit the four strictly conserved β-strands and one α-helix of the common core are shown in red. (Protein Data Bank, PDB:1ERI. Adapted from Pingoud, A. and Jeltsch, A. 2001. Structure and function of type II restriction endonucleases. *Nucleic Acids Research* 29:3705–3727. Copyright © 2001, with permission of the Oxford University Press.). (B) Catalytic centers of the *Eco*RI–DNA complex. (i) Coordination of Mg^{2+} by six ligands in the catalytic center: one carboxylate oxygen of the glutamic acid at position 111 (E111); two carboxylate oxygens of asparagine 91 (D91); the main-chain carbonyl of alanine 112 (A112); the O1P oxygen of the scissile phosphate GpAA (to polarize the phosphate and facilitate nucleophilic attack); and a water molecule, W_A, that forms the attacking nucleophile. (ii) Catalytic and recognition elements of the crystal structure of the Mg^{2+}-free *Eco*RI–DNA complex. The letters following the side chain numbers denote protein subunits A and B. Only one DNA strand (orange) is shown for part of the recognition site. (iii and iv) The *Eco*RI–DNA complexes in the absence (iii) and presence (iv) of Mg^{2+}. The black arrow in (iv) shows the direction of nucleophilic attack on phosphorus. The presence of Mg^{2+} causes a number of structural changes, including alteration of the position and orientation of the water molecules W_A and W_C, movement of D91, and movement of lysine 113 (K113) away from its hydrogen-bonding partner E111. (Reproduced from Kurpiewski, M.R., Engler, L.E., Wozniak, L.A., Kobylanska, A., Koziolkiewicz, M., Stec, W.J., Jen-Jacobsen, L. 2004. Mechanisms of coupling between DNA recognition and specificity and catalysis in *Eco*RI endonuclease. *Structure* 12:1775–1788. Copyright © 2004, with permission from Elsevier.)

*Eco*RI: kinking and cutting DNA

conserved four-stranded β-sheet surrounded on either side by α-helices (Fig. 1). The active site of the endonuclease lies at the C-terminus of this parallel β-sheet and forms a catalytic center, in which Mg^{2+} is bound by interaction with six amino acids (β2 and β3 contain the amino acid residues directly involved in catalysis). Upon specific DNA binding, about 150 water molecules are released; this expulsion of solvent molecules from the interface allows for close contact between the enzyme and the DNA. The N-terminus of the protein forms an arm that partially wraps around the DNA. A bundle of four parallel α-helices, two from each dimer, pushes into the major groove and directly recognizes the DNA base sequence. A major portion of the sequence specificity exhibited by this enzyme appears to be achieved through an array of 12 hydrogen bond donors and acceptors from protein side chains. These donors and acceptors are complementary to the donors and acceptors presented by the exposed edges of the base pairs in the hexanucleotide recognition sequence.

The binding of *Eco*RI to its recognition site induces a dramatic conformational change not only in the enzyme itself, but also in the DNA. A central kink (or bend) of about 20–40° in the DNA brings the critical phosphodiester bond between G and A deeper into the active site. The kink is accompanied by unwinding of the DNA. This unwinding of the top 6 bp relative to the bottom 6 bp results in a widening of the major groove by about 3.5 Å. The widening allows the two α-helices from each subunit of the dimer to fit (end on) into the major groove. Further, the realignment of base pairs produced by the kink creates sites for multiple hydrogen bonds with the protein not present in the undistorted DNA. Thus, the protein-induced distortions of the DNA are an intimate part of the recognition and catalysis process.

Choice of vector is dependent on insert size and application

The classic cloning vectors are plasmids, phages, and cosmids, which are limited to the size insert they can accommodate, taking up to 10, 20, and 45 kb, respectively (Table 8.2). The feature of plasmids and phages and their use as cloning vectors will be discussed in more detail in later sections. A cosmid is a plasmid carrying a

Table 8.2 Principal features and applications of different cloning vector systems.

Vector	Basis	Size limits of insert	Major application
Plasmid	Naturally occuring multicopy plasmids	≤ 10 kb	Subcloning and downstream manipulation, cDNA cloning and expression assays
Phage	Bacteriophage λ	5–20 kb	Genomic DNA cloning, cDNA cloning, and expression libraries
Cosmid	Plasmid containing a bacteriophage λ *cos* site	35–45 kb	Genomic library construction
BAC (bacterial artificial chromosome)	*Escherichia coli* F factor plasmid	75–300 kb	Analysis of large genomes
YAC (yeast artificial chromosome)	*Saccharomyces cerevisiae* centromere, telomere, and autonomously replicating sequence	100–1000 kb (1 Mb)	Analysis of large genomes, YAC transgenic mice
MAC (mammalian artificial chromosome)	Mammalian centromere, telomere, and origin of replication	100 kb to > 1 Mb	Under development for use in animal biotechnology and human gene therapy

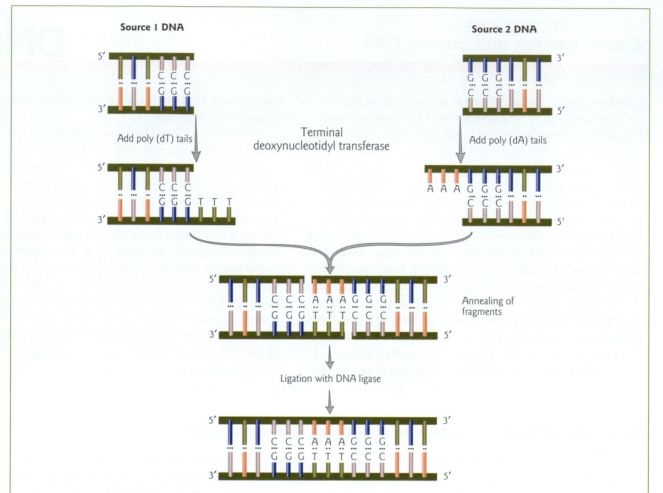

Figure 8.5 Modified blunt end ligation. Recombinant DNA molecules can be formed from DNA cut with restriction endonucleases that leave blunt ends, such as *Sma*I. Without end modification, blunt end ligation is of low efficiency. The efficiency is increased through using the enzyme terminal deoxynucleotidyl transferase to create complementary tails by the addition of poly(dA) and poly(dT) to the cleaved fragments. These tails allow DNA fragments from two different sources to anneal. "Source 1" DNA and "source 2" DNA are then covalently linked by treatment with DNA ligase to create a recombinant DNA molecule. Note that the *Sma*I site is destroyed in the process.

phage λ *cos* site, allowing it to be packaged into a phage head. Cosmids infect a host bacterium as do phages, but replicate like plasmids and the host cells are not lysed. Mammalian genes are often greater than 100 kb in size, so originally there were limitations in cloning complete gene sequences. Vectors engineered more recently have circumvented this problem by mimicking the properties of host cell chromosomes. This new generation of artificial chromosome vectors includes bacterial artificial chromosomes (BACs), yeast artificial chromosomes (YACs), and mammalian artificial chromosomes (MACs).

Plasmid DNA as a vector

Plasmids are naturally occurring extrachromosomal double-stranded circular DNA molecules that carry an origin of replication and replicate autonomously within bacterial cells (see Section 3.4). The plasmid vector pBR322, constructed in 1974, was one of the first genetically engineered plasmids to be used in

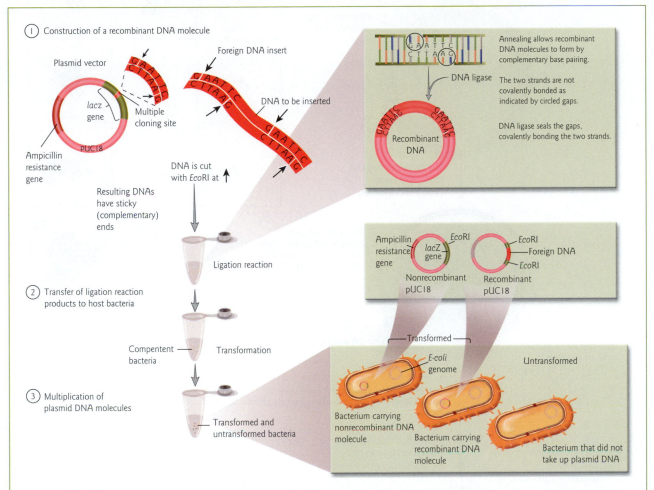

Figure 8.6 Molecular cloning using a plasmid vector. Molecular cloning using a plasmid vector involves five major steps. (1) Construction of a recombinant DNA molecule. In this example, vector DNA (the plasmid pUC18) and the foreign DNA insert are cleaved with *Eco*RI and mixed together in a ligation reaction containing DNA ligase. pUC18 carries the ampicillin resistance gene and has a large number of restriction sites comprising a multiple cloning site within a selectable marker gene. (2) Transfer of ligation reaction products to host bacteria. Competent *E. coli* are transformed with ligation reaction products. Any DNA that remains linear will not be taken up by the host bacteria. (3) Multiplication of plasmid DNA molecules. Within each transformed host bacterium, there is autonomous multiplication of plasmid DNA. Each bacterium may contain as many as 500 copies of pUC18. Some bacteria in the mixture will be untransformed (not carrying either recombinant or nonrecombinant plasmid DNA).

recombinant DNA. Plasmids are named with a system of uppercase letters and numbers, where the lowercase "p" stands for "plasmid." In the case of pBR322, the BR identifies the original constructors of the vector (Bolivar and Rodriquez), and 322 is the identification number of the specific plasmid. These early vectors were often of low copy number, meaning that they replicate to yield only one or two copies in each cell. pUC18, the vector shown in Fig. 8.6, is a derivative of pBR322. This is a "high copy number" plasmid (> 500 copies per bacterial cell).

Plasmid vectors are modified to contain a specific antibiotic resistance gene and a multiple cloning site (also called the polylinker region) which has a number of unique target sites for restriction endonucleases. Cutting the circular plasmid vector with one of these enzymes results in a single cut, creating a linear plasmid. A foreign DNA molecule, referred to as the "insert," cut with the same enzyme, can then be joined to the vector in a ligation reaction (Fig. 8.6). Ligations of the insert to vector are not 100% productive, because

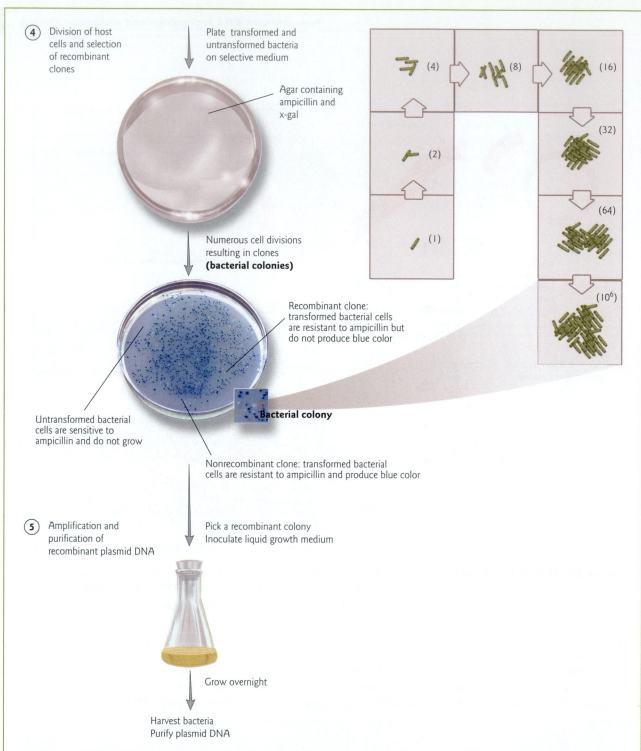

Figure 8.6 (cont'd) (4) Division of host cells and selection of recombinant clones by blue-white screening. Bacteria are plated on a selective agar medium containing the antibiotic ampicillin and X-gal (see Fig. 8.7). If foreign DNA is inserted into the multiple cloning site, then the *lac*Z′ coding region is disrupted and the N-terminal portion of β-galactosidase is not produced. Since there is no functional β-galactosidase in the bacteria, the substrate X-gal remains colorless, and the bacterial colony containing recombinant plasmid DNA appears white, thus allowing the direct identification of colonies carrying cloned DNA inserts. If there is no insertion of foreign DNA in the multiple cloning site, then the *lac*Z′ gene is intact and enzymatically active β-galactosidase is produced. The bacterial colonies containing nonrecombinant plasmid DNA thus appear blue. (Photograph courtesy of Vinny Roggero and the Spring 2006 Molecular Genetics Lab, College of William and Mary.) (5) Amplification and purification of recombinant plasmid DNA. A recombinant colony is used to inoculate liquid growth medium. After growing the bacteria overnight, the culture is harvested, bacterial cells are lysed, and the plasmid DNA is purified away from other cellular components.

the two ends of a plasmid vector can be readily ligated together, which is called self-ligation. The degree of self-ligation can be reduced by treatment of the vector with the enzyme phosphatase, which removes the terminal 5'-phosphate. When the 5'-phosphate is removed from the plasmid it cannot be recircularized by ligase, since there is nothing with which to make a phosphodiester bond. But, if the vector is joined with a foreign insert, the 5'-phosphate is provided by the foreign DNA. Another strategy involves using two different restriction endonuclease cutting sites with noncomplementary sticky ends. This inhibits self-ligation and promotes annealing of the foreign DNA in the desired orientation within the vector.

Transformation: transfer of recombinant plasmid DNA to a bacterial host

The ligation reaction mixture of recombinant and nonrecombinant DNA described in the preceding section is introduced into bacterial cells in a process called transformation (Fig. 8.6). The traditional method is to incubate the cells in a concentrated calcium salt solution to make their membranes leaky. The permeable "competent" cells are then mixed with DNA to allow entry of the DNA into the bacterial cell. Alternatively, a process called electroporation can be used that drives DNA into cells by a strong electric current.

Since bacterial species use a restriction-modification system to degrade foreign DNA lacking the appropriate methylation pattern, including plasmids, the question arises: why don't the transformed bacteria degrade the foreign DNA? The answer is that molecular biologists have cleverly circumvented this defense system by using mutant strains of bacteria, deficient for both restriction and modification, such as the common lab strain *E. coli* DH5α.

Successfully transformed bacteria will carry either recombinant or nonrecombinant plasmid DNA. Multiplication of the plasmid DNA occurs within each transformed bacterium. A single bacterial cell placed on a solid surface (agar plate) containing nutrients can multiply to form a visible colony made of millions of identical cells (Fig. 8.6). As the host cell divides, the plasmid vectors are passed on to progeny, where they continue to replicate. Numerous cell divisions of a single transfomed bacteria result in a clone of cells (visible as a bacterial colony) from a single parental cell. This step is where "cloning" got its name. The cloned DNA can then be isolated from the clone of bacterial cells.

Recombinant selection

What needs to be included in the medium for plating cells so that nontransformed bacterial cells are not able to grow at all? The answer depends on the particular vector, but in the case of pUC18, the vector carries a selectable marker gene for resistance to the antibiotic ampicillin. Ampicillin, a derivative of penicillin, blocks synthesis of the peptidoglycan layer that lies between the inner and outer cell membranes of *E. coli* (Table 8.3). Ampicillin does not affect existing cells with intact cell envelopes but kills dividing cells as they synthesize new peptidoglycan. The ampicillin resistance genes carried by the recombinant plasmids produce an enzyme, β-lactamase, that cleaves a specific bond in the four-membered ring (β-lactam ring) in the ampicillin molecule that is essential to its antibiotic action. If the plasmid vector is introduced into a plasmid-free antibiotic-sensitive bacterial cell, the cell becomes resistant to ampicillin. Nontransformed cells contain no pUC18 DNA, therefore they will not be antibiotic-resistant, and their growth will be inhibited on agar containing ampicillin. Transformed bacterial cells may contain either nonrecombinant pUC18 DNA (self-ligated vector only) or recombinant pUC18 DNA (vector containing foreign DNA insert). Both types of transformed bacterial cells will be ampicillin-resistant.

Blue-white screening

To distinguish nonrecombinant from recombinant transformants, blue-white screening or "*lac* selection" (also called α-complementation) can be used with this particular vector (Figs 8.6, 8.7). Bacterial colonies are grown on selective medium containing ampicillin and a colorless chromogenic compound called X-gal, for short (5-bromo-4-chloro-3-indolyl-β-D-galactoside). pUC18 carries a portion of the *lacZ* gene (called

Table 8.3 Some commonly used antibiotics and antibiotic resistance genes.

Antibiotic	Mode of action	Resistance gene
Ampicillin	Inhibits bacterial cell wall synthesis by disrupting peptidoglycan cross-linking	β–Lactamase (*amp*^{r}) gene product is secreted and hydrolyzes ampicillin
Tetracycline	Inhibits binding of aminoacyl tRNA to the 30S ribosomal subunit	*tet*^{r} gene product is membrane bound and prevents tetracycline accumulation by an efflux mechanism
Kanamycin	Inactivates translation by interfering with ribosome function	Neomycin or aminoglycoside phosphotransferase (*neo*^{r}) gene product inactivates kanamycin by phosphorylation

lacZ′) that encodes the first 146 amino acids for the enzyme β–galactosidase (see Section 10.5). The multiple cloning site resides in the coding region. If the *lacZ*′ region is not interrupted by inserted DNA, the amino-terminal portion of β-galactosidase is synthesized. Importantly, an *E. coli* deletion mutant strain is used (e.g. DH5α) that harbors a mutant sequence of *lacZ* that encodes only the carboxyl end of β-galactosidase (*lacZ*′ ΔM15). Both the plasmid and host *lacZ* fragments encode nonfunctional proteins. However, by α-complementation the two partial proteins can associate and form a functional enzyme. When present, the enzyme β-galactosidase catalyzes hydrolysis of X-gal, converting the colorless substrate into a blue-colored product (see Figs 8.6, 8.7).

Amplification and purification of recombinant plasmid DNA

Further screening of positive (white) colonies can be done by restriction endonuclease digest to confirm the presence and orientation of the insert (see Section 8.9). When a positive colony containing recombinant plasmid DNA is transferred aseptically to liquid growth medium, the cells will continue to multiply exponentially. Within a day or two, a culture containing trillions of identical cells can be harvested.

The final step in molecular cloning is the recovery of the cloned DNA. Plasmid DNA can be purified from crude cell lysates by chromatography (see Tool box 8.1) using silica gel or anion exchange resins that preferentially bind nucleic acids under appropriate conditions and allow for the removal of proteins and polysaccharides. The purified plasmid DNA can then be eluted and recovered by ethanol precipitation in the presence of monovalent cations. Ethanol precipitation of plasmid DNA from aqueous solutions yields a clear pellet that can be easily dissolved in an appropriate buffered solution.

Bacteriophage lambda (λ) as a vector

Bacteriophage lambda (λ) has been widely used in recombinant DNA since engineering of the first viral cloning vector in 1974. Phage λ vectors are particularly useful for preparing genomic libraries, because they can hold a larger piece of DNA than a plasmid vector (see Section 8.5). Today many variations of λ vectors exist. Insertion vectors have unique restriction endonuclease sites that allow the cloning of small DNA fragments in addition to the phage λ genome. These are often used for preparing cDNA expression libraries. Replacement vectors have paired cloning sites on either side of a central gene cluster. This central cluster contains genes for lysogeny and recombination, which are not essential for the lytic life cycle (see Fig. 8.1). The central gene cluster can be removed and foreign DNA inserted between the "arms." All phage vectors used as cloning vectors have been disarmed for safety and can only function in special laboratory conditions. A typical strategy for the use of a phage λ replacement vector is depicted in Fig. 8.8. The recombinant viral particle infects bacterial host cells, in a process called "transduction." The host cells lyse after phage

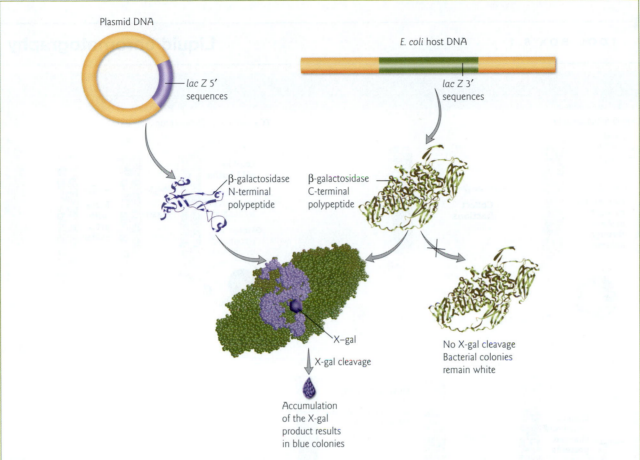

Figure 8.7 β-Galactosidase activity can be used as an indicator of the presence of a foreign DNA insert. Plasmids that express the N-terminal fragment of β-galactosidase (*lacZ* 5′) can be used in *E. coli* strains expressing the C-terminal fragment of the enzyme (*lacZ* 3′ sequences). The N-terminal and C-terminal fragments join and four subunits come together to form a functional tetrameric-enzyme. β-Galactosidase activity can be measured in live cells using a colorless chromogenic substrate called 5-bromo-4-chloro-3-indolyl-β-D-galactoside (X-gal). Cleavage of X-gal produces a blue-colored product that can be visualized as a blue colony on agar plates. If a foreign insert has disrupted the *lacZ* 5′ coding sequence, then only the C-terminal polypeptide will be produced in the bacterial cell. Thus, X-gal is not cleaved and bacterial colonies remain white.

reproduction, releasing progeny virus particles. The viral particles appear as a clear spot of lysed bacteria or "plaque" on an agar plate containing a lawn of bacteria. Each plaque represents progeny of a single recombinant phage and contains millions of recombinant phage particles. Most contemporary vectors carry a *lacZ′* gene allowing blue–white selection.

Artificial chromosome vectors

Bacterial artificial chromosomes (BACs) and yeast artificial chromosomes (YACs) are important tools for mapping and analysis of complex eukaryotic genomes. Much of the work on the Human Genome Project and other genome sequencing projects depends on the use of BACs and YACs, because they can hold greater than 300 kb of foreign DNA. BACs are constructed using the fertility factor plasmid (F factor) of *E. coli* as a starting point. The plasmid is naturally 100 kb in size and occurs at a very low copy number in

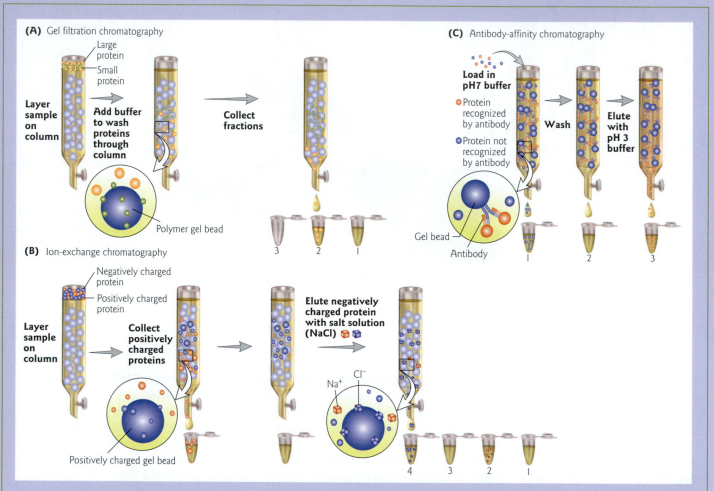

Figure 1 Liquid chromatography techniques. (A) Gel filtration chromatography is used to separate macromolecules that differ in size. For example, a protein mixture is layered on the top of a column packed with porous beads (agarose or polyacrylamide). Larger proteins flow around the beads. Because smaller proteins penetrate into the beads, they travel through the beads more slowly than larger proteins. Different proteins can be collected in separate liquid fractions. (B) Ion-exchange chromatography is used to separate macromolecules (such as proteins or nucleic acids) that differ in net charge. For example, proteins are added to a column packed with beads that are coated by amino (NH^{3+}) or carboxyl (COO^-) groups that carry either a positive charge (shown here) or a negative charge at neutral pH. Acidic proteins with the opposite charge (net negative charge) bind to the positively charged beads, while basic or neutral proteins with the same net charge flow through the column. Bound proteins, in this case negatively charged, are eluted by passing a salt gradient through the column. As the negatively charged salt ions bind to the beads, the protein is released. (C) Affinity chromatography relies on the ability of a protein or nucleic acid to bind specifically to another molecule. Columns are packed with beads to which ligand molecules are covalently attached that bind the protein or nucleic acid of interest. Ligands can be antibodies, enzyme substrates, or other small molecules that bind a specific macromolecule. For example, in antibody–affinity chromatography, the column contains a specific antibody covalently attached to beads. Only proteins with a high affinity for the antibody are retained by the column, regardless of mass or charge, while other proteins flow through. The bound protein can be eluted in an acidic solution, by adding an excess of ligand, or by changing the salt concentration.

Liquid chromatography

An important tool in molecular biology is chromatography. The technique of chromatography was first developed in the early 1900s by a botanist named Mikhail Semenovich Tswett. Tswett passed a leaf extract through a vertical tube packed with some absorbent resin. Through this procedure he was able to separate the main green and orange pigments from the leaves. The chlorophylls, xanthophylls, and carotenes appeared as distinct colored bands in the column. Based on these observations, Tswett name the technique "chromatography" (from the Greek word *khroma* for "color," and *graphein*, "to write").

Today, there are many variants of chromatography, but they all rely on the principles first observed by Tswett, that molecules dissolved in a solution will interact (bind and dissociate) with a solid surface. When the solution is allowed to flow across the surface, molecules that interact weakly with the solid surface will spend less time bound to the surface and will move more rapidly than molecules that interact strongly with the surface. Liquid chromatography is commonly used to separate mixtures of nucleic acids and proteins by passing them through a column packed tightly with spherical beads. The nature of these beads determines whether the separation of the nucleic acids or proteins depends on differences in mass (gel filtration chromatography), charge (ion-exchange chromatography), or binding affinity (affinity chromatography) (Fig. 1).

the host. The engineered BAC vector is 7.4 kb (including a replication origin, cloning sites, and selectable markers) and thus can accommodate a large insert of foreign DNA. The characteristics of YAC vectors are discussed below.

Immediately after the construction of the first YAC in 1983, efforts were undertaken to develop a mammalian artificial chromosome (MAC). From there on, it took 14 years until the first prototype MAC was described in 1997. Like YACs, MACs rely on the presence of centromeric sequences, sequences that can initiate DNA replication, and telomeric sequences. Their development is considered an important advance in animal biotechnology and human gene therapy for two main reasons. First, they involve autonomous replication and segregation in mammalian cells, as opposed to random integration into chromosomes (as for other vectors). Second, they can be modified for their use as expression systems of large genes, including not only the coding region but all control elements. A major drawback limiting application at this time, however, is that they are difficult to handle due to their large size and can be recovered only in small quantities. Two principal procedures exist for the generation of MACs. In one method, telomere-directed fragmentation of natural chromosomes is used. For example, a human artificial chromosome (HAC) has been derived from chromosome 21 using this method. Another method involves *de novo* assembly of cloned centromeric, telomeric, and replication origins *in vitro*.

Yeast artificial chromosome (YAC) vectors

Yeast, although a eukaryote, is a small single cell that can be manipulated and grown in the lab much like bacteria. YAC vectors are designed to act like chromosomes. Their design would not have been possible without a detailed knowledge of the requirements for chromosome stability and replication, and genetic analysis of yeast mutants and biochemical pathways. YAC vectors include an origin of replication (autonomously replicating sequence, ARS) (see Section 6.6), a centromere to ensure segregation into daughter cells, telomeres to seal the ends of the chromosomes and confer stability, and growth selectable markers in each arm (Fig. 8.9). These markers allow for selection of molecules in which the arms are joined and which contain a foreign insert. For example, the yeast genes *URA3* and *TRP1* are often used as markers. Positive selection is carried out by auxotrophic complementation of a *ura3-trp1* mutant yeast strain,

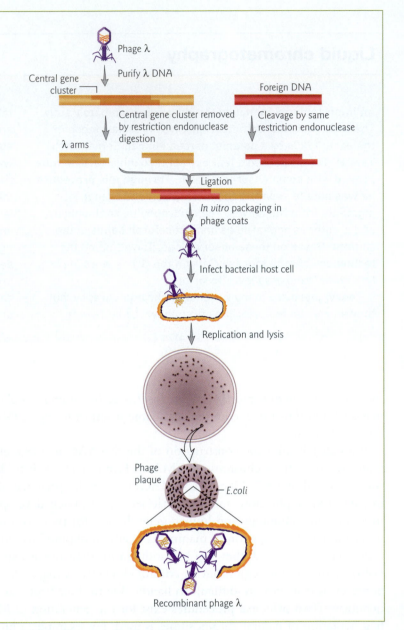

Figure 8.8 Use of bacteriophage lambda (λ) as a cloning vector. DNA is extracted from phage λ and the central gene cluster is removed by restriction endonuclease digestion. The foreign DNA to be cloned is cut with the same enzyme and ligated to the left and right "arms" of the phage λ DNA. The recombinant DNA is then mixed with phage proteins *in vitro*. The DNA is packaged into the phage head and tail fibers are attached via a self-assembly pathway. The recombinant viral particle is then able to infect bacterial cells on an agar plate. The phage replicates its genome, including the foreign DNA insert. Recombinant phage λ DNA directs the cell to make phage particles. The bacteria become filled with new phage particles, break open (lyse), and release millions of recombinant phages. The holes in the lawn of host bacteria, called plaques, are regions where phages have killed the bacteria. Each plaque represents progeny of a single recombinant phage.

which requires supplementation with uracil and tryptophan to grow. *URA3* encodes an enzyme that is required for the biosynthesis of the nitrogenous base uracil (orotidine-5′-phosphate decarboxylase). *TRP1* encodes an enzyme that is required for biosynthesis of the amino acid tryptophan (phosphoribosyl-anthranilate isomerase). YAC vectors are maintained as a circle prior to inserting foreign DNA. After cutting with restriction endonucleases *Bam*HI and *Eco*RI, the left arm and right arm become linear, with the end sequences forming the telomeres. Foreign DNA is cleaved with *Eco*RI and the YAC arms and foreign DNA are ligated and then transferred into yeast host cells. The yeast host cells are maintained as spheroplasts (lacking yeast cell wall). Yeast cells are grown on selective nutrient regeneration plates that lack uracil and tryptophan, to select for molecules in which the arms are joined bringing together the *URA3* and *TRP1* genes.

Red–white selection   In the example shown in Fig. 8.9, recombinant YACs are screened for by a "red–white" selection" process. Within the multiple cloning site of the YAC in this example, there is another marker,

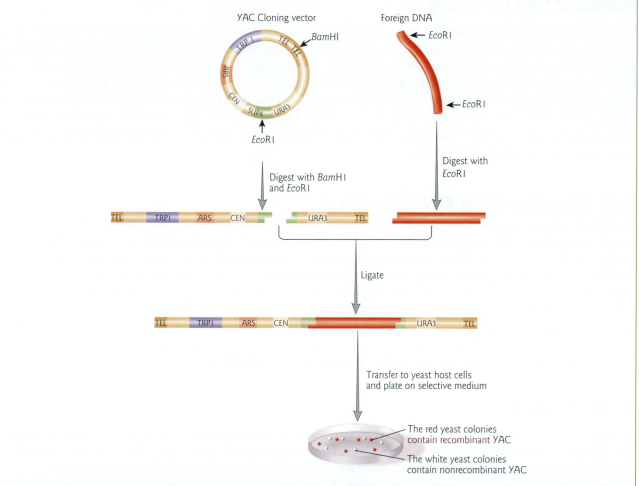

Figure 8.9 Use of yeast artificial chromosome (YAC) cloning vectors. YAC cloning vectors contain functional elements for chromosome maintenance in the yeast *Saccharomyces cerevisiae*. The YAC shown in this example contains an autonomously replicating sequence (ARS) to function as an origin of replication, centromere elements (CEN) for chromosome segregation during cell division, telomeric sequences (TEL) for chromosome stability, and growth selectable markers (URA3 and TRP1) to select positively for chromosome maintenance. Foreign DNA is partially digested with *Eco*RI and the material is then ligated to YAC vector DNA that has been digested with *Bam*HI to liberate telomeric ends and with *Eco*RI to create the insert cloning site. Yeast transformants containing recombinant YAC DNA can be identified by red–white color selection using a yeast strain that is Trp1⁻ and Ura3⁻ and contains the *Ade2-1* mutation, which is suppressed by the *SUP4* gene product. Inactivation of *SUP4* by DNA insertion into the *Eco*RI site results in the formation of a red colony.

SUP4. *SUP4* encodes a tRNA that suppresses the *Ade2-1* UAA mutation. *ADE1* and *ADE2* encode enzymes involved in the synthesis of adenine (phosphoribosylamino-imidazole-succinocarbozamide synthetase and phosphoribosylamino-imidazole carboxylase, respectively). In the absence of these critical enzymes, *Ade2-1* mutant cells produce a red pigment, derived from the polymerization of the intermediate phosphoribosylamino-imidazole. But *Ade2-1* mutant cells expressing *SUP4* are white (the color of wild-type yeast strains), because the *Ade2-1* mutation is suppressed. When foreign DNA is inserted in the multiple cloning site, *SUP4* expression is interrupted. In the absence of *SUP4* expression the red pigment reappears because the *Ade2-1* mutation is no longer suppressed. In contrast, the nonrecombinant YAC vectors retain the active *SUP4* suppressor. Thus, red colonies contain recombinant YAC vector DNA, whereas the white colonies contain nonrecombinant YAC vector DNA.

Sources of DNA for cloning

The cloning that has been described so far will work for any random piece of DNA. But since the goal of many cloning experiments is to obtain a sequence of DNA that directs the production of a specific protein, we need to first consider where to obtain such DNA. Sources of DNA for cloning into vectors may be DNA fragments representing a specific gene or portion of a gene, or may be sequences of the entire genome of an organism, depending on the end goal of the researcher. Typical "inserts" include genomic DNA, cDNA (Tool box 8.2), polymerase chain reaction (PCR) products (Tool box 8.3), and chemically synthesized oligonucleotides. When previously isolated clones are transferred into a different vector for other applications, this is called "subcloning."

8.5 Constructing DNA libraries

Vectors are used to compile a library of DNA fragments that have been isolated from the genomes of a variety of organisms. This collection of fragments can then be used to isolate specific genes and other DNA sequences of interest. DNA fragments are generated by cutting the DNA with a specific restriction endonuclease. These fragments are ligated into vector molecules, and the collection of recombinant molecules is transferred into host cells, one molecule in each cell. The total number of all DNA molecules makes up the library. This library is searched, that is screened, with a molecular probe that specifically identifies the target DNA. Once prepared the library can be perpetuated indefinitely in the host cells and is readily retrieved whenever a new probe is available to seek out a particular fragment. Two main types of libraries can be used to isolate specific DNAs: genomic and cDNA libraries.

Genomic library

A genomic library contains DNA fragments that represent the entire genome of an organism. The first step in creating a genomic library is to break the DNA into manageable size pieces (e.g. 15–20 kb for phage λ vectors), usually by partial restriction endonuclease digest. Under limiting conditions, any particular restriction site is cleaved only occasionally, so not all sites are cleaved in any particular DNA molecule. This generates a continuum of overlapping fragments. The second step is to purify fragments of optimal size by gel electrophoresis or centrifugation techniques. The final step is to insert the DNA fragments into a suitable vector. In humans, the genome size is approximately 3×10^9 bp. With an average insert size of 20 kb, the number of random fragments to ensure with high probability (95–99%) that every sequence is represented is approximately 10^6 clones for humans. The maths actually works out to 1.5×10^5 (i.e. $(3 \times 10^9$ bp$)/(2 \times 10^4$ bp$)$) but more clones are needed in practice, since insertion is random. Bacteriophage λ or cosmid vectors are typically used for genomic libraries. Since a larger insert size can be accommodated by these vectors compared with plasmids, there is a greater chance of cloning a gene sequence with both the coding sequence and the regulatory elements in a single clone.

cDNA library

The principle behind cDNA cloning is that an mRNA population isolated from a specific tissue, cell type, or developmental stage (e.g. embryo mRNA) should contain mRNAs specific for any protein expressed in that cell type or during that stage, along with "housekeeping" mRNAs that encode essential proteins such as the ribosomal proteins, and other mRNAs common to many cell types or stages of development. Thus, if mRNA can be isolated, a small subset of all the genes in a genome can be studied. mRNA cannot be cloned directly, but a cDNA copy of the mRNA can be cloned (see Tool box 8.2). Because a cDNA library is derived from mRNA, the library contains the coding region of expressed genes only, with no introns or regulatory regions. This latter point becomes important for applications of recombinant DNA technology to the production of transgenic animals and for human gene therapy (see Chapters 15 and 17).

8.6 Probes

Searching for a specific cloned DNA sequence in a library is called library screening. One of the key elements required to identify a gene during library screening is the probe. The term probe generally refers to a nucleic acid (usually DNA) that has the same or a similar sequence to that of a specific gene or DNA sequence of interest, such that the denatured probe and target DNA can hybridize when they are renatured together. The probe not only must have the same or a similar sequence to the gene of interest but the researcher must also be able to detect its hybridization. Thus, the probe is labeled; that is, it is chemically modified in some way which allows it, and hence anything it hybridizes to, to be detected. Specific enzymes are used that can add labeled nucleotides in a variety of ways. Typically the probe is made radioactive and added to a solution (Tool boxes 8.4, 8.5). Filters containing immobilized clones are then bathed in the solution. The principle behind this step is that the probe will bind to any clone containing sequences similar to those found on the probe. This binding step is called hybridization. In some cases a library is screened with a protein. For example, when a cDNA library is being screened an antibody can be used to identify the protein that is being expressed by the insert of the clone. In this case, the library is said to be "incubated" with the antibody probe, not hybridized. The use of antibodies in molecular biology research is discussed in more detail in Chapter 9 (Tool box 9.4).

Hybridization can occur between DNA and DNA, DNA and RNA, and RNA and RNA. There are three major types of probe: (i) oligonucleotide probes, which are synthesized chemically and end-labeled; (ii) DNA probes, which are cloned DNAs and may either be end-labeled or internally labeled during *in vitro* replication; and (iii) RNA probes (riboprobes), which are internally labeled during *in vitro* transcription from cloned DNA templates. RNA probes and oligonucleotide probes are generally single-stranded. DNA may be labeled as a double-stranded or single-stranded molecule, but it is only useful as a probe when single-stranded and therefore must be denatured before use. Oligonucleotide, cloned DNA, and RNA probes are of two major types: heterologous and homologous.

Heterologous probes

A heterologous probe is a probe that is similar to, but not exactly the same as, the nucleic acid sequence of interest. If the gene being sought is known to have a similar nucleotide sequence to a second gene that has already been cloned, then it is possible to use this known sequence as a probe. For example, a mouse probe could be used to search a human genomic library.

Homologous probes

A homologous probe is a probe that is exactly complementary to the nucleic acid sequence of interest. Homologous probes can be designed and constructed in a number of different ways. Examples include degenerate probes, expressed sequence tag (EST) based probes, and cDNA probes that are used to locate a genomic clone.

Use of degenerate probes: historical perspective

Before the advent of genome sequence databases, the classic method for designing a probe and screening a library relied on having a partial amino acid sequence of a purified protein. To generate an 18–21 nt oligonucleotide probe, all that was required was to know the sequence of about six to seven amino acids. Long before the advent of DNA sequencing, amino acid sequencing was a routine procedure for biochemists. In fact, in 1953, the same year that Watson and Crick proposed the double helix structure of DNA, Frederick Sanger – also at Cambridge University – worked out the sequence of amino acids in the polypeptide chains of the hormone insulin. This was a most important achievement, since it had long been thought that protein sequencing would be a nearly impossible task. Traditionally, protein sequencing was performed using the Edman degradation method (Fig. 8.10). Today, protein sequencing is more often

Complementary DNA (cDNA) synthesis

Most eukaryotic mRNAs are polyadenylated at the 3′ end to form a poly(A) tail (see Section 13.5). This has an important practical consequence that has been exploited by molecular biologists. The poly(A) region can be used to selectively isolate mRNA from total RNA by affinity chromatography (Fig. 1A). The purified mRNA can then be used as a template for synthesis of a complementary DNA (cDNA) (Fig. 1B,C).

Purification of mRNA

Total RNA is extracted from a specific cell type that expresses a specific set of genes. Of this total cellular RNA, 80–90% is rRNA, tRNA, and histone mRNA, not all of which have a poly(A) tail. These RNAs can be separated from the poly(A) mRNA by passing the total RNA through an affinity column of oligo(dT) or oligo(U) bound to resin beads. Under conditions of relatively high salt the poly(A) RNA is retained

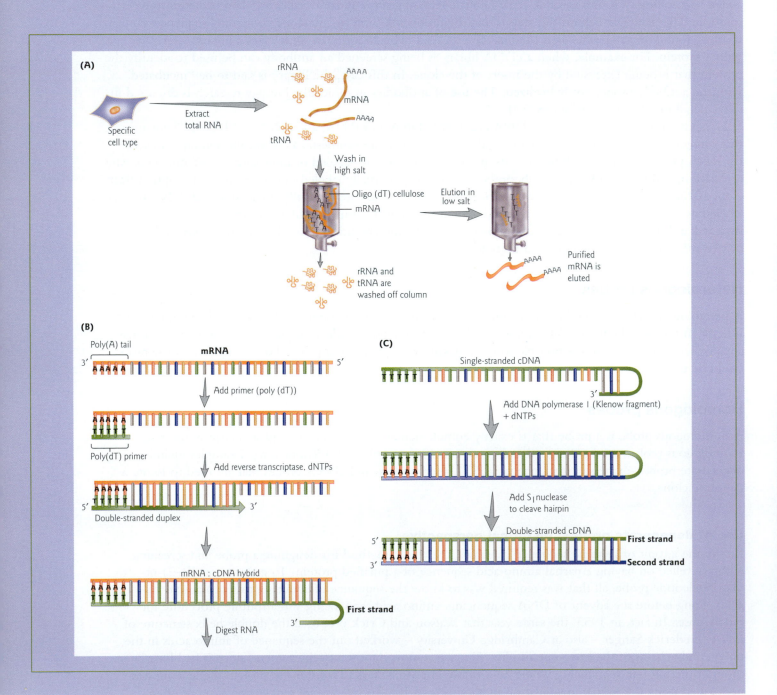

Complementary DNA (cDNA) synthesis

by formation of hydrogen bonds with the complementary bases, and the RNA lacking a poly(A) tail flows through. The salt conditions for hybridization are similar to the ion concentration in cells (e.g. 0.3–0.6 M NaCl). The poly(A) mRNA is then eluted from the column in low salt elution buffer (e.g. 0.01 M NaCl), which promotes denaturation of the hybrid (see Section 2.6).

First strand synthesis of cDNA

A number of strategies can be used to synthesize cDNA from purified mRNA. One strategy is as follows. In brief, cDNA is synthesized by the action of reverse transcriptase and DNA polymerase (Fig. 1B). The reverse transcriptase catalyzes the synthesis of a single-stranded DNA from the mRNA template. Like a regular DNA polymerase, reverse transcriptase also needs a primer to get started. A poly(dT) primer is added to provide a free 3′-OH end that can be used for extension by reverse transcriptase in the presence of deoxynucleoside triphosphates (dNTPs). Usually a viral reverse transcriptase is employed such as one from avian myeloblastosis virus (AMV). The reverse transcriptase adds dNTPs from 5′ to 3′ by complementary base pairing. This is called first strand synthesis. The mRNA is then degraded with a ribonuclease or an alkaline solution.

Second strand synthesis of cDNA

For most applications, including cloning of cDNAs, double-stranded DNA is required. The second DNA strand is generated by the Klenow fragment of DNA polymerase I from *E. coli* (Fig. 1C). The 5′ → 3′ exonuclease activity of

DNA polymerase I from *E. coli* (see Focus box 6.1) makes it unsuitable for many applications. However, this enzymatic activity can be readily removed from the holoenzyme by exposure to a protease. The large or Klenow fragment of DNA polymerase I generated by proteolysis has 5′ → 3′ polymerase and 3′ → 5′ exonuclease (proofreading) activity, and is widely used in molecular biology. Commercially available Klenow fragments are usually produced by expression in bacteria from a truncated form of the DNA polymerase I gene.

There is a tendency for the reverse transcriptase enzyme used in first strand synthesis to loop back on itself and start to make another complementary strand. This hairpin forms a natural primer for DNA polymerase and a second strand of DNA is generated. S1 nuclease (from *Aspergillus oryzae*) is then added to cleave the single-stranded DNA hairpin. Double-strand DNA linkers with ends that are complementary to an appropriate cloning vector are added to the double-strand DNA molecule before ligation into the cloning vector. The end result is a double-stranded cDNA in which the second strand corresponds to the sequence of the mRNA, thus representing the coding strand of the gene. The sequences that appear in the literature are the 5′ → 3′ sequences of the second strand cDNA (Fig. 1). Sequences corresponding to introns and to promoters and all regions upstream of the transcriptional start site are not represented in cDNAs. The library created from all the cDNAs derived from the mRNAs in the specific cell type forms the cDNA library of cDNA clones.

Figure 1 (*opposite*) Traditional cDNA synthesis. (A) Purification of mRNA. Total RNA is extracted from a specific cell type and loaded on an oligo(dT) affinity chromatography column under conditions (high salt buffer) that promote hybridization between the 3′ poly(A) tails of the mRNA and the oligo(dT) covalently coupled to the column matrix. After hybridization, the rRNAs and tRNAs are washed out of the column. The mRNA is eluted with a low salt buffer. The resulting purified mRNA contains many different mRNAs encoding different proteins. (B) First strand synthesis. Synthesis of the first strand of cDNA is carried out using the enzyme reverse transcriptase and a poly(dT) primer in the presence of dNTPs. An mRNA–cDNA hybrid is produced and the mRNA is then digested with an alkaline solution or the enzyme ribonuclease. (C) Second strand synthesis. Synthesis of double-stranded cDNA uses a self-priming method. The Klenow fragment of DNA polymerase I catalyzes synthesis of the second strand, using the natural hairpin of the first strand as a primer. The hairpin is cleaved with a single-strand DNA nuclease (S1 nuclease). The end result is a collection of double-strand cDNAs that correspond to the sequences of the many different mRNAs extracted from the cell.

Polymerase chain reaction (PCR)

The polymerase chain reaction (PCR) is the one of the most powerful techniques that has been developed recently in the area of recombinant DNA research. PCR has had a major impact on many areas of molecular cloning and genetics. With this technique, a target sequence of DNA can be amplified a billion-fold in several hours. Amplification of particular segments of DNA by PCR is distinct from the amplification of DNA during cloning and propagation within a host cell. The procedure is carried out entirely *in vitro*. In addition to its use in many molecular cloning strategies, PCR is also used in the analysis of gene expression (see Section 9.5), forensic analysis where minute samples of DNA are isolated from a crime scene (see Section 16.2), and diagnostic tests for genetic diseases (see Disease box 8.1).

PCR is a DNA polymerase reaction. As with any DNA polymerase reaction it requires a DNA template and a free 3′-OH to get the polymerase started. The template is provided by the DNA sample to be amplified and the free 3′-OH groups are provided by site-specific oligonucleotide primers. The primers are complementary to each of the ends of the sequence that is to be amplified. Note that *in vivo* DNA polymerase would use an RNA primer (see Section 6.4), but a more stable, more easily synthesized DNA primer is used *in vitro*. The three steps of the reaction are denaturation, annealing of primers, and primer extension (Fig. 1):

1 **Denaturation**. In the first step, the target sequence of DNA is heated to denature the template strands and render the DNA single-stranded.
2 **Annealing**. The DNA is then cooled to allow the primers to anneal, that is, to bind the appropriate complementary strand. The temperature for this step varies depending on the size of the primer, the GC content, and its homology to the target DNA. Primers are generally DNA oligonucleotides of approximately 20 bases each.
3 **Primer extension**. In the presence of Mg^{2+}, DNA polymerase extends the primers on both strands from 5′ to 3′ by its polymerase activity. Primer extension is performed at a temperature optimal for the particular

polymerase that is used. Currently, the most popular enzyme for this step is *Taq* polymerase, the DNA polymerase from the thermophilic (heat-loving) bacteria *Thermus aquaticus*. This organism lives in hot springs that can be near boiling and thus requires a thermostable polymerase.

These three steps are repeated from 28 to 35 times. With each cycle, more and more fragments are generated with just the region between the primers amplified. These accumulate exponentially. The contribution of strands with extension beyond the target sequence becomes negligible since these accumulate in a linear manner. After 25 cycles in an automated thermocycler machine, there is a 2^{25} amplification of the target sequence. PCR products can be visualized on a gel stained with nucleic acid-specific fluorescent compounds such as ethidium bromide or SYBR green. The error rate of *Taq* is 2×10^{-4}. If an error occurs early on in the cycles, it could become prominent. Other polymerases, such as *Pfu*, have greater fidelity. *Pfu* DNA polymerase is from *Pyrococcus furiosus*. Base misinsertions that may occur infrequently during polymerization are rapidly excised by the $3′ \rightarrow 5′$ exonuclease (proofreading) activity of this enzyme.

When Kary Mullis first developed the PCR method in 1985, his experiments used *E. coli* DNA polymerase. Because *E. coli* DNA polymerase is heat-sensitive, its activity was destroyed during the denaturation step at 95°C. Therefore, a new aliquot of the enzyme had to be added in each cycle. The purification, and ultimately the cloning, of the DNA polymerase from *T. aquaticus* made the reaction much simpler. In his first experiments, Mullis had to move the reaction manually between the different temperatures. Fortunately, this procedure has been automated by the development of thermal cyclers. These instruments have the capability of rapidly switching between the different temperatures that are required for the PCR reaction. Thus the reactions can be set up and placed in the thermal cycler, and the researcher can return several hours later (or the next morning) to obtain the products.

Polymerase chain reaction (PCR)

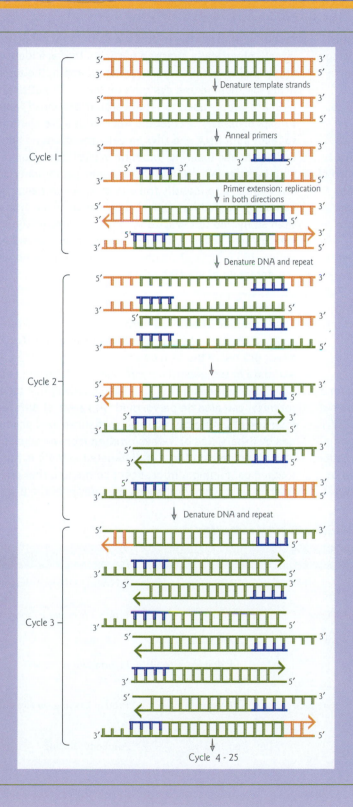

Figure 1 Polymerase chain reaction (PCR). PCR is an *in vitro* DNA replication method. The starting material is a double-stranded DNA. The target sequence to be amplified is indicated in green. Large numbers of two primers (blue) are added, each with a sequence complementary to that found in one strand at the end of the region to be amplified. A thermostable DNA polymerase (e.g. *Taq* polymerase) and dNTPs are also added. In the first cycle, heating to 95°C denatures the double-stranded DNA and subsequent cooling to 55–65°C then allows the primers to anneal to their complementary sequences in the target DNA. *Taq* polymerase extends each primer from 5′ to 3′, generating newly synthesized strands in both directions, which extend to the end of the template strands. The extension is performed at 72°C. In the second cycle, the original and newly made DNA strands are denatured at 95°C and primers are annealed to their complementary sequences at 55–65°C. Each annealed primer again is extended by *Taq* polymerase. In the third cycle, two double-strand DNA molecules are generated exactly equal to the target sequence. These two are doubled in the fourth cycle and are doubled again with each successive cycle.

TOOL BOX 8.4   # Radioactive and nonradioactive labeling methods

The world of cells and macromolecules is invisible to the unaided eye. Development of the tools of molecular biology has allowed researchers to make this world visible. Since World War II, when radioactive materials first became widely available as byproducts of work in nuclear physics, they have become indispensible tools for detecting biological molecules. Hundreds of biological compounds (e.g. nucleotides, amino acids, and numerous metabolic intermediates) are commercially available. The presence of a radioisotope does not change the chemical properties of a radioactively labeled precursor of a macromolecule. Enzymes, both *in vitro* and *in vivo*, catalyze reactions involving labeled substrates just as readily as those involving nonlabeled substrates. Because radioisotopes emit easily detected particles, the fate of radiolabeled molecules can be traced in cells and cellular extracts. For example, labeling nucleic acids is important for tracking their localization, for defining synthetic processes, and for labeling of hybridization probes.

Commonly used radioisotopes in molecular biology

Radioisotopes are unstable isotopes of an element. Isotopes of a given element contain the same number of protons but a different number of neutrons. During radioactive decay there is a change in the number of neutrons and protons from an unstable combination to a more stable combination. The nuclide has less mass after decay (mass converted to energy). For the radioisotopes used in molecular biology

research, this energy is emitted as beta (β) particles (small, electrically charged particles that are identical to electrons) or gamma (γ) rays. For example, the amino acids methionine and cysteine labeled with sulfur-35 (^{35}S) are widely used to label cellular proteins (Table 1). Phosphorus-32 (^{32}P) labeled nucleotides are routinely used to label both RNA and DNA in cell-free systems (*in vitro*). For metabolic labeling (labeling *in vivo*), compounds labeled with hydrogen-3 (^{3}H, tritium) are more commonly used. For example, to identify the site of RNA synthesis, cells can be incubated for a short period with ^{3}H-uridine and then subjected to a fractionation procedure to separate the various organelles or to autoradiography. The radioisotope iodine-125 (^{125}I) is often covalently linked to antibodies for the detection of specific proteins.

Detection techniques

The technique of autoradiography makes use of the fact that radioactive isotopes expose photographic film. The visible silver grains on the film can be counted to provide an estimate of the quantity of radioactive material present. Quantitative measurements of radioactivity in a labeled material can also be performed with several different instruments. A Geiger counter measures ions produced in a gas by β-particles or γ-rays emitted from a radioisotope. In a scintillation counter, a radiolabeled sample is mixed with a liquid containing a fluorescent compound that emits a flash of light when it absorbs the energy of the β-particles

Table 1 Some radioisotopes commonly used in molecular biology research.

Radioisotope	Symbol	Labeled macromolecule	Half-life*	Application
Tritium (hydrogen-3)	^{3}H	Nitrogenous bases (e.g. ^{3}H-uridine, ^{3}H-thymidine) ^{3}H-dNTPs, ^{3}H-NTPs	12.28 years	RNA and DNA labeling *in vivo* Probes for *in situ* hybridization
Carbon-14	^{14}C	^{14}C-chloramphenicol	5730 years	CAT assays
Phosphorus-32	^{32}P	^{32}P-NTPs ^{32}P-dNTPs	14.29 days	Hybridization probes
Sulfur-35	^{35}S	Amino acids (^{35}S-cysteine, ^{35}S-methionine)	87.4 days	Protein labeling (*in vivo* and *in vitro*)
Iodine-125	^{125}I	N/A	60.14 days	Antibody labeling

* The half-life is a means of classifying the rate of decay of radioisotopes according to the time it takes them to lose half their strength (intensity).

Radioactive and nonradioactive labeling methods

or γ-rays released during decay of the radioisotope; a phototube in the instrument detects and counts these light flashes. Phosphorimagers are used to detect radiolabeled compounds on a surface, storing digital data on the number of decays in disintegrations per minute (dpm) per small pixel of surface area. These instruments are commonly used to quantitate radioactive molecules separated by gel electrophoresis and are replacing photographic film for this purpose.

Nonradioactive labeling

As noted above, traditionally, nucleic acids have been labeled with radioisotopes. These radiolabeled probes are very sensitive, but their handling is subject to stringent safety precautions regulated in the US by the federal Nuclear Regulatory Commission, and, in the case of ^{32}P and ^{35}S, the signal decays relatively quickly. More recently, a series of nonradioactive labeling methods have been

developed that generate colorimetric or chemiluminescent signals. A widely used label is digoxygenin, a plant steroid isolated from foxglove, *Digitalis*. This can be conjugated to nucleotides and incorporated into DNA, RNA, or oligonucleotide probes and then detected using an antibody to digoxygenin. The antibody can be attached covalently (conjugated) to fluorescent dyes or enzymes that facilitate signal detection. For example, often anti-digoxygenin antibodies are conjugated to the enzyme alkaline phosphate. When a specific substrate is added, the attached enzyme catalyzes a chemical reaction producing light which exposes an X-ray film. Another system uses biotin, a vitamin, and the bacterial protein streptavidin, which binds to biotin with extremely high affinity. Biotin-conjugated nucleotides are incorporated as a label and detected using enzyme-conjugated streptavidin (see Focus box 10.1 for an application).

performed using mass spectrometry technology, such as matrix-assisted laser desorption/ionization–time of flight (MALDI-TOF) (see Fig. 16.15).

In the example shown in Fig. 8.11, all possible oligonucleotide combinations are synthesized as probes. This is based on the degeneracy of the genetic code (see Section 5.3). Some amino acids are coded for by more than one triplet combination. This can be optimized by choosing a region of the protein that has a high percentage of single or two codon amino acids. The oligonucleotides are made synthetically in the lab and then used to screen a library to identify the gene (or cDNA) encoding the purified protein. One of the oligonucleotides will be exactly the same as the cloned gene. For organisms with sequenced genomes (see Sections 16.4 and 16.5), the protein sequence of their gene products can be simply deduced from the DNA sequence of the respective genes. As a result of this, and the development of EST-based probes, degenerate probes are rarely used today.

Unique EST-based probes

The use of EST-based probes is a newer method than making degenerate probes. Although EST-based probes are no longer frequently used for organisms in which the whole genome sequence is available, they are still useful for organisms where only limited sequence information is available. ESTs are partial cDNA sequences of about 200–400 bp (because they represent just a short portion of the cDNA, they are called "tags"). This method uses cDNA sequence data and identifies a single oligonucleotide rather than a degenerate mixture. A computer program applies the genetic code to translate an EST into a partial amino acid sequence. If a match is found with the protein under study, the EST provides the unique DNA sequence of that portion of cDNA. A probe can then be synthesized and used to screen a library for the entire cDNA (or genomic) clone.

There are a variety of methods for labeling RNA and DNA. The choice of method depends on the application. Probes of the highest specific activity (proportion of incorporated label per mass of probes) are generated using internal labeling, where many labeled nucleotides are incorporated uniformly during DNA or RNA synthesis *in vitro*. End-labeling involves either adding a labeled nucleotide to the 3′-hydroxyl end of a DNA strand or exchanging the unlabeled 5′-phosphate group for a labeled phosphate. When deciding on a labeling method, one of the first questions to ask is whether internal labeling or end-labeling of the nucleic acid is desirable (Table 1). Internal labeling provides maximal labeling and is often used for probes to screen libraries or for Southern blots (see Tool box 8.7). End-labeling is used when precise definition of one end of the DNA (5′ or 3′) is required. An example of this is labeling a DNA fragment for use in DNase I footprinting (see Fig. 9.15), where the researcher wants to know the orientation of the fragments; that is, is a protein-binding site near the 5′ or 3′ end of the DNA sequence? Common methods of uniform labeling involve DNA or RNA synthesis reactions; for example, random primed labeling and synthesis of riboprobes. Some methods for end-labeling DNA fragments involve DNA synthesis reactions (e.g. Klenow fill-in), while oligonucleotides are generally end-labeled by other enzyme-mediated reactions.

Random primed labeling

Random primed labeling is a method of incorporating radioactive nucleotides along the length of a fragment of DNA. The DNA is denatured and random oligonucleotides are annealed to both strands. Each batch of random oligonucleotides contains all possible sequences (for hexamers, which are most commonly employed, this would be 4096 different oligonucleotides), so any DNA template can be used with this method. The odds are that some of these primers will anneal to the DNA of interest. The oligonucleotide primers provide the required free 3′-hydroxyl group for the initiation of DNA synthesis. A Klenow fragment of *E. coli* DNA polymerase I is then used to extend the oligonucleotides, using three unlabeled ("cold") nucleotides and one radioactively labeled ("hot") nucleotide (or a nonradioactively labeled nucleotide) provided in the reaction mixture, to produce a uniformly labeled double-stranded probe. Subsequently, the double-stranded probe is denatured and added to a hybridization solution (Fig. 1A).

In vitro transcription

RNA can be labeled by *in vitro* transcription from a DNA template. The DNA template (often a cDNA) is cloned into a special plasmid vector so that it can be transcribed under the control of a promoter specific for recognition by a bacteriophage RNA polymerase, typically SP6 RNA polymerase, T7 RNA polymerase, or T3 RNA polymerase. The transcription reaction is carried out *in vitro* by the addition of all four NTPs, with one or more labeled, and the appropriate phage RNA polymerase (Fig. 1B). Labeled RNA can be used for tracking the movement and localization of RNA transcripts in cells, analyzing RNA processing pathways, and as hybridization probes. RNA probes (riboprobes) may be either complementary to the sense or antisense strand of DNA depending on the purpose.

Klenow fill-in

A "fill-in" reaction is used to generate blunt ends on fragments created by cleavage with restriction endonucleases that leave 5′ single-stranded overhangs. The Klenow fragment of *E. coli* DNA polymerase I is used to fill in the gaps from 5′ to 3′, in the presence of dNTPs, including one labeled dNTP (Fig. 1C). The result is a double-stranded DNA with the 3′ ends labeled.

Table 1 Labeling of nucleic acids.

Labeling method	Type of labeling	Enzyme	Example of application
Random priming	Uniform	Klenow DNA polymerase	Hybridization probes
***In vitro* transcription**	Uniform	Phage SP6, T7, or T3 RNA polymerase	Hybridization probes, tracking RNA localization
Klenow fill-in	3′ end-labeling	Klenow DNA polymerase	DNase I footprinting
Oligonucleotides	5′ end-labeling 3′ end-labeling	T4 polynucleotide kinase Terminal transferase	Hybridization probes, EMSA

Oligonucleotide labeling

Methods for end-labeling oligonucleotides do not involve DNA synthesis reactions. Instead, they make use of other enzymes. In 5′ end-labeling, a γ-phosphate from ATP is added to the 5′ end of an oligonucleotide by the enzyme T4 polynucleotide kinase. In 3′ end-labeling, a labeled dNTP is added to the 3′ end by the enzyme terminal transferase.

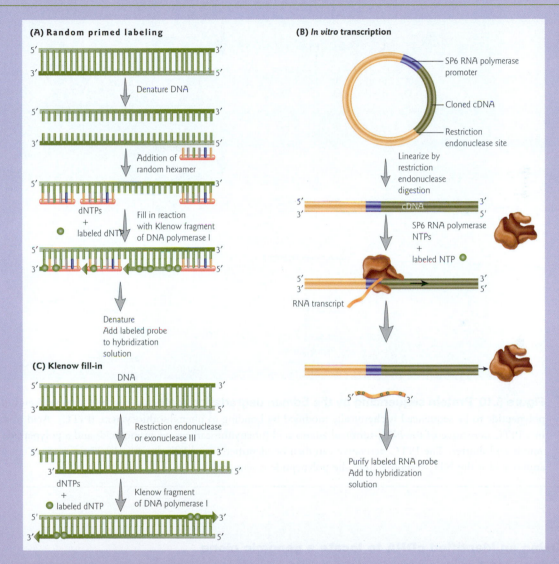

Figure 1 Some methods for labeling nucleic acids. Two methods of uniform labeling and one method of end-labeling are depicted that involve nucleic acid synthesis reactions. (A) In random primed labeling, the template DNA is denatured and random hexamer primers are annealed to both strands. Only one strand is shown for simplicity. The primers provide the 3′-OH for the initiation of DNA synthesis upon addition of the Klenow fragment of DNA polymerase I, and unlabeled and labeled dNTPs. The resulting labeled double-stranded DNA probe is denatured and added to a hybridization solution. (B) *In vitro* transcription generates a labeled RNA from a DNA template. The DNA template (cDNA) is cloned in a plasmid vector containing a promoter for bacteriophage SP6 RNA polymerase. The transcription reaction is carried out by the addition of all labeled and unlabeled NTPs and SP6 RNA polymerase. The labeled RNA probe can then be purified and added to a hybridization solution. (C) Klenow fill-in is used to create labeled blunt ends on fragments created by cleavage with a restriction endonuclease that leaves a 5′ overhang. The reaction is conceptually identical to the one described in (A), but with a long primer and a very short segment of single-stranded template.

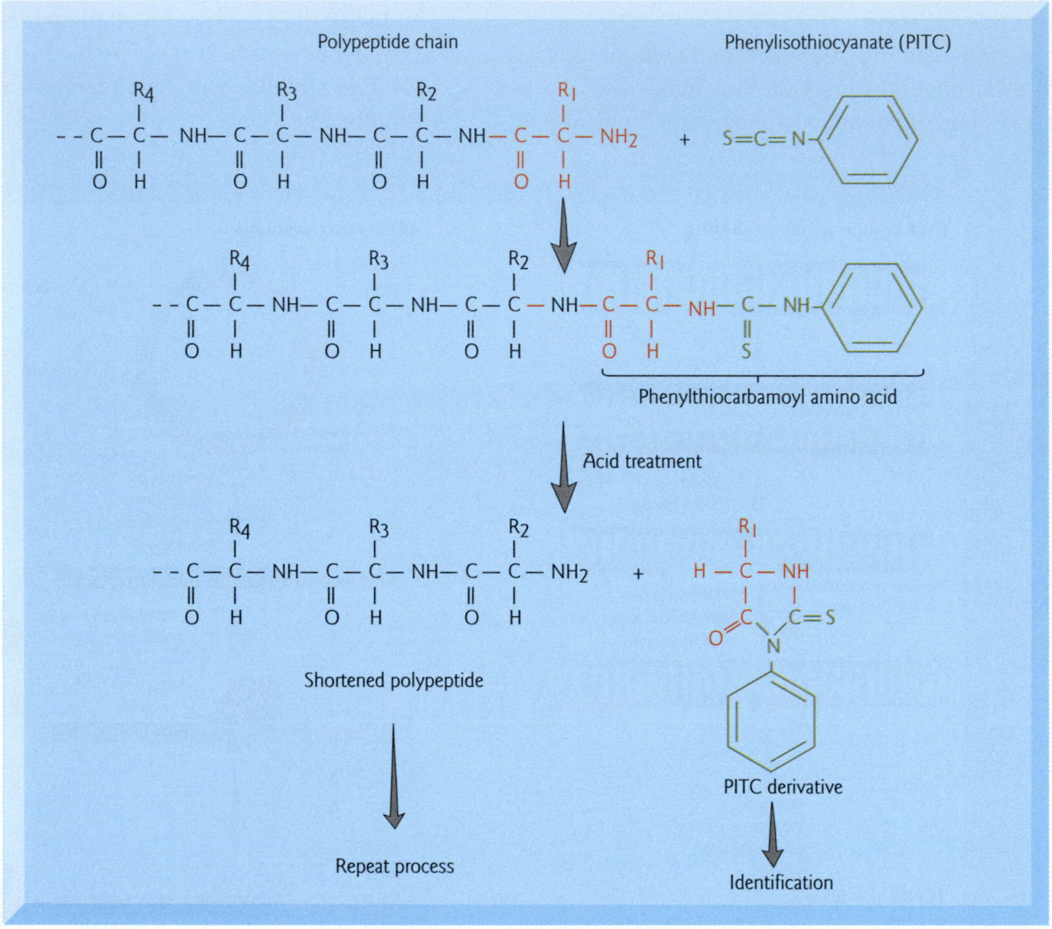

Figure 8.10 Protein sequencing by the Edman degradation method. The NH_2-terminal end of the polypeptide to be sequenced is chemically modified by bonding to phenylisothiocyanate (PITC). Acid treatment results in a PITC derivative of the NH_2-terminal amino acid (phenylthiocarbamoyl amino acid) and a polypeptide that is one amino acid shorter. The PITC derivative can then be identified by analytical methods. The process is repeated for each amino acid at the NH_2-terminus until the polypeptide is sequenced.

Using an identified cDNA to locate a genomic clone

Since a cDNA contains only the coding region of a gene, researchers often need to isolate a genomic clone for analysis of regulatory regions, introns, etc. Use of an identified cDNA to locate a genomic clone provides a highly specific probe for the gene of interest.

8.7 Library screening

Nowadays, because of the wealth of genomic sequence data available for many organisms, a DNA sequence of interest is more likely to be isolated by polymerase chain reaction (PCR) (see Tool box 8.3) than by a library screen. DNA cloning and PCR both amplify tiny samples of DNA into large quantities by repeated rounds of DNA duplication, either carried out by cycles of cell division in a host or cycles of DNA synthesis *in vitro*. In PCR, the pair of oligonucleotide primers limits the amplification process to the particular DNA sequence of interest from the beginning. In contrast, once prepared, a DNA library can be

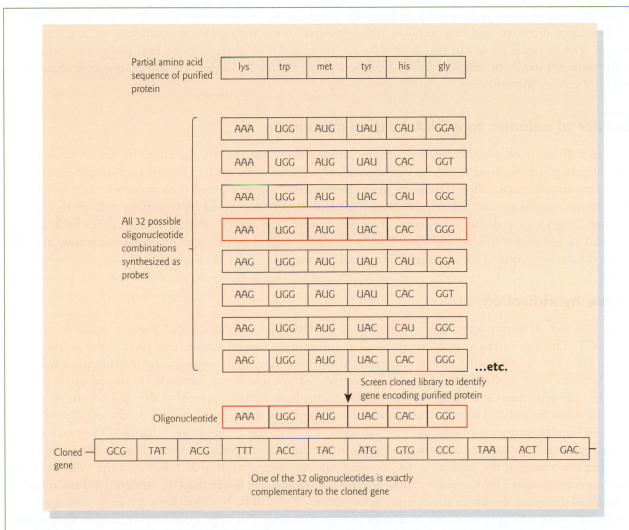

Figure 8.11 Generating degenerate oligonucleotide probes. The coding DNA sequence can be deduced from the partial amino acid sequence of a protein. In this example, two amino acids, tryptophan (Trp) and methionine (Met) have only one codon each. Three others, lysine (Lys), histidine (His), and tyrosine (Tyr) are encoded by two codons, while glycine (Gly) is encoded by codons that include all base combinations at the third position. For this amino acid sequence, 32 oligonucleotide sequences encompass all the possible combinations of codons. For simplicity, only eight of these are depicted in the diagram. The exact coding sequence of the gene of interest must be one of the base combinations. Using a mixture of radioactively labeled oligonucleotides as probes, a cloned library can be screened to isolate the gene for the complete protein.

perpetuated indefinitely in host cells, and can be readily retrieved whenever the researcher wants to seek out some particular fragment. Assuming a probe is available for the cloned sequence of interest, the library can be screened using the principles of hybridization. Complementary base pairing of single-stranded nucleic acids underlies some of the most important biological processes: replication, recombination, DNA repair, transcription, and translation. Library screening exploits this fundamental ability of double-stranded nucleic acids to undergo denaturation or melting (separation into single strands) and for complementary single strands to spontaneously anneal to a labeled nucleic acid probe to form a hybrid duplex (heteroduplex). The power of the technique is that the labeled nucleic acid probe can detect a complementary molecule in a complex mixture with exquisite sensitivity and specificity.

Like many procedures in molecular biology, the process of library screening sounds simple, but in practice it can be labor intensive, and may result in "false positives" if the hybridization conditions are not stringent enough. The example shown in Fig. 8.12 is for screening a cDNA library cloned into plasmid vectors. A similar protocol would be employed for screening a phage library, but would involve plaque hybridization instead of colony hybridization.

Transfer of colonies to a DNA-binding membrane

Bacterial colonies with recombinant vectors containing inserts representing the entire library are grown on nutrient agar plates, forming hundreds or thousands of colonies (Fig. 8.12). The colonies (members of the library) are transferred to nitrocellulose or nylon membranes by gently pressing to make a replica. Bacteria attach to the membrane and the cells are then lysed and the DNA purified by treatment with alkali and proteases. The DNA is denatured to make it single stranded, and fixed to the membrane either by heat treatment or ultraviolet irradiation. The DNA is covalently bound by its sugar–phosphate backbone and the unpaired bases are exposed for complementary base pairing.

Colony hybridization

In the next step of library screening, a radioactively labeled, single-stranded DNA probe is applied (Fig. 8.12). The hybridization step is performed at a nonstringent temperature that ensures the probe will bind to any clone containing a similar sequence. At the same time, some nonspecific hybridization will occur because some of the clones will contain limited, but not significant, similarity to the probe. A series of washes are performed at a stringent temperature that is high enough to remove the probe from all clones to which it has bound in a nonspecific manner. Heteroduplex stability is influenced by the number of hydrogen bonds between the bases and base stacking by hydrophobic interactions that hold the two single strands together. The number of hydrogen bonds is determined by various properties of the heteroduplex, including the length of the duplex, its GC content, and the degree of mismatch between the probe and the complementary target DNA sequence. The shorter the duplex, the lower the GC content, and the more mismatches there are, the lower the melting temperature (T_m) will be because there are fewer hydrogen bonds and base-stacking interactions to disrupt (see Section 2.6). The appropriate hybridization temperature is calculated according to the GC content and the percent homology of probe to target according to the following equation:

$$T_m = 49.82 + 0.41(\% \text{ G} + \text{C}) - (600/l)$$

where l is the length of the hybrid in base pairs.

It is important that the temperature is not so high that it removes the probe from clones that contain sequences that are similar (only a few mismatches) or identical to the probe itself (exact complementarity). Therefore, consideration about the source of the probe (homologous or heterologous) determines the temperature at which the high stringency washes are performed.

Detection of positive colonies

In the final phase of library screening, an X-ray film is applied to the membrane and is exposed by any remaining specifically bound radioactive probe. The resulting autoradiogram has a dark spot on the developed film where DNA–DNA hybrids have formed, by virtue of sequence complementarity to the radiolabeled probe. If the gene is large, it may be fragmented. Various fragments in different clones may need to be identified by finding overlapping fragments and reconstructing the order. The original plate is used to pick bacterial cells with recombinant plasmids that hybridized to the probe. Cells are transferred to medium for growth and further analysis.

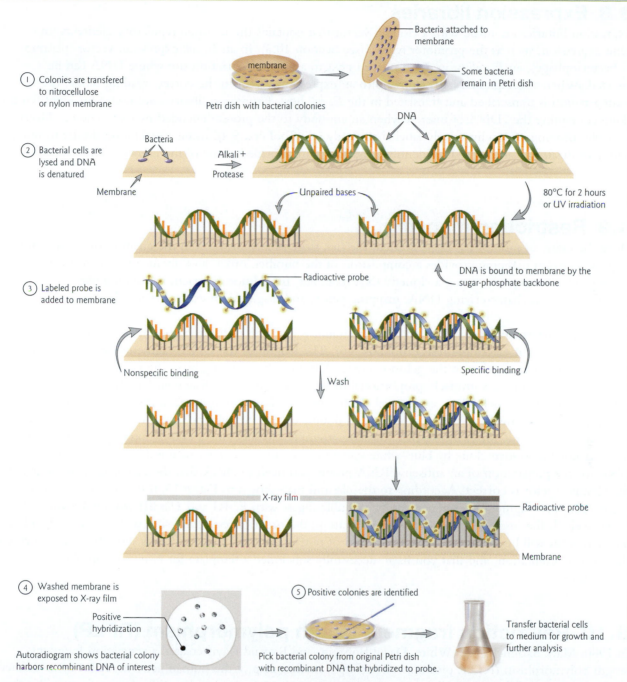

① Colonies are transfered to nitrocellulose or nylon membrane

membrane

Petri dish with bacterial colonies

Bacteria attached to membrane

Some bacteria remain in Petri dish

② Bacterial cells are lysed and DNA is denatured

Bacteria

Membrane

Alkali + Protease

DNA

Unpaired bases

80°C for 2 hours or UV irradiation

DNA is bound to membrane by the sugar-phosphate backbone

③ Labeled probe is added to membrane

Radioactive probe

Nonspecific binding

Specific binding

Wash

X-ray film

Radioactive probe

Membrane

④ Washed membrane is exposed to X-ray film

Positive hybridization

Autoradiogram shows bacterial colony harbors recombinant DNA of interest

⑤ Positive colonies are identified

Pick bacterial colony from original Petri dish with recombinant DNA that hybridized to probe.

Transfer bacterial cells to medium for growth and further analysis

Figure 8.12 Screening a library by nucleic acid hybridization. The example depicts a method for screening a cDNA library. Colony hybridization is used to identify bacterial cells that harbor a specific recombinant plasmid. (1) A sample of each bacterial colony in the library is transferred to a membrane. (2) The bacterial cells on the membrane are lysed and the DNA is denatured. Using heat or UV treatment, the DNA is covalently bound to the membrane. (3) The membrane is placed in a hybridization bag along with a labeled single-stranded DNA probe. After hybridization, the membrane is removed from the bag and washed to remove excess probe and nonspecifically bound probe. (4) Hybrids are detected by placing a piece of X-ray film over the membrane and exposing for a short time. The film is developed and the hybridization events are visualized as dark spots on the autoradiogram. (5) From the orientation of the film, a positive colony containing the insert that hybridized to the probe can be identified. Bacterial cells are picked from this colony for growth and further analysis.

8.8 Expression libraries

Expression libraries are made with a cloning vector that contains the required regulatory elements for gene expression, such as the promoter region (see Section 10.3). In an *E. coli* expression vector (plasmid or bacteriophage), an *E. coli* promoter is placed next to a unique restriction site where DNA can be inserted. When a foreign cDNA is cloned into an expression vector in the correct reading frame, the coding region is transcribed and translated in the *E. coli* host. Expression libraries are useful for identifying a clone containing the cDNA of interest when an antibody to the protein encoded by that gene or cDNA is available. Binding of a radioactively labeled antibody (see Tool box 9.4), using a technique similar to nucleic acid hybridization, can be used to identify a specific protein made by one of the clones of the expression library.

8.9 Restriction mapping

Once the clone of interest has been isolated, the first stage of analysis is often the creation of a restriction map. Restriction mapping provides a compilation of the number, order, and distance between restriction endonuclease cutting sites along a cloned DNA fragment. In addition, restriction mapping plays an important role in characterizing DNA, mapping genes, and diagnostic tests for genetic diseases (see Section 8.10).

To make a restriction map, a cloned DNA fragment is cut with restriction endonucleases and loaded on to an agarose gel for electrophoresis (Tool box 8.6). The lengths of the DNA fragments can be determined by comparing their position in the gel to reference DNAs of known lengths in the gel. A DNA fragment migrates a distance that is inversely proportional to the logarithm of the fragment length in base pairs over a limited range in the gel. Thus, agarose gel electrophoresis allows the restriction fragment lengths to be determined. The pattern of cutting in single and double digests indicates what the relationship is between the two sites. Assume, for example, that you have attempted to subclone a cDNA into a plasmid vector. You have obtained a positive clone by blue-white screening. Because you want to use the recombinant plasmid DNA for the preparation of an antisense RNA probe, you need to check that the orientation of the insert in the plasmid vector is correct. According to the plasmid map shown in Fig. 8.13, if the insert is in the desired orientation, the fragment sizes generated by a double digest with *Eco*RI and *Hin*dIII will be 4.5 and 1.3 kb, respectively. If the insert is in the opposite orientation, the order of restriction sites will be reversed and the fragment sizes will be 3.5 and 2.3 kb. The results of restriction endonuclease digests show that the insert is in the correct orientation, and that you have successfully subcloned a template for *in vitro* riboprobe preparation.

8.10 Restriction fragment length polymorphism (RFLP)

In 1980, Mark Skolnick, Ray White, David Botstein, and Ronald Davis created a restriction fragment length polymorphism (RFLP, pronounced "rif-lip") marker map of the human genome. A RFLP is defined by the existence of alternative alleles associated with restriction fragments that differ in size from each other. RFLPs are visualized by digesting DNA from different individuals with restriction endonucleases, followed by gel electrophoresis to separate fragments according to size, then Southern blotting (Tool box 8.7), and hybridization to a labeled probe that identifies the locus under investigation. A RFLP is demonstrated whenever the Southern blot pattern obtained with one individual is different from the one obtained with another individual. These variable regions do not necessarily occur in genes, and the function of most of those in the human genome is unknown. An exception is a RFLP that can be used to diagnose sickle cell anemia (Fig. 8.14). In individuals with sickle cell anemia, a point mutation in the β–globin gene has destroyed the recognition site for the restriction endonuclease *Mst*II. This mutation can be distinguished by the presence of a larger restriction fragment on a Southern blot in an affected individual, compared with a shorter fragment in a normal individual.

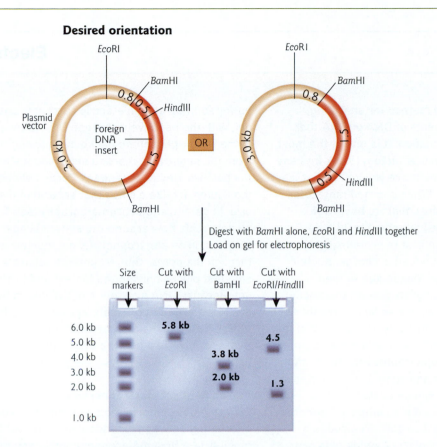

Figure 8.13 Analysis of recombinant DNA by restriction endonuclease digestion. Assume a 2.0 kb foreign DNA insert has been successfully ligated into the *Bam*HI site of a 3.8 kb plasmid vector. However, the orientation of the insert is unknown. Samples of the recombinant plasmid are digested with restriction endonucleases: one sample is digested with *Eco*RI, one with *Bam*HI, and one with both *Eco*RI and *Hin*dIII. The resulting fragments are separated by agarose gel electrophoresis. The sizes of the separated fragments can be measured by comparison with molecular weight standards in an adjacent lane. *Eco*RI linearizes the 5.8 kb plasmid which appears as a single band on the gel. *Bam*HI generates two fragments of 3.8 and 2.0 kb in size, representing the vector and the insert, respectively. Digestion with *Eco*RI and *Hin*dIII generates two fragments of 4.5 and 1.3 kb. These data indicate that the foreign DNA has been inserted in the desired orientation. If the DNA had been inserted in the opposite orientation, a double digest with *Eco*RI and *Hin*dIII would have generated fragment sizes of 3.5 and 2.3 kb.

RFLPs can serve as markers of genetic diseases

By carefully examining the DNA of members of families that carry genetic diseases, it has been possible to find forms of particular RFLPs that tend to be inherited with particular diseases. The simplest RFLPs are those caused by single base pair substitutions. However, RFLPs can also be generated by the insertion of genetic material such as transposable elements, or by tandem duplications, deletions, translocations, or other chromosomal rearrangements. In linkage analysis, families in which individuals are at risk for a genetic disease are identified (i.e. both parents are heterozygous for an autosomal recessive mutation associated with a particular disease). DNA samples from various family members are then analyzed to determine the frequency with which specific RFLP markers segregate with the mutant allele causing the disease. This frequency is a measure of the distance between the markers and the mutation-defined locus.

Generally, the fragment size differences occur not because a restriction site was created or disrupted by the diseased state itself, but rather because the nucleotide sequence differences just happen to be near the gene

Electrophoresis is the standard method for analyzing, identifying, and purifying fragments of DNA or RNA that differ in size, charge, or conformation. It is one of the most widely used techniques in molecular biology. Walk into any molecular biology lab, and the odds are you will see at least one gel apparatus in operation. When charged molecules are placed in an electric field, they migrate toward the positive (anode, red) or negative (cathode, black) pole according to their charge. In contrast to proteins, which can have either a net positive or net negative charge, nucleic acids have a consistent negative change due to their phosphate backbone, and they migrate toward the anode. Proteins and nucleic acids are separated by electrophoresis within a matrix or "gel." Most commonly, the gel is cast in the shape of a thin slab, with wells for loading the sample. The gel is immersed within an electrophoresis buffer that provides ions to carry a current and some type of buffer to maintain the pH at a relatively constant value.

The gels used for eletrophoresis are composed either of agarose or polyacrylamide. Agarose gels are used in a horizontal gel apparatus, while polyacrylamide gels are used in a vertical gel apparatus. These two differ in resolving power. Agarose gels are used for the analysis and preparation of fragments between 100 and 50,000 bp in size with moderate resolution, and polyacrylamide gels are used for the analysis and preparation of small molecules with single nucleotide resolution. This high resolution is required for applications such as for DNA sequencing (see Section 8.11), DNase I footprinting, and electrophoretic mobility shift assays (see Fig. 9.15). In contrast to agarose, polyacrylamide gels are also widely used for the electrophoresis of proteins (see Tool box 9.3).

Agarose gel electrophoresis

Agarose is a polysaccharide extracted from seaweed. Agarose gels are easily prepared by mixing agarose powder with buffer solution, boiling in a microwave to melt, and pouring the gel into a mold where the agarose (generally 0.5–2.0%) solidifies into a slab (Fig. 1). Agarose gels have a large range of separation, but relatively low resolving power. By varying the concentration of agarose, fragments of DNA from about 100 to 50,000 bp can be separated using standard electrophoretic techniques. A toothed comb forms wells in the agarose. The agarose slab is submerged in a

buffer solution and an electric current is passed through the gel, with the negatively charged DNA (due to the phosphate in the sugar–phosphate backbone) moving through the gel from the negatively charged electrode (cathode) towards the positive electrode (anode). Pores between the agarose molecules act like a sieve that separates the molecules by size. In gel filtration chromatography (see Tool box 8.1), nucleic acids flow around the spherical agarose beads. In contrast, in an electrophoretic gel, nucleic acids migrate through the pores; thus fragments separate by size with the smallest pieces moving the fastest and farthest through the gel. Because DNA by itself is not visible in the gel, the DNA is stained with a fluorescent dye such as ethidium bromide (EtBr). EtBr intercalates between the bases causing DNA to fluoresce orange when the dye is illuminated by ultraviolet light.

Pulsed field gel electrophoresis (PFGE)

DNA changes conformation as it moves through gels, alternating between extended and compact forms. The mobility of a DNA molecule depends upon the relationship between the pore size of the gel and the globular size of the DNA in its compact form, with larger molecules moving more slowly. Once DNA reaches a critical size, the molecule is too large to fit through any of the pores in an agarose gel and can move only as an extended molecule. The mobility of the DNA thus becomes independent of size, resulting in co-migration of all large molecules. To fractionate large DNA molecules such as YACs (see Section 8.4), agarose gel electrophoresis is carried out with a pulsed electric field. The periodic field causes the DNA molecule to reorient; longer molecules take longer to realign than shorter ones, thus delaying their progress through the gel and allowing them to be resolved. DNA molecules up to 200–400 Mb in size have been separated by various pulsed field-based methods.

Polyacrylamide gel electrophoresis (PAGE)

Polyacrylamide is a cross-linked polymer of acrylamide. The length of the polymer chains is dictated by the concentration of acrylamide used, which is typically between 3.5 and 20%. Polyacrylamide gels are significantly more cumbersome to prepare than agarose gels. Because oxygen inhibits the polymerization process, they must be

Electrophoresis

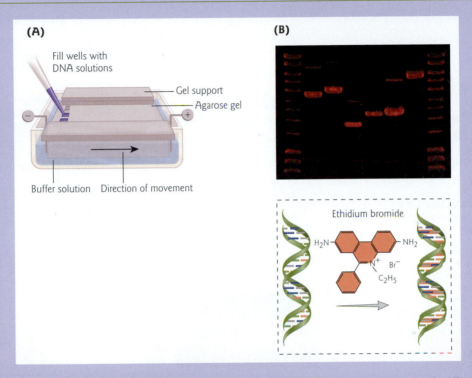

Figure 1 Agarose gel electrophoresis is used to separate DNA (and RNA) molecules according to size. (A) A pipet is used to load DNA samples on an agarose gel in a horizontal gel apparatus. The negatively charged nucleic acids move toward the positive electrode. Larger molecules move more slowly than smaller molecules, so the DNA (or RNA) molecules are separated according to size. (B) Photograph of an agarose gel stained with ethidium bromide (EtBr) to make the DNA bands visible. EtBr molecules intercalate between the bases (see inset) causing the DNA to fluoresce orange when the gel is illuminated with UV light. (Photograph courtesy of Vinny Roggero, College of William and Mary.)

poured between glass plates. Gels can be either nondenaturing or denaturing (e.g. contain 8 M urea). Denaturing gels are used, for example, when single-stranded DNA is being analyzed, as in DNA sequencing (see Fig. 8.15). Polyacrylamide gels have a rather small range of separation, but very high resolving power. In the case of

DNA, polyacrylamide is used for separating fragments of less than about 500 bp. However, under appropriate conditions, fragments of DNA differing in length by a single base pair are easily resolved. Bands in polyacrylamide gels are usually detected by autoradiography, although silver staining can also be used.

involved. A particular form of a polymorphism that is close to a diseased gene tends to stay with that gene during crossing-over (recombination) in meiosis. Relatively large segments of chromosomes are involved in crossing-over, so markers close together on a given chromosome are more likely to be transmitted together (not separated during recombination) than those that are far apart. Linkage thus refers to the likelihood of having one marker transmitted with another through meiosis. Markers that are transmitted together frequently are said to be closely linked. Thus RFLPs can serve as markers of disease, even when the RFLP is

Southern blot

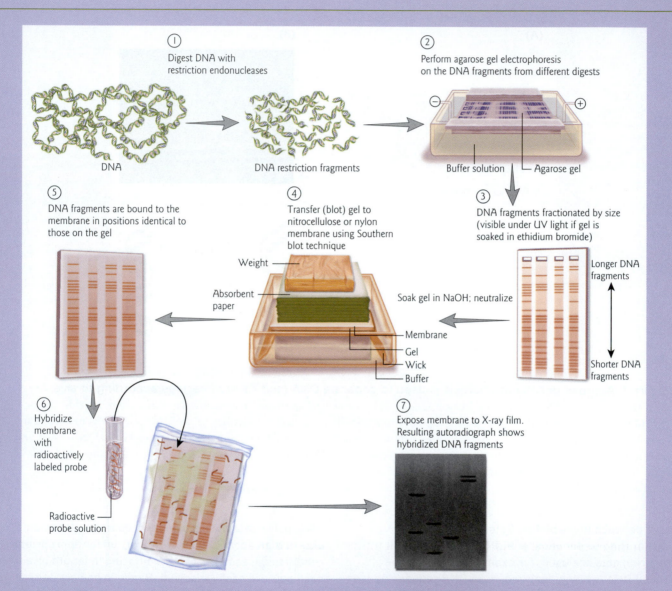

Figure 1 Southern blot. The steps involved in performing a Southern blot hybridization are depicted. (1–3) Samples of the DNA to be probed are cut with restriction endonucleases and the fragments are separated by gel electrophoresis. (4) After soaking the gel in an alkaline solution to denature the DNA, the single-stranded DNA is transferred to a DNA-binding membrane for hybridization by making a sandwich of the gel, membrane, filter paper, and absorbent paper. The blot is held in place with a weight. (5) Capillary action draws the buffer through the gel, transferring the pattern of DNA fragments from the gel to the membrane. (6) The membrane is hybridized with a radioactively labeled probe and then washed to remove any nonspecifically bound probe. (7) The membrane is overlaid with a piece of X-ray film for autoradiography. The hybridized fragments show up as bands on the X-ray film.

Southern blot

The capillary transfer of fragments of DNA separated on an agarose gel from the gel to a solid support was first carried out by Edward Southern. This technique thus bears his name and is called a Southern blot. Southern blots have many applications, but their primary purpose is to identify a specific gene fragment from the often many DNA bands on a gel.

Southern blot method

In the classic Southern blot method, DNA samples are first digested with one or more restriction endonucleases to reduce the size of the DNA molecules (Fig. 1). An endonuclease with a six base pair recognition sequence, statistically cuts once every 4^6 base pairs – producing many thousands of restriction fragments in the genomes of higher eukaryotes. For example, 732,422 restriction fragments will be generated in mouse genomic DNA by *Eco*RI. The digested DNA is then separated by agarose gel electrophoresis. The DNA fragments are denatured by an alkaline (high pH) buffer, and the single strands are transferred by capillary action to a nylon or nitrocellulose membrane. The membrane is a replica of the agarose gel. To identify the DNA fragment that contains a gene of interest, a specific DNA probe, such as a small region of the DNA of interest, can be used to hybridize to the membrane, as described earlier for colony hybridization (see Section 8.7). If the probe hybridizes to fragments on the membrane, then photographic film applied to the membrane will be exposed where the probe has hybridized to a specific DNA band or bands (Fig. 1).

Applications

Southern blots can be used to complement restriction mapping of cloned DNA and to identify overlapping fragments. For example, the exact location of a 2.0 kb cDNA within a 10 kb insert of foreign DNA can be identified so that it can be isolated for further analysis or sequencing. Alternatively, Southern blots can used to identify structural differences between genomes, through analysis of RFLPs, or to study families of related DNA sequences. To illustrate this point, assume that a gene of interest does not contain any internal *Hin*dIII sites. If a clone (e.g. a cDNA clone for some gene) is hybridized to a *Hin*dIII digestion of the DNA sample and two fragments are seen, then it can be concluded that the organism being analyzed has two copies of the gene. Normally, it is not known at the start of analysis what restriction endonuclease sites are located in the gene, so a series of digestions and hybridizations are performed. If four out of the five restriction endonucleases reveal two bands, and the probe hybridizes to three fragments of DNA digested with a fifth restriction endonuclease (e.g. *Eco*RI), then it can be concluded that two genes exist and one of these genes contains a restriction endonuclease site for *Eco*RI. In the case of transgenic animals and plants, Southern blots can be used to determine whether the foreign gene is present and is now part of the host chromosome (see Sections 15.2 and 15.6). The integration of foreign DNA can be visualized by an increase in restriction fragment sizes.

not within the disease gene. RFLPs have been useful for detecting such genetic diseases as cystic fibrosis, Huntington's disease, and hemophilia.

RFLPs were the predominant form of DNA variation used for linkage analysis until the advent of PCR. The main advantage of RFLP analysis over PCR-based protocols is that no prior sequence information or oligonucleotide synthesis is required. However, when a PCR assay for typing a particular locus is developed, it is generally preferable to RFLP analysis. In some cases, a combined approach of PCR and RFLP is used for analysis (Disease box 8.1).

8.11 DNA sequencing

Until 1977, determining the sequence of bases in DNA was a labor-intensive process that could be applied only to very short sequences, such as the template region for tRNA. With the development of techniques for rapid, large-scale DNA sequencing, today molecular biologists determine the order of bases in DNA as a matter of course. DNA sequencing is used to provide the ultimate characterization once a gene has

PCR-RFLP assay for maple syrup urine disease

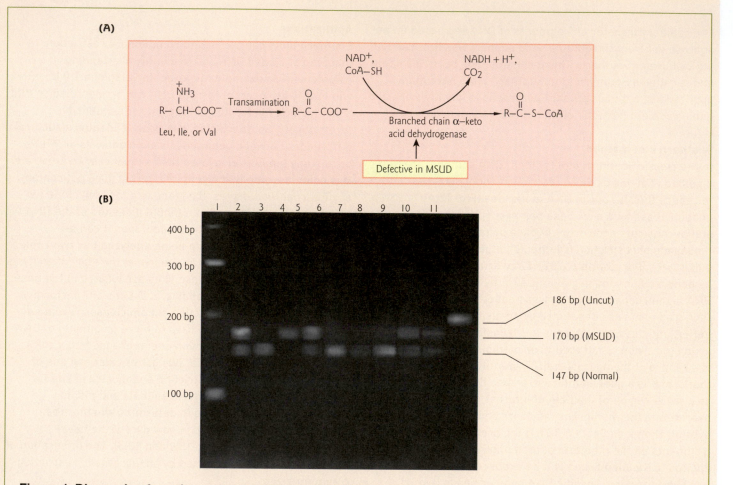

Figure 1 Diagnosis of maple syrup urine disease (MSUD). (A) Degradation of the amino acids leucine, isoleucine, and valine starts with transamination followed by oxidative decarboxylation of the respective keto acids. The latter reaction is carried out by a multienzyme complex, called the branched-chain α-keto acid dehydrogenase complex, which is defective in MSUD patients. (B) Genotype analysis of MSUD by PCR-RFLP assay. DNA samples were amplified by PCR with primers specific for the MSUD allele. Following digestion with restriction endonuclease *Sca*I, samples were visualized by agarose gel electrophoresis and staining with ethidium bromide. Heterozygous individuals are indicated by the presence of a 170 and 147 bp DNA fragment: lane 2 (father), lane 5 (sibling), lane 9 (mother), and lane 10 (maternal grandmother). Individuals homozygous for the normal allele are indicated by the presence of the 147 bp DNA fragment only: lane 3 (paternal grandmother) and lanes 6–8 (maternal great grandfather, maternal grandfather, maternal great grandmother, respectively). Lane 4 indicates the resulting 170 bp fragment from the family member homozygous for the MSUD allele. Lane 11 shows the undigested 186 bp PCR product. (Love-Gregory, L.D., Dyer, J.A., Grasela, J., Hillman, R.E., Philips, C.L. 2001. Carrier detection and rapid newborn diagnostic test for the common Y393N maple syrup urine disease allele by PCR-RFLP: culturally permissible testing in the Mennonite community. *Journal of Inherited and Metabolic Diseases* 24:393–403, Fig. 2. Copyright © 2001 SSIEM and Kluwer Academic Publishers. Reprinted with kind permission of Springer Science and Business Media.)

PCR-RFLP assay for maple syrup urine disease

Maple syrup urine disease (MSUD) was first described in 1954. It is a metabolic disorder inherited as an autosomal recessive that affects the metabolism of the three branched-chain amino acids, leucine, isoleucine, and valine. In a normal individual, when more protein is consumed than is needed for growth, the branched-chain amino acids are degraded to generate energy. Breakdown of these amino acids involves a series of chemical reactions. The second step of the degradation pathway is mediated by an enzyme system consisting of six components, called the multienzyme branched-chain α-keto acid dehydrogenase complex (Fig. 1). In MSUD, one or several of the genes encoding components of this complex are mutated. Because the degradation pathway is blocked, the α-keto acid derivatives of isoleucine, leucine, and valine accumulate in the blood and urine. This accumulation of keto acids gives the urine of affected children a sweet odor resembling maple syrup, and causes a toxic effect that interferes with brain function.

Infants with classic MSUD appear normal at birth and become symptomatic within 4–7 days after birth. They follow a progressive course of neurological deterioration, exhibiting lethargy, seizures, coma, and death within the first 2–3 weeks of life if untreated. Early diagnosis is essential for the child with MSUD to develop normally. Treatment involves a special, carefully controlled diet that requires detailed monitoring of protein intake. The diet centers around a synthetic formula that provides nutrients and all the amino acids except for leucine, isoleucine, and valine. These three amino acids are then added to the diet in carefully controlled amounts to provide enough of these essential amino acids for normal growth and development without exceeding the level of tolerance. In older patients,

this special metabolic product provides basic nutrients and takes the place of cow's milk in a normal diet. The remainder of the diet is essentially a vegetarian diet.

Worldwide, the incidence of MSUD is one in 185,000–225,000; however, in certain Old Order Mennonite communities, the incidence of classic MSUD is estimated to be one in 176 live births. The defect in Mennonite MSUD patients is caused by a single nucleotide change in one of the genes encoding a component of the enzyme complex. The missense mutation results in a tyrosine (Y) to asparagine (N) substitution, called the Y393N allele. MSUD in the non-Mennonite population is clinically and genetically heterogeneous; however the Y393N allele is present in a significant portion of the non-Mennonite MSUD population. A mismatch PCR-RFLP assay has been designed to identify the Y393N allele. Owing to religious and cultural preferences, prenatal testing is not permitted in the Mennonite community. Hence, neonatal testing is vital. Various tests are available to monitor the levels of the amino acids and their α-keto acid derivatives in the blood and urine. Traditional serum-based assays may not give results quickly enough, and there have been reports of infants dying before newborn screening results were reported. The mismatch PCR-RFLP assay provides rapid turnaround times. Since the assay is DNA-based it does not require time for the levels of keto acid derivatives to increase in the blood serum, which may take up to 72 hours. Buccal swabs or blood samples are used and results are available in a minimum of 8 hours. Identification of the normal allele (147 bp) results from the cleavage of a PCR product at a ScaI site. Identification of the Y393N allele (170 bp) results in the absence of this second ScaI site (Fig. 1).

been cloned or amplified by PCR. Although a sequence on its own is of limited value, it is the necessary stepping-stone to more informative analyses of the cloned gene. DNA sequencing is used to identify genes, determine the sequence of promoters and other regulatory DNA elements that control expression, reveal the fine structure of genes and other DNA sequences, confirm the DNA sequence of cDNA and other DNA synthesized *in vitro* (for example, after *in vitro* mutagenesis to confirm the mutation), and help deduce the amino acid sequence of a gene or cDNA from the DNA sequence. With the advent of automated DNA sequencing technology, large genome sequencing projects are yielding information about the evolution of genomes, the location of coding regions, regulatory elements, and other sequences, and the presence of mutations that give rise to genetic diseases (see Section 16.3).

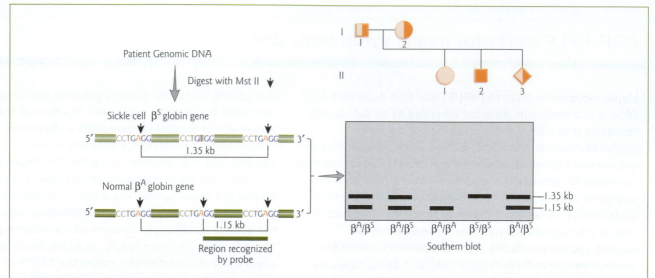

Figure 8.14 Diagnosis of sickle cell anemia by restriction fragment length polymorphism (RFLP) and Southern blot. Black arrows represent the location of recognition sites for the restriction endonuclease *Mst*II in the β-globin gene. In the mutant β-globin gene (β^S), a point mutation (GAG → GTG) has destroyed one *Mst*II recognition site. Digestion of patient genomic DNA with *Mst*II results in a 1.35 kb DNA fragment, compared with a 1.15 kb DNA fragment in normal individuals. For diagnosis, the restriction-digested DNA fragments are separated by gel electrophoresis and transferred to a nylon or nitrocellulose membrane (see Tool box 8.7). The fragments are visualized by hybridization using a probe that spans a portion of the β-globin gene where the 1.15 kb *Mst*II restriction fragment resides. In the pedigree, the family has one unaffected homozygous normal daughter (II-1), an affected homozygous son (II-2), and an unaffected heterozygous fetus (II-3). The genotypes of each family member can be read directly from the Southern blot. (Wilson, J.T., Milner, P.F., Summer, M.E., Nallaseth, F.S., Fadel, H.E., Reindollar, R.H., McDonough, P.G., Wilson, L.B. 1982. Use of restriction endonucleases for mapping the allele for β^S-globin. *Proceedings of the National Academy of Sciences USA* 79:3628–3631; and Chang, J.C., Alberti, A., Kan, Y.W. 1983. A β-thalassemia lesion abolishes the same *Mst*II site as the sickle mutation. *Nucleic Acids Research* 11:7789–7794.)

Manual DNA sequencing by the Sanger "dideoxy" DNA method

In 1977, Frederick Sanger, Allan Maxam, and Walter Gilbert pioneered DNA sequencing. The Maxam and Gilbert sequencing method uses a chemical method that involves selective degradation of bases. The most widely used method for DNA sequencing is the Sanger or "dideoxy" method, which is, in essence, a DNA synthesis reaction (Fig. 8.15). In this method, single-stranded DNA is mixed with a radioactively labeled primer to provide the 3'-OH required for DNA polymerase to initiate DNA synthesis. The primer is usually complementary to a region of the vector just outside the multiple cloning site. The sample is then split into four aliquots, each containing DNA polymerase, four dNTPs (at high concentration), and a low concentration of a replication terminator. The replication terminators are dideoxynucleoside triphosphates (ddNTPs) that are missing the 3'-OH. Because they lack the 3'-OH, they cannot form a phosphodiester bond with another nucleotide. Thus, each reaction proceeds until a replication-terminating nucleotide is added, and each of the four sequencing reactions produces a series of single-stranded DNA molecules, each one base longer than the last. The polymerase of choice for DNA sequencing is phage T7 DNA polymerase (called "Sequenase"). The sequencing mixtures are loaded into separate lanes of a denaturing polyacrylamide gel and electrophoresis is used to separate the DNA fragments. Autoradiography is used to detect a ladder of radioactive bands. The radioactive label (primer) is at the 5' end of each newly synthesized DNA molecule. Thus, the smallest fragment at the bottom of the gel represents the 5' end of the DNA. Reading the sequence of bases from the bottom up (5' → 3') gives the sequence of the DNA molecule synthesized in the sequencing reaction. The sequence of the original strand of DNA is complementary to the sequence read from the gel (3' → 5').

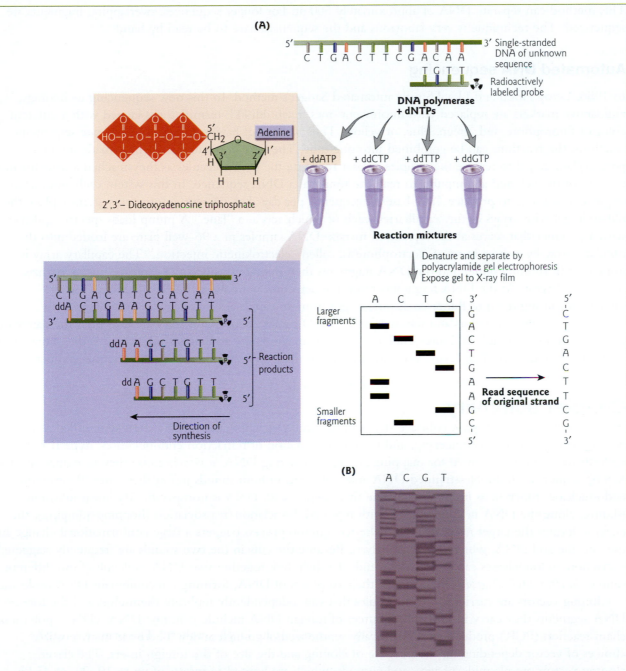

Figure 8.15 Sanger "dideoxy" DNA sequencing. (A) Four DNA synthesis reaction mixes are set up using template DNA, a labeled primer, DNA polymerase, and a mixture of dNTPs and one each of the four dideoxy NTPs (ddNTPs). The direction of synthesis is from 5′ to 3′. The radioactive products of each reaction mixture are separated by polyacrylamide gel electrophoresis and located by exposing the gel to X-ray film. The nucleotide sequence of the newly synthesized DNA is read directly from the autoradiogram in the 5′ → 3′ direction, beginning at the bottom of autoradiogram. The sequence in the original template strand is its complement (3′ → 5′). (Inset) The random incorporation of dideoxy ATP into the growing chain generates a series of smaller DNA fragments ending at all possible positions where adenine is found in the newly synthesized fragments. These correspond to positions where thymine occurs in the original template strand. (B) An exposed X-ray film of a DNA sequencing gel. The four lanes represent A, C, G, and T dideoxy reaction mixes, respectively. (Photograph courtesy of Jim Nicoll, College of William and Mary.)

This method can separate DNA of approximately 500 nt. For longer sequences, overlapping fragments are sequenced. The technique is very laborious and the sequences have to be read by hand.

Automated DNA sequencing

In 1986, Leroy Hood and Lloyd Smith automated Sanger's method. In this new sequencing technology, radioactive markers are replaced with fluorescent ones. Each ddNTP terminator is tagged with a different color of fluorophore: red, green, blue, or yellow. Thus, instead of having to run four separate sequencing reactions, the reactions can be combined into one tube. The first automated sequencer made use of a polyacrylamide gel to resolve the samples, a laser to excite the dye molecules as they reached a detector near the end of the gel, and a computer to read the results as a DNA sequence. In this system each automated sequencer was able to produce 4800 bases of sequence per day. The current automated systems replace the old-style gel with arrays of tiny capillaries, each of which acts as a "lane." A pump loads special capillaries with a polymer that serves as the separation matrix. DNA samples in a 96-well plate are loaded into the capillary array by a short burst of electrophoresis, called "electrokinetic injection." The capillary array is immersed in running buffer and the DNA fragments then migrate through the capillary matrix by size, smallest to largest. As the DNA fragments reach the detection window, a laser beam excites the dye molecules causing them to fluoresce. Emitted light from 96 capillaries is collected at once, spectrally separated into the four colors and focused onto a CCD camera. Computer software interprets the pattern of peaks to produce a graph of fluorescence intensity versus time (electropherogram), which is then converted to the DNA sequence (Fig. 8.16). With this system, as many as 2 million bases can be sequenced per day.

Chapter summary

Insights from bacteriophage λ cohesive sites and bacterial restriction and modification systems led to the development of genetic engineering, and the characterization of restriction endonucleases. Type II restriction endonucleases are widely used for mapping and reconstructing DNA *in vitro* because they recognize specific 4–8 bp sequences in double-stranded DNA and make cuts in both strands just at these sites. Restriction endonucleases function as homodimers. The first contact with DNA is nonspecific. By linear diffusion (sliding) along the DNA in combination with repeated dissociation/reassociation (hopping/jumping), the enzyme locates the target restriction site. The recognition process triggers a large conformational change in the enzyme and DNA, which leads to catalysis. Because the cuts in the two strands are frequently staggered, restriction endonucleases can create sticky ends that help link together two DNA molecules from different sources *in vitro*. DNA ligase is used to join the two pieces of DNA, forming a recombinant DNA molecule.

Cloning vectors are carrier DNA molecules that can independently replicate themselves and the foreign DNA segments they carry in host cells. Sources of foreign DNA include genomic DNA, cDNA, polymerase chain reaction (PCR) products, and chemically synthesized oligonucleotides. There are many possible choices of vector depending on the purpose of cloning and the size of the foreign insert. The classic cloning vectors are plasmids, phages, and cosmids, which are limited to inserts of up to 10, 20, or 45 kb, respectively. A new generation of artificial chromosome vectors that can carry much larger inserts include bacterial artificial chromosomes (BACs), yeast artificial chromosomes (YACs), and mammalian artificial chromosomes (MACs).

Among the first generations of plasmid cloning vectors were pUC plasmids that replicate autonomously in bacterial cells after transformation of the bacteria. These have an ampicillin resistance gene and a multiple cloning site that interrupts a partial *lacZ* (β–galactosidase) gene. The multiple cloning sites make it convenient to carry out directional cloning into two different restriction sites. Ampicillin-resistant clones are screened for those that do not make active β–galactosidase and therefore do not turn the indicator substrate, X–gal, blue. Positive bacterial colonies can be amplified in liquid growth medium, followed by purification of the amplified recombinant plasmid DNA by liquid chromatography methods. Ion-exchange chromatography can be used to separate substances according to their charges. Gel filtration chromatography

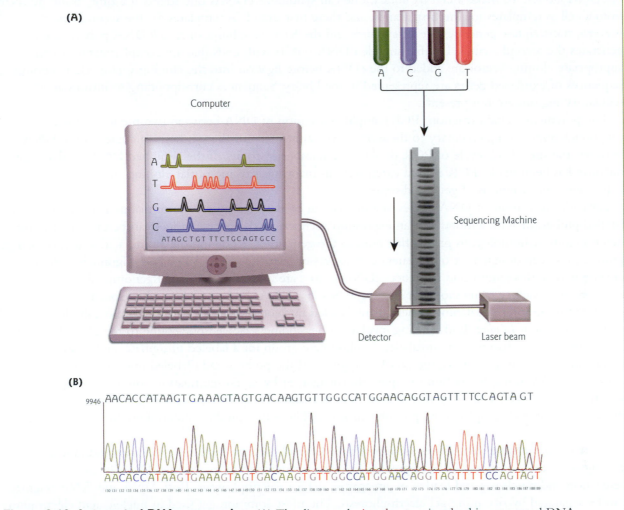

Figure 8.16 Automated DNA sequencing. (A) The diagram depicts the steps involved in automated DNA sequencing. This technique uses dideoxynucleotides, just as in the manual technique shown in Fig. 8.15, but the primers used in each of the four reactions are tagged with different fluorescent molecules. The products from each tube will emit a different color fluorescence when excited by light. (B) Sample computer printout of an automated DNA sequencing experiment. (Electropherogram courtesy of Ghislain Bonamy, College of William and Mary.)

uses columns filled with porous resins that let in smaller substances, but exclude larger ones. Thus, the smaller substances are slowed in their journey through the column, but larger substances travel relatively rapidly through the column.

Engineered phage λ from which certain nonessential genes have been removed to make room for inserts are useful for preparing genomic libraries, in which it is important to have large pieces of genomic DNA in each clone. Even more useful are YAC vectors that are designed to act like chromosomes in host yeast cells, and can accommodate up to 1 Mb of foreign DNA. YAC vectors included an origin of replication, a centromere, telomeres, and growth selectable markers in each arm. Positive selection is carried out by auxotrophic complementation for molecules in which the arms are joined together. Recombinant YACs are often screened for by a "red–white" selection process, in which insertion of foreign DNA leads to the expression of a red pigment in particular mutant strains of yeast.

A genomic library contains DNA fragments that represent the entire genome of an organism. The library created from all the cDNAs derived from the expressed mRNAs in a specific cell type forms a cDNA library of cDNA clones. To make a cDNA library, one can synthesize cDNAs one strand at a time, using mRNAs from a cell as templates for the first strands, and these first strands as templates for the second strands. Reverse transcriptase generates the first strands and the Klenow subunit of *E. coli* DNA polymerase I generates the second strands. Double-stranded DNA linkers with ends that are complementary to an appropriate cloning vector are added to the cDNA before ligation into the cloning vector. Gene coding sequences of expressed genes are represented in the library. Sequences corresponding to introns and regulatory regions are not present.

The polymerase chain reaction (PCR) amplifies a region of DNA between two predetermined sites. Oligonucleotides complementary to these sites serve as primers for the synthesis of copies of the DNA between the sites. Each cycle of PCR doubles the number of copies of the amplified DNA until a large quantity has been made. PCR is used extensively in many areas of molecular cloning, analysis of gene expression, and diagnosis of genetic diseases.

Particular recombinant DNA clones in a library can be detected by colony or plaque hybridization with labeled probes, or with antibodies if an expression vector is used. Specific clones can be identified using heterologous or homologous probes that bind to the gene itself. Knowing the amino acid sequence of a gene product, one can design a set of degenerate or expressed sequence tag (EST) based oligonucleotides that encode part of this amino acid sequence. cDNA probes are often used to screen genomic libraries.

There are a variety of methods for labeling RNA and DNA. Probes of the highest specific activity are generated using internal labeling, where many labeled nucleotides are incorporated uniformly during DNA or RNA synthesis *in vitro*. End-labeling involves either adding a labeled nucleotide to the 3'-OH end of a DNA strand or exchanging the unlabeled 5'-phosphate group for a labeled phosphate, and is used when precise definition of one end of the DNA is required. If the probe is radiolabeled one can detect it by autoradiography, using X-ray film or a phosphorimager, or by liquid scintillation counting. Some very sensitive nonradioactive labeling methods are now available. Those that employ chemiluminescence can be detected by autoradiography or by phosphorimaging. Those that produce colored products can be detected directly.

Restriction mapping determines the number, order, and distance between restriction endonuclease cutting sites along a cloned DNA fragment. To make a restriction map, a cloned DNA fragment is cut with restriction endonucleases and loaded on an agarose gel for electrophoresis. Both DNA and RNA fragments can be separated by size using gel electrophoresis. The most common gel used in nucleic acid electrophoresis is agarose, but polyacrylamide is used for the separation of smaller fragments such as in DNA sequencing. Some restriction fragment length polymorphisms (RFLPs) are used as markers for genetic diseases.

Labeled DNA (or RNA) probes can be used to hybridize to DNAs of the same, or very similar, sequence on a Southern blot. The number of bands that hybridize to a short probe gives an estimate of the number of closely related genes in an organism. Southern blots can be used to complement restriction mapping of cloned DNA and to identify overlapping fragments, or to identify structural differences between genomes, through analysis of RFLPs.

The Sanger DNA sequencing method uses a radiolabeled primer to initiate synthesis by DNA polymerase and dideoxynucleotides (ddNTPs) to terminate DNA synthesis, yielding a series of labeled DNA fragments, each one base longer than the last. These fragments can be separated according to size by electrophoresis. The last base in each of these fragments is known, because we know which ddNTP was used to teminate each of four reactions. Ordering these fragments by size tells us the base sequence of the DNA. The sequence of the original strand of DNA is complementary to the sequence read from the autoradiograph of the gel. In automated DNA sequencers, radioactive markers are replaced with fluorescent ones. An electropherogram – a graph of fluorescence intensity versus time – is converted to the DNA sequence.

Analytical questions

1 You have attempted to ligate a 1.5 kb fragment of foreign DNA into the *Eco*RI site in the multiple cloning site of the 4.0 kb plasmid vector shown below:

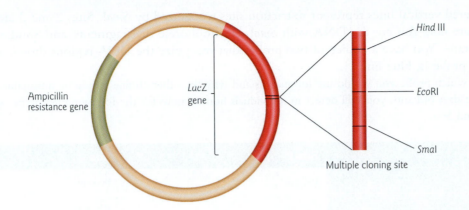

(a) After ligation you use the DNA in the ligation mixture to transform host bacteria. Why is it important to use host bacteria that are deficient for restriction modification?

(b) You screen the bacteria that supposedly have been transformed with recombinant plasmid DNA. Some of the bacterial colonies growing on the nutrient agar plate that contains ampicillin and X-gal are white and some are blue. Explain these results.

(c) To confirm the presence of the foreign DNA insert, you perform *Eco*RI restriction endonuclease digests on DNA extracted from bacterial colonies. Draw a diagram of an agarose gel showing the orientation of the positive and negative electrodes and the pattern of bands (label their size in kilobases) you would expect to see for *Eco*RI-digested recombinant plasmid and *Eco*RI-digested nonrecombinant plasmid vector, after electrophoresis and staining of the gel with ethidium bromide.

2 A chromatographic column in which oligo-dT is linked to an inert substance is useful in separating eukaryotic mRNA from other RNA molecules. On what principle does this column operate?

3 Starting with the nucleotide sequence of the human DNA ligase I gene, describe how you would search for a homologous gene in another organism whose genome has been sequenced, such as the pufferfish *Tetraodon nigroviridis*. Then, describe how you would obtain the protein and test it for ligase activity.

4 You plan to use the polymerase chain reaction to amplify part of the DNA sequence shown below, using oligonucleotide primers that are hexamers matching the regions shown in bold. (In practice, hexamers are too short for most purposes.) State the sequence of the primer oligonucleotides that should be used, including their polarity (5′ → 3′), and give the sequence of the DNA molecule that results from amplification.

5′-TAGGCAT**GCAATG**GTAATTTTTCAGGAACCAGGGCCCTT**AAGCCG**TAGGCAT-3′
3′-ATCGGTA**CGTTAC**CATTAAAAACTCCTTGGTCCCGGGAA**TTCGGC**ATCGGTA-5′

5 The following is a physical map of a region you are mapping for RFLP analysis:

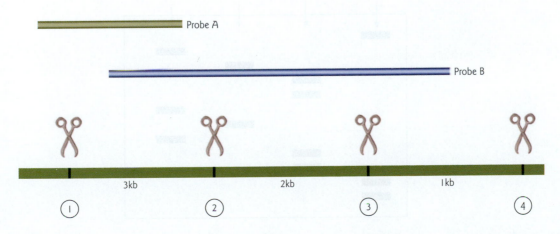

The numbered vertical lines represent restriction sites recognized by *Sma*I. Sites 2 and 3 are polymorphic, the others are not. You cut the DNA with *Sma*I, electrophorese the fragments, and Southern blot them to a membrane. You have a choice of two probes that recognize the DNA regions shown above: probe A, green line; probe B, blue line.

(a) Explain which probe you would use for analysis and why the other choice would be unsuitable.

(b) Give the sizes of bands you will detect in individuals homozygous for the following genotypes with respect to sites 2 and 3:

Haplotype	Site 2	Site 3
A	Present	Present
B	Present	Absent
C	Absent	Present
D	Absent	Absent

6 Complete the incomplete diagrams below to show the key structural difference between an NTP, dNTP, and ddNTP (e.g. CTP, dCTP, ddCTP):

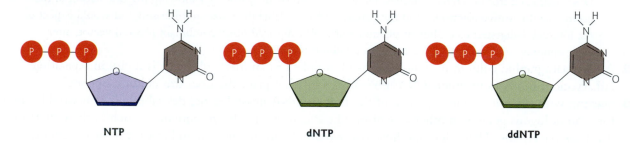

NTP　　　　　dNTP　　　　　ddNTP

(a) What do "d" and "dd" stand for?

(b) Explain why ddNTPs are called "chain terminators" in DNA sequencing reactions.

7 The nucleotide sequence of a DNA fragment was determined by the Sanger (dideoxy) DNA sequencing method. The data are shown below. What is the 5′ → 3′ sequence of the nucleotides in the original DNA fragment?

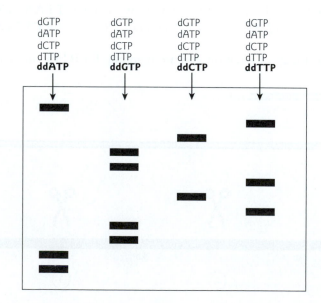

Suggestions for further reading

Ausubel, F.M., Brent, R., Kingston, R.E., Moore, D.D., Seidman, J.G., Smith, J.A., Struhl, K., eds (2002) *Short Protocols in Molecular Biology*, 5th edn. John Wiley & Sons, New York.

Echols, H. (2001) *Operators and Promoters. The Story of Molecular Biology and Its Creators*. University of California Press, Berkeley, CA.

Gowers, D.M., Wilson, G.G., Halford, S.E. (2005) Measurement of the contributions of 1D and 3D pathways to the translocation of a protein along DNA. *Proceedings of the National Academy of Sciences USA* 102:15883–15888.

Harrington, J.J., van Bokkelen, G., Mays, R.W., Gutashaw, K., Willard, H.F. (1997) Formation of *de novo* centromeres and construction of first generation human artificial microchromosomes. *Nature Genetics* 125:345–355.

Katoh, M., Ayabe, F., Norikane, S. et al. (2004) Construction of a novel human artificial chromosome vector for gene delivery. *Biochemical and Biophysical Research Communications* 321:280–290.

Kurpiewski, M.R., Engler, L.E., Wozniak, L.A., Kobylanska, A., Koziolkiewicz, M., Stec, W.J., Jen-Jacobsen, L. (2004) Mechanisms of coupling between DNA recognition and specificity and catalysis in *Eco*RI endonuclease. *Structure* 12:1775–1788.

Lipps, H.J., Jenke, A.C.W., Nehlsen, K., Scinteie, M.F., Stehle, I.M., Bode, J. (2003) Chromosome-based vectors for gene therapy. *Gene* 304:23–33.

Love-Gregory, L.D., Dyer, J.A., Grasela, J., Hillman, R.E., Philips, C.L. (2001) Carrier detection and rapid newborn diagnostic test for the common Y393N maple syrup urine disease allele by PCR-RFLP: culturally permissible testing in the Mennonite community. *Journal of Inherited and Metabolic Diseases* 24:393–403.

Morrow, J.F., Cohen, S.N., Chang, A.C.Y., Boyer, H.W., Goodman, H.M., Helling, R.B. (1974) Replication and transcription of eukaryotic DNA in *Escherichia coli*. *Proceedings of the National Academy of Science USA* 71:1743–1747.

Mullis, K.B. (1990) The unusual origin of the polymerase chain reaction. *Scientific American* 262:36–43.

Pingoud, A., Jeltsch, A. (2001) Structure and function of type II restriction endonucleases. *Nucleic Acids Research* 29:3705–3727.

Sambrook, J., Russell, D.W. (2001) *Molecular Cloning: a Laboratory Manual*, 3rd edn. Cold Spring Harbor Laboratory Press, New York.

Viadiu, H., Aggarwal, A.K. (2000) Structure of *Bam*HI bound to nonspecific DNA: a model for DNA sliding. *Molecular Cell* 5:889–895.

Watson, J. (1993) The human genome initiative. In: *Genetics and Society* (eds B. Holland, C. Kyriacou), pp. 13–26. Addison-Wesley Publishing Co., New York.

Chapter 9
Tools for analyzing gene expression

To discover for yourself what is already known can still be a source of wonder – which is why the study of nature can never disappoint.

Thomas Eisner, *For Love of Insects* (2003), p. 268.

Outline

9.1 Introduction

Since cloning of the first eukaryotic gene three decades ago, thousands of eukaryotic genes have now been isolated. Genome projects that are currently underway are leading to identification of entire gene complements for numerous organisms. After a new gene is cloned, the next step is to determine the structure of the gene, how its expression is regulated, and the biological functions of the encoded gene product. Gene expression is the production of a functional protein or RNA from the genetic information encoded in the genes. In its broadest sense, the term encompasses both transcription and translation, but often gene expression is used to refer to the process of transcription only. It is the differential expression of genes that regulates the remarkable development of a single cell into a multicellular organism, and that distinguishes a normal cell from a cancer cell, or a skin cell from a liver cell. Thus, understanding the molecular mechanisms regulating gene expression is an important field with far-reaching implications.

The world of living cells is extraordinarily complex. Understanding how the molecular processes of the cell work requires a variety of experimental approaches, ranging from biochemical assays to genetic analysis to microscopic visualization. In this chapter, tools for analyzing gene regulation and function are described, including analysis of gene expression at the level of both transcription and translation. In addition, techniques for analyzing DNA–protein and protein–protein interactions are outlined (Table 9.1). It is not intended that this chapter be read continuously from start to finish. Instead, individual sections may be better used as a resource when they become relevant for understanding experiments referred to in subsequent or previous chapters.

Table 9.1 Summary of some commonly used tools for analyzing gene expression.

Level	Methods
Inhibition	*In vitro* mutagenesis Antisense oligonucleotides Expression of antisense RNA RNA interference (RNAi)
Transcription	Northern blot *In situ* hybridization RNase protection assay (RPA) Reverse transcriptase–PCR (RT-PCR)
Translation	Reporter gene enzyme activity Western blot *In situ* analysis Enzyme-linked immunosorbent assay (ELISA)
DNA–protein interactions	Electrophoretic mobility shift assay (EMSA) DNase I footprinting Chromatin immunoprecipitation (ChIP) assay
Protein–protein interactions	Pull-down assays Yeast two-hybrid assay Coimmunoprecipitation assay Fluorescence resonance energy transfer (FRET)
Protein structure	X-ray crystallography Nuclear magnetic resonance (NMR) spectroscopy Cryoelectron microscopy Atomic force microscopy (AFM)

Table 9.2 Various cell transfection methods.

Method	Comments
Chemical transfection	Many methods, e.g. calcium phosphate or DEAE-dextran transfection of animal cells. DNA is internalized by endocytosis
Lipofection	DNA complexed with cationic liposomes and taken up by endocytosis. Highly efficient method for transfecting animal cells, yeast spheroplasts, and plant protoplasts*
Electroporation	Naked DNA taken into cells through transient pores created by brief pulses of high voltage. Very efficient method for the transfection of yeast, plant, and animal cells
Direct injection	Labor-intensive but 100% efficient. Routinely applied to animal oocytes, eggs, and zygotes (see Fig. 15.2)
Microballistics (biolistics)	The use of microprojectiles, tungsten, or gold particles coated with DNA, which are fired into cells at high velocity using a gene gun. Gives efficient transfection of plant cells without removing cells walls. Can also be used to transfect whole plant and animal tissues (see Section 15.6 and Table 17.4)

★ A spheroplast is a yeast cell from which the cell wall has been removed. A protoplast is a plant cell with the cell wall removed.

9.2 Transient and stable transfection assays

One of the most important basic techniques for molecular cloning and analysis of gene expression is the introduction of DNA into cells. As described in Section 8.4, a highly efficient procedure called "transformation" is used routinely to introduce DNA into bacterial cells. A range of techniques also allows the introduction of DNA into eukaryotic cells. Most involve getting cells to take up naked DNA in a process called "transfection." Methods for uptake of naked DNA include chemical transfection, lipofection, electroporation, direct injection, and microballistics (Table 9.2). Other gene transfer techniques are based on the uptake of DNA packaged in viral capsids.

In bacteria and yeast, plasmid vectors can be replicated and maintained episomally (extrachromosomally, i.e. separate from the host genome) in the host cell. However, in mammalian cells, plasmid DNA is not replicated and is eventually lost by dilution (as the cells divide) and degradation of the recombinant vector. In many experiments it is unnecessary for the DNA to be stably maintained. In this case, "transient transfection" – the introduction of DNA into cells for a short duration – is sufficient (Fig. 9.1). For example, experiments such as reporter gene assays (see Section 9.3) can be carried out relatively quickly (over 24–72 hours). For techniques such as protein overexpression where longer term analysis is desirable, "stable transfection" is required. Plasmid DNA introduced into eukaryotic cells frequently integrates randomly into the genome. Cells in which this has happened can be selected for when the plasmid vector contains a drug resistance gene. After drug selection, only cells in which the plasmid has stably integrated into a chromosome will survive. Typically, multiple copies of plasmid DNA integrate randomly into a chromosomal site in a tandem array. Each stably transfected cell has a unique integration site. Dosage effects need to be considered in analyzing results because overexpression can sometimes lead to mislocalization or abnormal function of gene products within the cell. Methods for gene targeting to a specific chromosomal site by homologous recombination are addressed in Chapter 15 (Section 15.3).

9.3 Reporter genes

Reporter genes are widely used to analyze gene expression. A reporter gene is a known gene whose RNA or protein levels can be measured easily and accurately. Thus, they offer a simple means to monitor changes in gene expression. These genes are often used to replace other coding regions whose protein products are

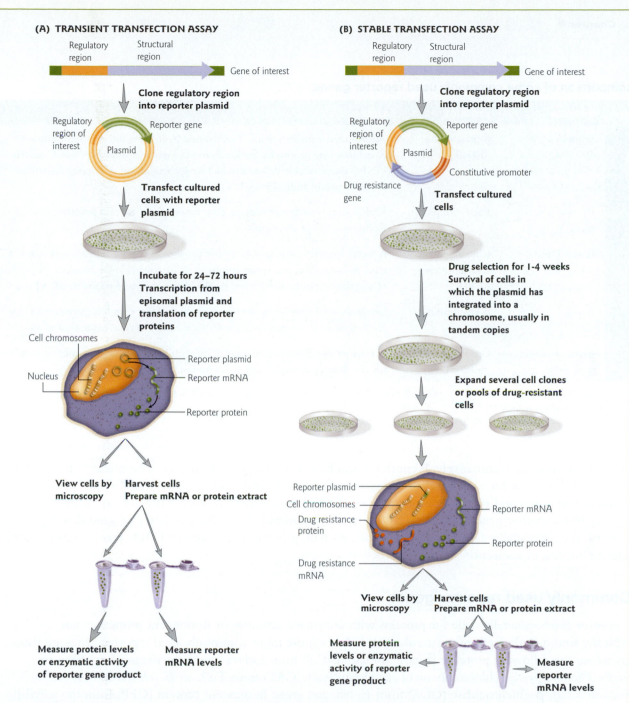

(A) TRANSIENT TRANSFECTION ASSAY

Regulatory region
Structural region
Gene of interest

Clone regulatory region into reporter plasmid

Regulatory region of interest
Reporter gene
Plasmid

Transfect cultured cells with reporter plasmid

Incubate for 24–72 hours Transcription from episomal plasmid and translation of reporter proteins

Cell chromosomes
Nucleus
Reporter plasmid
Reporter mRNA
Reporter protein

View cells by microscopy

Harvest cells Prepare mRNA or protein extract

Measure protein levels or enzymatic activity of reporter gene product

Measure reporter mRNA levels

(B) STABLE TRANSFECTION ASSAY

Regulatory region
Structural region
Gene of interest

Clone regulatory region into reporter plasmid

Regulatory region of interest
Reporter gene
Plasmid
Constitutive promoter
Drug resistance gene

Transfect cultured cells

Drug selection for 1-4 weeks Survival of cells in which the plasmid has integrated into a chromosome, usually in tandem copies

Expand several cell clones or pools of drug-resistant cells

Reporter plasmid
Cell chromosomes
Drug resistance protein
Drug resistance mRNA
Reporter mRNA
Reporter protein

View cells by microscopy

Harvest cells Prepare mRNA or protein extract

Measure protein levels or enzymatic activity of reporter gene product

Measure reporter mRNA levels

Figure 9.1 Comparison of transient and stable transfection assays. A recombinant DNA plasmid is constructed in which a reporter gene is attached to a regulatory region of interest (e.g. a promoter or enhancer). The reporter gene is a known gene whose mRNA or protein levels can be measured conveniently and accurately. The method of analysis depends on the reporter gene used, and may involve viewing the cells by fluorescence microscopy (e.g. GFP), measuring reporter mRNA or protein levels, or measuring the enzymatic activity of the reporter gene product (e.g. CAT, luciferase, β–Gal, or GUS). For stable transfection assays, the reporter plasmid also contains a drug resistance gene (e.g. neomycin resistance gene) under the control of a constitutively active promoter. The reporter plasmid is introduced into cultured cells by chemical transfection, lipofection, or electroporation. (A) For transient transfection assays, the cells are incubated for 24–72 hours to allow time for transcription of mRNA from the plasmid and translation of the reporter protein. The assay is considered transient because the plasmids remain episomal (separate from the host cell chromosomes) and rarely integrate into the host genome. (B) For stable transfection assays, cells that have stably integrated the plasmid into a chromosome are selected by supplementing the growth medium with the appropriate drug (e.g. neomycin), which kills cells that do not stably express the drug resistance gene. After selection for 1–4 weeks, several cell clones or pools of drug-resistant cells are expanded and analyzed for reporter gene expression.

Table 9.3 A comparison of some commonly used reporter genes.

Reporter gene	Species	Product	Use
lacZ (*lac* operon)	*Escherichia coli* (bacteria)	β-galactosidase (β-Gal)	Widely used reporter system. The enzyme hydrolyzes the colorless substrate X-gal to a blue precipitate for localization of gene expression *in situ*. Converts ONPG into a soluble yellow product for quantification of enzyme activity in cellular extracts. Inducible by IPTG
luc	*Photinus pyralis* (firefly)	Luciferase (Luc)	Highly sensitive reporter enzyme that oxidizes luciferin and generates a bioluminescent product (photons)
cat	*E. coli* Tn9	Chloramphenicol acetyltransferase (CAT)	A useful reporter for *in vitro* assays but protein gives poor resolution *in situ*. CAT transfers nonradioactive acetyl groups to radioactive chloramphenicol; the extent of acetylation can be determined by thin-layer chromatography in a CAT assay
GUS	*E. coli*	β-glucuronidase (GUS)	Generally used reporter in plant systems; hydrolyzes colorless glucuronides (e.g. X-gluc) to yield colored products for localization of gene expression *in situ*
gfp	*Aequorea victoria* (jellyfish)	Green fluorescent protein (GFP)	A reporter that fluoresces on irradiation with UV; because it is autofluorescent it has the distinct advantage that it can be used for imaging live cells

IPTG, isopropyl-β-D-thiogalactopyranoside; ONPG, O-nitrophenyl-β-D-galactopyranoside; X-gal, 5-bromo-4-chloro-3-indolyl-β-D-galactopyranoside; X-gluc, 5-bromo-4-chloro-3-indolyl-β-D-glucuronide.

difficult to measure quantitatively. Reporter genes have several major advantages. The researcher does not need a separate assay for each regulatory region being studied, and reporter genes can "report" many different properties and events. For example, reporter genes are used to analyze the activity of the regulatory regions from another gene in different tissues or developmental stages, the efficiency of gene delivery systems, the intracellular fate of a gene product, protein–protein interactions or DNA–protein interactions, and the success of molecular cloning efforts.

Commonly used reporter genes

Reporter genes generally code for proteins with enzymatic activities or fluorescent properties not typically found in the cells of most eukaryotes. Among the more commonly used reporter genes are those encoding the following proteins: β-galactosidase (β-Gal) from *Escherichia coli*, luciferase (Luc) from the firefly *Photinus pyralis*, chloramphenicol acetyltransferase (CAT) from Tn9, an *E. coli* transposon (see Section 12.5), β-glucuronidase (GUS) from *E. coli*, and green fluorescent protein (GFP) from the jellyfish *Aequorea victoria* (Table 9.3). The choice of reporter gene depends on a variety of factors, including the cell system being used, the sensitivity required, and the desired method of analysis (whether *in vitro*, *in vivo*, or *in situ*).

Beta-galactosidase catalyzes the hydrolysis of β-galactosides (see Fig. 8.7 and Section 10.5), while GUS hydrolyzes glucuronides. Standard substrates used in β-Gal and GUS assays generate reaction products that can be quantified using a spectrophotometer. Luciferase is an enzyme that oxidizes luciferin, yielding a fluorescent product that can be quantified by measuring the released light. The CAT enzyme catalyzes the acetylation of chloramphenicol, with the acetyl group donated by acetyl CoA. The acetylated chloramphenicol can be monitored in a variety of ways. Most commonly, ^{14}C-chloramphenicol is used as the reaction substrate, with acetylation monitored by autoradiography following thin-layer chromatography (TLC) to separate the acetylated from the unacetylated forms (Fig. 9.2). Alternatively, the presence of the CAT protein can be measured by enzyme-linked immunosorbent assay (ELISA, see Section 9.6). GFP is a naturally fluorescent protein. It requires no substrate or cofactors for light production. The emitted light can usually be measured in intact cells, resulting in a simple assay for reporter gene expression (see below).

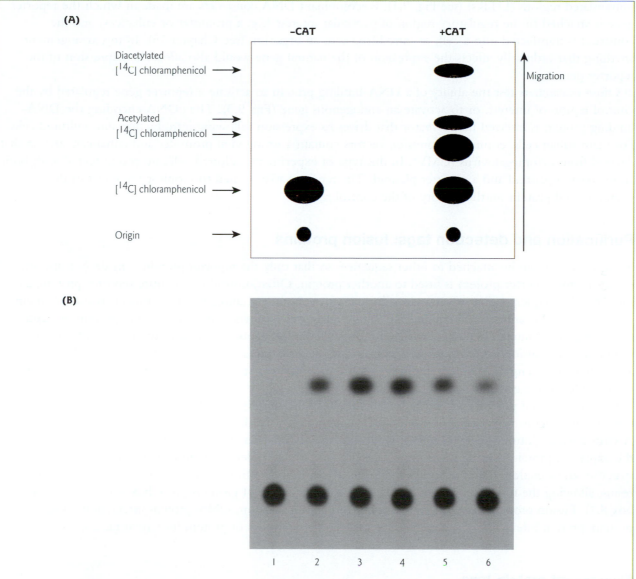

Figure 9.2 CAT reporter gene assay. (A) A chloramphenicol acetyltransferase (CAT) reporter gene assay using thin-layer chromatography (TLC). The CAT enzyme catalyzes the acetylation of [^{14}C] chloramphenicol, with the acetyl group donated by acetyl CoA. TLC is used to separate the acetylated from the unacetylated forms. The percent conversion of [^{14}C] chloramphenicol to acetyl [^{14}C] chloramphenicol can be measured by phosphorimager analysis of the TLC plate, by densitometry of an autoradiograph, or by excising the radioactive spots from the TLC plate and counting in a scintillation counter. (B) Actual experimental results from a transient transfection assay. Lane 1 is a negative control with no cell extract. Lanes 2–6 have varying levels of CAT activity. (Photograph courtesy of Patty Zwollo, College of William and Mary.)

Analysis of gene regulation

One of the most common uses of a reporter gene is to analyze how the expression of a gene is regulated. For example, a typical protein-coding gene is composed of a regulatory region lying upstream of the transcription start point and a structural region, including the open reading frame (ORF) and any 5′ or 3′

untranslated regions (UTRs) (see Fig. 9.1). Recombinant DNA constructs are made in which the reporter gene is attached to the regulatory region of particular interest (e.g. a promoter or enhancer) and the construct is transfected into a cell, or introduced into an organism (see Chapter 15). In this arrangement, anything that ordinarily affects the expression of the natural gene would also affect the expression of the reporter gene.

Often researchers test the ability of a DNA-binding protein to activate a reporter gene regulated by the control region of interest, or to activate an endogenous gene (Fig. 9.3). The cDNA encoding the DNA-binding protein is inserted into a vector that drives its expression following introduction into cultured cells. For mammalian cells, common expression vectors contain a strong viral promoter and enhancer, such as that derived from cytomegalovirus (CMV). In this type of experiment, cultured cells are cotransfected with both an expression plasmid and a reporter plasmid. The reporter assay is used to monitor the effect of the overexpressed protein on the activity of the control region.

Purification and detection tags: fusion proteins

Reporter genes can be attached to other sequences so that only the reporter protein is made (see above), or so that the reporter protein is fused to another protein. Often, instead of an entire reporter protein, a short peptide sequence that serves as an affinity tag or epitope tag (antigenic determinant) is used. Protein expression vectors are typically engineered with a nucleotide sequence that encodes the protein or peptide tag. The gene of interest is cloned in-frame relative to the tag. Upon expression, the protein of interest is synthesized as a fusion protein with the reporter protein or peptide tag. The availability of highly specific antibodies to the engineered tags eliminates the time-consuming step of making antibodies to proteins from each newly cloned gene.

Commonly used protein or peptide tags include 6-histidine (His), glutathione-*S*-transferase (GST), the transcription factor c-Myc, FLAG (amino acid sequence Asp-Tyr-Lys-Asp-Asp-Asp-Asp-Lys), and influenza A virus haemagglutinin (HA) (Fig. 9.4). Generally, these sequences are fused to the N- or C-terminus of the expressed protein making them more accessible for antibody detection and less likely to disrupt protein structure and function. The tags either bind tightly to antibody affinity resins or to other types of affinity resins, allowing the fusion protein to be purified from a mixture of proteins (Tool box 9.1; see also Tool box 8.1). Fusion proteins are used for studies of protein localization, DNA–protein interactions, and protein–protein interactions, and are used to make large quantities of protein for structural studies.

Fluorescent protein tags

As noted earlier, the gene coding for green fluorescent protein (GFP) was originally isolated from the jellyfish *Aequorea victoria*. *GFP* is widely used as a reporter gene for investigation of tissue-specific gene expression and cellular localization of proteins for two main reasons. First, the fluorescence of GFP can be detected directly in living cells (Tool box 9.2). In contrast, detection of other reporter gene products requires fixation or lysis of cells. Second, GFP can be artificially expressed effectively in every cell type and organism tested so far, including bacteria, yeast, plants, *Caenorhabditis elegans*, *Drosophila melanogaster*, zebrafish, frogs, and mice.

GFP contains an intrinsic peptide fluorophore that arises from an autocatalytic post-translational modification (Fig. 9.5). A fluorophore is a group of atoms in a molecule responsible for absorbing light energy and producing the color of the compound. The GFP fluorophore consists of a cyclic tripeptide derived from Ser65–Tyr66–Gly67 in the 238 amino acid polypeptide. Following excitation with ultraviolet or blue light, the fluorophore only emits green light when embedded within the complete GFP protein. X-ray crystallographic analysis of GFP has revealed a particularly striking and elegant three-dimensional structure. The fluorophore is buried in the center of a nearly perfect cylinder formed by an intricately folded, 11-stranded β-barrel (Fig. 9.5B). This structure provides the proper environment for fluorophore activity by excluding solvent and oxygen.

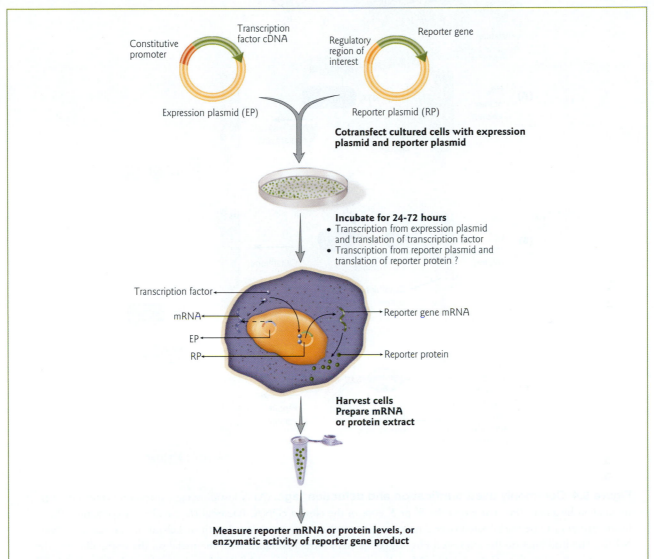

Figure 9.3 Cotransfection assay. (A) The activation of reporter gene expression by overexpression of a transcription factor. Cultured cells are transiently cotransfected with an expression plasmid that contains the transcription factor cDNA under control of a constitutively active promoter, and a reporter plasmid that contains the reporter gene under control of the regulatory region of interest. Cells are incubated for 24–72 hours, to allow transcription of mRNA from the expression plasmid and translation of the transcription factor. If the transcription factor binds the regulatory region of interest and activates transcription, then transcription from the reporter plasmid will occur, followed by translation of the reporter gene product. The activity of the control region of interest is determined by measuring reporter gene activity. The method of analysis depends on the reporter gene used.

Plasmid vectors have been constructed that include a strong promoter, the coding region for *GFP*, and a multiple cloning site for insertion of the cDNA coding for the protein of interest. The plasmid DNA is introduced into a cell (or used to make a transgenic animal, see Fig. 15.2). An RNA transcript including the coding region for GFP and the protein of interest is transcribed as one unit. This RNA is then translated, resulting in a chimeric protein with the protein of interest attached to either the C-terminus or the N-terminus of GFP, depending on the vector design. An example of expression of a GFP fusion protein in cells is shown in Fig. 9.6.

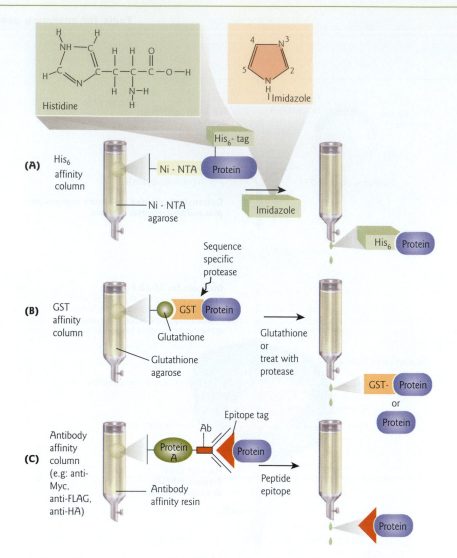

Figure 9.4 Commonly used purification and detection tags. (A) A histidine tag expression vector encodes six tandem histidines (His$_6$) at either the 5′ or 3′ ends of the cloned cDNA. Bacterial (*E. coli*) lysates containing the fusion protein can be fractionated over a resin (e.g. agarose) column complexed with nickel–nitrilotriacetic acid (Ni–NTA). The histidines on the expressed His$_6$-tagged protein form a complex with the metal on the resin, allowing the protein to adhere to the column matrix. Most *E. coli* proteins do not bind Ni-NTA and pass through the column. The His$_6$-tagged protein that has adhered is then eluted from the column using an imidazole-containing buffer. Because of its structural similarity with histidine (top left inset), imidazole competes with the histidines for the metal and disrupts the interaction, thereby eluting the protein from the resin. (B) GST fusions. A GST expression vector encodes glutathione-*S*-transferase, a 26 kD protein, at the 5′ end of the cloned cDNA of interest. Bacterial lysates containing the fusion protein are passed over a glutathione–agarose column. The GST portion of the fusion protein adheres to the matrix and can be eluted by the addition of excess free glutathione. GST vectors often include a cleavage site for a sequence-specific protease. If it is necessary to remove the tag to facilitate biochemical analysis of the protein, protease treatment can be used to cleave the protein of interest from the GST tag. Cleavage allows its elution from the column. (C) Immunotags. Some types of expression vectors encode epitope tags at the 5′ or 3′ end of the cloned cDNA of interest; e.g. Myc, FLAG, and influenza virus hemagglutinin (HA). Specific antibodies are covalently linked to protein A-sepharose (or agarose) or other types of activated affinity resins. Protein A is a bacterial protein from *Staphylococcus aureus* that binds to most antibodies nonspecifically. The cellular extracts containing the fusion protein are then passed over the antibody affinity resin via column or batch chromatography. The resin is washed extensively with high salt buffers to remove nonspecifically bound proteins. The bound protein is eluted with excess peptide corresponding to the epitope and moderate to high salt concentrations. The FLAG antibody requires Ca^{2+} for binding; removal of Ca^{2+} by EGTA causes antibody dissociation from the FLAG tag, providing an easy method of elution.

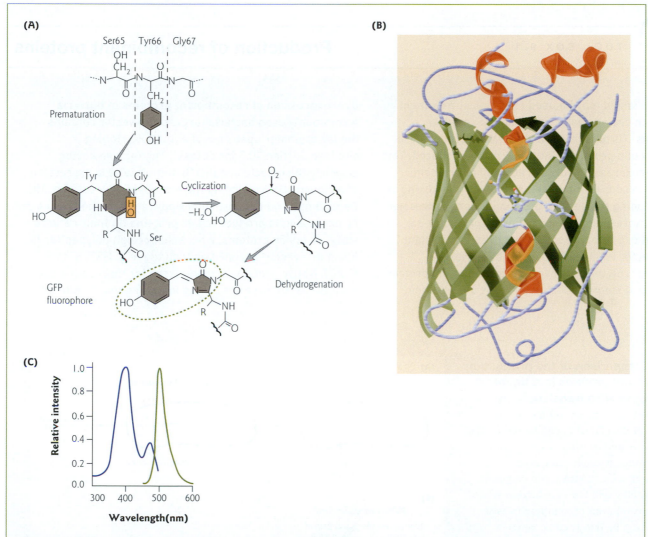

Figure 9.5 Properties of green fluorescent protein (GFP). (A) The GFP fluorophore forms by an oxygen-dependent post-translational modification involving three consecutive amino acids in its primary sequence, a serine at position 65 (Ser65), a tyrosine at position 66 (Tyr66), and a glycine at position 67 (Gly67). A reaction between the amino nitrogen of Gly67 and the carboxyl carbon of Ser65 (nucleophilic attack) leads to cyclization of the Ser–Tyr–Gly tripeptide and 1,2-dehydrogenation of the Tyr. (B) Ribbon diagram of the structure of GFP. The α-helices are shown in red, the β-strands are shown in green, and the fluorophore is shown as a ball-and-stick model. The 11-stranded β-barrel forms a cylinder 4.2 nm long and 2.4 nm in diameter with a single α-helix running through its center. The fluorophore is positioned in the middle of this central helix where it is protected from bulk solvent by the surrounding β-strands. (Protein Data Bank, PDB:1EMB. Reprinted with permission from Brejc, K., Sixma, T.K., Kitts, P.A., Kain, S.R., Tsien, R.Y., Ormö, M., Remington, S.J. 1997. Structural basis for dual excitation and photoisomerization of the *Aequorea victoria* green fluorescent protein. *Proceedings of the National Academy of the Sciences USA* 94:2306–2311. Copyright ©1997 National Academy of Sciences, USA). (C) The excitation spectrum of native GFP from *Aequorea victoria* (blue) has two excitation maxima at 395 nm and at 470 nm. The fluorescence emission spectrum (green) has a peak at 509 nm and a shoulder at 540 nm.

Since the introduction of wild–type GFP, many mutant forms such as enhanced GFP (EGFP), cyan fluorescent protein (CFP), and yellow fluorescent protein (YFP) have been created by altering the amino acid sequence of the fluorophore. These mutants display different spectra of fluorescence. For example, the red-shifted variant EGFP contains the double amino acid substitutions Phe64 to Leu and Ser65 to Thr. The

TOOL BOX 9.1

Production of recombinant proteins

Many proteins which may be used for medical treatment or for research are normally expressed at very low concentrations. Through recombinant DNA technology, a large quantity of a protein of interest can be produced. This is called "overexpression" of the recombinant protein. Production of recombinant proteins involves the cloning of the cDNA encoding the desired protein into an "expression vector." The expression vector contains a promoter so that the cDNA can be expressed. Next, the recombinant expression vector is introduced into a bacterial or eukaryotic host cell to allow overexpression of the protein. There are also systems available for *in vitro* translation.

Overexpression of recombinant proteins in bacteria

A commonly used bacterial expression vector contains the *lac* promoter upstream of a multiple cloning site (see Section 10.5 for details). The lactose analog isopropylthiogalactoside (IPTG) stimulates the expression of a cloned cDNA encoding the protein of interest (Fig. 1A). Depending upon the particular application, cloned genes can be expressed to produce native proteins (in their natural state) or fusion proteins, where the foreign polypeptide is fused to a vector-encoded epitope tag (e.g. GST, His, or FLAG). Native proteins are preferred for therapeutic use because fusion proteins can be immunogenic in humans.

Figure 1 Comparison of the production of recombinant proteins in a bacterial host and by *in vitro* translation. (A) Production of recombinant proteins involves the cloning of the cDNA encoding the protein of interest into an expression vector. The expression vector typically contains a *lac* promoter so that the cDNA can be expressed in the *E. coli* host cell. Upon induction with IPTG the recombinant protein is expressed. (B) Proteins can be translated *in vitro* in a rabbit reticulocyte lysate. The lysate is a cell-free extract made from red blood cell precursors. The lysate contains the cellular components necessary for protein synthesis (tRNA, ribosomes, initiation, elongation, and termination factors). In this system, the cDNA is cloned into an expression vector under control of a phage RNA polymerase promoter, such as SP6, T3, or T7. Upon addition of the appropriate phage polymerase to the cell lysate, transcription of an mRNA transcript occurs, followed by translation. If a radioactively labeled amino acid, such as 35S-methionine, is included in the reaction mixture, then a labeled protein is produced.

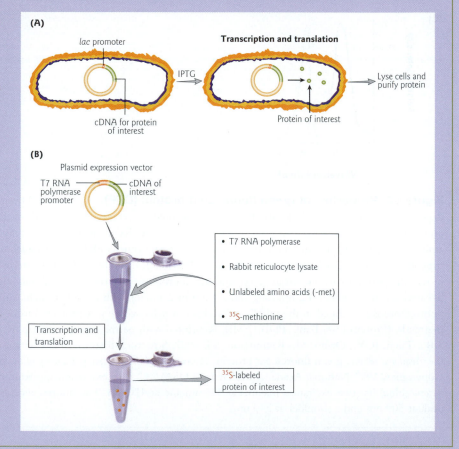

Production of recombinant proteins

However, fusion proteins are usually more stable in *E. coli* because they resemble endogenous proteins, whereas native proteins can be targeted for degradation. Fusion proteins also offer another advantage: they can be easily purified (e.g. by affinity chromatography). It is sometimes possible to cleave the vector-derived polypeptide from the fusion protein using specific proteases, to yield native protein.

Although the overexpression of cloned genes in *E. coli* has facilitated the industrial-scale synthesis of many prokaryotic and eukaryotic proteins, there are a number of problems associated with this system. Overexpressed foreign proteins often can form insoluble inclusion bodies which must be broken up by harsh chemical treatments that denature proteins. *E. coli* often fails to fold and process eukaryotic proteins properly, probably because it lacks the molecular chaperones present in eukaryotic cells. For example, post-translational modification such as cleavage or glycosylation does not take place, and correct disulfide bonds do not form.

Overexpression of recombinant proteins in eukaryotic cells

Where bacterial cells fail to process expressed proteins correctly, eukaryotic cells may be used as alternative expression hosts. Three types of eukaryotic hosts are used for protein overexpression: yeast, insect cells, and mammalian cells. The insect cell system involves a baculovirus expression system where foreign genes are overexpressed from the strong promoter of a nonessential polyhedrin gene in baculovirus-infected insect cells. Mammalian cell systems are the least efficient of the three; however, sometimes they are the only option for producing recombinant protein if processing or post-translational modification does not occur correctly in other cells.

***In vitro* translation of recombinant proteins**

In vitro translation in a cell-free extract can be used to produce small quantities of labeled protein for analysis from purified mRNA or from a DNA template (Fig. 1B). If a radioactively labeled amino acid is included in the reaction mixture, then a labeled protein is produced.

mutant fluoresces 35-fold more intensely than wild-type GFP when excited at 488 nm. CFP contains six amino acid substitutions. One of these mutations, Tyr66 to Trp, shifts the fluorophore's excitation and emission properties, while the other five substitutions enhance the brightness and solubility of the protein. Recently, a new fluorescent protein gene, red fluorescent protein (*RFP* or *DsRed*) was isolated from a tropical coral (*Discosoma striata*). The availability of these different fluorescent proteins makes it possible to carry out multiple labeling of different organelles or structures within the same cells or different tissues or cells in the same organism (Fig. 9.6, see p. 248).

9.4 *In vitro* mutagenesis

Once a DNA molecule has been cloned, *in vitro* mutagenesis techniques can be used to introduce sequence changes. These can be specific mutations that allow functional comparison between mutant and wild-type clones; e.g. to identify regulatory elements or critical amino acid codons. Alternatively, they can be random mutations at a defined region that allow the screening of many variants; e.g. to identify those with improved or diminished performance. The analysis of gene regulation may involve comparing the activity of a series of reporter plasmids in which the regulatory element of interest has been modified by *in vitro* mutagenesis. Such analysis can be carried out by *in vitro* transcription using different cell lysates, by transient transfection of reporter plasmids into cells, or, for multicellular organisms, by introducing the plasmid into the germline (see Fig. 15.2).

There are three main types of *in vitro* mutagenesis: deletion, scanning, and site-directed (Fig. 9.7). Although a number of different strategies are available for all three types, most molecular biologists use PCR-based methods. Deletion mutagenesis removes segments of DNA from a clone (Fig. 9.7A).

Fluorescence, confocal, and multiphoton microscopy

One of the ultimate goals of microscopy is to be able to locate and observe the dynamics of single molecules in chemically and biologically relevant environments in "real time." Advances in technology and new fluorescent probes such as GFP have revolutionized light microscopy imaging in living biological samples, and have opened entire new worlds to the microscopic eye. Fluorescence, confocal, and multiphoton microscopy have become some of the most commonly used techniques by molecular and cellular biologists to visualize gene expression and the localization of cellular components, and for scanning microarrays (see Fig. 16.13). The major application of these types of microscopy in molecular biology are for imaging either fixed or living tissues that have been labeled with one or more fluorescent probes. When samples are imaged using a conventional fluorescence microscope, the fluorescence in the specimen away from the region of interest interferes with resolution of the structures in focus, especially for those specimens that are thicker than approximately 2 µm. The development of confocal and multiphoton microscopy has enabled the imaging of discrete regions of tissues virtually free of out-of-focus fluorescence (Fig. 1).

Fluorescence terminology

A fluorochrome is a natural or synthetic dye or molecule that is capable of exhibiting fluorescence. Fluorochromes (also termed fluorescent molecules, fluorescent probes, fluorescent dyes, or fluorescent tags) are usually heterocyclic molecules containing nitrogen, sulfur, and/or oxygen with delocalized electron systems and reactive moieties that enable the compounds to be attached to proteins and nucleic acids. One of the most commonly

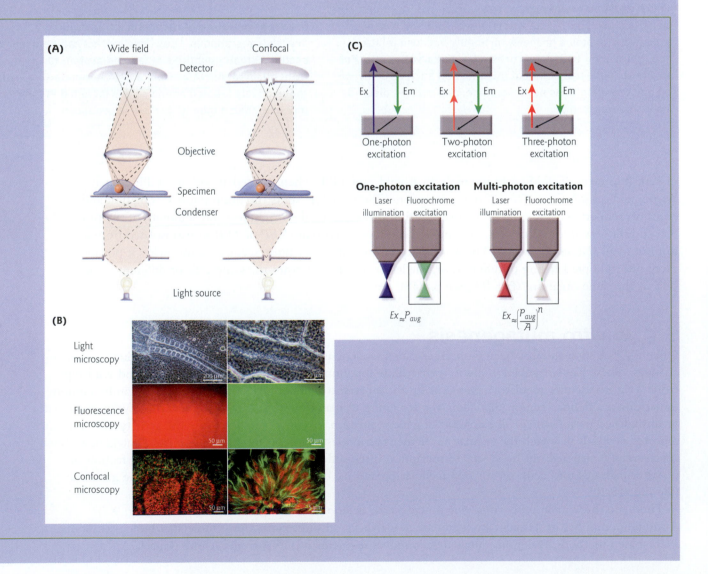

$$Ex \approx P_{avg}$$

$$Ex \approx \left(\frac{P_{avg}}{A}\right)^n$$

used fluorochromes is fluorescein isothiocyanate (FITC, pronounced "fit-see"). FITC fluoresces a yellow-green light when excited with ultraviolet light.

Another term that has increased in popularity is "fluorophore." A fluorophore is defined as the region of a molecule capable of exhibiting fluorescence. In other words, a fluorophore is a conjugated fluorochrome. For example, when the fluorochrome FITC is bound to a nucleic acid or protein, FITC is then referred to as a "fluorophore." Similarly, GFP is referred to as a fluorophore when used as a reporter protein. Not surprisingly, all of these terms tend to get used interchangeably in the scientific literature.

Confocal microscopy

The method of image formation in a confocal microscope is fundamentally different from that in a conventional wide-field microscope in which the entire specimen is bathed in light from a mercury source, and the image can be viewed directly by eye. In contrast, illumination in a confocal microscope is achieved by scanning one or more focused beams of light from a laser across the specimen (Fig. 1). Confocal microscopy uses the resolving power of the objective lens twice: first the illumination light is focused to a diffraction-limited spot; second, the signal photons are focused onto a detector pinhole that rejects scattered and out-of-focus light. The images produced by scanning the specimen in this way are called "optical sections." This term refers to the noninvasive method of image collection by the instrument, which uses light rather than physical means to "section" the specimen. By collecting a series of such optical sections or "slices" (called a Z series) researchers

Figure 1 (*opposite*) **Comparison of conventional fluorescence microscopy, confocal microscopy, and multiphoton microscopy.** (A) Schematic illustration of the operating principles of a conventional, or wide-field, fluorescence microscope and a confocal microscope. In a wide-field fluorescence microscope (left), the specimen is illuminated over an extended region by a light source and condenser. The detector forms an image from the sum of all the simultaneously arriving light rays. Light rays from three points in the specimen are shown. The darker dashed lines show a centrally located point, the lighter dashed lines show an off-axis point, and the dotted lines indicated a point that is on-axis but is located below the plane of focus and gives a blurred image at the detector. In a confocal microscope (right), two pinhole apertures are present. The upper aperture allows only the focused light rays from the on-axis, in-focus point of the specimen to pass to the detector (darker dashed lines). The lower aperture restricts the illumination so that it is focused on the point seen by the upper pinhole aperture. Light rays arising from the other two points are not detected. (B) Comparison of images of a thick (0.5 μm) fluorescent specimen from a conventional and confocal microscope. The sample is a chick embryo labeled with propidium iodide to stain the cell nucleus (red) and a FITC-labeled antibody against tubulin (a component of the cell cytoskeleton) (green). (Top left) Low magnification, wide-field, light microscope image of the entire embryo. (Top right) Light microscope image at the same magnification as the fluorescence microscope images. (Middle row) Conventional fluorescence microscope images showing the same field of view for propidium iodide (red) and tubulin (green) distribution. The large amount of out-of-focus light makes it impossible to see the fine cellular structures. (Bottom left) Optical sections obtained by confocal microscopy of exactly the same field and focal plane as in the middle row. (Bottom right) Higher magnification view of a portion of the same field. Bundles of tubulin (green) and nuclei with condensed chromatin (red) can be seen (dotted white ellipse) (Parts A and B reprinted with permission from Murray, J.M. 2005. Confocal microscopy, deconvolution, and structured illumination methods. Chapter 14 in R.D. Goldman and David L. Spector, eds. *Live Cell Imaging. A Laboratory Manual.* Cold Spring Harbor Laboratory Press, Cold Spring Harbor, NY, pp. 239–279. Copyright © 2005 Cold Spring Harbor Laboratory Press. Images kindly provided by John Murray, University of Pennsylvania, and Camille DiLullo, Philadelphia College of Osteopathic Medicine) (C) Comparison of the principles of one-photon, two-photon, and three-photon excitation. (Top) The three diagrams show the theoretical differences between one-, two-, and three-photon excitation with respect to the energy levels of the fluorochrome. (Bottom) Diagram showing the difference between single- and multiphoton excitation in the sample. For one-photon excitation, fluorochrome excitation is directly proportional to the photon flux of the incident light. For two-photon excitation, excitation depends on the square of the intensity of the incident light, and on the intensity cubed for three-photon excitation. Because of these principles, multiphoton excitation is limited to the focal plane. *Ex*, excitation; P_{avg}, average power of the incident beam of the sample; *A*, numerical aperture of the lense; *n*, number of photons. (Reprinted with permission from Dickinson, M.E. 2005. Multiphoton and multispectral laser-scanning microscopy. Chapter 15 in R.D. Goldman and David L. Spector, eds. *Live Cell Imaging. A Laboratory Manual.* Cold Spring Harbor Laboratory Press, Cold Spring Harbor, NY, pp. 239–279. Copyright © 2005 Cold Spring Harbor Laboratory Press.)

Fluorescence, confocal, and multiphoton microscopy

can create, with the help of sophisticated computer algorithms, high-resolution, three-dimensional images of a sample.

There are two current ways to obtain a confocal image: laser scanning and spinning disk systems. Laser scanning confocal microscopes use motorized mirrors to scan a single laser-produced spot across the sample. Photomultiplier tubes (PMTs) are then used to detect emitted light. In spinning disk confocal microscopy, a disk containing multiple pinholes rotates so that the sample is scanned at multiple points simultaneously at a rate of 1000 scans per second. Spinning disk systems use CCD cameras rather than PMTs. One of the main advantages of spinning disk systems is that they can more easily image live cells in real time. Standard laser scanning confocal technology concentrates light at a single point, potentially causing photobleaching or cell death. In contrast, spinning disk confocal microscopes dissipate the same amount of light over the entire focal area, allowing the investigator to scan for longer periods of time with less damage to the sample.

Multiphoton microscopy
A more recent development is multiphoton excitation

microscopy (also commonly known as two-photon microscopy) (Fig. 1). Typically, the laser used in multiphoton microscopy is a Ti:sapphire mode-locked laser (sapphire crystals embedded with titanium ions) that emits light in the near-infrared range (700–1000 nm). This excitation wavelength is in the region of the spectrum in which there is virtually no absorption in cells or most chemical systems. Excitation of the fluorophore to a higher energy state occurs via the simultaneous absorption of two photons of excitation light. Normally the fluorophore would be excited by a single photon with a shorter wavelength. The nonlinear optical absorption property of two-photon excitation limits the fluorophore excitation to the point of focus. The infrared excitation light has lower scattering and offers better depth penetration and the absence of background fluorescence. The sensitivity of detection is much higher than for confocal microscopy because no aperture is required in the emission path and a greater number of photons reach the photodetector. Limiting the excitation light to the point of focus rather than exposing the entire sample considerably reduces total photobleaching and photodamage of samples. These features are important for live cell analysis, particularly in thick tissue.

Unidirectional deletions by exonucleases can be used to create a nested set of deletions (where one end is common and the other variable). These are useful for analyzing regulatory sequences in DNA, such as a gene promoter (referred to as "promoter bashing"), or as a starting point to provide information on the position of functional domains within a regulatory protein (e.g. the DNA-binding domain or activation domain). Scanning mutagenesis is the systematic replacement (substitution) of each part of a gene clone to determine its function. For example, in linker-scanning mutagenesis, small blocks of DNA sequence are deleted and then replaced by oligonucleotide linkers at each position along the gene clone (Fig. 9.7B). By replacing the deleted DNA, this preserves the spatial relationship of the remaining DNA sequence. Often, the spacing of DNA regulatory elements or protein domains is critical for their function (see Sections 11.3 and 11.5).

The introduction of specific base substitutions or small insertions at defined sites in a cloned DNA molecule is called site-directed mutagenesis (Fig. 9.7C). A widely used variation of site-directed mutagenesis in proteins is alanine substitution. The codon for alanine, a small hydrophobic amino acid with a methyl group side chain, is introduced in place of the pre-existing amino acid codon. When placed on the solvent-exposed surface of a protein, the alanine is assumed to have little overall effect on the structure. Used in conjunction with X-ray crystallography, site-directed mutagenesis can be employed to validate the importance of critical amino acids in protein structure and function.

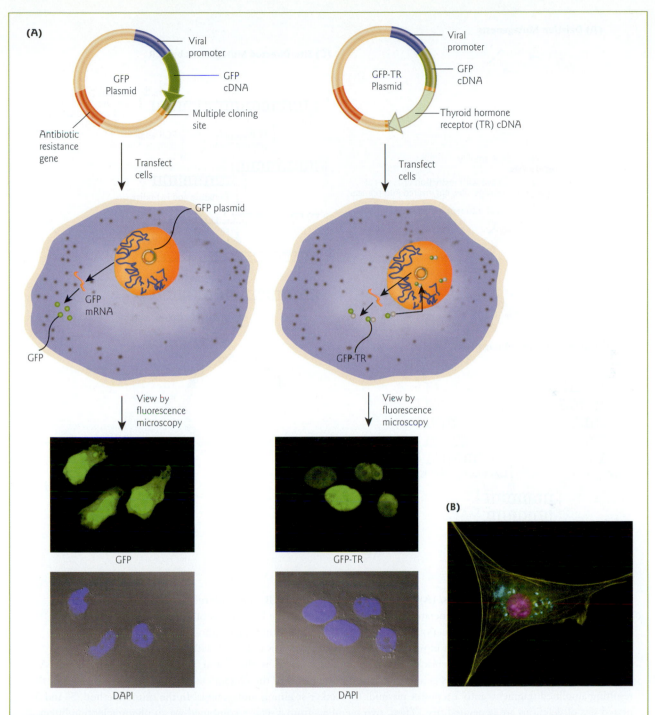

Figure 9.6 Use of Green fluorescent protein (GFP) fusion proteins. (A) Tracking the intracellular localization of a protein of interest. In this example, cells were transfected either with a GFP expression plasmid, or an expression plasmid for GFP-tagged thyroid hormone receptor (GFP-TR). The thyroid hormone receptor is a transcription factor that alters gene expression in response to thyroid hormone. Cells were incubated for 24 hours to allow time for transcription and translation, and then viewed by fluorescence microscopy and Nomarski interference microscopy. Nomarski optics use light refraction to highlight cell structure. In the fluorescence microscope image on the left, it can be seen the GFP diffuses throughout the whole cell. The image on the right shows that the GFP-tagged thyroid hormone receptor (green) becomes localized to the cell nucleus. The nucleus is stained blue with DAPI, a DNA-specific stain. The two panels show the same field of view, but viewed separately using fluorescence detection filters specific for GFP and DAPI. (B) Double labeling with cyan fluorescent protein (CFP) and yellow fluorescent protein (YFP). HeLa cells were transiently transfected with expression vectors for a CFP-tagged Golgi-specific protein and YFP-tagged actin. The nucleus was stained with DAPI (pseudocolored magenta). (Photographs in Parts A and B courtesy of Ghislain Bonamy, College of William and Mary.)

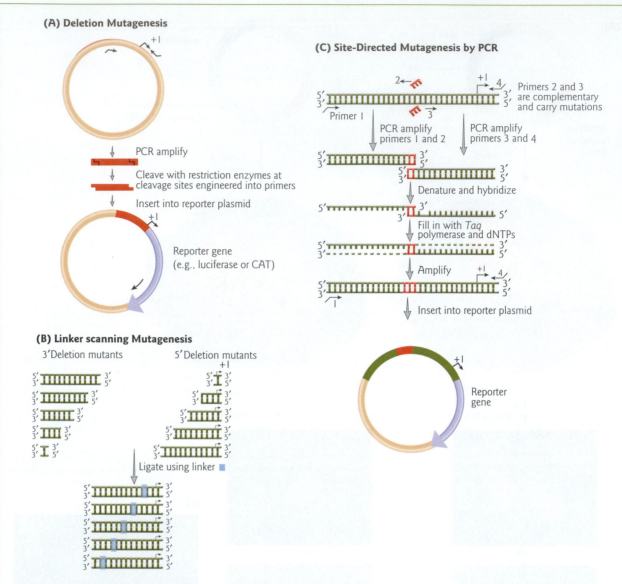

Figure 9.7 *In vitro* mutagenesis. (A) Deletion mutagenesis by PCR. The polymerase chain reaction (PCR) provides a convenient method for generating deletions to determine the boundaries of a regulatory region. The schematic diagram depicts the construction of a simple promoter deletion mutant using this method. The cloned regulatory region of interest is shown in red. The start of transcription is indicated with a short arrow labeled +1. Two oligonucleotide primers (longer black arrows) are used that represent the 5′ and 3′ end points of the deletion. A restriction endonuclease cutting site is generally added onto the ends of the oligonucleotides to allow subcloning of the resulting amplified fragments into a reporter plasmid. (B) Linker-scanning mutagenesis. In the classic method, 5′ and 3′ nested sets of deletions are generated first. Then, two members from a set are combined via an oligonucleotide linker in such a way as to replace the missing 9 bp of wild-type sequence. By placing such linkers throughout the regulatory region, the entire region can be scanned for functional elements without altering the relative spacing of the remaining regulatory sites. (C) Site-directed mutagenesis by PCR. In the example shown, PCR is used to generate a mismatch within a regulatory region of interest. Using this approach, mutations of any size, from one to dozens of base pairs, can be generated with the appropriate primers. The mutant is synthesized in three steps. In the first step, the internal primers that carry the mutations (primers 2 and 3), and the two flanking primers (primers 1 and 4), are used to generate two mutant PCR fragments. The internal primers are designed so that there is a 15 bp region of overlap between primers 2 and 3 (red). Because of this, the downstream end of the first fragment contains a 15 bp region of overlap with the upstream end of the second fragment. In the second step, the two PCR products are denatured and the top strand of the first product hybridizes with the bottom strand of the second product. *Taq* polymerase is then used to "fill in" the ends, linking the two fragments. In the third step, the fragment is amplified using the flanking primers (primers 1 and 4). The mutant PCR product can then be inserted into a reporter plasmid for analysis.

9.5 Analysis at the level of gene transcription: RNA expression and localization

Gene expression levels within a cell, as well as the complex mechanisms that regulate differential expression, are of great interest to researchers. In general, changes in cellular mRNA levels directly correlate to changes in their corresponding protein levels, although there are exceptions to this rule. In general, genes are expressed constitutively, temporally, or spatially. Constitutive expression implies that the gene is expressed at all times. mRNA for genes that exhibit spatial expression are only found in specific tissues of an organism. If a gene is only expressed during a specific time in development, it is said to exhibit temporal expression. Combinations of expression patterns are possible. For example, a gene that is always expressed in the liver is exhibiting constitutive spatial expression. Whereas, if the probe only hybridizes to RNA from liver tissue after an individual reaches adult age, it can be concluded that the gene is only expressed in the adult liver and thus exhibits temporal spatial expression.

Monitoring mRNA levels can be accomplished by using a number of different techniques, such as Northern blotting, *in situ* hybridization, ribonuclease (RNase) protection assays, and reverse transcription–polymerase chain reaction (RT-PCR) (Fig. 9.8).

Northern blot

A method similar to the DNA blotting and hybridization method described in Chapter 8 (see Tool box 8.7) can be used to probe RNA molecules. Since the DNA-based method was named "Southern blotting" after Edward Southern, the RNA-based technique was rather humorously named "Northern" blot hybridization (Fig. 9.8A). The first step in a Southern blot is typically digestion of the DNA with restriction endonucleases. Because mRNAs are relatively short (typically less than 5 kb), there is no need for them to be digested with any enzymes prior to gel electrophoresis. Northern blot hybridization is used to measure the quantity and determine the size of specific transcribed RNAs. Thus, this method is useful for studying the expression of specific genes. For example, RNA can be isolated and analyzed from different tissues and from different developmental stages of an organism. RNA splicing variants that differ in molecular weight can also be characterized by Northern blot analysis. Northern blotting provides a good approximation of gene transcript size in kilobases; however, the technique is not very sensitive and low abundance mRNAs can be difficult to detect.

In situ hybridization

In situ hybridization allows an investigator to look at the precise localization of RNA within a cell. This technique is often used to confirm and extend the results of a Northern blot. In essence, the same principles of nucleic acid hybridization are used, but hybridization with a labeled probe is carried out on sample tissue instead of isolated RNA (Fig. 9.8B). For radioactive labeling, tritium (^{3}H) is often used because it has lower energy emissions than phosphorus-32 (^{32}P) (see Tool box 8.4) and thus allows more precise localization. After incubation in tritium-labeled probe, the slide is dipped in photographic emulsion and developed. Silver grains represent the location of mRNA complementary to the probe (either DNA or antisense RNA). Nonradioactive-detection methods use fluorescently labeled probes or probes tagged with an antigen. When a fluorescently labeled probe is used the technique is called fluorescent *in situ* hybridization (FISH). The samples are visualized by fluorescence microscopy. FISH can, of course, also be used for detecting DNA *in situ* and is a technique commonly used for analysis of the location of specific gene sequences on metaphase chromosomes. Alternatively, when a probe is tagged with an antigen, an enzyme-conjugated antibody reaction is used to produce a colored product where hybridization has occurred (Fig. 9.8B). The colored product, representing the localization of the RNA of interest, is visualized by light microscopy.

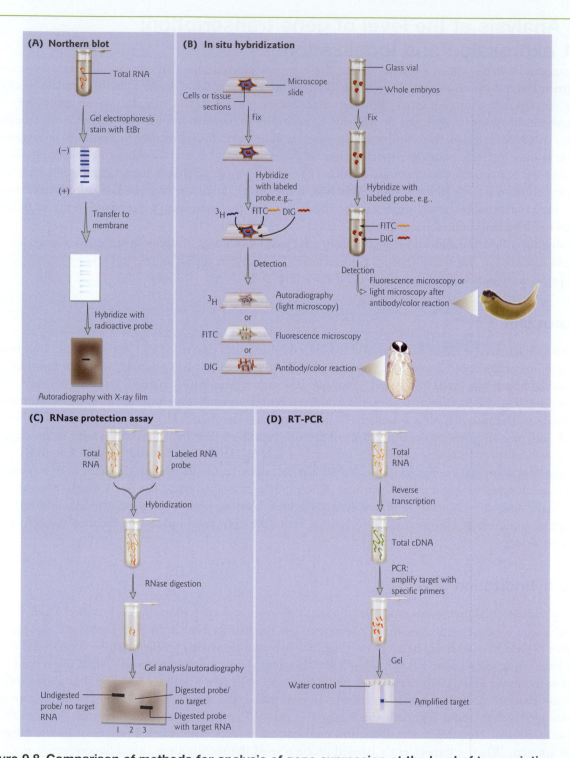

Figure 9.8 Comparison of methods for analysis of gene expression at the level of transcription.
(A) Northern blot. This method is similar to a Southern blot, except that instead of DNA, RNA is separated by a denaturing agarose gel. The RNA can be visualized by staining with ethidium bromide (EtBr). The RNA is then transferred from the gel to a membrane and hybridized with a single-stranded DNA or antisense RNA probe. Hybridization of a radioactively labeled probe to a complementary RNA can be visualized by autoradiography. The density of a band is proportional to the amount of mRNA present. (B) *In situ* hybridization. In this method a labeled DNA or RNA probe is hybridized to cells or sections of tissue fixed on a slide, or to whole embryos fixed in a glass

RNase protection assay (RPA)

RPA is a sensitive method for detecting and quantifying specific mRNA transcripts in a complex mixture of total RNA or mRNA molecules (Fig. 9.8C). The basic principle underlying RPA is as follows. Hybridization of a labeled RNA probe to a transcript protects part of the probe from digestion with an RNase that specifically degrades single-stranded RNA. The length of the section of probe protected by the transcript locates the end of the transcript, relative to the known location of an end of the probe. Because the amount of probe protected by the transcript is proportional to the concentration of transcription, RPA can also be used as a quantitative method. Because of their high sensitivity and resolution, RPAs are well suited for mapping internal and external boundaries in mRNA, such as mapping the start of transcription.

Reverse transcription-PCR (RT-PCR)

RT-PCR is a highly sensitive technique for the detection and quantitation of mRNA (Fig. 9.8D). Compared to the two other commonly used techniques for quantifying mRNA levels – Northern blot analysis and RPA – RT-PCR can be used to quantify mRNA levels from much smaller samples. The technique is sensitive enough to quantify RNA levels in a single cell. By the incorporation of fluorescent dyes in the PCR mixture, real-time measurement of PCR products can be made, allowing for more accurate quantitation of the amounts of the original mRNA. Fluorescence detection filters in the PCR machine are used to measure the fluorescence signal at the end of each PCR extension step and the data are plotted graphically. This approach is called quantitative RT-PCR (qRT-PCR) or "real time" PCR. In addition to its use for analysis of gene expression at the level of mRNA, RT-PCR is routinely used as a diagnostic test for the presence of RNA viruses, such as the agents causing acquired immune deficiency syndrome (AIDS), measles, and mumps.

9.6 Analysis at the level of translation: protein expression and localization

Protein expression can be analyzed in a variety of ways using protein gel electrophoresis (Tool box 9.3) and the tools of immunology (Tool box 9.4), including Western blots, *in situ* analyses, and ELISA, or constructing fusion proteins with an easy to detect tag (see Section 9.3).

Western blot

A method similar to the nucleic acid blotting and hybridization described earlier can be used to probe proteins. In keeping with the naming tradition started with the Southern (DNA) and Northern (RNA)

vial. The detection method depends on how the probe is labeled. (Insets) A whole-mount *in situ* hybridization is compared with the expression pattern in histological sections of *Xenopus* embryos. Neural β-tubulin expression in the whole *Xenopus* embryo can be observed throughout the developing nervous system (upper inset). A transverse section of a *Xenopus* embryo shows neural β-tubulin expression localized in the hindbrain and/or neural tube (lower inset). (Photographs courtesy of Matthew R. Wester, College of William and Mary.) (C) RNase protection assay (RPA). A labeled RNA probe complementary to the gene sequence of interest is synthesized by an *in vitro* transcription reaction. The labeled antisense RNA probe is then hybridized to samples of total RNA (or mRNA). After hybridization, the mixture of single-stranded RNA and double-stranded probe–target hybrid is treated with a single-strand-specific ribonuclease (RNase), which digests all single-stranded RNA but not double-stranded RNA. Any RNA that remains undigested is complementary to the antisense probe, and therefore transcribed from the gene of interest. The sample is analyzed by gel electrophoresis and autoradiography. (D) Reverse transcription-PCR (RT-PCR). In the first step, cDNA copies of the total RNA (or mRNA) sample are synthesized using the enzyme reverse transcriptase. In the second step, the specific cDNA of interest is amplified by PCR. The PCR products are analyzed by gel electrophoresis.

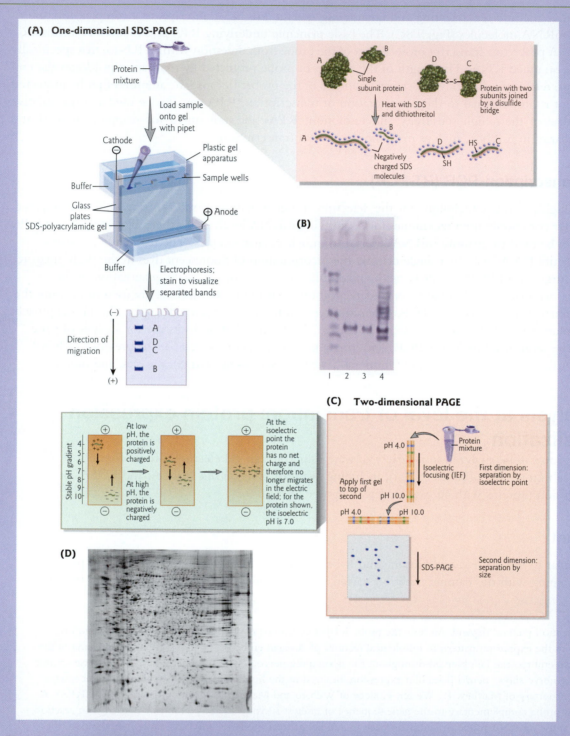

Figure 1 Protein gel electrophoresis. (A) One-dimensional (1D) polyacrylamide gel electrophoresis: SDS–PAGE. An electrophoresis apparatus is depicted. The protein mixture is first treated with SDS, a negatively charged detergent that binds to proteins (inset). This binding dissociates proteins with more than one subunit and forces all polypeptide chains into denatured conformations. In addition, a reducing agent such as 2-mercaptoethanol or dithiothreitol is included in the sample buffer to break any disulphide linkages in the proteins, to ensure that all of the polypeptides in the multisubunit complexes are analyzed separately.

Gel electrophoresis of proteins is a useful analytical method. For protein analysis, polyacrylamide is used instead of agarose because it gives better resolution. Proteins can be visualized as well as separated, permitting a researcher to estimate quickly the number of proteins in a mixture or the degree of purity of a particular protein preparation. Also, gel electrophoresis allows determination of important properties of a protein such as its isoelectric point and approximate molecular weight. A protein's isoelectric point or pI is the pH at which the protein has an equal number of positive and negative charges. Proteins can be separated by one-dimensional (1D) or two-dimensional (2D) polyacrylamide gel electrophoresis (PAGE). 1D-PAGE separates proteins by size, while 2D-PAGE separates proteins by both charge and size.

One-dimensional PAGE: SDS-PAGE

Like nucleic acid electrophoresis, the charge to mass ratio of each protein determines its migration rate through a polyacrylamide gel. Because the carbon backbone of protein molecules is not negatively charged, negative charge is provided by including the anionic detergent sodium dodecyl sulfate (SDS) in the loading, gel, and electrophoresis buffers (Fig. 1). Because of this, 1D-PAGE is generally referred to as SDS-PAGE. The amount of SDS bound to each protein is proportional to its molecular weight, and the rate of migration through the gel is inversely proportional to the logarithm of molecular weight. Thus, in SDS-PAGE separations, migration is determined not by the intrinsic electrical charge of the polypeptide, but by its molecular weight. SDS-PAGE can be used to purify specific components of a mixture that contains more than one protein.

Two-dimensional PAGE

Separating proteins in two dimensions achieves greater resolution of the molecules. Proteins are separated on the basis of two properties (isoelectric point and molecular weight), thus thousands of different proteins can be resolved from each other in a single experiment (Fig. 1).

During electrophoresis, the negatively charged SDS–protein complexes migrate through the porous polyacrylamide gel. Small proteins are able to move through the pores more easily, and faster, than larger proteins. Thus, the proteins separate into bands according to their size as they migrate through the gel. After electrophoresis on the vertical slab gel, proteins are visualized as bands by a dye (e.g. Coomassie blue or silver staining) that binds to the proteins in the gel. Radiolabeled protein samples can be detected by autoradiography. (B) Analysis of protein samples by SDS-PAGE. The photograph shows a Coomassie-stained gel that has been used to detect the proteins present at successive stages in the purification of a GST-tagged fusion protein. The leftmost lane (lane 1) contains a mixture of proteins of known molecular weight used as a marker for protein size. Lane 4 contains the complex mixture of proteins in the starting bacterial cell extract, and lanes 2–3 represent elutions from a glutathione affinity column. Individual proteins appear as sharp, dye-stained bands; a band broadens, however, when it contains too much protein (Credit: Allison Lab, College of William and Mary.) (C) Two-dimensional (2D) PAGE. Preparation of a 2D protein gel by isoelectric focusing (IEF) followed by SDS-PAGE is depicted. First, the proteins are separated according to their isoelectric point (inset) by isoelectric focusing. The net charge of a protein depends on its amino acid composition. If it has more positively charged amino acids such that the sum of the positive charges exceeds the sum of the negative charges, the protein will have an overall positive charge. Proteins with a variation of even one amino acid will have a different overall charge, and thus are distinguishable by electrophoresis. Since proteins are amphoteric compounds (possessing both acidic and basic properties), their net charge is determined by the pH of the medium in which they are suspended. In a solution with a pH above its isoelectric point, a protein has a net negative charge and migrates towards the anode in an electrical field. Below its isoelectric point, the protein is positively charged and migrates towards the cathode. The net charge carried by a protein is independent of its size; that is, the charge carried per unit length of the protein differs from protein to protein. The isoelectric point is the pH at which a protein exhibits no net charge and hence becomes stationary (focuses) in the pH gradient. Thus, when a tube containing a fixed gradient from pH 4.0 to pH 10.0 is subjected to a strong electric field in the appropriate direction, each protein present migrates until it forms a sharp band at its isoelectric pH. For the second dimension separation, the tube is placed horizontally along the top of an SDS-containing vertical slab gel, and proteins move by electrophoresis out of the tube and into the gel, which separates them by size, as in 1D PAGE. Proteins are visible as spots rather than bands in a 2D separation. (D) An image of a two-dimensional gel of a protein extract of antisense oligonucleotide-treated melanoma cells. Each spot corresponds to a different polypeptide chain. The intensity and size of the spot corresponds to the amount of protein present. The proteins were first separated on the basis of their isoelectric points by IEF from left to right. They were then further fractionated according to their molecular weights by electrophoresis from top to bottom in the presence of SDS. (Photograph courtesy of Johannes Winkler, University of Vienna, Austria).

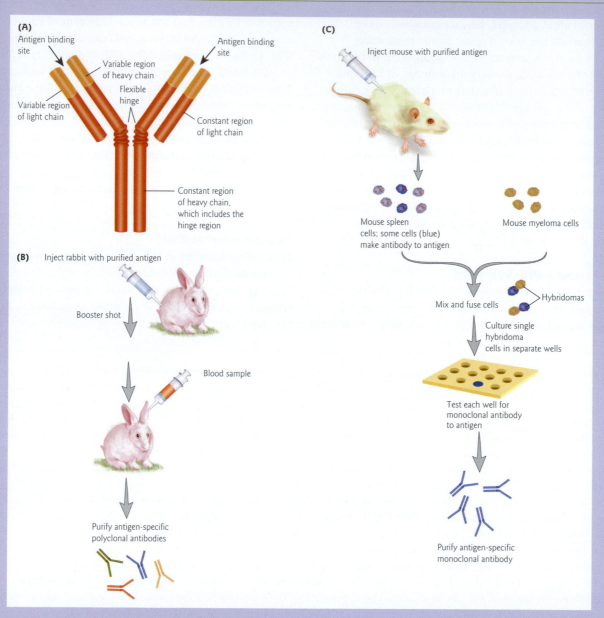

Figure 1 Polyclonal and monoclonal antibodies. (A) Structure of an antibody. An antibody is composed of four polypeptide chains (two heavy chains and two light chains) joined in the shape of a Y. The F_c fragment is formed by the constant region of the heavy and light chains and is present in all antibodies in exactly the same form in a given animal. The F_{ab} fragment is formed by the variable region of the heavy and light chains and recognizes and binds to an epitope of the antigen. (B) Production of a polyclonal antibody. The purified antigen (e.g. a protein) is injected into a rabbit. A booster shot is given a couple of weeks later. A mixture of antibodies is produced by the B cells of the rabbit's immune system, each antibody recognizing a specific epitope within the antigen. A sample of blood is drawn from the rabbit's ear and the antibodies are purified. (C) Production of a monoclonal antibody. A mouse is injected with the purified antigen of interest. B cells from the mouse's spleen are isolated and fused with myeloma cells (a cancerous form of B cells) that are able to divide indefinitely in culture. Single hybridoma cells are cultured in separate wells in a plastic microplate. Each well is tested for the monoclonal antibody to the antigen. The fusion product, called a hybridoma, secretes one type of antibody against a specific antigen. The cells derived from the hybridoma are all clones and thus produce the same monoclonal antibody.

Antibody production

Antibodies are used extensively as tools for molecular biology research and as pharmaceuticals. They are proteins made by B cells of the immune system (see Fig. 12.23). They consist of two fragments, the F_c fragment and the F_{ab} fragment (Fig. 1). The F_c fragment is present in all antibodies in exactly the same form in a given animal. In other words, the F_c fragment says, for example, "this is a rabbit antibody." The F_{ab} fragment is variable and recognizes and binds to an epitope of the antigen. An antigen is a substance that will induce an immune response. An epitope is the name given to a region on an antigen to which an antibody can bind. One antibody will recognize and bind to one and only one epitope. Depending on their use, antibodies are classified as primary or secondary antibodies. Primary antibodies are either polyclonal or monoclonal.

Primary antibodies

Polyclonal antibodies
When an antigen such as a protein is injected into an animal (usually mice, rabbits, or goats), a mixture of antibodies is produced and isolated. Hundreds and possibly thousands of different parts of the protein (epitopes) are involved in activating hundreds and thousands of B cell clones, each producing a different antibody. These are called polyclonal antibodies because they have different specificities to different epitopes (Fig. 1B). Thus, each antibody in the mixture recognizes a specific epitope within the protein molecule. The same process occurs when our immune system responds to infection by a foreign antigen, such as a bacterium.

Monoclonal antibodies
In contrast to polyclonal antibodies, monoclonal antibodies (MAbs) are identical antibodies to a specific epitope of a protein produced by a clone originating from one cell (Fig. 1C). Monoclonal antibodies provide researchers with an unlimited amount of very specific antibodies.

Secondary antibodies
If the primary antibody (the first set of antibodies made for a specific epitope) is conjugated to a fluorochrome, the target protein will be labeled with fluorescent antibodies for easy identification by fluorescence microscopy (see Tool box 9.2). Alternatively, primary antibodies may be linked to enzymes that convert colorless substrates to a colored precipitate. In practice, however, primary antibodies are usually left unlabeled, and a second set of antibodies (secondary antibodies) are created to target the F_c fragment of the primary antibodies.

There are two main advantages to using secondary antibodies that are covalently bonded to detectable tags. First, secondary antibodies provide an additional step for signal amplification, increasing the overall sensitivity of the assay. Second, since the labeled secondary antibody is directed against all primary antibodies of a given species (e.g. anti-rabbit), it can be used with a wide variety of primary antibodies. This means the investigator does not have to label each primary antibody to be used. This is an advantage because antibody labeling can be a time-consuming and expensive process. In addition, certain primary antibodies may be unsuitable for direct labeling. To make it even more convenient, the appropriate secondary antibodies are commercially available. For example, if an investigator is using a primary antibody that was made in rabbit, then an anti-rabbit secondary antibody would be purchased that was made, for example, in goat.

blots, protein blotting from electrophoresis gels is called "Western" blotting. Because the protein of interest is detected with a labeled antibody, Western blotting is also, more descriptively, called "immunoblotting." There is no "eastern" blotting, although there is a technique called a "southwestern" that analyzes the ability of specific proteins to bind DNA.

Western blots are used to detect a specific protein, often present in very low concentration and mixed with other proteins. The steps in the procedure are depicted in Fig. 9.9A. There are several advantages to working with a blot instead of the original gel. These include: (i) a blotting membrane is much easier to handle than a floppy gel that is prone to sticking and tearing; (ii) membrane staining and destaining is faster;

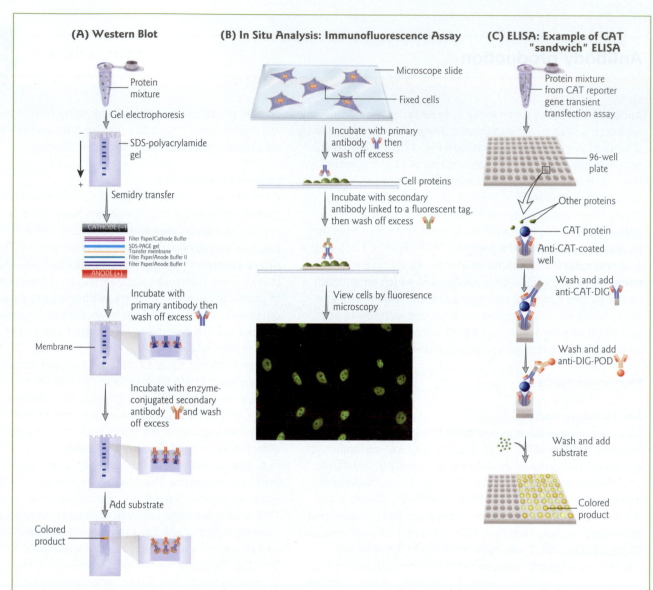

Figure 9.9 Methods for analyzing gene expression at the level of translation. (A) Western blot. The first step in Western blotting, or immunoblotting, is to separate a protein mixture through an SDS–polyacrylamide gel. The proteins migrate into the gel, and the distance they move is proportional to the logarithm of size. After separation of the protein sample on the polyacrylamide gel, the proteins are blotted by electrophoretic transfer to a membrane in a pattern replicating the separation seen on the gel. A semidry transfer system is depicted. In this method, the gel and membrane are saturated with transfer buffer and stacked horizontally between buffer–saturated blotting paper, then sandwiched between solid electrodes. At this point in the procedure, the proteins are not visible. After blotting, a labeled antibody is used as a probe to identify its target protein (the antigen). Only the band containing this protein binds the antibody, forming a layer of antibody molecules (their position cannot be seen at this point). After sufficient time for binding, the membrane is washed to remove any excess unbound primary antibody. In the next step, the membrane is incubated with a secondary antibody that is covalently linked to an enzyme, such as alkaline phosphatase. The secondary antibody binds specifically to the primary antibody. When a chromogenic substrate is added, the enzyme catalyzes a reaction and a deep purple precipitate forms marking the band containing the protein of interest. Alternatively, a chemiluminescent substrate can be added, leading to the production of light. In this case, the membrane is exposed to an X–ray film. (B) *In situ* analysis. An example of an indirect immunofluorescence assay is shown. Cells or tissues are fixed on a microscope slide and the protein of interest is detected by a primary antibody against the target protein and a fluorescently tagged secondary antibody to the primary antibody. (Inset) Micrograph of

(iii) the blot provides a permanent record of the gel; and (iv) small amounts of protein are more readily detected since they are concentrated on the membrane surface instead of being spread throughout the thickness of the gel. Western blotting can be performed by a variety of methods, ranging from capillary diffusion to electrophoretic transfer. The latter is the most widely used because of its speed and precision. In "tank" electrophoretic transfer, the gel and membrane are mounted in a cassette and suspended vertically in a buffer-filled tank between electrode panels. This method is used when longer transfer times are required, since tank transfer has greater buffering capacity and temperature control capability. "Semidry" electrophoretic transfer, as shown in Fig. 9.9A, is a popular method because it provides rapid transfer at relatively low voltages and uses very little transfer buffer.

After blotting, a labeled antibody is used as a probe to identify its target protein (the antigen). Usually, the membrane is incubated in a solution of specific monoclonal or polyclonal antibodies against the protein of interest, that are then detected by using a labeled secondary antibody that binds to the primary antibody probe (Tool box 9.4). The membrane is washed and the places where proteins have bound with antibody are visible. Depending on the method, the bands are visible using autoradiography, chemiluminescence, or the enzymatic production of a colored precipitate. Western blotting allows a researcher to detect the presence, quantity, and size of specific proteins in a particular preparation.

In situ analysis

The expression and localization of fluorescent protein-tagged (e.g. GFP-tagged) fusion proteins can be analyzed directly within living cells. In addition, immunoassays performed *in situ* allow investigators to look at the precise localization of proteins that are not themselves fluorescently tagged. *In situ* immunoassays are often used to confirm and extend the results of a Western blot (Fig. 9.9B). When fluorescently tagged primary antibodies are used for detection, the technique is called direct immunofluorescence assay. When fluorescently tagged secondary antibodies to the primary antibody against the target protein are used for detection, the technique is called indirect immunofluorescence assay. When enzyme-conjugated secondary antibodies are used, two different terms are used, depending on the type of sample. The technique is called "immunohistochemistry" when organs are tagged with antibodies. The term "immunocytochemistry" is reserved for analysis of intracellular material.

Enzyme-linked immunosorbent assay (ELISA)

Immunoassays quantify antigen–antibody reactions. One of the most commonly used immunoassays is ELISA. ELISA combines the specificity of antibodies with the sensitivity of simple enzyme assays. There are two main variations on this method: ELISA can be used to detect the presence of antigens that are recognized by an antibody or it can be used to test for antibodies that recognize an antigen.

cultured cells showing localization of a nuclear protein. The cells were fixed and stained with an antibody specific for a nuclear protein; the secondary antibody was labeled with the dye FITC, which fluoresces green in ultraviolet light. (Photograph courtesy of Vinny Roggero, College of William and Mary.) (C) Enzyme-linked immunosorbent assay (ELISA). An example of a CAT "sandwich" ELISA is depicted. First, antibodies to CAT (anti-CAT) are prebound to the surface of wells in a 96-well plastic microplate. Second, cell extracts from a CAT reporter gene transient transfection assay are added to the wells; any CAT present in the extract will bind to the anti-CAT antibodies prebound to the microplate surface. Third, a digoxigenin-labeled antibody to CAT (anti-CAT-DIG) is added that will bind to any CAT protein bound to the microplate surface. Fourth, an antibody to digoxigenin conjugated to the enzyme peroxidase (anti-DIG-POD) is added to bind to digoxigenin. Finally, a peroxidase substrate is added. The peroxidase enzyme catalyzes the cleavage of the substrate yielding a colored reaction product. The absorbance of the sample is determined using a microplate (ELISA) reader and is directly correlated to the level of CAT present in the cell extract.

One of the most common types of ELISA is "sandwich" ELISA. The steps in this procedure are depicted in Fig. 9.9C. ELISA is a highly sensitive and relatively inexpensive method compared to other types of immunoassay and is regularly used as a clinical diagnostic tool. The assay is often used to detect viral infections such as hepatitis, human immunodeficiency virus type 1 (HIV-1), rubella, and herpes simplex.

9.7 Antisense technology

In 1978, molecular biologists demonstrated that the specificity of bonding between complementary nucleic acid strands could be exploited to create specific inhibitors of gene expression. This marked the beginning of a field that was called "antisense-mediated inhibition of gene expression." Today, this powerful technology is providing major insights into gene function and is being tested for its therapeutic applications. Several types of antisense methods are used to inhibit the expression of a target gene. These include antisense oligonucleotides and RNA interference (RNAi).

Antisense oligonucleotides

Early work in the antisense field primarily used antisense oligonucleotides that were introduced into cells. Antisense oligonucleotides of 15–25 nt are designed to selectively bind to a specific mRNA by Watson–Crick base pairing. The DNA–RNA duplex that is formed inhibits translation of the mRNA into the corresponding protein by altering mRNA splicing, translation, and/or degradation. Typically, either the mRNA strand in the hybrid duplex is cleaved by RNase H, or translation arrest is mediated by blocking of read-through by the ribosome (Fig. 9.10). In general, antisense oligonucleotides suppress

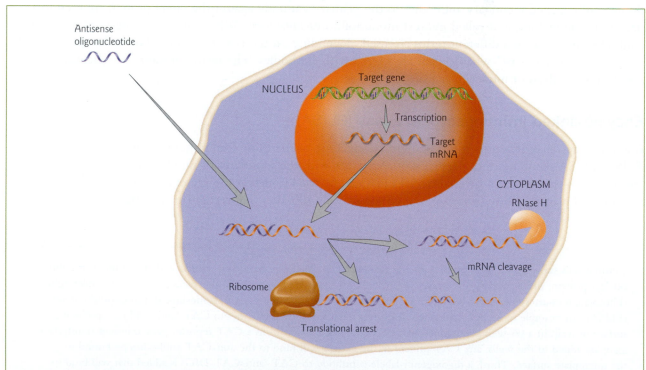

Figure 9.10 Antisense oligonucleotide-mediated inhibition of gene expression. An antisense oligonucleotide with a sequence complementary to that of the target mRNA is introduced into the cell. It hybridizes with the target mRNA and blocks mRNA translation through translational arrest or mRNA cleavage by ribonuclease (RNase) H.

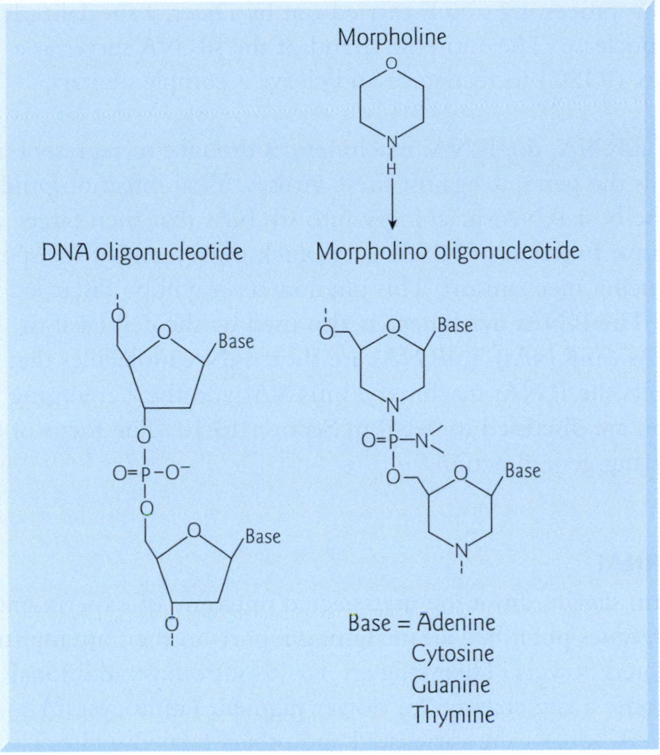

Figure 9.11 Structures of DNA and morpholino oligonucleotides.
Morpholino oligonucleotides are assembled from four different morpholine subunits, each of which contains one of the four bases (adenine, cytosine, guanine, and thymine) linked to a six-membered morpholine ring. Typically, 18–25 subunits are joined in a specific order by nonionic phosphorodiamidate intersubunit bonds to give a morpholino oligonucleotide, but they can be any length. The structure is compared with a DNA oligonucleotide in which the nucleotide subunits are joined by phosphodiester bonds.

gene expression relatively poorly, in part because of their inefficient delivery to cells and to the target mRNA.

Another antisense strategy is to introduce a recombinant expression vector encoding antisense RNA into a host cell by standard methods. The antisense RNA is transcribed from the template DNA, and then forms a duplex with the complementary mRNA. This prevents translation of the corresponding protein. Because the antisense construct is expressed within the cell, it lasts longer compared to antisense oligonucleotides. However, the widespread use of antisense RNA expression has been limited by low transfection efficiencies.

A major improvement in the chemical properties of oligonucleotides came with the development of various modified oligonucleotides, such as antisense morpholinos. These have been particularly useful in studies of zebrafish development. Morpholino oligonucleotides are modified DNA analogs with an altered backbone linkage that lacks a negative charge (Fig. 9.11). They are readily delivered into cultured cells or embryos. Despite their modifications, they bind to complementary RNA sequences by Watson–Crick base pairing. The target RNA–morpholino hybrids are not substrates for RNase H, so the mRNA is not degraded. Morpholinos are usually targeted to the 5′ UTR or start codon of a target mRNA to prevent the ribosome from binding, thereby blocking protein synthesis.

RNA interference (RNAi)

The phenomenon of RNA silencing – or RNA interference (RNAi) as it is now known – is one of the most exciting and revolutionary discoveries of the last decade. In 2002, the journal *Science* designated RNAi as the "breakthrough of the year." RNAi is a sequence-specific gene-silencing process that occurs at the post-transcriptional level. The triggers of RNAi are double-stranded RNA (dsRNA) molecules.

These are processed into short RNAs of ~21–26 nt in length with two-nucleotide 3′ overhangs called small interfering RNAs (siRNAs) (Fig. 9.12). This processing step is carried out by Dicer, a specialized cytoplasmic ribonuclease III (RNase III) family nuclease. The antisense strand of the siRNA serves as a template for the RNA-induced silencing complex (RISC) to recognize and cleave a complementary mRNA, which is then rapidly degraded.

Because some viral genomes are composed of dsRNA, the RNAi machinery is thought to represent an ancient, highly conserved mechanism that defends the genome against these viruses. Viral infection produces dsRNA intermediates, which are processed by the host RNAi machinery into siRNAs that then target viral RNAs for destruction. Another significant defensive function of RNAi is to block movement of transposons within the genome through epigenetic gene-silencing mechanisms. This phenomenon will be discussed in a later chapter on epigenetics (see Section 12.6). The RNAi machinery is also used by the cell itself to regulate gene activity through cellular microRNAs (miRNAs). miRNAs are short RNA molecules that fold into a hairpin to create a dsRNA that then triggers the RNAi machinery. miRNAs and their emerging central role in post-transcriptional gene regulation are discussed in detail in Section 13.10. The focus of this section is on the use of RNAi as a tool for analyzing gene function.

Historical perspective: the discovery of RNAi

The discovery of RNAi arose by a circuitous path, starting from the unexpected outcome of experiments in petunias. In 1990, Richard Jorgensen and colleagues published an intriguing report on their attempt to engineer petunias with deeper purple or red-colored flowers. Their strategy was to introduce additional copies of the chalcone synthase gene, which encodes a key enzyme for flower pigment (anthocyanin) biosynthesis. Instead of darker purple or red flowers, some were variegated and others turned white, indicating that expression of both the introduced transgene and the endogenous gene had been knocked down. RNase protection assays showed that levels of endogenous chalcone synthase mRNA were reduced 50-fold from wild-type levels. A few years later plant virologists made a similar observation of virus-induced gene silencing, but the mechanisms remained unknown at the time.

After the initial observations of gene silencing in plants, many researchers began investigating whether a similar phenomenon occurred in other organisms. RNAi was first described and so named by Andrew Fire, Craig Mello, and colleagues in 1998. They discovered that injecting dsRNAs into the body cavity of *Caenorhabditis elegans* worms silenced only mRNAs containing a complementary sequence. The dsRNA had to include exon regions; dsRNA corresponding to introns and promoter sequences did not cause RNAi. Introducing dsRNAs into *C. elegans* became even easier, when it was shown later that the worm can take up dsRNAs by ingestion of bacteria expressing dsRNA molecules (Fig. 9.13).

Particularly remarkable was the observation by Fire, Mello, and colleagues that the effect of dsRNA crossed cell boundaries and the effect spread throughout the whole organism. It turns out that *C. elegans* has an enzyme called RNA-directed RNA polymerase (RdRP) that uses antisense siRNAs as primers and the target mRNA as a template to make many copies of full-length dsRNA (see Fig. 9.12). This new dsRNA is then digested by Dicer into siRNAs. Although RdRP activity is also present in fruitflies, plants, and fungi, it is not found in mammalian cells.

Before RNAi was well characterized it was referred to by a number of different names: post-transcriptional gene silencing (PTGS) in plants, quelling in fungi, and RNAi in animals. Only after these phenomena were characterized at the molecular level was it obvious that they were the all using the same machinery.

RNAi machinery

Gene silencing through RNAi is carried out by RISC, the RNA-induced silencing complex. RISC is formed through an ordered assembly process in which the RISC loading complex acts in an ATP-dependent manner to place one strand of the siRNA duplex into the RISC complex (Fig. 9.12). Helicase activity is required to unwind the duplex siRNA. When associated with an siRNA that is fully complementary to the

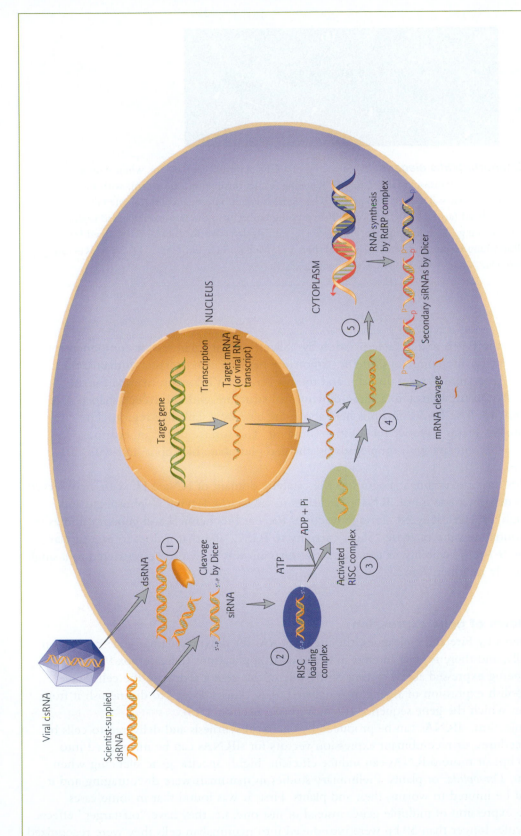

Figure 9.12 Mechanism of RNA interference. The diagram depicts the RNAi pathway triggered by the introduction into cells of either viral double-stranded RNA (dsRNA) or scientist-supplied dsRNA. (1) The ribonuclease Dicer processes long dsRNA into double-stranded small interfering RNAs (siRNAs), with two-nucleotide 3' overhangs. (2) The siRNAs trigger the formation of an RNA-induced silencing complex (RISC). (3) The ATP-dependent unwinding of the siRNA duplex by helicase activity in the RISC loading complex (blue) leads to activated RISC (green). (4) The single-stranded siRNA is used as a guide for target RNA (viral RNA or cellular mRNA) recognition. The complex targets RNAs of complementary sequence for cleavage by "Slicer" activity at the site where the antisense siRNA strand is bound. (5) In worms, flies, plants, and fungi, RNA-directed RNA polymerase (RdRP) uses the siRNA antisense strands as primers and targets RNA as a template to make new dsRNA. Dicer can then process the dsRNA to make more siRNA. This starts a new round of priming and siRNA amplification, and mRNA or viral RNA cleavage.

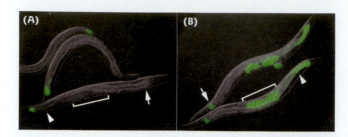

Figure 9.13 RNAi in *Caenorhabditis elegans*. (A) Silencing of a GFP reporter gene occurs when transgenic nematode worms are fed on bacteria expressing GFP dsRNA. Silencing occurs throughout the worm, with the exception of a few cells in the tail that still express some GFP. The signal is lost in intestinal cells near the tail (arrowhead) as well as near the head (arrow). The lack of GFP-expressing embryos in the uterus (bracketed region) demonstrates inheritance of silencing. (B) Silencing does not occur in animals defective for RNAi. (Reprinted by permission from Nature Publishing Group and Macmillan Publishers Ltd: Mello, C.C. and Conte, D. 2004. Revealing the world of RNA interference. *Nature* 431:338–342. Copyright © 2004).

target RNA, RISC cleaves the RNA at a discrete position by "Slicer" activity. The major component of RISC is a member of the Argonaute family of proteins. Argonaute has two characteristic domains, the RNA-binding PAZ (Piwi Argonaute Zwille) domain, and the nuclease PIWI domain (named for the protein Piwi) (Fig. 9.14). Argonaute folds into a three-dimensional structure with a crescent-shaped base made up of the PIWI domains. The PAZ domain is held above the base by a stalk-like region. The placement of the two domains forms a groove for substrate binding. The surface of the inner groove is lined with positive charges that interact with the negatively charged phosphate backbone and with the 2′-OH of the ribose sugar.

The PIWI domain folds into a structure analogous to the catalytic domain of RNase H, with a conserved active site aspartate-aspartate-glutamate motif. RNase H cleaves its substrates, leaving 5′-phosphate and 3′-hydroxyl groups through a metal-catalyzed cleavage reaction. Recent biochemical and genetics studies suggest that Argonaute contributes the catalytic "Slicer" activity to RISC, and that it does so in a similar manner to RNase H (Fig. 9.14). The ability to form an active enzyme was shown to be restricted to a single mammalian Argonaute family member, Ago2.

Applications: knockdown of gene expression

Since the initial experiments by Fire and colleagues, the use of RNAi for downregulating gene expression has increased exponentially, and there is great interest in therapeutic applications (Disease box 9.1). Repressing a gene from being expressed allows testing of the role of the gene product in the cell. Since RNAi may not totally abolish expression of a gene, it is referred to as "knockdown" to distinguish it from "knockout" procedures in which the gene sequence is removed (see Section 15.3).

In addition to direct injection, siRNAs can be produced by chemical synthesis and delivered to cells by standard transfection procedures, or recombinant expression vectors for siRNAs can be introduced into cells. Long (typically 500 bp) or more dsRNAs can induce efficient, highly specific gene silencing when introduced into *C. elegans*, *Drosophila*, or plants. Preliminary studies in mammals were discouraging and it seemed that RNAi might be limited to worms, flies, and plants. First, it was found that in some cases siRNAs knock down the expression of multiple genes instead of just one, i.e. they have "off-target" effects. Second, if dsRNA molecules longer than 30 bp were introduced into mammalian cells they were recognized by the RNA-dependent protein kinase (PKR) and initiated an interferon-based inflammatory response

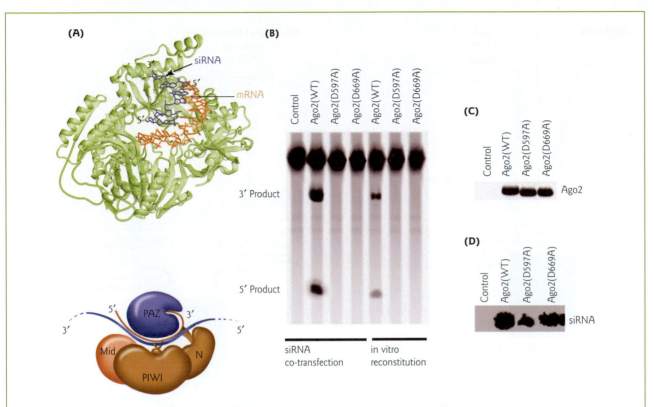

Figure 9.14 siRNA-guided mRNA cleavage by Argonaute. (A) Crystal structure of the Argonaute protein from the archaebacterium *Pyrococcus furiosus*, with siRNA (blue) and mRNA (orange) inserted by model building (upper panel). The phosphate between nucleotides 11 and 12 from the 5′ end of the mRNA falls near the active site residues. A schematic model for siRNA-guided mRNA cleavage is shown below. The siRNA (red) binds with its 3′ end in the PAZ cleft (blue) and the 5′ end near the other end of the cleft. The mRNA (blue) comes in between the N-terminal domain (brown) and PAZ domains and out between the PAZ and middle domains (red). The active site in the PIWI domain (brown) (shown as scissors) cleaves the mRNA opposite the middle of the siRNA guide. (Protein Data Bank, PDB: 1U04. Reprinted with permission from Song, J.J., Smith, S.K., Hannon, G.J., Joshua-Tor, L. 2004. Crystal structure of Argonaute and its implications for RISC Slicer activity. *Science* 305:1434–1437. Copyright © 2004 AAAS.). (B) Mammalian cells were cotransfected with vectors encoding Myc-tagged wild-type Argonaute2 (Ago2WT) or two Ago2 mutants (Ago2D597A and Ago2D669A) along with an siRNA that targets firefly luciferase mRNA. After assembly into RISC *in vivo*, the complexes were immunoprecipitated and tested for siRNA-directed cleavage against labeled firefly luciferase mRNA. Alternatively they were assembled by *in vitro* reconstitution, mixing affinity-purified proteins with single-stranded siRNAs. 5′ and 3′ cleavage products of the mRNA are indicated. The control lane represents cells that were cotransfected with a control vector that did not express Argonaute. Only wild-type Argonaute was able to direct cleavage of the mRNA. (C) Both mutant proteins were expressed at levels similar to wild-type Ago2 as shown by Western blot using an anti-Myc antibody. (D) Both mutant proteins bound siRNAs as readily as wild-type Ago2. siRNA binding was examined by Northern blotting of immunoprecipitates. (Parts B, C, and D reprinted with permission from Liu, J., Carmell, M.A., Rivas, F.V. et al. 2004. Argonaute2 is the catalytic engine of mammalian RNAi. *Science* 305:1437–1441. Copyright © 2004 AAAS.)

that caused general nonspecific inhibition of protein translation. Then, short (< 30 bp) synthetic dsRNAs were shown to trigger sequence-specific knockdown of gene expression without PKR activation. This breakthrough discovery affirmed RNAi as a powerful tool for analyzing the function of mammalian genes, leading to genome-wide phenotypic screens in cell culture and the creation of knockdown mice.

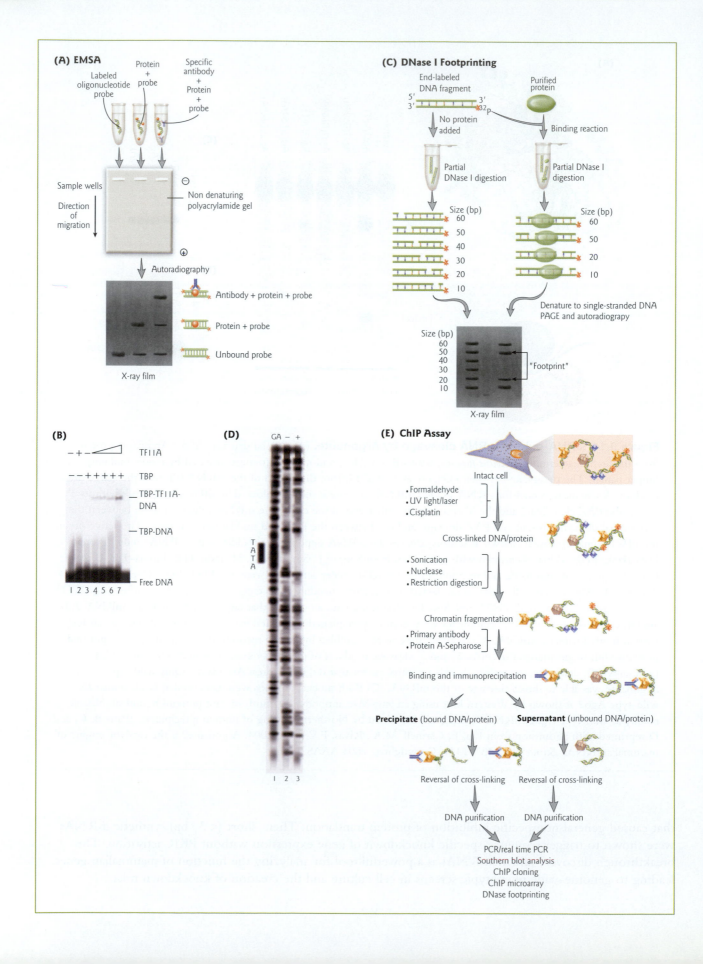

(A) EMSA

Labeled oligonucleotide probe

Protein + probe

Specific antibody + Protein + probe

Sample wells

⊖

Direction of migration

Non denaturing polyacrylamide gel

⊕

Autoradiography

Antibody + protein + probe

Protein + probe

Unbound probe

X-ray film

(B)

− + − ⟋ TFIIA

− − + + + + + TBP

TBP-TFIIA-DNA

TBP-DNA

Free DNA

1 2 3 4 5 6 7

(C) DNase I Footprinting

End-labeled DNA fragment

5'   3'
3'   *32P

Purified protein

No protein added

Binding reaction

Partial DNase I digestion

Partial DNase I digestion

Size (bp)
60
50
40
30
20
10

Size (bp)
60
50
20
10

Denature to single-stranded DNA PAGE and autoradiograpy

Size (bp)
60
50
40
30
20
10

"Footprint"

X-ray film

(D)

GA − +

T
A
T
A

1 2 3

(E) ChIP Assay

Intact cell

• Formaldehyde
• UV light/laser
• Cisplatin

Cross-linked DNA/protein

• Sonication
• Nuclease
• Restriction digestion

Chromatin fragmentation

• Primary antibody
• Protein A-Sepharose

Binding and immunoprecipitation

Precipitate (bound DNA/protein)

Supernatant (unbound DNA/protein)

Reversal of cross-linking

Reversal of cross-linking

DNA purification

DNA purification

PCR/real time PCR
Southern blot analysis
ChIP cloning
ChIP microarray
DNase footprinting

9.8 Analysis of DNA–protein interactions

DNA–protein interactions play a key role in the regulation of a diversity of critical cellular functions, including DNA replication and gene transcription. Often researchers want to know whether a protein binds to a particular region of DNA; for example, does a candidate transcription factor bind to the promoter region of a gene? Three methods are commonly used to demonstrate the association between proteins and DNA. Electrophoretic mobility shift assay (EMSA) is a technique for analysis of DNA–protein interactions *in vitro*, i.e. not within the context of the living cell. Deoxyribonuclease I (DNase I) footprinting is typically used for analysis of DNA–protein interactions *in vitro*, but the first step of DNase treatment can be carried out *in vivo*. Chromatin immunoprecipitation (ChIP) assays allow DNA–protein interactions to be studied within the context of the living cell (Fig. 9.15).

Figure 9.15 (*opposite*) Methods for analyzing DNA–protein interactions. (A) Electrophoretic mobility shift assay (EMSA). The mobility of a ^{32}P-labeled DNA probe with the sequence of interest is compared with the mobility of the probe prebound with protein on a nondenaturing polyacrylamide gel. The source of test protein may be a purified protein, either native or recombinant, or a cell extract containing a mixture of proteins. After electrophoresis, the gel is exposed to X-ray film. Protein–^{32}P–DNA complexes and ^{32}P–DNA alone appear as bands on an autoradiogram. Protein alone is not visible, because it is not labeled. One control for binding specificity in EMSA when using cell extracts is the addition of protein-specific antibodies to the binding reaction, which results in a "supershift." (B) Autoradiogram of an EMSA performed with two components of the general RNA polymerase II transcription machinery, TFIIA and TBP (see Section 11.4), and with a ^{32}P-labeled oligonucleotide encoding a high-affinity binding site for TBP. The amount of TBP was kept constant, while TFIIA was added in increasing amounts. TFIIA did not bind DNA on its own (lane 2), but in the presence of TBP it formed a TBP–TFIIA–DNA complex (lanes 4–7). (Reproduced with permission from Biswas, D., Imbalzano, A.N., Eriksson, P., Yu, Y., Stillman, D.J. 2004. Role for Nhp6, Gcn5, and the Swi/Snf complex in stimulating formation of the TATA-binding protein–TFIIA–DNA complex. *Molecular and Cellular Biology* 24:8312–8321. Copyright © 2004 American Society for Microbiology. Photograph courtesy of David Stillman.) (C) DNase I footprinting. An end-labeled double-stranded DNA fragment with the gene regulatory sequence of interest is incubated in binding buffer with or without a purified DNA-binding protein. Usually a restriction fragment is chosen that can be selectively radiolabeled at the end of one strand. Knowing which is the 5′ end allows for orientation and matching to sequence information. Partial cleavage of the end-labeled DNA with DNase I creates fragments of unique length; any particular phosphodiester bond is broken in some, but not all, DNA molecules. Usually there is one cut per DNA molecule. In samples incubated with the protein of interest, the same region is protected from cleavage in all the DNA molecules. After electrophoresis, samples are visualized by autoradiography. The footprint – the region protected from DNase I cleavage – appears as a gap in the banding pattern. The exact position in the DNA sequence of the footprint can be determined by comparison with the separate sequencing reaction run on the labeled DNA. (D) Autoradiogram of a DNase I footprinting gel showing specific binding of TBP (+) to a gene regulatory sequence (the TATA box, see Chapter 11). The markers (GA) correspond to DNA sequencing reactions performed with the same end-labeled DNA fragment. (Reproduced with permission from Nature Publishing Group and Macmillan Publishers Ltd: Imbalzano, A.N., Kwon, H., Green, M.R., Kingston, R.E. 1994. Facilitated binding of TATA-binding protein to nucleosomal DNA. *Nature* 370:481–485. Copyright © 1994; and from Biswas, D. et al. 2004. *Molecular and Cellular Biology* 24:8312–8321. Copyright © 2004 American Society for Microbiology. Photograph courtesy of David Stillman.) (E) Chromatin immunoprecipitation (ChIP) assay. The first step is usually fixation (cross-linking) of proteins bound to DNA by formaldehyde or UV light treatment. The second step is to shear the high molecular weight DNA into small fragments by sonication (ultrasound waves), nuclease, or restriction digestion. In the final step, the DNA fragments are incubated with specific primary antibodies to the protein of interest in an immunoprecipitation assay. The primary antibody is then bound to a secondary antibody. The secondary antibody has been prebound to a resin (protein A-sepharose), which can be pelleted in a centrifuge. If a particular DNA fragment binds the protein of interest, then it will be recovered in the pellet (immunoprecipitate) by virtue of this association. Any unbound DNA will remain in the supernatant after centrifugation. The immunoprecipitated DNA can then be analyzed after its release from the protein by a variety of methods.

Medical researchers hope that RNAi might be used to knockdown disease-related gene expression in humans. The goal is to deliver chemically synthesized siRNAs complexed with lipids to human cells, thereby triggering sequence-specific silencing of complementary mRNAs. The first clinical trial of RNAi therapy for use in the eye disease macular degeneration was launched in October 2004. Because treatments can be restricted to the eye, the risk of off-target effects was considered of less concern. Time will tell whether these treatments will be successful.

In another trial using an animal model for human disease, it was shown that RNAi can lower cholesterol levels in mice. The targeted mRNA encoded apolipoprotein B (ApoB), a molecule involved in the metabolism of cholesterol. The concentration of this protein in mouse (and human) blood samples correlates with those of cholesterol. Higher levels of both compounds are associated with an increased risk of coronary heart disease. The siRNA was joined to a cholesterol group and injected directly into the bloodstream. The presence of the cholesterol group allowed uptake of the siRNA into mouse tissues, including liver, heart, kidneys, lung, fat, and jejunum (part of the small intestine). siRNA treatment was shown to reduce levels of *apoB* mRNA by more than 50% in the liver and 70% in the jejunum. In addition, the levels of cholesterol in the blood were lower, and were comparable to those observed in mice with a deletion of the *apoB* gene. siRNA treatment did not interfere with expression of any unrelated genes that were analyzed. However, only a small subset of unrelated genes were looked at and the mice were studied over a relatively short period of time. A more comprehensive analysis would need to be carried out before this type of therapy is moved to the clinic.

Electrophoretic mobility shift assay (EMSA)

The principle of EMSA is that protein–DNA complexes migrate more slowly (have "shifted mobility") in an electrophoretic field than unbound (naked) DNA (Fig. 9.15A, B). This type of assay is referred to by a variety of abbreviated names, including gel shift assay and gel retardation assay. EMSA provides an indication of whether a protein binds to a specific fragment of DNA, but it does not determine the exact nucleotides sequence with which the protein interacts.

DNase I footprinting

DNase I "footprinting" was named with the "fingerprinting" methods of chromatographic analyses in mind. This technique is used to precisely map protein-binding sites on DNA (Fig. 9.15C, D). There is, in fact, a technique called "toeprinting" (also known as primer extension inhibition) which is used to map the position of an arrested ribosome on an mRNA transcript. DNase I footprinting involves partial digestion of a DNA fragment of interest with DNase I, in the presence and absence of a protein of interest (see Tool box 14.1). A lack of nuclease cutting shows that the protein binds within a particular region of the DNA. This region where the phosphodiester bonds are protected from cleavage is the "footprint."

Chromatin immunoprecipitation (ChIP) assay

ChIP assay is an important technique for studying DNA–protein (or RNA–protein) interactions *in vivo*, within the context of the living cell (Fig. 9.15E). Cells are treated with a cross-linking agent to form covalent bonds between the DNA and any proteins bound to it. Cell extracts are made and the protein–DNA complexes are immunoprecipitated with an antibody against the protein of interest.

9.9 Analysis of protein–protein interactions

One powerful method for figuring out the function of a newly characterized gene product is to identify its interacting partners. For example, proteins that interact with one another or are part of the same complex are generally involved in the same cellular process. The majority of cellular processes are carried out by protein "machines" or aggregates of 10 or more proteins. Recently, there have been intensive efforts to identify protein–protein interactions on a large scale (see Section 16.6). Four methods are commonly used to demonstrate the interaction between proteins. A pull-down assay is a technique for the analysis of DNA–protein interactions *in vitro*, i.e. not within the context of the living cell. The yeast two-hybrid system and coimmunoprecipitation assays are techniques for analysis of protein–protein interactions *in vivo*, i.e. within the context of the living cell (Fig. 9.16). However, the yeast two-hybrid system involves artificial constructs and coimmunoprecipitation requires cell lysis for analysis, so the precise intracellular localization of protein–protein interactions cannot be determined. In contrast, fluorescence resonance energy transfer (FRET) allows protein–protein interactions to be studied *in situ*, at the precise location in the cell where they normally occur.

Pull-down assay

Pull-down assays are used for the analysis of protein–protein interactions *in vitro*. For example, a GST pull-down assay tests interactions between a GST-tagged protein (the "bait") and another protein (the "prey") (Fig. 9.16A). The bait serves as the secondary affinity support for identifying new protein partners or for confirming a previously suspected protein partner to the bait.

Yeast two-hybrid assay

The yeast two-hybrid assay is used to identify and analyze protein–protein interactions *in vivo*. The basic principle underlying the development of this technique was the observation by molecular biologists that many eukaryotic transcription factors are modular in nature. They contain both a site-specific DNA-binding domain and a transcriptional activation domain that recruits the transcriptional machinery (see Section 11.5). The binding domain and activation domain do not necessarily have to be on the same polypeptide. In fact, a protein with a DNA-binding domain can activate transcription when simply bound to another protein containing an activation domain. It is this principle that forms the basis for the yeast two-hybrid technique (Fig. 9.16B). Typically, a protein of interest fused to the DNA-binding domain (the so-called "bait") is screened against a library of activation domain hybrids ("prey") to select interacting partners. The two-hybrid system has been scaled up for high-throughput screening by the construction of ordered arrays of strains expressing either DNA-binding domain or activation domain fusion proteins.

Coimmunoprecipitation assay

Coimmunoprecipitation assays are used to analyze protein–protein interactions *in vivo*. Cell extracts are made and protein–protein complexes are immunoprecipitated with an antibody against the protein of interest (Fig. 9.16C). The immunoprecipitates are then analyzed by gel electrophoresis or Western blot to identify interacting protein partners.

Fluorescence resonance energy transfer (FRET)

Another method to detect protein–protein interactions *in vivo* involves the use of FRET between fluorescent tags on interacting proteins (Fig. 9.16D). FRET is a particularly powerful, elegant technique because it can be used to make measurements directly in living cells. Protein–protein interactions can be studied *in situ*, at the precise location in the cell where they normally occur. Second, transient interactions between proteins can be followed in real time in single cells.

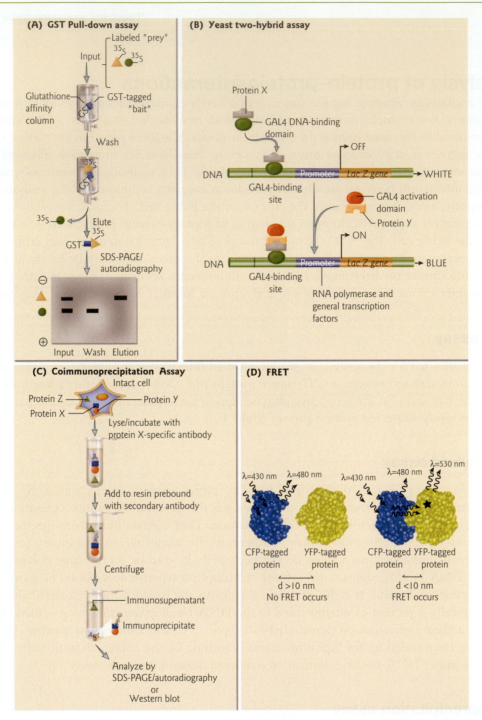

(A) GST Pull-down assay

Labeled "prey"

Input

Glutathione affinity column — GST-tagged "bait"

Wash

Elute

GST

SDS-PAGE/autoradiography

⊖

⊕

Input   Wash   Elution

(B) Yeast two-hybrid assay

Protein X

GAL4 DNA-binding domain

DNA   GAL4-binding site   Promoter   *Lac Z gene* → WHITE   → OFF

GAL4 activation domain

Protein Y

DNA   GAL4-binding site   Promoter   *Lac Z gene* → BLUE   → ON

RNA polymerase and general transcription factors

(C) Coimmunoprecipitation Assay

Intact cell

Protein Z   Protein Y

Protein X

Lyse/incubate with protein X-specific antibody

Add to resin prebound with secondary antibody

Centrifuge

Immunosupernatant

Immunoprecipitate

Analyze by SDS-PAGE/autoradiography or Western blot

(D) FRET

$\lambda = 430$ nm $\lambda = 480$ nm

$\lambda = 430$ nm $\lambda = 480$ nm $\lambda = 530$ nm

CFP-tagged protein   YFP-tagged protein

CFP-tagged protein   YFP-tagged protein

d >10 nm
No FRET occurs

d <10 nm
FRET occurs

Figure 9.16 Methods for analyzing protein–protein interactions. (A) GST pull–down assay. The GST–tagged bait is purified from an appropriate expression system (e.g. bacterial host). It is then immobilized on a glutathione affinity gel in a column. Prey protein can be obtained from multiple sources including recombinant purified proteins, cell lysate, or *in vitro* translation reactions. In this example, if the prey is associated with the GST–tagged bait, then it will elute with the bait from the column, rather than being washed off in a prior step. The eluted proteins can be visualized by SDS-PAGE. The detection method will depend on the sensitivity requirements of the interacting proteins. These methods include Coomassie or silver staining, Western blotting, and ^{35}S radioisotopic detection. (B) Yeast two-hybrid assay. One example of a common strategy is depicted. A DNA sequence encoding a protein of interest (protein X) is fused to the DNA sequence encoding the DNA–binding domain of a well–characterized transcription factor such as GAL4. Another DNA sequence encoding its potential binding partner (protein Y) is fused to the sequence encoding the transcriptional activation domain of GAL4. The activation domain hybrid is introduced to a yeast strain already expressing the DNA–binding domain hybrid. The resulting yeast strain is

9.10 Structural analysis of proteins

Ultimate understanding of protein function requires a detailed knowledge of protein structure. Different information can be gained by studies using X-ray crystallography, nuclear magnetic resonance (NMR) spectroscopy, cryoelectron microscopy, and atomic force microscopy. For example, cyroelectron microscopy can provide a picture of the overall assembly of a large protein complex with multiple subunits, while X-ray crystallography can provide a detailed look at each subunit. NMR spectroscopy and atomic force microscopy provide information about dynamic changes in protein conformation under specific conditions.

X-ray crystallography

Pioneered by Max Perutz and John Kendrew in the 1950s, X-ray crystallography is the oldest of the techniques in structural biology, and still one of the most challenging. Because it provides the most detailed structures, it is the most prolific technique for determining the three-dimensional structure of proteins. More than 8000 proteins have been characterized by X-ray crystallography so far. However, because it requires that dynamic macromolecules be transformed into static crystals, the method cannot capture the dynamics of protein assembly.

In this technique, beams of X-rays that have a wavelength short enough to resolve atoms (0.1–0.2 nm) are passed through a crystal of protein (Fig. 9.17A). Atoms in the protein crystal scatter the X-rays, which produce a diffraction pattern of discrete spots when they are intercepted by photographic film. As many as 25,000 diffraction spots can be obtained from a small protein, so solving the structure of the protein from this pattern is a challenging and complicated process. The process has been likened to reconstructing the precise shape of a rock from the ripples it creates in a pond. Extensive calculations and modifications of the protein (such as binding of heavy metals) must be made to interpret the diffraction pattern.

Nuclear magnetic resonance (NMR) spectroscopy

While X-ray crystallography has provided the majority of structures in the database of biomolecular structures, the fraction determined by NMR spectroscopy is also significant (currently 14%). NMR spectroscopy allows the study of protein dynamics in solution at atomic resolution. In this technique a concentrated protein solution is placed in a magnetic field and the effects of different radio frequencies on the resonances of different atoms are measured. Because NMR is carried out in solution, it has an inherent ability to detect intramolecular motion. The behavior of any atom is influenced by neighboring atoms in

assayed for association of the two fusion proteins. Interaction between protein X and protein Y restores the activity of the transcription factor by bringing together the DNA-binding domain and the activation domain. This leads to expression of a reporter gene with a recognition site for the GAL4 DNA-binding domain. Typically *lacZ* is used as the reporter gene, since its product (β-galactosidase) can be easily detected and measured by conversion of the chromogenic indicator X-Gal from colorless to blue. The amount of the reporter produced can be used to measure the interaction between protein X and its potential partner protein Y. The assay can also use a reporter that is essential for growth, such as histidine synthesis. (C) Coimmunoprecipitation assay. In this technique, cells are lysed and the lysate is incubated with a primary antibody specific for protein X. The primary antibody is then bound to a secondary antibody. The secondary antibody is prebound to a resin (e.g. protein A-agarose, see Fig. 9.4C) which can be pelleted in a centrifuge. If protein Y binds protein X, then it will be recovered in the immunoprecipitate (pellet) by virtue of this association. Unbound proteins in the cell lysate (protein Z) will remain in the supernatant. The immunoprecipitated proteins can be analyzed by SDS-PAGE and autoradiography, or Western blot to confirm their identity. (D) Fluorescence resonance energy transfer (FRET). FRET is a nonradiative process whereby energy from an excited donor fluorophore is transferred to an acceptor fluorophore that is within approximately 10 nm of the excited fluorophore. After excitation of the first fluorophore, FRET is detected by emission from the second fluorophore using appropriate filters, or by alteration of the fluorescence lifetime of the donor. Two fluorophores commonly used are variants of green fluorescent protein (GFP): cyan fluorescent protein (CFP) and yellow fluorescent protein (YFP).

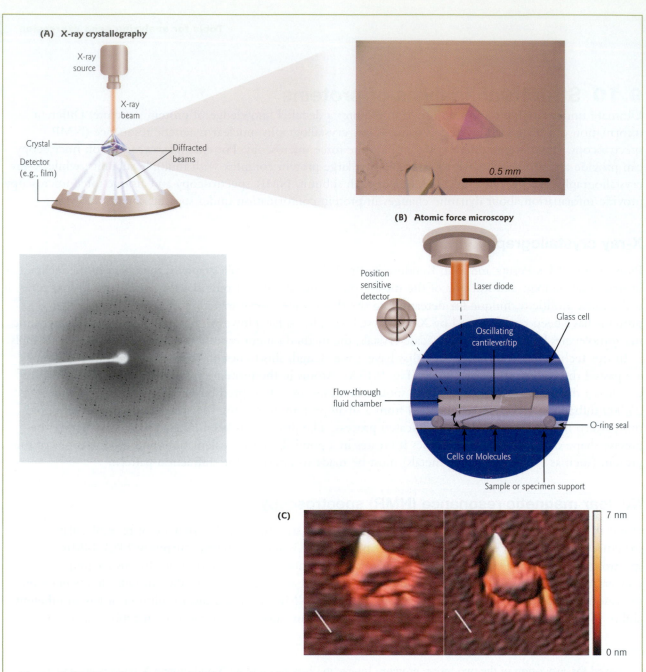

Figure 9.17 Structural analysis of proteins. (A) X-ray crystallography. The basic components of X-ray diffraction analysis are shown in the upper panel. When a narrow beam of X-rays strikes a protein crystal, part of it passes straight through and the rest is diffracted (scattered) in various directions. The intensity of the diffracted wave is recorded on an X-ray film or with a solid-state electronic detector. The electron density is obtained from the diffraction patterns and then a model of the structure is fitted to the electron density. The inset shows a *Haemophilus influenzae* carbonic anhydrase crystal. The bottom panel shows the X-ray diffraction pattern for the carbonic anhydrase crystal. (Photographs courtesy of Roger Rowlett, Colgate University.) (B) Atomic force microscopy (AFM). The diagram depicts an atomic force microscope, which collects topographical images by scanning a cantilever tip across a surface and measuring the cantilever's deflection. (Hohlbauch, S., Ohler, B. 2004. Atomic force microscopy. Unfolding new biological applications. *Biophotonics International* 11:38–43). (C) High-resolution AFM images of the DNA-repair complex Mre11–Rad50–Nbs1 (see Fig. 7.12). From the globular domain, two coiled coils extend that are either not connected (open conformation, left panel) or overlapping (closed conformation, right panel) near the apices of the coiled coils. Scale bars, 25 nm. (Reprinted with permission from Nature Publishing Group and Macmillan Publishers Ltd: Moreno-Herrero, F., de Jager, M., Dekker, N.H., Kanaar, R., Wyman, C., Dekker, C. 2005. Mesoscale conformational changes in the DNA-repair complex Rad50/Mre11/Nbs1 upon binding DNA. *Nature* 437:440–443. Copyright © 2005.)

adjacent residues; the closely spaced residues are more perturbed than distant residues. From the magnitude of the effect, the distances between residues can be calculated. These distances then are used to generate a model of the three-dimensional structure of the protein. Because NMR is limited to proteins smaller than about 20 kDa, it is often used to analyze protein domains, for example to define which parts of the molecule are flexible and which parts are rigid. NMR can provide insight into which domains might be good candidates for crystallization, if the whole protein itself is too unstructured to form crystals. Finally, NMR can be used to measure binding affinities.

Cryoelectron microscopy

Cryoelectron microscopy is particularly useful for determining the structure of large protein complexes that are difficult to crystallize. For example, major progress on spliceosome (see Fig. 13.11) and ribosome (see Section 14.2) structure and function has come from combining data from cryoelectron microscropic analysis with X-ray crystallography results.

In cryoelectron microscopy, the protein sample is snap-frozen at extremely low temperatures in liquid helium (-180 to $-260°C$). When frozen so quickly, the protein is fully hydrated without water crystals, embedded in vitreous ice. The structure is examined in a cryoelectron microscope. The pictures are recorded on film using a low dose of electrons to prevent radiation-induced damage to the structure. Computer algorithms are used to analyze the images and reconstruct the protein's structure in three dimensions. This allows analysis of a protein in its native form, as opposed to fixed material in conventional electron microscopy.

Atomic force microscopy (AFM)

Atomic force microscopy is a technique for analyzing the surface of a rigid material all the way down to the level of the atom. AFM uses a mechanical probe – a cantilever tip with a radius of about 10 nm – to magnify surface features up to 100 million times. By measuring the cantilever's deflection as it scans the surface, AFM produces a topographic (three-dimensional) image of the surface (Fig. 9.17B,C). Amongst its many applications, this technology can be used to characterize the mechanical and structural properties of proteins, and to monitor dynamic events in response to various stimuli by exchanging buffer composition *in situ*.

9.11 Model organisms

The fundamental molecular mechanisms of DNA replication, RNA transcription, protein translation, and gene regulation were first worked out in bacteriophage lambda (λ) and the bacterium *E. coli*. These early studies provided the foundation to understanding the more complex processes in eukaryotes. Many species could claim to be model organisms for eukaryotes. Here, we will concentrate on seven of the most widely used in the study of eukaryotic molecular biology: yeast, *Caenorhabditis elegans*, *Drosophila*, zebrafish, *Arabidopsis*, mouse, and *Xenopus*. Other organisms such as the slime mold *Dictyostelium discoideum* have provided important insights into cell–cell communication and morphogenesis, and the ciliate *Tetrahymena* is the model organism for the study of telomere biology (see Section 6.9). Model organisms share common attributes as well as unique characteristics. In general, they are relatively cheap and plentiful, inexpensive to house, straightforward to propagate, have short gestation periods that produce large numbers of offspring, and are easy to manipulate in the lab. Some have the added attribute of a fairly small and relatively uncomplicated genome.

Each model organism is distinctively suited, as a simplified model, to the study of particular complex aspects of biology. Members of the general public (and even other scientists) sometimes question how findings in a worm or fly can be relevant to human biology. In fact, many discoveries in these organisms have led to important medical breakthroughs. Even where differences exist, they provide important insights to understanding both normal cell physiology and the pathology of diseased cells.

Yeast: *Saccharomyces cerevisiae* and *Schizosaccharomyces pombe*

Outside of scientific circles, yeasts are well known for their important contribution to beer and bread making. Within scientific circles, they are equally well known for their contribution to understanding cell cycle control (see Focus box 6.2). Two species of yeast have become model organisms, the budding yeast *Saccharomyces cerevisiae* and the fission yeast *Schizosaccharomyces pombe*. Both have the advantage of being unicellular organisms that grow very rapidly in the lab. They are about 2 μm in diameter, only one-tenth the size of a white blood cell. Their cell cycle is 90 minutes, and their lifespan is 38 hours from first division to senescence. Yeasts grow readily on a fermentable carbon source and a nitrogen source. Functional genomic and proteomic approaches (see Section 16.3) are being applied to yeast because of its well-characterized genetics. Researchers have individually knocked-out almost 96% of *S. cerevisiae* genes, thus providing 5916 gene-disruption mutants for functional analysis of the yeast genome. Both species have nearly 200 genes homologous to human genes involved in disease, with 23 for cancer alone.

Worm: *Caenorhabditis elegans*

The nematode *Caenorhabditis elegans* is a tiny 1 mm long worm with an unsegmented cylindrical body tapered at both ends. Easy to rear in the lab, the worm eats a diet of bacteria and has a lifespan of 2–3 weeks. It is the most primitive animal to exhibit characteristics that are important in the study of human biology and disease. The versatile genetics and complete map of the cell fate of all 1090 cells of *C. elegans* has led to key discoveries in the genetic regulation of organ development and gave rise to the field of apoptosis (programmed cell death) research. *C. elegans* contains a full set of differentiated tissues, including a nervous system which allows the study of behavior. *C. elegans* is ideal for the study of functional genomics. The genome sequence is complete and well annotated, and systematic gene function can be studied by RNAi.

Fly: *Drosophila melanogaster*

The ubiquitous fruitfly, *Drosophila melanogaster*, is the classic organism for the study of animal genetics. It was introduced to the lab in the early 20th century by Thomas Hunt Morgan, because of its short life cycle (2 weeks), ease of culture, and high fecundity (females can lay up to 100 eggs in 1 day). Mutant flies, with defects in any of several thousand genes, are now used for the study of genetics, development, behavior, and other topics. Major advances have been made in the genetic control of early embryonic development, and in the determination of *Drosophila* genes that control the body plan, formation of body segments, and how genes control the further development of these body segments into specialized organs.

Fish: *Danio rerio*

Danio rerio is a small, tropical zebrafish, native to South East Asia. These fish have gained model organism status because they are easy to look after and breed prodigiously, with an average clutch size of 200. External fertilization allows easy genetic manipulation and analysis, and the transparent embryos develop to adults in 3 months. Their lifespan is approximately 5 years. Developmental defects can be tracked in the transparent embryo by light microscopy. Zebrafish share many structures with their higher (but opaque) vertebrate cousins, thus they can model both development and pathology. Many zebrafish mutants were first identified by observing the color and number of blood cells flowing through their veins, or by noting gross defects such as brain and heart deformities.

Plant: *Arabidopsis thaliana*

Arabidopsis thaliana is commonly known as thale cress or mouse-ear cress. This weed has been known since the 16th century when first identified by Johannes Thal. The first mutant *Arabidopsis* plant was reported by

Alexander Braun in 1873. Ease of cultivation, rapid life cycle (6 weeks from germination to production of mature seeds), and high seed production (5000 seeds per plant), along with small genome size and ease of transformation, have made this organism the undisputed model plant. Many findings of fundamental importance in *Arabidopsis* have been applied to the improvement of crops such as corn, wheat, and rice.

Mouse: *Mus musculus*

For the last century, *M. musculus* has been virtually indispensable in the research lab, although it has a longer generation time compared to other model organisms. Gestation is 19–21 days, with an average of 1–10 pups per litter. Mice reach sexual maturity after 4 weeks and have a lifespan of 1.3–3 years. Although evolution separates mouse from human by an estimated 75 to 100 million years, approximately 85% of the DNA in mice is the same as in humans. Less than 1% of mouse genes have no detectable homolog in humans. As the only mammalian model organism, mice remain the preferred model of human disease because their physiology is at least similar to ours.

Frog: *Xenopus laevis* and *Xenopus tropicalis*

The South African clawed frog, *Xenopus laevis*, is used extensively for studies on development. The fertilized egg is 1.2 mm in diameter and the huge embryos develop externally. In a single mating, *X. laevis* can rapidly produce thousands of embryos. Researchers can manipulate the embryos surgically, chemically, and by microinjection to observe the developmental effects of the manipulation and to study induction. Induction is a process in development whereby a tissue or cell directs neighboring tissues or cells to develop in a particular way. One major limitation for the use of this model organism is that genetic analysis is virtually impossible because *X. laevis* is tetraploid. Also, it takes years to reach sexual maturity. Because of these two features, frog researchers have now turned to another species of clawed frog, *X. tropicalis*, for some studies. This smaller frog is diploid and reaches sexual maturity in less than 3 months.

Chapter summary

Gene expression can be studied in eukaryotic cells using transient and stable transfection assays. To measure the activity of a promoter or other regulatory region, one can link it to a reporter gene, such as the genes encoding β-galactosidase, CAT, luciferase, GUS, or GFP, and let the easily assayed reporter gene products indicate the activity of the regulatory region. Reporter genes can be attached to other sequences so that the reporter protein or peptide tag is fused to the protein of interest. The peptide tag allows recombinant fusion proteins to be purified from a mixture of proteins (such as after overexpression in bacterial cells) by affinity chromatography. Fusion proteins can be used for studies of protein localization, DNA–protein interactions, protein–protein interactions, and for structural studies. For studies in living cells, GFP is the reporter protein of choice. GFP and its spectral variants are naturally fluorescent, containing an intrinsic peptide fluorophore buried in the center of a barrel-shaped protein. GFP and other fluorescent tags can be visualized in cells by fluorescence, confocal, or multiphoton microscopy.

In vitro mutagenesis techniques can be used to introduce sequence changes to allow functional comparison between mutant and wild-type clones. For example, regulatory elements of a gene or critical amino acids in a protein can be identified by deletion, scanning, and/or site-directed mutagenesis. Deletion mutagenesis removes segments of DNA from a clone. Scanning mutagenesis is the systematic replacement of each part of a gene clone to determine its function. The introduction of specific base substitutions or small insertions at defined sites in a cloned DNA molecule is called site-directed mutagenesis.

Genes are expressed all of the time (constitutive expression), at a specific time in development (temporal expression), or are found only in specific tissues of an organism (spatial expression). Gene expression can be quantified by measuring the accumulation of mRNA transcripts of genes. Northern blotting, *in situ* hybridization, RNase protection assays (RPA), and reverse transcription–PCR (RT-PCR) are common ways of accomplishing this task. A Northern blot is similar to a Southern blot, but it contains electrophoretically separated RNAs instead of DNAs. The RNAs on the blot can be detected by hybridizing them to a labeled probe. The intensities of the bands reveal the relative amounts of specific RNA in each. One can hybridize labeled probes to cells or tissues to locate RNA transcripts within a cell. This type of procedure is called *in situ* hybridization; if the probe is fluorescently labeled the technique is called fluorescence *in situ* hybridization (FISH). One can also hybridize labeled probes to whole chromosomes to locate genes or other specific DNA sequences. In RPA, a labeled RNA probe is used to detect the 5′ or 3′ end of a transcript. Hybridization of the probe to the transcript protects a portion of the probe from digestion with RNase, which specifically degrades single-stranded RNA. The length of the section of probe protected by the transcript locates the end of the transcript relative to the known location of the end of the probe. Because the amount of probe protected by the transcript is proportional to the concentration of the transcript, RPA can also be used as a quantitative method. RT-PCR is a highly sensitive technique for the detection and quantitation of RNA. In the first step, cDNA copies of the total RNA sample are synthesized using reverse transcriptase. In the second step, the specific cDNA of interest is amplified by PCR.

Sodium dodecyl sulfate (SDS) polyacrytamide gel electrophoresis (PAGE) is used to separate polypeptides according to their masses. High-resolution separation of polypeptides can be achieved by two-dimensional gel electrophoresis, which uses isoelectric focusing in the first dimension and SDS-PAGE in the second.

Gene expression can be quantified by measuring the accumulation of the protein products of genes. Western blotting (immunoblotting), *in situ* analyses, and enzyme-linked immunosorbent assay (ELISA) are common ways of accomplishing this task. In a Western blot, proteins are separated by electrophoresis, transferred to a membrane, and probed with a labeled antibody. *In situ* immunoassays can be used to analyze protein localization within cells or tissue. When fluorescently tagged secondary antibodies against the target protein are used for detection, the technique is called indirect immunofluorescence assay. The expression of GFP-tagged fusion proteins can be analyzed directly within living cells. ELISA combines the specificity of antibodies with the sensitivity of simple enzyme assays in a microplate format. By using primary antibodies against an antigen of interest in combination with enzyme-conjugated secondary antibodies, ELISA can be used to detect the presence of antigens that are recognized by an antibody, or it can be used to test for antibodies that recognize an antigen.

Antisense-mediated inhibition of gene expression is a powerful technology for the analysis of gene function. Antisense oligonucleotides are designed to selectively bind a specific mRNA. The mRNA in the DNA–RNA duplex that is formed is either cleaved by RNase H or translation arrest is mediated by blocking of read-through by the ribosome. Antisense oligonucleotides are generally inefficient, although some modified oligonucleotides such as morpholino oligonucleotides have proved more effective in gene silencing.

Post-transcriptional gene silencing, or RNA interference (RNAi), occurs when a cell encounters dsRNA. RNAi is an ancient, highly conserved mechanism that defends the genome against viruses and transposons. First discovered in plants and nematode worms, this mechanism has been exploited by molecular biologists as a tool for analyzing gene function by "knockdown" of gene expression. dsRNA, whether introduced into a cell by the scientist or by a virus, is degraded into 21–26 nt fragments (siRNAs) by a nuclease called Dicer. The siRNA duplex is unwound by helicase activity and the antisense strand serves as a template for the RNA-induced silencing complex (RISC) to recognize and cleave a complementary mRNA, which is then rapidly degraded. The major component of RISC is Argonaute, the protein that contributes the catalytic "Slicer" activity which cleaves the mRNA. In plants, nematode worms, and fruitflies (but not in mammalian cells), the siRNA is amplified during RNAi. This occurs when antisense siRNAs hybridize to target mRNAs and serve as primers for the synthesis of full-length antisense RNA by an RNA-dependent RNA polymerase. This new dsRNA is then digested by Dicer into new pieces of siRNA.

DNA–protein interactions play a key role in gene expression. An electrophoretic mobility shift assay (EMSA) detects the interaction between a protein and DNA, by the slower migration of a labeled DNA fragment in an electrophoretic field that occurs on binding to a protein. Footprinting is a means of finding the target DNA sequence, or binding site, of a DNA-binding protein. DNase I footprinting is performed by binding the protein to its end-labeled DNA target, then incubating the DNA–protein complex with DNase. When the resulting DNA fragments are separated by gel electrophoresis, the protein-binding site shows up as a gap, or "footprint," in the pattern where the protein protected the DNA from degradation. Chromatin immunoprecipitation (ChIP) assays allow the study of DNA–protein interactions *in vivo*. Cells are treated with a cross-linking agent to form covalent bonds between the DNA and any proteins bound to it. Cell extracts are made and the protein–DNA complexes are immunoprecipitated with an antibody against the protein of interest.

Gene expression is mediated by multiprotein complexes or "machines." There are a number of methods for analyzing protein–protein interactions. *In vitro* pull-down assays, such as GST pull-down assays, test for interactions between a peptide-tagged protein (the so-called "bait") and another protein (the "prey"). The bait, bound to an affinity column, serves as the secondary affinity support for identifying protein partners. The yeast two-hybrid assay is used to identify protein–protein interactions *in vivo*. A protein of interest fused to a DNA-binding domain of a transcription factor (the bait) is screened against a library of activation-domain hybrids (the prey). Interacting partners are identified by assaying for the activation of reporter gene expression. Interaction between protein X and protein Y restores the activity of the transcription factor by bringing together the DNA-binding and activation domains. In coimmunoprecipitation assays, cell extracts are made and protein–protein complexes are immunoprecipitated with an antibody against the protein of interest, and analyzed by gel electrophoresis or Western blot to identify the interacting protein partners. Fluorescence resonance energy transfer (FRET) allows protein–protein interactions to be studied *in situ*. Energy from an excited donor fluorophore (e.g. CFP) is only transferred to an acceptor fluorophore (e.g. YFP) when the two tagged proteins are adjacent.

Information on protein structure can be gained from a variety of methods. In X-ray crystallography, beams of X-rays that are of a short enough wavelength to resolve atoms are passed through a crystal of protein and the resulting diffraction pattern is used to solve the structure. In NMR spectroscopy, a concentrated protein solution is placed in a magnetic field and the effects of different radio frequencies on the resonances of different atoms are measured. The behavior of any atom is influenced by its neighboring atoms. From the magnitude of effect, the distances between residues can be calculated and used to generate a model of the three-dimensional structure of the protein. NMR is limited to proteins smaller than about 20 kDa. Cryoelectron microscopy is useful for determining the structure of large protein complexes that are difficult to crystallize. The protein is snap-frozen at extremely low temperatures to preserve its native structure and examined in a cryoelectron microscope. Atomic force microscopy uses a mechanical probe to produce a topographic image of the surface of a protein, and to monitor dynamic changes in its conformation.

Many advances in eukaryotic molecular biology have come through the study of gene expression in model organisms such as yeast, the nematode worm *C. elegans*, the fruitfly *Drosophila*, the zebrafish *Danio rerio*, the plant *Arabidopsis*, the mouse, and the frog *Xenopus*. Each of these organisms is distinctively suited, as a simplified model, to the study of particular complex molecular and cellular processes. Typically, model organisms are relatively cheap and plentiful, have a short generation time, produce large numbers of offspring, and are easy to manipulate in the lab. Some have the added feature of a fairly small and relatively uncomplicated genome.

Analytical questions

1 You have purified a protein. When you subject it to SDS–PAGE, two bands are seen. Provide a possible explanation and describe how you could test your hypothesis.
2 Consider a Nothern blot:
   (a) Which strand of a DNA probe will hybridize to mRNA on the blot: the antisense strand, the sense strand, or both strands?
   (b) On the diagram of the DNA–RNA hybrid shown below, which bonds form when the DNA probe hybridizes to the mRNA?

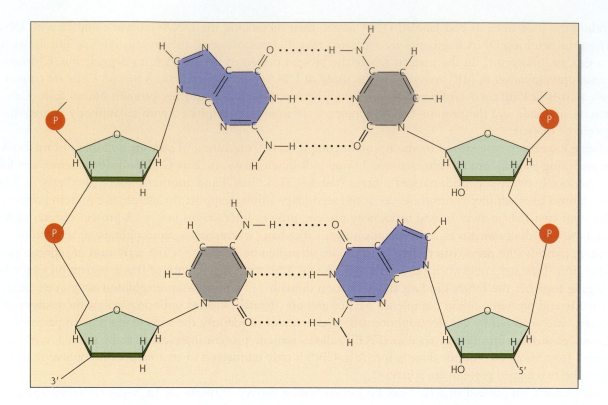

3 Consider a cotransfection assay with expression vectors for two putative transcription factors that you suspect work together to activate transcription, and with a regulatory region of interest attached to a CAT reporter gene. You also test two deletion (Δ) mutants of the regulatory region.

(a) How could you confirm that the transcription factors are transcribed and translated in the transfected cells?

(b) Interpret the results of the CAT assay below. The plus and minus symbols indicate the presence or absence of expression vectors for the listed components in the transfection assay.

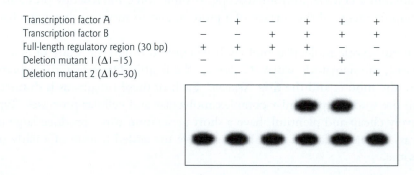

Transcription factor A	–	+	–	+	+	+
Transcription factor B	–	–	+	+	+	+
Full-length regulatory region (30 bp)	+	+	+	+	–	–
Deletion mutant I (Δ1–15)	–	–	–	–	+	–
Deletion mutant 2 (Δ16–30)	–	–	–	–	–	+

(c) How would you demonstrate protein–protein interaction between the two transcription factors *in vivo* and *in vitro*? Show sample positive results.

(d) Describe protocols to further characterize direct binding of the proteins to DNA *in vitro* and *in vivo*. Show sample results.

4 You have transfected siRNA against transcription factor A from question 3 into cells that normally express the transcription factor.

(a) What would be the effect on CAT reporter gene expression?

(b) How could you determine whether RNAi acts at the level of mRNA or translation in this case?

Suggestions for further reading

Ausubel, F.M., Brent, R., Kingston, R.E., Moore, D.D. Seidman, J.G., Smith, J.A., Struhl, K., eds (2002) *Short Protocols in Molecular Biology*, 5th edn. John Wiley & Sons, New York.

Bahls, C., Weitzman, J., Gallagher, R. (2003) Biology's models. *Scientist* 17(Suppl. 1):S5–S34.

Biswas, D., Imbalzano, A.N., Eriksson, P., Yu, Y., Stillman, D.J. (2004) Role for Nhp6, Gcn5, and the Swi/Snf complex in stimulating formation of the TATA-binding protein–TFIIA–DNA complex. *Molecular and Cellular Biology* 24:8312–8321.

Brejc, K., Sixma, T.K., Kitts, P.A., Kain, S.R., Tsien, R.Y., Ormö, M., Remington, S.J. (1997) Structural basis for dual excitation and photoisomerization of the *Aequorea victoria* green fluorescent protein. *Proceedings of the National Academy of the Sciences USA* 94:2306–2311.

Carey, M., Smale, S.T. (2000) *Transcriptional Regulation in Eukaryotes. Concepts, Strategies, and Techniques.* Cold Spring Harbor Laboratory Press, Cold Spring Harbor, NY.

Das, P.M., Ramachandran, K., vanWert, J., Singal, R. (2004) Chromatin immunoprecipitation assay. *Biotechniques* 37:961–969.

Day, R.N., Periasamy, A., Schaufele, F. (2001) Fluorescence resonance energy transfer microscopy of localized protein interactions in the living cell nucleus. *Methods* 25:4–18.

Eccleston, A., Eggleston, A.K., eds (2004) RNA interference. Nature insight. *Nature* 431:337–378.

Fire, A., Xu, S., Montgomery, M.K., Kostas, S.A., Driver, S.E., Mello, C.C. (1998) Potent and specific genetic interference by double-stranded RNA in *Caenorhabditis elegans*. *Nature* 391:806–811.

Goldman, R.D., Spector, D.L. (2005) *Live Cell Imaging. A Laboratory Manual.* Cold Spring Harbor Laboratory Press. Cold Spring Harbor, NY.

Hohlbauch, S., Ohler, B. (2004) Atomic force microscopy. Unfolding new biological applications. *Biophotonics International* 11:38–43.

Jackson, A.L., Linsley, P.S. (2004) Noise amidst the silence: off-target effects of siRNAs? *Trends in Genetics* 20:521–524.

Lee, L.K., Roth, C.M. (2003) Antisense technology in molecular and cellular bioengineering. *Current Opinion in Biotechnology* 14:505–511.

Liu, J., Carmell, M.A., Rivas, F.V. et al. (2004) Argonaute2 is the catalytic engine of mammalian RNAi. *Science* 305:1437–1441.

Mello, C.C., Conte, D. (2004) Revealing the world of RNA interference. *Nature* 431:338–342.

Miesfeld, R.L. (1999) *Applied Molecular Genetics*. Wiley-Liss, New York.

Moreno-Herrero, F., de Jager, M., Dekker, N.H., Kanaar, R., Wyman, C., Dekker, C. (2005) Mesoscale conformational changes in the DNA-repair complex Rad50/Mre11/Nbs1 upon binding DNA. *Nature* 437:440–443.

Napoli, C., Lemieux, C., Jorgensen, R. (1990) Introduction of a chimeric chalcone synthase gene into petunia results in reversible co-suppression of homologous genes *in trans*. *Plant Cell* 2:279–289.

Sambrook, J., Russell, D.W. (2001) *Molecular Cloning: a Laboratory Manual*, 3rd edn. Cold Spring Harbor Laboratory Press, Cold Spring Harbor, NY.

Sandy, P., Ventura, A., Jacks, T. (2005) Mammalian RNAi: a practical guide. *BioTechniques* 39:215–224.

Song, J.J., Smith, S.K., Hannon, G.J., Joshua-Tor, L. (2004) Crystal structure of Argonaute and its implications for RISC Slicer activity. *Science* 305:1434–1437.

Soutschek, J., Akinc, A., Bramlage, B. et al. (2004) Therapeutic silencing of an endogenous gene by systemic administration of modified siRNAs. *Nature* 432:173–178.

Williams, T., Fried, M. (1986) A mouse locus at which transcription from both DNA strands produces mRNAs complementary to their 3′ ends. *Nature* 322:275–279.

Wong, M.L., Medrano, J.F. (2005) Real-time PCR for mRNA quantitation. *BioTechniques* 39:75–85.

Chapter 10

Transcription in prokaryotes

Few proteins have had such strong impact on a field as the *lac* repressor has had in Molecular Biology.

Mitchell Lewis, *Comptes Rendus Biologies* (2005) 328:521.

Outline

10.1 Introduction

A central event in gene expression is the copying of the sequence of the template strand of a gene into a complementary RNA transcript. All cells have at least one kind of RNA polymerase – the enzyme that transcribes RNA from DNA – and the machinery that translates the mRNA into protein. The biochemistry of transcript formation is straightforward, but the regulatory mechanisms that have been developed by organisms to control transcription are complex and highly variable. This chapter focuses on the process of transcription in prokaryotes. Some of the similarities and differences between prokaryotes and eukaryotes are pointed out in this chapter, but the details of transcriptional regulation in eukaryotes are reserved for Chapter 11. The lactose operon, a classic example of the process of induction, is highlighted because of the many important principles it illustrates, and for its relevance to molecular biology research. Other key features of transcriptional regulation in bacteria are illustrated by the arabinose and tryptophan operons. The mode of action of transcriptional regulators is addressed, focusing in particular on the cooperative binding of proteins to DNA, allosteric modification of protein activity, and how distant DNA regulatory sites are brought together by protein–protein interactions that cause DNA looping. Regulation by differential folding of RNA and riboswitches is also discussed.

10.2 Transcription and translation are coupled in bacteria

In bacteria, transcription and translation are said to be "coupled," since they occur within a single cellular compartment. As soon as transcription of the mRNA begins, ribosomes attach and initiate protein synthesis. The whole process of transcription and translation occurs within minutes (Fig. 10.1). In contrast, in eukaryotes, mRNA and protein synthesis are separated between two cellular compartments. Transcription takes place in the nucleus and translation takes place in the cytoplasm. The whole process may take hours or, in some cases, months for developmentally regulated genes. Transcription initiation is the point at which most genes are regulated in both prokaryotes and eukaryotes. However, since transcription and translation are "uncoupled" in eukaryotes, the control of gene expression can be exerted at many additional levels including processing of the RNA transcript, transport of RNA to the cytoplasm, and mRNA stability and translation into protein (Fig. 10.2). These additional levels of control are discussed in Chapters 11–14.

10.3 Mechanism of transcription

By 1967 it was clear the RNA polymerase (often abbreviated as RNAP) is the enzyme that catalyzes RNA synthesis. Using DNA as a template, this enzyme polymerizes (joins) nucleoside triphosphates (NTPs) by phosphodiester bonds from 5′ to 3′. In bacteria such as *Escherichia coli*, there is one type of RNA polymerase, while eukaryotes have multiple nuclear DNA-dependent RNA polymerases and organelle-specific polymerases (Table 10.1).

RNA polymerase binds to a region on DNA called a promoter. Promoters are located right before the start of transcription of the gene(s) physically connected on the same stretch of DNA. For this reason, a promoter is classified as a *cis*-acting sequence or element. A promoter differs from DNA sequences whose role is to be transcribed or translated. It serves exclusively as a sequence of DNA whose function is to be recognized by a regulatory protein. A unifying theme in the control of gene transcription is that a protein complex binds to DNA with specific motifs (stretches of amino acids) recognizing a particular sequence of DNA. A regulatory protein that binds a *cis*-acting sequence is called a *trans*-acting factor. The DNA sequence coding for a *trans*-acting factor is transcribed and translated. The *trans*-acting gene product diffuses to its target, the *cis*-acting DNA sequence (usually) upstream of the gene, and regulates its transcription. The minimal requirements for gene transcription are the gene promoter and the RNA polymerase, while additional factors are required for the regulation of transcription.

Bacterial promoter structure

If nucleotide sequences of DNA are aligned with each other and each has exactly the same series of nucleotides in a given region, the sequence is said to be conserved. If there is some variation in the sequence but certain nucleotides are present at a high frequency, those nucleotides are referred to as making up a consensus sequence. The sequencing of numerous bacterial promoters has shown that they are not absolutely conserved, but they do have a consensus sequence (Fig. 10.3). Promoters rarely match the consensus perfectly. In general, the more closely regions within a promoter resemble the consensus sequences, the

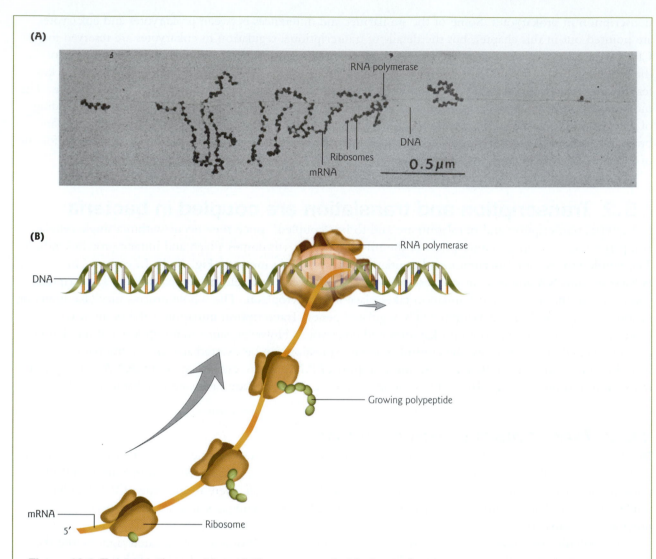

(A)

RNA polymerase

Ribosomes

mRNA

DNA

0.5 µm

(B)

RNA polymerase

DNA

Growing polypeptide

mRNA

5′

Ribosome

Figure 10.1 Transcription and translation are coupled in bacteria. (A) Electron micrograph of bacterial genes in action. The growing polypeptide chain cannot be seen in this view. Magnification × 44,000. (Reprinted with permission from Miller, O.L., Hamkalo, B.A., Thomas, C.A. 1970. Visualization of bacterial genes in action. *Science* 169:392–395. Copyright © 1970 AAAS.) (B) Schematic diagram of the events shown in (A). Translation of mRNA by ribosomes begins when the 5′ end is accessible. When the first ribosome moves from the 5′ end, a second ribosome can attach.

greater the strength of the promoter. Promoter strength refers to the relative frequency of transcription initiation and is related to the affinity of RNA polymerase for the promoter region. In addition, some very strong promoters in *E. coli* rRNA genes have an additional element further upstream called a UP element that increases their ability to attract RNA polymerase.

For the majority of *E. coli* genes, the promoter consensus sequence consists of two hexamer sequences. These sequences are separated by about 17 base pairs and are located upstream of the first base transcribed (Fig. 10.3). It is convention to indicate the start of transcription by the number +1 and to use positive numbers to count farther down the DNA in the direction of transcription, the direction referred to as downstream (3′ to the start site). If transcription is proceeding to the right, then the direction to the left (5′ to the start site) is called upstream with the bases indicated by negative numbers. Centered at −10 is

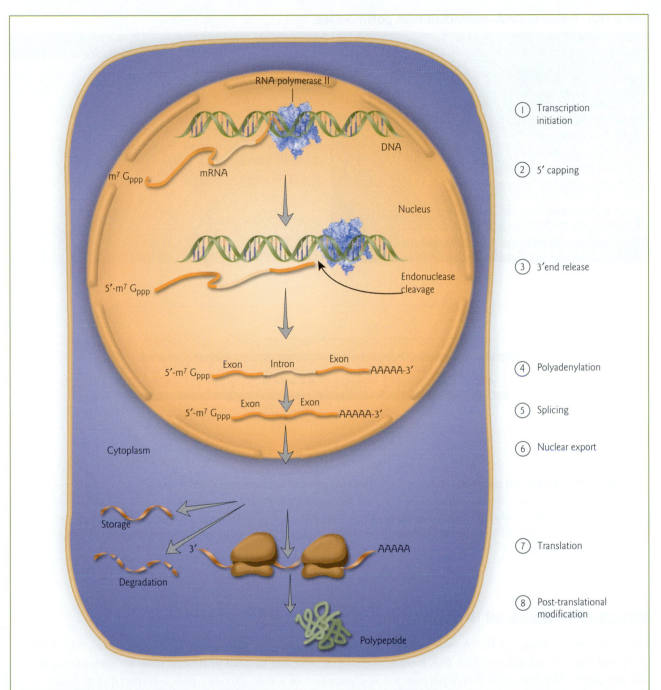

Figure 10.2 Transcription and translation are uncoupled in eukaryotes. This diagram depicts transcriptional, post-transcriptional, translational, and post-translational levels of control of gene expression for a eukaryotic protein-coding gene (see Chapters 11–14).

the consensus sequence TATAAT, which is also known as the TATA box or Pribnow box after one of its discoverers, David Pribnow (who marked a box around the sequence). Now, it is usually called the −10 sequence (or −10 region or element). Another region with similar sequences among many promoters is centered at −35. The consensus sequence at −35 is TTGACA. The spacing between the −10 and −35 sequences and the start point for transcription is important, and deletions or insertions that change the spacing are deleterious.

Table 10.1 Types of DNA-dependent RNA polymerases.

Polymerase	Genes transcribed
Bacteria	
RNA polymerase	All bacterial genes
Eukaryotic nuclei	
RNA polymerase I	Ribosomal RNA (rRNA) genes
RNA polymerase II	mRNA, snoRNAs, some snRNAs, miRNAs
RNA polymerase III	tRNA, 5S rRNA, some snRNAs
RNA polymerase IV (plants only?)	siRNAs involved in heterochromatin formation
Eukaryotic organelles	
Mitochondrial RNA polymerase	Mitochondrial genes
Chloroplast RNA polymerases (multiple;	Chloroplast genes
both chloroplast and nuclear encoded)	

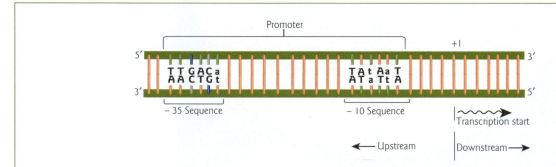

Figure 10.3 Bacterial promoters have two distinct consensus sequences. The −35 and −10 sequences for a typical *E. coli* promoter are shown. The first base transcribed is indicated by the number +1. The upstream and downstream directions relative to the start of transcription are indicated. Capital letters indicate bases found in those positions in more than 50% of promoters analyzed; small letters denote bases found in those positions in 50% or fewer of promoters analyzed.

Structure of bacterial RNA polymerase

The structure of bacterial RNA polymerase has been determined at the atomic resolution level. Much of the biochemical and genetic data that have accumulated over the past 20 years can now be interpreted by the structure. The bacterial RNA polymerase is comprised of a core enzyme plus a transcription factor called the sigma factor (σ). Together, they form the complete, fully functional enzyme complex called the holoenzyme (from the Greek word *holos* for "whole").

The core enzyme

The core enzyme is the component of the holoenzyme that catalyzes polymerization. It is 400 kD in size and has five subunits: two copies of the α-subunit (αI, αII), and one copy each of the β-, β′-, and ω-subunits (Fig. 10.4). The sequence, structure, and function of the core polymerase are evolutionarily conserved from bacteria to humans. The core enzyme has high affinity for most DNA. In the absence of σ, it can initiate synthesis anywhere on a DNA template *in vitro*. The σ factor is responsible for decreasing the nonspecific binding affinity of the RNA polymerase.

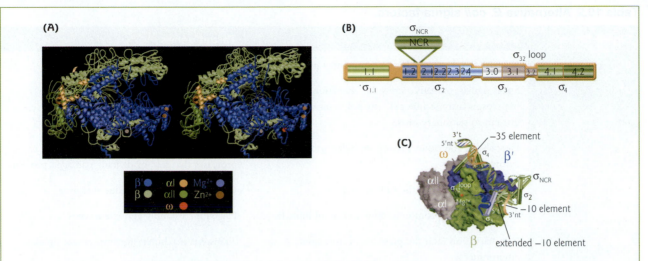

Figure 10.4 Structure of bacterial sigma factor and the RNA polymerase holoenzyme. (A) Crystal structure of the *Thermus aquaticus* RNA polymerase core enzyme. One stereo view is shown. The core enzyme has a crab claw shape. Subunits αI, αII, and ω form the base of the claw, and subunits β and β′ form the pincers. The metal ions are depicted as small colored spheres. The larger green and red spheres denote unstructured regions in the β- and β′-subunits that are missing from the structure. (Reprinted from Zhang, G., Campbell, E.A., Minakhin, L., Richter, C., Severinov, K., Darst, S.A. 1999. Crystal structure of *Thermus aquaticus* core RNA polymerase at 3.3 Å resolution. *Cell* 98:811–824. Copyright © 1999, with permission from Elsevier). (B) Sigma factor domains (based on σ^{70}). The black bar represents the primary amino acid sequence of σ. The conserved domains are numbered and color coded to correspond to regions shown in (C). NCR, nonconserved region. (C) Structure of the RNA polymerase holoenzyme bound to promoter DNA. The holoenzyme is composed of the core enyzme and the σ factor. The α-helices of the σ factor are shown as cylinders; conserved domains are colored as in (A). The template strand (t) of the DNA double helix is dark green and the nontemplate (nt) strand is light green. The −35 and −10 promoter elements are shown in yellow and the extended −10 element (present in some promoters that lack a −35 region) is shown in red. (Parts B and C adapted from from Murakami, K.S., Darst, S.A. 2003. Bacterial RNA polymerases: the wholo story. *Current Opinion in Structural Biology* 13:31–39. Copyright © 2003, with permission from Elsevier).

X-ray crystallographic analysis of the core RNA polymerase from the bacterium *Thermus aquaticus* has revealed a complex with a shape likened to a crab claw, with the β- and β′-subunits forming the pincers (Fig. 10.4A). These pincers form a 27 Å (2.7 nm) wide internal channel. In the holoenzyme, the globular domains of σ are spread out across the face of the crab claw. Binding of σ factor to the core results in closing of the pincers, and a change in shape of the internal channel where transcription takes place. The enzyme active site is located at the back of this internal channel where an essential magnesium ion (Mg^{2+}) is bound. Closure of the pincers presumably contributes to the highly processive nature of the enzyme during transcript elongation.

Sigma factor

The σ factor is primarily involved in recognition of gene promoters. The −35 and the −10 sequences are necessary for recognition by the σ^{70} factor, while the −10 sequence is the region of contact for the core enzyme. In addition, the −10 sequence is necessary for the initial melting of the DNA to expose the template strand. A domain of the σ^{70} factor binds to the nontemplate strand of the −10 sequence in a sequence-specific manner that stabilizes the initial transcription bubble.

Most of the σ factors share four regions of amino acid sequence homology that play a role in recognizing the promoter (Fig. 10.4). These four regions are further divided into subdomains with specific functions. Among the many different σ factors expressed in bacteria, the most abundant is a protein of 70 kDa, referred

Table 10.2 Alternative *E. coli* sigma factors.

Sigma factor	Inducer	Target genes
σ^{70}	Normal requirements during exponential growth.	General "housekeeping" genes
σ^S	Stress signals: oxidative stress, UV radiation, heat shock, hyperosmolarity, acidic pH, ethanol, transition from growth to stationary phase	General stress regulator genes (> 70 genes)
σ^{32}	Heat shock and other stresses: unfolded proteins in the cytoplasm	Heat shock proteins: chaperone proteins and proteases that fold or degrade damaged proteins
σ^E	Unfolded proteins in the cell envelope	Genes that restore envelope integrity
σ^F (σ^{28})	Conditions that promote production of multiple flagella	Flagellum assembly and chemotaxis
FecI	Iron starvation (and the presence of iron citrate in the environment)	Transport machinery for iron citrate uptake
σ^{54}	Nitrogen starvation (absence of ammonia)	Metabolism of alternative nitrogen sources

to as σ^{70}. It has a higher binding affinity for the RNA polymerase core enzyme than other σ factors. σ^{70} is required for specific binding of RNA polymerase to the promoter of the majority of genes in *E. coli*, and stimulates tight binding of the enzyme to template DNA. This property of the σ factor was characterized over 30 years ago, using nitrocellulose filter-binding assays to measure RNA polymerase–DNA complex dissociation rates. The holoenzyme containing σ dissociates more slowly from template DNA compared with the core polymerase alone (Fig. 10.5).

For expression of some genes, bacterial cells use alternative σ factors, specific to different subsets of promoters. The number of σ factors varies from one in *Mycoplasma genitalia* to more than 63 in *Streptococcus coelicolour*. In general, organisms with more varied lifestyles contain more σ factors. *E. coli* uses seven alternative σ factors to respond to some environmental changes (e.g. elevated temperatures induce the expression of heat shock proteins) and for the expression of flagellar genes (Table 10.2). These alternative σ factors recognize a different promoter sequence than that recognized by σ^{70}. Another bacterium, *Bacillus subtilis*, has 18 different σ factors, five of which orchestrate the process of sporulation, transcribing sporulation-specific genes in a cascade-like fashion.

Stages of transcription

The transcription process consists of three stages: initiation, elongation, and termination. Initiation is further divided into three stages: formation of a closed promoter complex, formation of an open promoter complex, and promoter clearance (Fig. 10.6).

Initiation

The RNA polymerase holoenzyme initially binds to the promoter at nucleotide positions −35 and −10 relative to the transcription start site (+1) to form a closed promoter complex (Fig. 10.6). The term "closed" indicates that the DNA remains double-stranded and the complex is reversible. The complex then undergoes a structural transition to the "open" form in which approximately 18 bp around the transcription start site are melted to expose the template strand of the DNA. Transcription is aided by negative supercoiling of the promoter region of some genes. Strand separation relieves the strain of supercoiled structures, thus less free energy is required for the initial melting of DNA in the initiation complex. AT-rich promoters like the −10 sequence also require less energy to melt, because of the differences in stacking interactions (hydrophobicity) and hydrogen bonding compared with GC-rich DNA (see Section 2.6). Formation of the open complex is generally irreversible

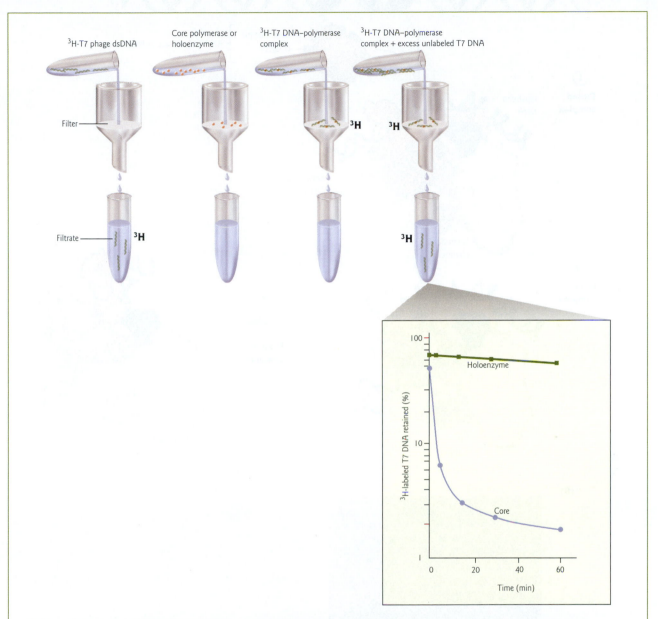

Figure 10.5 The sigma factor stimulates tight binding of RNA polymerase to the promoter. A nitrocellulose filter-binding assay measures DNA–protein interactions. Double-stranded DNA (dsDNA) does not bind to nitrocellulose filters. Protein does bind, however, so if dsDNA is bound to a protein, the protein–DNA complex will bind the nitrocellulose filter. In this experiment, *E. coli* core RNA polymerase (lacking σ) or holoenzyme (containing σ) were isolated from bacteria and allowed to bind to ³H-labeled T7 phage DNA, whose early promoters are recognized by *E. coli* RNA polymerase. Next, an excess of unlabeled T7 DNA was added so that any polymerase that dissociated from the labeled DNA would be more likely to rebind to unlabeled DNA. The mixture was passed through nitrocellulose filters at various time points to monitor the dissociation of the labeled T7 DNA–polymerase complexes. The radioactivity on the filter and in the filtrate was monitored by liquid scintillation counting. Since the labeled DNA only binds to the filter if it is still bound to RNA polymerase, this assay measures the dissociation rate of the polymerase–DNA complex. The much slower dissociation rate of the holoenzyme (green) relative to the core polymerase (blue) shows much tighter binding between T7 DNA and holoenzyme. The most tightly bound polymerase dissociates from the labeled DNA last. (Reprinted from Hinkle, D.C., Chamberlin, M.J. 1972. Studies of the binding of *Escherichia coli* RNA polymerase to DNA. I. The role of sigma subunit in site selection. *Journal of Molecular Biology* 70:157–185. Copyright © 1972, with permission from Elsevier.)

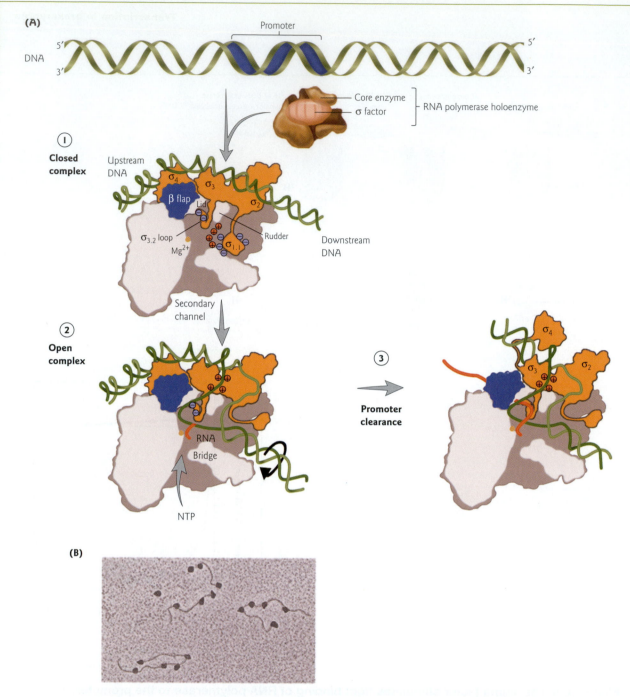

Figure 10.6 Conformational changes during the steps of transcription initiation. (A) Cross-sectional views of the RNA polymerase holoenzyme (β flap, blue; σ, orange; rest of RNA polymerase, gray; catalytic Mg²⁺, yellow sphere), promoter DNA (template strand, dark green; nontemplate strand, light green), and the RNA transcript (red). The view is looking down on top of the β-subunit, but with most of β removed, revealing the inside of the RNA polymerase active site channel (compare with Fig. 10.4). (1) With the help of the σ factor, RNA polymerase core enzyme binds the promoter in a closed complex, in which the DNA strands remain base paired. (2) The RNA polymerase holoenzyme separates the DNA strands around the start site of transcription, creating the "open complex." An incoming nucleoside triphosphate (NTP) is shown. (3) In the promoter clearance step, RNA polymerase initiates transcription, and moves off the promoter. The interaction with σ is altered (see text for details). (Adapted from Murakami, K.S., Darst, S.A. 2003. Bacterial RNA polymerases: the wholo story. *Current Opinion in Structural Biology* 13:31–39. Copyright © 2003, with permission from Elsevier) (B) Electron micrograph of RNA polymerase molecules from *E. coli* bound to several promoter sites of phage T7 DNA. Magnification × 200,000. (Reprinted with permission from: Fisher, H.W., Williams, R.C. 1979. Electron microscopic visualization of nucleic acids and of their complexes with proteins. *Annual Review of Biochemistry* 48:649–679. Copyright © 1979, with permission from Annual Reviews.)

and transcription is initiated in the presence of NTPs. In contrast to most DNA polymerases, no primer is required for initiation by RNA polymerase. RNA polymerase can initiate RNA synthesis *de novo*.

During the "promoter clearance" step, there is a staged disruption of σ factor–core enzyme interaction. The classic model proposes that σ dissociates from the core as the polymerase undergoes promoter clearance and switches from initiation to elongation mode. In this model, the σ factor then joins with a new core polymerase to initiate another RNA chain. However, recent experiments and structural studies provide evidence that σ does not completely dissociate from the core polymerase. Instead, there is sequential displacement of some domains of σ that would otherwise act as a barrier to the extension of the nascent RNA as it emerges from the RNA exit channel. The displaced portions include region 3.2, which is positioned within the RNA exit channel, and region 4, which interacts with the β-flap of RNA polymerase and is positioned adjacent to the end of the RNA exit channel (see Figs 10.4 and 10.6). Mutations that alter the strength of the interaction between σ^{70} region 4 and the β-flap affect not only transcription initiation but also transcription elongation.

Elongation

After about 9–12 nt of RNA have been synthesized, the initiation complex enters the elongation stage. The transition from initiation to elongation is marked by a significant conformational change in the core enzyme. This leads simultaneously to the modification or loss of RNA polymerase–DNA contacts, disruption of some σ contacts (as described above), and formation of a highly processive elongation complex. As the RNA polymerase moves, it holds the DNA strands apart forming a characteristic transcription "bubble" as it unwinds the strands at the front, and rewinds them at the back (Fig. 10.7). The moving polymerase protects a "footprint" of about 30 bp along the DNA against nuclease digestion. This footprint includes the transcription bubble as well as some double-stranded DNA on either side.

Within the transcription bubble, one strand of DNA acts as the template for RNA synthesis by complementary base pairing. The catalytic site of the polymerase has both a substrate-binding subsite, at which the incoming NTP is bound to the polymerase and to the complementary nucleotide residue of the template, and a product-binding subsite, at which the 3′ terminus of the growing RNA chain is positioned. A pyrophosphate is removed from the NTP and a phosphodiester bond forms with the 3′-OH group of the last nucleotide in the RNA chain. The Mg^{2+} bound in the active site helps to increase the nucleophilicity of the attacking 3′-OH and/or to stabilize the negative charge on the pyrophosphate leaving group. Transcription, like DNA replication, always proceeds in the 5′ → 3′ direction.

Completion of the single nucleotide addition cycle is accompanied by a shift of the active site of the RNA polymerase forward by one position along the DNA template. As a result, the 9–12 bp RNA–DNA hybrid retains a constant length but becomes one base pair longer at the downstream end and one base pair shorter at the upstream end. Transcription continues in a processive manner as nucleotides are added to the growing RNA strand by RNA polymerase according to the rules of complementary base pairing. Whether it is the RNA polymerase that moves along the DNA or vice versa remains a subject of debate (Focus box 10.1).

Termination

The RNA polymerase core enzyme moves down the DNA until a stop signal or terminator sequence is reached by the RNA polymerase. There are two types of terminators recognized, Rho-dependent and Rho-independent terminators. As the names suggest, the difference between the two types lies in their dependency on the Rho protein (Greek letter ρ). Rho-independent terminators are also called "intrinsic terminators" because they cause termination of transcription in the absence of any external factors. In contrast, Rho-dependent terminators require the Rho protein; without it RNA polymerase continues to transcribe past the terminator, a process known as readthrough. *E. coli* uses both kinds of transcript terminators. This is not true for all bacteria; a few such as *Mycoplasma* lack a *rho* gene. There is no evidence yet for a *rho*-like gene in eukaryotes.

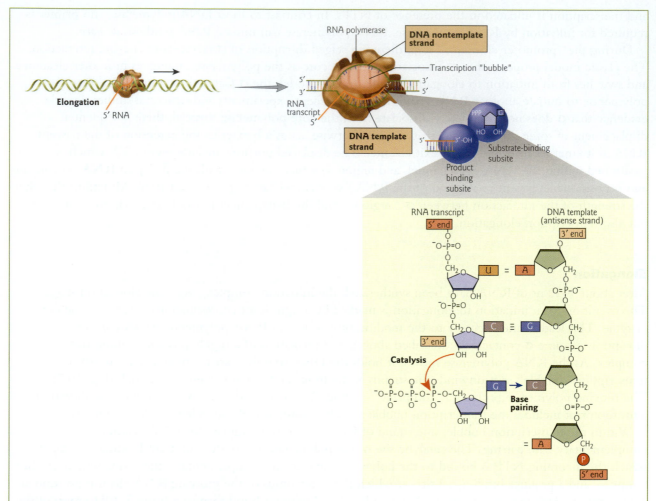

Figure 10.7 Transcription elongation. A schematic representation of a transcription bubble and the essential structural features of the elongation complex. The active (catalytic) site of the polymerase at the downstream end of the transcription complex includes the substrate-binding subsite for the next NTP and the product-binding subsite for the 3′ end of the newly synthesized RNA transcript. The bottom inset shows how RNA synthesis occurs from 5′ to 3′. The red arrow marks the polymerization site where a new phosphodiester bond forms. A cytosine (C) in the DNA template base pairs with an incoming GTP (blue arrow). GMP is then linked to the 3′ end of the nascent transcript. Pyrophosphate (PPi) is released.

Rho-independent termination   Rho-independent terminators are characterized by a consensus sequence that is an inverted repeat (Fig. 10.8A). Stem-loop structures can form within the mRNA just before the last base transcribed, by the pairing of complementary bases within the inverted repeat. The stem-loop structure may destabilize the transcription bubble, causing it to collapse. The inverted repeat sequence in the mRNA is followed by seven to eight uracil-containing nucleotides. A hybrid helix of U in the RNA base paired with A in the DNA is less stable than other complementary base pairs (e.g. GC, CG, or AT). This property, combined with formation of the stem loop in the exit channel of RNA polymerase, is sufficient to cause the enzyme to pause, resulting in transcript release.

Rho-dependent termination   Rho-dependent termination is controlled by the ability of the Rho protein to gain access to the mRNA. Because of the presence of a ribosome translating the mRNA at the same rate

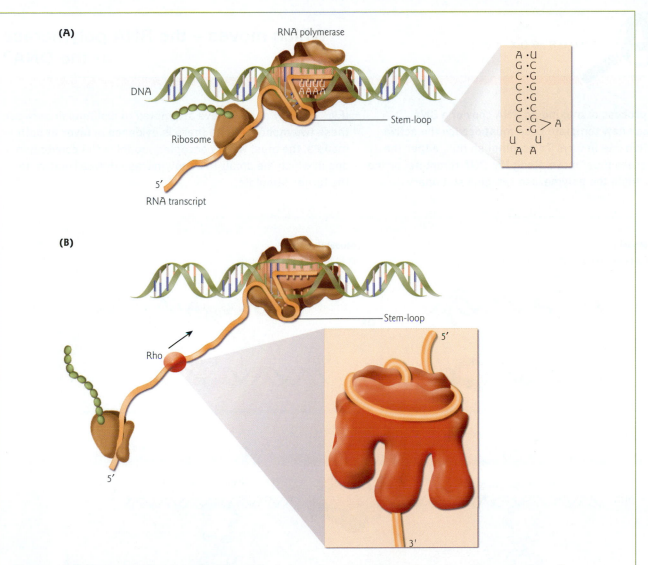

(A)

RNA polymerase

DNA

Ribosome

Stem-loop

5′

RNA transcript

A · U
G · C
C · G
C · G
C · G
G · C
C · G > A
C · G
U   U
A   A

(B)

Stem-loop

5′

Rho

5′

3′

Figure 10.8 Transcription termination. Rho-independent (A) and Rho-dependent (B) termination of transcription are preceded by a pause of the RNA polymerase at a termination sequence. (Top inset) The inverted repeat in the tryptophan (*trp*) attenuator is depicted. The repeat is not perfect, but 8 bp are still possible and seven of these are strong GC pairs. (Bottom inset) A topological model of the hexameric Rho bound to an mRNA. The 5′ end of the RNA is bound in the continuous cleft that extends around the upper periphery of Rho. The 3′ segment of the RNA passes through the center of the Rho hexamer, ending in the active site of the RNA polymerase. (Adapted from Richardson, J.P. 2003. Loading Rho to terminate transcription. *Cell* 114:157–159. Copyright © 2003, with permission from Elsevier.)

with which the mRNA is being transcribed, Rho is prevented from loading onto the newly formed RNA until the end of the gene or operon. At that point the ribosome is no longer moving along the mRNA, so the segment of newly formed RNA emerging from RNA polymerase becomes accessible to Rho. Rho-dependent terminators are very different from Rho-independent terminators. Although they consist of an inverted repeat, there is no string of Ts in the nontemplate strand, and they are not definable by a simple consensus sequence. Rho binds specifically to a C-rich site called a *Rho utilization* or *rut* site at the 5′ end of the newly formed RNA, as it emerges from the exit site of RNA polymerase (Fig. 10.8B). Rho

FOCUS BOX 10.1

Which moves – the RNA polymerase or the DNA?

During the process of making an RNA copy of a DNA template each new template base must occupy the active polymerization site in turn. To accomplish this, either the RNA polymerase must move along the DNA template, or the DNA moves while the polymerase remains stationary (Fig. 1). Many studies have attempted to distinguish between these two models. While there is evidence in favor of both models, the most widely accepted model is the conventional one in which the small RNA polymerase moves relative to the larger template.

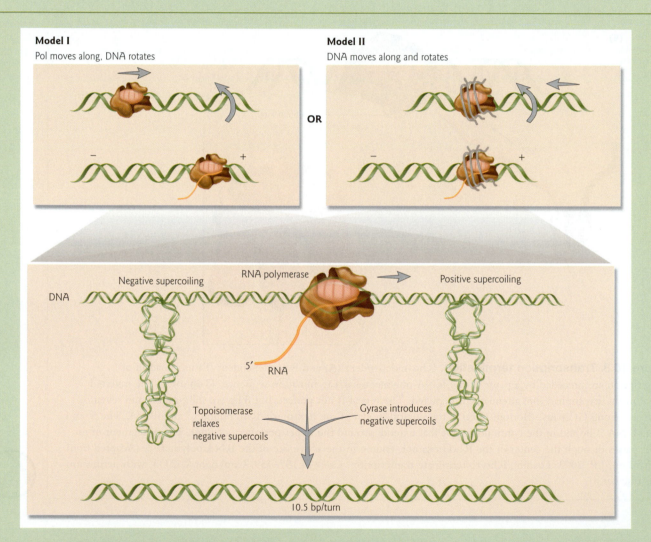

Figure 1 Possible movement of RNA polymerase and template. Two models are depicted. In model 1, RNA polymerase moves along and the DNA rotates. In model 2, the RNA polymerase remains stationary, and the DNA moves along and rotates. In both models, as the RNA polymerase moves along the torsionally constrained DNA, the DNA ahead of the RNA polymerase is wound more tightly, leading to the formation of positive supercoils (+). Behind the polymerase, the DNA becomes less tightly wound, leading to the formation of negative supercoils (−). Topoisomerase I and gyrase (bacterial topoisomerase II) resolve this supercoiling and restore the DNA to its relaxed form with 10.5 base pairs per turn.

Which moves – the RNA polymerase or the DNA?

Regardless of the model, the overall process of transcription itself has a significant local effect on DNA structure. As RNA polymerase pushes forward on the double helix, or as the DNA moves along and rotates, the DNA becomes more tightly wound leading to the formation of positive supercoils. Behind the polymerase, the DNA is underwound leading to the formation of negative supercoils. Topoisomerases are required to resolve the situation (see Chapter 2). Topoisomerase I relaxes the negative supercoils, while gyrase (bacterial topoisomerase II) introduces negative supercoils to counteract the positive supercoils in front of the polymerase.

Experiments *in vitro* have addressed whether the RNA polymerase precisely follows the DNA double helix as it moves along. To answer this question, real-time optical microscopy was used to catch RNA polymerase in the act of transcribing (Fig. 2). In this experiment, a DNA template was constructed containing one strong promoter. An 850 nm magnetic bead decorated with smaller fluorescent beads and coated with streptavidin was attached to the 3′ end of the DNA where nine nucleotides were biotinylated. The biotin/streptavidin system is widely used in molecular biology. The vitamin biotin can be conjugated to nucleotides which can then be incorporated as a label into DNA.

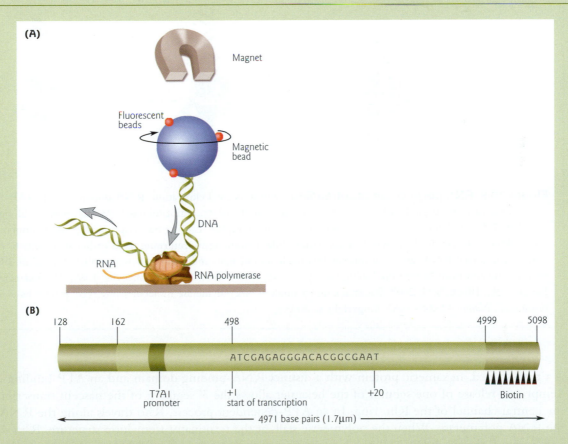

Figure 2 RNA polymerase – a molecular motor? (A) Observation system (not to scale) for DNA rotation by RNA polymerase (see text for details). (B) The DNA template. The numbers above are from the T7 phage DNA sequence. Rotation assay started from position +20. The magnetic bead was attached to nine biotins. (Adapted by permission from Nature Publishing Group and Macmillan Publishers Ltd: Harada, Y., Ohara, O., Takatsuki, A., Itoh, H., Shimamoto, N., Kinosita, K. 2001. Direct observation of DNA rotation during transcription by *Escherichia coli* RNA polymerase. *Nature* 409:113–115. Copyright © 2001.)

FOCUS BOX 10.1 (*cont'd*)

Which moves – the RNA polymerase or the DNA?

The bacterial protein streptavidin binds to biotin with extraordinary affinity, allowing the capture and detection of the biotin-labeled (biotinylated) DNA.

Transcription was initiated in the absence of UTP so that the RNA polymerase would pause at the adenine at the +20 position. The stalled polymerase was then attached to a glass surface. All four NTPs were added to allow further transcription. By monitoring the movement of the

fluorescent beads by microscopy, rotation of DNA could be observed directly. The experiment showed that RNA polymerase acts like a molecular motor and rotates DNA with a rate consistent with high-fidelity tracking. The polymerase tracks the DNA right-handed helix over thousands of base pairs, producing measurable torque. However, whether RNA polymerase rotates around DNA or vice versa *in vivo* remains to be determined.

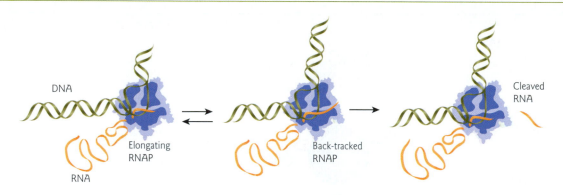

Figure 10.9 RNA polymerase proofreading. During normal elongation, RNA polymerase (RNAP, blue) moves downstream on the DNA (green) as it elongates the RNA transcript (orange). At each position along the DNA template, RNA polymerase may slide backward along the template, causing transcription to pause temporarily. From the backtracked state, RNA polymerase can either slide forward again, returning to its earlier state (left) or cleave the newly synthesized RNA removing mismatched nucleotides (right) and resume transcriptional elongation. (Reprinted with permission from Nature Publishing Group and Macmillan Publishers Ltd: Shaevitz, J.W., Abbondanzieri, E.A., Landick, R., Block, S.M. 2003. Backtracking by single RNA polymerase molecules observed at near-base-pair resolution. *Nature* 426:684–687. Copyright © 2003.).

is a ring-shaped, hexameric protein with a distinct RNA-binding domain and an ATP-binding domain. Temporary release of one subunit of the hexamer allows the 3′ segment of the nascent transcript to enter the central channel of the Rho ring. In an ATP-dependent process, Rho travels along the RNA, "chasing" the RNA polymerase. When the polymerase stalls at the terminator stem-loop structure, Rho catches up and unwinds the weak DNA–RNA hybrid. This causes termination of RNA synthesis and release of all the components.

Proofreading

E. coli RNA polymerase synthesizes RNA with remarkable fidelity *in vivo*. Its low error rate may be achieved, in part, by a proofreading mechanism similar to that found in DNA polymerases (see Section 6.6).

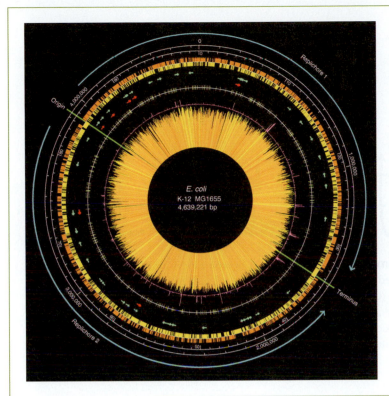

Figure 10.10 Direction of transcription around the *E. coli* circular genome. The origin and terminus of replication are shown as green lines, with blue arrows indicating replichores 1 and 2. The distribution of genes is depicted on the two outer rings: The orange boxes are genes located on the presented strand, and the yellow boxes are genes on the opposite strand. Red arrows show the location and direction of transcription of rRNA genes, and tRNA genes are shown as green arrows. The central yellow sunburst depicts the extent to which codon usage agrees with an *E. coli* reference set from highly expressed genes. Long yellow rays highlight regions of the genome with unusual codon usage (Reprinted with permission from Blattner, F.R., Plunkett, G. III, Bloch, C.A. et al. 1997. The complete genome sequence of *Escherichia coli* K-12. *Science* 277:1453–1462. Copyright © 1997 AAAS.)

Proofreading involves two key events. The first event is a short backtracking motion of the enzyme along the DNA template through several (~5) base pairs (Fig. 10.9). The movement is directed upstream in the opposite direction to transcriptional elongation (3′ → 5′). This backwards motion carries the 3′ end of the nascent RNA transcript away from the enzyme active site. The second event is nucleolytic cleavage, which occurs after a variable "pause" of the polymerase. This delay can last anywhere from 20 seconds to 30 minutes *in vitro*. In its backtracked state, the polymerase is able to cleave off and discard the most recently added base(s) by nuclease activity. In this process, a new 3′ end is generated at the active site, ready for subsequent polymerization onto the nascent RNA chain.

Direction of transcription around the *E. coli* chromosome

In a schematic representation of a gene sequence, it is a general convention to show the nontemplate strand on top and the template strand on the bottom. This is because the nontemplate strand and the transcribed RNA have the same sequences, substituting U for T in RNA, and they are both read from left to right (5′ → 3′). However, if one draws out the entire double-stranded circular DNA genome of *E. coli*, transcription is not always from the "top" strand in the drawing. Of the 50 operons or genes whose transcription direction is known, 27 are transcribed clockwise and 23 in the counterclockwise direction around the circle (i.e. the opposite strand is used as a template). In all cases, only one strand of a given operon's DNA is used as a template for transcription.

Many features of *E. coli* are oriented with respect to replication. The origin and terminus of replication divide the genome into oppositely replicated halves, or "replichores." Replichore 1 is replicated clockwise and replichore 2 is replicated counterclockwise. All seven ribosomal RNA operons and 53 of 86 tRNA genes are transcribed in the direction of replication (Fig. 10.10). Approximately 55% of protein-coding genes are also aligned with the direction of replication. This arrangement most likely leads to fewer collisions of DNA and RNA polymerase, and less topological strain from opposing supercoils generated during replication and transcription (Focus box 10.1).

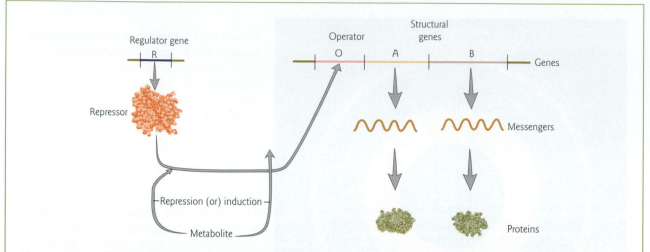

Figure 10.11 The Jacob–Monod operon model. The Jacob–Monod operon model for the control of the synthesis of sugar-metabolizing enzymes predicted the existence of a repressor molecule that is produced from a regulator gene (R) and binds to an operator site (O) (called an operator "gene" in the original model), thereby stopping the expression of the structural genes (A, B) that follow the operator site. The repressor also binds an inducer (metabolite) that lowers the affinity of the repressor for the operator and allows expression of the structural genes. The model also predicted the existence of an RNA intermediate (messenger) in protein synthesis.

10.4 Historical perspective: the Jacob–Monod operon model of gene regulation

Some of our first real far-reaching insights into how genes can be regulated came from the 1959 operon model of François Jacob and Jacques Monod. The operon model introduced the novel concept of regulatory genes that code for products that control other genes (Fig. 10.11). At the time scientists were still thinking only in terms of the one gene–one enzyme hypothesis of Beadle and Tatum, where genes code for enzymes involved in metabolic processes.

Jacob and Monod's model arose from their experimental observations in bacteria and phages. Monod was investigating how the enzyme β-galactosidase was produced in bacterial cells only when bacteria needed this enzyme to use the sugar lactose. Jacob was working on a similar problem – how phage lambda (λ) could be induced to switch from the lysogenic (quiescent) state, during which the λ genome replicates along with that of its bacterial host cell, to the lytic state, during which λ multiplies rapidly and lyses the host cell. Their collaborative work showed that regulation of the three enzymes involved in lactose metabolism occurs at the level of gene expression and that the inducer (lactose) acts on a repressor of transcription.

The essential features of gene regulation as described for *lac* operon induction are also represented in the λ switch. The *lac* operon provides an example of negative control of the enzymes involved in lactose metabolism. Initially Jacob and Monod proposed that all gene regulation occurred by negative control. Now it is known that, in fact, the *lac* operon is also regulated by positive control under certain environmental conditions.

The operon model led to the discovery of mRNA

The operon model was determined by genetic studies long ago, although the molecular mechanisms were unknown at the time. The model put forth a number of testable hypotheses. Jacob and Monod had noted that the pathway from gene to protein was very rapid when the *lac* operon was induced. From this

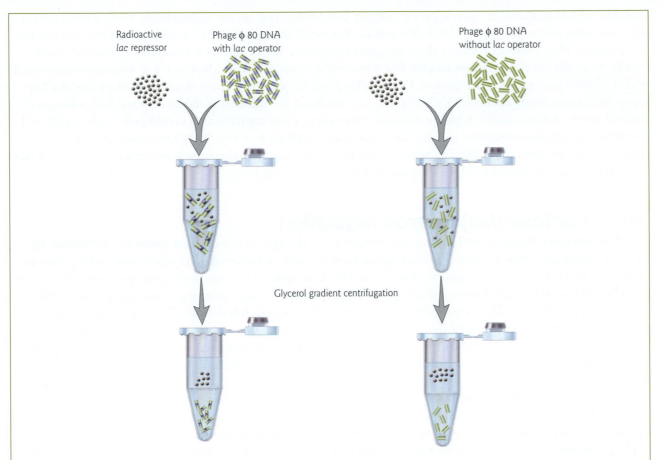

Figure 10.12 Gilbert and Müller-Hill's experiment demonstrating that the Lac repressor binds operator DNA. After glycerol gradient centrifugation, radioactively labeled Lac repressor sedimented with phage ϕ80 DNA, which had the *lac* operator, but not with ϕ80 DNA, which lacked the *lac* operator. (Redrawn from Gilbert, W., Müller-Hill, B. (1967) The lac operator is DNA. *Proceedings of the National Academy of Sciences USA* 58:2415–2421.)

observation, Jacob postulated, in 1959, the existence of an unstable RNA as an intermediate in protein synthesis. Few really took notice at the time. But a year and a half later, different experimental approaches in a number of research labs, including those of François Jacob, Sydney Brenner, James Watson, Charles Kurland, and François Gros, led to the discovery of messenger RNA and demonstrated its role as an informational molecule.

Characterization of the Lac repressor

The Jacob–Monod model for gene regulation also proposed the existence of a repressor protein. Between 1966 and 1972, it was shown that both Lac and λ repressors are indeed proteins. They bind to operator DNA adjacent to the promoter and inhibit the capacity of RNA polymerase to transcribe. Since a bacterial cell contains only 10–20 copies of the *lac* operon repressor, its detection and isolation in 1966 by Walter Gilbert and Benno Müller-Hill was a remarkable accomplishment, when many sensitive technologies we use today were not available.

To determine whether Lac repressor bound to operator DNA, Gilbert and Müller-Hill carried out an *in vitro*-binding assay. Before the advent of gene cloning techniques, studies of bacterial genes had to rely on bacteriophage variants that had incorporated pieces of bacterial DNA. Conveniently, a phage strain was available which included *lac* operon DNA. Gilbert and Müller-Hill mixed radioactively labeled purified Lac repressor with phage phi (φ) 80 DNA that had incorporated the *lac* operator, or with phage φ80 lacking the *lac* operator. They then centrifuged these mixtures to separate the large DNA molecules that sedimented rapidly from the small protein molecules that sedimented more slowly. Radioactive Lac repressor sedimented with *lac* DNA but not with the control DNA lacking the *lac* operator. These findings showed that the Lac repressor protein binds operator DNA (Fig. 10.12). At about the same time, Mark Ptashne and colleagues isolated the λ cI repressor for λ phage operons. Since then, more sophisticated techniques, such as DNase I footprinting and electrophoretic gel mobility shift assays (EMSA) (see Fig. 9.15) have confirmed the sequence-specific binding of the Lac repressor to the *lac* operator, and there are now crystal structures of the Lac repressor and many other DNA-binding proteins (see Section 10.6).

10.5 Lactose (*lac*) operon regulation

A major difference between prokaryotes and eukaryotes is the way in which their genes are organized. In bacteria, genes are organized into operons. An operon is a unit of bacterial gene expression and regulation, including structural genes and control elements in DNA recognized by regulatory gene products. The genes in an operon are transcribed from a single promoter to produce a single primary transcript (pre-mRNA) or "polycistronic mRNA." The nematode worm *Caenorhabditis elegans* differs from all other eukaryotes in having ~15% of its genes grouped in operons. However, unlike in prokaryotes, each *C. elegans* pre-mRNA is processed into a separate mRNA for each gene rather than being translated as a unit. Some of the genes in *C. elegans* operons appear to be involved in the same biochemical function, but this may not be the case for most.

Bacteria need to respond swiftly to changes in their environment, to switch from metabolizing one substrate to another quickly and energetically efficiently. When glucose is abundant, bacteria use it exclusively as their food source, even when other sugars are present. However, when glucose supplies are depleted, bacteria have the ability to rapidly take up and metabolize alternative sugars, such as lactose. The process of induction – the synthesis of enzymes in response to the appearance of a specific substrate – is a widespread mechanism in bacteria and single-celled eukaryotes, such as yeast. The lactose (*lac*) operon in *E. coli* is regarded as a paradigm for understanding bacterial gene expression. Features of the *lac* operon illustrate basic principles of gene regulation that are universal. There is a constitutively active RNA polymerase that alone works with a certain frequency. The transcriptional activator increases the frequency of initiation by recruiting the RNA polymerase to the gene promoter, and the transcriptional repressor decreases the frequency of initiation by excluding the polymerase. Both the repressor and activator are DNA-binding proteins that undergo allosteric modifications.

Lac operon induction

The *lac* operon consists of three structural genes, *lacZ*, *lacY*, and *lacA* (Fig. 10.13). The *lacZ* gene encodes β-galactosidase. This enzyme cleaves lactose into galactose and glucose, both of which are used by the cell as energy sources. The active form of β-galactosidase is a tetramer of approximately 500 kD. *LacY* codes for lactose permease, a 30 kDa membrane-bound protein that is part of the transport system to bring β-galactosides such as lactose into the cell. *LacA* codes for a transacetylase which rids the cell of toxic thiogalactosides that get taken up by the permease. *LacY* may not be absolutely required for lactose metabolism; mutations in *lacZ* or *lacY* create cells which cannot use lactose, while *lacA* mutants still can metabolize lactose. Upstream of the *lac* operon is the regulatory gene that codes for the 38 kDa Lac repressor. By convention, the gene name is given in lower case and italics, whereas the protein name is in plain type with the first letter in capitals. The Lac repressor is constitutively transcribed under control of its own promoter. In the absence of lactose, the Lac repressor binds as a tetramer to the operator DNA sequence. Because the *lac* operator sequence overlaps with the promoter region, the Lac repressor blocks RNA polymerase from binding to the promoter. As a consequence, transcription of the *lac* operon structural genes is repressed.

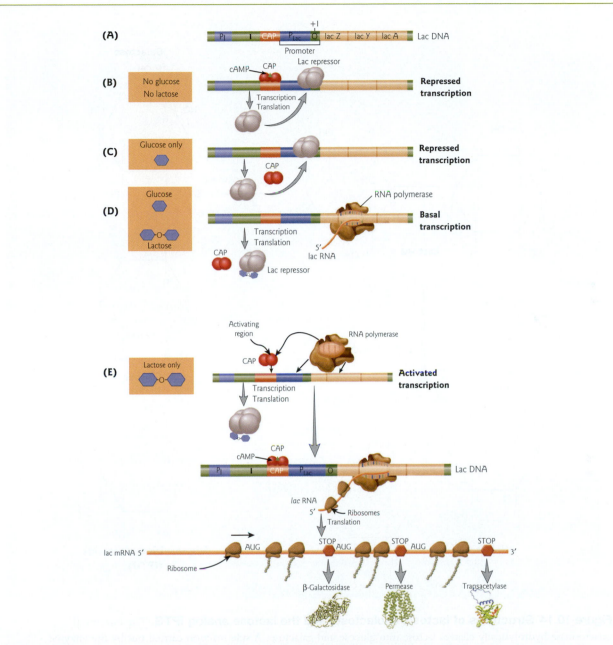

Figure 10.13 *Lac* operon regulation by glucose and lactose. (A) The components of the *lac* operon. The Lac repressor protein is encoded by the repressor gene (I), which is under control of its own promoter (P_I). The Lac repressor binds to the *lac* operator (O) as a tetramer. The start of transcription (+1) is indicated. The catabolic activator protein (CAP) site is the DNA-binding site for the activator protein CAP. The CAP protein is encoded by a separate gene distant from the *lac* operon. It binds DNA as a dimer. The *lac* operon structural genes are under control of the *lac* promoter (P_{Lac}). (B and C) Transcription of the *lac* operon is repressed in the absence of lactose, whether glucose is absent (B) or present (C). Under both conditions, the Lac repressor protein binds the operator and excludes RNA polymerase. (D) In the presence of both glucose and lactose, RNA polymerase binds the *lac* promoter very poorly, resulting in a low (basal) level of transcription. (E) When lactose is present and glucose is absent the *lac* operon is induced. Binding of the inducer allolactose (see Fig. 10.14) changes the conformation of the Lac repressor and alters its operator–binding domain. CAP, along with its small molecule effector cAMP, recruits RNA polymerase and binds the CAP site and transcription is stimulated 20–40-fold. The structural genes are transcribed as a polycistronic mRNA that is then translated using the start and stop codon for each individual protein.

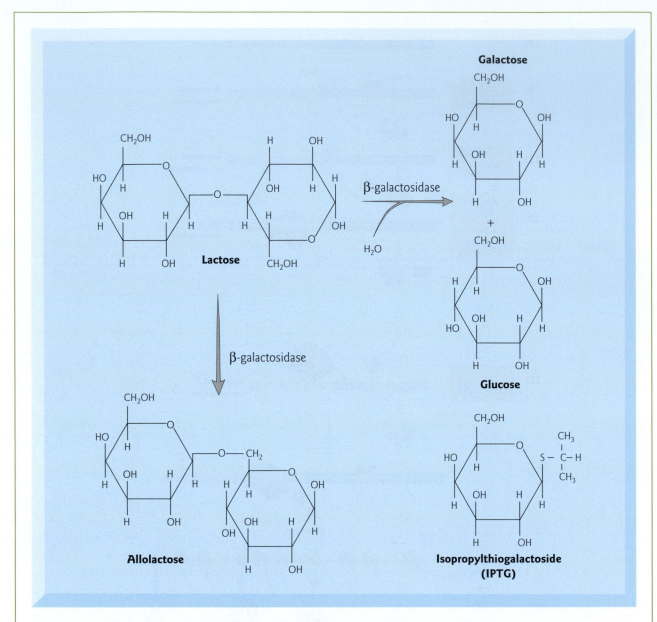

Figure 10.14 Structures of lactose, allolactose, and the lactose analog IPTG. The enzyme β-galactosidase hydrolytically cleaves lactose into glucose and galactose. A side reaction carried out by the enzyme rearranges lactose to form the inducer, allolactose. Note the change in the galactosidic bond from β–1,4 in lactose, to β–1,6 in allolactose. β-galactosidase cannot metabolize isopropylthiogalactoside (IPTG), a sulfur-containing analog of lactose that is used in molecular biology research.

In the presence of lactose (and the absence of glucose) the *lac* operon is induced. The real inducer is an alternative form of lactose called allolactose. Because repression of the *lac* operon is not complete, there is always a very low level of the *lac* operon products present (< 5 molecules per cell of β–galactosidase), so some lactose can be taken up into the bacterium and metabolized. When β-galactosidase cleaves lactose to galactose plus glucose it rearranges a small fraction of the lactose to allolactose (Fig. 10.14). Even a small amount of the inducer is enough to start activating the *lac* operon. Upon binding allolactose, the Lac repressor undergoes a conformational (allosteric) change, which alters its operator-binding domain. This allosteric change reduces its DNA-binding affinity to nonspecific levels, thereby relieving *lac* repression (see Section 10.6). Not only is there release from repression, there is also activation of transcription. At a site distance from the *lac* operon is the gene that encodes catabolic activator protein (CAP). The same protein is often called CRP, for cyclic AMP receptor protein. CAP binds to the DNA sequence within the *lac* operon called the CAP site. Recruitment of RNA polymerase requires the formation of a complex of CAP, polymerase, and DNA (see Fig. 10.13). The formation of this complex is an example of cooperative binding of proteins to DNA (see Section 10.6).

When the *lac* operon is activated, RNA polymerase begins transcription from the promoter and transcribes a common mRNA for the three structural genes, from 5′ to 3′. The mRNA has a start (AUG) codon and stop codon for each protein. The ribosome binds to the 5′ end of the mRNA and begins translation. When it reaches the stop codon at the end of the β-galactosidase-coding region the ribosome may detach, but most continue on to the next coding region, to synthesize the permease, followed by the transacetylase. Within 8 minutes after induction, approximately 5000 molecules of β-galactosidase per cell are produced.

Basal transcription of the *lac* operon

The *lac* operon is subject to both positive and negative regulation. As described above, the *lac* operon is transcribed if and only if lactose is present in the medium; however, this signal is almost entirely overridden by the simultaneous presence of glucose, a more efficient energy source than lactose. When provided with a mixture of sugars, including glucose, the bacteria use glucose first. So long as glucose is present, operons such as lactose are not transcribed efficiently. Only after exhausting the supply of glucose does the bacterium fully turn on expression of the *lac* operon. Glucose exerts its effect by decreasing synthesis of cAMP, which is required for the activator CAP to bind DNA. Without the cooperative binding of CAP, RNA polymerase transcribes the *lac* genes at a low level, called the basal level (see Fig. 10.13). This basal level of transcription is determined by the frequency with which RNA polymerase spontaneously binds the promoter and initiates transcription. The basal level is some 20–40-fold lower than activated levels of transcription.

Regulation of the *lac* operon by Rho

The *E. coli lac* operon contains latent Rho-dependent terminators within the early part of the operon. Rho has been shown to terminate the synthesis of transcripts when the cells are starved of amino acids. The intragenic terminators do not function under conditions of normal expression, presumably because a ribosome is present translating the RNA as it emerges from the exit site of the RNA polymerase. However, when the movement of the ribosome is slowed or blocked because of the absence of an amino acid, a segment of a transcript containing a Rho utilization (*rut*) site becomes exposed, allowing Rho to bind and terminate the partial transcript. This is advantageous to the cell since it prevents the loss of energy in making a transcript that will not be translated.

The *lac* promoter and *lacZ* structural gene are widely used in molecular biology research

Knowledge of the *lacZ* gene and the function of the enzyme it encodes has been exploited by molecular biologists in the lab (see Sections 8.4 and 9.3). Because its activity is easily detected by color reactions and its expression is inducible, β-galactosidase has become an important enzyme in DNA biotechnology. It is often used in screening strategies for bacterial colonies that have been transformed with recombinant DNA during gene cloning procedures, as a reporter gene for studies of gene regulation in transgenic animals or in cultured cells. The *lac* operon transcriptional machinery is widely used to induce expression of heterologous proteins in *E. coli*. In the lab, isopropylthiogalactoside (IPTG), a sulfur-containing analog of lactose, is used as an inducer of the *lac* operon (see Fig. 10.14). The advantage of IPTG over lactose is that IPTG interacts with the Lac repressor and induces the *lac* operon but is not metabolized by β-galactosidase. Thus, IPTG can continue inducing the operon for longer periods of time in the laboratory.

10.6 Mode of action of transcriptional regulators

Study of the *lac* operon and many other operons in bacteria has revealed many fundamental aspects of the mode of action of transcriptional regulatory proteins. The proteins are modular, consisting of domains with distinct functions, such as for DNA binding and protein–protein interactions. In many cases, these regulatory proteins bind to DNA in a cooperative fashion with other proteins. Allosteric modification plays a key role in regulation of their activities. Distant DNA regulatory sites are brought in close proximity through cooperative protein–protein interactions that cause DNA looping.

Cooperative binding of proteins to DNA

Cooperative binding of proteins to DNA plays a central role in gene regulation in both prokaryotes and eukaryotes. Cooperative binding does not require allosteric changes. The effects are mediated by protein–protein and protein–DNA interactions. For example, CAP has two major functional domains: a DNA-binding domain and an "activating region," which contacts the RNA polymerase. The distinct functions of these domains have been characterized by mutagenesis studies. Chemical cross-linking shows that the CAP-activating region interacts directly with the C-terminal domain of one of the α-subunits of RNA polymerase. Through this interaction, CAP recruits RNA polymerase to the promoter (see Fig. 10.13). When CAP and RNA polymerase are both present their binding sites are much more likely to be occupied, even at very low concentrations, because they help each other bind to DNA. One protein might dissociate from the DNA, but due to its continued interaction with the other DNA-bound protein, it does not diffuse away and is more likely to rebind to its DNA site. Through this interaction, CAP helps RNA polymerase bind tightly to the promoter until the polymerase changes from the closed to open complex and transcription begins.

Allosteric modifications and DNA binding

Both CAP (Fig. 10.15) and the Lac repressor (Fig. 10.16) bind to their DNA sites using a similar structural motif, called a helix-turn-helix (HTH). Each HTH has one α-helix called the recognition helix that inserts into the major groove of DNA. The side chains of amino acids exposed along the recognition helix make sequence-specific contacts with functional groups exposed on the base pairs. A second α-helix lies across the DNA. It helps position the recognition helix and strengthens the binding affinity. Differences in the residues along the outside of the recognition helix largely account for differences in the DNA-binding specificities of regulators. The HTH motif is the predominant DNA recognition motif found among *E. coli* transcriptional regulatory proteins. A somewhat modified form is found in eukaryotes in homeodomain proteins (see Fig. 11.15 and Focus box 11.4).

The allosteric change undergone by CAP upon binding cAMP increases its ability to bind DNA (Fig. 10.15). In contrast, the allosteric change in the Lac repressor upon binding the inducer allolactose (or the lactose analog, IPTG) decreases its ability to bind DNA (Fig. 10.16). The addition of IPTG has been shown to cause a conformational change in the N-terminal domain of the Lac repressor dimer, leading to separation of the hinge helices. The HTH DNA-binding motifs become disordered and dissociate from the major groove binding site.

The hinge region of the Lac repressor between the α-helices also acts as a structural switch between nonspecific and specific binding modes. In general, interaction of regulatory DNA-binding proteins with their target sites is preceded by binding to nonspecific DNA. The proteins then translocate to their specific sites by a "random walk" (see Fig. 8.4) along the DNA (Fig. 10.16B). When bound to DNA nonspecifically, the hinge region of the Lac repressor remains disordered. In its unfolded state it makes no contacts with the DNA minor groove. The DNA remains in canonical B form, instead of becoming bent, as observed for the specific complex. When bound to specific DNA sequences of the operator, the Lac repressor hinge forms an α-helix. The DNA bends approximately 36°, resulting in a central kink within the operator, and the protein contacts both the major and minor grooves by the HTH motif. Specific

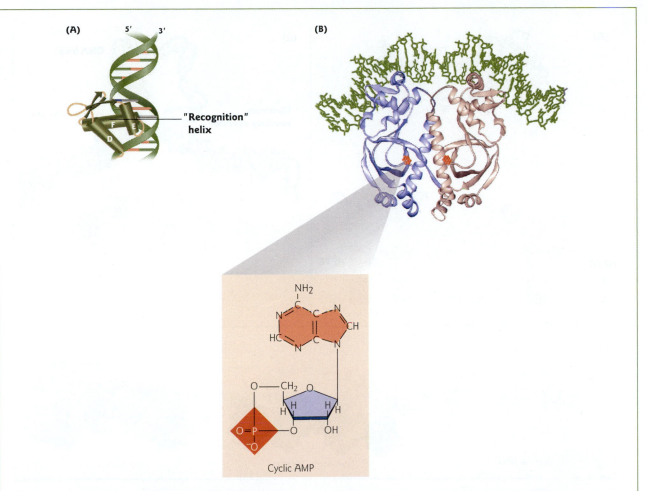

Figure 10.15 The CAP–DNA complex. (A) Model showing the helix-turn-helix (HTH) DNA-binding motif of one subunit of CAP. The recognition helix (F) contacts the DNA in the major groove. The three α-helices are depicted as cylinders and β–pleated sheets as flat ribbons. (B) Ribbon model of the CAP–DNA complex model derived from a co-crystal structure. The inset shows the location of the two bound cAMP molecules (red). The DNA (green) is bent by about 90° overall. The protein dimer (blue and gray subunits) is held together through interaction between two long α-helices. (Protein Data Bank, PDB:1CGP, 1O3R, 1O3S). (Lower inset) The structure of cAMP.

interactions between the hydrogen bond donors and acceptors of the protein-binding site and those of the base pairs in the major and minor grooves of the DNA double helix provide the molecular basis for binding specificity and target recognition. Binding is supported and stabilized by electrostatic interactions between negatively charged phosphates of the sugar–phosphate backbone of the DNA and the basic amino acid residues that surround the binding site of the Lac repressor.

DNA looping

DNA looping is a mechanism now known to be widely used in gene regulation. DNA looping allows multiple proteins to interact with RNA polymerase, some from adjacent sites and some from distant sites. As discussed earlier, the cooperative binding of proteins to multiple DNA-binding sites increases their effective binding constants and allows regulatory proteins to function at very low concentrations within the cell.

A classic example of DNA looping (as well as allosteric modification) is found in the operon controlling use of the sugar arabinose. In this operon, the regulatory protein AraC acts both as a repressor and activator

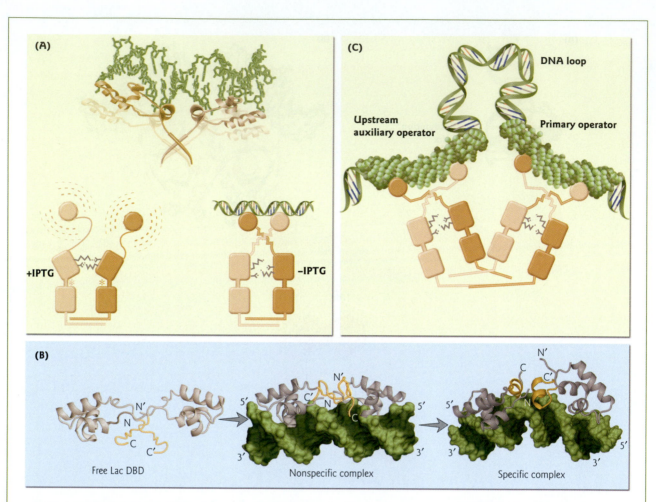

Figure 10.16 Lac repressor–DNA recognition. (A) Allosteric changes in the Lac repressor. A ribbon diagram of the Lac dimer–DNA complex is shown in the darker brown shade; the Lac–IPTG complex is shown in the lighter brown shade. The addition of IPTG (a lactose analog, see Fig. 10.14) causes the hinge helices in the repressor to move apart. The helix-turn-helix (HTH) DNA-binding motifs become disordered and move out of the major groove binding site. The cartoons below the structures summarize these changes. The left side shows a dimer of Lac repressor bound to IPTG (asterisk). A number of salt bridges (gray symbols) exist between the dimers but the HTH domains are far apart and the hinge helices are not formed. The right side shows a dimer of the Lac repressor–DNA complex. The salt bridges are broken, the hinge helices form, and the HTH domain becomes ordered and binds DNA. (Redrawn from Lewis, M. 2005. The lac repressor. *Comptes Rendus Biologies* 328:521–548.) (B) The hinge region of the DNA-binding domain (DBD) of the Lac repressor (colored orange) remains unstructured in both the free state and the nonspecific complex state. It folds up into a HTH motif in the specific complex with *lac* operator DNA. In the nonspecific complex, the DNA double helix adopts a B-DNA conformation. In the specific complex the DNA is bent by ~36°. (Reprinted with permission from Kalodimos, C.G., Biris, N., Bonvin, A.M., Levandoski, M.M., Guennuegues, M., Boelens, R., Kaptein, R. 2004. Structure and flexibility adaptation in nonspecific and specific protein–DNA complexes. *Science* 305:386–389. Copyright © 2004 AAAS.) (C) Cartoon of the Lac tetramer bound to an upstream auxiliary operator and the primary operator DNA sequence (space-filling representation), forming a DNA loop in between (not drawn to scale). The Lac repressor is a tethered dimer of dimers.

of transcription, rather than dividing the two functions among two proteins as in the *lac* operon (i.e. Lac repressor and CAP). Arabinose binds to AraC, changing the shape of the activator so that it binds as a dimer to two regulatory sequence half sites (Fig. 10.17). This places one monomer of AraC close to the promoter from which it can activate transcription. The promoter is also further activated by CAP recruitment of

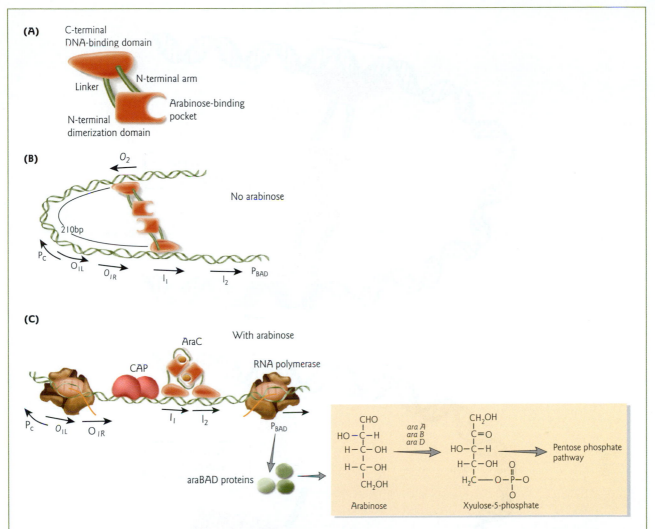

Figure 10.17 Regulation of the arabinose operon. (A) The domain structure is shown for one subunit of the dimeric regulatory protein AraC. AraC acts as a repressor in the absence of arabinose (B) and as an activator in the presence of arabinose (C). AraC binds to 17 bp half sites of similar sequence called O_2, I_1, and I_2. Another regulatory element O_1 is formed of two half sites (O_{1L} and O_{1R}) that bind two subunits of AraC. AraC and the arabinose operon structural genes (*ara*B, *ara*A, and *ara*D) are transcribed in opposite directions from the p_C and p_{BAD} promoters, respectively (arrows show the direction of RNA polymerase movement). At p_C and O_1, the RNA polymerase and AraC compete for binding. There is a single CAP-binding site required for the activation of the p_{BAD} promoter, but not the p_C promoter. The *ara*BAD structural genes are only expressed in the presence of arabinose. The *ara*BAD operon encodes the enzymes responsible for converting arabinose into xylulose-5-phosphate. In the absence of arabinose, two AraC proteins bind to both O_2 and I_1 and then to each other. This results in the formation of a loop in the DNA. The presence of this loop blocks activation of the p_{BAD} promoter by RNA polymerase and thus no *ara* operon expression occurs. (Adapted from Schleif, R. 2000. Regulation of the L-arabinose operon of *Escherichia coli*. *Trends in Genetics* 16:559–565. Copyright © 2000, with permission from Elsevier.)

RNA polymerase. In the absence of arabinose, the AraC dimer folds into a different conformation and one monomer binds to a half site 194 bp upstream. When AraC binds in this manner, the DNA between the two sites forms a loop. The DNA loop sterically blocks access of RNA polymerase to the promoter.

DNA looping was first predicted upon the discovery that the negative control element *ara*O_2 was nearly 200 bp upstream of all the sequences required for positive control. Helical–twist experiments confirmed this model. In these experiments, half a turn of the DNA helix was added to the arabinose operon sequence

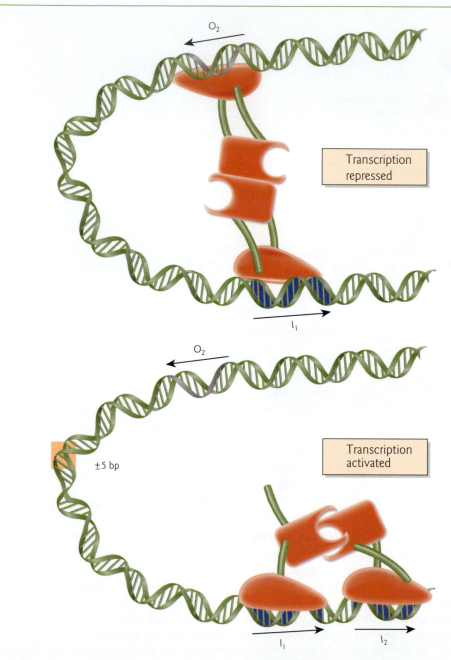

O_2

Transcription
repressed

I_1

O_2

±5 bp

Transcription
activated

I_1 I_2

Figure 10.18 Schematic representation of the helical-twist experiment that demonstrated DNA looping. When half-integral turns were introduced in the DNA between the O_2 and I_1 half sites in the arabinose operon (see Fig. 10.17), this interfered with repression of p_{BAD} in the absence of arabinose. There was a 5–10-fold elevation of the basal level of transcription from the operon. Introduction of integral numbers of turns did not interfere with this repression. (Adapted from Schleif, R. 2000. Regulation of the L-arabinose operon of *Escherichia coli*. *Trends in Genetics* 16:559–565. Copyright © 2000, with permission from Elsevier.)

anywhere between the *ara*O_2 site required for full repression and the downstream site (I_1) required for induction (Fig. 10.18). The added half-turn rotated one of the two sites to which AraC binds to the opposite side of the DNA, thereby interfering with protein–protein interactions and hindering loop formation. The introduction of half-rotations decreased repression of the arabinose operon, whereas the introduction of integral numbers of rotation that maintained the AraC-binding sites on the same side of the DNA did not (see Fig. 10.17).

After the discovery of DNA looping in the arabinose operon system, looping was found to occur in a number of other prokaryotic systems, including the *lac* operon. In addition, DNA looping is now known to explain how eukaryotic enhancers act from a distance (see Sections 11.3 and 11.7). The Lac repressor binds DNA as a tetramer, but functions as "a dimer of dimers" (Fig. 10.16C). Each operator is contacted by only two of the four subunits. The other two subunits within the tetramer can bind to one of two other *lac* operators, located 400 bp downstream and 90 bp upstream of the primary operator adjacent to the promoter. In each case, the intervening DNA loops out to accommodate the reaction.

10.7 Control of gene expression by RNA

An important concept has emerged from gene regulation research over the past several years – RNA frequently plays a more direct role in controlling gene expression than previously thought. It has long been known that differential folding of RNA plays a major role in transcriptional attenuation in bacteria. Other RNA-based regulatory mechanisms have been discovered more recently, including pathways involving riboswitches.

Differential folding of RNA: transcriptional attenuation of the tryptophan operon

Regulation of the tryptophan (*trp*) operon in bacteria is a classic example of transcriptional attenuation. During transcription of the leader region of the *trp* operon, a domain of the newly synthesized RNA transcript can fold to form either of two competing hairpin structures, an antiterminator or a terminator (Fig. 10.19A). The leader RNA preceding the antiterminator contains a 14 nt coding region, *trpL*, which includes two tryptophan codons. When bacterial cells have adequate levels of tryptophan-charged tRNA^{Trp} for protein synthesis, the leader peptide (*trpL*) is synthesized, the terminator forms in the leader transcript, and transcription is terminated. When cells are deficient in charged tRNA^{Trp}, the ribosome translating *trpL* stalls at one of these tryptophan codons. This stalling allows the downstream sequence to fold, forming an antiterminator structure that prevents formation of the competing terminator. Termination is blocked, allowing transcription of sequences encoding the structural genes involved in tryptophan biosynthesis. In addition to post-transcriptional control by differential RNA folding, transcription initiation in the *trp* operon is also controlled by more conventional means. A tryptophan-activated repressor binds to operator sites located within the *trp* promoter region, blocking access of RNA polymerase to the *trp* promoter (Fig. 10.19B). The major effect of transcription is through *trp* repression; the effect of attenuation on transcription is about 10-fold. Combined, these two regulatory mechanisms allow about a 600-fold range of transcription levels of the *trp* operon structural genes.

Riboswitches

The expression of the majority of genes is controlled by protein factors. However, specialized domains within certain mRNAs act as switchable "on–off" elements or "riboswitches," which selectively bind metabolites and control gene expression without the need for protein transcription factors. RNA can function as a sensor for signals as diverse as temperature, salt concentration, metal ions, amino acids, and other small organic metabolites. Riboswitches are widespread in bacteria. However, only one type of riboswitch – a thiamine pyrophosphate-sensing riboswitch present in plants and fungi – has been found in eukaryotes so far.

Riboswitches are typically found in the 5′ untranslated regions of mRNAs. Most riboswitches can be divided roughly into two structural domains: an aptamer (RNA receptor) and an "expression platform" (Fig. 10.20A). The aptamer domain selectively binds to the target metabolite. The expression platform converts metabolite-binding events into changes in gene expression via changes in RNA folding that are

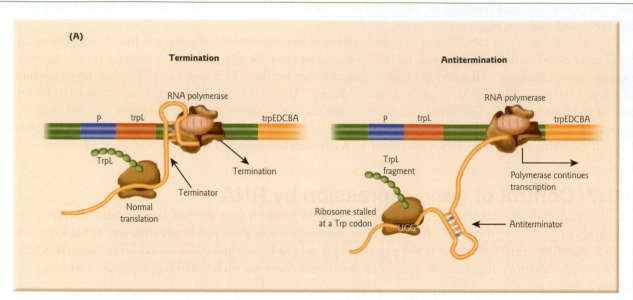

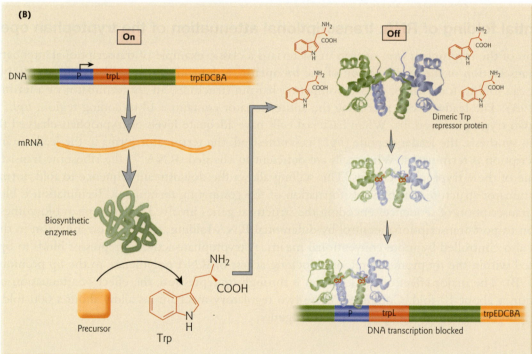

Figure 10.19 Transcriptional attenuation of the *E. coli trp* operon. (A) Repression mediated by differential folding of RNA. Termination: when the amino acid tryptophan is abundant in the cell, the ribosome translating *trpL* does not stall at the tandem Trp codons in *trpL* and rapidly reaches the *trpL* stop codon. The terminator hairpin forms in the transcript, resulting in the termination of transcription. The structural genes for the enzymes involved in tryptophan biosynthesis (*trpEDCBA*) are not transcribed. Antitermination: a deficiency in charged tRNA^{Trp} stalls the translating ribosome at one of the two tandem Trp codons in *trpL*. This stalling allows the antiterminator hairpin to form, which prevents terminator formation. Transcription of the structural gene coding regions takes place. (B) Protein-mediated repression. In the absence of tryptophan, the genes encoding the biosynthetic enzymes for tryptophan are transcribed and translated. When enough tryptophan has been produced, the tryptophan binds to the dimeric Trp repressor protein, inducing a conformational change that enables it to bind the *trp* operator, thereby blocking access of RNA polymerase to the *trp* promoter.

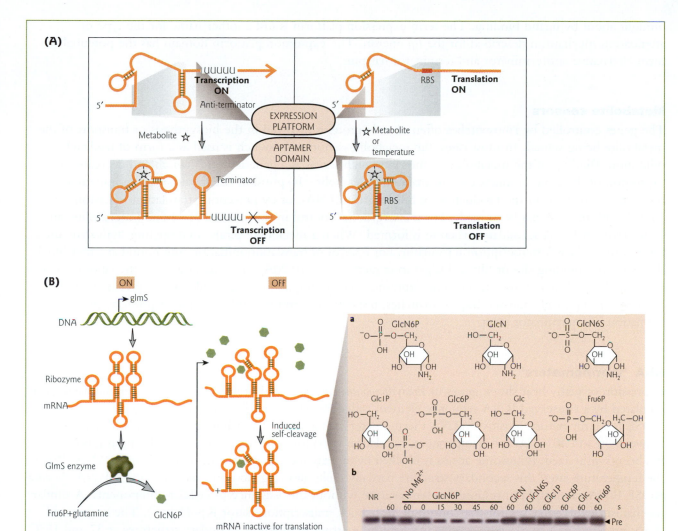

Figure 10.20 Mechanisms of riboswitch gene control. (A) Most riboswitches are comprised of a < 100 nt aptamer (RNA receptor) domain and an expression platform which has inducible secondary structures. Transcription control involves metabolite binding and stabilization of a specific conformation of the aptamer domain that precludes formation of a competing antiterminator stem in the expression platform. This allows formation of a terminator stem, which prevents the full-length mRNA from being synthesized. In contrast, control of translation is accomplished by metabolite- or temperature-induced structural changes that sequester the ribosome-binding site (RBS), thereby preventing the ribosome from binding to the mRNA. (Adapted from Tucker, B.J., Breaker, R.R. 2005. Riboswitches as versatile gene control elements. *Current Opinion in Structural Biology* 15:342–348. Copyright © 2005, with permission from Elsevier.) (B) A ribozyme riboswitch. The GlmS enzyme, which is involved in the synthesis of GlcN6P (green hexagon), is encoded by the *glmS* mRNA, which also contains a ribozyme sequence. In the presence of GlcN6P, the ribozyme cleaves its own mRNA. Self-destruction of the *glmS* mRNA inhibits further production of GlcN6P. (Inset) (i) Chemical structures and names of various compounds used to explore the substrate specificity of the *glmS* ribozyme. (ii) Ribozyme cleavage assays using the *glmS* RNA and various compounds showed that significant cleavage (Clv) only occurred in the presence of GlcN6P and Mg^{2+}. 5′-[^{32}P]-labeled precursor (Pre) RNAs were incubated for the times indicated in the absence (−) or presence of 200 μM effector as noted for each lane. Self-cleaved RNA was separated from precursor RNA by polyacrylamide gel electrophoresis and visualized by autoradiography. (Reprinted with permission from Nature Publishing Group and Macmillan Publishers Ltd: Winkler, W.C., Nahvi, A., Collins, J.A., Breaker, R.R. 2004. Control of gene expression by a natural metabolite-responsive ribozyme. *Nature* 428:281–286. Copyright © 2004.)

brought about by ligand binding. The term expression platform is just another name for the type of attenuation mechanism described for the *trp* operon. The expression platform domain has the potential to form alternative antiterminator and terminator hairpins.

Metabolite sensors

The genes controlled by riboswitches often encode proteins involved in the biosynthesis or transport of the metabolite being sensed. In most cases, the metabolite-sensing riboswitch is used as a form of feedback inhibition. Binding of the metabolite to the riboswitch serves as a genetic "off" switch and decreases the expression of the gene products used to make the metabolite. Repression occurs either by terminating transcription to prevent the production of full-length mRNAs, or by preventing translation initiation once a full-length mRNA has been made. For transcriptional control in the absence of a bound metabolite, an antiterminator RNA secondary structure is formed. When a metabolite binds, a competing stem structure is formed that acts as a transcription terminator. For control of translation initiation, the bound aptamer blocks the ribosome-binding site or Shine–Dalgarno sequence (Fig. 10.20A). In some rare cases, the riboswitch acts as a genetic "on" switch to activate gene expression. For example, binding of adenine or glycine to the adenine-sensing or glycine-sensing riboswitches, respectively, promotes mRNA transcription by preventing formation of a terminator stem.

RNA "thermometers"

Expression of many heat shock genes in *Bradyrhizobium japonicum* and other rhizobia (root nodule bacteria) is regulated by a conserved RNA sequence element called ROSE (repression of heat shock gene expression). ROSE is a temperature-sensitive "RNA thermometer" riboswitch. The riboswitch responds to temperatures between 30 and 40°C. At low temperature, translation initiation is prevented once a full-length mRNA has been synthesized. Ribosome access is blocked by an extended secondary structure in the mRNA. When the temperature rises, this secondary structure partially melts allowing ribosome access to the mRNA. *In vitro* ROSE RNA can unfold in response to temperature without the assistance of cellular components. A similar mechanism exists in *E. coli* for the control of heat shock transcription factor RpoH (σ^{32}). The major cold shock protein CspA works by a reverse mechanism, existing in different secondary structures at 37 and 15°C.

Riboswitch ribozymes

The riboswitches described above do not use a catalytic RNA as part of their mechanism of genetic control. In 2004, Ronald Breaker and colleagues reported the discovery of a metabolite-responsive ribozyme (Fig. 10.20B). The *glmS* gene in *Bacillus subtilis* encodes the enzyme glutamine fructose-6-phosphate amidotransferase. This enzyme generates glucosamine-6-phosphate (GlcN6P) from fructose-6-phosphate and glutamine. The *glmS* ribozyme responds to the addition of GlcN6P by self-cleaving its mRNA. Self-cleavage generates a 2′,3′-cyclic phosphate and a 5′-OH product, as in other small ribozymes (see Table 4.2). Self-destruction of the *glmS* mRNA inhibits further production of GlcN6P.

Chapter summary

The process of transcription and translation are coupled in bacteria. Transcription initiation is the level at which most genes are regulated. RNA polymerase binds to a *cis*-acting sequence on the DNA called a promoter. The majority of *E. coli* genes have promoter consensus sequences at −10 and −35 upstream of the start of transcription. The fully functional bacterial RNA polymerase (holoenzyme) consists of a core enzyme plus a regulatory protein called the sigma (σ) factor. For expression of some genes, bacterial cells use alternative σ factors. The σ factor is required for specific binding of the polymerase. RNA polymerase has a crab–claw-like structure. Binding of σ results in closing of the pincers and formation of the enzyme active site.

The process of transcription consists of three stages: initiation, elongation, and termination. Initiation is further divided into three stages: formation of a closed promoter complex where the DNA is still double-stranded, formation of an open complex in which the DNA melts to expose the template strand of DNA, and promoter clearance. During promoter clearance, certain domains of σ are displaced to allow exit of the nascent RNA from the RNA exit channel. As RNA polymerase moves from 5′ to 3′ it unwinds the strands at the front and rewinds them at the back, forming a characteristic transcription "bubble." Transcription continues in a processive manner as nucleotides are added to the growing RNA chain, unless an error is made. In this case, the polymerase backtracks and cleaves the most recently added nucleotides. When a terminator stem-loop sequence is reached, either Rho-independent or Rho-dependent, the RNA polymerase pauses and the nascent transcript is released. Transcription of most *E. coli* genes are aligned with the direction of replication. Some operons or genes are transcribed clockwise and some counterclockwise around the circular chromosome.

The operon model of Jacob and Monod introduced the novel concept at the time of regulatory genes that code for products that control other genes. Their model led to the discovery of mRNA and the characterization of the first repressor protein. Bacterial genes are organized into operons. An operon is a unit of bacterial expression and regulation, including structural genes and control elements in DNA recognized by regulatory gene products. The lactose (*lac*) operon encodes enzymes involved in the regulation of lactose metabolism. In the absence of lactose (or the presence of glucose, the preferred sugar), the *lac* operon is repressed. The Lac repressor binds to the operator DNA site and blocks RNA polymerase from binding the promoter. In the presence of lactose, the *lac* operon is induced. Upon binding of allolactose (the actual inducer), the Lac repressor undergoes a conformational change which alters its operator-binding domain. Lac repression is relieved and CAP binds its DNA site and recruits RNA polymerase to the promoter. Without the cooperative binding of CAP, only basal levels of transcription occur. When cells are starved of amino acids, a Rho-dependent terminator stops synthesis of mRNA. The *lac* promoter and the *lacZ* structural gene are widely used in molecular biology research; for example, as a reporter gene and for expression of heterologous proteins in bacteria.

Transcriptional regulators are modular proteins consisting of domains with distinct functions, such as for DNA binding and protein–protein interactions. A common DNA-binding motif is the helix-turn-helix motif, in which an α-helix interacts with the major groove of the DNA. In many cases, regulatory proteins bind to DNA in a cooperative manner with other proteins, as illustrated by CAP recruiting RNA polymerase to the promoter. Allosteric modification plays a role in the regulation of their activities. Such modification is exemplified by the shape changes in the Lac repressor upon inducer binding. These changes disrupt the helix-turn-helix DNA-binding motif and an important hinge region that holds the dimers together. In contrast, binding of cAMP to CAP increases its affinity for DNA. Through cooperative binding, distant DNA regulatory sites are brought in close proximity by protein–protein interactions that cause DNA looping. DNA looping was first discovered in the arabinose operon. Arabinose binds to the regulatory protein AraC, changing its shape so that it binds as a dimer to regulatory sequences near the promoter. In the absence of arabinose, one monomer of AraC binds to a distant site, forming a DNA loop that blocks access of RNA polymerase to the promoter.

Differential folding of RNA to form antiterminator or terminator structures leads to transcriptional attenuation of the tryptophan operator. When cells have adequate levels of tryptophan for protein synthesis, the terminator forms and transcription is terminated. Modulation by a tryptophan-activated repressor that binds an operator site within the promoter region also blocks access of RNA polymerase to the promoter. When cells are deficient in tryptophan, the repressor protein cannot bind the operator. In addition, the antiterminator forms and allows transcription of the structural genes involved in tryptophan biosynthesis.

Riboswitches are domains within certain mRNAs that act as switchable "on–off" elements that selectively bind metabolites and control gene expression. Riboswitches can function as a sensor for signals as diverse as temperature, salt concentration, metal ions, amino acids, and other small organic metabolites. Riboswitches have two structural domains: an aptamer that binds the target metabolite, and an expression platform that has the potential to form alternative antiterminator and terminator hairpins. Repression either occurs by terminating transcription or preventing translation initiation. One riboswitch ribozyme, *glmS*, has been characterized that responds to the addition of glucosamine-6-phosphate by self-cleaving its mRNA, which encodes the enzyme that generates this metabolite.

Analytical questions

1 A particular sequence containing six base pairs is located in 10 different organisms. The observed sequences are: 5′-ACGCAC-3′, ATACAC, GTGCAC, ACGCAC, ATACAC, ATGTAT, ATGCGC, ACGCAT, GTGCAT, and ATGCGC. What is the consensus sequence?

2 Draw a diagram of a prokaryotic gene being transcribed and translated. Show the nascent mRNA with ribosomes attached. With an arrow, indicate the direction of transcription.

3 Show how you could use DNase footprinting to demonstrate that a sigma (σ) factor is required for specific binding of RNA polymerase to a bacterial gene promoter.

4 Consider *E. coli* cells, each having one of the following mutations:
   (a) A mutant *lac* operator sequence that cannot bind Lac repressor.
   (b) A mutant Lac repressor that cannot bind to the *lac* operator.
   (c) A mutant Lac repressor that cannot bind to allolactose.
   (d) A mutant *lac* promoter that cannot bind CAP plus cAMP.

   What effect would each mutation have on the function of the *lac* operon (assuming no glucose is present)?

5 You are studying a new operon in bacteria involved in tyrosine biosynthesis.
   (a) You sequence the operon and discover that it contains a short open reading frame at the 5′ end of the operon that contains two codons for tyrosine. What prediction would you make about this leader sequence, the RNA transcript, and the peptide that it encodes?
   (b) How would you predict this operon is regulated; i.e. is it inducible or repressible by tyrosine? Why?
   (c) Would this kind of regulation work in a eukaryotic cell? Why or why not?

6 You are studying a repressor protein that you suspect forms a DNA loop between two operator sites, one located very near the promoter and the other located at a distant site upstream. Describe an experiment to test your hypothesis.

Suggestions for further reading

Blattner, F.R., Plunkett, G. III, Bloch, C.A. et al. (1997) The complete genome sequence of *Escherichia coli* K-12. *Science* 277:1453–1462.

Borukhov, S., Nudler, E. (2003) RNA polymerase holoenzyme: structure, function and biological implications. *Current Opinion in Microbiology* 6:93–100.

Cech, T.R. (2004) RNA finds a simpler way. *Nature* 428:263–264.

Chowdhury, S., Ragaz, C., Kreuger, E., Narberhaus, F. (2003) Temperature-controlled structural alterations of an RNA thermometer. *Journal of Biological Chemistry* 278:47915–47921.

Cook, P.R. (2001) *Principles of Nuclear Structure and Function*. John Wiley & Sons, New York.

Echols, H. (2001) *Operators and Promoters. The Story of Molecular Biology and its Creators*. University of California Press, Berkeley, CA.

Fedor, M.J., Williamson, J.R. (2005) The catalytic diversity of RNAs. *Nature Reviews Molecular Cell Biology* 6:399–412.

Fisher, H.W., Williams, R.C. (1979) Electron microscope visualization of nucleic acids and their complexes with proteins. *Annual Review of Biochemistry* 48:649–679.

Gruber, T.M., Gross, C.A. (2003) Multiple sigma subunits and the partioning of bacterial transcription space. *Annual Reviews of Microbiology* 57:441–466.

Harada, Y., Ohara, O., Takatsuki, A., Itoh, H., Shimamoto, N., Kinosita, K. (2001) Direct observation of DNA rotation during transcription by *Escherichia coli* RNA polymerase. *Nature* 409:113–115.

Henkin, T.M., Yanofsky, C. (2002) Regulation by transcription attenuation in bacteria: how RNA provides instructions for transcription termination/antitermination decisions. *BioEssays* 24:700–707.

Herr, A.J., Jensen, M.B., Dalmay, T., Baulcombe, D.C. (2005) RNA polymerase IV directs silencing of endogenous DNA. *Science* 308:118–120.

Hinkle, D.C., Chamberlin, M.J. (1972) Studies of the binding of *Escherichia coli* RNA polymerase to DNA. I. The role of sigma subunit in site selection. *Journal of Molecular Biology* 70:157–185.

Jacob, F., Monod, J. (1961) Genetic regulatory mechanisms in the synthesis of proteins. *Journal of Molecular Biology* 3:318–356.

Kalodimos, C.G., Biris, N., Bonvin, A.M.J.J., Levandoski, M.M., Guennuegues, M., Boelens, R., Kaptein, R. (2004) Structure and flexibility adaptation in nonspecific and specific protein–DNA complexes. *Science* 305:386–389.

Kuznedelov, K., Minakhin, L., Niedziela-Majka, A. et al. (2002) A role for interaction of the RNA polymerase flap domain with the σ subunit in promoter recognition. *Science* 295:855–857.

Lewis, M. (2005) The lac repressor. *Comptes Rendus Biologies* 328:521–548.

Ma, D., Cook, D.N., Pon, N.G., Hearst, J.E. (1994) Efficient anchoring of RNA polymerase in *Escherichia coli* during coupled transcription-translation of genes encoding integral inner membrane polypeptides. *Journal of Biological Chemistry* 269:15362–15370.

Masters, B.S., Stohl, L.L., Clayton, D.A. (1987) Yeast mitochondrial RNA polymerase is homologous to those encoded by bacteriophages T3 and T7. *Cell* 51:89–99.

Miller, O.L., Hamkalo, B.A., Thomas, C.A. (1970) Visualization of bacterial genes in action. *Science* 169:392–395.

Murakami, K.S., Darst, S.A. (2003) Bacterial RNA polymerases: the wholo story. *Current Opinion in Structural Biology* 13:31–39.

Nickels, B.E., Garrity, S.J., Mekler, V., Minakhin, L., Severinov, K., Ebright, R.H., Hochschild, A. (2005) The interaction between σ70 and the β-flap of *Escherichia coli* RNA polymerase inhibits extension of nascent RNA during early elongation. *Proceedings of the National Academy of Sciences USA* 102:4488–4493.

Onodera, Y., Haag, J.R., Ream, T., Nunes, P.C., Pontes, O., Pikaard, C.S. (2005) Plant nuclear RNA polymerase IV mediates siRNA and DNA methylation-dependent heterochromatin formation. *Cell* 120:613–622.

Ptashne, M., Gann, A. (2002). *Genes and Signals*. Cold Spring Harbor Laboratory Press, Cold Spring Harbor, NY.

Richardson, J.P. (2002) Rho-dependent termination and ATPases in transcript termination. *Biochimica et Biophysica Acta* 1577:251–260.

Schleif, R. (2000) Regulation of the L-arabinose operon of *Escherichia coli*. *Trends in Genetics* 16:559–565.

Shaevitz, J.W., Abbondanzieri, E.A., Landick, R., Block, S.M. (2003) Backtracking by single RNA polymerase molecules observed at near-base-pair resolution. *Nature* 426:684–687.

Shiina, T., Tsunoyama, Y., Nakahira, Y., Khan, M.S. (2005) Plastid RNA polymerases, promoters, and transcription regulators in higher plants. *International Review of Cytology* 244:1–68.

Steitz, T.A. (1993) *Structural Studies of Protein–Nucleic Acid Interaction: the Sources of Sequence-Specific Binding*. Cambridge University Press, Cambridge, UK.

Tucker, B.J., Breaker, R.R. (2005) Riboswitches as versatile gene control elements. *Current Opinion in Structural Biology* 15:342–348.

Von Hippel, P.H. (1998) An integrated model of the transcription complex in elongation, termination, and editing. *Science* 281:660–665.

Von Hippel, P.H. (2004) Completing the view of transcriptional regulation. *Science* 305:350–353.

Winkler, W.C., Nahvi, A., Collins, J.A., Breaker, R.R. (2004) Control of gene expression by a natural metabolite-responsive ribozyme. *Nature* 428:281–286.

Zhang, G., Campbell, E.A., Minakhin, L., Richter, C., Severinov, K., Darst, S.A. (1999) Crystal structure of *Thermus aquaticus* core RNA polymerase at 3.3 Å resolution. *Cell* 98:811–824.

Chapter 11
Transcription in eukaryotes

. . . the modern researcher in transcriptional control has much to think about.

James T. Kadonaga, *Cell* (2004) 116:247.

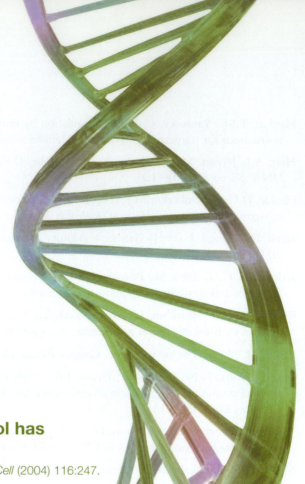

Outline

11.1 Introduction

Many eukaryotes are estimated to have 20,000–25,000 genes (see Table 16.4). Some of these are expressed (transcribed) in all cells all of the time, while others are expressed as cells enter a particular pathway of differentiation or as conditions in and around the cells change. In the early 1980s, transcription researchers primarily explored DNA–protein interactions *in vitro*. Research focused on the purification of sequence-specific DNA-binding proteins by affinity chromatography, analysis of the transcriptional activity of promoters by reporter gene assays, *in vitro* transcription assays that allowed the fractionation of the general transcription machinery, and assays such as electrophoretic gel mobility shift assays (EMSA) and DNase I footprinting for analysis of *cis*-acting DNA elements with *trans*-acting factors (see Chapter 9 for methods). By the late 1980s, many sequence-specific DNA-binding proteins had been identified, purified, and their genes cloned. Upon further study, it became clear that in addition to DNA–protein interactions, protein–protein interactions were of critical importance for regulating gene transcription. This insight was followed closely by the realization in the early 1990s that chromatin structure, nuclear architecture, and cellular compartmentalization must also be taken into account. Sections within this chapter will cover protein-coding gene regulatory elements, transcription factors and their DNA-binding motifs, the general transcription machinery and the mechanism of RNA polymerase II transcription, transcriptional coactivators and corepressors, including chromatin modification and remodeling complexes, and signal-mediated nuclear import and export of proteins involved in regulating gene transcription.

11.2 Overview of transcriptional regulation

The most important and widely used strategy for regulating gene expression is altering the rate of transcription of a gene. However, the control of gene expression can be exerted at many other levels, including processing of the RNA transcript, transport of RNA to the cytoplasm, translation of mRNA, and mRNA and protein stability. These additional levels of control are discussed in Chapters 13 and 14. There are also instances where genes are selectively amplified during development and, as a consequence, there is an increase in the amount of RNA transcript synthesized. The ribosomal RNA genes of *Xenopus* are an example of this form of gene regulation (see Fig. 6.17).

In this chapter, the regulation of transcription of protein-coding genes by RNA polymerase II (RNA pol II) will be highlighted. RNA pol II is located in the nucleoplasm and is responsible for transcription of the vast majority of genes including those encoding mRNA, small nucleolar RNAs (snoRNAs), some small nuclear RNAs (snRNAs), and microRNAs. Gene transcription is a remarkably complex process. The synthesis of tens of thousands of different eukaryotic mRNAs is carried out by RNA pol II. During the process of transcription, RNA pol II associates transiently not only with the template DNA but with many different proteins, including general transcription factors. The initiation step alone involves the assembly of dozens of factors to form a preinitiation complex. Transcription is mediated by the collective action of sequence-specific DNA-binding transcription factors along with the core RNA pol II transcriptional machinery, an assortment of coregulators that bridge the DNA-binding factors to the transcriptional machinery, a number of chromatin remodeling factors that mobilize nucleosomes, and a variety of enzymes that catalyze covalent modification of histones and other proteins. Not surprisingly, the transcription literature is replete with a sometimes bewildering array of acronyms such as TBP, CBP, HDAC, LSD1, and SWI/SNF, to name a few (Table 11.1).

There are two other important eukaryotic polymerases – RNA polymerase I and RNA polymerase III (see Table 10.1). RNA polymerase I resides in the nucleolus and is responsible for synthesis of the large ribosomal RNA precursor. RNA polymerase III is also located in the nucleoplasm and is responsible for synthesis of transfer RNA (tRNA), 5S ribosomal RNA (rRNA), and some snRNAs. Plants have a fourth nuclear polymerase, named RNA polymerase IV, which is an RNA silencing-specific polymerase that mediates synthesis of small interfering RNAs (siRNAs) involved in heterochromatin formation (see Section 12.6). A full treatment of transcriptional regulation by all of the polymerases is beyond the scope of this chapter.

11.3 Protein-coding gene regulatory elements

Expression of protein-coding genes is mediated in part by a network of thousands of sequence-specific DNA-binding proteins called transcription factors. Transcription factors interpret the information present in gene promoters and other regulatory elements, and transmit the appropriate response to the RNA pol II transcriptional machinery. Information content at the genetic level is expanded by the great variety of regulatory DNA sequences and the complexity and diversity of the multiprotein complexes that regulate gene expression. Many different genes and many different types of cells in an organism share the same transcription factors. What turns on a particular gene in a particular cell is the unique combination of regulatory elements and the transcription factors that bind them.

Protein-coding sequences make up only a small fraction of a typical multicellular eukaryotic genome. For example, they account for less than 2% of the human genome. The typical eukaryotic protein-coding gene consists of a number of distinct transcriptional regulatory elements that are located immediately 5′ of the transcription start site (termed +1). The regulatory regions of unicellular eukaryotes such as yeast are usually only composed of short sequences located adjacent to the core promoter (Fig. 11.1A). In contrast, the regulatory regions in multicellular eukaryotes are scattered over an average distance of 10 kb of genomic DNA with the transcribed DNA sequence only accounting for just 2 or 3 kb (Fig. 11.1B). Genes range in size from very small, such as a histone gene that is only 500 nt long with no introns, to very large. The largest known human gene encodes the protein dystrophin, which is missing or nonfunctional in the disease muscular dystrophy. The transcribed sequence is 2.5 million nucleotides in length, including 79 introns. It takes over 16 hours to produce a single transcript, of which more than 99% is removed during splicing to generate a mature mRNA.

Gene regulatory elements are specific *cis*-acting DNA sequences that are recognized by *trans*-acting transcription factors (see Section 10.3 for more discussion of the terms "*cis*" and "*trans*"). *Cis*-regulatory elements in multicellular eukaryotes can be classified into two broad categories based on how close they are to the start of transcription: promoter elements and long-range regulatory elements. In comparing the regulatory region of a particular gene with another in multicellular eukaryotes, there will be variation in whether a particular element is present or absent, the number of distinct elements, their orientation relative to the transcriptional start site, and the distance between.

Structure and function of promoter elements

The "gene promoter" is loosely defined as the collection of *cis*-regulatory elements that are required for initiation of transcription or that increase the frequency of initiation only when positioned near the transcriptional start site. The gene promoter region includes the core promoter and proximal promoter elements. Proximal promoter elements are also sometimes designated as "upstream promoter elements" or "upstream regulatory elements."

Table 11.1 Proteins that regulate transcription.

Category	Acronym	Derivation of name	Function
Transcription factors (activators or repressors)	Some examples mentioned in text:		
	CBF	CAAT binding factor	Binds CAAT box
	C/EBP	CAAT/enhancer-binding protein	Binds CAAT box
	CREB	cAMP response element-binding protein	Binds the cAMP response element
	CTCF	CCCTC-binding factor	Binds insulator element (CCCTC) and mediates enhancer blocking activity
	FOG-1	Friend of GATA-1	Required for developmental expression of β-globin genes
	GATA-1	GATA-binding protein	Required for developmental expression of β-globin genes
	NF-E2	Nuclear factor erythoid-derived 2	Required for developmental expression of β-globin genes
	NF-κB	Nuclear factor of kappa light polypeptide enhancer in B cells	Central mediator of human stress and immune responses
	USF1, USF2	Upstream stimulatory factor 1 and 2	Bind insulator element, recruit chromatin-modifying enzymes
	SATB1	Special AT-rich binding protein 1	Matrix attachment region (MAR) binding protein required for T-cell-specific gene regulation
	Sp1	SV40 early and late promoter-binding protein 1	Binds GC box
General transcription machinery	RNA pol II (pol II, RNAPII)	RNA polymerase II	Catalysis of RNA synthesis
	General transcription factors:		
	TFIIB	Transcription factor for RNA polymerase II B	Stabilization of TBP–DNA interactions, recruitment of RNA pol II–TFIIF, start site selection by RNA pol II
	TFIID:		
	TBP	TATA-binding protein	Core promoter recognition, TFIIB recruitment
	TAF	TBP-associated factor	Core promoter recognition/selectivity
	TFIIE	Transcription factor for RNA polymerase II E	TFIIH recruitment
	TFIIF	Transcription factor for RNA polymerase II F	Recruitment of RNA pol II to promoter DNA–TBP–TFIIB complex
	TFIIH	Transcription factor for RNA polymerase II H	Promoter melting, helicase, RNA pol II CTD kinase
	Mediator	Mediator	Transduces regulatory information from activator and repressor proteins to RNA pol II
Coactivators and corepressors	Chromatin modification complexes:		
	HAT	Histone acetyltransferase	Acetylates histones
	HDAC	Histone deacetylase	Deacetylates histones
	CBP	CREB-binding protein	HAT activity
	HMT	Histone methyltransferase	Methylates histones
	LSD1	Lysine-specific demethylase 1	Demethylates histones
	Chromatin remodeling complexes:		
	SWI/SNF	Mating-type switching defective/sucrose nonfermenters	ATP-dependent chromatin remodeling (sliding and disassembly)
	ISWI	Imitation Swi2	ATP-dependent chromatin remodeling (sliding)
	SWR1	Swi2/Snf2 related 1	ATP-dependent chromatin remodeling (histone replacement)
Elongation factors	FACT	Facilitates chromatin transcription	Transcription-dependent nucleosome alterations
	Elongator	Elongator	Exact function in elongation unknown
	TFIIS	Transcription factor for RNA polymerase II S	Facilitates RNA pol II passage through regions that cause transcriptional arrest

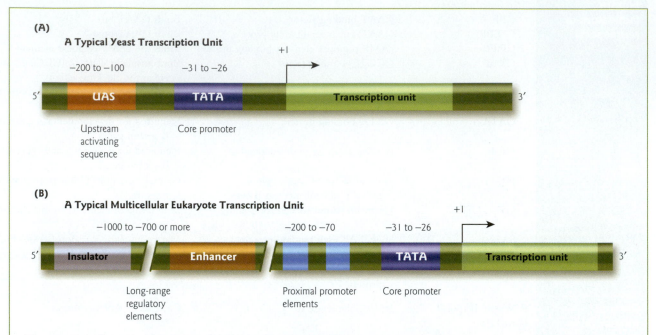

(A)

A Typical Yeast Transcription Unit

−200 to −100   −31 to −26   +I

5′ UAS TATA Transcription unit 3′

Upstream activating sequence

Core promoter

(B)

A Typical Multicellular Eukaryote Transcription Unit

−1000 to −700 or more   −200 to −70   −31 to −26   +I

5′ Insulator Enhancer TATA Transcription unit 3′

Long-range regulatory elements

Proximal promoter elements

Core promoter

Figure 11.1 Comparison of a simple and complex RNA pol II transcription unit. (A) A typical yeast (unicellular eukaryote) transcription unit. The start of transcription (+1) of the protein-coding gene (transcription unit) is indicated by an arrow. (B) A typical multicellular eukaryote transcription unit with clusters of proximal promoter elements and long-range regulatory elements located upstream from the core promoter (TATA). There is variation in whether a particular element is present or absent, the number of distinct elements, their orientation relative to the transcriptional start site, and the distance between them. Although the figure is drawn as a straight line, the binding of transcription factors to each other draws the regulatory DNA sequences into a loop.

Core promoter elements

The core promoter is an approximately 60 bp DNA sequence overlapping the transcription start site (+1) that serves as the recognition site for RNA pol II and general transcription factors (see Section 11.4). Promoter elements become nonfunctional when moved even a short distance from the start of transcription or if their orientation is altered. The general transcription factor TFIID is responsible for the recognition of all known core promoter elements, with the exception of the BRE which is recognized by TFIIB. Some of the known core promoter elements are the TATA box, the initiator element (Inr), the TFIIB recognition element (BRE), the downstream promoter element (DPE), and the motif ten element (MTE) (Fig. 11.2,

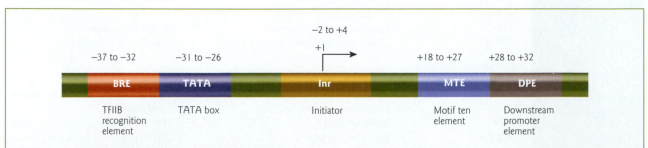

−2 to +4

−37 to −32   −31 to −26   +I   +18 to +27   +28 to +32

BRE TATA Inr MTE DPE

TFIIB recognition element

TATA box

Initiator

Motif ten element

Downstream promoter element

Figure 11.2 RNA pol II core promoter motifs. Sequence elements that can contribute to basal transcription from the core promoter. A particular core promoter may contain some, all, or none of these motifs. The locations of the TFIIB recognition element (BRE), TATA box (TATA), initiator (Inr), motif ten element (MTE), and downstream promoter element (DPE) motifs are indicated relative to the start of transcription (+1).

Table 11.2 Eukaryotic promoter elements.

Promoter	Position	Transcription factor	Consensus sequence
Upstream core promoter elements			
TFIIB recognition element (BRE)	−37 to −32	TFIIB	(G/C)(G/C)(G/A)CGCC
TATA box	−31 to −26	TBP	TATA(A/T)AA(G/A)
Initiator (Inr)	−2 to +4	TAF1 (TAF$_{II}$250)	PyPyA$_{+1}$N(T/A)PyPy
		TAF2 (TAF$_{II}$150)	
Downstream core promoter elements			
Motif ten element (MTE)	+18 to +27	TFIID	C(G/A)A(A/G)C(G/C)
			(C/A/G)AACG(G/C)
Downstream promoter element (DPE)	+28 to +32	TAF9 (TAF$_{II}$40)	(A/G)G(A/T)(C/T)(G/A/C)
		TAF6 (TAF$_{II}$60)	
Proximal promoter elements			
CAAT box	−200 to −70	CBF, NF1, C/EBP	CCAAT
GC box	−200 to −70	Sp1	GGGCGG

Most, but not all, CAAT and GC boxes are located between −200 and −70.
CBF, CAAT-binding protein; C/EBP, CAAT/enhancer-binding protein; N, any (A, T, C, or G); Py, pyrimidine (C or T).

Table 11.2). Each of these sequence motifs is found in only a subset of core promoters. A particular core promoter may contain some, all, or none of these elements.

The TATA box (named for its consensus sequence of bases, TATAAA) was the first core promoter element identified in eukaryotic protein-coding genes. A key experiment by Pierre Chambon and colleagues demonstrated that a viral TATA box is both necessary and sufficient for specific initiation of transcription by RNA pol II. When they cloned a viral promoter into the plasmid pBR322, it was able to promote specific initiation of transcription *in vitro* (Fig. 11.3).

From early studies, it seemed that the TATA box was present in the majority of protein-coding genes. However, recent sequence database analysis of human genes found that TATA boxes are present in only 32% of potential core promoters. Thus, it has become increasingly important for molecular biologists to be aware of the full repertoire of possible promoter elements, and to continue the search for novel regulatory elements. The TATA box is the binding site for the TATA-binding protein (TBP), which is a major subunit of the TFIID complex (see Section 11.4). The TATA box can function in the absence of BRE, Inr, and DPE motifs. The Inr element was defined as a discrete core promoter element that is functionally similar to the TATA box. The Inr element is recognized by two other subunits of TFIID, TBP-associated factor 1 (TAF1) and TAF2 (TAF$_{II}$250 and TAF$_{II}$150 in the old TAF nomenclature). Inr can function independently of the TATA box, but in TATA-containing promoters, it acts synergistically to increase the efficiency of transcription initiation. Synergistic means that they act together, often to produce an effect greater than the sum of the two promoter elements acting separately. The Inr consensus sequence is shown in Table 11.2.

The DPE is a distinct seven nucleotide element that is conserved from *Drosophila* to humans. It functions in TATA-less promoters and is located about +30 relative to the transcription start site (see Fig. 11.2). The DPE consensus sequence is shown in Table 11.2. In contrast to the TATA box, the DPE motif requires the presence of an Inr. The DPE is bound by two specific subunits of the TFIID complex, TAF9 and TAF6 (TAF$_{II}$40 and TAF$_{II}$60, respectively, in the old TAF nomenclature). The recently identified MTE is located at positions +18 to +27 relative to the start of transcription (see Fig. 11.2 and Table 11.2). It promotes transcriptional activity and binding of TFIID in conjunction with the Inr. Although it can function independently of the TATA box or DPE, it exhibits strong synergism with both of these elements. Other downstream promoter motifs that contribute to transcriptional activity have been described that appear to be distinct from DPE and MTE. For example, the downstream core element (DCE) was first identified in the

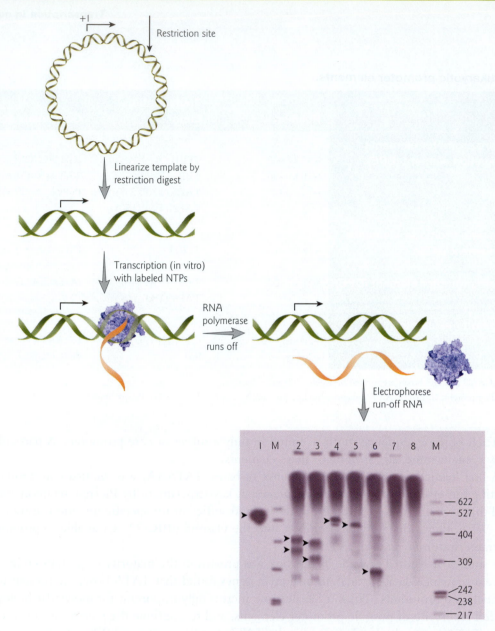

Figure 11.3 A TATA box-containing region promotes specific initiation of transcription *in vitro*. A 21 or 43 bp DNA fragment containing the adenovirus type 2 major late promoter TATA box was cloned into plasmid pBR322. Plasmid "C" contains two copies of the 21 bp region, cloned in the same orientation. Plasmid "F" contains the 21 and 43 bp region cloned in the opposite orientation. Transcription was measured using an *in vitro* run-off assay, in which the plasmid template is linearized with a restriction endonuclease, then incubated in a cell-free extract with radiolabeled nucleoside triphosphates (NTPs). When RNA pol II (blue) reaches the end of the linear template, it falls off and releases the labeled run-off transcript (orange). Run-off transcripts are then separated by polyacrylamide gel electrophoresis and visualized by autoradiography. The size of the transcript corresponds to the distance between the start of transcription and the end of the template. The more actively the template is transcribed, the stronger the transcript signal. Lane 1 and lane M, size markers; lane 2, plasmid C cut with *Eco*RI; lane 3, plasmid C cut with *Hin*dIII; lane 4, plasmid F cut with *Eco*RI; lane 5, plasmid F cut with *Hin*dIII; lane 6, plasmid F cut with *Sal*I; lane 7, pBR322 wild-type plasmid (lacks a TATA box) cut with *Eco*RI; lane 8, pBR322 wild-type plasmid cut with *Sal*I. Arrowheads point to the run-off transcripts. No bands were produced by pBR322 lacking a TATA box. In all cases the TATA region directed RNA pol II to initiate about 30 bp downstream from the first T. Transcript sizes were similar to predicted sizes, indicating specific initiation from the TATA box (about 380 and 335 nt for plasmid C, and 515, 485, and 300 nt for plasmid F). (Reprinted from Sassone-Corsi, P., Corden, J., Kédinger, C., Chambon, P. 1981. Promotion of specific *in vitro* transcription by excised "TATA" box sequences inserted in a foreign nucleotide environment. *Nucleic Acids Research* 9:3941–3958, by permission of Oxford University Press).

human β-globin promoter. It consists of three sub-elements located at approximately +10, +20, and +30 of a subset of TATA-containing promoters. The DCE is bound by TAF1 and contributes to transcriptional activity of TATA-containing promoters.

Proximal promoter elements

The regulation of TFIID binding to the core promoter element in yeast depends on an upstream activating sequence (UAS) located within a few hundred base pairs of the promoter (see Fig. 11.1A). The vast majority of yeast genes contain a single UAS, which is usually composed of two or three closely linked binding sites for one or two different transcription factors. In contrast, a typical multicellular eukaryote gene is likely to contain several proximal promoter elements. Promoter proximal elements are located just 5′ of the core promoter and are usually within 70–200 bp upstream of the start of transcription. Recognition sites for transcription factors tend to be located in clusters. Examples include the CAAT box and the GC box (see Table 11.2). The CAAT box is a binding site for the CAAT-binding protein (CBF) and the CAAT/enhancer-binding protein (C/EBP). The GC box is a binding site for the transcription factor Sp1. Sp1 was initially identified as one of three components required for the transcription of *SV40* early and late promoters. Promoter proximal elements increase the frequency of initiation of transcription, but only when positioned near the transcriptional start site. The transcription factors that bind promoter proximal elements do not always directly activate or repress transcription. Instead, they might serve as "tethering elements" that recruit long-range regulatory elements, such as enhancers, to the core promoter.

Structure and function of long-range regulatory elements

Protein-coding genes of multicellular eukaryotes typically contain additional regulatory DNA sequences that can work over distances of 100 kb or more from the gene promoter. These long-range regulatory elements are instrumental in mediating the complex patterns of gene expression in different cells types during development. Such long-range regulation is not generally observed in yeast, although a few genes have regulatory sequences located further upstream than the UAS (e.g. silencers of the mating-type locus, see Section 12.7). The function of many long-range regulatory elements was confirmed by their effect on gene expression in transgenic animals. These elements tend to protect transgenes from the negative or positive influences exerted by chromatin at the site of integration (Focus box 11.1). Long-range regulatory elements in multicellular eukaryotes include enhancers and silencers, insulators, locus control regions (LCRs), and matrix attachment regions (MARs).

Enhancers and silencers

A typical protein-coding gene is likely to contain several enhancers which act at a distance. These elements are usually 700–1000 bp or more away from the start of transcription. The hallmark of enhancers is that, unlike promoter elements, they can be downstream, upstream, or within an intron, and can function in either orientation relative to the promoter (Fig. 11.4). A typical enhancer is around 500 bp in length and contains in the order of 10 binding sites for several different transcription factors. Each enhancer is responsible for a subset of the total gene expression pattern. Enhancers increase gene promoter activity either in all tissues or in a regulated manner (i.e. they can be tissue-specific or developmental stage-specific). Similar elements that repress gene activity are called silencers.

Insulators

Eukaryotic genomes are separated into gene-rich euchromatin and gene-poor, highly condensed heterochromatin. Because heterochromatin has a tendency to spread into neighboring DNA, natural barriers to spreading are critical when active genes are nearby. A mutation in *Drosophila* affecting a chromatin

Over the past 15–20 years of making transgenic animals and plants, a major problem has been a lack of expression of the transgene, or inappropriate expression (see Chapter 15). This has been attributed to "position effect." Position effect is a phenomenon in which expression of the transgene is unpredictable; it varies with the chromosomal site of integration. Because integration is random when transgenic animals are made by pronuclear microinjection of foreign DNA (see Fig. 15.2), it is possible for the transgene to be integrated into either inactive or active chromatin. Because heterochromatin has a tendency to spread into neighboring DNA, transgenes that are integrated near heterochromatin tend to undergo inactivation. Traditionally, transgenes were constructed by fusing a complementary DNA (cDNA) coding for the protein of interest to a strong promoter. When a combination of long-range regulatory elements was included in the transgene construct, research found that position-independent expression units could be established, regardless of where the transgene integrated into the chromatin. Two classic examples of experiments showing how enhancers and MARs can protect transgenes from position effects are described below.

Intron enhancers contribute to tissue-specific gene expression

Introns were long considered to be "junk" DNA. We now know that they can include important coding DNA sequences (see Focus box 13.1), as well as regulatory elements such as enhancers. The importance of intron enhancers has been demonstrated for a number of genes *in vivo*. For example, Beatriz Levy-Wilson and colleagues showed that an enhancer located in the second intron of the human apolipoprotein B (*apoB*) gene is essential for tissue-specific gene expression. The gene product is a protein responsible for clearance of low-density lipoproteins (LDLs) by the LDL receptor and is involved in cholesterol homeostasis. The *apoB* gene is transcribed primarily in the liver and intestine in humans. In cell culture, a gene construct that contained only the *apoB* promoter linked to the *lacZ* reporter gene (see Section 9.3) was efficiently expressed. However, the addition of a sequence representing the second intron enhancer stimulated β-galactosidase activity 5–7-fold (Fig. 1A). In contrast, in transgenic mice there was no expression of the promoter-only construct. Integration of the construct was confirmed by Southern blot analysis (see Tool box 8.7) and transcription was assessed by RNase protection assay (see Fig. 9.8). In addition, β-galactosidase activity was assayed in

tissue sections of livers from transgenic and control mice. Further experiments showed that the second intron enhancer was absolutely required for expression of the reporter gene in the liver. Neither the promoter-only nor the promoter-enhancer construct were expressed in the small intestine, suggesting that additional tissue-specific regulatory elements are required.

MARs promote formation of independent loop domains

The existence of matrix attachment regions (MARs) was considered to be without question; however, their biological significance was originally uncertain. Researchers thus set out to test whether MARs are essential for appropriately regulated gene expression *in vivo*. In one of the first studies of its kind, Lothar Hennighausen and colleagues examined transcriptional regulation of the whey acidic protein (WAP) gene, which codes for a major milk protein in mice. This gene provided an excellent model system because its expression is tissue-specific and developmentally and hormonally regulated. WAP is only expressed in the mammary gland during lactation, under control of the insulin–hydrocortisone–prolactin signaling pathway. Transgenic mice were generated with the 7 kb WAP-coding region and its associated promoter but no other flanking regions, or with the inclusion of chicken lysozyme gene MARs (Fig. 1B). The reason a chicken MAR was used was because at the time it was the best characterized MAR. A *Hind*III linker in the 5′ untranslated region was included to distinguish the transgene from the endogenous gene. Polymerase chain reaction (PCR) and Southern blot analysis were carried out to identify transgenic mice and to map the site of integration. Northern blot analysis was used to analyze gene transcription. The results showed that in the absence of the MARs, WAP mRNA expression was position-dependent. In other words, expression was unpredictable and depended on where the transgene was integrated. Expression was mammary-specific in 50% of the mouse lines, but it was hormone-independent and the levels were variable. In some mouse lines, WAP was activated early during pregnancy and then turned off during lactation. In contrast, in the presence of MARs, there was position-independent regulation. That is, when MARs were included, the transgene was unaffected by neighboring chromatin regardless of the site of integration. All transgenic mouse lines showed mammary-specific expression and four out of five lines showed accurate hormonal and developmental regulation.

This experiment and other similar ones (e.g. chicken lysozyme locus MARs have even been shown to reduce the

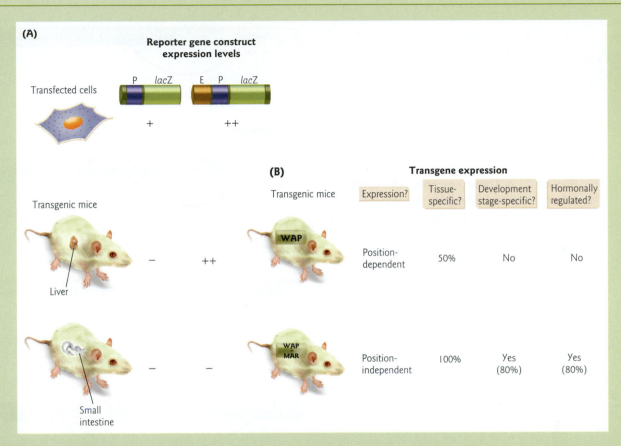

Figure 1 Long-range regulatory elements can protect transgenes from position effect. (A) Intron enhancer. In transiently transfected cells, a gene construct that contained only the apolipoprotein B gene promoter (P) linked to the *lacZ* reporter gene was efficiently expressed (indicated by "+" symbol) (see Fig. 9.1 and Section 9.3). The addition of a sequence representing the second intron enhancer (E) stimulated β-galactosidase activity 5—fold (++). In contrast, in transgenic mice (see Fig. 15.2) there was no expression of the promoter-only construct (indicated by "−" symbol). The second intron enhancer was absolutely required for expression of the reporter gene in the liver. Neither the promoter-only nor the promoter-enhancer construct were expressed in the small intestine. (Brand, M., Ranish, J.A., Kummer, N.T. et al. 1994. Sequences containing the second-intron enhancer are essential for transcription of the human apolipoprotein B gene in the livers of transgenic mice. *Molecular and Cellular Biology* 14:2243–2256. (B) Matrix attachment regions (MARs). Transgenic mice were generated with the whey acid protein (WAP) coding region and its associated promoter but no other flanking regions, or with the inclusion of chicken lysozyme gene MARs (WAP + MAR). In the absence of MARs, WAP mRNA expression was position-dependent. Expression was mammary-specific in 50% of the mouse lines, but expression was not developmentally or hormonally regulated. In the presence of MARs, there was position-independent regulation. All transgenic mouse lines showed mammary-specific expression and four out of five lines showed accurate hormonal and developmental regulation. (McKnight, R.A., Shamay, A., Sankaran, L., Wall, R.J., Hennighausen, L. 1992. Matrix-attachment regions can impart position-independent regulation of a tissue-specific gene in transgenic mice. *Proceedings of the National Academy of Sciences USA* 89:6943–6947.)

variability in transgene expression in rice plants) lend support for the following model of gene regulation. The association of MARs with the nuclear architecture allows formation of an independent DNA loop domain. This loop domain can adopt an altered chromatin structure distinct from the structure of neighboring chromatin. In this altered configuration, the gene promoter and other regulatory elements become accessible to tissue-specific and/or developmental stage-specific transcription factors.

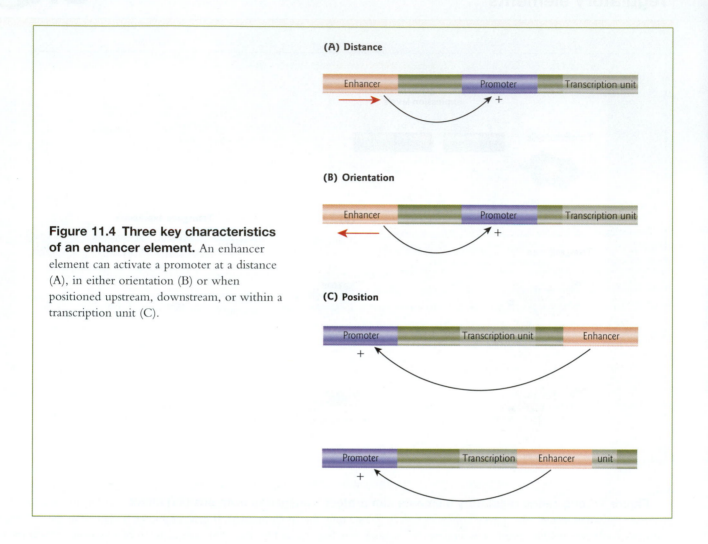

Figure 11.4 Three key characteristics of an enhancer element. An enhancer element can activate a promoter at a distance (A), in either orientation (B) or when positioned upstream, downstream, or within a transcription unit (C).

boundary was first observed by Ed Lewis in the 1970s. However, the general concept of boundary elements functioning as "insulators" was not fully established until the early 1990s. An insulator is a DNA sequence element, typically 300 bp to 2 kb in length, that has two distinct functions (Fig. 11.5):

1 Chromatin boundary marker: an insulator marks the border between regions of heterochromatin and euchromatin.
2 Enhancer blocking activity: an insulator prevents inappropriate cross-activation or repression of neighboring genes by blocking the action of enhancers and silencers.

Typically, insulators contain clustered binding sites for sequence-specific DNA-binding proteins. The exact molecular mechanism by which they block enhancers and silencers is not clear. One model proposes that insulators tether the DNA to subnuclear sites, forming loops that separate the promoter of one gene from the enhancer of another.

The vertebrate β-globin locus is an excellent model system for examining the interaction between insulator elements and chromatin structure. Gary Felsenfeld and co-workers first identified a chromatin boundary separating adjacent heterochromatin from β-globin genes within the locus. The boundary is located at a deoxyribonuclease I (DNase I) hypersensitive (HS) site called HS4. DNase I hypersensitive sites are a hallmark of active genes and regulatory elements (Disease box 11.1). Felsenfeld and colleagues went on to show that HS4 insulator also provides enhancer blocking activity from neighboring genes.

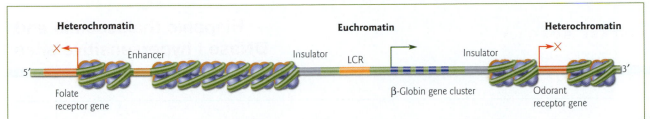

Figure 11.5 Insulators function as chromatin boundary markers and have enhancer blocking activity. An insulator (gray) marks the border between regions of heterochromatin (depicted as nucleosome-bound DNA) and euchromatin. In erythrocytes it separates the actively transcribed β-globin genes (indicated by the green arrow) from the inactive folate receptor gene (indicated by the red X symbol). Another insulator prevents the inactive odorant receptor gene from cross-activation by the locus control region (LCR) (orange) upstream of the β-globin genes, and vice versa in other cell types. (Adapted from Bell, A.C., West, A.G., Felsenfeld, G. 2001. Insulators and boundaries. Versatile regulatory elements in the eukaryotic genome. *Science* 291:447–450.)

The insulator elements are recognized by at least three different DNA-binding proteins, CCCTC-binding factor (CTCF), and upstream stimulatory factor (USF) 1 and 2. CTCF mediates the enhancer blocking activity, while USF proteins bind to the insulator and recruit several chromatin-modifying enzymes (see Section 11.6).

Locus control regions (LCRs)

LCRs are DNA sequences that organize and maintain a functional domain of active chromatin and enhance the transcription of downstream genes. Although sometimes referred to as "enhancers" of transcription, LCRs, unlike classic enhancer elements, operate in an orientation-dependent manner. The prototype LCR was characterized in the mid-1980s as a cluster of DNase I-hypersensitive sites upstream of the β-globin gene cluster (Fig. 11.6A). At the time, DNase I-hypersensitive sites were known to be important elements in the control of chromatin structure and transcriptional activity (Disease box 11.1). This series of hypersensitive sites became known collectively first as a "locus activation region" and later as the "locus control region." Subsequently, LCRs have been shown to be present in other loci, including gene clusters encoding the α-globins, visual pigments, major histocompatibility proteins, human growth hormones, serpins (a family of structurally related proteins that inhibit proteases), and T-helper type 2 cytokines (involved in the immune response).

Beta-globin gene LCR is required for high-level transcription   Hemoglobin is the iron-containing oxygen transport metalloprotein in the red blood cells of mammals and other animals. In adult humans, the most common hemoglobin is a tetramer composed of two α-like globin polypeptides and two β-like polypeptides plus four heme groups (an organic molecule with an iron atom) (see Fig. 5.10). Given their critical function, it is not surprising that the α- and β-globin genes are highly regulated. In particular, the LCR of the β-like globin gene cluster provides an excellent illustration of the complexity of regulatory regions (Fig. 11.6A). The β-like globin-coding regions are each 2–3 kb in size and the entire cluster spans approximately 100 kb. The genes are expressed in erythroid cells in a tissue- and developmental stage-specific manner: the epsilon (ε) globin gene is activated in the embryonic stage, the gamma (γ) globin is activated in the fetal stage, and the β-globin gene is expressed in adults. Physiological levels of expression of each of these genes can be achieved only when they are downstream of the LCR. The DNase hypersensitive sites contain clusters of transcription factor-binding sites and interact via extensive protein–DNA and protein–protein interactions.

Early studies of β-globin gene expression *in vivo* were often inconclusive. Proper developmental regulation and high-level expression could not be achieved coordinately in transgenic mice carrying an artificial

DISEASE BOX 11.1

Hispanic thalassemia and DNase I hypersensitive sites

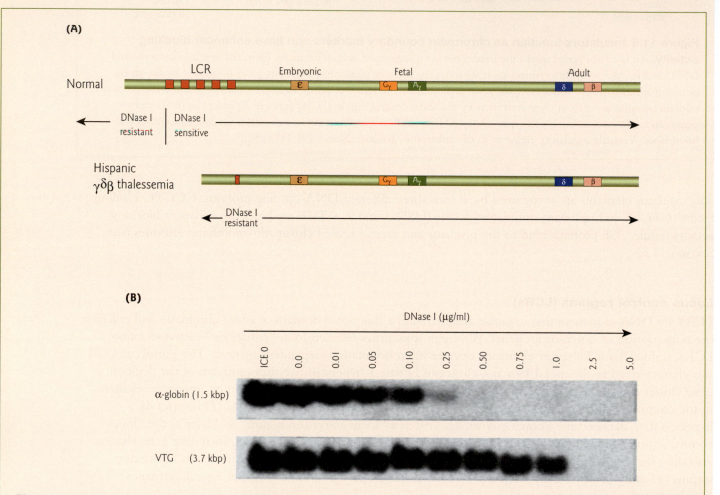

Figure 1 Hispanic thalassemia and DNase I sensitivity. (A) An approximately 35 kb deletion of the β-globin LCR causes Hispanic γδβ-thalassemia. The Hispanic locus is transcriptionally silent and the entire gene cluster is DNase I-resistant. In normal individuals the locus is transcriptionally active; chromatin upstream of the LCR is DNase I-resistant, whereas downstream of the LCR it is DNase I-sensitive. (B) Method for showing that transcriptionally active genes are more susceptible than inactive genes to DNase I digestion. Chick embryo erythroblasts at 15 days actively synthesize α-globin, but not vitellogenin. Nuclei were isolated and exposed to increasing concentrations of DNase I. The nuclear DNA was extracted and digested with the restriction endonuclease *Bam*HI, which cleaves the α-globin gene resulting in a 1.5 kb fragment and the vitellogenin gene resulting in a 3.7 kb fragment. The digested DNA was analyzed by Southern blot with a probe of labeled α-globin DNA. The transcriptionally active α-globin DNA from the 15-day erythroblasts was sensitive to DNase I digestion, indicated by the absence of the 1.5 kb band at higher nuclease concentrations. In contrast, the inactive vitellogenin DNA (VTG) was resistant to DNase I digestion, indicated by the presence of the 3.7 kb band at higher nuclease concentrations. (Reprinted with permission from Conklin, K.F., Groudine, M. 1986. Varied interactions between proviruses and adjacent host chromatin. *Molecular and Cellular Biology* 6:3999–4007. Copyright © 1986 American Society for Microbiology.)

Hispanic thalassemia and DNase I hypersensitive sites

γδβ-Thalassemia is a rare disorder characterized by partial or complete deletions of the most 5′ sequences of the β-like globin gene cluster, leading to reduced amounts of hemoglobin in the blood. Usually babies are diagnosed with the disease between the ages of 6 and 18 months. Depending on the extent of the deletion, and whether a patient is heterozygous or homozygous for the mutation, symptoms range from severe to more mild forms of anemia. Regular transfusion with red blood cells may be necessary to sustain life. Analysis of patients with this disease has led to significant advances in understanding of the locus control region (LCR) of the β-globin gene locus.

Hispanic thalassemia

In 1989, a naturally occurring ~35 kb deletion of the LCR was found in a Hispanic patient with a form of γδβ-thalassemia now called "Hispanic thalassemia." The LCR deletion was shown to result in drastic changes in activity of the β-globin locus. It quickly became apparent that deletions of the LCR in the β-globin gene cluster result in silencing of the genes, even though the genes themselves are intact. Analysis of the wild-type and Hispanic deletion alleles in mouse erythroid cells (following transfer of these chromosomes from the heterozygous patient into the cells) showed that the Hispanic locus was transcriptionally silent and the entire region of the β-like globin gene cluster was DNase I-resistant (Fig. 1). These findings led researchers to propose that normally the LCR maintains an "open" chromatin structure and enhances transcription by establishing an independent domain. Deletion of the LCR leads to "closed" chromatin and inactive β-like globin genes.

Analysis of DNase I sensitivity

When chromatin is digested *in situ* with a low concentration of DNase I, certain regions are particularly sensitive to the nuclease. Such DNase I sensitivity is one feature of genes that are able to be transcribed. The nuclease introduces double-strand breaks in transcriptionally active chromatin over 100 times more frequently than in inactive chromatin. In addition to this general sensitivity to nucleases, there are also short DNA sequences (100–200 bp) called DNase I hypersensitive sites. These sites are the first place DNase I introduces a double-strand break in chromatin, and are > 2 orders of magnitude more accessible to cleavage compared with neighboring active chromatin. These sites are typically composed of clusters of recognition sites for sequence-specific DNA-binding proteins. DNase I hypersensitive sites are not necessarily nucleosome free. They may represent the stable association of a transcription factor or complex on the surface of the nucleosome. Figure 1 shows the relative levels of nuclease sensitivity in chromatin from 15-day-old chicken embryo erythrocytes. A comparison was made between the rates of DNase I digestion of the α-globin gene which is expressed in erythrocytes, and the vitellogenin gene which is only expressed in the liver. Results show that the inactive vitellogenin gene is DNase I-resistant and the transcriptionally active globin genes are DNase I-sensitive.

construct, such as a plasmid-based β-globin gene construct. Advances in understanding LCR function came with the development of yeast artificial chromosome (YAC) vectors (see Section 8.4). Unlike plasmid vectors, which have an insert size limit of less than 10 kb, YACs can stably maintain a large enough insert of foreign DNA to encompass the entire β-globin gene locus in its natural configuration, which spans about 200 kb. Transgenic mouse experiments revealed that the LCR is required for high-level transcription of all the β-like globin genes, but the regulation of chromatin "loop" formation is the main mechanism controlling developmental expression of the β-globin genes.

Both developmentally proper regulation and physiological levels of expression of the β-globin genes can be recaptured in transgenic mice carrying a YAC construct. For example, a transgenic mouse line was constructed carrying a β-globin locus YAC that lacked the LCR. RNase protection assays showed no detectable levels of ε-, γ-, or β-globin gene transcripts at any stage of development. These results demonstrated conclusively that the LCR is a minimum requirement for globin gene expression. Additional *in vivo* mutagenesis studies have shown the developmental regulation of the β-like globin genes is mediated

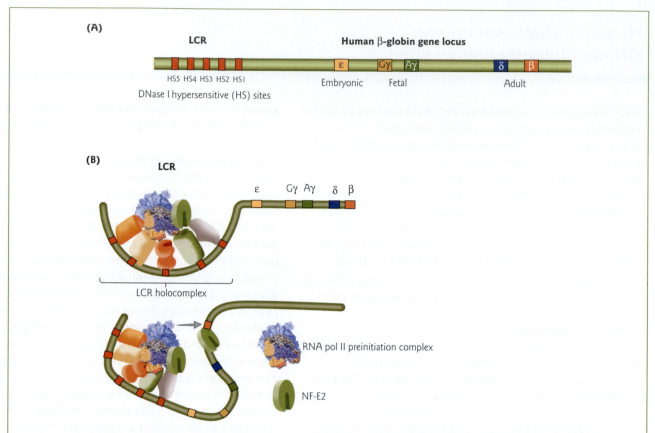

(A)

LCR   Human β-globin gene locus

HS5 HS4 HS3 HS2 HS1   Embryonic   Fetal   Adult

DNase I hypersensitive (HS) sites

(B)

LCR

ε   Gγ Aγ   δ β

LCR holocomplex

RNA pol II preinitiation complex

NF-E2

Figure 11.6 The human β-like globin gene locus. (A) Diagrammatic representation of the human β-like globin gene locus on chromosome 11, which encodes embryonic ε-globin, the two fetal γ-globins, and the adult δ- and β-globins. The locus control region (LCR) upstream of the ε-globin gene has five DNase I hypersensitive (HS) sites (HS1–HS5) separated from each other by 2–4 kb. (B) Model for transcription complex recruitment. The general transcription machinery (RNA pol II and the preinitiation complex) and other transcriptional regulatory proteins are recruited to the LCR to form a "holocomplex." A developmental stage-specific chromatin loop forms and transcription complexes are then transferred from the LCR to the appropriate globin gene promoter. The transfer is facilitated by the transcription factors NF-E2, GATA-1, and FOG-1. (Adapted with permission of The American Society for Biochemistry and Molecular Biology, Inc. from Vieira, K.F., Levings, P.P., Hill, M.A., Cruselle, V.J., Kang, S.H.L., Engel, J.D., Bungert, J. 2004. Recruitment of transcription complexes to the β-globin gene locus *in vivo* and *in vitro*. *Journal of Biological Chemistry* 279:50350–50357; permission conveyed through the Copyright Clearance Center, Inc.)

by the regulatory elements within their individual promoters. Targeted deletions in mice reveal an absolute requirement of the LCR for high-level transcription of all the β-like globin genes, but the LCR interacts with only one gene promoter at any one time.

Chromatin "loop" formation controls developmental expression of the β-globin genes   Chromatin immunoprecipitation (ChIP) assays (see Fig. 9.15E) were used to demonstrate that RNA pol II is first recruited to LCR DNase I hypersensitive sites sites *in vivo*. The transfer of RNA pol II from the LCR to the β-globin gene promoter is stimulated by the erythroid transcription factor NF-E2 (Fig. 11.6B). Researchers showed that the transcription factor GATA-1 and its cofactor FOG-1 (*friend of* GATA-1) are required for the physical interaction between the β-globin LCR and the β-globin promoter. GATA-1 has a zinc finger DNA-binding motif (see Section 11.5) and binds to the DNA sequence 5′-GATA-3′. ChIP assays were used to show that direct interaction with FOG-1 is required for GATA to induce formation of a tissue-specific chromatin "loop." In this context, the term "loop" means a chromatin conformation where two distant regions of DNA located in *cis* along the chromatin are physically close to one another, but not to intervening DNA sequences. Because GATA-1 induced loop formation correlates with the onset of β-globin gene transcription, regulation of loop formation is thought to be the main mechanism controlling developmental expression of the β-globin genes. This model provides an explanation of how the LCR can enhance the rate of transcription over such large distances. As we saw in Section 10.6, such DNA looping is an equally important transcriptional regulatory mechanism in bacteria.

Matrix attachment regions (MARs)

Much of the initial work on the basic molecular mechanisms of gene expression was done in simple test tube systems. However, it is now becoming clear that the three-dimensional organization of chromatin within the cell nucleus plays a central role in transcriptional control. There is increasing evidence that eukaryotic chromatin is organized as independent loops. Following histone extraction, these loops can be visualized as a DNA halo anchored to a densely stained matrix or chromosomal scaffold (Focus boxes 11.2 and 11.3). The formation of each loop is dependent on specific DNA sequence elements that are scattered throughout the genome at 5–200 kb intervals. These DNA sequences are termed either scaffold attachment regions (SARs) when prepared from metaphase cells, or matrix attachment regions (MARs) when prepared from interphase cells. MARs are thought to organize the genome into approximately 60,000 chromatin loops with an average loop size of 70 kb. Active genes tend to be part of looped domains as small as 4 kb, whereas inactive regions of chromatin are associated with larger domains of up to 200 kb.

Greater than 70% of characterized MARs are AT rich. The particular mode of MAR–matrix interaction indicates that binding is not directly correlated to the primary DNA sequence. Instead, the secondary structure-forming potential of DNA that tends to unwind in the AT-rich patches is of greater importance. In addition some MARs are GT rich and have the potential to form Z-DNA (see Fig. 2.8). MARs are typically located near enhancers in 5′ and 3′ flanking sequences. They are thought to confer tissue specificity and developmental control of gene expression by recruiting transcription factors and providing a "landing platform" for several chromatin-remodeling enzymes. Some MARs include recognition sites for topoisomerase II.

A model for the organization of transcription in active loop domains which localize transcription factors and actively transcribed genes is shown in Fig. 11.7. Two types of nuclear matrix-binding sites are proposed to exist within the loops, structural and functional MARs (also known as constitutive and facultative, respectively). Structural MARs serve as anchors, wheareas functional MARs are more dynamic and help to bring genes onto the nuclear matrix. MARs reversibly associate with ubiquitous factors. These contacts can be altered by specific interactions with components of enhancers and LCRs. For example, SATB1 (special AT-rich-binding protein 1) is one of the best-characterized MAR-binding proteins. SATB1 is preferentially expressed in thymocytes, the precursors of T cells in the immune system. The protein binds to the base of the chromatin loop and is thought to play a key role in T cell-specific gene regulation.

11.4 General (basal) transcription machinery

Five to 10% of the total coding capacity of the genome of multicellular eukaryotes is dedicated to proteins that regulate transcription. The yeast genome encodes a total of approximately 300 proteins involved in the regulation of transcription, while there may be as many as 1000 in *Drosophila* and 3000 in humans. These proteins fall into three major classes (see Table 11.1):

1 The general (basal) transcription machinery (this section): general, but diverse, components of large multiprotein RNA polymerase machines required for promoter recognition and the catalysis of RNA synthesis.

DN🔍A

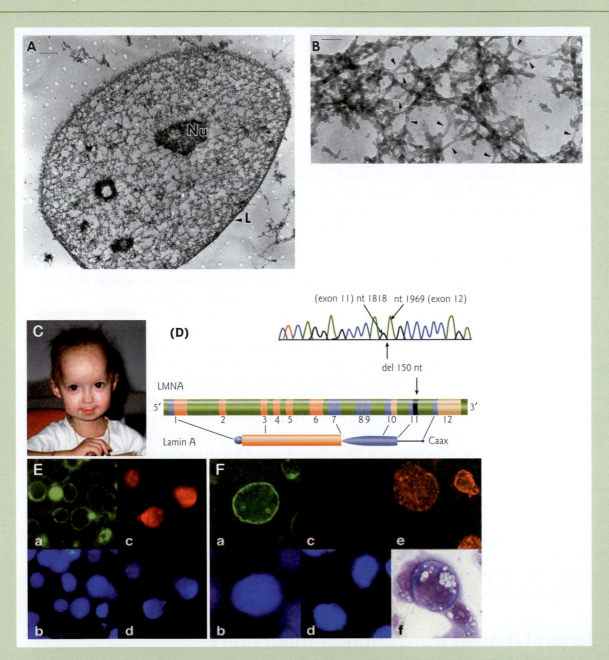

Figure 1 The nuclear matrix. (A) Resinless section electron micrograph of the core filaments of a CaSki cell (human cell line) nuclear matrix. Soluble proteins and chromatin have been removed from this nucleus. The core filament network, shown at high magnification, is connected to the nuclear lamina (L). Remnants of nucleoli (Nu) remain and are connected to fibers of the internal nuclear matrix. Bar, 1.0 μm. (B) Higher magnification showing the fibrous network structure in more detail. The underlying 10-nm filaments are seen most clearly when they are free of covering material (arrowheads). Bar, 100 nm. (Reprinted with permission from:

Is there a nuclear matrix?

Prior to the mid-1970s, the nucleus was viewed by many scientists as "a bag of chromatin floating in a sea of nucleoplasm." This was despite many observations to the contrary. For example, as early as 1925, Edmund B. Wilson in his classic textbook *The Cell in Development and Heredity* described the nucleus as containing a "nuclear framework" that was a "net-like or sponge-like reticulum," and in 1948 researchers observed that extraction of nuclei with high salt solutions produced a residual structure. In 1974, Ronald Berezny and D.S. Coffey also extracted a salt-insoluble residual structure from nuclei that they called the "nuclear matrix."

At first the general scientific community was skeptical that this meshwork was an artefact because of the harsh methods used to extract it from nuclei. After all, the nuclear matrix is operationally defined as "a branched meshwork of insoluble filamentous proteins within the nucleus that remains after digestion with high salt, nucleases, and detergent." In the mid-1980s techniques for visualizing the meshwork in whole cells were developed that provided evidence that such an architectural component exists (Fig. 1). At the same time, observations highlighted its dynamic nature. Because of its dynamic elusive qualities, this structure has been referred to at various times as the nuclear skeleton, nuclear scaffold, or even "chickenwire." The current preferred term is "matrix" (a dimensional field of variables) over the alternative terms that imply rigidity.

What does the nuclear matrix do?

The nuclear matrix is proposed to serve as a structural organizer within the cell nucleus. For example, direct interaction of matrix attachment regions (MARs) with the nuclear matrix is proposed to organize chromatin into loop domains and to maintain chromosomal territories (see Focus box 11.3). Active genes are found associated with the nuclear matrix only in cell types in which they are expressed. In cell types where they are not expressed, these genes no longer associate with the nuclear matrix.

What are the components of the nuclear matrix?

There are over 200 types of proteins associated with the nuclear matrix, of which most have not been characterized. What forms the framework for the branching filaments remains unknown. Some nuclear matrix proteins are common to all cell types and others are tissue-specific. For example, steroid receptors, such as the glucocorticoid receptor, have been shown to be associated with the nuclear matrix in hormone-responsive cells. General components of the matrix include the heterogeneous nuclear ribonucleoprotein (hnRNP) complex proteins and the nuclear lamins; hnRNP proteins are involved in transcription, transport, and processing of hnRNA.

The nuclear lamina is a protein meshwork underlying the nuclear membrane that is primarily composed of the intermediate filament proteins lamins A, B, and C. Internal

Nickerson, J.A., Krockmalnic, G., Wan, K.M., Penman, S. 1997. *Proceedings of the National Academy of Sciences USA* 94:4446–4450. Copyright ©1997 National Academy of Sciences, USA. Photographs courtesy of Jeffrey Nickerson, University of Massachusetts Medical Center.) (C–F) Lamin A truncation in Hutchinson–Gilford progeria, a premature aging syndrome. (C) Hutchinson–Gilford progeria affecting a 6-year-old female. (D) Schematic representation of the *LMNA* gene, which encodes both lamin A and lamin C, and of the lamin A protein, correlated by blue (globular domains) and red (rod domains) colors. The deleted LMNA transcript junction sequence is shown. The 150 bp deletion (indicated by a black bar) extends from G1819 to the end of exon 11. The CaaX (cysteine-aliphatic-aliphatic-any amino acid) sequence mediates farnesylation of the protein. (E) Nuclei from normal lymphocytes: (a) detection of lamin A/C and (c) lamin A by immunostaining; (b and d) DAPI staining of DNA. (F) Nuclei from the patient: (a) detection of lamin A/C by immunostaining; (c) absence of lamin A; (b and d) DAPI staining of DNA; (e) lamin B1 localizes both at the nuclear envelope (normal distribution) and nucleoplasm (abnormal distribution); (f) Giemsa staining shows nuclear deformities and cytoplasmic vacuoles. Width of fluorescent images, 80 µm; width of brightfield images, 120 µm. (Parts C–F reprinted with permission from: De Sandre-Giovannoli, A., Bernard, R., Cau, P. et al. 2003. Lamin A truncation in Hutchinson–Gilford progeria. *Science* 300:2055. Copyright © 2003 AAAS).

lamins form a "veil" that appears to branch throughout the interior of the nucleus (Fig. 1A,B). The importance of the lamina is highlighted by the wide variety of human disorders resulting from mutations in the *LMNA* locus which encodes lamin A/C proteins. For example, a splicing mutation in the lamin A gene leads to expression of a truncated form of the protein and causes Hutchinson–Gilford progeria syndrome – a premature aging syndrome in which patients have an average life expectancy of ~13 years. Patient cells have altered nuclear sizes and shapes, with disrupted nuclear membranes and extruded chromatin (Fig. 1C–F).

There are other filament-forming protein families present in the cell nucleus, but none seems to account for the long-range filament system visualized after high salt extraction. Actin was considered a candidate for the meshwork; however, although β-actin is present in the nucleus, it remains in monomer form. Two nuclear pore complex-associated proteins, Nup153 and Tpr, are organized into filaments that may play a role in mRNA export. But these filaments only extend 100–350 nm into the nucleus, ruling them out as the component of the long-range filament system.

Is there a nuclear matrix?

Whether the matrix is viewed as a stable, rigid, scaffold or a dynamic, transient, transcription-dependent structure depends to a large extent on the method applied for its characterization. The biological reality of the nuclear matrix remains in question. An alternative model suggests that nothing contributes as much to nuclear structure as does the chromatin itself. In other words, the "matrix" may be established by particular nuclear functions, as opposed to being present as a structural framework which then promotes function.

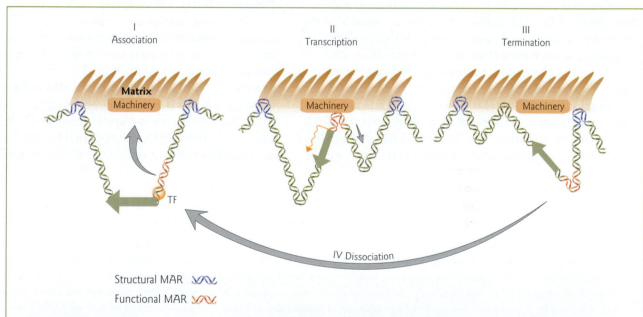

Structural MAR

Functional MAR

Figure 11.7 A model for transcriptional regulation by matrix attachment regions (MARs). (I) A gene (gray arrow) is located within a chromatin loop domain with structural MARs (blue) at its termini and a functional MAR (red) located near the gene promoter. (II) When specific association of the gene with the transcriptional machinery is required, the functional MAR moves the gene to the nuclear matrix, initiating transcription (indicated by wavy arrow) in association with transcription factors (TF). Following the initiation of transcription, the gene is pulled through the transcriptional machinery. (III) Transcription is terminated by release of the functional MAR from the nuclear matrix. (IV) Transcription is terminated by dissociation, which restores the silent but transcriptionally-ready chromatin state. (Adapted from Bode, J., Goetze, S., Heng, H., Krawetz, S.A., Benham, C. 2003. From DNA structure to gene expression: mediators of nuclear compartmentalization and dynamics. *Chromosome Research* 11:435–445.)

The chromatin fiber in a typical human chromosome is long enough to pass many times around the nucleus, even when condensed into loops (see Fig. 3.2). Chromosome "painting" – *in situ* hybridization with chromosome-specific probes – has shown that in the nucleus, each chromosome occupies its own distinct region or "territory" (Fig. 1). The territories do not generally intermingle. Specific genes do not always occupy the same relative position in three-dimensional space; however, in many vertebrates, chromosomes with low gene density reside at the nuclear periphery, whereas chromosomes with high gene density are located in the nuclear interior. Transcription appears to drive decondensation of chromatin territories. One model proposes that the DNA loops that form in the decondensed regions are associated with transcription "factories," containing a number of actively transcribed genes, RNA polymerase, and associated factors. It is estimated that ~16 loops would be associated with each factory. The loops often seem to surround the factory in a "cloud." These factories are associated with the underlying nuclear matrix (see Focus box 11.2). Transcriptionally active genes also appear to be preferentially associated with nuclear pore complexes (see Focus box 11.6). This may promote direct entry of pre-mRNAs into the processing and nuclear export pathways.

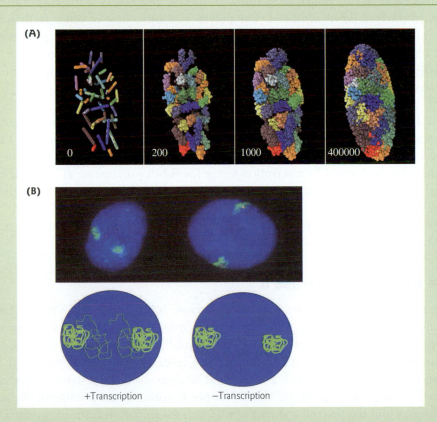

Figure 1 Chromosome territories and transcription factories. (A) Simulation of a human model nucleus based on results from 24-color chromosome fluorescent *in situ* hybridization (FISH). The first image shows 46 statistically placed rods representing the 46 human chromosomes. The next images simulate the decondensation process and show the resulting chromosome territory arrangement. (Reprinted from Bolzer, A., Kreth, G., Solovei, I. et al. 2005. Three-dimensional maps of all chromosomes in human male fibroblast nuclei and prometaphase rosettes. *PLoS Biology* 3:826–842.) (B) Transcription decondenses chromatin. (Upper panel) Loci (red) in areas of very high transcriptional activity are frequently found outside chromosome territories (green). When transcription is blocked (e.g. with an inhibitor), loci are now found more frequently within chromosome territories. (Lower panels) A model is shown suggesting that transcription decondenses chromosome territories, extruding large chromatin loop domains in a "cloud." These loops may associate with transcription "factories" containing a number of actively transcribed genes. The loops collapse back into condensed territories when transcription ceases. (Reprinted from Chubb, J.R., Bickmore, W.A. 2003. Considering nuclear compartmentalization in the light of nuclear dynamics. *Cell* 112:403–406. Copyright © 2003, with permission from Elsevier.)

2 Transcription factors (see Section 11.5): sequence-specific DNA-binding proteins that bind to gene promoters and long-range regulatory elements and mediate gene-specific transcriptional activation or repression.

3 Transcriptional coactivators and corepressors (see Section 11.6): proteins that increase or decrease transcriptional activity through protein–protein interactions without binding DNA directly. Coactivators are operationally defined as components required for activator-directed ("activated") transcription, but dispensable for activator-independent ("basal") transcription. Coactivators and corepressors either serve as scaffolds for the recruitment of proteins containing enzymatic activities, or they have enzymatic activities themselves for altering chromatin structure. They include chromatin remodeling and modification complexes that assist the transcriptional apparatus to bind and move through chromatin.

Components of the general transcription machinery

RNA polymerase II (RNA pol II) is recruited to specific promoters at the right time by extremely elaborate machineries to catalyze RNA synthesis. The polymerase and a host of other factors, including the general transcription factors, work together to form a preinitiation complex on core promoters and to allow subsequent transcription initiation (Fig. 11.8). Over the last 20 years the three major components of the general (basal) transcription machinery have been identified by techniques such as biochemical fractionation of cell extracts and *in vitro* transcription assays:

1 RNA pol II: a 12-subunit polymerase capable of synthesizing RNA and proofreading nascent transcript.

2 General transcription factors: a set of five general transcription factors, denoted TFIIB, TFIID, TFIIE, TFIIF, and TFIIH, is responsible for promoter recognition and for unwinding the promoter DNA. The nomenclature denotes "*transcription factor for RNA polymerase II*," with letters designating the individual factors. RNA pol II is absolutely dependent on these auxiliary transcription factors for the initiation of transcription.

3 Mediator: a 20-subunit complex, which transduces regulatory information from activator and repressor proteins to RNA pol II.

Structure of RNA polymerase II

Among the three nuclear eukaryotic RNA polymerases, RNA pol II is the best characterized. The most detailed analysis has been completed for RNA pol II from the budding yeast *Saccharomyces cerevisiae*. The 0.5 MDa enzyme complex consists of 12 subunits (Rpb1 to 12), numbered according to size, that are highly conserved among eukaryotes. Crystal structures have revealed that yeast RNA pol II has two distinct structures and can be dissociated into a 10-subunit catalytic core and a heterodimer of subunits Rpb4 and Rpb7 (Rpb4/7 complex) (Fig. 11.9). The RNA pol II core enzyme is catalytically active but requires the Rpb4/7 complex and the general transcription factors for initiation from promoter DNA. Rpb4/7 functions at the interface of the transcriptional and post-transcriptional machinery, playing a part in mRNA nuclear export and transcription-coupled DNA repair (see Section 7.6). An additional component of RNA pol II, the mobile C-terminal domain (CTD) of Rpb1, is not seen in crystals because it is unstructured.

RNA polymerase II catalytic core

The three-dimensional structure of the 10-subunit core enzyme of yeast RNA pol II was reported in 2001. The structure has been determined both alone and with DNA and RNA, in the form of a transcribing complex (Fig. 11.9). The two large subunits Rpb1 and Rbp2 form the central mass of the enzyme and a positively charged "cleft." The nucleic acids occupy this deep cleft, with 9 bp of RNA–DNA hybrid at the center. One side of the cleft is formed by a massive, mobile protein element termed the "clamp." The active center is formed between the clamp, a "bridge helix" that spans the cleft, and a "wall" of protein density

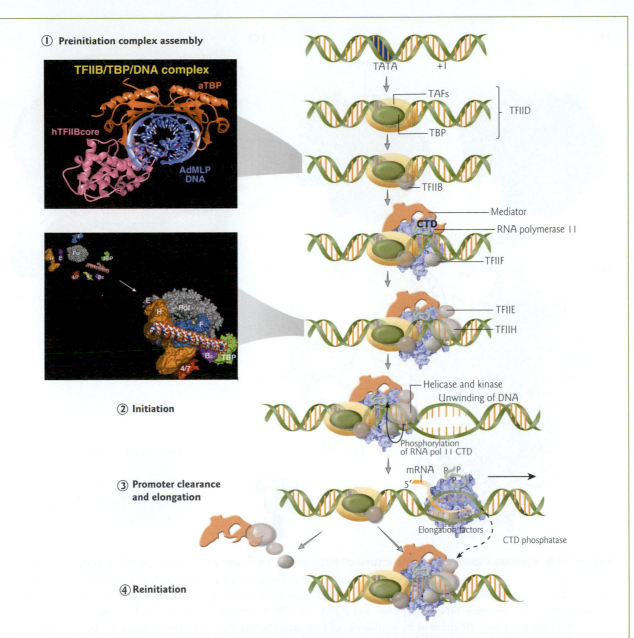

① Preinitiation complex assembly

② Initiation

③ Promoter clearance and elongation

④ Reinitiation

Figure 11.8 Preinitiation complex formation and initiation of transcription. (1) Assembly of a stable preinitiation complex for RNA pol II transcription. Binding of TFIID to the promoter provides a platform to recruit TFIIB, TFIIF together with RNA pol II (in a complex with Mediator), and then TFIIE and TFIIH. (Upper inset) The saddle-like structure of the TATA-binding protein (TBP) bound to a TATA-containing sequence in DNA, which it unwinds and bends sharply. TAF, TBP-associated factor. (Image courtesy of Song Tan, Pennsylvania State University). (Lower inset) A "minimal" initiation complex of RNA pol II and the general transcription factors from combined results of X-ray diffraction and electron microscopy. (Reprinted by permission of Federation of the European Biochemical Societies from: Boeger, H., Bushnell, D.A., Davis, R. et al. 2005. Structural basis of eukaryotic gene transcription. FEBS Letters 579:899–903.) (2) Initiation. The helicase activity of TFIIH unwinds the DNA allowing its transcription into RNA, and its kinase activity phosphorylates the C-terminal domain (CTD) of RNA pol II. (3) Promoter clearance and elongation. As the polymerase moves away from the promoter to transcribe the gene, TFIID remains bound at the TATA box allowing the formation of a new stable complex and further rounds of transcription. (4) Reinitiation requires dephosphorylation of the RNA pol II CTD.

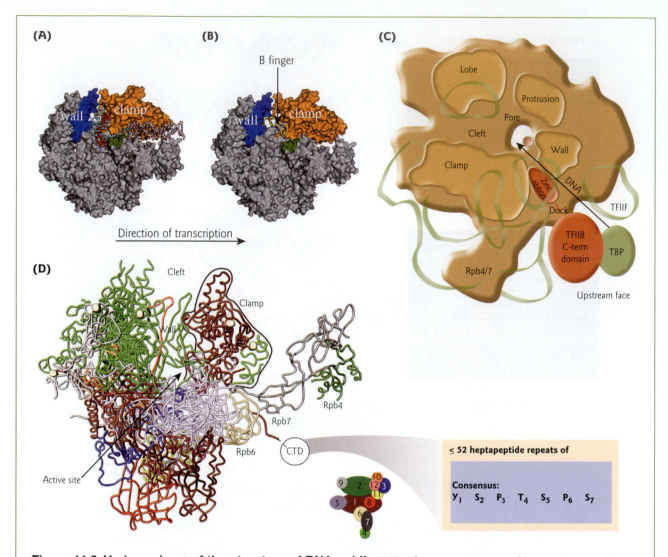

Figure 11.9 Various views of the structure of RNA pol II. (A) Surface representation of an atomic model with a cutaway view to show the active center cleft of the transcribing RNA pol II complex. Features described in the text are colored as follows: clamp, orange; wall, blue; bridge helix, green; active center Mg^{2+} ion, pink; remainder of polymerase, gray; template DNA, turquoise; and RNA, red. (B) RNA pol II–TFIIB complex, with a backbone model of TFIIB in yellow. (Reprinted by permission of Federation of the European Biochemical Societies from: Boeger, H., Bushnell, D.A., Davis, R. et al. 2005. Structural basis of eukaryotic gene transcription. *FEBS Letters* 579:899–903.) (C) Model of a minimal RNA pol II initiation complex. The structure of RNA pol II is shown schematically in brown with the locations of general transcription factors TBP, TFIIB, and TFIIF shown in dark green, red, and light green, respectively. The zinc ribbon domain of TFIIB extends into an "opening" of RNA pol II. A black arrow indicates the presumed direction of upstream promoter DNA towards the active site (magenta sphere). (D) Ribbon model of the complete yeast RNA pol II, viewed from the front. Beige spheres depict zinc ions and an active Mg^{2+}, respectively. The linker to the unstructured RNA pol II C-terminal domain (CTD) is indicated by a dashed line. A key to subunit color is provided, with subunits Rpb1–Rpb12 numbered 1–12. (Inset) The consensus sequence of the heptapeptide repeat. Serines 2 and 5 are the major phosphorylated residues. (Adapted from: Cramer, P. 2004. RNA polymerase II structure. *Current Opinion in Genetics & Development* 14:218–226. Copyright © 2004, with permission from Elsevier.)

that blocks the end of the cleft. Because the "wall" prevents straight passage of nucleic acids through the cleft, the axis of the RNA–DNA hybrid is at nearly 90° to that of the entering DNA duplex. A "pore" beneath the active site widens towards the outside, like an inverted funnel. The rim of the pore includes a loop of Rpb1 that binds a metal ion (Mg^{2+}). A second metal ion can bind weakly further in the pore. A linker that connects to the CTD of Rpb1 extends from the base of the clamp. Rpb4/7 is located on the core enzyme surface, below the clamp.

RNA polymerase II C-terminal domain (CTD)

The CTD is a unique tail-like feature of the largest RNA pol II subunit, which consists of up to 52 heptapeptide repeats of the amino acid consensus sequence Tyr-Ser-Pro-Thr-Ser-Pro-Ser (see Fig. 11.9 inset). The CTD is required for mRNA processing *in vivo* and has been shown to bind to processing factors *in vitro* (see Fig. 13.6). During the transcription cycle, the CTD undergoes dynamic phosphorylation of serine residues at positions 2 and 5 in the repeat (see Fig. 11.8). Phosphorylation is catalyzed by TFIIH and other kinases, and dephosphorylation is catalyzed by the phosphatase Fcp1. Transcription initiation requires an unphosphorylated CTD, whereas elongation requires a phosphorylated CTD. When phosphorylated, the CTD can bind mRNA processing factors during termination events (see Section 13.5). For recycling of RNA pol II and reinitiation of transcription, the CTD must again be dephosphorylated (see Fig. 11.8).

General transcription factors and preinitiation complex formation

General transcription factors have been defined biochemically as factors required for the correct initiation of RNA pol II transcription *in vitro* on a promoter with a classic TATA box and a strong initiator (Inr) element. Crystallographic studies suggest that small domains of the general transcription factors can enter "openings" in RNA pol II to modulate its function during transcription initiation. Assembling the general transcription apparatus involves a series of highly ordered steps (see Fig. 11.8). Binding of TFIID provides a platform to recruit other general transcription factors and RNA pol II to the promoter. *In vitro*, these proteins assemble at the promoter in the following order: TFIIB, TFIIF together with RNA pol II (in a complex with yet more proteins such as Mediator), and then TFIIE and TFIIH which bind downstream of RNA pol II. In some cases, TFIIA is recruited prior to TFIIB and contributes to complex stability. Because TFIIA (and its subunit TFIIJ) is not absolutely required for preinitiation complex formation and transcription initiation *in vitro*, it is not typically considered a general transcription factor. TFIID and the other general transcription factors are not sufficient to reconstitute DPE-dependent transcription. Two additional factors, a protein kinase (CK2) and a coactivator protein (PC4), are necessary to initiate DPE-dependent transcription.

TFIID recruits the rest of the transcriptional machinery

The first general transcription factor to associate with template DNA is TFIID (see Fig. 11.8). TFIID is a complex composed of the TATA-binding protein (TBP) and 14 TBP-associated factors (TAFs). TBP is a sequence-specific DNA-binding protein (see Fig. 9.15B,D) that recognizes the TATA box, as well as some other core promoter elements. TBP is highly conserved through evolution and appears to be required for transcription by all three eukaryotic RNA polymerases. TBP contains an antiparallel β-sheet that sits on the DNA like a saddle in the minor groove and bends the DNA (see Fig. 11.8). The most common recognition motif between a transcription factor and DNA is the α-helix domain of the protein and base pairs within the major groove (see Section 11.5). TBP is a notable exception to this general rule. The seat of TBP's "saddle" is the site of many protein–protein interactions both with the TAFs and the general transcription factors.

Binding of TFIID to the core promoter is a critical rate limiting step at which activators and/or chromatin remodeling factors can control transcription (see section on "Mediator" below). Transcription initiation can also be regulated at the level of preinitiation complex formation at core promoters.

Multicellular eukaryotes have evolved multiple related TFIID complexes that can function at distinct promoters through the use of several distinct TBP-like factors and tissue-specific TAFs. These TAFs interact with the transactivation domains of specific transcription factors (see Section 11.5). For example, the glutamine-rich transactivation domain of Sp1 binds to TAF4 (TAF$_{II}$110 in the old nomenclature).

TFIIB orients the complex on the promoter

Upon TFIID binding, TFIIB joins the growing assemblage (see Fig. 11.8). The TATA box promoter element to which TBP binds has a rough two-fold rotational symmetry (see Table 11.2). TBP could in principle align with it in either of two directions on either strand of the DNA double helix. Instead, binding of TFIIB orients the complex via specific protein–protein and protein–DNA contacts. TFIIB binds to only one end of TBP and simultaneously to a GC-rich DNA sequence that follows the TATA motif. The TFIIB–TBP–DNA complex signposts the direction for the start of transcription by RNA pol II and indicates which strand of the DNA double helix acts as the template (see Fig. 11.8). The N-terminal "zinc ribbon" domain (a cysteine-rich, zinc-binding region) of TFIIB binds RNA pol II and is essential for its recruitment (see Fig. 11.9). This brings the initiation complex to a point on the surface of RNA pol II from which the DNA need only follow a straight path to the active site. Because of the conserved spacing from the TATA box to the transcription start site (+1), the start site is positioned in the polymerase active center.

TFIIE, TFIIF, and TFIIH binding completes the preinitiation complex formation

Promoter loading onto RNA pol II also requires TFIIF, which forms a tight complex with the polymerase (see Fig. 11.8). Entry of TFIIB, core promoter DNA, and TFIIF in the growing assemblage leads to binding of TFIIE which, in turn, recruits TFIIH. TFIIH is the most complex of all the general transcription factors. It contains at least nine polypeptide subunits with diverse functions. These include a cyclin-dependent protein kinase (CDK7), cyclin H, a $5' \rightarrow 3'$ helicase, a $3' \rightarrow 5'$ helicase, two zinc finger proteins, and a ring finger protein. Danny Reinberg and co-workers demonstrated that TFIIH is the protein kinase that phosphorylates the CTD of RNA pol II, by showing that purified TFIIH could convert the unphosphorylated polymerase to its phosphorylation form in vitro (Fig. 11.10). ATP-dependent helicase activity was demonstrated for yeast TFIIH (encoded by the RAD25 gene) by Satya Prakash and colleagues. They overexpressed the protein in yeast, purified it almost to homogeneity, and then showed that the protein had helicase activity in vitro (Fig. 11.11).

TFIIH functions not only in transcription but also in DNA repair (see Fig. 7.11). Upon binding of these factors, the complex of core promoter, general transcription factors, and RNA pol II (in association with Mediator, see below) is called the preinitiation complex (PIC). Promoter melting (unwinding of the double-stranded DNA) requires hydrolysis of ATP. This reaction is mediated by the ATPase/helicase subunit of TFIIH, with the help of TFIIE. Unwinding is followed by "capture" of the nontemplate strand by TFIIF. The template strand descends to the active site of RNA pol II (see Fig. 11.9).

Mediator: a molecular bridge

In vitro transcription assays can be performed using a minimum set of general transcription factors and purified core RNA pol II. In such assays, RNA pol II and associated factors can stimulate low levels of transcription (referred to as basal transcription). However, under these conditions, the core RNA pol II is not responsive to transcriptional activators that can increase the frequency of initiation or can increase gene activity in vivo. Following up on these observations, Roger Kornberg and colleagues provided the first evidence that an additional factor is required for activator-responsive transcription (Fig. 11.12). This led to the characterization of a protein complex called "Mediator" that serves as a molecular bridge that connects transcriptional activators bound at enhancers, or other long-range regulatory elements, with RNA pol II (Fig. 11.13). Mediator is not required for basal RNA pol II transcription from a core promoter in vitro and

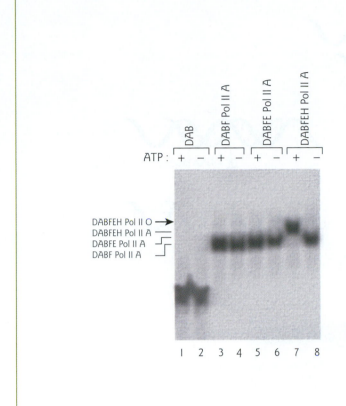

Figure 11.10 RNA pol II phosphorylation. Electrophoretic mobility shift assays were performed with a ^{32}P-end-labeled DNA fragment containing a viral promoter, and preinitiation complexes containing the unphosphorylated form of RNA polymerase II (Pol IIA) and various combinations of TFIID, TFIIA, TFIIB, TFIIF, TFIIE, and TFIIH (DABFEH) in the presence and absence of ATP as indicated at the top (see Fig. 9.15A for methods). Only when TFIIH was present did ATP shift the mobility of the complex (compares lane 7 and 8). The simplest explanation is that TFIIH promotes phosphorylation of the input polymerase from its unphosphorylated form to its phosphorylated form (Pol IIO). Additional experiments by Reinberg and colleagues demonstrated directly that TFIIH phosphorylates RNA pol II. (Reprinted by permission from Nature Publishing Group and Macmillan Publishers Ltd: Lu, H., Zawel, L., Fisher, L., Egly, J.M., Reinberg, D. 1992. Human general transcription factor IIH phosphorylates the C-terminal domain of RNA polymerase II. *Nature* 358:641–645. Copyright © 1992.)

hence is not classified as a general transcription factor. It is often referred to as a coactivator; however, this classification is a misnomer because Mediator is a component of the preinitiation complex and several Mediator subunits are required for transcription of almost all genes. In addition to promoting preinitiation complex assembly, Mediator also stimulates the kinase activity of TFIIH.

RNA polymerase II and the general transcription factors have counterparts in bacteria (see Section 10.3); however, Mediator is unique to eukaryotes. Mediator is expressed ubiquitously in eukaryotes from yeast to mammals. While yeast has one such complex, multicellular eukaryotes contain several related complexes. At least seven different mammalian Mediator complexes consisting of 25–30 proteins similar to yeast Mediator were subsequently identified in many labs. They include previously characterized complexes such as TRAP–SMCC (thyroid hormone receptor-associated protein/SRB-Med-containing cofactor) and DRIP (vitamin D receptor-interacting proteins) complexes. A unified nomenclature has been proposed, renaming these complexes as Mediator T/S and Mediator D, respectively, to more clearly designate their function in the cell. The majority of Mediator complexes act as transcriptional coactivators (e.g. Mediator D). However, there is a Mediator complex that represses transcription and Mediator T/S can act as both a repressor and an activator depending on the conditions used in the cell-free transcription assay.

11.5 Transcription factors

The regulation of gene activity at the transcriptional level generally occurs via changes in the amounts or activities of transcription factors. Of course, the genes encoding the transcription factors themselves may be transcriptionally induced or repressed by other regulatory proteins, or the transcription factors may be activated or deactivated by proteolysis, covalent modification, or ligand binding. Transcription factors

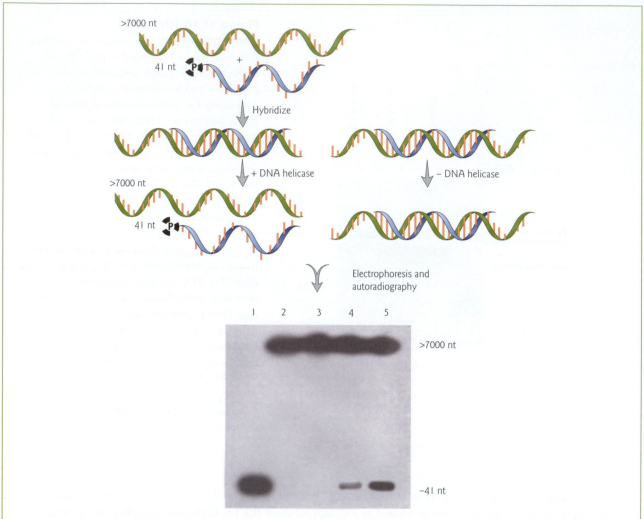

Figure 11.11 TFIIH helicase activity. A labeled 41 nt piece of single-stranded DNA (blue) was hybridized to its complementary region in a much larger (> 7000 nt), unlabeled M13 phage DNA. This substrate was then incubated with or without RAD25 protein (the yeast TFIIH) in the presence and absence of ATP. DNA helicase activity unwinds the partial duplex and releases the labeled 41 nt DNA from the phage DNA. The 41 nt DNA is distinguished from the hybrid (> 7000 nt) by gel electrophoresis. The results of the helicase assay are shown: lane 1, heat-denatured 41 nt DNA (size marker); lane 2, no protein (negative control); lane 3, 20 ng RAD25 with no ATP; lane 4, 10 ng of RAD25 plus ATP; lane 5, 20 ng of RAD25 plus ATP. (Reprinted by permission from Nature Publishing Group and Macmillan Publishers Ltd: Guzder, S.N., Sung, P., Bailly, V., Prakash, L., Prakash, S. 1994. RAD25 is a DNA helicase required for DNA repair and RNA polymerase II transcription. *Nature* 369:578–581. Copyright © 1994.)

influence the rate of transcription of specific genes either positively or negatively (activators or repressors, respectively) by specific interactions with DNA regulatory elements (see Section 11.3) and by their interaction with other proteins.

Transcription factors mediate gene-specific transcriptional activation or repression

Transcription factors that serve as repressors block the general transcription machinery, whereas transcription factors that serve as activators increase the rate of transcription by several mechanisms:

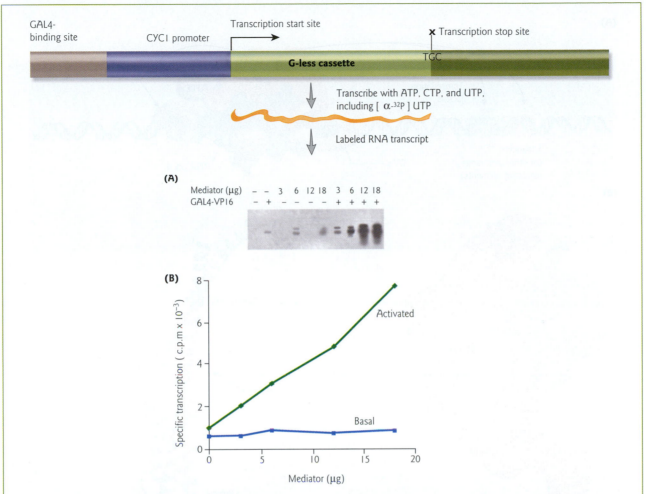

Figure 11.12 The discovery of Mediator. The yeast *CYC1* promoter was placed downstream of a GAL4-binding site and upstream of a G-less cassette, so transcription of the G-less cassette depended on both the *CYC1* promoter and GAL4. The G-less cassette assay is a variation of a run-off transcription assay (see Fig. 11.3). Instead of cutting the template DNA with a restriction endonuclease, a stretch of nucleotides lacking guanine in the nontemplate strand is inserted downstream of the promoter. This template is transcribed *in vitro* with CTP, ATP, and UTP, one of which is labeled, but no GTP. Transcription will stop at the end of the cassette where the first G is required, yielding a transcript of predictable size (based on the size of the G-less cassette, which is usually a few hundred base pairs long) on a polyacrylamide gel. The more transcript produced, the stronger will be the corresponding band on the autoradiograph. In this experiment, the construct was transcribed in the presence of a Mediator-containing fraction from yeast cells in the amounts indicated, and in the absence (−) or presence (+) of the activator protein GAL4-VP16 as indicated. (A) Phosphorimager scan of the gel. (B) Graphic presentation of the results in (A). Mediator greatly stimulates transcription in the presence of the activator (lanes 7–10), but has no effect on unactivated (basal) transcription (lanes 3–6). (Reprinted by permission from Nature Publishing Group and Macmillan Publishers Ltd: Flanagan, P.M., Kelleher, R.J. III, Sayre, M.H., Tschochner, H., Kornberg, R.D. 1991. A mediator required for activation of RNA polymerase II transcription *in vitro*. *Nature* 350:436–438. Copyright © 1991.)

1 Stimulation of the recruitment and binding of general transcription factors and RNA pol II to the core promoter to form a preinitiation complex.

2 Induction of a conformational change or post-translational modification (such as phosphorylation) that stimulates the enzymatic activity of the general transcription machinery.

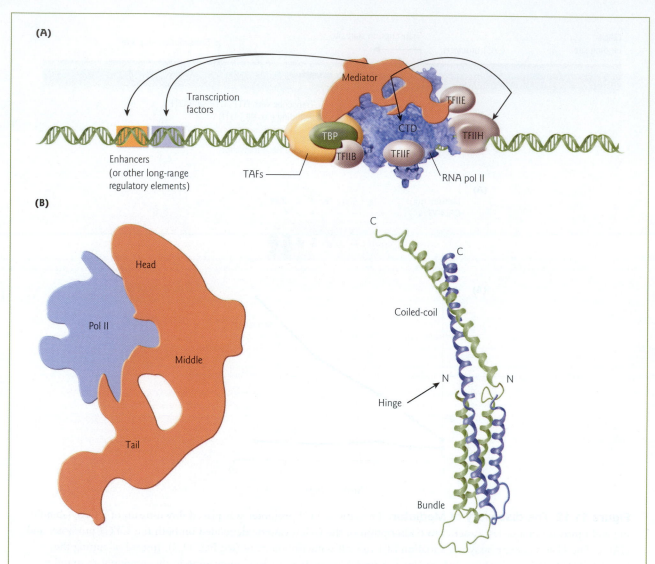

Figure 11.13 Mediator: a molecular bridge. (A) Mediator serves as a molecular bridge between the transactivation domains of various transcription factors and RNA pol II. (B) Structure of the Mediator complex. (Left panel) Outline of Mediator bound to RNA pol II according to electron microscopy. The suggested head, middle, and tail modules are indicated (Redrawn from Baumli, S., Hoeppner, S., Cramer, P. 2005. A conserved Mediator hinge revealed in the structure of the MED7/MED21 (Med7/Srb7) heterodimer. *Journal of Biological Chemistry* 280:18171–18178.) (Right panel) A ribbon model of the MED7 (blue)/MED21 (green) heterodimer is shown, based on the 3.0 Å crystal structure. The heterodimer structure has a very extended four helix-bundle domain and a coiled-coil protrusion, connected by a flexible hinge. It spans one-third the length of Mediator and almost the diameter of RNA pol II. The hinge may account for changes in Mediator structure upon binding RNA pol II or transcription factors (Protein Data Bank, PDB:1YKH).

3 Interaction with chromatin remodeling and modification complexes to permit enhanced accessibility of the template DNA to general transcription factors or specific activators.

These different roles can be promoted directly via protein–protein interaction with the general transcription machinery (see Section 11.4) or via interactions with transcriptional coactivators and corepressors (see Section 11.6).

Many transcription factors are members of multiprotein families. For example, nuclear receptors are members of a superfamily of related proteins, including the receptors for steroid hormones, thyroid hormone, and vitamin D. NF-κB is yet another family of proteins (see Section 11.10), and Sp1 – one of the first transcription factors to be isolated – is a member of the Sp family of proteins. Within each family, the members often display closely related or essentially identical DNA-binding properties but distinct activator or repressor properties.

Transcription factors are modular proteins

Transcription factors are modular proteins consisting of a number of domains (Fig. 11.14). Recognition of this feature triggered the development of a powerful technique for analyzing protein–protein interactions *in vivo* – the yeast two-hybrid assay (see Fig. 9.16B). The three major domains are a DNA-binding domain, a transactivation domain, and a dimerization domain. In addition, transcription factors typically have a nuclear localization sequence (NLS), and some also have a nuclear export sequence (NES) (see Section 11.9). Some transcription factors also have ligand-binding (regulatory) domains, such as hormone-binding domains, which are essential for controlling their activity.

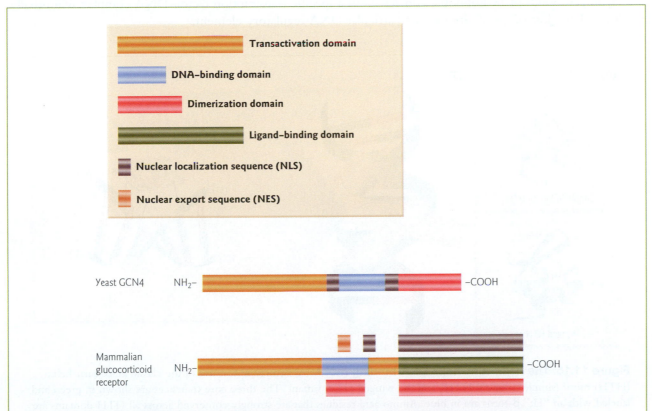

Figure 11.14 Transcription factors are composed of modular domains. Transcription factors are composed of separable, functional components and typically include a DNA-binding domain, a transactivation domain, and a dimerization domain. Transcription factors may contain more than one transactivation domain, but rarely contain more than one DNA-binding domain. In addition, transcription factors usually have a nuclear localization sequence (NLS), and some also have a nuclear export sequence (NES) (see Section 11.10). Some transcription factors also have ligand-binding (regulatory) domains, such as hormone-binding domains, which are essential for controlling their activity. The domain structure of the yeast GCN4 transcription factor and the mammalian glucocorticoid receptor are compared. Both GCN4 and the glucocorticoid receptor have two NLS motifs. Functional domains may overlap, as shown by stacked colored boxes for the glucocorticoid receptor.

DNA-binding domain motifs

The DNA-binding domain positions a transcription factor on a specific DNA sequence. Hundreds of protein–DNA complexes have now been analyzed by X-ray crystallography. In addition, NMR spectroscopy has been used to study complexes in solution (see Section 9.10 for methods). These studies have provided a detailed picture of how the DNA-binding domain interacts specifically with the bases of DNA. High-affinity binding is dependent on the overall three-dimensional shape and formation of specific hydrogen bonds. The amino acids of a protein can make specific hydrogen bonds with exposed atoms on the sides of the base pairs or along the "floor" of the major or minor groove in the DNA. The most common recognition pattern between transcription factors and DNA is an interaction between an α-helical domain of the protein and about five base pairs within the major groove of the DNA double helix (Fig. 11.15). The α-helical domain is complementary in its shape to the surface of the DNA formed by base pairs and phosphates. For high-affinity binding, both surfaces must match closely in terms of hydrogen bonds and hydrophobic contacts. Before the transcription factor and DNA come together, their "polar groups" (e.g. N-H, O-H, N or O) form hydrogen bonds to surrounding water molecules. The hydrogen bonds to water are mostly replaced in the transcription factor–DNA complex by hydrogen bonds made directly between protein and DNA. The replacement of these water molecules lends stability to the complex. Loss of just a few hydrogen bonds or hydrophobic contacts from a specific transcription factor–DNA complex will usually result in a large loss of specificity for that particular DNA regulatory element.

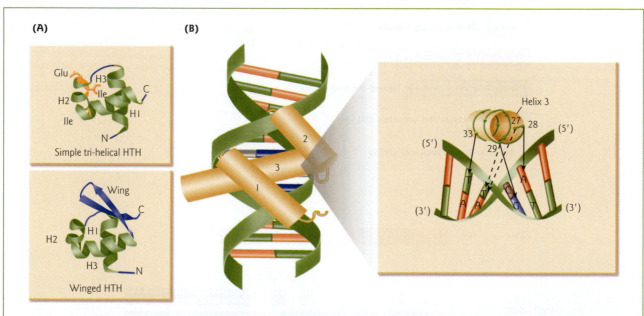

Figure 11.15 The helix-turn-helix DNA-binding motif. (A) Structural features of the classic helix–turn–helix (HTH) motif (simple trihelical HTH) and the winged HTH variant. The three core α-helices are shown in green and labeled with an "H;" β-turns are in blue. Amino acid residues that are strongly conserved across all HTH domains are shown in stick representation. (Reprinted with permission from: Aravind, L., Anantharaman, V., Balaji, S., Babu, M.M., Iyer, L.M. 2005. The many faces of the helix-turn-helix domain: transcription regulation and beyond. *FEMS Microbiology Reviews* 29:231–262. Copyright © 2005, with permission from Elsevier.) (B) The third helix, or "recognition helix," typically forms the principal DNA–protein interface by inserting itself into the major groove of the DNA. (Inset). A detailed view of how amino acids 27, 28, 29, and 33 from the "recognition helix" (α-helix 3) of the bacteriophage 434 repressor protein contact different base pairs within the major groove at the sequence TGTT. Hydrogen bonds between amino acids and base pairs are drawn as continuous arrows, while hydrophobic contacts are drawn as dashed arrows. (Modified from Calladine, C.R., Drew, H.R., Luisi, B.F., Travers, A.A. 2004. *Understanding DNA. The Molecule and How it Works.* Third Edition. Elsevier Academic Press, San Diego, CA).

When the DNA-binding domains are compared, many transcription factors fall into groups defined by related "motifs." A motif is defined as a cluster of amino acid residues that has a characteristic three-dimensional folding pattern and carries out a specific function. The following is an overview of some of the more common, well described, sequence-specific DNA-binding motifs: the helix-turn-helix, zinc finger, basic leucine zipper, and basic helix-loop-helix motifs.

Helix-turn-helix (HTH)

The HTH motif was the first DNA-binding domain to be well characterized. It was initially identified in 1982 by comparison of the structures of CAP (*E. coli* catabolite activator protein) and Cro (repressor protein from bacteriophage λ). The HTH motif is now known to be present in most prokaryotic regulatory proteins, including the *E. coli* Lac repressor protein and CAP (see Figs 10.15 and 10.16). The classic HTH domain is a simple amino acid fold composed of three core α-helices that form a right-handed helical bundle with a partly open configuration (Fig. 11.15). The third helix, or "recognition helix," typically forms the principal DNA–protein interface by inserting itself into the major groove of the DNA. A characteristic sharp turn of several amino acids, called the β-turn, separates the second and third helix.

The homeodomain is a variant of the classic HTH that is present in many transcription factors that regulate development. It is a conserved 60 amino acid domain that is encoded by a 180 bp "homeobox" sequence in the DNA (Focus box 11.4). Other variant forms of the trihelical HTH may contain additional elaborations, such as the winged HTH motif, which was discovered in 1993. This variant is distinguished by the presence of the "wing," which is a C-terminal β-strand hairpin that folds against the shallow cleft of the partially open trihelical core (Fig. 11.15A). At least 80 genes with this motif are known, many with developmentally specific patterns of expression. The winged HTH is also termed a fork head domain after the founding member of this group, the *Drosophila fork head* gene. *Fork head* mutations cause homeotic transformation of portions of the gut; e.g. the foregut and hindgut are replaced by head structures of the fruitfly. A homeotic mutation is a mutation that transforms one part of the body into another part (Focus box 11.4).

Zinc finger (Zif)

The zinc finger structural motif is one of the most prevalent DNA-binding motifs. It was first described in 1985 for *Xenopus laevis* TFIIIA – a transcription factor essential for 5S ribosomal RNA (rRNA) gene transcription by RNA polymerase III. TFIIIA binds to the 5S rRNA gene promoter, as well as to the 5S rRNA itself. The name zinc finger was coined because the two-dimensional diagram of its structure resembles a finger (Fig. 11.16). A "finger" is formed by interspersed cysteines (Cys) and/or histidines (His) that covalently bind a central zinc (Zn^{2+}) ion, folding a short length of the amino acid chain into a compact loop domain. When the three-dimensional structure was solved, it was shown that the left side of the finger folds back on itself to form a β-sheet. The right side twists into an α-helix. Binding of zinc by cysteines in the β-sheet and histidines in the α-helix draws the halves together near the base of the finger. It also brings hydrophobic amino acids close to one another at the fingertip where their mutual attraction helps to stabilize the motif. The finger inserts its α–helical portion into the major groove of the DNA. Generally, there is a linker region of 7–8 amino acids in between each zinc finger module. The number of fingers is variable between different zinc finger-containing transcription factors.

There are a number of different types of zinc finger motifs. For example, TFIIIA has nine fingers and GLI3 (Disease box 11.2) has five of the classic Cys_2-His_2 (C_2H_2) pattern described above, while nuclear receptors have two fingers of a Cys_2-Cys_2 (C_2C_2) pattern. Figure 11.16 illustrates the zinc finger DNA-binding domain of the glucocorticoid receptor. Three to four amino acids at the base of the first finger confer specificity of binding. A dimerization domain near the base of the second finger is the region that interacts with another glucocorticoid receptor to form a homodimer. Each protein in the pair recognizes half of a two-part DNA regulatory element called a glucocorticoid response element (GRE). The binding

Homeoboxes and homeodomains

In the late 1970s developmental biologists began to understand the elegant simplicity of pattern formation in the developing fruitfly embryo. Clusters of genes are turned on in sequence, determining the overall body pattern – anterior to posterior, and dorsal to ventral – and then successive gene cascades send signals telling the cells in the various segments of a developing embryo what kind of structures to make, whether it be legs, wings, or antennae. A major advance was the discovery of the homeobox genes first in *Drosophila* in the late 1970s, and later in many other organisms from humans to the plant *Arabidopsis*. The name "homeobox" is derived from the fact that mutations in some of these genes result in a homeotic transformation, or a situation in which one body or plant part is replaced by another or is duplicated. Homeobox genes encode transcription factors containing a 60 amino acid DNA-binding motif called a homeodomain, a variant form of the classic helix-turn-helix motif. Homeodomain transcription factors regulate many embryonic developmental programs, including axis formation, limb development, and organogenesis. Some homeodomain transcription factors are also expressed in adult tissues such as liver, kidney, and intestine, where they are thought to play a role in the regenerative differentiation of cells. The best known homeobox gene subclass is the *Hox* family. Disruption of a *Drosophila Hox* gene can lead to a phenotype known as *Antennapedia* in which fly legs develop in place of the antennae (Fig. 1A). In humans, a mutation in the *Hoxd13* gene results in the duplication of a digit (polydactyly) resulting in the development of six fingers.

The *Hox* genes are present in large clusters. These clusters tend to contain fewer chromosomal breakpoints between human and mouse than expected by chance, which suggests that they are being held together by natural selection. Many *Hox* genes display a colinear pattern of expression determined by their position in the cluster. Clusters of homeobox genes are regulated, in part, by the Polycomb group proteins.

Colinear expression of homeobox genes

The clustering of homeobox genes allows a pattern of organized, differential gene transcription in which an expression gradient is achieved either spatially, temporally, or quantitatively, depending on the location of each gene within the cluster. Spatial colinearity means the position of

a gene in a cluster correlates with its expression domain, e.g. 5′ and 3′ *Hox* genes are typically expressed in the posterior and anterior portions of developing embryos, respectively (Fig. 1A). In temporal colinearity, genes at one end of the cluster are turned on first and genes at the opposite end of the cluster are turned on last. Quantitative colinearity refers to a situation where the first gene in a cluster displays the maximum level of mRNA expression and downstream genes exhibit progressively lower expression. Colinear expression of *Hox* genes was recently reported to be regulated by global enhancer elements that establish a long-range chromosomal domain.

Almost 20 years after the initial discovery of the homeobox DNA element, a new cluster of 12 homeobox genes has been reported (Fig. 1B). The cluster is on the X chromosome and is a subfamily of the reproductive homeobox X-linked (*Rhox*) genes, which appear to play an important regulatory role in reproduction. They are selectively expressed in the testis, epididymus, ovary, and placenta. Their expression pattern is colinear and corresponds to their chromosomal position. Two of the three subclusters show temporal colinearity and all three show quantitative colinearity.

Polycomb group proteins silence homeobox genes

The *Polycomb* gene was discovered nearly 60 years ago through observations of *Drosophila* mutants. Normal male fruitflies have structures called sex combs on their front legs, used for grasping females during mating. Mutant flies were observed that also had sex combs on the second and third pairs of legs, and so were named "Polycomb." Later work suggested that the Polycomb protein normally represses expression of the bithorax gene, and prevents structures such as sex combs or wings from forming in the wrong body segments. The Polycomb group (PcG) proteins act in conjunction with DNA sequences termed "Polycomb response elements" to maintain lineage-specific "off" transcriptional states of homeotic genes. They mediate gene silencing by altering the higher order structure of chromatin. The founding member of the PcG family, Polycomb, contains a chromodomain – a 37 amino acid motif similar to a known chromatin-binding domain in a protein called heterochromatin-associated protein 1 (HP1). Recent work shows that gene silencing in the fruitfly requires two distinct complexes of various PcG proteins. PRC2

Homeoboxes and homeodomains

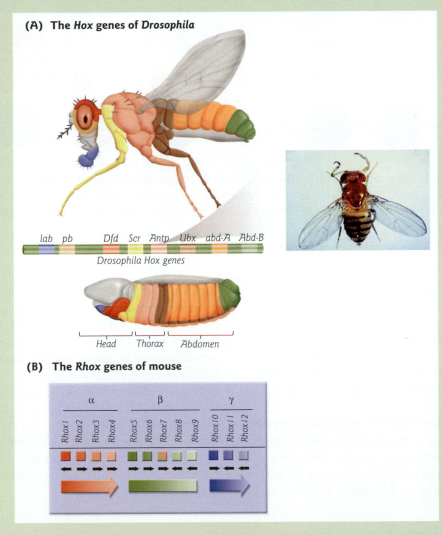

(A) The *Hox* genes of *Drosophila*

lab pb   Dfd Scr Antp Ubx abd-A Abd-B

Drosophila Hox genes

Head   Thorax   Abdomen

(B) The *Rhox* genes of mouse

α   β   γ

Rhox1 Rhox2 Rhox3 Rhox4 Rhox5 Rhox6 Rhox7 Rhox8 Rhox9 Rhox10 Rhox11 Rhox12

Figure 1 Homeobox genes. (A) The *Hox* genes of *Drosophila*. Eight *Hox* genes regulate the identity of regions within the adult fruitfly and the embryo. The color coding represents the segments and structures that are affected by mutations in *Hox* genes (Adapted with permission by Nature Publishing Group and Macmillan Publishers Ltd: Carroll S.B. 1995. Homeotic genes and the evolution of arthropods and chordates. *Nature* 376:479–485. Copyright © 1995. (Inset) *Antennapedia* homeotic mutant in which the antennae are transformed into legs. (Reprinted from Carroll S.B., Grenier, J.K., Weatherbee, S.D. 2005. *From DNA to Diversity. Molecular Genetics and the Evolution of Animal Design*. Second Edition. Blackwell Publishing, Malden, MA). (B) The *Rhox* genes of mouse. Twelve *Rhox* genes on the X chromosome are specifically expressed in male and female reproductive tissues. The genes of each subcluster (α, β, γ) are expressed in the order that they occur on the chromosome during sperm differentiation, to different degrees (graded coloured arrows) and, in the α and γ subclusters, at different times. The first genes in each subcluster are expressed earlier and to a greater maximal level than the next ones. Black arrows denote the direction of gene transcription. The pink shading for the *Rhox7* gene shows that, based on its DNA sequence, it is more closely related to the α- than the β-cluster. (Adapted with permission by Nature Publishing Group and Macmillan Publishers Ltd: from Spitz, F., Duboule, D. 2005. Developmental biology: reproduction in clusters. *Nature* 434:715–716. Copyright © 2005.)

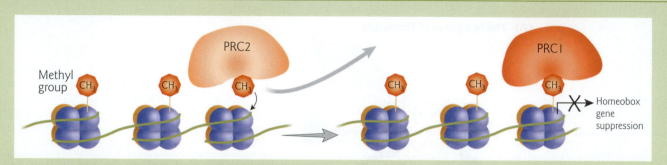

Figure 2 Model for how Polycomb group proteins silence developmental control genes. The methyl groups added to histone H3 of chromatin by the Polycomb group protein complex PRC2 attract PRC1, which then silences the activity of neighboring genes, such as homeobox genes. (Francis N.J., Kingston, R.E., Woodcock, C.L. 2004. Chromatin compaction by a polycomb group protein complex. *Science* 306:1574–1577.)

(Polycomb repressive complex 2) first marks the genes to be silenced. One of its components, known as E(Z) for enhancer of zeste, adds methyl groups to the amino acid lysine 27 in histone H3. The second set of proteins, PRC1, then is recruited by this mark and blocks transcription of the marked genes, possibly by blocking the effects of the SWI/SNF chromatin-remodeling complex (see Section 11.6) (Fig. 2).

PcG proteins are now known to play a much wider role in gene silencing. Humans have structurally similar proteins, which form complexes with similar activity. For example, a human PcG complex, the EED–EZH2 (embryonic ectoderm development/enhancer of zeste 2 homolog 2) complex

contains methyltransferase activity and is required for histone methylation and heterochromatin formation during X chromosome inactivation (see Section 12.4). Further, PcG proteins appear to play a role in the repression of various genes involved in controlling cell growth and division. Consistent with this, the expression of EZH2 is much higher in prostate cancers that have spread (metastasized) to other tissues than it is in localized tumors or normal tissues. This may be due to the inhibition of tumor-suppressor genes. EZH2 overproduction also leads to the formation of a Polycomb protein complex that differs in protein composition from PRC2, which could also lead to changes in gene expression.

site is a palindrome with two half sites and three intervening base pairs. Other hormone receptors share the same sequence of half sites, but have different spacing in between. These intervening base pairs are of critical importance for receptor–DNA interaction. The glucocorticoid receptor distinguishes both the base sequence and the spacing of the half sites.

Basic leucine zipper (bZIP)

The bZIP motif is not as common as the HTH or zinc finger motifs. The motif was first described in 1987 for the CAAT/enhancer-binding protein (C/EBP), which recognizes both the CCAAT box found in many viral and mammalian gene promoters and the "core homology" sequence common to many enhancers. The structure of the DNA-binding domain was solved by comparison with two other related proteins, the GCN4 regulatory protein from yeast and the transcription factor Jun. The latter protein is encoded by the

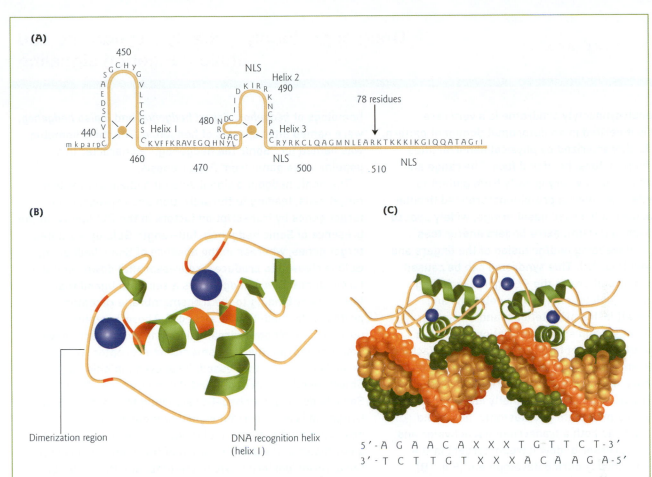

Figure 11.16 The zinc finger DNA-binding domain. (A) A schematic of the two zinc finger motifs in the DNA-binding domain of the glucocorticoid receptor (GR). The orange spheres depict the zinc ions coordinating the cysteine residues. The positions of two nuclear localization sequences (NLSs) are shown. (B) The three-dimensional structure of the GR DNA-binding domain. (Reprinted from Low, L.Y., Hernández, H., Robinson, C.V., O'Brien, R., Grossmann, J.G., Ladbury, J.E., Luisi, B. 2002. Metal-dependent folding and stability of nuclear hormone receptor DNA-binding domains. *Journal of Molecular Biology* 319:87–106. Copyright © 2002, with permission from Elsevier.) (C) The three-dimensional structure of a GR homodimer bound to DNA. (Protein Data Bank, PDB:1GDC, 1LAT). The base sequence of the glucocorticoid response element (GRE) in the DNA recognized by GR is given below. The half sites in the elements are alike if their base pairs are read along opposite strands of the DNA, in the 5' → 3' direction. One GR monomer binds to each half site.

proto-oncogene *c-jun*. Proto-oncogenes are genes that have a normal function in the cell but can become cancer-causing when they undergo certain mutations (see Section 17.2).

The bZIP motif is not itself the DNA-binding domain of the transcription factor and does not directly contact the DNA. Instead, it plays an indirect structural role in DNA binding by facilitating dimerization of two similar transcription factors. The joining of two bZIP-containing transcription factors results in the correct positioning of the two adjacent DNA-binding domains in the dimeric complex (Fig. 11.17A). The bZIP motif is a stretch of amino acids that folds into a long α-helix with leucines in every seventh position. The leucines form a hydrophobic "stripe" on the face of the α-helix. Two polypeptide chains with this motif coil around each other to form a dimer in a "coiled coil" arrangement. The amino acids protrude like knobs from one α-helix and fit into complementary holes between the knobs on the partner helix, like a zipper. The dimer forms a Y-shaped structure and one end of each α-helix protrudes into the major groove of the DNA. This allows the two basic binding regions (rich in arginine and lysine) to contact the DNA.

Greig cephalopolysyndactyly syndrome and Sonic hedgehog signaling

Greig cephalopolysyndactyly syndrome is a very rare disorder that is inherited in an autosomal dominant pattern. The disorder is characterized by physical abnormalities affecting the fingers, toes, head and face. The range and severity of symptoms may vary greatly from patient to patient. Patients often have a prominent forehead (frontal bossing), an abnormally broad nasal bridge, widely spaced eyes (ocular hypertelorism), extra fingers and/or toes (polydactyly), and webbing and/or fusion of the fingers and toes (syndactyly) (Fig. 1A). This syndrome can be caused by many types of mutations in the *GLI3* gene, including translocations, large deletions, exonic deletions and duplications, small in-frame deletions, and missense, frameshift/nonsense, and splicing mutations. *GLI3* is a member of the GLI-Kruppel family of transcription factors. The *GLI1*, *GLI2*, and *GLI3* genes all encode transcription factors that share five highly conserved tandem Cys_2-His_2 zinc fingers. The *GLI1* gene was originally identified as an amplified gene in a malignant *gli*oblastoma, hence the acronym. *GLI3* acts as both a transcriptional activator and transcriptional repressor of downstream targets in the Sonic hedgehog pathway during development (Fig. 1B).

Embryogenesis is regulated by a number of complex signaling cascades that are critical for normal development. One such pathway begins with the secreted protein "Sonic hedgehog." Sonic hedgehog is one of three vertebrate homologs to the *Drosophila hedgehog* gene. The original *hedgehog* gene was named for the appearance of the mutant phenotype in which a *Drosophila* embryo is covered with pointy denticles resembling a hedgehog. The first two homologs of *hedgehog*, *desert hedgehog* and *indian hedgehog*, were named after species of hedgehog. The third homolog was named for "Sonic the hedgehog," a character in a popular video game from Sega Genesis.

The Sonic hedgehog signal sets off a chain of events in target cells, leading to the activation and repression of target genes by transcription factors in the GLI family. In the presence of Sonic hedgehog, full-length GLI3 upregulates target genes, whereas in the absence of Sonic hedgehog, GLI3 is cleaved to produce a repressor that downregulates target genes. Sonic hedgehog is a secreted ligand that functions by binding to the transmembrane receptor proteins, Patched-1 and Patched-2. These receptors normally inhibit downstream signaling by interaction with Smoothened, another transmembrane receptor. When Sonic hedgehog is expressed, it relieves inhibition of Smoothened by Patched-1 and Patched-2, and allows Sonic hedgehog–Patched–GLI signaling to progress. Sonic hedgehog is an important signaling molecule for the developmental patterning of many tissues and organs. It is a key factor in the determination of the ventral neural tube, the anterior/posterior axis of the limb, and the cartilage-producing region of the somites. Studies of human GLI3 protein have shown that it interacts with CBP (CREB-binding protein), a common transcriptional coactivator. Mutations in CBP can cause Rubinstein–Taybi syndrome (Disease box 11.3). The similarity between the Rubinstein–Taybi phenotype and Greig cephalopolysyndactyly syndrome suggests that the biochemical interaction between CBP and GLI3 plays a role in Sonic hedgehog signaling.

The dimer can be either a homodimer – two of the same polypeptide – or a heterodimer – two different polypeptides zipped together. For example, C/EBP forms both homodimers and heterodimers composed of mixed pairs of different variants. The transcription factor AP-1 is a combination of members of two different families of transcription factors, Fos and Jun. Tony Kouzarides and Edward Ziff showed that at low concentrations of protein Jun and Fos bind to their DNA target better together than either one does separately (Fig. 11.18). They also demonstrated that the bZIP domains of each protein are essential for binding. Jun can form both homodimers and heterodimers, whereas Fos can only form heterodimers.

Basic helix-loop-helix (BHLH)

The BHLH motif is distinct from the HTH motif (Fig. 11.17B). It forms two amphipathic helices, containing all the charged amino acids on one side of the helix, which are separated by a nonhelical

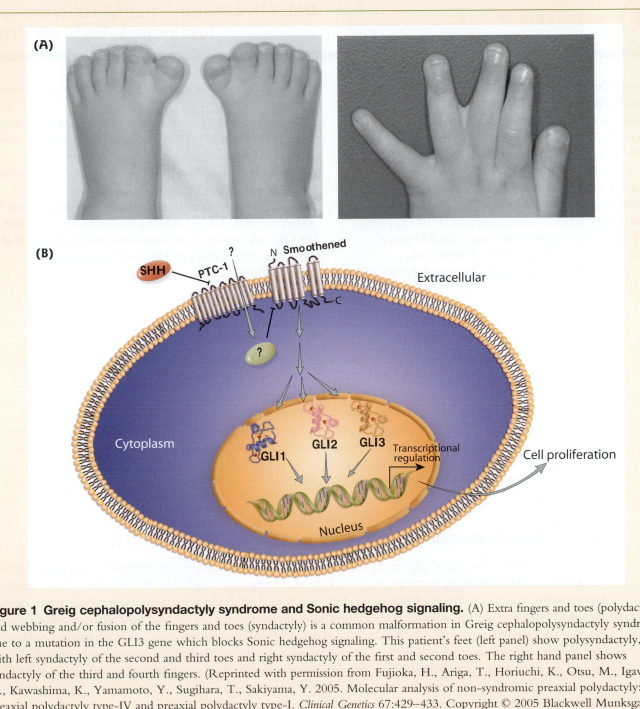

Figure 1 Greig cephalopolysyndactyly syndrome and Sonic hedgehog signaling. (A) Extra fingers and toes (polydactyly) and webbing and/or fusion of the fingers and toes (syndactyly) is a common malformation in Greig cephalopolysyndactyly syndrome due to a mutation in the GLI3 gene which blocks Sonic hedgehog signaling. This patient's feet (left panel) show polysyndactyly, with left syndactyly of the second and third toes and right syndactyly of the first and second toes. The right hand panel shows syndactyly of the third and fourth fingers. (Reprinted with permission from Fujioka, H., Ariga, T., Horiuchi, K., Otsu, M., Igawa, H., Kawashima, K., Yamamoto, Y., Sugihara, T., Sakiyama, Y. 2005. Molecular analysis of non-syndromic preaxial polydactyly: preaxial polydactyly type-IV and preaxial polydactyly type-I. *Clinical Genetics* 67:429–433. Copyright © 2005 Blackwell Munksgaard.) (B) The Sonic hedgehog signal transduction pathway. The Sonic hedgehog signal sets off a chain of events in target cells, leading to the activation and repression of target genes by transcription factors in the GLI family, including GLI3. Sonic hedgehog (SHH) binds to the transmembrane receptor protein, Patched-1 (PTC-1). This receptor normally inhibits downstream signaling by interaction with Smoothened, another transmembrane receptor. When SHH is expressed, it relieves inhibition of Smoothened by PTC-1, and allows SHH–PTC-1–GLI signaling to progress. Not all the steps in the signaling pathway have been defined.

Defective histone acetyltransferases in Rubinstein–Taybi syndrome

Rubinstein–Taybi syndrome is a rare disease that was first described by Rubinstein and Taybi in 1963. It is inherited in an autosomal dominant manner, and occurs approximately once in every 125,000 live births. The disease is characterized by facial abnormalities, broad digits, stunted growth, and mental retardation with impairments in learning and long-term memory formation. The disease is due to mutations in the gene coding for CBP which is a coactivator with histone acetyltransferase (HAT) activity. Patients have a variety of mutations in the CBP gene, including point mutations and 5′ or 3′ deletions. Patients are typically heterozygous for a mutation in CBP.

Homozygous mutations are likely to be embryonic lethal in humans.

The defect in long-term memory formation in patients is linked to the role of CBP in mediating transcription from some cAMP-responsive genes. The cAMP response element-binding protein (CREB) is a transcription factor that is activated by phosphorylation. Once activated, it recruits CBP and binds to a cAMP-responsive element (CRE) within the DNA. CBP acetylates the histones associated with the CRE, disrupting chromatin within the promoter and facilitating the binding of RNA pol II and the general transcription machinery.

loop. Like the bZIP motif, the BHLH motif is not itself the DNA-binding domain of the transcription factor and does not directly contact the DNA. Instead, it plays an indirect structural role in DNA binding by facilitating dimerization of two similar transcription factors. The joining of two BHLH-containing transcription factors results in the correct positioning of the two adjacent DNA-binding domains in the dimer. The DNA-binding domains are rich in basic amino acids and can interact directly with the acidic DNA. As an example, efficient DNA binding by the mouse transcription factor Max requires dimerization with another BHLH protein. Max binds DNA as a heterodimer with either with Myc or Mad. The Myc–Max complex is a transcriptional activator, whereas the Mad–Max complex is a repressor.

Transactivation domain

The transactivation domain of a transcription factor is involved in activating transcription via protein–protein interactions. Transactivation domains may work by recruiting or accelerating the assembly of the general transcription factors on the gene promoter, but their mode of action remains unclear. Some transcription factors do not contact the general transcription machinery directly but instead bind coactivators that in turn contact the general apparatus. Unlike the well-defined DNA-binding domains, transactivation domains are structurally more elusive. They are often characterized by motifs rich in acidic amino acids, so-called "acid blobs." In addition to acid blobs, there are other distinct motifs. For example, transcription factor Sp1 contains a nonacidic transactivation region with multiple glutamine-rich motifs. Other motifs associated with transactivation include proline-rich regions and hydrophobic β-sheets.

Dimerization domain

The majority of transcription factors bind DNA as homodimers or heterodimers. Accordingly, they have a domain that mediates dimerization between the two identical or similar proteins. In contrast to our detailed knowledge of protein–DNA interactions, far less is known about the exact molecular characteristics of these protein–protein contacts. As described above, two dimerization domains that are relatively well characterized structurally are the helix-loop-helix and leucine zipper motifs (Fig. 11.17).

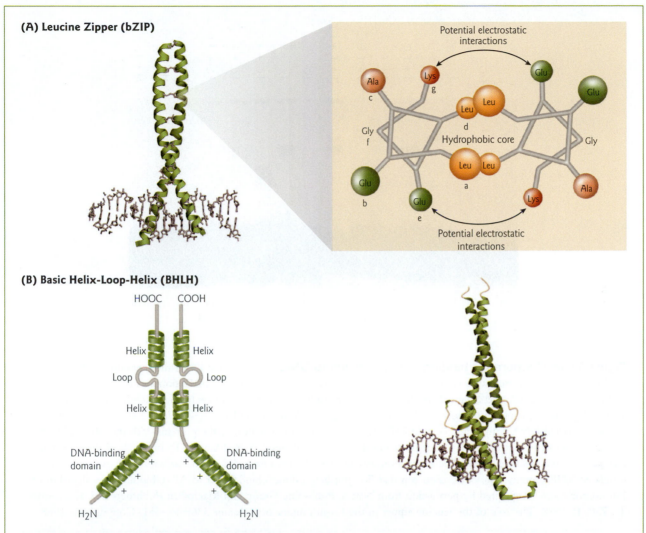

(A) Leucine Zipper (bZIP)

Potential electrostatic interactions

Ala c

Lys g

Glu

Glu

Leu Leu d

Gly f

Hydrophobic core

Gly

Leu Leu a

Glu b

Glu e

Lys

Ala

Potential electrostatic interactions

(B) Basic Helix-Loop-Helix (BHLH)

HOOC   COOH

Helix   Helix

Loop   Loop

Helix   Helix

DNA-binding domain

DNA-binding domain

H$_2$N   H$_2$N

Figure 11.17 The basic DNA-binding domain. (A) The leucine zipper and the basic DNA-binding domain are illustrated by the transcription factor AP-1–DNA complex. AP-1 is a dimer formed by Jun and Fos. Two helices "zipper" together by their leucine residues (inset) (Protein Data Bank, PDB: 1FOS.) (B) (Left) Schematic representation of the basic helix-loop-helix motif and the basic DNA-binding domain. (Right) Three-dimensional image of a Myc–Max–DNA complex. (Protein Data Bank, PDB: 1NKP).

11.6 Transcriptional coactivators and corepressors

Gene transcription is a multistep process involving a very large number of proteins functioning in discrete complexes. As described above, transcription factors bind to DNA in a sequence-specific manner. They mark a gene for activation or repression through the recruitment of coactivators or corepressors. Coactivators and corepressors are proteins that increase or decrease transcriptional activity, respectively, without binding DNA directly. Instead they bind directly to transcription factors and either serve as scaffolds for the recruitment of other proteins containing enzymatic activities, or they have enzymatic activities themselves for altering chromatin structure. Coactivators and corepressors have been much harder to study compared with transcription factors. In general, assays for protein–protein interactions are more difficult to perform than techniques for studying DNA–protein interactions (see Sections 9.8 and 9.9). In addition, techniques for determining which coactivator is docking on a particular transcription factor *in vivo* were not available until recently.

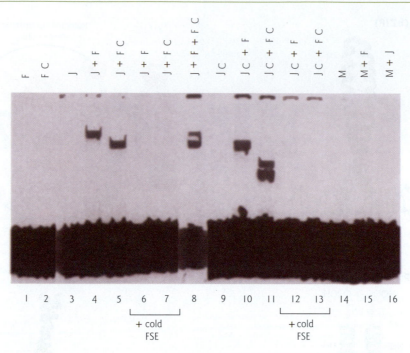

Figure 11.18 Cooperative binding of Fos and Jun to DNA. A ^{32}P-labeled DNA that contains an AP-1 binding site was incubated with *in vitro*-translated products of Fos (F), Jun (J), the bZIP domains of Fos (Fos core, FC) and Jun (Jun core, JC), and Myc (M) in the combinations indicated at the top of each lane. Samples were then analyzed by an electrophoretic mobility shift assay (see Fig. 9.15A for methods). Lanes 6, 7, 12, and 13 contained excess unlabeled DNA competitor (+ cold FSE), which prevented interaction of the proteins with the labeled DNA. Combinations of Fos or FC with Jun or JC bound to the DNA (lanes 4, 5, 8, 10, and 11), but each of these proteins did not bind when added alone. The interaction was also specific because it did not occur with Myc, which is an unrelated bZIP protein. The conclusion was that Fos–Jun heterodimers bind better to AP-1-binding sites than Fos or Jun homodimers. (Reprinted by permission from Nature Publishing Group and Macmillan Publishers Ltd: Kouzarides, T., Ziff, E. 1988. The role of the leucine zipper in the fos-jun interaction. *Nature* 336:646–651. Copyright © 1988.)

Coactivators, in the broadest sense, can be divided into two main classes:

1 Chromatin modification complexes: multiprotein complexes that modify histones post-translationally, in ways that allow greater access of other proteins to DNA.
2 Chromatin remodeling complexes: multiprotein complexes of the yeast SWI/SNF family (or their mammalian homologs BRG1 and BRM) and related families that contain ATP-dependent DNA unwinding activities.

Corepressors have the opposite effect on chromatin structure, making it inaccessible to the binding of transcription factors or resistant to their actions. Mechanisms for transcriptional silencing will be discussed in more detail in Chapter 12.

Chromatin modification complexes

The assembly of the eukaryotic genome into chromatin is essential for the compaction of DNA into the relatively tiny nucleus (see Section 3.2). Until less than a decade ago, the histones were widely regarded as inert building blocks that package DNA into nucleosomes. Nucleosomes were viewed as general repressors

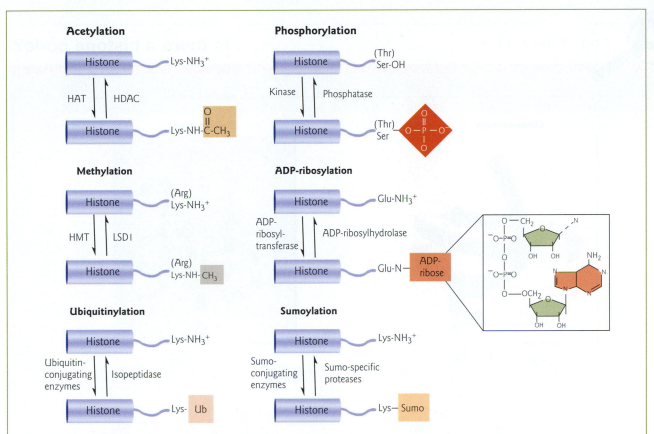

Figure 11.19 Post-translational modification of histone N-terminal tails. Six types of modifications to histone tails (wavy lines) known to play a role in gene regulation are depicted: acetylation, methylation, ubiquitinylation, phosphorylation, ADP–ribosylation, and sumoylation.

of transcription that make DNA sequences inaccessible to transcription factors. However, recent progress in the chromatin field has markedly changed our perspective. Chromatin modification complexes are now known to play a central role in both gene activation and repression.

The N-terminal tails of histones H2A, H2B, H3, and H4 stick out of the core octamer and are subject to a wide range of post-translational modifications (see Fig. 3.5). Because these covalent modifications alter the accessibility of chromatin to the general transcriptional machinery they have been proposed to function as master on/off switches that determine whether a gene is active or inactive. There are four main types of modification to histone tails known to play a role in regulating gene expression: (i) acetylation of lysines; (ii) methylation of lysines and arginine; (iii) ubiquitinylation of lysines; and (iv) phosphorylation of serines and threonines. Other less common modifications are ADP–ribosylation of glutamic acid and sumoylation of lysine residues. Levels of specific histone modifications or "marks" are maintained by the balanced activities of modifying and demodifying enzymes (Fig. 11.19). The activities of these two sets of enzymes may be shifted by changes in their intracellular distribution, their targeting to chromatin, or the action of inhibitors. In addition, some histone modifications may act as molecular switches, enabling or blocking the setting of other covalent marks. These modifications are used as recognition landmarks by other proteins that bind chromatin and initiate downstream processes, such as chromatin compaction or transcriptional regulation (Focus box 11.5). Besides modification, linker histone subtypes can also promote selective gene expression.

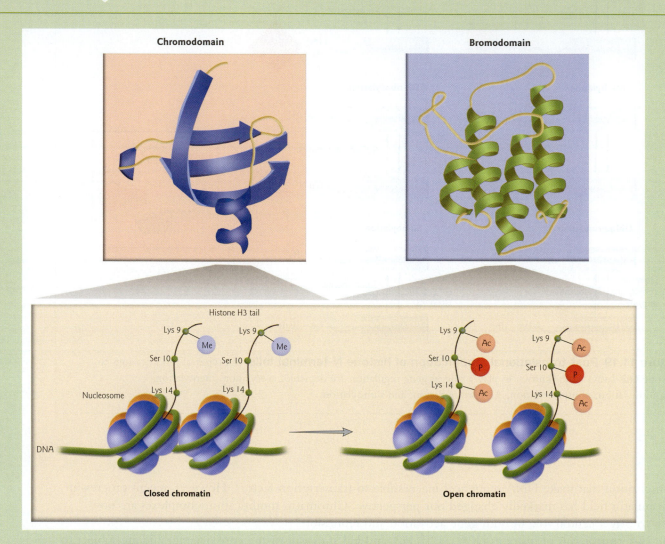

Figure 1 Histone modifications are recognition landmarks for chromatin-binding proteins. Proteins containing a chromodomain (left inset; Protein Data Bank, PDB: 1N72) are targeted to methylated lysines in histone tails, leading to tightly packed "closed" chromatin. Demethylation of the lysine amino acid at position 9 in histone H3 facilitates phosphorylation of serine 10 and the acetylation of lysines 9 and 15 leading to a more open chromatin structure. The bromodomain protein motif (right inset; PDB: 1PFB) forms a four-helix bundle with loops between pairs of helices. Conserved residues in these two loops form a hydrophobic pocket which serves as the binding site for an acetylated lysine.

Histone acetyltransferases

The enzyme histone acetyltransferase (HAT) directs acetylation of histones. When a histone is acetylated, an acetyl group (CH_3CO) is added to one or more of its lysine residues at the ε-amino group, thereby removing the positive charge (Fig. 11.19). The addition of the negatively charged acetyl group reduces the

Is there a histone code?

The histone code hypothesis proposes that covalent post-translational modifications of histone tails are read by the cell and lead to a complex, combinatorial transcriptional output. In this model, the chromatin structure of a particular gene is specified by this combinatorial pattern of modifications of the histones that package it. Histone modifications are proposed to provide binding sites for proteins that can change the chromatin state to either active or repressed. Some protein domains present in chromatin-binding proteins and histone-modifying enzymes are known to interact with covalent marks in the histone tails. For example, the ~110 amino acid bromodomain recognizes specific acetylated lysines in histone tails. Another protein motif called the chromodomain (the name is derived from "chromatin organization modifier") is targeted to methylated lysines and arginines in histone tails (Fig. 1).

An increasing amount of experimental data has provided support for different aspects of the histone code hypothesis, but the hypothesis continues to be the subject of much debate. To date, acetylation of individual lysines has not been shown to be combinatorial or consistent from gene to gene. Genome-wide analysis of histone acetylation patterns shows evidence for groups of genes with consistent acetylation patterns involving combinations of 11 lysines. However, some researchers argue that the use of histone modifications individually or sequentially cannot be considered a code since the total number of modifications does not necessarily contain more information than the sum of the individual modifications. Only if acetylation patterns are read combinatorially can they provide the language of a true code. Single or sequential use of modifications is proposed to be more analogous to a protein-signaling pathway.

Recently, a detailed study of the complexity of histone H4 acetylation patterns was performed that provides some insight into the nature of the "code." Researchers constructed yeast strains carrying all possible combinations of mutations among four lysines in the histone H4 tail that are involved in transcriptional silencing when acetylated. They then characterized the resulting genome-wide changes in gene expression by using DNA microarrays (see Fig. 16.13). They found that the four H4 lysines give rise to eight transcriptional states, rather than the full complement of 16 possible states. Cumulative effects were seen as changes in gene expression correlating with an increase in the total number of mutations; the consequences of H4 acetylation are simple and cumulative. The researchers concluded that, at least for these particular histone modifications, if there is indeed a "code" it is a simple one and not a "complex, multimark code."

overall positive charge of the histones. This results in a decreased affinity of the histone tails for the negatively charged DNA. Histone deacetylase (HDAC) catalyzes the removal of acetyl groups from lysines near the N-termini of histones. Because of the decreased affinity of acetylated histone tails for DNA, hyperacetylation of histones was generally assumed to be associated with gene activation. This view has been modified by two lines of evidence. First, genome-wide analysis indicates that both acetylation and deacetylation of specific lysines are correlated with gene activity. Second, histone deacetylation by a particular HDAC (Hos2) was shown to be required for the activity of specific genes. The current model proposes that the acetylation state of lysine residues provides a specific binding surface that can either serve to recruit repressors or activators of gene activity, depending on the context.

Histone methyltransferases

The enzyme histone methyltransferase (HMT) directs methylation of histones. Unlike histone acetylation which takes place only on lysine (K), methylation occurs both on lysine and arginine (R) residues. Either one, two, or three methyl groups (CH_3) are added at lysine residues. The methyl groups increase their bulk but do not alter the electric charge (Fig. 11.19). Arginine can be modified by the addition of one methyl group. Histone methylation is linked to both activation and repression. The site specificity of lysine or arginine methylation, as well as the number of methyl groups attached to a particular lysine or arginine, can have distinct effects on transcription. For example, dimethylation of histone H3 at lysine 9 (abbreviated as dimethylated H3-K9) and trimethylation of histone H3 at lysine 27 (trimethylated H3-K27) are both largely associated with gene silencing and heterochromatin formation. In contrast, methylation of H3-K4, H3-K36, or H3-K79 is associated with active chromatin. Demethylation of H3-K9 facilitates phosphorylation at serine 10 and acetylation at lysine 14. In addition, serine 10 phosphorylation facilitates acetylation of lysine 9, thereby preventing the setting of the repressive lysine 9 methylation mark (see Focus box 11.5). These modifications lead to tightly packed, "closed" chromatin becoming "open" chromatin. Histone methylation is generally associated with the recruitment of other proteins, which may then combine to form an altered chromatin structure.

Histone acetylation was known to be dynamically regulated by HATs and HDACs, but histone methylation had been considered to be a "permanent" modification, because no histone demethylase had been isolated. However, in 2004, scientists finally isolated an enzyme that acts as a histone demethylase. The enzyme – lysine-specific demethylase 1 (LSD1) – represses specific genes by maintaining unmethylated histones. LSD1 is conserved from *Schizosaccharomyces pombe* to humans. Discovery of this enzyme suggests that histone methylation is also a dynamic process and is subject to regulation by both methylases and demethylases (see Fig. 11.19).

Ubiquitin-conjugating enzymes

Usually, the addition of polyubiquitin chains targets a protein for degradation by the proteasome (see Fig. 5.17). However, the addition of one ubiquitin (monoubiquitinylation) can alter the function of a protein without signaling its destruction. A conjugating enzyme catalyzes the formation of a peptide bond between ubiquitin and the side chain-NH_2 of a lysine residue in a target protein. For example, monoubiquitinylation of histone H2B is, depending on the gene, associated with activation or silencing of transcription as well as transcription elongation. Monoubiquitinylation of linker histone H1 leads to its release from the DNA. In the absence of the linker histone, chromatin becomes less condensed, leading to gene activation.

Kinases

When a histone is phosphorylated by a specific kinase, a phosphate group is added, usually to one or more of its serine or threonine amino acids, thereby adding a negative charge (see Fig. 11.19). Phosphorylation of the linker histone H1 either removes H1 from the DNA, or makes it bind DNA less tightly. Either case is generally associated with gene activation. As noted above, phosphorylation of histone H3 is also associated with the activation of specific genes. Phosphorylation is a dynamic process, subject to regulation by phosphatases.

ADP-ribosyltransferases

Mono-ADP-ribosylation is the enzymatic transfer of an ADP-ribose residue from NAD^+ to a specific amino acid of the acceptor protein by an ADP-ribosyltransferase (see Fig. 11.19). Histones are modified at glutamic acid residues; however, in other proteins cysteine, asparagine, and arginine residues can be modified. Removal of the ADP-ribose residue is mediated by an ADP-ribosylhydrolase. MacroH2A, a histone variant associated with X chromosome inactivation (see Section 12.4), has an N-terminal region with high sequence homology to H2A, but it also contains a 25 kDa nonhistone macro domain. A recent crystallographic study points to the possibility that macro domains may function in ADP-ribosylation of histones. This was an exciting finding because it suggests for the first time that a histone variant may have inherent enzymatic potential.

SUMO-conjugating enzymes

Small ubiquitin–like modifier (SUMO) has 20% identity with ubiquitin. SUMO conjugation involves the same set of enzymatic steps as ubiquitin conjugation (see Fig. 5.17). SUMO is a 97 amino acid protein that is added through its carboxyl end to the amino functional group on lysines in a target protein (see Fig. 11.19). Typically, lysine residues subject to sumoylation are found within a consensus motif, ψ–Lys-x-Glu (where ψ is a large hydrophobic residue and x is any residue). Deconjugation is mediated by SUMO-specific proteases. In most cases reported to date, sumoylation is associated with transcriptional repression. Once thought to be a relatively rare post-translational modification, sumoylation is now known to play a role in regulating protein transport, segregation of chromosomes during mitosis, and repair of damaged DNA. Novel functions of SUMO continue to be discovered. Although SUMO can operate throughout the cell, its main actions are concentrated in the nucleus.

Linker histone variants

In some cases, the selection of a different histone linker protein is important for gene regulation. Mammals contain eight histone H1 subtypes including H1a through H1e and H1° found in somatic cells, as well as two germ cell-specific subtypes, $H1_t$ and $H1_{oo}$. The expression of each mammalian histone H1 subtype depends on the tissue type, phase of the cell cycle, and developmental stage. For example, when the Msx1 homeodomain protein forms a complex with linker histone H1b, the complex binds to an enhancer element in the *MyoD* gene and inhibits gene expression. This prevents the differentiation of muscle progenitor cells because MyoD (a helix–loop–helix transcription factor) is a master regulator of skeletal muscle cell differentiation. Decreased production of histone H1b and Msx1 during development relieves the repression of the *MyoD* gene, allowing skeletal muscle differentiation to proceed (Fig. 11.20).

Chromatin remodeling complexes

The recent discovery of ATP-dependent chromatin remodeling complexes marked an exciting advance in the chromatin field. This new class of molecular motors uses the energy derived from ATP hydrolysis to change the contacts between histones and DNA, thereby allowing transcription factors to bind to DNA regulatory elements. Chromatin remodeling complexes can mediate at least four different changes in chromatin structure (Fig. 11.21):

1 Nucleosome sliding: the position of a nucleosome changes on the DNA.
2 Remodeled nucleosomes: the DNA becomes more accessible but the histones remain bound.
3 Nucleosome displacement: the complete dissociation of DNA and histones.
4 Nucleosome replacement: replacement of a core histone with a variant histone.

There are a number of different families of ATP-dependent remodeling complexes that have been characterized, including the SWI/SNF, ISWI, and SWR1 complex families. Each family is defined by a unique subunit composition and the presence of a distinct ATPase. The precise mechanism of remodeling appears to be distinct from one family member to another.

SWI/SNF chromatin remodeling complex family

The SWI/SNF complex (pronounced "switch-sniff") from the budding yeast *Saccharomyces cerevisiae* was the first chromatin remodeling complex to be characterized. It is a massive 2 MDa complex composed of at least 11 different polypeptides (Fig. 11.21). The name derives from the components of the catalytic ATPase subunit, which were first described in mutants that were mating-type *swi*tching defective (Swi2) (see Section 12.7) and *s*ucrose *n*on*f*ermenters (Snf2). Many other chromatin remodeling factors have since been identified. The human cell homolog of SWI/SNF is called BAF (*B*RG-1 *a*ssociated *f*actor) and

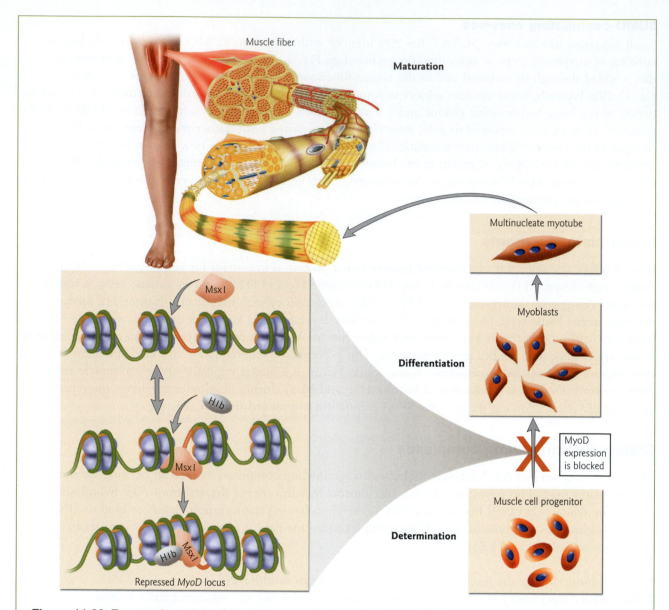

Figure 11.20 Repression of the *MyoD* gene by the linker histone H1b. The Msx1 homeodomain protein forms a complex with linker histone H1b, which binds to the enhancer of the *MyoD* gene. This blocks expression of MyoD and prevents the differentiation of muscle progenitor cells. (Lee, H., Habas, R., Abate-Shen, C. 2004. Msx1 cooperates with histone H1b for inhibition of transcription and myogenesis. *Science* 304:1675–1678).

contains two ATPase subunits, human Brahma (hBRM) and *Brahma-related gene 1* (BRG1). The *Drosophila* homolog BAP (*B*RM complex *a*ssociated *p*rotein) has a single ATPase subunit named Brahma.

RSC (*r*emodel the *s*tructure of *c*hromatin) is another yeast chromatin remodeling complex closely related to SWI/SNF. This complex contains about 15 subunits, sharing two identical subunits and at least four homologs with the subunits of SWI/SNF. PBAF and PBAP are RSC-like complexes found in mammals and *Drosophila*, respectively. The "P" stands for "polybromo associated." In polybromo-associated proteins there is an unusually high number of bromodomains (Focus box 11.5). Typically bromodomains are only present as single or double domains, but in PBAP there are six bromodomains. The bromodomain is a distinguishing feature of the SWI/SNF–like family that is absent in ISWI and other remodeling complexes.

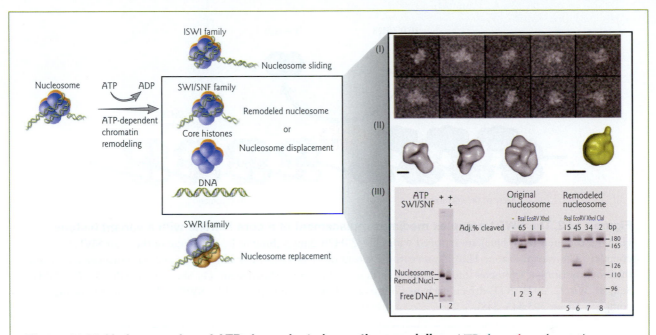

Figure 11.21 Various modes of ATP-dependent chromatin remodeling. ATP-dependent chromatin remodeling complexes alter histone–DNA contacts. Nucleosomes either slide to another position (ISWI family), or are remodeled or displaced (SWI/SNF family), or core histones are replaced with a variant (SWR1 family). (Inset) (i) Raw images of the yeast SWI/SNF complex obtained by scanning transmission electron microscopy. (ii) Rendered surface three-dimensional structures of the SWI/SNF complex (gray) and the nucleosome core particle (yellow) for comparison. The SWI/SNF complex has a cone-shaped depression at the top which may serve as a nucleosome-binding pocket. (Reprinted with permission from: Smith, C.L., Horowitz-Scherer, R., Flanagan, J.F., Woodcock, C.L., Peterson, C.L. 2003. Structural analysis of the yeast SWI/SNF chromatin remodeling complex. *Nature Structural Biology* 10:141–145. Copyright © 2003 Nature Publishing Group). (iii) Remodeled nucleosomes have faster mobility in an electrophoretic mobility shift assay (EMSA) (see Fig. 9.15A) and differential accessibility to restriction endonuclease digest (see Section 8.3). Core octamers were assembled *in vitro* on DNA radiolabeled at the 5′ end, and incubated with or without SWI/SNF. After remodeling, SWI/SNF was removed from the labeled nucleosomes by the addition of excess unlabeled competitor DNA. Samples were separated by native polyacrylamide gel electrophoresis (PAGE). A small proportion of nucleosomes were released as a slightly slower migrating band, identified as a hexamer–DNA complex. Nucleosomes were purifed from the gel and digested with various restriction endonucleases as indicated. The resulting restriction fragments were separated by denaturing PAGE. (Reprinted from: Kassabov, S.R., Zhang, B., Persinger, J., Bartholomew, B. 2003. SWI/SNF unwraps, slides, and rewraps the nucleosome. *Molecular Cell* 11:391–403. Copyright © 2003, with permission from Elsevier.)

Mode of action of SWI/SNF: nucleosome sliding and disassembly

The SWI/SNF and RSC-type chromatin remodeling complexes cause nucleosome sliding and significant perturbation of nucleosome structure. A general model of the mode of action of SWI/SNF is shown in Fig. 11.21. The ATPase activity of SWI/SNF hydrolyzes ATP and the energy is used for modification of the path of DNA around the histone octamer. Some 50 bp of DNA is unwrapped from the edges of the nucleosome. This results in the exposure of DNA on the nucleosomal surface and the sliding of nucleosomes to new positions. In fact, in some highly active genes, such as the yeast *PHO5* gene, the nucleosomes dissociate completely from the DNA. *PHO5* encodes a secreted acid phosphatase and is induced in response to a lack of inorganic phosphate in the growth medium. After the remodeling complex alters the path of DNA around the histone octamer, transcription factors and the general transcription machinery can gain access to DNA. Often there is cooperation between SWI/SNF and histone acetyltransferase (HAT) in remodeling chromatin structure.

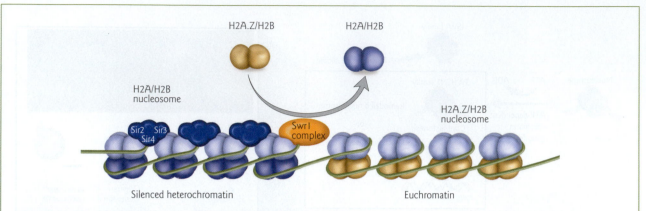

Figure 11.22 The SWR1 complex mediates replacement of a core histone with a variant histone.
H2A–H2B dimers (light blue) are replaced with H2A.Z–H2B dimers (brown) by the action of the yeast SWR1 complex. The variant histone H2A.Z prevents the spread of Sir proteins (dark blue) associated with silenced chromatin (heterochromatin) at telomeres into adjacent regions of euchromatin. (Mizuguchi, G., Shen, X., Landry, J., Wu, W.H., Sen, S., Wu, C. 2004. ATP-driven exchange of histone H2AZ variant catalyzed by SWR1 chromatin remodeling complex. *Science* 303:343–348).

ISWI chromatin remodeling complex family

In contrast to SWI/SNF and RSC–type chromatin remodeling complexes, members of the ISWI (*imitation Swi2*) family (such as NURF from *Drosophila*) relocate the nucleosomes by sliding the histone octamers along the DNA without apparent perturbation of their structure (see Fig. 11.21). This process yields nucleosomes at different positions on the DNA without their disassembly, or the altered nucleosome forms that are typical of SWI/SNF action. Evidence for this comes from experiments demonstrating that the ISWI complexes do not alter the nuclease sensitivity of nucleosome core particles.

SWR1 chromatin remodeling complex family

The SWR1 complex is named after its ATPase subunit Swr1 (for *Swi2/Snf2 r*elated). SWR1 may be targeted to the appropriate genomic regions by the Bdf1 subunit (*bromo*domain *factor 1*), which contains two bromodomains. The SWR1 complex adds a new theme to chromatin remodeling: histone replacement with a variant histone in the core octamer (Fig. 11.22). The disruption of the core octamer was demonstrated by studies showing that the SWR1 complex can transfer H2A.Z–H2B dimers in exchange for H2A–H2B dimers to nucleosomal arrays but not to free DNA. The reaction is dependent on the ATPase subunit of SWR1. Histone variant H2A.Z is strongly enriched in active genes that are adjacent to transcriptionally repressed regions, such as near telomeres. The presence of this variant histone prevents the spread of silent heterochromatin into neighboring euchromatin.

11.7 Transcription complex assembly: the enhanceosome model versus the "hit and run" model

Knowledge of the identity and roles of transcription factors and coactivators is well advanced. However, the dynamic process by which these proteins interact on DNA to activate transcription is the subject of much study. There is clearly an interplay between chromatin remodeling and modification complexes. Both types of coactivators can be recruited to the same gene promoters, but there appears to be no general rule for the order of recruitment of the many different proteins involved in gene transcription. Instead there is a gene-specific order of events.

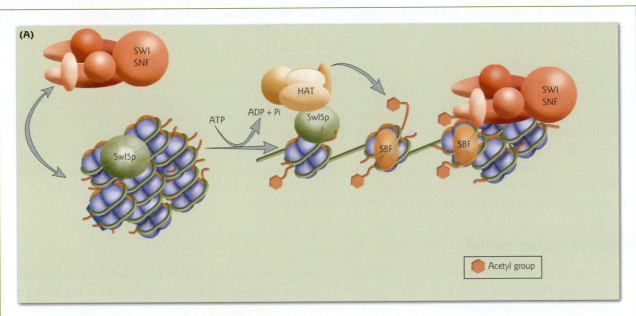

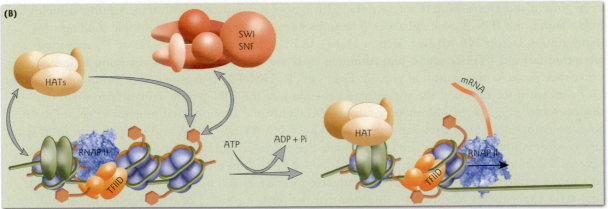

Figure 11.23 The order of recruitment of transcriptional regulatory proteins is gene-specific. (A) At the yeast *HO* gene promoter, the Swi5p transcription factor recruits SWI/SNF and a histone acetyltransferase (HAT) complex, followed by a second transcription factor, SBF, well before assembly of the preinitiation complex. (B) At the human α-antitrypsin gene promoter, multiple HAT complexes and the SWI/SNF complex are recruited after preinitiation complex assembly. (Reprinted with permission from: Fry, C.J., Peterson, C.L. 2002. Unlocking the gates to gene expression. *Science* 295:1847–1848. Copyright © 2002 AAAS.)

Order of recruitment of various proteins that regulate transcription

The order of recruitment depends on chromatin structure of the gene promoter, the phase of the cell cycle, and many other factors. Two examples are shown in Fig. 11.23. In the first example, SWI/SNF is recruited by an activator protein bound to an upstream regulatory region in the yeast *HO* gene (see Section 12.7). After SWI/SNF remodels the *HO* gene promoter, a histone acetyltransferase (HAT) complex is recruited. The cooperative action of SWI/SNF and HAT facilitates the binding of a gene-specific transcription factor, followed by the general transcription factors and RNA pol II. Unexpectedly, in the second example, the entire preinitiation complex is assembled on the α_1-antitrypsin gene promoter prior to recruitment of chromatin remodeling complexes or HATs. The human interferon-β (IFN-β) gene promoter follows a different series of events upon induction by viral infection. A DNA–activator complex (enhanceosome, see below) binds to a nucleosome-free region upstream of the IFN-β gene promoter. This leads to rapid recruitment of HAT and acetylation of nucleosomes at the TATA box in the gene promoter. Subsequently, the SWI/SNF complex slides a nucleosome array from the TATA box, allowing access to the general transcription machinery.

Individual transcription factors generally bind to DNA regulatory elements such as promoters and enhancers with relatively low specificity both *in vitro* and *in vivo*. However, a high degree of specificity is achieved through combinatorial interactions that occur when multiple transcription factors bind to clustered DNA regulatory elements. Cooperative binding of regulatory proteins to DNA was introduced as an important mechanism for bacterial gene regulation in Chapter 10 (Section 10.6), and is an equally important mechanism in eukaryotic gene regulation. Two models for binding of transcription factors and assembly of transcription complexes have been proposed, the enhanceosome model and the "hit and run" model. A recent study suggests that these models are not mutually exclusive.

Enhanceosome model

The enhanceosome model proposes that interactions among transcription factors promote their cooperative, step-wise assembly on DNA and give the complex exceptional stability. The context in which the elements are organized within three-dimensional space is essential for transcriptional regulation, as are extensive protein–protein and DNA–protein interactions. This model is compatible with transcription complex assembly within the IFN-β gene enhancer. The inducible enhancer consists of multiple binding sites for transcription factors NF-κB (see Section 11.10), IRF1 (*interferon regulatory factor 1*), activating transcription factor 2 (ATF2), and c-Jun. Although each site will not activate transcription independently, together they facilitate the activation process. For example, HMG-I/Y (*high mobility group protein isoform I and Y*) is essential for directing the appropriate unbending of the IFN-β gene enhancer to form the "enhanceosome" (Fig. 11.24). Members of the HMG family are "architectural proteins" that are the major nonhistone component of mammalian chromatin. They bind AT-rich DNA within the minor groove and induce conformational changes in the DNA double helix that facilitate the formation of multiprotein complexes. By binding to DNA near the NF-κB- and ATF2/c-Jun-binding sites, HMG-I/Y makes direct protein–protein contacts with the two transcription factors and increases their affinity for DNA. Helical-twist experiments (e.g. Fig. 10.18) show that the relative phasing of transcription factor and HMG-I/Y-bindings sites cannot be changed without interfering with the induction of transcription.

Hit and run model

The classic enhanceosome model at first glance seems incompatible with the observation that transient and dynamic binding is a common property of all chromatin proteins with the exception of core histones. Fluorescence recovery after photobleaching (FRAP) experiments (Fig. 11.25A) have shown that most nuclear proteins are highly mobile and the interaction of proteins with chromatin and nuclear compartments is highly dynamic. For example, the glucocorticoid receptor was shown by FRAP to bind and unbind to chromatin in cycles of only a few seconds. Even such transient interactions, however, are sufficient to promote large-scale remodeling of chromatin lasting a few hours and to correlate with transcriptional activation. In the hit and run model, transcriptional activation reflects the probability that all components required for activation will meet at a certain chromatin site (the "hit") – i.e. transcription complexes are assembled in a stochastic fashion from freely diffusible proteins – and that their binding is transient (the "run").

Merging of models

A more recent study suggests that the two models are not mutually exclusive. Each protein in a complex influences the binding kinetics of its partners. The bottom line is that the principles of combinatorial interaction and complex stability apply to hit and run models even if the complex itself has a very limited

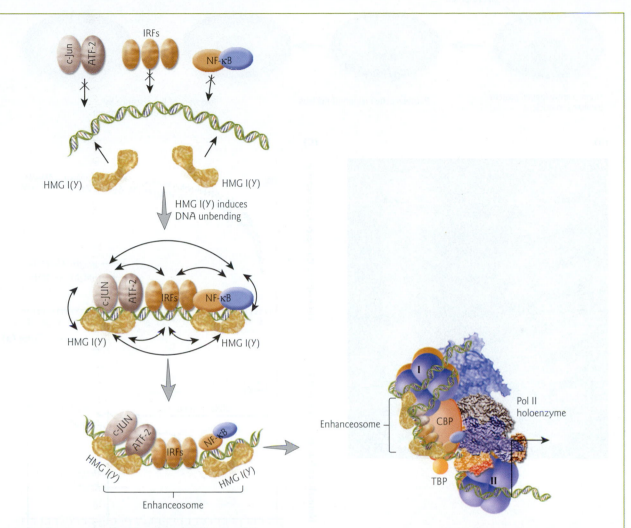

Figure 11.24 The INF-β enhanceosome. A two-step model for enhanceosome assembly. Transcription factors recognize their binding sites in the INF-β promoter region with low affinity due to unfavorable intrinsic DNA curvature. Binding of HMG-I/Y to the promoter region unbends DNA, lowering the free energy required for binding of the transcription factors c-Jun, ATF2, IRF1, and NF-κB. Enhanceosome assembly is completed by protein–protein interactions (arrows) between all the components leading to a highly stable structure. The enhanceosome interacts with the general transcription machinery at the core promoter through protein–protein interactions. Multiple HATs are recruited during assembly of the preinitiation complex, which includes RNA polymerase II (Pol II holoenzyme). The SWI/SNF complex joins the assemblage and facilitates TBP binding to the TATA box. Nucleosomes I and II have been remodeled by the SWI/SNF complex. CBP is a histone acetyltransferase complex. The black arrow indicates the start of transcription. (Yie, J., Merika, M., Munshi, N., Chen, G., Thanos, D. 1999. The role of HMG I(Y) in the assembly and function of the INF-β enhanceosome. *EMBO Journal* 18:3074–3089; and Agalioti, T., Lomvardas, S., Parekh, B., Yie, J., Maniatis, T., Thanos, D. 2000. Ordered recruitment of chromatin modifying and general transcription factors to the IFN-β promoter. *Cell* 103:667–678.)

lifetime. High mobility group box 1 protein (HMGB1) and the glucocorticoid receptor (GR) were analyzed in living cells, by tagging them with yellow fluorescent protein (YFP) and cyan fluorescent protein (CFP), respectively (see Section 9.3). The interaction between CFP–GR and HMGB1–YFP in living cells was followed via fluorescence resonance energy transfer (FRET) (see Fig. 9.16D) and FRAP. The two proteins were shown to only interact within chromatin and to lengthen each other's residence time on chromatin

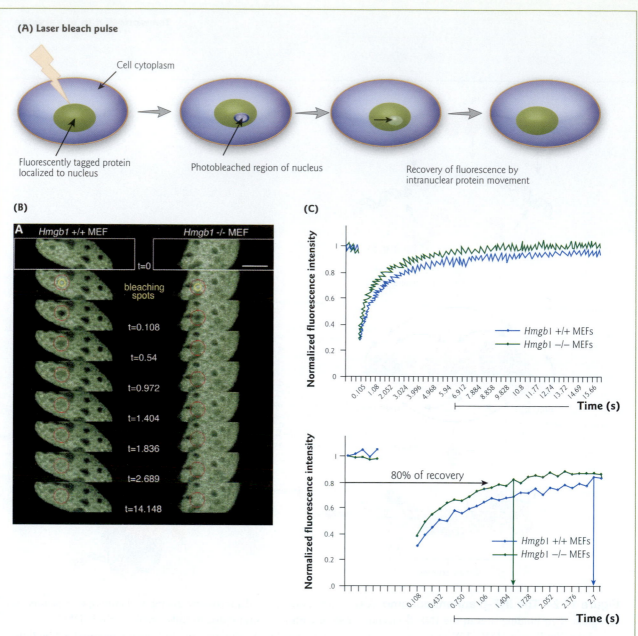

Figure 11.25 Protein dynamics within the nucleus. (A) In a fluorescence recovery after photobleaching (FRAP) experiment, a small area of the nucleus in a cell expressing a fluorescently tagged protein is rapidly and irreversibly bleached using a targeted laser pulse. Photobleaching results in a small region devoid of fluorescence signal. The recovery of the fluorescence signal is measured as a function of time using confocal microscopy. Recovery of fluoresence is due to the movement of unbleached proteins into the bleached area. The kinetics of recovery correlates with the mobility of the fluorescently tagged proteins. (B) GFP-tagged glucocorticoid receptor (GR) shows increased mobility in $Hmg1^{-/-}$ knockout mouse cells (see Fig. 15.4 for methods) as compared to $Hmgb1^{+/+}$ wild-type mouse cells. Yellow circles indicate the photobleaching areas in the nucleus (green), and the red contours highlight the region of interest. Images were collected every 108 ms. The scale bar represents 10 μm. (C) FRAP analysis of GFP-GR in $Hmgb1^{-/-}$ (green dots) and $Hmgb1^{+/+}$ (blue dots) mouse cells. (Upper graph) The dots indicate mean values, and error bars indicate standard error, from 10 cells of a representative experiment. (Lower graph) An enlargement of the recovery kinetics of the two cells closest to the mean in their group, in order to show the calculation of 80% recovery. Recovery is faster in knockout cells than in wild-type cells: 80% recovery is reached in 2.7 and 1.4 seconds respectively; complete recovery in knockout cells is reached in 9 seconds; while it is beyond 16 seconds in wild-type cells. (Reprinted from: Agresti, A., Scaffidi, P., Riva, A., Caiolfa, V.R., Bianchi, M.E. 2005. GR and HMGB1 interact only within chromatin and influence each other's residence time. *Molecular Cell* 18:109–121. Copyright © 2005, with permission from Elsevier.)

(Fig. 11.25B). Mutants that were unable to bind chromatin showed no FRET, and GR was more mobile in cells devoid of HMGB1 (e.g. $Hmbg1^{-/-}$ cells from knockout mice).

11.8 Mechanism of RNA polymerase II transcription

Crystallographic models at 2.3 Å of the complete RNA pol II, together with new biochemical and electron microscopic data, have begun to provide mechanistic insights into the process of initiation and subsequent synthesis of the RNA transcript. After assembly of the preinitiation complex, a period of abortive initiation follows before the polymerase escapes the promoter region (promoter clearance) and enters the elongation phase. The basic process of RNA synthesis is subdivided into multiple stages: selection of a nucleoside triphosphate (NTP) complementary to the DNA template, catalysis of phosphodiester bond formation, and translocation of the RNA and DNA (Fig. 11.26). In essence, the overall process is the same as that described

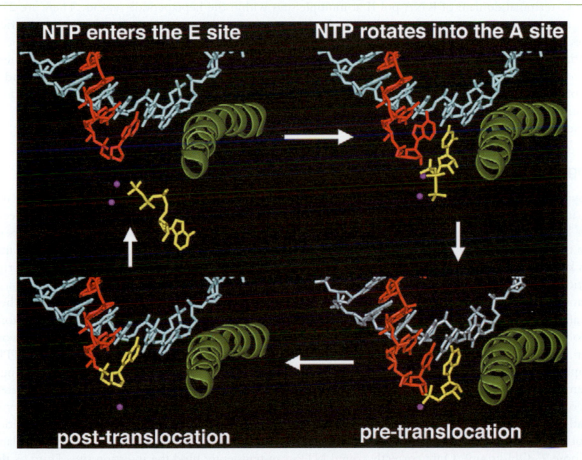

Figure 11.26 RNA transcript synthesis by RNA pol II. Four crystal structures of RNA pol II transcribing complexes are shown illustrating the four step cycle of RNA transcript synthesis. (1) A nucleotide (NTP) enters the entry (E) site beneath the active center. (2) The NTP rotates into the nucleotide addition (A) site and is checked for mismatches. (3) Pretranslocation: phosphodiester bond formation. (4) Translocation and post-translocation: the NTP just added to the RNA transcript moves into the next position, leaving the A site open for the entry of another NTP. Only nucleic acids in the active center region (template DNA, turquoise; RNA transcript, red), bridge helix (green), and Mg^{2+} ions (purple) are shown. The incoming NTP is shown in yellow. (Reprinted by permission of Federation of the European Biochemical Societies from: Boeger, H., Bushnell, D.A., Davis, R. et al. 2005. Structural basis of eukaryotic gene transcription. *FEBS Letters* 579:899–903.)

for bacterial RNA transcription in Section 10.3; however, the details differ because of all the additional factors involved in RNA pol II-mediated transcription.

Promoter clearance

During abortive initiation, the polymerase synthesizes a series of short transcripts. As RNA pol II moves, it holds the DNA strands apart forming a transcription "bubble." The upstream edge of the transcription bubble forms 20 bp from the TATA box. TFIIB contacts both DNA strands within the transcription bubble and stabilizes the transcribing complex until a complete 8 or 9 bp DNA–RNA hybrid is formed. The bubble expands downstream until 18 base pairs are unwound and the RNA is at least seven nucleotides long. When this point is reached, the upstream approximately eight bases of the bubble reanneal. This so-called "bubble collapse" marks the end of the need for the TFIIH helicase for transcript elongation. Synthesis of RNA greater than about 10 residues in length leads to displacement of TFIIB from RNA pol II, because the TFIIB-binding site overlaps the RNA exit site (see Fig. 11.9). Promoter clearance requires phosphorylation of the C-terminal domain of RNA pol II at multiple sites within the heptapeptide repeats. There are 27 of these repeats in the yeast RNA pol II CTD. Each repeat contains sites for phosphorylation by specific kinases including one that is a subunit of TFIIH. Hyperphosphorylation of the CTD tail is essential for activating the polymerase and allowing it to begin the elongation phase. During elongation, the polymerase starts to move away from the promoter. Addition of these phosphates helps RNA pol II to leave behind most of the general transcription factors used for initiation. TFIID remains bound and allows the rapid formation of a new preinitiation complex. Once phosphorylated, RNA pol II can unwind DNA, polymerize (synthesize) RNA, and proofread.

Elongation: polymerization of RNA

The synthesis of mRNA precursors by eukaryotic RNA pol II is a multistage process consisting of four major steps: initiation, promoter clearance, elongation, and termination. So far, the focus of this chapter has been on events leading up to the initiation of transcription and promoter clearance, and how transcription factors gain access to promoters and enhancers. Although much has been learned about steps involved in initiation, much less is known about later stages in transcription, especially chain elongation. The transition from transcription initiation to elongation is marked by a change in the factors that are associated with RNA pol II.

Transcription elongation is the process by which RNA pol II moves through the coding region of the gene after it initiates mRNA synthesis at the promoter. Incoming (downstream) DNA is unwound before the polymerase active site and is rewound beyond it to form the exiting (upstream) duplex. In the unwound region, the DNA template strand forms a hybrid duplex with growing mRNA. RNA pol II selects NTPs in a template-directed manner. Binding of a NTP to a RNA pol II-transcribing complex with base pairing to the template DNA has been observed by X-ray crystallography. The results are consistent with a four step cycle of nucleotide selection and addition (Fig. 11.26). First, the incoming nucleotide binds to an entry site beneath the active center in an inverted orientation. Second, the NTP rotates into the nucleotide addition site for sampling of correct pairing with the template DNA and for discrimination from dNTPs (that lack the ribose 2′-OH group). Only correctly paired NTPs can transiently bind the insertion site. Third, is the pretranslocation step in which phosphodiester bond formation occurs. Fourth, translocation occurs to repeat the cycle. In the post-translocation complex, the nucleotide just added to the RNA has moved to the next position, leaving the A site open for the entry of a nucleotide. At the upstream end of the hybrid, RNA pol II separates the nascent RNA from the DNA.

Proofreading and backtracking

In DNA polymerase, the growing DNA shuttles between widely separated active sites for DNA synthesis and cleavage during proofreading (see Fig. 6.13). Surprisingly, in RNA pol II the growing RNA remains at

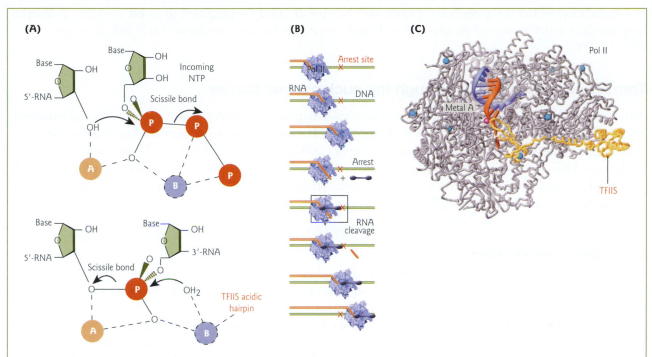

Figure 11.27 RNA pol II polymerization and cleavage. (A) Model of the tunable RNA pol II active site. RNA polymerization and cleavage both require metal ion "A" (e.g. Mg^{2+}) in the active center region. The differential positioning of metal ion "B" switches activity from polymerization to cleavage. (Upper panel) Proposed mechanism of nucleotide (NTP) incorporation during RNA polymerization. (Lower panel) Proposed mechanism of TFIIS-mediated RNA cleavage during backtracking. A metal ion "B" interacts either with the phosphates of the incoming NTP (RNA polymerization) or with the TFIIS acidic hairpin and a nucleophilic water molecule (RNA cleavage). (B) A model for the rescue of arrested RNA pol II by TFIIS-mediated mRNA cleavage. When transcribing, RNA pol II (blue) encounters an arrest site (red cross) on DNA (green), it then pauses and backtracks leading to transcriptional arrest. Cleavage of the extruded RNA (orange) is induced by TFIIS (blue). Transcription then continues on past the arrest site. (C) Cut-away view of the crystallographic model of the RNA pol II (gray)–TFIIS (yellow) complex, showing metal ion A in the active site. Template DNA, blue; RNA, red; zinc ions, cyan spheres. (Protein Data Bank, PDB: 1Y1Y. Adapted from Cramer, P. 2004. RNA polymerase II structure. *Current Opinion in Genetics & Development* 14:218–226. Copyright © 2004, with permission from Elsevier.)

a single active site that switches between RNA synthesis and cleavage. X-ray crystallographic analysis of RNA pol II with the elongation factor TFIIS (see below) supports the idea that the polymerase has a "tunable" active site that switches between mRNA synthesis and cleavage (Fig. 11.27). RNA polymerization and cleavage both require a metal ion "A" (e.g. Mg^{2+}), but differential coordination of another metal ion "B" switches RNA pol II activity from polymerization to cleavage. In this model, for RNA polymerization, metal B binds the phosphates of the substrate NTP. For cleavage during proofreading, metal B is bound by an additional unpaired nucleotide located in the active center. The tunable polymerase active site allows efficient mRNA proofreading because RNA cleavage creates a new RNA 3′ end at metal A, from which polymerization can continue. Two types of proofreading reactions may occur: removal of a misincorporated nucleotide directly after its addition, or cleavage of a dinucleotide after misincorporation and backtracking by one nucleotide.

During mRNA elongation, RNA pol II also can encounter DNA sequences that cause reverse movement or "backtracking" of the enzyme. During backtracking the register of the RNA–DNA base pairing is maintained but the 3′ end of the transcript is unpaired and extruded from the active center. This can lead to transcriptional arrest. Escape from arrest requires cleavage of the extruded RNA with the help of TFIIS.

For stimulation of the weak nuclease activity of RNA pol II, TFIIS is proposed to insert an acidic β–hairpin loop into the active center to position metal B and a nucleophilic water molecule for RNA cleavage (Fig. 11.27).

Transcription elongation through the nucleosomal barrier

Although chromatin is remodeled upon transcription initiation, the DNA remains packaged in nucleosomes in the coding region of transcribed genes. As RNA pol II moves through the gene-coding region, the enzyme encounters a nucleosome approximately every 200 bp. How does RNA pol II overcome this nucleosomal barrier? Recent studies suggest the existence of two distinct mechanisms for the progression of RNA polymerases through chromatin: (i) nucleosome mobilization or "octamer transfer" (i.e. movement of the octamer on the DNA); and (ii) H2A-H2B dimer depletion (Fig. 11.28).

Nucleosome mobilization

Polymerases such as RNA polymerase III and bacteriophage SP6 RNA polymerase appear to use the nucleosome mobilization or "octamer transfer" mechanism for overcoming the nucleosome barrier. In this mode of action, nucleosomes are translocated without release of the core octamer into solution. In other words, DNA is displaced from the nucleosomes and the nucleosomes are transferred to a region of DNA already transcribed by RNA pol II (Fig. 11.28A). This process may be facilitated by the elongation factor FACT (see below). Currently, the mechanism of RNA pol I-dependent chromatin remodeling is unknown.

H2A-H2B dimer removal

Nucleosomes are disrupted on active RNA pol II transcribed genes by H2A–H2B dimer removal (Fig. 11.28B). This mode of overcoming the nucleosome barrier requires a number of auxiliary elongation factors. The first protein factor that has been firmly established to promote RNA pol II elongation *in vitro* is FACT (*f*acilitates *c*hromatin *t*ranscription). Other complexes that facilitate chromatin transcription *in vitro* are Elongator and TFIIS. Even in the presence of FACT, Elongator, or TFIIS, the rate of RNA pol II transcription through nucleosomes *in vitro* is much slower than the rate of transcript elongation in the cell. *In vitro*, a significant number of RNA pol II molecules are blocked by the nucleosome, suggesting that overcoming the nucleosome barrier requires the coordinated action of additional, unidentified factors.

FACT promotes nucleosome displacement

FACT plays a role in the elongation of transcripts through nucleosome arrays by promoting transcription-dependent nucleosome alterations. FACT is composed of two protein subunits, designated hSpt16 and SSRP1 in humans. Spt6 was first discovered by a genetic screen in yeast called SPT (suppressor of TY) performed more than two decades ago. This screen uncovered transcription factor mutations that overcome the consequences of inserting a foreign piece of DNA (a TY transposon element) in the promoter of a reporter gene. The FACT complex has a DNA-binding domain and a domain that interacts with histones H2A and H2B. The large subunit (hSpt16) also interacts with the catalytic subunit of a histone acetyltransferase (HAT). This interaction provides a possible mechanism explaining the extent of acetylation observed at transcriptionally active regions. A model for FACT's mode of action has been developed, based on an *in vitro* assay system that allows the study of transcription through positioned nucleosomes. In this model, FACT enables the displacement of a dimer of H2A-H2B in front of RNA pol II, leaving a histone "hexamer" at the same location as the initial octamer. FACT does not require ATP hydrolysis for its mode of action. After passage of the polymerase, FACT enables the immediate reassembly of the H2A-H2B dimer (Fig. 11.28B).

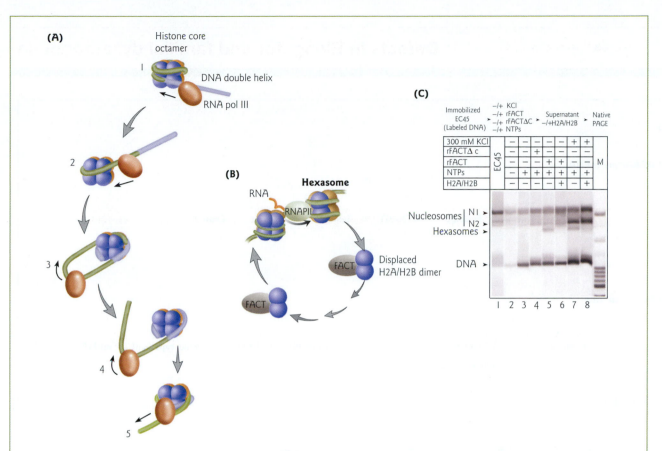

Figure 11.28 Mechanisms of transcription through the nucleosome by RNA polymerases. (A) Nucleosome mobilization or "octamer transfer" model for RNA pol III. (1) RNA pol III rapidly transcribes the first ~25 bp of nucleosomal DNA causing (2) partial dissociation of the DNA from the core octamer (blue). DNA that has been transcribed is shown in blue. DNA that has not been transcribed is green. (3) The DNA behind RNA pol III transiently binds to the exposed surface of the octamer forming a loop. When the polymerase has moved ~60 bp into the core, the downstream portion of the DNA dissociates from the octamer (4), completing octamer transfer. (5) Transcription proceeds to the end of the template. (Studitsky, V.M., Walter, W., Kireeva, M., Kashlev, M., Felsenfeld, G. 2004. Chromatin remodeling by RNA polymerases. *Trends in Biochemical Sciences* 29:127–135.) (B) Histone H2A-H2B dimer depletion model for RNA pol II. FACT mediates displacement of the H2A-H2B dimer from the core octamer, leaving a "hexasome" on the DNA. The histone chaperone activity of FACT helps to redeposit the dimer on the DNA after passage of RNA pol II (RNAPII). (C) Experiment demonstrating that FACT-facilitated transcription through the nucleosome results in the formation of hexasomes. Immobilized nucleosomes containing end-labeled DNA were transcribed *in vitro* in the presence or absence of NTPs, excess H2A-H2B dimer, recombinant FACT (rFACT), a nonfunctional mutant lacking the C-terminus (rFACTÄC), or 300 mM KCl as a control. This control was used as it had been previously shown that nucleosomes transcribed in high salt release the H2A-H2B dimer, forming a hexasome. Transcribed nucleosomes were analyzed on a native gel. The positions of polymerase-free nucleosomes N1 and N2, hexasomes, and DNA are indicated by arrowheads. M, molecular weight markers. (Reprinted with permission from: Belotserkovskaya, R., Oh, S., Bondarenko, V.A., Orphanides, G., Studitsky, V.M., Reinberg, D. 2003. FACT facilitates transcription-dependent nucleosome alteration. *Science* 301:1090–1093. Copyright © 2003 AAAS.)

Elongator facilitates transcript elongation

The Elongator complex was first isolated in yeast, but its exact function in transcript elongation remains elusive (Disease box 11.4). Elongator is composed of six major subunits, designated ELP1 to ELP6. ELP3 has been shown to have HAT activity. Subsequently, the human Elongator complex was purified from a human

Defects in Elongator and familial dysautonomia

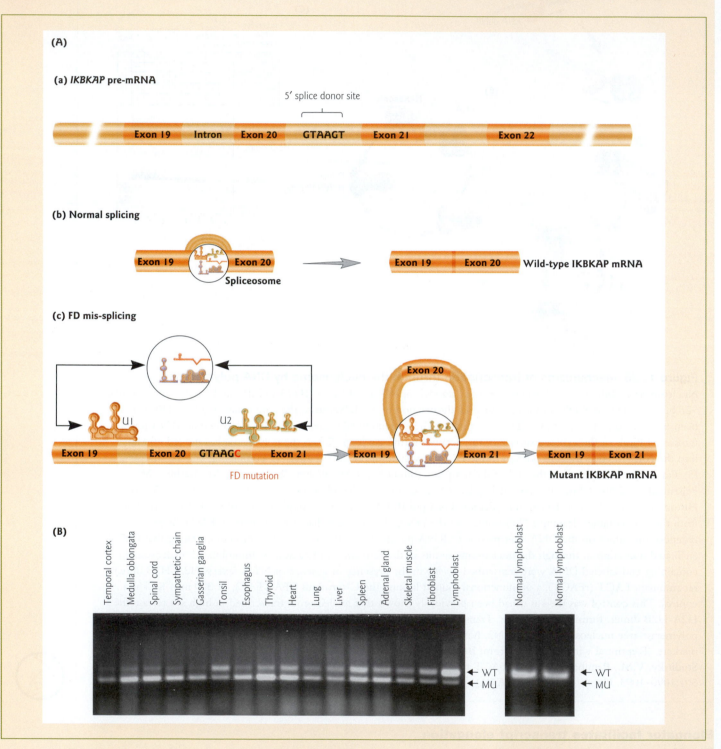

Familial dysautonomia is a disorder of the sensory and autonomic nervous system, which involves progressive depletion of unmyelinated sensory and autonomic neurons. It is inherited as an autosomal recessive and is very common in the Ashkenazi Jewish population; there is a carrier frequency of one in 30 descendents of Jews from Germany, Poland, and Austria, and Eastern Europe.

Symptoms of familial dysautonomia

The autonomic nervous system controls such involuntary functions as swallowing, digestion, and regulation of body temperature and blood pressure. The most distinctive feature of the disorder is the absence of tears when crying. The loss of neuronal function also results in decreased perception of heat, pain, and taste. For example, a person with familial dysautonomia leaning on a pot of boiling water may not feel it and could be seriously burned. Other symptoms include breath-holding episodes, vomiting in response to stress, profuse sweating, and spinal curvature (scoliosis) in 90% of patients by age 13. The average life expectancy is approximately 30 years, and patients frequently have long hospital stays. Treatment currently is mainly preventative and supportive; for example, use of artificial tears, special feeding therapies, and protection from injury.

A defect in the *IKBKAP* gene causes familial dysautonomia

Two mutations in the gene *IKBKAP* are responsible for familial dysautonomia. *IKBKAP* was originally identified as the gene encoding a protein that was thought to function as an I-κB kinase-associated protein (see Section 11.10), hence the name "IKAP." Subsequent reports determined that, in fact, IKAP is a subunit of the Elongator complex, homologous to the yeast ELP1 protein.

Neither of the two mutations in *IKBKAP* result in failure to express the IKAP protein. The most common mutation (> 99.5% of cases) is a splice site mutation that results in tissue-specific skipping of one exon during mRNA splicing. The *IKBKAP* gene contains 37 exons. A single T → C change at base pair 6 of the intron 20 donor splice site leads to symptoms in homozygotes. There is a decrease in splicing efficiency with sporadic skipping of exon 20, reducing the level of wild-type *IKBKAP* mRNA. Every familial dysautonomia cell type expresses both wild-type and mutant IKAP, the ratio of which is variable (Fig. 1). *IKBKAP* mRNA is primarily wild type in fibroblasts and in lymphoblast cell lines. However, in the brain, mutant mRNA with exon 20 missing is primarily expressed. It is not clear why brain cells are particularly sensitive to exon skipping. A second extremely rare mutation is a single G → C change in an exon, which causes an arginine to proline missense mutation that disrupts a threonine phosphorylation site in the protein. This mutation has only been seen in heterozygous patients and has never been detected in the homozygous state.

Researchers are still puzzling over whether a defect in transcription elongation leads to the degeneration of specific sensory and autonomic neurons in humans. Alternatively, the disease symptoms could be caused by disruption of an as yet unidentified cellular pathway in which IKAP alone plays a role. There is some evidence that IKAP may have multiple roles in the cell. A recent study suggests that ELP1P, the yeast homolog of IKAP, negatively regulates polarized exocytosis of post-Golgi secretory vesicles. Since neurons are highly dependent on polarized exocytosis for their development and function, this could make them especially sensitive to dysregulation of IKAP function.

Figure 1 (*opposite*) **Familial dysautonomia.** (A) Schematic representation of the aberrant splicing of *IKBKAP* pre-mRNA seen in familial dysautonomia (FD). The wild-type splice site sequence (GTAAGT) in between exons 20 and 21 of *IKBKAP* pre-mRNA is shown above. In normal splicing, the 5′ splice donor site is recognized by the spliceosome (see Section 13.5), resulting in joining of exons 19 and 20. In FD missplicing, the major FD mutation at base pair 6 of intron 20 is shown in red. This mutation decreases the efficiency of splicing and sometimes results in the skipping of exon 20. (B) Expression of wild-type (WT) and mutant (MU) *IKBKAP* mRNA in postmortem FD tissue samples and cell lines. All samples were assayed by reverse transcription–polymerase chain reaction (RT-PCR) and PCR products were fractionated by electrophoresis on an agarose gel stained with ethidium bromide (see Tool box 8.6 and Fig. 9.8D for methods). The relative WT : MU transcript ratio was consistently observed to be highest in FD lymphoblast lines, lowest in FD nervous system tissues, and intermediate in other tissues. Mutant *IKBKAP* transcript is never detected in RNA from normal cells or tissues. (Reprinted from Cuajungco, M.P., Leyne, M., Mull, J. et al. 2003. Tissue-specific reduction in splicing efficiency of *IKBKAP* due to the major mutation associated with familial dysautonomia. *American Journal of Human Genetics* 72:749–758. Copyright © 2003 by The American Society of Human Genetics, with permission of the University of Chicago Press.)

cell line (HeLa cells) by column chromatography. Like the yeast complex, human Elongator is also composed of six subunits, including a HAT with specificity for histone H3 and to a lesser extent for histone H4. Elongator directly interacts with RNA pol II and facilitates transcription, but it does not appear to interact directly with any other elongation factors, including FACT.

TFIIS relieves transcriptional arrest

The elongation factor, TFIIS, facilitates passage of RNA pol II through regions of DNA that can cause transcription arrest. These sites include AT-rich sequences, DNA-binding proteins, or lesions in the transcribed DNA strand. TFIIS rescues transcriptional arrest by a backtracking mechanism that stimulates endonucleolytic cleavage of the nascent RNA by the RNA pol II active center (see Fig. 11.27).

11.9 Nuclear import and export of proteins

One of the hallmarks of eukaryotic cells is the compartmentalization of the genome into a separate organelle called the nucleus. The sections in this chapter so far have focused on proteins involved in regulating gene transcription, including transcription factors, the general transcription machinery, coactivators, and corepressors. Since protein synthesis occurs in the cytoplasm, this means that transcriptional regulatory proteins must be delivered to their site of activity in the nucleus. The control of nuclear localization of transcriptional regulatory proteins represents a level of transcriptional regulation in eukaryotes that is not present in prokaryotes.

Trafficking between the nucleus and the cytoplasm occurs via the nuclear pore complexes (NPCs). NPCs are large multiprotein complexes embedded in the nuclear envelope – the double membrane that surrounds the nucleus (Focus box 11.6). The NPCs allow bidirectional passive diffusion of ions and small molecules. In contrast, nuclear proteins, RNAs, and ribonucleoprotein (RNP) particles larger than ~9 nm in diameter (and greater than ~40–60 kD) selectively and actively enter and exit the nucleus by a signal-mediated and energy-dependent mechanism (Fig. 11.29). Proteins are targeted to the nucleus by a specific amino acid sequence called a nuclear localization sequence (NLS). In some cases, a nuclear protein without a NLS dimerizes with an NLS-bearing protein and rides "piggyback" into the nucleus. In addition, some nuclear proteins shuttle repeatedly between the nucleus and cytoplasm. Their exit from the nucleus requires a nuclear export sequence (NES). Nuclear import and export pathways are mediated by a family of soluble receptors referred to as importins or exportins, and collectively called karyopherins (Table 11.3). The presence of several different NLSs and NESs and multiple karyopherins suggests the existence of multiple pathways for nuclear localization.

Karyopherins

Karyopherins are proteins composed of helical molecular motifs called HEAT repeats (the acronym is derived from the four name-giving proteins: *H*untington, *E*longation factor 3, the "*A*" subunit of protein phosphatase 2, and *T*OR1 kinase) or Armadillo repeats (so named for their discovery in the *Drosophila* Armadillo protein). The repeats are stacked on top of each other to form highly flexible superhelical or "snail-like" structures (Fig. 11.29). The largest class of soluble receptors is the karyopherin-β family, which is involved in the transport of proteins and RNP cargoes (Table 11.3). A second group of receptors, which is structurally unrelated to the karyopherin-β family, is the family of nuclear export factors (NXFs) that are involved in the export of many mRNAs. A third class is represented by the small nuclear transport factor 2 (NTF2), which imports the small GTPase Ran into the nucleus.

Of the more than 20 members of the karyopherin-β family in vertebrates, 10 of these play a role in nuclear import, whereas seven function in nuclear export (Table 11.3). Importin-β1 is one of the predominant karyopherins that drives import. Although a small number of cargo proteins may bind importin-β1 directly, most cargoes require the adaptor protein importin-α. Seven importin-α adapters have

Table 11.3 Nuclear import and export factors of the karyopherin-β family.

Name	Cargo
Import receptors	
Importin-β1	Many cargoes with basic NLSs via an importin-α adapter, snRNPs via snurportin
Karyopherin-β2	mRNA-binding proteins, histones, ribosomal proteins
Transportin SR	mRNA-binding (phospho-SR domain) proteins
Transportin SR2	HuR
Importin 4	Histones, ribosomal proteins
Importin 5	Histones, ribosomal proteins
Importin 7	Glucocorticoid receptor, ribosomal proteins
Importin 8	SRP19
Importin 9	Histones ribosomal proteins
Importin 11	UbcM2, ribosomal protein L12
Export receptors	
Crm1	Leucine-rich NES cargoes
Exportin-t	tRNA
CAS	Karyopherin α
Exportin 4	eIF-5A
Exportin 5	microRNA precursors
Exportin 6	Profilin, actin
Exportin 7	p50Rho-GAP, 14-3-3δ
Import/export	
Importin 13	Rbm8, Ubc9, Pax6 (import) eIF-1A (export)

NES, nuclear export sequence; NLS, nuclear localization sequence.

been characterized in mammals (importin α1–α7). Although interchangeable for many cargoes *in vitro*, there are reports of preferential use of specific importin-α adapters *in vivo*. These receptors bind to signals in their cargo and to a subset of nucleoporins containing repeats of the amino acids phenylalanine and glycine (designated as FG repeats). Binding to FG repeats mediates passage through the NPCs. There is some redundancy between certain transport pathways. For example, *in vitro* import assays in mammalian cells have shown that five different importins can mediate nuclear entry of histones, and at least four importin-β–like factors are able to transport ribosomal proteins into the nucleus for assembly into ribosomes.

Nuclear localization sequences (NLSs)

Unlike signal sequences targeting proteins to the endoplasmic reticulum or mitochondrion, which are generally removed from proteins during transit, nuclear proteins retain their NLS. This may be to ensure

Over the past few years there has been much progress in knowledge of the structure and molecular architecture of the nuclear pore complexes (NPCs). NPCs are highly dynamic, modular machines embedded in the nuclear membrane or "nuclear envelope" (Fig. 1). The nuclear envelope that surrounds the nucleus is a double lipid bilayer. There is an outer membrane, a lumen (perinuclear space of 20–40 nm), and an inner membrane. Electron microscopy thin sections show that the outer membrane is continuous with the membrane of the endoplasmic reticulum, and its outer surface is studded with ribosomes like the cytoplasmic face of the rough endoplasmic reticulum. The NPCs serve as a gateway for the exchange of material between the nucleus and cytoplasm. The nucleus of a human cell has several thousand NPCs connected by the nuclear lamina. The whole NPC structure is ~50 MDa in size in yeast. In mammals, estimates range from 60 to 125 MDa. The overall architecture appears to be well conserved between species despite differences in mass.

The most striking feature of the NPCs is their eight-fold radial symmetry. They are composed of eight globular subunits that form a central spoke–ring complex, including a cylindrical structure which surrounds the central translocation channel (Fig. 1). Ring-like structures flank the spoke–ring complex on both its cytoplasmic and nuclear side. Eight fibrils are attached to each of these rings. On the cytoplasmic face of the NPC the fibrils have free ends that extend into the cytoplasm. On the nucleoplasmic face the fibrils form a basket-like structure that ends in a terminal ring. The NPCs are composed of multiple copies of a set of proteins that are collectively referred to as nucleoporins. Proteomic analysis of NPCs from both yeast and mammals

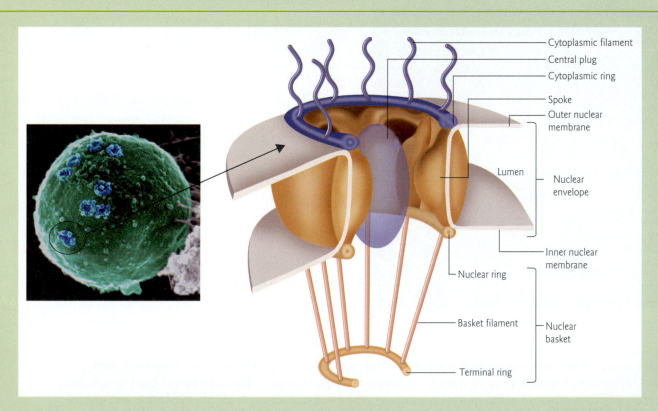

Figure 1 The nuclear pore complex (NPC). Field emission scanning electron microscope image of the yeast nucleus. Blue pseudocoloring highlights the NPCs; green pseudocoloring highlights the nuclear envelope together with attached ribosomes. Scale bar, 100 nm. (Reprinted by permission of Nature Publishing Group and Macmillan Publishers Ltd: Kiseleva, E. 2004. Cell of the month. *Nature Cell Biology* 6:497. Copyright © 2004. Photograph kindly provided by Elena Kiseleva, Institute of Cytology and Genetics, Novosibirsk-90, Russia). (Inset) A cut-away model of the NPC showing the key structural features. The NPC consists of eight spokes that form a cylinder embedded in the nuclear envelope. Several ring-like structures connect the spokes. Additional filamentous structures point towards the lumen of the nucleus and cytoplasm.

The nuclear pore complex

has shown that there are about 30 different nucleoporins. Because of the eight-fold symmetry of the NPC, each nucleoporin is present at a copy number of either eight or an integer multiple of eight. Individual nucleoporins have unique roles in regulating NPC function and nuclear import and export. Immunogold electron microscopy has shown that most of the nucleoporins are located symmetrically on both faces of the NPC, whereas only a few are located asymmetrically on either the nuclear face or the cytoplasmic face.

The central channel in the NPC connecting the cytoplasm to the nucleus has a length of about 90 nm. The nuclear

basket and the central channel appear to dilate in response to the translocation of large cargoes, increasing the resting 9–11 nm channel to an effective diameter of ~45–50 nm. When viewed in projection, the central pore often appears to be obstructed by a particle that varies greatly in size and shape and is commonly referred to as the "central plug" or "transporter" (Fig. 1). Three-dimensional reconstruction of native NPC and several studies using atomic force microscopy suggest that the central plug represents a view of the distal ring of the nuclear basket and/or cargo translocating through the central pore of the NPC.

that they can reaccumulate in the nucleus after each mitotic cell division. The best studied NLSs are basic amino acid sequences, typically rich in lysine and arginine. Despite the impressive number of receptor–cargo interactions that have been studied, the prediction of NLSs in candidate proteins remains extremely difficult. There is no real consensus sequence for an NLS. Some proteins, such as nucleoplasmin, have a bipartite NLS (Table 11.4, Focus box 11.7). NLSs interact with members of the karyopherin β family either directly or through the adapter importin-α.

Nuclear export sequences (NESs)

Of the nuclear export signals found in eukaryotes, the best characterized are the small hydrophobic leucine-rich NESs, first described in the human immunodeficiency virus 1 (HIV-1) Rev protein (Table 11.4). These classic Rev-type NESs function via interaction with the export factor CRM1. CRM1 was named for a mutant of the fission yeast *Schizosaccharomyces pombe* that was shown to be essential for proliferation and for "*c*hromosome *r*egion *m*aintenance." Although nuclear transport of RNP complexes by members of the

Table 11.4 Nuclear localization sequences and nuclear export sequences.

Signal type	Protein	Amino acid sequence*
Nuclear localization sequence (NLS)	SV-40 T antigen (monopartite NLS)	P**KKKRK**V
	Nucleoplasmin (bipartite NLS)	**KR**PAATKKAGQA**KKKKLD**
Nuclear export sequence (NES)	HIV-1 Rev	L-PPL-ERLTL
	IκBα	MVKEL-QEIRL
	TFIIIA	L-PVL-ENLTL
	Consensus	$\phi x_{2-3}\phi x_{2-3}\phi x\phi$ (ϕ = conserved hydrophobic residues L, I, V, F, or M; x is any amino acid)

* See Fig. 5.3 for standard one-letter amino acid designations.

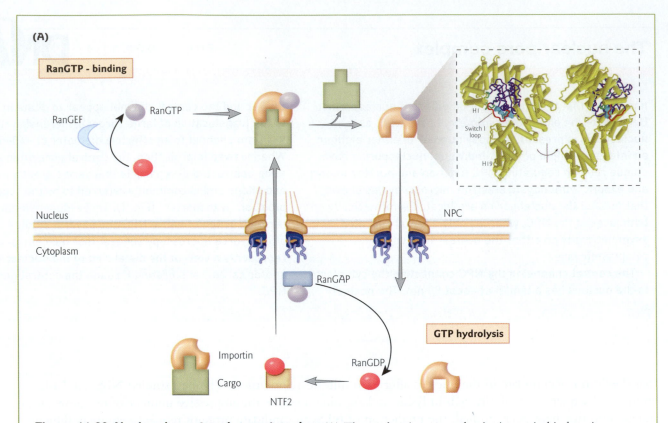

Figure 11.29 Nuclear import and export cycles. (A) The nuclear import cycle. An importin binds to its NLS-bearing cargo in the cytoplasm and translocates through the nuclear pore complex (NPC) into the nucleus. In the nucleus the importin binds RanGTP, resulting in cargo release. The importin–RanGTP complex recycles back to the cytoplasm. After translocation to the cytoplasm, GTP hydrolysis on Ran by RanGAP dissociates the importin–RanGTP complex. A gradient of RanGTP exists in the cell, with RanGDP at a high concentration in the cytoplasm and RanGTP at a high concentration in the nucleus. To maintain the high nuclear concentration of Ran, a dedicated transporter protein, NTF2, functions to recycle Ran continuously back to the nucleus. RanGTP is generated in the nucleus by chromatin-bound RanGEF. (Inset) Two views of the importin-β-RanGTP complex, rotated by 90° relative to one another. Importin-β is shown in yellow and RanGTP is shown in dark blue, with the switch I loop in red and the switch II loop in green. The bound GDP is shown in space filling format. The switch I and II loops of Ran change conformation with nucleotide state and regulate its interactions with importin-β (Reprinted by permission from Nature Publishing Group and Macmillian Publishers Ltd: Lee, S.J., Matsuura, Y., Liu, S.M., Stewart, M. 2003. Structural basis for nuclear import complex dissociation by RanGTP. *Nature* 435:693–696. Copyright © 2003.)

karyopherin-β family is often specified by the signals present on their protein components, two exportins, exportin-t and exportin 5, interact with tRNA directly (see Table 11.3).

Nuclear import pathway

Factors involved in nuclear import have been well studied biochemically. Recent advances into regulatory mechanisms have been made by combining computer simulation and real-time assays to test the predictions in intact cells. The process of nuclear import involves three main steps: (i) cargo recognition and docking; (ii) translocation through the nuclear pore complex (NPC); and (iii) cargo release and receptor recycling (see Fig. 11.29A).

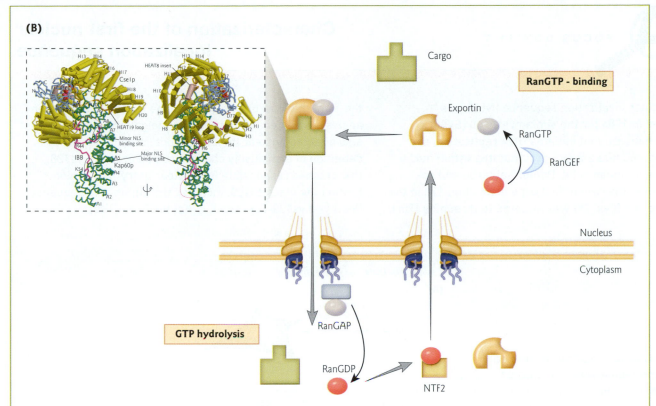

Figure 11.29 (cont'd) (B) Nuclear export cycle. Exportins bind to their cargo in the nucleus in the presence of RanGTP. In the cytoplasm, GTP hydrolysis causes disassembly of the export complex and recycling of the export receptor. (Inset) Two orthogonal views showing the structure of an export complex containing Cse1p (yellow), Kap60p (importin-α), and RanGTP (blue). Cse1p envelopes RanGTP and the C-terminal region of importin-α (green) and its IBB (importin-β-binding) domain. The "spring-loaded" complex disassembles spontaneously upon RanGTP hydrolysis. The Cse1p α-helices are represented by cylinders. The ARM repeats of Kap60p and HEAT repeats of Cse1p are labeled A1–A10 and H1–H20, respectively. GTP is shown as space-filling spheres. (Reprinted by permission from Nature Publishing Group and Macmillian Publishers Ltd: Matsuura, Y., Stewart, M. 2004. Structural basis for the assembly of a nuclear export complex. *Nature* 432:872–877. Copyright © 2004.)

Cargo recognition and docking

The import receptor for the classic, lysine/arginine-rich NLS is a complex of an importin-α adapter and importin-β1. Members of the importin-α family contain a C-terminal region that binds directly to cargo proteins containing an NLS. Importin-β1 binds to both importin-α and to nuclear pore complex (NPC) proteins. Binding occurs at the cytoplasmic filaments of the NPC. Cargo recognition and docking does not require energy from either ATP or GTP hydrolysis.

Translocation through the nuclear pore complex

The exact molecular mechanism by which cargo translocation through the NPC occurs is poorly understood. However, weak hydrophobic interactions between importins and the FG repeat domains of nucleoporins seem to be essential. For many years it was assumed that translocation through the NPC would occur by some type of ATP-driven motor. However, when it became evident through experimentation that neither ATP nor GTP hydrolysis were required for translocation, different models had to be developed. An older "affinity gradient" model suggested that karyopherins travel along a gradient of nucleoporin-binding

FOCUS BOX 11.7

Characterization of the first nuclear localization sequence

The first nuclear localization sequence (NLS) was characterized in 1984 for the simian virus 40 (SV40) T antigen (the helicase that initiates DNA replication, see Fig. 6.7). The > 60 kDa antigen accumulates within nuclei of infected mammalian cells. Daniel Kalderon and co-workers showed that when the lysine at position 128 in the polypeptide chain (Lys128) was mutated to threonine (Thr), the T antigen was unable to enter the nucleus. Subsequent analysis revealed that mutation of neighboring amino acids also affected nuclear import. This observation led to the definition of a positively charged region around Lys128 that seemed responsible for nuclear entry. The region, termed the classic NLS, included the amino acid sequence: Pro-Lys-Lys128-Lys-Arg-Lys-Val (or PKKKRKV).

Figure 1 Characterization of nuclear localization sequences and import receptors. (A) When bovine serum albumin (BSA) tagged with fluorescein (green) was microinjected into the cytoplasm of cells, as expected the serum protein did not enter the nucleus. Attaching the short amino acid sequence "PKKKRKV" from the SV40 T antigen to the fluorescein-tagged BSA allowed nuclear entry. When threonine was substituted for one of the lysines (PTKKRKV), BSA was unable to enter the nucleus. These experiments demonstrated that the SV40 T antigen nuclear localization sequence (NLS) was necessary and sufficient for nuclear import. (Source: Kalderon D., Richardson W.D., Markham, A.F., Smith, A.E. 1984. Sequence requirements for nuclear location of simian virus 40 large-T antigen. *Nature* 311:33–38.) (B) Biochemical fractionation of cytosol led to the discovery of the various soluble receptors that bind specifically to NLS-containing. Permeabilized cells were incubated with fluorescein-tagged BSA-NLS and cytosol fractions, and then assayed for nuclear entry of the protein by fluorescence microscopy.

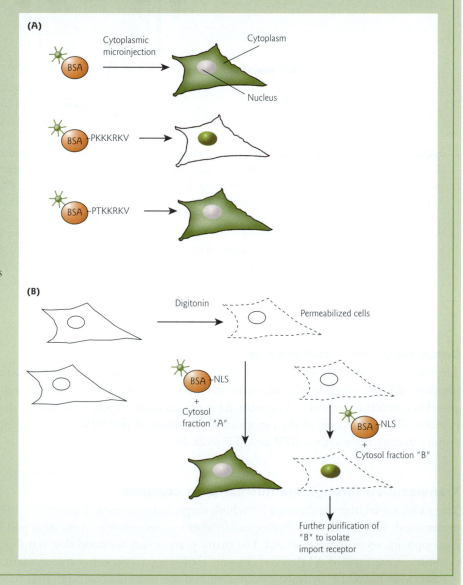

Characterization of the first nuclear localization sequence

FOCUS BOX 11.7

Decisive evidence that the classic NLS sequence was necessary and sufficient for nuclear entry was provided as follows. When bovine serum albumin (BSA) tagged with fluorescein was microinjected into the cytoplasm, as expected the serum protein did not enter the nucleus (Fig. 1). However, attaching the short polypeptide sequence "PKKKRKV" to the fluorescein-tagged BSA (fluorescein-BSA:NLS) allowed nuclear entry. Injecting more of the fluorescein-BSA:NLS soon saturated the pathway. These early experiments predicted that NLSs within nuclear proteins bind to saturable receptors at the NPCs. Thus, molecular biologists began to search for such receptors.

An *in vitro* assay was developed, in which the plasma membrane of mammalian cells is permeabilized by digitonin, a weak detergent that leaves the nuclear membrane intact. Cells were then incubated with fluorescein-BSA:NLS. Nuclear entry of fluorescein-BSA:NLS was monitored by fluorescence microscopy. Entry in buffer alone was inefficient but could be greatly enhanced by adding "cytosol" (contents of cytoplasm, excluding the various membrane-bound organelles) from another cellular source. Biochemical fractionation of the cytosol led to the discovery of the various soluble receptors that bind specifically to NLS-containing proteins.

sites, encountering nucleoporins of increasing affinity during their translocation. However, the cytoplasmic filaments of the NPC can be removed with no obvious effect on transport. Also, experiments have shown that the direction of transport can be inverted in the absence of the RanGTP gradient, and deletion of all asymmetric FG repeats has only modest effects on the nuclear import pathway. These findings suggest that the NPC has no intrinsic directionality. The two most widely supported models are the Brownian affinity or "virtual gating" model and the selective phase or "sieve" model (Fig. 11.30). Both of these models (and other variations on the general theme) agree that transport through the central channel occurs via diffusion, but differ in the basis for the permeability barrier in the NPC and the structure of the central channel.

Nucleocytoplasmic transport through the NPC occurs against a concentration gradient, and because this cycle is bidirectional, an energy source and a directional cue are needed. Both are provided by the small GTPase Ran (see Fig. 11.29). The 24 kDa protein Ran (for *Ra*s-related *n*uclear) belongs to a superfamily of GTP-binding proteins that act as molecular switches cycling between GDP- and GTP-bound states. The conversion from the GDP- to GTP-bound state involves nucleotide exchange. In contrast, the conversion of RanGTP to RanGDP occurs by removal of the terminal phosphate from the bound GTP. RanGTP is in a high concentration within the nucleus and a low concentration in the cytoplasm. The concentration difference between free nuclear and cytoplasmic RanGTP is estimated to be at least 200-fold. This gradient is created by an asymmetric distribution of the Ran guanosine-nucleotide exchange factor (RanGEF; also known as RCC1) and the Ran-specific GTPase-activating protein (RanGAP). RanGEF is a resident nuclear protein that binds to nucleosomes through an interaction with histones H2A and H2B and promotes nucleotide exchange, replacing the GDP bound to Ran with GTP. RanGAP, on the other hand, is excluded from the nucleus and acts to maintain Ran in the GDP-bound state in the cytoplasm. RanGAP is found at its highest concentration at the outer face of the NPC where it associates with the nucleoporin RanBP2 via its SUMO modification. The actual process of NPC translocation itself does not require RanGTP hydrolysis. None of the ~30 proteins in the NPC proteome encodes motor or ATPase domains, providing further evidence that the complex does not actively pump cargoes across the nuclear membrane.

Cargo release and receptor recycling

Once the cargo–import receptor complex reaches the nuclear side of the NPC, RanGTP binds to the importin and dislodges it from the cargo (see Fig. 11.29A). Cargo dissociation occurs by an allosteric

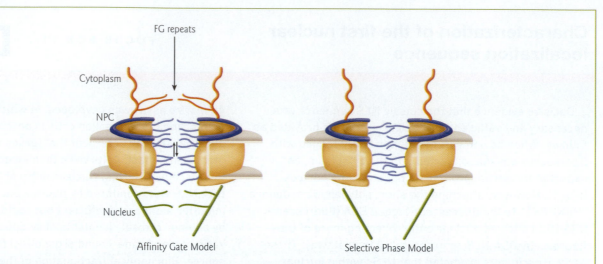

Figure 11.30 Schematic illustrations of the nuclear pore complex (NPC) translocation models. Two models are depicted for the selectivity of translocation through the NPC. (Left) In the affinity gate model the FG repeat-containing nucleoporins form multiple waving filaments at both ends of the NPC. The FG repeats function as a selective barrier to diffusion of non-nuclear proteins. The affinity of importins or exportins for these flexible filaments, moved by Brownian motion, increases the residence time of karyopherin–cargo complexes at the nuclear periphery and the probability of their access to the NPC central channel. Proteins gain access to the channel and diffuse through the NPC by "hopping" of the karyopherin–cargo complex from FG repeat to FG repeat. High-affinity nucleoporin-binding sites present in the destination compartment support the directionality of transport. (Right) As in the affinity gate model, the selective phase model assumes that the karyopherin–cargo complex moves randomly by diffusion through the central channel. The hydrophobic FG repeats are proposed to form a selective semiliquid phase or meshwork into which the karyopherins partition. These interactions enable exclusion of non-nuclear proteins that cannot dissolve into the sieve. The hydrophobic karyopherins in association with cargo enter (dissolve into) the sieve structure and cross the NPC, via their transient and low-affinity interactions with FG repeats.

mechanism in which binding of the importin to RanGTP results in a conformational change that is transmitted to its cargo-binding domain. At this point in the pathway, the cargo is free to carry out its nuclear function. The RanGTP–importin complex subsequently translocates back to the cytoplasm. For the importin-α/importin-β1–mediated pathway, the exportin CAS (also known as Cse1p) binds to importin-α and mediates its nuclear export in association with RanGTP.

The receptor recycling step is the only energy-requiring step in nuclear import – at least two GTP molecules are consumed per NLS import cycle. GTP hydrolysis occurs with the aid of the accessory proteins RanGAP and RanBP1 converting RanGTP to RanGDP. After GTP hydrolysis, the export complexes dissociate and the importins are recycled for another round of import (see Fig. 11.29A). RanGDP is rapidly imported into the nucleus by transport factor NTF2, where it is converted to RanGTP with the aid of RanGEF. Several million molecules of Ran have to be imported every minute into the nucleus of an actively dividing mammalian cell to keep up with the demands of nuclear import and export cycles.

Nuclear export pathway

As described above, RanGTP has very different functions in import and export. In the nuclear import pathway it causes disassembly of import complexes. But, in nuclear export it is required for the assembly of a cargo–exportin complex. The cargo may be importin-α, or a nuclear protein that shuttles between the nucleus and cytoplasm. Upon binding, RanGTP induces a conformational change in the export receptor that results in energy storage. The exportin is twisted from an S shape into a horseshoe–like conformation in which both arches of the "horseshoe" interact with RanGTP. When the export complex translocates through the central channel of the NPC and arrives on the outer (cytoplasmic) face it is exposed to proteins that cause GTP hydrolysis (see above). This releases the energy stored in the "spring–loaded" exportin which "pops" open and releases its cargo (see Fig. 11.29B).

11.10 Regulated nuclear import and signal transduction pathways

Any protein with both an NLS and NES has the potential to shuttle back and forth between the nucleus and cytoplasm. If such a protein is a transcription factor, this has clear implications for post-translational regulation. For example, a transcription factor may be sequestered in the cytoplasm at a particular developmental stage or in unstimulated cells, and may remain cytoplasmic until an extracellular signal induces its nuclear import. Spatial separation of the transcription factor from its DNA target (by exclusion from the nucleus) acts as a potent inhibitor of function. Regulation of NLS and NES activity can occur by several mechanisms, including post-translational modifications (e.g. phosphorylation and dephosphorylation) that mask (or unmask) the NLS or NES. Where a protein is localized at "steady state" depends on the balance between import, retention, and export, and which signal is dominant. Nuclear retention may be mediated by domains of the protein that interact with components of the nucleus, such as chromatin or the nuclear matrix.

There are various ways that signals are detected by a cell and communicated to a gene. The effect may be direct, where a small molecule, such as a sugar or steroid hormone, enters the cell and binds the transcriptional regulator directly. Or the effect of the signal may be indirect where the signal induces a kinase that phosphorylates the transcriptional regulator or an associated inhibitory protein. This type of indirect signaling is an example of a signal transduction pathway. The signal-mediated nuclear import of transcription factors NF-κB and the glucocorticoid receptor are used to illustrate the wide variety of mechanisms for controlling gene activity.

Regulated nuclear import of NF-κB

NF-κB (nuclear factor of kappa light polypeptide gene enhancer in B cells) is a dimeric transcription factor that is a central mediator of the human stress response. It plays a key role in regulating cell division, apoptosis, and immune and inflammatory responses. The discovery of NF-κB attracted widespread interest because of the variety of extracellular stimuli that activate it, the diverse genes and biological responses that it controls, and the striking evolutionary conservation of structure and function among family members. NF-κB was first identified in 1986 as a transcription factor in the nuclei of mature B lymphocytes that binds to a 10 bp DNA element in the kappa (κ) immunoglobulin light-chain enhancer. Since then a family of various distinct subunits has been identified. The events leading to signal-mediated nuclear import of NF-κB involve three main stages: (i) cytoplasmic retention by I-κB; (ii) a signal transduction pathway that induces phosphorylation and degradation of I-κB; and (iii) I-κB degradation resulting in exposure of the NLS on NF-κB, allowing nuclear import of NF-κB (Fig. 11.31).

Cytoplasmic retention by I-κB

In a resting B lymphocyte, NF-κB subunits form homodimers or heterodimers in the cytoplasm. The dimers are composed of a DNA-binding subunit (e.g. p50) and a transcription-activating subunit (e.g. p65); p65 also has a DNA-binding domain and binds DNA as a homodimer. In cells that have not received an external cue, the dimers are retained in the cytoplasm in an inactive form. Typically, they are held by an anchor protein called I-κB (inhibitor of κB) (Fig. 11.31). The I-κB family includes at least eight structurally related proteins, of which I-κBα is the most abundant inhibitor protein. I-κB characteristically contains a

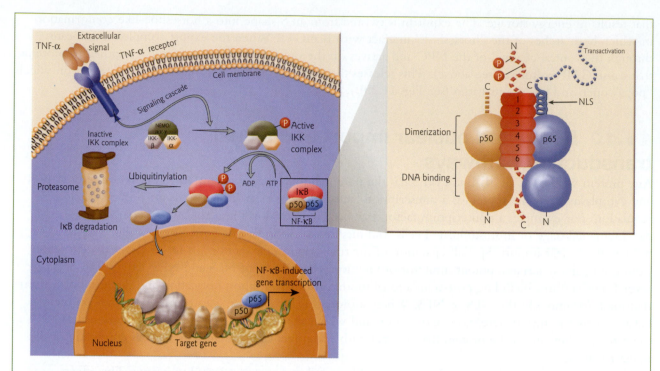

Figure 11.31 Signal-mediated nuclear import of NF-κB. The pathway of NF-κB activation occurs by the following steps: TNF-α receptor-mediated transduction of an extracellular signal, activation of the I-κB kinase (IKK) complex, phosphorylation of I-κB, ubiquitinylation of I-κB and subsequent degradation by proteosome activity, release of NF-κB from I-κB, nuclear import of NF-κB, and activation of target genes. Phosphorylation is indicated by a red circled "P." (Inset) Schematic representation of the I-κB–NF-κB complex showing the functional domains of the p50 and p65 subunits of NF-κB. Ankyrin repeats 1 and 2 (red cylinders) of I-κB mask the p65 nuclear localization sequence (NLS).

stretch of 5–7 ankyrin repeat domains that mask the NLS within NF-κB. These repeats (named for their discovery in the cytoskeletal protein ankyrin) consist of a pair of antiparallel α-helices stacked side by side, which are connected by a series of intervening β-hairpin turn motifs.

Signal transduction pathways induce phosphorylation and degradation of I-κB

Upon receipt of an extracellular signal, a signal transduction pathway is triggered that ultimately leads to transient activation of the serine-specific I-κB kinase (IKK) complex (Fig. 11.31). There are over 150 different possible signals or initiating ligands, including interleukins, bacterial lipopolysaccharides, lymphotoxin-β, tumor necrosis factor α (TNF-α), viruses, various stress stimuli, and chemotherapeutic agents. Whatever the signal, it is detected by a specific cell surface receptor. The ligand binds to an extracellular domain of the receptor and this binding is communicated to the intracellular domain to initiate a signaling cascade (or second messenger system). There are many complex steps involved in this cascade depending on the particular signal and receptor. For example, the TNF-α signal is relayed through a cascade of kinases ending with phosphorylation and activation of IKK. IKK is composed of two catalytic subunits, IKKα and IKKβ, and a structural/regulatory subunit IKKγ (also called NEMO, not after the clownfish in the Disney movie, but for "NF-κB essential modulator"). The IKKβ subunit phosphorylates I-κB at two conserved serines (position 32 and 36) in the N-terminus. This post-translational modification leads to release of I-κB from NF-κB.

I-κB degradation results in exposure of the NLS on NF-κB

Originally, it was postulated that phosphorylation simply caused a conformational change leading to release of I-κB from NF-κB. In fact, phosphorylation of I-κB also triggers another pathway, that of ubiquitinylation and proteasome-mediated degradation (Fig. 11.31; see also Fig. 5.17). Upon release from I-κB, the NLS of NF-κB is exposed. Once the NLS is unmasked, NF-κB can interact with importin-α/β1 and translocate through the NPC. In the nucleus, NF-κB activates target genes by binding to specific DNA regulatory elements. There is an impressively broad range of over 150 target genes with NF-κB-binding sites, including the enhancer of the interferon-β gene (see Fig. 11.24) and the kappa (κ) chain gene involved in the immune response.

Regulated nuclear import of the glucocorticoid receptor

In the preceding example of NF-κB, gene expression was induced in response to a signal received by a cell surface receptor, and transmitted by a signal transduction cascade. In contrast, the glucocorticoid receptor mediates a highly abbreviated signal transduction pathway: the receptor for the extracellular signal is cytoplasmic and carries the signal directly into the nucleus (Fig. 11.32).

Steroid hormone receptors, such as the glucocorticoid receptor, activate gene expression in response to hormones. Steroid hormones (e.g. cortisol) do not require a cell surface receptor. Because of their lipophilic nature, they can pass through the cell membrane by diffusion. Once in the cytoplasm, they bind the cytoplasmic receptor. In the absence of hormone, the glucocorticoid receptor is bound to heat shock protein 90 (Hsp90) and a 59 kD protein in a complex in the cytoplasm. Hsp90 was first identified as a protein synthesized in response to stress, hence its name. Hsp90 binds to the glucocorticoid receptor via a C-terminal region of the receptor that also binds steroid hormone and masks the region of the glucocorticoid receptor required for dimerization or DNA binding. In the classical model, activation of the glucocorticoid receptor involves ligand-induced conformational changes that result in rapid dissociation of the inhibitory Hsp90 protein. Subsequently, two glucocorticoid receptors join together to form a homodimer. The NLS interacts with importins and the receptor is translocated through the nuclear pore complex into the nucleus. However, some recent studies suggest that continued association with Hsp90 is critical for efficient nuclear import. Thus GR interactions with Hsp90 after ligand binding requires further investigation. Once in the nucleus, the glucocorticoid receptor dimer binds DNA at a glucocorticoid responsive element (GRE) (see Fig. 11.16). Activation of hormone-responsive target genes leads to many diverse cellular responses, ranging from increases in blood sugar to anti-inflammatory action.

Chapter summary

The synthesis of tens of thousands of different eukaryotic mRNAs is carried out by RNA polymerase II (RNA pol II). During the process of transcription, RNA pol II associates transiently not only with the template DNA but with many different proteins, including general transcription factors. Transcription factors are sequence-specific DNA-binding proteins that bind to gene promoters and other regulatory elements, interpret the information present in these regulatory elements, and transmit the appropriate response to the RNA pol II transcriptional machinery. The gene promoter consists of core promoter elements and proximal promoter elements that are required for the initiation of transcription or that increase the frequency of initiation only when positioned near the transcriptional start site. The initiation step alone involves the assembly of dozens of factors to form a preinitiation complex. The general transcription factor TFIID is responsible for the recognition of most of the known core promoter elements, the best characterized of which is the TATA box. Long-range regulatory DNA elements act over distances of 100 kb or more from the gene promoter. Enhancers increase gene promoter activity downstream, upstream, or in either orientation relative to the promoter. Similar elements that repress gene activity are called silencers. Insulators act as barriers between regions of heterochromatin and euchromatin and block enhancer or silencer activity of neighboring genes. Locus control regions (LCRs) maintain functional, independent domains of active

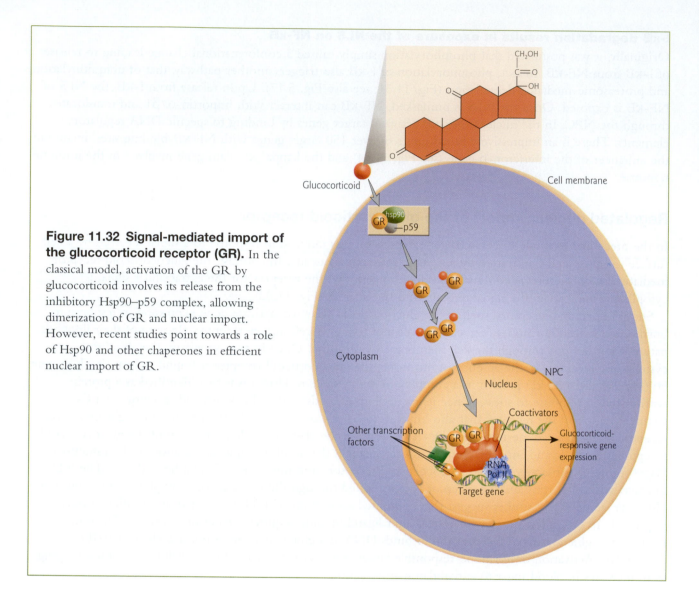

Figure 11.32 Signal-mediated import of the glucocorticoid receptor (GR). In the classical model, activation of the GR by glucocorticoid involves its release from the inhibitory Hsp90–p59 complex, allowing dimerization of GR and nuclear import. However, recent studies point towards a role of Hsp90 and other chaperones in efficient nuclear import of GR.

chromatin. They are often characterized by DNase I hypersensitive sites which are indicative of an "open" chromatin conformation. The LCR of the β-globin gene cluster is required for high-level transcription of all genes in the cluster, but chromatin loop formation controls their developmental expression. Matrix attachment regions (MARs) organize chromatin into independent loop domains, and localize transcription factors and actively transcribed genes to the nuclear matrix.

Transcription is mediated by the collective action of sequence-specific DNA-binding transcription factors along with the general RNA pol II transcriptional machinery, an assortment of coregulators that bridge the DNA-binding factors to the transcriptional machinery, a number of chromatin-remodeling factors that mobilize nucleosomes, and a variety of enzymes that catalyze covalent modification of histones and other proteins.

RNA pol II is a 12-subunit polymerase, capable of synthesizing RNA and proofreading nascent transcript. A set of five general transcription factors, denoted TFIIB, TFIID, TFIIE, TFIIF, and TFIIH, is responsible for promoter recognition and for unwinding the promoter DNA. RNA pol II is absolutely dependent on these auxiliary transcription factors for the initiation of transcription. Assembling the transcription preinitiation complex involves a series of highly ordered steps. Binding of TFIID (TBP and associated TAFs) to the TATA box provides a platform to recruit other general transcription factors and RNA pol II to the gene promoter. *In vitro*, these proteins assemble at the promoter in the following order: TFIIB (orients the complex on the promoter), TFIIF together with RNA pol II, and then TFIIE and TFIIH. TFIIH has both kinase and helicase activity. Phosphorylation of the C-terminal domain (CTD) of RNA pol II is required for transcript elongation. For recycling of RNA pol II and reinitiation of transcription, the CTD must again be dephosphorylated. Mediator is a 20-subunit complex, which acts as a molecular bridge to transmit regulatory information from activator and repressor proteins to RNA pol II.

Transcription factors mediate gene-specific transcriptional activation or repression. They are modular proteins consisting of a number of domains, including a DNA-binding domain, a transactivation domain that activates transcription via protein–protein interactions, and a dimerization domain. The majority of transcription factors bind DNA as homodimers or heterodimers. Some of the more common DNA-binding domain motifs are the helix–turn–helix, zinc finger, basic leucine zipper, and basic helix–loop–helix motifs. Despite differences in overall three-dimensional structure, all these common motifs involve interaction between an α-helical domain of the protein and about five base pairs within the major groove of the DNA double helix. The homeodomain is a variant of the helix-turn-helix motif found in many transcription factors that are important in development. The zinc finger is formed by interspersed cysteines and/or histidines that covalently bind a central zinc ion, folding a short length of the amino acid chain into a compact loop domain, one side of which is an α-helix. The leucine zipper and helix–loop–helix motifs do not directly contact the DNA, instead they play an indirect structural role in DNA binding by facilitating dimerization of two similar transcription factors with basic α-helical DNA-binding domains.

Transcriptional coactivators and corepressors are proteins that increase or decrease transcriptional activity through protein–protein interactions without binding DNA directly. There are two main classes: (i) chromatin modification complexes that modify histones post-translationally, in ways that allow greater access of other proteins to DNA; and (ii) chromatin remodeling complexes, such as the yeast SWI/SNF family and related families (ISWI and SWR1) that contain ATP-dependent DNA unwinding activities. Chromatin modification complexes include histone acetyltransferases that direct acetylation of lysine residues in histones, and histone methyltransferases that direct methylation of lysines and arginines. These processes are dynamic and are subject to regulation by histone deacetylases and demethylases. Histone modifications provide binding sites for proteins that can change the chromatin state to either active or repressed. The histone code hypothesis proposes that covalent post-translational modifications of histone tails are read by the cell and lead to a complex, combinatorial transcriptional output. Other important histone modifications include ubiquitinylation, phosphorylation, ADP-ribosylation, and sumoylation. In some cases, the selection of linker histone variants is important for gene regulation. Chromatin remodeling complexes can mediate at least four different changes in chromatin structure: (i) nucleosome sliding in which the position of a nucleosome on the DNA changes; (ii) remodeled nuclesomes in which the DNA becomes more accessible but the histones remain bound; (iii) complete dissociation of the DNA and histones; or (iv) the replacement of a core histone with a variant histone. Each family of ATP-dependent chromatin remodeling complexes is defined by a unique subunit composition, the presence of a distinct ATPase, and their precise mechanisms for remodeling chromatin.

The particular order of recruitment of the preinitiation complex, the transcription factors, and the chromatin remodeling and modification complexes is gene-specific. The enhanceosome model proposes that interactions among transcription factors promote their cooperative, step-wise assembly on DNA and give the complex exceptional stability. The hit and run model proposes that transcriptional activation reflects the probability that all components required for activation will meet at a certain chromatin site; i.e. transcription complexes are assembled in a stochastic fashion from freely diffusible proteins, and their binding is highly transient and dynamic. Most likely, transcription complex formation is a combination of elements of both models. The principles of combinatorial interaction and complex stability apply even if the complex itself has a very limited lifetime.

After assembly of the preinitiation complex and the rest of the transcriptional machinery, a period of abortive initiation follows before the polymerase escapes the promoter region and enters the elongation phase. Promoter clearance requires phosphorylation of the CTD of RNA pol II at multiple sites within a series of heptapeptide repeats. Most of the general transcription factors are released. TFIID remains bound to the gene promoter to allow rapid formation of a new preinitiation complex. Once phosphorylated, RNA pol II can unwind DNA without the help of the helicase activity of TFIIH. The elongation phase is subdivided into multiple stages: the selection of a NTP complementary to the DNA template, the catalysis of phosphodiester bond formation, and the translocation of the RNA transcript and DNA template. The growing RNA remains at a single "tunable" active site in RNA pol II that switches between RNA synthesis and cleavage during proofreading of misincorporated nucleotides. The two reactions are mediated by differential coordination of one of the two metal ions held in the active site.

Movement of RNA pol II through the array of nucleosomes on the DNA template occurs by histone H2A-H2B dimer removal. RNA pol III uses a different mechanism which involves nucleosome mobilization – the transfer of the histone core octamer along the DNA without displacement. H2A-H2B dimer removal is mediated by a number of auxiliary elongation factors, including FACT, Elongator, and TFIIS.

Since protein synthesis occurs in the cytoplasm, transcriptional regulatory proteins must be delivered to their site of activity in the nucleus. Trafficking between the nucleus and the cytoplasm occurs via the nuclear pore complexes (NPCs). The NPC is a large multiprotein complex which consists of eight spokes that form a cylinder embedded in the nuclear envelope. The NPCs allow bidirectional passive diffusion of ions and small molecules. Nuclear proteins, RNAs, and RNPs selectively enter and exit the nucleus by a signal-mediated and energy-dependent mechanism. Proteins are targeted to the nucleus by a basic amino acid sequence (typically lysine and arginine rich) called a nuclear localization sequence (NLS). Nuclear proteins that shuttle between the nucleus and cytoplasm also have a nuclear export sequence (NES); the most common of these has a leucine-rich consensus sequence. Nuclear import and export pathways are mediated by a family of soluble receptors referred to as importins or exportins, and collectively called karyopherins. There are multiple pathways for nuclear localization mediated by several different classes of NLSs and NESs and multiple types of karyopherins. Bidirectional nucleocytoplasmic transport through the NPC occurs against a concentration gradient. The energy source and the directional cue are both provided by the small GTPase Ran.

In the nuclear import cycle, an importin binds to its NLS-bearing cargo in the cytoplasm and translocates through the NPC into the nucleus. The exact mechanism by which cargo translocation occurs is poorly understood. In the nucleus, the importin binds RanGTP, resulting in cargo release. The importin–RanGTP complex recycles back to the cytoplasm. After translocation to the cytoplasm, RanGAP mediates GTP hydrolysis on Ran, resulting in dissociation of the complex. The transporter NTF2 carries RanGDP into the nucleus. The conversion of RanGDP to RanGTP is mediated by chromatin-bound RanGEF. In the nuclear export cycle, exportins bind to their cargo in the nucleus in the presence of RanGTP. In the cytoplasm, GTP hydrolysis causes disassembly of the export complex and recycling of the export receptor.

Regulation of transcription factor nuclear import by signal transduction pathways is a powerful level of gene regulation. The events leading to the signal-mediated nuclear import of NF-κB, a transcription factor that is a central mediator of the human stress response, involve three main stages: (i) cytoplasmic retention by the inhibitor protein I-κB; (ii) a signal transduction pathway triggered by an extracellular signal (e.g. bacterial lipopolysaccharides or TNF-α) that induces phosphorylation and degradation of I-κB; and (iii) I-κB degradation resulting in exposure of the NLS on NF-κB, allowing nuclear import of NF-κB. Once in the nucleus, NF-κB activates target genes by binding to specific DNA regulatory elements.

The glucocorticoid receptor (GR) mediates a highly abbreviated signal transduction pathway leading to many diverse cellular responses, ranging from increases in blood sugar to anti-inflammatory action. In the absence of hormones, GR is bound in a complex with Hsp90 and p59. Hsp90 masks the regions of GR necessary for formation of homodimers and DNA binding. In the classical model, activation of GR by a glucocorticoid involves a ligand-induced conformational change that releases it from the Hsp90–p59 complex, allowing dimerization and nuclear import. Recent evidence suggests, however that Hsp90 may play a role in efficient nuclear entry. Once in the nucleus, the GR activates target genes by binding to specific DNA regulatory elements.

Analytical questions

1 You suspect that a sequence upstream of a transcriptional start site is acting as an enhancer and not as a promoter. Describe an experiment you would run to test your hypothesis. Predict the results.

2 A team of molecular biologists has made transgenic mice using a viral promoter and a cDNA coding for a human protein called p45. In most of the transgenic mice, they observe position-dependent expression of the human gene. They seek your advice. Suggest regulatory regions that you could link to the promoter–p45 DNA construct that might confer position-independent expression. Explain your choice of regulatory regions.

3 You have purified a transcription factor that has a leucine-rich region. You perform an electrophoretic mobility shift assay (EMSA) using a double-stranded oligonucleotide that you know from other studies contains the site recognized by this transcription factor *in vivo*. However, the transcription factor does not bind to the labeled oligonucleotide in your EMSA. Provide an explanation for this result.

4 You are studying a new class of eukaryotic promoters recognized by a novel RNA polymerase. You discover two general transcription factors that are required for transcription of these promoters. You suspect that one has helicase activity and that the other is required to recruit the helicase and the RNA polymerase to the promoter. Describe experiments you would perform to test your hypothesis. Provide sample results of your experiments.

5 The figure below shows an electrophoretic mobility shift assay (EMSA). The plus and minus symbols indicate which of the following components were included in each shift assay:

Naked DNA = ^{32}P-labeled DNA fragment containing a promoter element
Mononucleosome = same ^{32}P-labeled DNA wrapped around the histone core octamer
ZF = zinc finger protein
ΔZF = ZF with the zinc finger deleted
SWI/SNF = SWI/SNF complex
HAT = histone acetyltransferase
HDAC = histone deacetylase

Naked DNA	+	+	+	−	−	−	−	−	−
Mononucleosome	−	−	−	+	+	+	+	+	+
ZF	−	+	−	−	+	+	+	+	+
ΔZF	−	−	+	−	−	−	−	−	−
SWI/SNF	−	−	−	−	−	+	+	−	+
HAT	−	−	−	−	−	−	+	+	−
HDAC	−	−	−	−	−	−	−	−	+

Interpret the results for each lane.

6 You include an inhibitor of the protein kinase activity of TFIIH in an *in vitro* transcription assay. What step in transcription would you expect to see blocked? Describe an experiment you would run to test your hypothesis. Predict the results.

7 Assume that you are studying a transcription factor (TF). You have constructed a plasmid vector for the expression of a GFP-tagged TF. You discover that the TF is sometimes in the nucleus and sometimes in the cytoplasm. Describe a possible mechanism for its cytoplasmic retention and subsequent signal-mediated import. Design an experiment to test your hypothesis and provide sample results.

Suggestions for further reading

Adelman, K., Lis, J.T. (2002) How does pol II overcome the nucleosome barrier? *Molecular Cell* 9:451–452.

Agalioti, T., Lomvardas, S., Parekh, B., Yie, J., Maniatis, T., Thanos, D. (2000) Ordered recruitment of chromatin modifying and general transcription factors to the IFN-β promoter. *Cell* 103:667–678.

Agresti, A., Scaffidi, P., Riva, A., Caiolfa, V.R., Bianchi, M.E. (2005) GR and HMGB1 interact only within chromatin and influence each other's residence time. *Molecular Cell* 18:109–121.

Aravind, L., Anantharaman, V., Balaji, S., Babu, M.M., Iyer, L.M. (2005) The many faces of the helix-turn-helix domain: transcription regulation and beyond. *FEMS Microbiology Reviews* 29:231–262.

Armache, K.J., Kettenberger, H., Cramer, P. (2005) The dynamic machinery of mRNA elongation. *Current Opinion in Structural Biology* 15:197–203.

Baumli, S., Hoeppner, S., Cramer, P. (2005) A conserved Mediator hinge revealed in the structure of the MED7/MED21 (Med7/Srb7) heterodimer. *Journal of Biological Chemistry* 280:18171–18178.

Bell, A.C., West, A.G., Felsenfeld, G. (2001) Insulators and boundaries. Versatile regulatory elements in the eukaryotic genome. *Science* 291:447–450.

Belotserkovskaya, R., Oh, S., Bondarenko, V.A., Orphanides, G., Studitsky, V.M., Reinberg, D. (2003) FACT facilitates transcription-dependent nucleosome alteration. *Science* 301:1090–1093.

Bird, R.C., Stein, G.S., Lian, J.B., Stein, J.L. (1997) *Nuclear Structure and Gene Expression*. Academic Press, New York.

Bode, J., Goetze, S., Heng, H., Krawetz, S.A., Benham, C. (2003) From DNA structure to gene expression: mediators of nuclear compartmentalization and dynamics. *Chromosome Research* 11:435–445.

Boeger, H., Bushnell, D.A., Davis, R. et al. (2005) Structural basis of eukaryotic gene transcription. *FEBS Letters* 579:899–903.

Boeger, H., Griesenbeck, J., Strattan, J.S., Kornberg, R.D. (2004) Removal of promoter nucleosomes by disassembly rather than sliding *in vivo*. *Molecular Cell* 14:667–673.

Bolzer, A., Kreth, G., Solovei, I. et al. (2005) Three-dimensional maps of all chromosomes in human male fibroblast nuclei and prometaphase rosettes. *PLoS Biology* 3:826–842.

Brand, M., Ranish, J.A., Kummer, N.T. et al. (1994) Sequences containing the second-intron enhancer are essential for transcription of the human apolipoprotein B gene in the livers of transgenic mice. *Molecular and Cellular Biology* 14:2243–2256.

Calladine, C.R., Drew, H.R., Luisi, B.F., Travers, A.A. (2004) *Understanding DNA. The Molecule and How it Works*, 3rd edn. Elsevier Academic Press, New York.

Carroll, S.B. (2005) *From DNA to Diversity. Molecular Genetics and the Evolution of Animal Design*, 2nd edn. Blackwell Publishing, Malden, MA.

Chubb, J.R., Bickmore, W.A. (2003) Considering nuclear compartmentalization in the light of nuclear dynamics. *Cell* 112:403–406.

Cirillo, L., Zaret, K. (2004) Developmental biology. A linker histone restricts muscle development. *Science* 304:1607–1609.

Cook, P.R. (2001) *Principles of Nuclear Structure and Function*. Wiley-Liss, New York.

Corda, D., Di Girolamo, M. (2003) Functional aspects of protein mono-ADP-ribosylation. *EMBO Journal* 22:1953–1958.

Crabtree, G.R., Aebersold, R., Groudine, M. (2004) Dynamic changes in transcription factor complexes during erythroid differentiation revealed by quantitative proteomics. *Nature Structural Molecular Biology* 11:73–80.

Cramer, P. (2004) RNA polymerase II structure: from core to functional complexes. *Current Opinion in Genetics and Development* 14:218–226.

Cuajungco, M.P., Leyne, M., Mull, J. et al. (2003) Tissue-specific reduction in splicing efficiency of *IKBKAP* due to the major mutation associated with familial dysautonomia. *American Journal of Human Genetics* 72:749–758.

de la Cruz, X., Lois, S., Sánchez-Molina, S., Martínez-Balbás, M.A. (2005) Do protein motifs read the histone code? *BioEssays* 27:164–175.

De Sandre-Giovannoli, A., Bernard, R., Cau, P. et al. (2003) Lamin A truncation in Hutchinson–Gilford progeria. *Science* 300:2055.

Dion, M.F., Altschuler, S.J., Wu, L.F., Rando, O.J. (2005) Genomic characterization reveals a simple histone H4 acetylation code. *Proceedings of the National Academy of Sciences USA* 102:5501–5506.

Flanagan, P.M., Kelleher, R.J. III, Sayre, M.H., Tschochner, H., Kornberg, R.D. (1991) A mediator required for activation of RNA polymerase II transcription *in vitro*. *Nature* 350:436–438.

Fry, C.J., Peterson, C.L. (2002). Unlocking the gates to gene expression. *Science* 295:1847–1848.

Green, M.R. (2005) Eukaryotic transcription activation: right on target. *Molecular Cell* 18:399–402.

Goldfarb, D.S., Corbett, A.H., Mason, D.A., Harreman, M.T., Adam, S.A. (2004) Importin α: a multipurpose nuclear-transport receptor. *Trends in Cell Biology* 14:505–514.

Guzder, S.N., Sung, P., Bailly, V., Prakash, L., Prakash, S. (1994) RAD25 is a DNA helicase required for DNA repair and RNA polymerase II transcription. *Nature* 369:578–581.

Hay, R.T. (2005) SUMO: a history of modification. *Molecular Cell* 18:1–12.

Heng, H.H.Q., Goetze, S., Ye, C.J. et al. (2004) Chromatin loops are selectively anchored using scaffold/matrix-attachment regions. *Journal of Cell Science* 117:999–1008.

Kadonaga, J.T. (2004) Regulation of RNA polymerase II transcription by sequence-specific DNA-binding factors. *Cell* 116:247–257.

Kalderon, D., Richardson, W.D., Markham, A.F., Smith, A.E. (1984) Sequence requirements for nuclear location of simian virus 40 large-T antigen. *Nature* 311:33–38.

Kassabov, S.R., Zhang, B., Persinger, J., Bartholomew, B. (2003) SWI/SNF unwraps, slides, and rewraps the nucleosome. *Molecular Cell* 11:391–403.

Kettenberger, H., Armache, K.J., Cramer, P. (2004) Complete RNA polymerase II elongation complex structure and its interactions with NTP and TFIIS. *Molecular Cell* 16:955–965.

Kim, J.H., Lane, W.S., Reinberg, D. (2002) Human Elongator facilitates RNA polymerase II transcription through chromatin. *Proceedings of the National Academy of Sciences USA* 99:1241–1246.

Korber, P., Hörz, W. (2004) SWRred not shaken: mixing the histones. *Cell* 117:5–7.

Kouzarides, T., Ziff, E. (1988) The role of the leucine zipper in the fos–jun interaction. *Nature* 336:646–651.

Kurdistani, S.K., Tavazoie, S., Grunstein, M. (2004) Mapping global histone acetylation patterns to gene expression. *Cell* 117:721–733.

Kutay, U., Güttinger, S. (2005) Leucine-rich nuclear export signals: born to be weak. *Trends in Cell Biology* 15:121–124.

Latchman, D. (2004) *Eukaryotic Transcription Factors*, 4th edn. Elsevier Academic Press, New York.

Laudet, V., Gronemeyer, H. (2002) *The Nuclear Receptor Facts Book*. Academic Press, New York.

Levine, M., Tjian, R. (2003) Transcription regulation and animal diversity. *Nature* 424:147–151.

Lewis, B.A., Sims, R.J. III, Lane, W.S., Reinberg, D. (2005) Functional characterization of core promoter elements: DPE-specific transcription requires the protein kinase CK2 and the PC4 coactivator. *Molecular Cell* 18:471–481.

Lim, C.Y., Santoso, B., Boulay, T., Dong, E., Ohler, U., Kadonaga, J.T. (2004) The MTE, a new core promoter element for transcription by RNA polymerase II. *Genes and Development* 18:1606–1617.

Low, L.Y., Hernández, H., Robinson, C.V., O'Brien, R., Grossmann, J.G., Ladbury, J.E., Luisi, B. (2002) Metal-dependent folding and stability of nuclear hormone receptor DNA-binding domains. *Journal of Molecular Biology* 319:87–106.

Lu, H., Zawel, L., Fisher, L., Egly, J.M., Reinberg, D. (1992) Human general transcription factor IIH phosphorylates the C-terminal domain of RNA polymerase II. *Nature* 358:641–645.

MacLean, J.A. II, Chen, M.A., Wayne, C.M. et al. (2005) *Rhox*: a new homeobox gene cluster. *Cell* 120:369–382.

Marx, J. (2005) Combing over the Polycomb group proteins. *Science* 308:624–626.

Matsuura, Y., Stewart, M. (2004) Structural basis for the assembly of a nuclear export complex. *Nature* 432:872–877.

McKnight, R.A., Shamay, A., Sankaran, L., Wall, R.J., Hennighausen, L. (1992) Matrix-attachment regions can impart position-independent regulation of a tissue-specific gene in transgenic mice. *Proceedings of the National Academy of Sciences USA* 89:6943–6947.

McKnight, S.L. (1991) Molecular zippers in gene regulation. *Scientific American* 265:32–39.

Mohrmann, L., Verrijzer, C.P. (2005) Composition and functional specificity of SWI2/SNF2 class chromatin remodeling complexes. *Biochimica et Biophysica Acta* 1681:59–73.

Mosammaparast, N., Pemberton, L.F. (2004) Karyopherins: from nuclear-transport mediators to nuclear-function regulators. *Trends in Cell Biology* 14:547–556.

Müller, F., Tora, L. (2004) The multicolored world of promoter recognition complexes. *EMBO Journal* 23:2–8.

Navas, P.A., Li, Q., Peterson, K.R., Swank, R.A., Rohde, A., Roy, J., Stamatoyannopoulos, G. (2002) Activation of the β-like globin genes in transgenic mice is dependent on the presence of the β-locus control region. *Human Molecular Genetics* 11:893–903.

Owen-Hughes, T., Bruno, M. (2004) Breaking the silence. *Science* 303:324–325.

Pal, M., Ponticelli, A.S., Luse, D.S. 2005. The role of the transcription bubble and TFIIB in promoter clearance by RNA polymerase II. *Molecular Cell* 19:101–110.

Pederson, T. (2000) Half a century of "the nuclear matrix." *Molecular Biology of the Cell* 11:799–805.

Pemberton, L.F., Paschal, B.M. (2005) Mechanisms of receptor-mediated nuclear import and nuclear export. *Traffic* 6:187–198.

Rachez, C., Freedman, L.P. (2001) Mediator complexes and transcription. *Current Opinion in Cell Biology* 13:274–280.

Rahl, P.B., Chen, C.Z., Collins, R.N. (2005) Elp1p, the yeast homolog of the FD disease syndrome protein, negatively regulates exocytosis independently of transcriptional elongation. *Molecular Cell* 17:841–853.

Renault, L., Kuhlmann, J., Henkel, A., Wittinghofer, A. (2001) Structural basis for guanine nucleotide exchange on Ran by the regulator of chromosome condensation (RCC1). *Cell* 105:245–255.

Rhoades, D., Klug, A. (1993) Zinc fingers. *Scientific American* 268:32–39.

Riddick, G., Macara, I.G. (2005) A systems analysis of importin-α-β mediated nuclear protein import. *Journal of Cell Biology* 168:1027–1038.

Sassone-Corsi, P., Corden, J., Kédinger, C., Chambon, P. (1981) Promotion of specific *in vitro* transcription by excised "TATA" box sequences inserted in a foreign nucleotide environment. *Nucleic Acids Research* 9:3941–3958.

Schwartz, T.U. (2005) Modularity within the architecture of the nuclear pore complex. *Current Opinion in Structural Biology* 15:221–226.

Semenza, G.L. (1999) *Transcription Factors and Human Disease*. Oxford University Press, New York.

Shank, L.C., Paschal, B.M. (2005) Nuclear transport of steroid hormone receptors. *Critical Reviews in Eukaryotic Gene Expression* 15:49–73.

Shi, Y., Lan, F., Matson, C., Mulligan, P., Whetstine, J.R., Cole, P.A., Casero, R.A., Shi, Y. (2004) Histone demethylation mediated by the nuclear amine oxidase homolog LSD1. *Cell* 119:941–953.

Slaugenhaupt, S.A., Gusella, J.F. (2002) Familial dysautonomia. *Current Opinion in Genetics and Development* 12:307–311.

Smale, S.T., Kadonaga, J.T. (2003) The RNA polymerase II core promoter. *Annual Review of Biochemistry* 72:449–479.

Smith, C.L., Horowitz-Scherer, R., Flanagan, J.F., Woodcock, C.L., Peterson, C.L. (2003) Structural analysis of the yeast SWI/SNF chromatin remodeling complex. *Nature Structural Biology* 10:141–145.

Spiegelman, B.M., Heinrich, R. (2004) Biological control through regulated transcriptional coactivators. *Cell* 119:157–167.

Spitz, F., Duboule, D. (2005) Developmental biology: reproduction in clusters. *Nature* 434:715–716.

Studitsky, V.M., Walter, W., Kireeva, M., Kashlev, M., Felsenfeld, G. (2004) Chromatin remodeling by RNA polymerases. *Trends in Biochemical Sciences* 29:127–135.

Svejstrup, J.Q. (2003) Histones face the FACT. *Science* 301:1053–1055.

Tolhuis, B., Palstra, R.-J., Splinter, E., Grosveld, F., de Laat, W. (2002) Looping and interaction between hypersensitive sites in the active β-globin locus. *Molecular Cell* 10:1453–1465.

Vakoc, C.R., Letting, D.L., Gheldof, N. et al. (2005) Proximity among distant regulatory elements at the β-globin locus requires GATA-1 and FOG-1. *Molecular Cell* 17:453–462.

Vieira, K.F., Levings, P.P., Hill, M.A., Cruselle, V.J., Kang, S.H.L., Engel, J.D., Bungert, J. (2004) Recruitment of transcription complexes to the β-globin gene locus *in vivo* and *in vitro*. *Journal of Biological Chemistry* 279:50350–50357.

Villavicencio, E.H., Walterhouse, D.O., Iannaccone, P.M. (2000) The Sonic hedgehog-Patched-Gli pathway in human development and disease. *American Journal of Human Genetics* 67:1047–1054.

Wang, G., Balamotis, M.A., Stevens, J.L., Yamaguchi, Y., Handa, H., Berk, A.J. (2005) Mediator requirement for both recruitment and postrecruitment steps in transcription initiation. *Molecular Cell* 17:683–694.

Weeber, E.J., Levenson, J.M., Sweatt, J.D. (2002) Molecular genetics of human cognition. *Molecular Interventions* 2:376–391.

Weis, K. (2003) Regulating access to the genome: nucleocytoplasmic transport throughout the cell cycle. *Cell* 112:441–451.

Wilson, E.B. (1925) *The Cell in Development and Heredity*. Macmillan Co., New York.

Wolffe, A. (1998) *Chromatin Structure and Function*, 3rd edn. Academic Press, New York.

Yie, J., Merika, M., Munshi, N., Chen, G., Thanos, D. (1999) The role of HMG I(Y) in the assembly and function of the INF-β enhanceosome. *EMBO Journal* 18:3074–3089.

Yusufzai, T.M., Tagami, H., Nakatani, Y., Felsenfeld, G. (2004) CTCF tethers an insulator to subnuclear sites, suggesting shared insulator mechanisms across species. *Molecular Cell* 13:291–298.

Zhou, J., Berger, S.L. (2004) Good fences make good neighbors: barrier elements and genomic regulation. *Molecular Cell* 16:500–502.

Chapter 12

Epigenetics and monoallelic gene expression

When you know you're right, you don't care what others think. You know sooner or later it will come out in the wash.

Barbara McClintock, quoted in Claudia Wallis, "Honoring a modern Mendel,"
Time (1983) 24 October:43–44.

Outline

Table 12.1 Monoallelic expression of mammalian genes.

Gene	Chromosome	Selection of allele
Imprinted genes	Autosomal	Nonrandom
X-inactivated genes	X	Random
Immunoglobulin genes	Autosomal	Random
T cell receptor genes	Autosomal	Random
Natural killer cell receptor genes	Autosomal	Random
Interkeukin-2 gene	Autosomal	Random

12.1 Introduction

Cells normally have two copies (alleles) of autosomal genes on chromosomes other than the X and Y. One allele is inherited from the mother (maternal allele) and one is inherited from the father (paternal allele). For most genes, both copies are expressed by the cell. However, a small class of genes is "monoallelically" expressed," i.e. transcribed preferentially from a single allele in each cell (Table 12.1). In most cases of monoallelic gene expression, cells randomly select only one allele to encode RNA and protein for that gene. This is typical in cells of the immune system and in olfactory neurons, and is considered to be a way of ensuring that a single kind of receptor is displayed on the cell surface. An exception is genomic imprinting where selection of the active allele is nonrandom and based on the parent of origin.

Monoallelic expression in mammals is exemplified by genomic imprinting, X chromosome inactivation, and allelic exclusion. Epigenetic silencing mechanisms play a role in all three systems. In addition, programmed gene rearrangements – rearrangements of DNA that regulate the expression of some genes – play a central role in some forms of allelic exclusion. Classic examples include mating-type switching in yeast, antigen switching in trypanosomes, and V(D)J recombination in the mammalian immune system. The primary function of eukaryotic DNA methylation may be defense of the genome from the potentially detrimental effects of transposable elements. An overview of the types of transposable elements and the phenotypic consequences of their movement within the genome is followed by a discussion of their epigenetic silencing.

12.2 Epigenetic markers

Scientists have long puzzled over the question of how a single cell can differentiate into the many different cell types of a multicellular organism. The general conclusion is that additional regulatory information – epigenetic information – must exist beyond the level of the genetic code. The developmental biologist Conrad H. Waddington is often credited with coining the term "epigenetics" (where "epi" means "outside of" or "in addition to"). However, the term was already in use at least as early as 1896 by the German biologist, Wilmelm August Oscar Hertwig. In 1942, Waddington defined epigenetics as "the branch of biology which studies the causal interaction between genes and their products which bring the phenotype into being."

Today, epigenetic research is primarily concerned with the study of mitotically and/or meiotically heritable changes in gene expression without changes in DNA sequence. This shift in research focus traces back to Barbara McClintock's discovery of transposable elements in maize in the 1940s and 1950s. The study of epigenetic phenomena in mammals developed independently of the plant genetics field, beginning with the genetic interpretation of X chromosome inactivation by Mary Lyon in 1961. Most cell differentiation processes are initiated or maintained through epigenetic processes. The epigenetic changes that take place during cell differentiation are normally erased in the germline, but there are examples in both plants and mammals of epigenetic variants that are transmitted through meiosis.

Modification of chromatin structure is a classic example of an epigenetic characteristic. The major type of DNA modification present in most animals and plants is methylation of the base cytosine. DNA methylation patterns are the best studied and best understood epigenetic markers but there are other important sources of epigenetic regulation, such as the modification of histones.

Cytosine DNA methylation marks genes for silencing

Cytosine DNA methylation is a covalent modification of DNA. In this reaction a methyl group is transferred from *S*-adenosylmethionine to the carbon-5 position of cytosine by a family of cytosine DNA methyltransferases (Fig. 12.1A). 5-Methyl-cytosine is the only modified base commonly found in eukaryotes, although *Caenorhabditis elegans*, *Drosophila*, and yeast contain little or no 5-methyl-cytosine. DNA methylation occurs almost exclusively at the dinucleotide CG in mammals. The CG dinucleotide is often denoted as "CpG", where p stands for the phosphate group. In plant DNA, cytosine methylation occurs at either CG or CNG, where N can be any base. In both mammals and plants, the C residues on both strands of the DNA are methylated. Methylation is maintained through DNA replication by a semiconservative process, like DNA replication itself (Fig. 12.1B). After replication, the DNA double helix is "hemimethylated," i.e. the old template strand is methylated and the newly synthesized strand is unmethylated. A maintenance DNA methyltransferase recognizes only hemimethylated sites and methylates the new strand of DNA appropriately.

Methylation is a way of marking genes for silencing. In many genes, the methylation pattern is constant at most CG dinucleotide sites, but varies at some sites. As a general rule, genes in which the majority of CG dinucleotide sites are methylated (hypermethylation) tend to be inactive, while genes in which the minority of CG sites are methylated (hypomethylation) tend to be active. This is not a universal rule, since the loss of DNA methylation has been shown in some cases to correlate with increased gene activity.

One way to demonstrate whether DNA methylation correlates with gene activity or repression is to treat cells in culture with the modified base "5-aza-cytosine," usually in the form of the nucleoside 5-aza-deoxycytidine. 5-Aza-cytosine is an analog of normal cytosine that can be incorporated into DNA during replication. However, it cannot be methylated because the 5-carbon of normal cytosine is replaced by nitrogen (Fig. 12.1A). The nitrogen acts as a strong inhibitor of most of the cellular enzymes that add methyl groups to cytosine in that same 5 position. Drug treatment leads to the loss of methyl groups from many different cytosine bases across the whole genome. Despite these global effects, the use of 5-aza-deoxycytidine

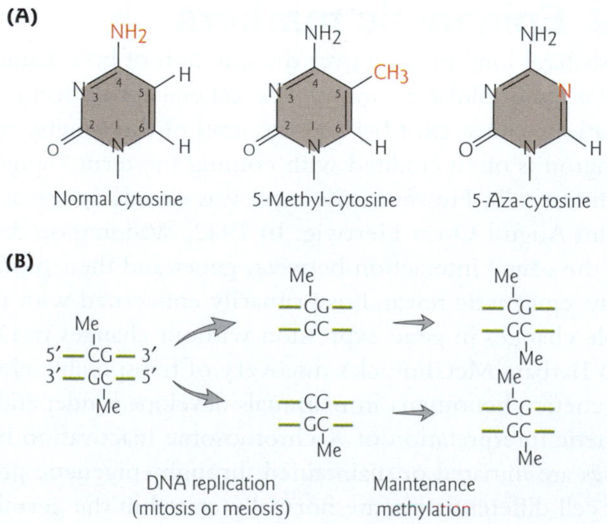

Figure 12.1 Inheritance of methylation states. (A) Structure of the normal DNA base cytosine compared with 5-methyl-cytosine and the drug 5-aza-cytosine. (B) Replication of methylated DNA results in hemimethylated progeny DNA. DNA methyltransferases methylate cytosine based on the hemimethylated states of symmetric CG motifs.

in cancer therapy has been considered (Disease box 12.1). *In vitro*, this type of treatment has proven effective for reversing the repression of important genes. For example, *BNIP3* (*BCL2*/adenovirus E1B interacting protein 3) expression is silenced in some pancreatic cancer cells by methylation of its promoter region. When Tamaki Abe and colleagues treated these cancer cells with 5-aza-deoxycytidine, *BNIP3* expression was restored and induced hypoxia-mediated cell death (Fig. 12.2). A similar type of experiment was first used to show that epigenetic information modulates gene expression without modifying actual DNA sequence, as in classic mutations. In plant and mammalian tissue culture a significant percentage of mutant variants were demonstrated to not be true mutations because they could be efficiently reverted by treatment with 5-aza-kdeoxycytidine.

Most of the 5-methyl-cytosine in the genome lies within retrotransposons or other repetitive sequences (see Section 12.5). Because of this observation, scientists have proposed that methylation evolved as a host defense mechanism to prevent the mobilization of these elements and to reduce the occurrence of chromosomal rearrangements. Normally, about 70–80% of all CG dinucleotides in the mammalian genome are methylated. The remainder of CG dinucleotides occur in clusters known as "CpG islands" near the 5′ end of genes and are protected from methylation (Fig. 12.3).

CpG islands are found near gene promoters

CpG islands are small regions of CG-rich DNA (1–2 kb in size). Approximately 50,000 CpG islands punctuate the human genome. They are found associated with the promoters of ~40–50% of housekeeping genes. CpG islands were first detected by their sensitivities to the restriction endonuclease *Hpa*II, which cuts only unmethylated CCGG. When genomic DNA is digested with *Hpa*II, short pieces result from cutting within the unmethylated CG regions (Fig. 12.3A). These fragments separate on an agarose gel like an "island" from the long uncut pieces of methylated DNA found between unmethylated regions. Unlike cytosine, 5-methyl-cytosine is highly susceptible to spontaneous C → T deamination that results in the generation of a TG dinucleotide (Fig. 12.3B). Consequently, the mammalian genome has become progressively depleted of CG dinucleotides through evolution. Because they are normally unmethylated, the CpG islands are protected from spontaneous deamination (Fig. 12.3C). Important exceptions to the unmethylated status of CpG islands include those that are associated with imprinted genes, genes subject to X chromosome inactivation, and transposable elements.

Stable maintenance of histone modifications

Histone modifications such as acetylation and methylation are important in transcriptional regulation (see Section 11.6). Many of these post-translational modifications are stably maintained during cell division, although epigenetic inheritance is not yet well understood. Proteins that mediate these modifications are often associated within the same complexes as those that regulate DNA methylation. Typically, histone hypoacetylation and hypermethylation are characteristic of DNA sequences that are methylated and inactive in normal cells.

The study of the epigenome – the genome-wide pattern of methyl groups and other epigenetic markers – has led to important insights into differences in gene expression between normal and diseased cells. For example, aberrant DNA methylation patterns and histone modifications are found in human cancer cells (Disease box 12.1) and in patients with fragile X mental retardation (Disease box 12.2).

12.3 Genomic imprinting

Genomic imprinting affects a small subset of genes and results in the expression of those genes from only one of the two parental chromosomes. This is brought about by epigenetic instructions – imprints – that are laid down in the parental germ cells. These instructions are in the form of differential methylation of the two parental alleles of the imprinted gene.

DISEASE BOX 12.1

Cancer and epigenetics

CpG island hypermethylation and genome-wide hypomethylation are common epigenetic features of cancer cells. Too little methylation across the genome or too much methylation in the CpG islands can cause problems, the former by activating nearby oncogenes, and the latter by silencing tumor suppressor genes (see Section 17.2). Global hypomethylation may stimulate oncogene expression. However, since it takes places primarily in DNA repetitive sequences its main effect may be linked to chromosomal instability. Genome-wide demethylation appears to progress with age. This loss of methylation may, in part, explain the higher incidence of cancer among the elderly. Loss of methylation has also been linked to poor nutrition. For example, *S*-adenosylmethionine, a derivative of folic acid, is the primary methyl donor in the cell (Fig. 1). A lack of folic acid in the diet has been shown to predispose cells of an organism to cancer.

In addition to having aberrant DNA methylation patterns, cancer cells also have a histone H4 "cancer signature." This signature is found in many types of cancer, ranging from leukemia to cancer of the bone, cervix, prostate, and testis, and is characterized by a loss of monoacetylation at lysine 16 and trimethylation at lysine 20. The deacetylation occurs mainly in regions of repetitive DNA that also undergo hypomethylation in cancer cells. Because of this loss of acetylated histones in cancer cells, there is great interest in using histone deacetylase (HDAC) inhibitors alone, or in combination with DNA demethylating agents, to reactivate silenced tumor suppressor genes. A large number of clinical trials are underway to determine whether these inhibitors are safe and mediate the desired effect in treatment of various types of cancer.

Figure 1 Folic acid and related pathways for synthesis of *S*-adenosylmethionine (SAM). SAM is the methyl donor for the methylation of DNA and histones. DHFR, dihydrofolate reductase; MTHF, methyltetrahydrofolate; MTHFD, methylenetetrahydrofolate dehydrogenase; SAH, *S*-adenosylhomocysteine; THF, tetrahydrofolate.

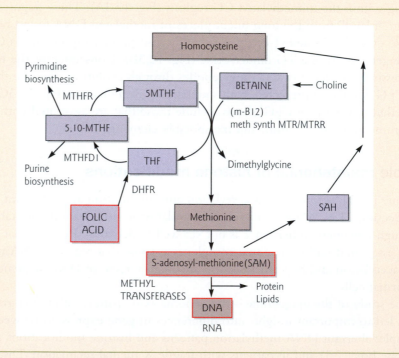

Imprinting occurs in mammals, but no other vertebrates looked at so far. There are around 80 different genes currently known to be imprinted in mammals. These genes tend to have important roles in development and the loss of imprinting is implicated in a number of genetic diseases and types of cancer in humans (Table 12.2, Disease box 12.3). Nearly all imprinted genes are organized in clusters in the genome.

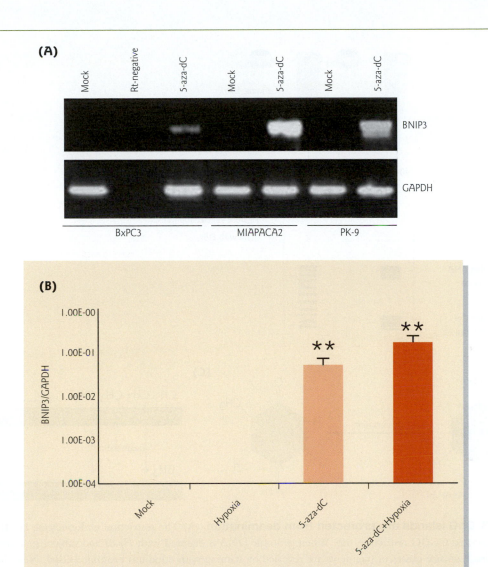

Figure 12.2 Induction of *BNIP3* gene expression by treatment with 5-aza-deoxycytidine (5-aza-dC).
(A) RT-PCR analysis of *BNIP3* expression in three different pancreatic cancer cell lines (B×PC3, MIAPACA2, PK-9) treated with 5-aza-dC compared with phosphate-buffered saline-treated cells (mock) (see Fig. 9.8D for methods). (B) Real-time PCR analysis of *BNIP3* expression in pancreatic cancer cells. Cells were incubated under the conditions indicated, after which total RNA was extracted for analysis. Cells subjected to hypoxic conditions (reduction in the normal level of tissue oxygen tension) were incubated at an atmosphere containing only 1% O_2 balanced by CO_2 and nitrogen. The *BNIP3* signals were normalized to levels of the housekeeping protein GAPDH (glyceraldehyde-3-phosphate dehydrogenase). Values are expressed as means ± SEM. There was a significant increase in *BNIP3* expression in 5-aza-dC-treated cells compared with mock-treated cells (★★, $P < 0.01$). (Reprinted from Abe, T., Toyota, M., Suzuki, H. et al. 2005. Upregulation of BNIP3 by 5-aza-2′-deoxycytidine sensitizes pancreatic cancer cells to hypoxia-mediated cell death. *Journal of Gastroenterology* 40:504–510. Copyright © 2005 Springer-Verlag, with kind permission of Springer Science and Business Media).

Within these clusters, typically there is at least one maternally expressed and one paternally expressed gene. However, there appear to be no particular rules governing the direction of transcription or the distribution of maternally versus paternally imprinted genes. Clusters of imprinted genes are regulated in a coordinated manner by DNA methylation of large (up to 100 kb) specific intergenic regions called imprinting control

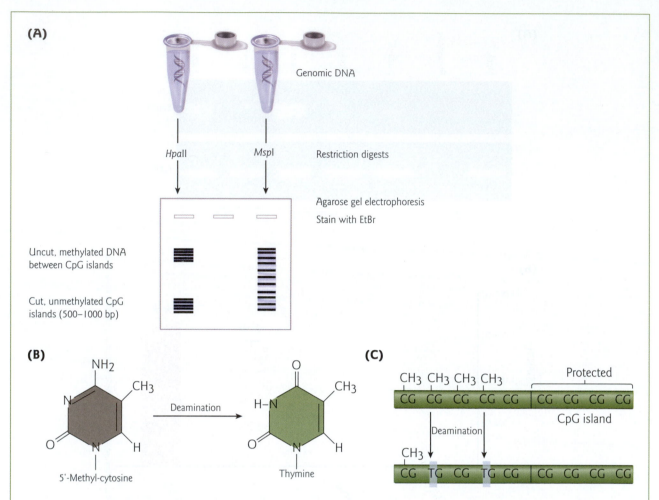

Figure 12.3 CpG islands are protected from deamination. (A) The restriction endonuclease *Hpa*II cannot cut at a methylated CCGG restriction site. When genomic DNA is digested with *Hpa*II and subject to agarose gel electrophoresis, two size classes of fragments are revealed by staining with ethidium bromide (EtBr). Near the top of the gel, there are long fragments of uncut, methylated DNA. Near the bottom of the gel, there are short pieces of 500–1000 bp that result from cutting within the unmethylated CG regions. Restriction endonuclease *Msp*I can cut at both unmethylated and methylated CCGG sequences. Digesting genomic DNA with *Msp*I results in a continuum of fragment sizes ranging from long (at the top of the gel) to short (at the bottom). (B) 5-Methyl-cytosine undergoes spontaneous deamination to thymine. (C) CpG islands are protected from methylation by bound proteins. All methylated CG dinucleotides undergo slow deamination to TG, but unmethylated CG dinucleotides in the nearby CpG islands remain unchanged.

regions (ICRs) (also referred to as "differentially methylated regions" or "imprinting centers") (Fig. 12.4). The ICRs are responsible for establishing the differential imprint and for the maintenance of this imprint during development. They are rich in CG dinucleotides and many correspond to CpG islands. Silent alleles of imprinted loci are targeted for methylation of cytosine residues during gametogenesis, while expressed alleles generally remain relatively undermethylated. Silenced alleles are then protected from global demethylation during early embryogenesis to achieve monoallelic expression later in development (see below).

Imprinting also occurs in flowering plants. However, imprinting in flowering plants involves a different mechanism for controlling DNA methylation compared with imprinting in mammals – no global demethylation has been detected during the plant life cycle and imprinting apparently results from the

Table 12.2 Genomic imprinting and neurodevelopmental disorders.

Disorder and prevalence	Symptoms	Genetic mechanism	Genes suspected or known to be affected
Prader–Willi syndrome (PWS): one in 25,000 births	Feeding difficulties as infants due to poor motor skills Gross obesity Obsessive-compulsive behavior (think about food 24 hours a day) Hypogonadism (underdeveloped genitals) Behavioral problems Mild mental retardation Characteristic facial appearance (small narrow mouth) Life expectancy may be normal if weight is controlled	Defect in genomic imprinting of PWS/AS locus on chromosome 15 Missing paternal allele(s) due to: Deletion (65–75% of cases) Uniparental disomy (20–30%) ICR mutation (≤ 5%)	Numerous paternally expressed genes present in imprinted region; role in PWS remains uncertain: *MKRN3* *MAGEL2* *NDN* *SNURF-SNRPN* *HBII* snoRNA genes
Angelman syndrome (AS) (also known as "happy puppet" syndrome): one in 15,000–20,000 births	Seizures Mental retardation (lack of speech) Slowing of head growth Puppet-like gait Unusually wide mouth Bouts of inappropriate laughter Autistic-like behavior Hyperactivity Sleep disorders Life expectancy may be normal	1) Defect in genomic imprinting of PWS/AS locus on chromosome 15 Missing maternal allele(s) due to: Deletion (65–75% of cases) Uniparental disomy (≤ 5%) ICR mutation (≤ 5%) 2) Mutation in *UBE3A* gene (~10% of cases)	*UBE3A*: maternally expressed in a brain-region specific manner; encodes a protein degradation regulator
Rett syndrome: one in 15,000 females	Continuous, rhythmic handwringing and clasping Mental retardation Seizures Decreased growth Breathing problems Autism Neurodegeneration and motor destruction (generally confined to a wheelchair by school age)	Mutation in *MeCP2* gene on X chromosome: Leads to loss of imprinting of *UBE3A* antisense gene, and reduced UBE3A protein expression in the brain	*MeCP2* (methyl-CpG-binding protein 2)

removal of the methylation mark from one of the parental alleles. The imprinted status of plants is not inherited and appears to be confined to the endosperm, which does not contribute to the next generation.

Establishing and maintaining the imprint

Imprinting is reprogrammed or "reset" in the germline by erasure of the DNA methylation marks in the primordial germ cells (Fig. 12.4). This appears to be an active demethylation process involving yet unknown enzymatic activities. Imprinted genes subsequently acquire different marks in the sperm and the egg, and these methylation marks are heritable through subsequent cell divisions. Methylation is an ideal marker for imprinting, since it can be established by *de novo* methylation in one of the gametes. Once this occurs, the differential pattern will be automatically preserved by means of the maintenance methylase, DNA methyltransferase 1 (DNMT1). After fertilization, DNMT1 acts preferentially on hemimethylated DNA substrates during DNA replication to maintain methylation patterns throughout cell division. At most

Fragile X mental retardation and aberrant DNA methylation

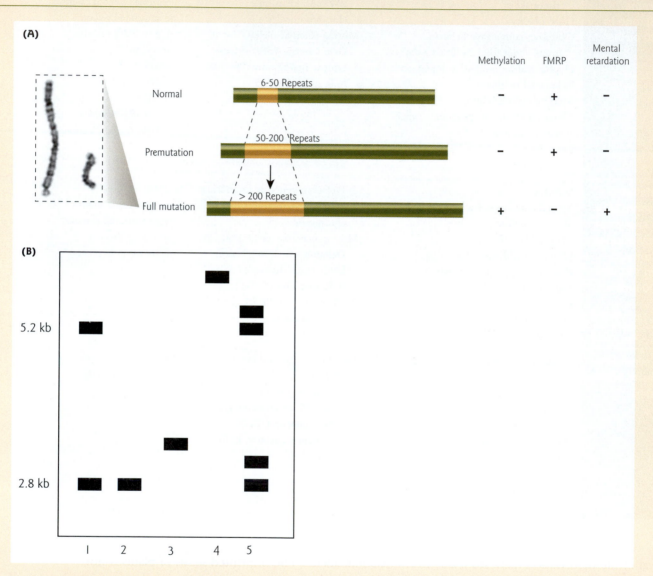

Figure 1 Trinucleotide repeat expansion in fragile X syndrome. (A) Schematic representation of the different *FRM1* alleles showing the size ranges of the CGG repeat. For the different alleles, it is shown whether they are methylated, whether they are translated into protein (FMRP), and whether individuals with such an allele are mentally retarded. (Inset) X and Y chromosomes from a boy with fragile X syndrome, showing the "fragile" point (arrowed). (Photograph courtesy of Barbara Jackson and Rebecca Galczynski, Kennedy Krieger Institute, Baltimore, Maryland). (B) Diagram illustrating results of fragile X analysis by Southern blot. Lane 1, normal female; Lane 2, normal male; Lane 3, premutation male; Lane 4, full mutation male; Lane 5, female with 18 and ~80 repeats, with equal X-inactivation. (Modified from Maddalena, A., Richards, C.S., McGinniss, M.J., Brothman, A., Desnick R.J., Grier, R.E., Hirsch, B., Jacky, P., McDowell, G.A., Popovich, B., Watson, M., Wolff, D.J. 2001. Technical standards and guidelines for fragile X: The first of a series of disease-specific supplements to the standards and guidelines for clinical genetics laboratories of the American College of Medical Genetics. *Genetics in Medicine* 3:200–205).

Fragile X mental retardation and aberrant DNA methylation

Fragile X mental retardation syndrome, as the name denotes, is an X-linked disorder. It is the most commonly inherited form of mental retardation and affects approximately one in 4000 males. Male phenotypic characteristics include a tendency towards large head size (macrocephaly), long and narrow facial features with unusually large ears, and decreased "floppy" muscle tone (hypotonia). Mental retardation severity can range from moderate to severe. In females, fragile X is seen in about one in 7000 live births. Only approximately 30% of females show any physical characteristics and all affected females exhibit a milder mental handicap. Symptoms are less severe because in heterozygous females there is a 50/50 chance that the defective X will be inactive (see Section 12.4). Patients with fragile X have a trinucleotide repeat expansion (CGG) within the 5'-untranslated region of the *FRM1* gene located on the X-chromosome (Fig. 1).

Consequences of trinucleotide repeat expansion

Trinucleotide repeat expansion often does not appear all at once, but rather occurs over several generations. If the repeat number is between ~50 and 200, normal protein is produced and no symptoms are seen. However, if the expansion exceeds 200 CGG repeats, affected individuals will exhibit disease symptoms. In some cases more than 1000 repeats have been observed. Trinucleotide repeat expansion in the *FMR1* gene results in hypermethylation of DNA and hypoacetylation of histones in the promoter region. The altered chromatin structure then leads to a loss of *FRM1* gene expression. The *FRM1* gene encodes the fragile X mental retardation protein (FMRP), a cytoplasmic RNA-binding protein that is involved in neuronal maturation and/or development.

Diagnostic tests for fragile X

The name "fragile X" was given to this syndrome because the majority of patients have a fragile piece that appears to be hanging off one end of the X chromosome (the fragile site) (Fig. 1A). The original test used to identify individuals with fragile X was cytogenetic (chromosome) analysis. However, there are many fragile X carriers whose X chromosomes appear to be normal. The most definitive testing method used is direct DNA analysis to determine the length of the CGG repeat in the patient's DNA. Two main approaches are used: polymerase chain reaction (PCR) and Southern blot analysis (Fig. 1B) (see Tool boxes 8.3 and 8.7 for methods). PCR analysis uses flanking primers to amplify a fragment of DNA spanning the repeat region. The sizes of the PCR products indicate the approximate number of repeats present in each allele of the individual being tested. However, the efficiency of the PCR reaction decreases as the number of CGG repeats increases, and large repeat expansions may fail to yield a detectable PCR product. In this case, Southern blot analysis can be used to accurately detect alleles in all size ranges. An added advantage is that both the size of the repeated region and methylation status can be assayed simultaneously. A methylation-sensitive restriction enzyme that fails to cleave methylated sites is used to distinguish between methylated and unmethylated alleles. The disadvantages of Southern blotting are that precise sizing of alleles is not possible, and it is more labor-intensive and requires larger quantities of genomic DNA than PCR.

ICRs, the allelic methylation originates in the egg; at a few only, it is established during spermatogenesis. Significant advances in understanding how this methylation pattern is established and maintained have come from the analysis of conditional knockout mice (see Fig. 15.7). Conditional knockouts are required, since disruption of *DNMT* genes is embryonic lethal.

The *de novo* methyltransferase enzyme DNMT3a, together with its cofactor the DNMT3L protein, is required to establish imprinting *de novo*. Exactly how the differential imprints in sperm and oocytes are established is not fully understood but it is possible that the binding of other protein factors to individual ICRs could prevent methylation in either the sperm or egg. For example, a testis-specific zinc finger DNA-binding protein was recently identified named BORIS (*b*rother *o*f the *r*egulator of *i*mprinted *s*ites) that may play a role in *de novo* establishment of the methylation imprint in sperm. BORIS binds to the same DNA sequences as the CCCTC-binding factor (CTCF), a protein that functions in maintaining and reading

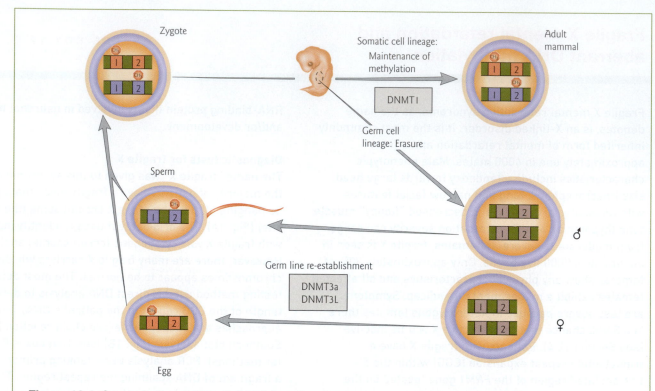

Figure 12.4 Genomic imprinting throughout development. Imprinted genes are silenced by DNA methylation (red symbol). During gametogenesis, the imprint marks present on the maternal (pink) and paternal (blue) chromosomes are erased (gray) by demethylation. Gender-specific imprints are re-established in gametes by differential (*de novo*) methylation at sequence elements called imprinting control regions (ICRs) (pink or blue boxes) that control imprinted gene expression. Methylation is mediated by DNA methyltransferase DNMT3a, its cofactor DNMT3L, and other oocyte- or sperm-specific factors. After fertilization of the egg by the sperm, these imprints are maintained throughout development by DNMT1. Two examples of ICRs are depicted: an ICR with maternally derived methylation (ICR1) and one with paternally-derived methylation (ICR2).

imprint marks. BORIS and CTCF have identical zinc finger domains, likely the result of a gene duplication event. Their expression is mutually exclusive, suggesting that they carry out similar functions that are subject to strict developmental regulation.

Mechanisms of monoallelic expression

DNA methylation and associated differences in chromatin are "read" after fertilization to ensure that the correct allele is expressed. This occurs in different ways depending on the gene. Adding to the complexity, reading mechanisms may vary in different tissues resulting in tissue-specific imprinting of some genes. There are three main mechanisms for ensuring monoallelic expression: (i) altered chromatin structure in the gene promoter; (ii) differential expression of an antisense RNA transcript; and (iii) blocking of an enhancer by an insulator (Fig. 12.5).

Altered chromatin structure in the gene promoter

In the simplest mechanism, DNA methylation occurs and repressive chromatin structure forms within one allele's promoter region, silencing that allele. This is exemplified by a large imprinted domain that is

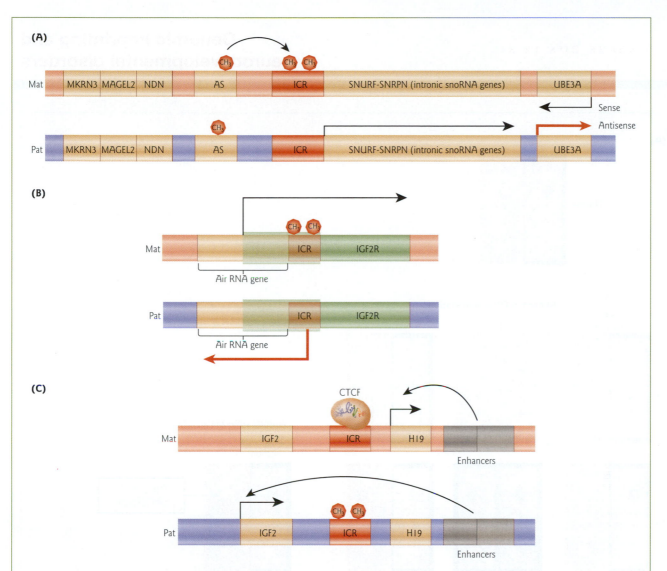

Figure 12.5 Regulation of imprinting. Schematic representation of regulatory mechanisms for three different imprinted domains (not drawn to scale). (A) The imprinting region (ICR) (red) at the *SNURF-SNRPN* gene of the Prader–Willi syndrome locus regulates imprinting along a large chromosomal domain. The ICR is unmethylated on the paternal chromosome (blue, Pat) and directs expression bidirectionally to all imprinted genes in this region, including the *UBE3A* antisense gene. Arrows indicate the direction of transcription (black, sense; red, antisense). Methylation (red symbol) at the ICR on the maternal chromosome (pink, Mat) requires a region located > 30 kb further upstream, called the Angelman syndrome (AS) element. The AS element is not differentially methylated, but it has an open chromatin structure that directs the ICR element to become methylated, and this silences all the paternally expressed genes on the maternal chromosome. In the absence of antisense *UBE3A* expression, the maternal *UBE3A* allele is expressed (see Disease box 12.3 and Table 12.2 for details). (B) *Igf2r* locus. When the ICR controlling the *Igf2r* imprinted region is unmethylated on the paternal chromosome, an antisense noncoding RNA (*Air* RNA) is expressed and the paternal *Igf2r* allele is silenced. On the maternal chromosome, the ICR is methylated. This promotes expression of the maternal *Igf2r* allele and silences the maternal *Air* allele. (C) *Igf2-H19* locus. On the maternal chromosome, the ICR is protected against methylation by CTCF binding. This creates a chromatin boundary that prevents interaction of the *Igf2* gene and enhancers (green) located downstream of *H19*. Methylation of the ICR on the paternal chromosome prevents CTCF binding. The *Igf2* promoter can interact with the enhancers, but the paternal *H19* allele is repressed because of ICR methylation spreads into the nearby *H19* promoter.

Genomic imprinting and neurodevelopmental disorders

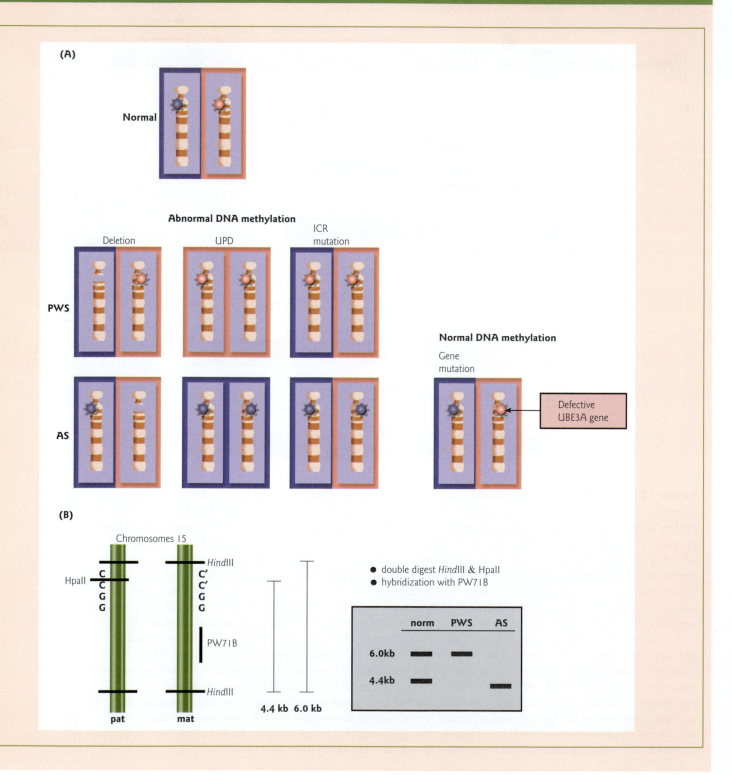

Genomic imprinting and neurodevelopmental disorders

DISEASE BOX 12.3

Three neurodevelopmental disorders, Prader–Willi syndrome, Angelman syndrome, and Rett syndrome (all named after the physicians who first described the disorders) are the result of either direct or indirect deregulation of imprinted genes (see Table 12.2). Recently, there have been major advances in understanding the molecular genetics of these diseases and in their diagnosis, but not in their treatment.

Defects in genomic imprinting lead to Prader–Willi syndrome and Angelman syndrome

Genes associated with both Prader–Willi syndrome and Angelman syndrome map to the long (q) arm of chromosome 15 (15q11-13). The entire 2 Mb imprinted domain is under the coordinated control of an imprinting control region (ICR) composed of two DNA sequence elements. The imprinting center regulates imprinted gene expression bidirectionally (see Fig. 12.5). In a normal child the Prader–Willi syndrome allele(s) are only expressed from the paternal chromosome 15, and the Prader–Willi syndrome allele(s) on the maternal chromosome 15 are inactive. Prader–Willi syndrome occurs when the paternal allele(s) that would normally be expressed are missing (see Table 12.2). In a normal child the Angelman syndrome allele(s) are only expressed from maternal chromosome 15, and the Angelman syndrome allele(s) on the paternal chromosome 15 are inactive. Angelman syndrome occurs when the maternal allele(s) that would normally be expressed are missing. There are several mechanisms leading to "missing" alleles, including *de novo* deletion of a large region of chromosome 15 encompassing the Prader–Willi syndrome/Angelman syndrome locus, uniparental disomy, where both chromosomes are inherited from the same parent, and ICR mutations (Fig. 1). In 24% of cases, Angelman syndrome results from classic gene mutation. Classic gene mutation has not been seen in any patients with Prader–Willi syndrome so far.

Clinical diagnosis of Prader–Willi syndrome and Angelman syndrome

Because imprinted genes show differential patterns of DNA methylation, clinical diagnosis of Prader–Willi syndrome and Angelman syndrome can be carried out by detection of 5-methyl-cytosine in DNA samples. One type of diagnostic test is shown in Fig. 1. More sensitive techniques for analyzing 5-methyl-cytosine content in clinical DNA samples are also available, such as the bisulfite-PCR method (Fig. 2).

Figure 1 (*opposite*) **DNA methylation testing for Prader–Willi syndrome (PWS) and Angelman syndrome (AS).** (A) Schematic representation of maternal (pink box) and paternal (blue box) chromosome 15, depicting the common mechanisms for PWS and AS: deletion of chromosomal region 15q11-13, uniparental disomy (UPD), imprinting control region (ICR) mutation, and classic gene mutation. The normal methylation pattern is indicated by a blue (paternal ICR) or pink (maternal ICR) sun symbol. If a paternal chromosome is present that has an ICR with a mutation resulting in a maternal imprint at this site, this leads to PWS. Conversely, if a maternal chromosome is present that has an ICR with a mutation resulting in a paternal imprint at this site, this leads to AS. (B) DNA methylation analysis. Genomic DNA is extracted from a patient blood sample, double digested with restriction endonucleases *Hind*III and *Hpa*II, and analyzed by Southern blot. *Hind*III is a rare cutting enzyme and *Hpa*II only cuts at a nonmethylated restriction site in the paternal chromosome (CCGG), and does not cut the methylated restriction site on the maternal chromosome (C′C′GG). The probe is complementary to a DNA sequence located in between the two restriction sites (PW71B). A 4.4 kb *Hpa*II–*Hind*III fragment from the paternal chromosome hybridizes with the probe on a Southern blot, but only a 6.0 kb *Hind*III–*Hind*III fragment from the maternal chromosome hybridizes with the probe. A normal child has both the maternal and paternal fragments (6.0 and 4.4 kb bands). A child with PWS is lacking a functional paternal contribution to the imprinted region, so only the maternal (6.0 kb) fragment is visible. When AS is due to biparental deletion, uniparental disomy, or ICR mutations, the child lacks a functional maternal contribution to the imprinted region. In this case, only the paternal (4.4 kb) fragment is visible. When AS is due to a mutation in the *UBE3A* gene, the methylation pattern would be identical to that of a normal child. (Buchholz, T., Jackson, J., Robson, L., Smith, A. 1998. Evaluation of methylation analysis for diagnostic testing in 258 referrals suspected of Prader-Willi or Angelman syndromes. *Human Genetics* 103:535–539.)

Genomic imprinting and neurodevelopmental disorders

Paternally and maternally expressed genes present at the imprinted locus

There is one protein-encoding gene that is maternally expressed, and four protein-encoding and several small nucleolar RNA (snoRNA) genes that are paternally expressed in the imprinted region (see Fig. 12.5, Table 12.2). In 1997, the gene responsible for Angelman syndrome was identified as encoding an E6-AP ubiquitin-protein ligase (UBE3A), an enzyme involved in the binding of substrate prior to

ubiquitinylation (see Fig. 5.17). In Angelman cases that result from classic mutations, four loss-of-function mutations in the *UBE3A* gene have been identified. This was the first example of a genetic disease associated with the ubiquitin-mediated protein degradation pathway. E6-AP ubiquitin-protein ligase is very selective in its binding and only four substrates have so far been identified. The possible molecular mechanisms underlying the human Angelman syndrome mental retardation state remain to be determined.

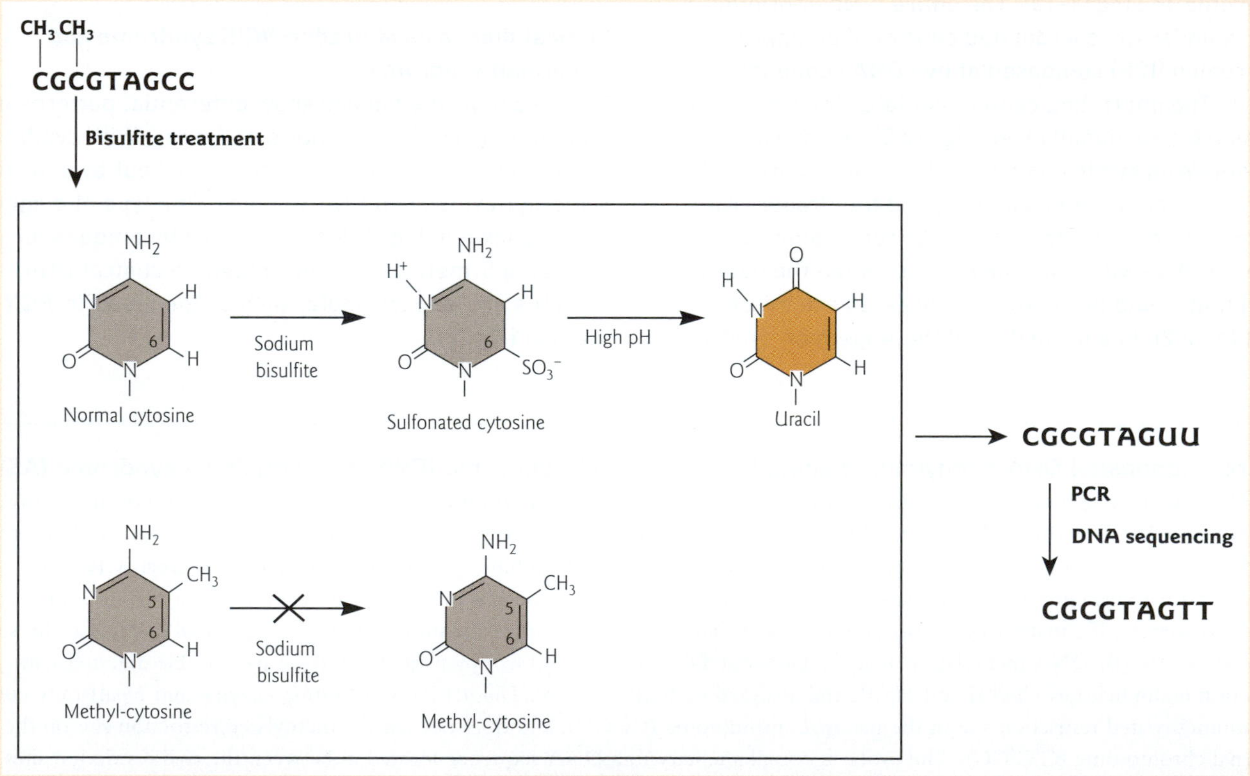

Figure 2 Bisulfite-PCR method for distinguishing normal cytosine from 5-methyl-cytosine. When genomic DNA samples are denatured and treated with sodium bisulfite ($NaHSO_3$), a sulfite group (SO_3) is added to the 6-position of cytosine only if the cytosine is unmethylated. Steric hindrance between the 6-ring position and the large 5-methyl group prevents reaction of sodium bisulfite with 5-methyl-cytosine. When the treated DNA is then subjected to high pH, the sulfonated cytosines are converted to uracil. Subsequently, the converted sample is amplified by PCR, using two specific primers for the region of interest, followed by sequencing of the PCR product. Every methylated cytosine will continue to read as C. Every unmethylated cytosine will be converted by bisulfite from C to U, which is then read as T in the PCR and sequencing reactions.

Genomic imprinting and neurodevelopmental disorders

The contribution of the paternally expressed protein-encoding genes and the snoRNAs to Prader–Willi syndrome remains unclear. MKRN3 (makorin ring finger protein 3) mediates ubiquitin conjugating enzyme E2-dependent ubiquitinylation and is expressed in all tissues. MAGEL2 (melanoma antigen-like gene 2) is expressed in the brain. NDN (necdin), a nuclear protein expressed in neurons and at high levels in the hypothalamus, is involved in post-mitotic neuron-specific growth suppression. SNURF (SNRPN upstream reading frame proteins) and SNRPN (small nuclear ribonucleoprotein-associated polypeptide N) form an extremely complex locus which encodes many novel human brain (HBII) snoRNAs. The HBII snoRNAs are expressed within introns of the SNURF-SNRPN large transcript, which has > 148 possible exons that undergo alternative splicing. In addition to encoding multiple snoRNA species, the spliced transcript serves as an antisense RNA for *UBE3A*, controlling expression of the paternal *UBE3A* allele. The *UBE3A* gene is expressed from both paternal and maternal alleles in most human tissues, yet maternally expressed in various parts of the brain. The paternal silencing of the *UBE3A* gene occurs in a brain region-specific manner; the maternal allele is active almost exclusively in neurons present in the hippocampus and cerebellum.

Mutations in the *MeCP2* gene cause Rett syndrome

Patients with Rett syndrome have mutations in the gene coding for methyl-CpG-binding protein 2 (MeCP2) on the X chromosome. It is not lethal in females since there is a 50/50 chance that the mutant X will be inactive (see Section 12.4). MeCP2 is part of a corepressor complex including histone deacetylase (HDAC) that controls transcription through the deacetylation of core histones (see Section 11.6). MeCP2 has a methyl-CpG-binding domain and a transcriptional repressor domain that associates with the corepressor complex. MeCP2 was originally believed to be a general transcriptional repressor that binds to methylated DNA with no obvious sequence preference. However experiments using microarray analysis have questioned this concept.

Rett syndrome and Angelman syndrome share some phenotypic characters (see Table 12.2), and are sometimes misdiagnosed prior to genetic testing. It turns out that a deficiency of MeCP2 causes epigenetic alterations at the Prader–Willi syndrome/Angelman syndrome imprinting control region (ICR), including changes in the patterns of histone acetylation and methylation. Altered histone modifications lead to loss of imprinting of the *UBE3A* antisense gene in the brain, an increase in *UBE3A* antisense RNA levels, and consequently a reduction in *UBE3A* expression. As the changes have no effect on paternally expressed alleles, there is no phenotypic similarity with Prader–Willi syndrome. The reduced expression of UBE3A protein in the brain of Rett syndrome patients provides the first example of a direct link between gene repression and MeCP2 deficiency.

deregulated in two neurodevelopmental diseases, Prader–Willi syndrome and Angelman syndrome (Fig. 12.5A; see also Disease box 12.3).

Differential expression of an antisense RNA transcript

In this indirect mechanism, the ICR contains the promoter of a nonprotein-coding gene. On the chromosome in which the ICR is unmethylated, the gene is expressed to produce an antisense RNA. This antisense RNA then, by an undetermined mechanism, silences the protein-coding imprinted gene on the same chromosome. On the other chromosome, methylation of the ICR ensures that the antisense RNA is not expressed. In the absence of the antisense RNA, the protein-coding gene is active (Fig. 12.5B).

This mechanism is exemplified by the insulin-like growth factor 2 receptor (*Igf2r*) gene, which encodes an important regulator of fetal growth. *Igf2r* is expressed exclusively from the maternal allele on chromosome 17. Deletion analysis has shown that a small (3 kb) CpG island located in the first intron of *Igf2r* functions as the ICR. This region also contains the promoter for the *Air* gene. There are allele-specific differences in histone modification that work together with DNA methylation to ensure imprinting. The silenced (DNA methylated) maternal ICR is associated with histone H3 methylation at lysine 9 (H3-K9), while the active paternal ICR is marked by histone H3 acetylation and H3-K4 methylation. On the paternal chromosome where the ICR is unmethylated, the *Air* gene is transcribed and the antisense RNA inhibits the expression of the paternal *Igf2r* allele. Methylation of the maternal ICR blocks *Air* expression, allowing the maternal *Igf2r* allele to be actively expressed.

Blocking of an enhancer by an insulator

In a third mechanism, the ability of shared enhancers to activate one or other imprinted genes is determined by a chromatin boundary element (insulator) present on the unmethylated allele between the two genes. This situation is exemplified by the *Igf2-H19* imprinted locus (Fig. 12.5C). The *Igf2* and *H19* genes are located in one cluster on mouse chromosome 7 and are oppositely imprinted with *Igf2* paternally expressed and *H19* maternally expressed. The insulin-like growth factor 2 (Igf2) is a highly conserved, potent growth factor that stimulates placental and fetal growth. The *H19* gene is transcribed by RNA polymerase II as a capped, spliced, and polyadenylated RNA. These are all features suggestive of a protein-coding mRNA (see Section 13.5) but, to date, all attempts to characterize an H19 protein have failed. The *H19* gene product is proposed to function as a "ribomodulator" or regulatory RNA.

Several elements are involved in the imprinted expression of *Igf2* and *H19*. In 1993, M. Azim Surani and colleagues demonstrated that hypermethylation and condensed chromatin in the region of the *H19* promoter region are associated with repression of the paternally inherited copy of the gene (Fig. 12.6). In contrast most of these sites are unmethylated in sperm, with methylation of the paternal promoter region occurring after fertilization. Studies since then have shown that both genes require a set of enhancers located downstream of *H19* for their expression (see Fig. 12.5). An ICR that is essential for imprinted expression is located upstream of the *H19* promoter. The CTCF protein regulates imprinted expression by binding to the unmethylated maternal ICR but not to the methylated paternal ICR. CTCF was first characterized for its role in regulating β-globin gene expression, where it functions as an insulator-binding protein and forms a chromosomal boundary (see Section 11.3). When bound to the unmethylated maternal DNA, CTCF blocks the downstream enhancer elements from interacting with the promoter of the *Igf2* gene and only the *H19* gene is expressed. Methylation of the paternal ICR prevents the binding of CTCF. In this case, the promoter of the *Igf2* paternal allele is activated by the downstream enhancers. Because the enhancer can no longer interact with the *H19* promoter, the paternal *H19* allele is silenced. Silencing is proposed to be mediated by long-range chromatin interactions, involving chromatin-loop formation.

Genomic imprinting is essential for normal development

Normal embryogenesis cannot proceed without the machinery of epigenetic regulation. Experiments in mice have shown that knockouts of DNA methyltransferases and histone modifiers are embryonic lethal. Not surprisingly, imprinted regions of the genome are associated with a number of diverse developmental disorders that result from impaired regulation, altered dosage, or mutation of these domains.

Since imprinted genes often occur in clusters coordinately regulated by an ICR, single alterations in these key regions can lead to disruption of many genes resulting in the formation of multiple disorders. Epigenetic inheritance has been associated with a number of human diseases, including Silver–Russell syndrome (asymmetry dwarfism), Prader–Willi and Angelman syndromes (Disease box 12.3), retinoblastoma (a tumor of the retina, see Disease box 17.1), and Beckwith–Wiedemann syndrome. Beckwith–Wiedemann syndrome results from a rare birth defect in which *Igf2* is biallelically expressed. It has been suggested that the double dosage of *Igf2* does its damage by inhibiting apoptosis and promoting cell proliferation. Infants born with Beckwith–Wiedemann syndrome are more likely to develop abdominal wall defects, various types of malignant tumors, and macroglossia (enlarged tongue). Loss of imprinting of *Igf2* occurs in many common cancers in humans, including in ~30% of patients with colorectal cancer. In a transgenic mouse

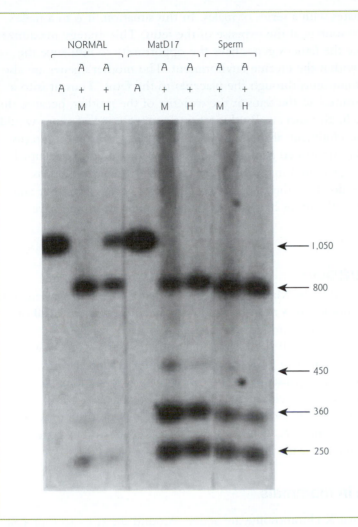

Figure 12.6 Differential methylation of *H19* in normal mouse embryos, MatDi7 embryos, and sperm. MatDi7 mouse embryos carry a maternal duplication/paternal deficiency on chromosome 7 and express a double dose of *H19*. DNA samples were digested with *Apa*I (A), *Msp*I (M), and *Hpa*II (H), a methylation-sensitive restriction endonuclease which only cuts at a nonmethylated restriction site, and were analyzed by Southern blot with an *H19* promoter-specific probe. In normal embryos the methylated maternal copy of *H19* remains undigested, so there are two distinct higher molecular weight bands (1050 and 800 bp) on the blot for the *Apa*I-*Hpa*II (A + H) double digest. In MatDi7 embryos, there is digestion of both maternal chromosomes, so only the 800 bp band appears on the blot (along with lower molecular weight bands of 450, 350, and 250 bp). The blot shows that the promoter region is unmethylated in sperm DNA. Both copies of *H19* are cut and only the 800 bp band is visible. (Reprinted by permission from Nature Publishing Group and Macmillan Publishers Ltd: Ferguson-Smith, A.C., Sasaki, H., Cattanach, B.M., Surani, M.A. 1993. Parental-origin-specific epigenetic modification of the mouse H19 gene. *Nature* 362:751–755. Copyright © 1993.)

model of loss of imprinting of *Igf2*, the mice developed twice as many intestinal tumors as did control littermates.

Some studies conclude that the incidence of diseases associated with imprinting defects is higher among babies born with the aid of assisted reproduction technology. However, the *in vitro* approach is a powerful one and the benefits are considered to far outweigh any potential risks. Assisted reproduction technology involves fertility treatments in which both egg and sperm are manipulated in the laboratory; i.e. *in vitro* fertilization and related procedures. There is strong circumstantial evidence for an association between epigenetic defects and assisted reproduction technology, but further in–depth study is required to confirm this association and to determine the mechanism. Epigenetic alterations could arise directly from some aspect of *in vitro* culture. Alternatively, epigenetic alterations could be a significant cause of infertility, rather than a consequence of the procedure used to treat it. The difficulties associated with reprogramming all the chromatin, histones, and methylation patterns along the entire length of the DNA sequence may also explain why cloned embryos have so many developmental failures (see Section 15.5).

Origins of genomic imprinting

Multiple hypotheses have been proposed to explain the origins of imprinting in early mammalian evolution. The most popular of these is the "conflict hypothesis." Imprinting is theorized to have evolved in early mammals with polyandry, where each female mates with a series of males. In this situation, it is to a male's benefit to silence genes that conserve maternal resources at the expense of the fetus. This strategy maximizes the father's immediate reproductive success, since the father's genes have the opportunity to influence the growth and competitive fitness of his offspring within the uterine environment. The mother's genes are also capable of regulating energy and distribution of nutrients through the placenta to the fetus. Thus, it is to a female's benefit to silence genes that allocate resources to the fetus at the expense of the mother, because this strategy maximizes the female's long-term reproductive success. By this rationale, imprinting of genes would only be relevant to mammalian species where development of the fetus occurs primarily within the uterus, and should be absent in egg-laying species. So far, imprinted genes have only been identified in marsupial and placental mammals but not in monotreme (egg-laying) mammals. The "conflict hypothesis" also is supported by the fact that some imprinted genes do affect the allocation of resources between mother and fetus in the direction that would be predicted. On the other hand, many genes that are imprinted have no obvious connection to maternal–fetal conflict.

12.4 X chromosome inactivation

X chromosome inactivation was first described in 1961 by Mary Lyon. From studying coat color variegation in mice, she concluded that one of the two X chromosomes must be randomly inactivated in each cell of female mice. She hypothesized that this process of whole-chromosome silencing evolved to balance transcriptional dosage between XX females and XY males. If X-linked genes were expressed equally well in each sex, the female would have twice as much of each gene product as males. Lyon's insights are now referred to as either the Lyon hypothesis or "dosage compensation." The calico cat provides a familiar visual example of the epigenetic linkage between genetics and developmental biology (see Focus box 15.4). Different organisms deal with dosage compensation in different ways. In *Drosophila*, genes on the male X chromosome are expressed at twice the level of the female X, whereas in *C. elegans* expression of genes on the female X chromosome is reduced by half compared with the male X.

Random X chromosome inactivation in mammals

Within the zygote, both the maternal and paternal X chromosome are active. Around the two- to four-cell stage, the paternal X is preferentially inactivated (Fig. 12.7). In marsupials, the paternal X remains inactivated in all tissues. In contrast, in placental mammals (e.g. humans and mice), the paternal imprint is erased at a later stage in cells that subsequently give rise to the fetus, but remains inactive in placental tissues. At the onset of gastrulation around embryonic day 6.5, there is random inactivation of one X chromosome in the fetal tissues. From this point on, there is stable maintenance of X chromosome inactivation. The same X is inactive in all progeny of a given cell. This leads to cellular mosaicism in adult tissues (Fig. 12.7). The inactive X appears as dark-staining heterochromatin, forming a highly condensed object in the interphase nucleus called the "Barr body".

Molecular mechanisms for stable maintenance of X chromosome inactivation

X inactivation is controlled by one regulatory domain, the X inactivation center. A key player in the function of the X inactivation center is the X inactivation specific transcript (XIST, pronounced "exist"), transcribed from the *XIST* gene located in the X inactivation center. Prior to the onset of X inactivation, *XIST* is expressed at low levels from both X chromosomes, together with an overlapping antisense transcript, *Tsix*. When initiation of X inactivation is induced during differentiation, *XIST* transcript levels are upregulated from the chromosome that will become the inactive X, while *Tsix* expression, which has a repressive role, is downregulated. Activity of the active X requires persistent expression of *Tsix* in *cis* (Fig. 12.8).

The mechanism for initiation of silencing of the inactive X chromosome is not fully understood. What is known is that the coating of the X chromosome destined for inactivation with *XIST* RNA is central to

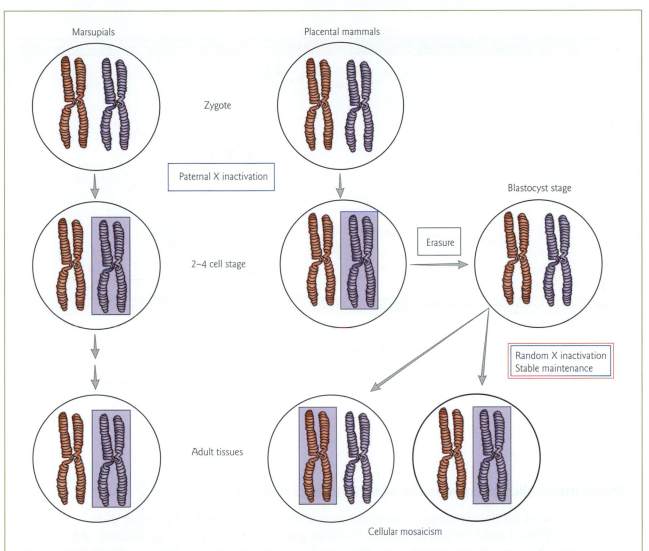

Figure 12.7 X chromosome inactivation in marsupials and placental mammals. The paternal X chromosome (blue) undergoes preferential inactivation (indicated by light blue box) around the 2–4 cell stage, and remains inactive in all tissues in the marsupial. In placental mammals, the mark is erased in cells of the blastocyst that give rise to the fetus. During subsequent development there is random inactivation of either the paternal or maternal (pink) X chromosome. X chromosome inactivation is stably maintained in all progeny of a given cell, resulting in cellular mosaicism in adult tissues.

the process. The inactive X is characterized by a series of epigenetic chromatin modifications including histone H3 methylation, histone H4 hypoacetylation, enrichment of variant histone macroH2A, and DNA methylation (many CpG islands are heavily methylated on the inactive X). One of the functions of XIST is to recruit, either directly or indirectly, the EED–EZH2 Polycomb group protein complex (see Focus box 11.4) that contains histone modifying enzymes. Subsequent maintenance of the inactive state is independent of XIST. How *XIST* RNA and other epigenetic modifications are directed to sites along the inactive X chromosome, and how inactivation spreads over the 155 Mb X chromosome, still remain poorly understood. Over one-third of the X chromosome is composed of *L1* retrotransposon repeat elements (see Section 12.5). Their distribution is consistent with the proposal that they function as "way stations" in the process of X chromosome inactivation.

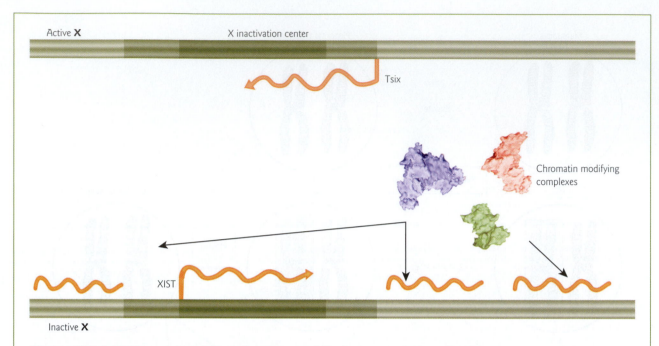

Figure 12.8 Initiation and stable maintenance of X chromosome inactivation. Random inactivation of one X chromosome is brought about by "coating" of the chromosome by the untranslated XIST transcript and subsequent recruitment of chromatin modifying complexes. Upregulation of the antisense Tsix transcript represses expression of XIST from the active X chromosome.

Is there monoallelic expression of all X-linked genes?

Because of the need for dosage compensation, the dogma for over 40 years has been that there is monoallelic expression of X-linked genes. It was presumed that most genes on the inactive X are inactive, although a few involved in the silencing process were known to be active. However, in March 2005 results of a large-scale survey of gene activity on the inactive X chromosome were reported by Laura Carrel and Huntington F. Willard that altered this view. Their work shows that genes once thought to be silenced on the inactive X are sometimes expressed (Fig. 12.9). Although 75% of genes are permanently silent as was expected, about 15% were shown to escape inactivation. In an additional 10% of genes, the level of expression differs from woman to woman. The genes that escape inactivation are clustered, suggesting that control is at the level of chromatin domains. The clusters are primarily towards the end of the X chromosome short arm, in the "pseudoautosomal" regions. Mammalian X and Y chromosomes are proposed to have diverged from an identical pair of ancestral autosomes. Genes within the pseudoautosomal regions are shared between the sex chromosomes. These findings have a number of important implications. Since these genes are active on both X chromosomes they are expressed at twice the level in females as in males. Such genes are potential contributors to sexually dimorphic traits, and may explain phenotypic variability among females heterozygous for X-linked conditions.

12.5 Phenotypic consequences of transposable elements

Mobile, or transposable, genetic elements and their derivatives are abundant in the genomes of bacteria, plants, and animals. In mammals, transposable elements may account for nearly half of the genome and in some higher plants they make up as much as 90% of the genome. Transposable elements are DNA sequences

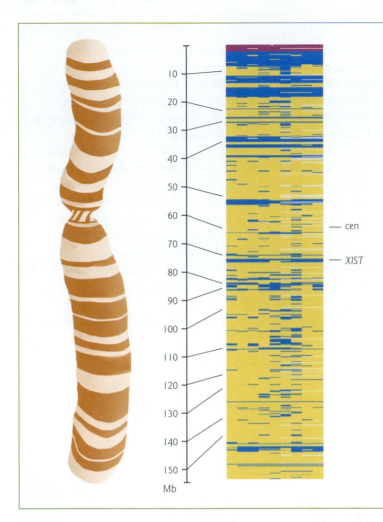

Figure 12.9 X inactivation profile of the human X chromosome. Gene expression from human active X (X_a) and inactive X (X_i) chromosomes was analyzed using a panel of rodent/human somatic cell hybrids. Reverse transcription–polymerase chain reaction (RT-PCR) was used to measure RNA transcript levels and to determine whether gene expression only occurred in hybrid lines retaining X_a chromosomes (and is therefore a gene subject to X inactivation) or from both X_a hybrid lines and X_i hybrid lines, indicating a gene that escapes X inactivation. A total of 624 genes were tested in nine X_i hybrids. Each gene is linearly displayed. Blue denotes signficant X_i gene expression, yellow shows silenced genes, pseudoautosomal genes are purple, and untested hybrids remain white. Positions of the centromere (cen) and *XIST* are indicated. (Reprinted by permission from Nature Publishing Group and Macmillan Publishers Ltd: Carrel, L., Willard, H.F. 2005. X-inactivation profile reveals extensive variability in X-linked gene expression in females. *Nature* 434:400–404. Copyright © 2005.)

that have the ability to integrate into the genome at a new site within their cell of origin. Because they appear to propagate like a parasite in host DNA, transposable elements are sometimes referred as "selfish DNA." Movement (transposition) from one place in the genome to another may disrupt genetic function and result in phenotypic variation. Considering the abundance of transposable elements in the genome of plants and animals, it is surprising that only a low proportion of spontaneous mutations are caused by them. In fact, transposable elements seem to have achieved a fine balance between detrimental effects on the individual and long-term beneficial effects on a species through the potential for genome modification. Mechanisms appear to exist that suppress uncontrolled transposition of these elements.

Eukaryotic genomes with a high proportion of repeated sequences generally have high methylated cytosine content. In the mammalian genome, most of the methylated cytosine is found in transposable elements and other types of repetitive DNA. Some eukaryotic species with less genomic DNA methylation, such as *Drosophila*, suffer from a high frequency of transposon-mediated mutations compared with vertebrates and higher plants. These observations led to the proposal that the primary function of eukaryotic DNA methylation may be defense of the genome from the potentially deleterious effects of transposable elements. From a practical viewpoint, mobile genetic elements have also provided a powerful tool for insertional mutagenesis studies (Tool box 12.1). This section provides an overview of the types of transposable elements, their mechanism of insertion, and the phenotypic consequences of their movement within the genome. Section 12.6 focuses on the epigenetic control of transposable element activity.

Transposons are often used by molecular biologists as a powerful tool for insertional mutagenesis. They provide a method to link phenotype with genomic sequence. For example, in a technique called "transposon tagging," a transposon carrying an antibiotic resistance gene can be introduced into pathogenic bacteria. Bacteria are then screened for nonfunctional mutants. If the bacteria are no longer pathogenic, this indicates that insertion of the transposon disrupted a gene important in pathogenicity. The disease gene which is "tagged" by the antibiotic-resistant phenotype can then be easily cloned. P transposable elements of *Drosophila* are used in a similar way for insertional mutagenesis studies in transgenic fruitflies. This technique has been used to identify many important genes involved in development and behavior.

More recently, transposons have been used to determine the function of mammalian genes after gene knockout by insertional mutagenesis. When a *Tc1/mariner*-type DNA transposon called Sleeping Beauty is inserted into the genome of mouse strains that already contain the Sleeping Beauty transposase, it is mobilized in the subsequent generation from one genomic location to another genomic site (Fig. 1). Insertions tend to be concentrated relatively close to the original transposon site, since transposons typically undergo "local hopping." LINEs, on the other hand, offer the potential for generating retrotranspositions at random, distant sites throughout the genome. However, so far, efficiencies are low in transgenic mice. Retrotransposition of human LINE transgenes occurs in only approximately one in every 15–20 offspring.

Figure 1 Sleeping Beauty transposition activates green fluorescent protein (GFP) in transgenic mice. (A) An overview of the experimental strategy. Multiply integrated *GFP* genes are epigenetically repressed at the original integration site in the mouse chromosome by DNA methylation of the flanking Sleeping Beauty transposon elements (gray arrows). Excision of transposons by the Sleeping Beauty transposase releases *GFP* genes from their repressed state. Transposition events are detected by GFP activity at the novel site. (B) *ITRA-GFP:SB* doubly transgenic mice were obtained by the mating of *ITRA-GFP* mice (carrying the *GFP* gene flanked by Sleeping Beauty transposon elements) and *SB* transgenic mice (carrying the Sleeping Beauty transposase gene). Neither *ITRA-GFP* transgenic mice carrying the *GFP* gene only or *ITRA-GFP:SB* doubly transgenic mice showed GFP signal. However, some of the progeny from *ITRA-GFP:SB* doubly transgenic mice mated with wild-type mice were GFP positive, indicating efficient chromosomal transposition of GFP by the Sleeping Beauty transposase and subsequent activation of GFP at its new chromosomal location. (Reprinted with permission from Horie, K., Kuroiwa, A., Ikawa, M., Okabe, M., Kondoh, G., Matsuda, Y., Takeda, J. 2001. Efficient chromosomal transposition of a *Tc1/mariner*-like transposon *Sleeping Beauty* in mice. *Proceedings of the National Academy of Sciences USA* 98:9191–9196. Copyright © 2001 National Academy of Sciences, USA.)

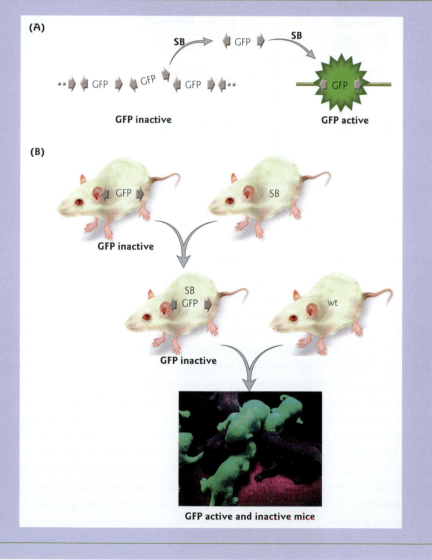

Historical perspective: Barbara McClintock's discovery of mobile genetic elements in maize

In the 1940s to 1950s, Barbara McClintock discovered mobile genetic elements. While studying chromosome breakage events in maize, she noticed a high level of phenotypic variegation; for example, spots of purple pigment in white corn kernels (Fig. 12.10A). These pigmented sectors were genetically

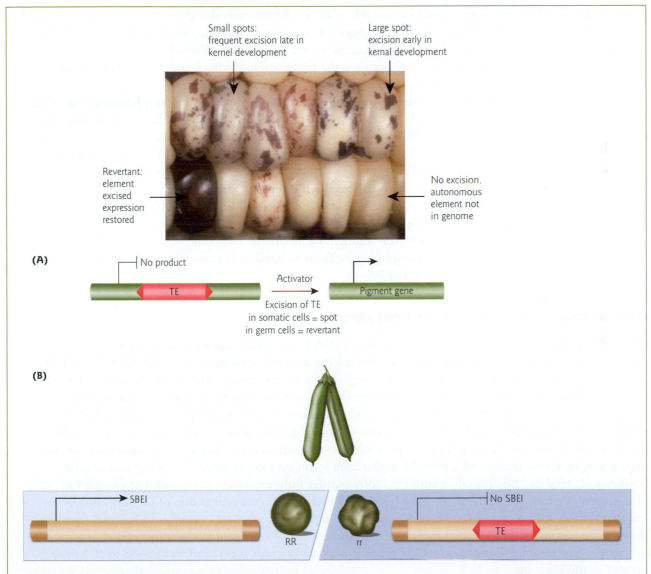

Figure 12.10 Transposable elements in plants. (A) Using maize kernel phenotypes to study transposable elements. Kernels on a maize ear show unstable phenotypes due to the interaction between a transposable element (TE) and a gene that encodes an enzyme involved in anthocyanin (pigment) production. Sectors of revertant (pigmented) tissue result from the transposition (excision) of the TE in a single cell. The size of the sector reflects the time in kernel development at which excision occurred. (Adapted by permission from Nature Publishing Group and Macmillan Publishers Ltd: Feschotte, C., Jiang, N., Wessler, S.R. 2002. Plant transposable elements: where genetics meets genomics. *Nature Reviews Genetics* 3:329–341. Copyright © 2002.) (B) The wrinkled seed (*rr*) character of the garden pea described by Mendel is caused by the insertion of a transposable element (red) in a gene encoding starch-branching enzyme (SBEI).

distinct from the surrounding normal tissue, and the genetic changes often correlated with the appearance of chromosomal breaks. These points of breakage were found in the same locations in different maize plants of the same strain, but McClintock found that sites of breakage could disappear or move to a new chromosomal location. After such a "transposition" event, chromosomal breaks could occur selectively at the new location. McClintock named such a mobile hot spot for breakage the "Dissociation" (*Ds*) element. Further genetic crosses showed that *Ds* elements did not act alone, but required another element present at a different chromosomal location. This element was named "Activator" (*Ac*) for its effect on *Ds*. McClintock found that not only could the *Ds* elements break chromosomes, but they could also induce new mutations in maize genes. These mutations could be unstable if an *Ac* element was also present in the strain, but stable in the absence of *Ac*. Furthermore, an active transposable element frequently lost the ability to mobilize and this inactive state was transmitted and maintained in later generations. Kernels on a maize ear show variegated phenotypes due to the interplay between the transposable element and a gene that encodes an enzyme involved in anthocyanin (pigment) production.

Similar transposable element families were later found to underlie unstable mutant phenotypes in other plants, such as the variegated patterns of flower pigmentation in snapdragon, petunia, and morning glory, and the color of grapes. In the early 1990s the surprise discovery was made that Mendel's wrinkled peas (see Fig. 1.3) result from a disruption of the starch-branching enzyme I gene by a transposable element (Fig. 12.10B). The element is similar to McClintock's *Ac* transposable element. Because of the 0.8 kb insertion, a defective transcript that encodes a nonfunctional enzyme is produced. This leads to sucrose accumulation, lack of starch synthesis, and higher water content. Because of the osmotic pressure in developing seeds, wrinkled seeds lose more water than smooth seeds, and thus appear wrinkled when dry.

For a long time, transposable elements were thought to be unique to plants, but transposable elements have now been described in many organisms ranging from bacteria to humans. For her pioneering work in this field, McClintock won the 1983 Nobel Prize.

DNA transposons have a wide host range

DNA transposons are transposable elements with a DNA intermediate during transposition. This distinguishes them from the retrotransposons which have an RNA intermediate (Table 12.3, Fig. 12.11). DNA transposons are abundant in bacteria and include the bacteriophage Mu and transposable elements called insertion sequences. Insertion sequences are defined as mobile genetic elements that are known to encode only functions involved in insertion events. In contrast, bacterial transposons, such as Tn7, also contain within their internal DNA sequence one or more bacterial genes whose functions are unrelated to the insertion process. As an example, one antibiotic-resistant strain of *Staphylococcus aureus* appears to have arisen by lateral transfer of a plasmid with a transposon containing genes encoding vancomycin resistance. Lateral or "horizontal" transfer is DNA transfer from organism to organism as opposed to inheritance of genes by "vertical" descent from one's parents. Vancomycin is one of the few antibiotics left that reliably kills *S. aureus*. *S. aureus* ("Staph") is a pathogenic microbe common in hospital settings, thus the development of a resistant strain is cause for concern. The plasmid originally resided in a strain of vancomycin-resistant *Enterococcus faecalis* present in a patient's foot ulcer, and was passed to *S. aureus* present at the same site.

DNA transposons are also found in many multicellular eukaryotes including insects, nematode worms, and humans. They include *Drosophila* P elements and the *Tc1/mariner* superfamily (Table 12.3). P elements do little harm in *Drosophila* because expression of their transposase gene is usually repressed. However, when male flies with P elements mate with female flies lacking them, the transposase becomes active in the germline producing so many mutations that their offspring are sterile (hybrid dysgenesis). The *Tc1/mariner* superfamily (named for the most extensively studied family members) of transposable elements is one of the most diverse and widespread. They are abundant in invertebrates and found even in some vertebrate species. Because of their wide host ranges, it is speculated that they have evolved to switch hosts by lateral transfer to avoid extinction. DNA transposons are the least abundant of the transposable elements in mammals. In

Table 12.3 Classes of transposable elements.

Class	Transposition intermediate	Examples

Class I
LTR retrotransposons
	RNA	Yeast:
		  Ty elements
		Human:
		  Human endogenous retroviruses (HERV)
		Mouse:
		  Intracisternal A particles (IAPs)

Non-LTR retrotransposons
  LINES (autonomous)
  SINES (nonautonomous)
	RNA	Human:
		  L1 elements
		  Alu elements

Class II
DNA transposons
	DNA	Bacteria:
		  Insertion sequence
		  Bacteriophage Mu
		  Transposons (e.g. Tn7)
		Drosophila:
		  P elements
		Maize:
		  Ac and *Ds* elements
		Invertebrates and vertebrates:
		  Tc1/mariner superfamily

humans, transposons make up approximately 3% of the genome and most appear to have become completely inactive. In the case of *mariner*, however, one element on chromosome 17 has been implicated as a "hot spot" for recombination (Disease box 12.4).

DNA transposons move by a "cut and paste" mechanism

When active, DNA transposons are excised from one genomic site and integrated into another by a "cut and paste" mechanism (Fig. 12.12A). Many of the DNA transposons, such as the bacterial insertion sequences and the *Tc1/mariner* family elements, have a quite simple design. These elements consist of a transposase gene flanked by inverted terminal repeats that bind transposase and thereby mediate transposition. The transposase enzyme has a number of functional domains, including a catalytic domain and a DNA-binding domain. The *Tc1* transposase, for example has two DNA-binding domains, one related to the fruitfly paired DNA-binding domain (two helix–turn–helix motifs and a β–hairpin turn structure) and the other related to the homeodomain DNA-binding motif (see Section 11.5). The enzyme can catalyze the whole set of reactions necessary for DNA transposition. The transposase binds at or near the inverted repeats and to the target DNA. Three critical amino acids (two aspartic acid and one glutamic acid) fold together to form a three-dimensional structure called the "DDE" motif that binds two metal atoms (Mg^{2+} or Mn^{2+}) which are central to the chemistry of DNA breaking and joining. The transposase then catalyzes hydrolysis of phosphodiester bonds between the transposon and flanking DNA to remove the transposon from its old site (Fig. 12.12A). The free 3′-OH residues generated in the DNA breakage reaction carry out the attack at the new site and a joining reaction inserts the transposon into its new site. Sequence specificity of integration is limited to a small number of nucleotides, thus insertions can occur at a large number of genomic sites. The *Tc1/mariner*-type transposons have specificity for TA dinucleotides. Because the two strands of the new

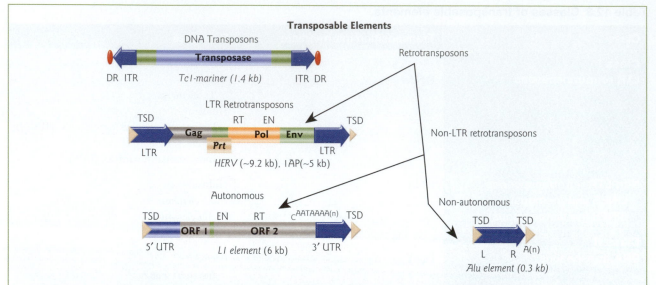

Figure 12.11 Classes of transposable elements. DNA transposons, e.g. *Tc1/mariner*, have inverted terminal inverted repeats (ITRs), a single open reading frame (ORF) that encodes a transposase, and are flanked by short direct repeats (DRs). Retrotransposons are divided into autonomous and nonautonomous classes depending on whether they have ORFs that encode proteins required for retrotransposition. An example of a LTR retrotransposon is the human endogenous retroviruses (HERV). These elements have long terminal repeats (LTRs) and slightly overlapping ORFs for group-specific antigen (*gag*), protease (*prt*), polymerase (*pol*), and envelope (*env*) genes, and produce target site duplications (TSDs) upon insertion. Also shown are the reverse transcriptase (RT) and endonuclease (EN) domains. *L1* is an example of a non-LTR retrotransposon. *L1* elements consist of a 5′-untranslated region (5′ UTR) containing an internal promoter, two ORFs, a 3′ UTR, a poly(A) signal followed by a poly(A) tail (A_n), and 7–20 bp TSDs. The RT, EN, and a conserved cysteine-rich domain (C) are shown. An *Alu* element is an example of a nonautonomous retrotransposon. *Alu* elements contain two similar monomers, the left (L) and the right (R), and end in a poly(A) tail. Approximate full-length element sizes are given in parentheses. (Adapted with permission from: Ostertag, E.M., Kazazian, H.H. 2001. Biology of mammalian L1 retrotransposons. *Annual Review of Genetics* 35:501–538. Copyright © 2001 by Annual Reviews.)

DNA are attacked at staggered sites, the inserted transposon is flanked by small gaps which, when filled in by host DNA repair enzymes, lead to short duplications of sequence at the target sites. The empty donor site is filled in by DNA polymerase and rejoined by blunt-end ligation, often leaving an "excision footprint" of a few nucleotides left behind by the transposable element.

There are also nonautonomous transposons that do not encode transposase, but consist of a pair of inverted repeats that function as transposase-binding sites. Nonautonomous family members are usually derived from an autonomous family member by internal deletion. For example, the *Ds* element in maize is derived from *Ac* elements.

Retrotransposons move by a "copy and paste" mechanism

Retrotransposons are thought to be remnants of ancestral retroviral infections that became fixed in the germline DNA and subsequently increased in copy number. They move by a "copy and paste" mechanism that duplicates the element (Fig. 12.12B). They are first transcribed into RNA, and then reverse transcribed into a cDNA and reintegrated into the genome. Retrotransposons can be divided into two groups based on their transposition mechanism and structure. They either contain long terminal repeats at both ends (LTR retrotransposons) or lack long terminal repeats and have a polyadenylate (poly(A)) sequence at their 3′ end (non-LTR retrotransposons) (see Table 12.3 and Fig. 12.11).

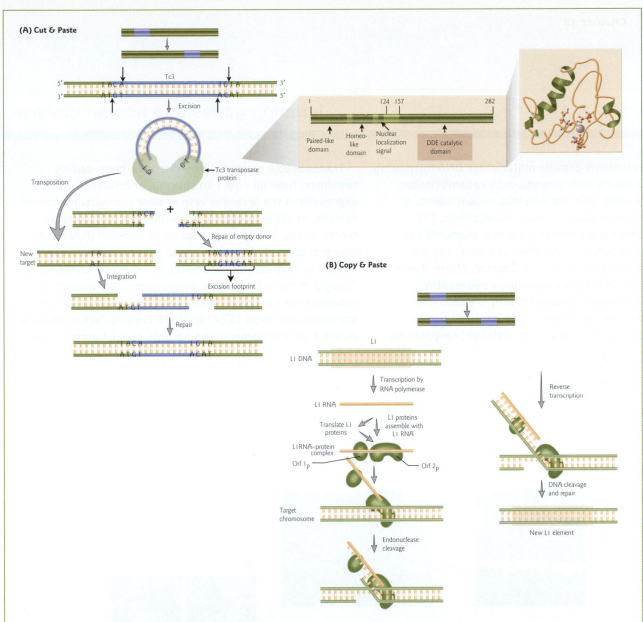

Figure 12.12 Mechanisms of transposition. (A) "Cut and paste." A cycle of excision and integration is shown for *Tc3*, a member of the *Tc1/mariner* family of DNA transposons. The element is cleaved by the transposase (arrows) to release the transposon from the flanking host DNA. The transposon DNA remains associated with the transposase, and the complex then carries out integration at a new TA target site. Because the 5′ cleavage sites are within the transposon DNA, two bases derived from the 5′ end of the transposon remain at the empty transposon donor site. The empty donor site is repaired leaving an excision footprint, in which both single-stranded regions are filled in by DNA polymerase prior to end-joining. The transposon 3′ ends are joined to 5′ protruding ends in the target DNA. Repair of the resulting gaps completes integration. (Inset) Structure of a *Tc1/mariner* transposase, showing the DDE motif and functional regions, and predicted structure of a typical DDE motif. The DDE amino acid residues are shown using ball–and–stick rendering and the central ball represents the metal ion Mg^{2+}. (Adapted with permission from: Lin, T.H., Tsai, K.C., Lo, T.C. 2003. Homology modeling of the central catalytic domain of insertion sequence IS*LC3* isolated from *Lactobacillus casei* ATCC 393. *Protein Engineering, Design & Selection* 16:819–829. Copyright © 2003, Oxford University Press.) (B) "Copy and paste." Retrotransposition of the *L1* element is depicted. The DNA sequence encoding an *L1* element is transcribed into RNA and translated into the L1 proteins ORF1p and ORF2p. The RNA and proteins form a complex which binds to target DNA. The endonuclease makes a nick in the DNA, generating a free end. The reverse transcriptase uses this free end as a primer for synthesis of a DNA copy of the *L1* RNA. The RNA is degraded (not shown) and the single-stranded DNA copy is used as a template to make double-stranded DNA, which is then integrated into the target site by DNA cleavage and repair processes. (Redrawn from Bushman, F. (2004) *Nature* 429:253–255.)

Jumping genes and human disease

Transposable elements provide material for DNA mispairing and unequal crossing-over (homologous recombination) between the repeats, and are potential causal agents of human disease through insertional mutagenesis. For example, recombination between *mariner* elements can yield either a deletion or duplication of a short region. Two human neurological diseases, Charcot–Marie–Tooth disease type 1A (CMT1A) and hereditary neuropathy with liability to pressure palsies (HNPP) are both due to such rearrangements in this region of chromosome 17 – a small duplication and a small deletion, respectively.

Transposable elements may disrupt a gene-coding sequence, have an effect on splicing, or influence gene expression if the insertion is in or near promoter/enhancer regions. In addition, promoters within the transposable element may initiate transcription of adjacent genes, and they are susceptible to epigenetic silencing which may spread to neighboring chromatin. In 1988, two LINEs were "caught in the act" of retrotransposition. Two unrelated patients with hemophilia were shown to have *L1* element insertions into exon 14 of the blood-clotting factor VIII gene on the X chromosome. In the parents, this *L1* element was

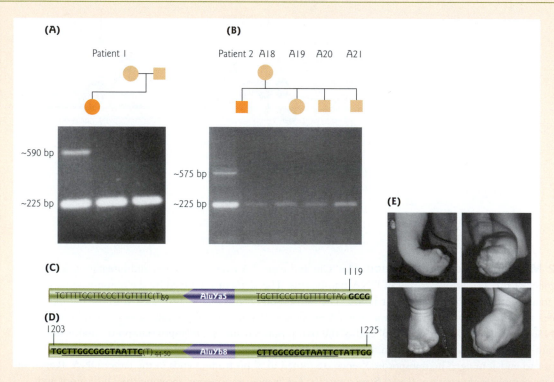

Figure 1 *Alu* **element insertion in patients with Apert syndrome.** (A, B) Amplification of the fibroblast growth factor receptor 2 gene (*FGFR2*) exon 9 and flanking intron sequence by PCR. PCR products from patients 1 (A) and 2 (B) and their available family members were electrophoresed on an agarose gel. All samples contained the expected 225 bp product, but patients 1 and 2 also showed DNA fragments of ~590 and ~575 bp, respectively, containing the inserted *Alu* elements. (C) Position and orientation of the *Alu* element inserted into intron 8 of patient 1, determined by DNA sequencing of the PCR products. (D) Position and orientation of the *Alu* element inserted into exon 9 of patient 2. The repeat flanking sequence is underlined, and the exon sequence is in bold. The numbering indicates the position within the *FGFR2* gene. (E) Appearance of hands and feet in a patient with Apert syndrome caused by *Alu* element insertions. (Reprinted by permission from the University of Chicago Press: Oldridge, M., Zackai, E.H., McDonald-McGinn, D.M. et al. 1999. De novo *Alu*-element insertions in *FGFR2* identify a distinct pathological basis for Apert syndrome. *American Journal of Human Genetics* 64:446−461. Copyright © 1999, The American Journal of Human Genetics.)

Jumping genes and human disease

DISEASE BOX 12.4

present on chromosome 22, an autosome. This suggests that the retrotransposon moved from one chromosome to another in the gamete-forming cells of the mother. Fukuyama-type congenital muscular dystrophy (FCMD) is another example of a disease caused by insertional mutation. In this disease there is a 3 kb insertion in the 3′-untranslated region of the FCMD gene. The insert has characteristics of a LINE, but it also has SINE-like repeats. FCMD is one of the most common autosomal recessive disorders in Japan, effecting approximately one in 10,000 births. The FCMD gene product, a secreted protein named fukutin, is not expressed because the insertion affects mRNA stability.

It has been estimated that one out of every 100–200 human births has a *de novo* *Alu* insertion, making the insertions a significant factor in human mutagenesis. *Alu* elements vary in different tissues of individuals and in the DNA of normal and diseased individuals. *Alu* insertions have accounted for over 20 cases of human genetic disease, including types of X-linked severe combined immunodefiency (X-SCID, see Section 17.3), hemophilia, leukemia, breast cancer, Apert's syndrome, and X-linked agammaglobulinemia. Generally, Apert syndrome results from a missense mutation in exon 7 of one allele of the fibroblast growth factor receptor II gene on chromosome 10; however, in two reported cases the disorder results from insertion of an *Alu* element in or near exon 9 of one allele (Fig. 1). This disorder is characterized by craniofacial abnormalities (premature fusion of the cranial sutures) and syndactyly of the hands and feet. X-linked agammaglobulinemia is a disease characterized by a lack of mature B cells that, in some patients, is due to an insertion in the Bruton agammaglobulinemia tyrosine kinase (Btk) gene.

Some LTR retrotransposons are active in the mammalian genome

Because LTR retrotransposons encode activities necessary for their retrotransposition, they are called autonomous, even though they may require host proteins to complete the process. LTR retrotransposons and retroviruses are very similar in structure (see Focus box 17.5). They both contain *gag* and *pol* genes that encode a viral protein coat, a reverse transcriptase, ribonuclease H, and integrase. These gene products provide the enzymatic activities for making cDNA from RNA and for inserting it into the host genome. Retrotransposons differ from retroviruses in that they lack a complete *env* gene. They can only reinsert into their current host's genome and cannot move between hosts. The complete retroviral *env* gene encodes an envelope protein that facilitates viral particle movement from one cell to another. Reverse transcription of LTR–retrotransposon RNA occurs within a viral-like particle in the cytoplasm in a multistep process. Many LTR retrotransposons target their insertions to relatively specific genomic sites. For example, *Ty3* elements of the yeast *Saccharomyces cerevisiae* specifically target a few conserved nucleotides present in RNA polymerase III promoters.

Although a variety of LTR retrotransposons exist, only the vertebrate-specific endogenous retroviruses appear to be active in the mammalian genome (e.g. human endogenous retrovirus, HERV, see Fig. 12.11). There are three distinct types of LTR retrotransposons in the mouse and human, with nearly all being inactive in humans and all three types being active in the mouse. Most is known about the active LTR-containing intracisternal A particles (IAPs), which account for approximately 15% of disease-producing mutations in the mouse. These include four different IAP insertions in the *agouti* gene. The insertion of IAPs into genes can cause disruption of coding sequence and subsequent dysfunctional proteins or reduced mRNA stability, or can affect the patterns of gene expression of neighboring genes (see Section 12.6).

Non-LTR retrotransposons include LINEs and SINEs

Non-LTR retrotransposons lack long terminal repeats and have a polyadenylate (poly(A)) sequence at their 3′ end. They are further divided into two major types: long interspersed nuclear elements (LINEs) and short interspersed nuclear elements (SINEs). LINEs are autonomous retrotransposons and SINEs are nonautonomous (see Table 12.3 and Fig. 12.11).

LINEs are widespread in the human genome

Human non-LTR retrotransposons that are currently active belong to the LINE1 family of repeats, whose members are commonly referred to as "*L1* elements." LINEs are derived from RNA polymerase II transcripts. The *L1* family has expanded to 20,000–50,000 copies and makes up about 21% of human DNA. There are two other distantly related LINE families found in humans, L2 and L3, but they are inactive. Full-length non-LTR retrotransposons are 4–6 kb in length and usually have two open reading frames, one encoding a RNA-binding protein, and the other encoding an endonuclease and a reverse transcriptase (see Fig. 12.11). Mammalian LINEs can integrate at a very large number of sites in the genome. This is because their endonuclease cleaves DNA at a short consensus sequence that is common in the genome (5'-TTT/A-3', where / designates the site of cleavage). The steps leading to retrotransposition of LINEs are not well understood, except for the reverse transcription process which is known to take place within the nucleus. The frequency of *L1* element transposition events per individual has been estimated to be one insertion in every 2–30 individuals, which means we are particularly vulnerable to this type of transposable element (see Disease box 12.4).

It was recently suggested that *L1* elements may play a role in X inactivation in mammals (see Fig. 12.8). The *L1* elements, which are clustered around the X inactivation center, are proposed to act as "boosters" or "way stations" which promote the spread of XIST RNA along the X chromosome. Genes subject to monoallelic expression (random monoallelic genes and imprinted genes) have also been shown to be flanked by high densities of *L1* sequences. Thus, LINE elements may act not only as insertional mutagens and regulatory control elements of neighboring genes, but also as long-distance modifiers of chromatin.

Alu elements are active SINEs

Nonautonomous retrotransposons are typified by SINEs. These are transposable elements that do not encode the enzymatic activities necessary for their retrotransposition (see Fig. 12.11). The only active SINEs in the human genome are members of the abundant *Alu* element family. *Alu* elements contain specific sequences that are cleaved by the restriction endonuclease *Alu* I, hence the name. These ~300 bp elements are made up of two similar but nonidentical DNA sequences that are proposed to be independently derived from 7SL RNA some 100 million years ago in a common ancestor of primates and rodents. The 7SL RNA is a core component of the signal recognition particle – an RNA–protein complex that interacts with ribosomes and is involved in the cotranslational transport of proteins across cell membranes (see Fig. 14.18). *Alu* elements do not encode proteins, yet have expanded to 1.1 million copies, or 11% of the human genome. The human *Alu* family is composed of numerous subfamilies, some of which continue to be actively retrotransposed (see Disease box 12.4).

Since *Alu* elements do not encode the enzymes required for transposition, the question arises, how do they actively move throughout the genome? *Alu* elements are usually flanked by 7–20 p target site duplications that share homology with *L1* elements (see Fig. 12.11). The reverse transcriptase encoded by some LINEs is proposed to interact with the shared 3' end sequence to mobilize the *Alu* elements in *trans*. *Alu* repeats and other mammalian SINEs preferentially insert at DNA target sites of about 15 bp that contain the conserved consensus sequence 5'-TTAAAA-3'.

12.6 Epigenetic control of transposable elements

Although transposable elements are continuously entering new genomic sites, phenotype-altering mutations caused by their insertions are much less frequent than are point mutations in most organisms, with the exception of fruitflies, corn, and wheat. Although many genomes contain a large number of active elements, they remain reasonably stable. At least two control methods are known that silence transposition: (i) methylation of transposable elements; and (ii) heterochromatin formation mediated by RNA interference (RNAi) and RNA-directed DNA methylation.

Methylation of transposable elements

Cytosine methylation of transposable elements has been proposed to function as a host defense mechanism. In fact, it is postulated that this is the primary reason for the existence of DNA methylation in higher eukaryotes. Because methylated cytosine residues are susceptible to transition mutation to thymine residues (see Fig. 12.3), cytosine methylation may aid in the mutation and inactivation of transposable elements. In addition, methylation may contribute to silencing of transposable element transcriptional activity and transposition.

Effect of transposable element methylation on plant pigmentation

The first evidence that the activity of transposable elements might be subject to methylation control came from studies of the maize *Ac* and *Spm* (suppressor-mutator) elements. Inactive elements were found to have methylated cytosines, whereas active elements were nonmethylated. These changes in methylation pattern were heritable over generations. More recently, genetics and epigenetics in flower pigmentation associated with transposable elements in morning glories (*Ipomoea*) have been studied extensively. The majority of spontaneous mutations were found to be caused by the insertion of DNA transposons in the genes for anthocyanin pigmentation in flowers (Fig. 12.13).

A diet lacking folic acid can activate a retrotransposon in mice

During the brief period of global demethylation in the early embryo (see Section 12.3), mammalian retrotransposons are demethylated and thus could become active and retrotranspose. In contrast, in somatic cells they are typically hypermethylated. As a consequence, their expression is not detectable and they cannot be mobilized. Recently, it was shown that diet can alter heritable phenotypic change in mice by changing the DNA methylation pattern of the mouse genome and thereby activating a retrotransposon.

A particular strain of yellow *agouti* mice with the A^{vy} allele (*agouti viable yellow*) has a LTR retrotransposon insertion (Table 12.3, IAP) upstream of the *agouti* gene (Fig. 12.14A). This is a striking case where the pattern of gene expression in adjacent genes is affected by the promoter in a transposable element. The *agouti* gene encodes a signaling protein that is expressed in the middle of the hair growth cycle. Transcription is initiated from a hair growth cycle-specific promoter in exon 2 of the *agouti* (*A*) allele. *Agouti* regulates a transient switch in pigment synthesis between black (eumelanin) and yellow (phaeomelanin) in hair follicle melanocytes to produce a yellow band on an otherwise black hair. The retrotransposon insertion places the *agouti* gene under the control of a cryptic promoter in the 5′ LTR of the retrotransposon. This causes constitutive expression of the *agouti* gene, resulting in mice with completely yellow fur. CpG methylation in the inserted retrotransposon varies dramatically among individual A^{vy} mice. This epigenetic variability causes a wide variation in individual coat color. Normally the fur of A^{vy} mice is yellow, brown, or a calico-like mixture of the two, depending on the number of attached methyl groups in the 5′ LTR promoter (Fig. 12.14B). Each mottled mouse is different owing to stochastic silencing of the retrotransposon.

Early nutrition can influence DNA methylation because methyl donors and essential cofactors required for methyl metabolism must be obtained from the diet (see Disease box 12.1, Fig. 1). For example, dietary methionine and choline are major sources of one-carbon units, and folic acid, vitamin B_{12}, and pyridoxal phosphate are critical cofactors in methyl metabolism. In a landmark experiment, Robert A. Waterland and Randy L. Jirtle fed folic acid and other methyl-rich supplements to pregnant *agouti* mice with the A^{vy} allele. The mice gave birth to offspring with mostly brown fur (pseudoagouti), indicating that the retrotransposon

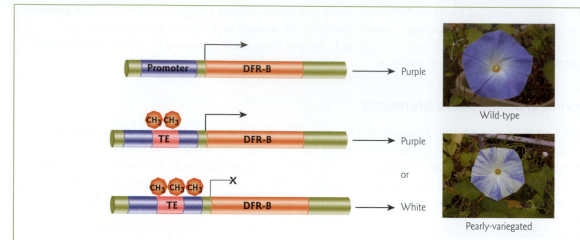

Figure 12.13 Methylation of a DNA transposons leads to flower variegation in morning glory. The pearly variegated mutant of *Ipomoea tricolor* results from insertion of a 0.4 kb *Mu*-related transposable element (TE) (magenta box) in the promoter region of the *DFR-B* gene, about 0.2 kb upstream of the start of transcription. *DFR-B* encodes dihydroflavonol-4-reductase, an enzyme involved in anthocyanin (purple) pigment biosynthesis. In the wild-type morning glory, *DFR-B* is expressed (indicated by black arrow) and flowers are solid bluish-purple in color. The insertion of the DNA transposon alone into the promoter region has little effect on gene expression; however, the transposon is heavily methylated (red sun symbols) and occasionally DNA methylation spreads into the adjacent promoter region, blocking *DFR-B* transcription (indicated by X). These heritable changes in methylation patterns result in a variegated flower, with sectors of bluish-purple and white. (Photographs courtesy of Atsushi Hoshino; kindly provided by Shigeru Iida, National Institute for Basic Biology, Japan.)

was inactive (Fig. 12.14B). In these mice, the retrotransposon promoter region was hypermethylated, thereby shutting off expression of the downstream *agouti* gene. What came as a surprise to researchers was that the *agouti* gene was turned off not only in the mother fed the supplements but in her offspring as well. Mother mice without methyl-rich supplements gave birth to mostly yellow pups with a high susceptibility to obesity, diabetes, and cancer. In these mice, the retrotransposon was active.

Extensive human epidemiological data have indicated that prenatal and early postnatal nutrition influence adult susceptibility to diet-related chronic diseases, including cardiovascular disease, type 2 diabetes, obesity, and cancer. The findings in mice suggest that it is the impact of nutrition on epigenetic gene regulation that provides the link with later metabolism and chronic disease susceptibility.

Heterochromatin formation mediated by RNAi and RNA-directed DNA methylation

A primary line of host defense against transposable elements is to target them for heterochromatin formation, which acts to suppress their transcription, transposition, and recombinational activity. Heterochromatin domains are in general inaccessible to DNA-binding factors and are transcriptionally silent. Domains of heterochromatin are stably maintained and inherited through many cell divisions. In addition to histone modifications, DNA methylation contributes to the stability of heterochromatin. Transposable elements and satellite repeats preferentially form heterochromatin. How they do this is not known, but the repetitive nature of these elements seems to be important in the process.

In plants, heterochromatin targeting involves RNA-directed DNA methylation. RNA-directed DNA methylation is linked with RNA interference (RNAi), another RNA-based silencing mechanism (see Section 9.7). The RNAi pathway has a well-known role in post-transcriptional gene silencing. In RNAi, double-stranded RNA (dsRNA) provides a trigger for the degradation of transcripts that have complementary sequences. The dsRNA is first diced into small interfering RNAs (siRNAs) of

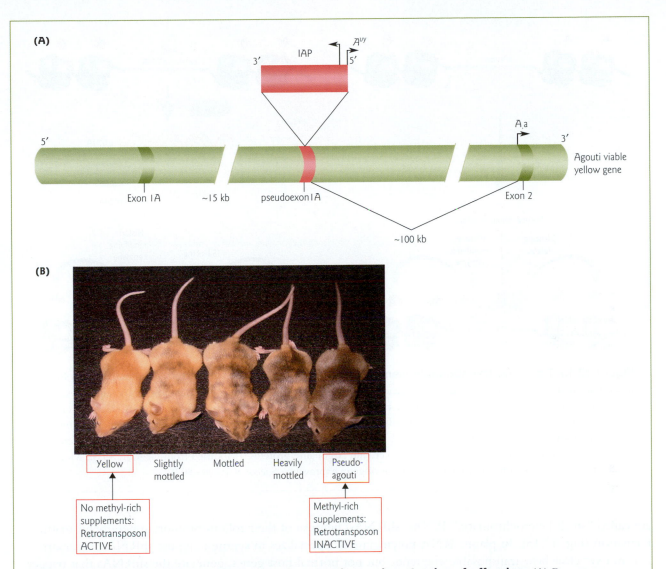

Figure 12.14 Maternal dietary methyl supplementation and coat color of offspring. (A) Retrotransposon (IAP) insertion site within pseudo-exon 1A (PS1A) of the mouse *agouti viable yellow* (A^{vy}) allele. PS1A arose from an inverted duplication that encompasses exon 1A of the *agouti* gene. A cryptic promoter in the long terminal repeat (LTR) proximal to the *agouti* gene (short arrowhead labeled A^{vy}) drives constitutive *agouti* expression in A^{vy} animals. In mice carrying the normal *agouti* allele (*A*) or the nonagouti (black) allele (*a*) transcription starts from a hair cycle-specific promoter in exon 2 (short arrowhead labeled *A, a*). (B) A^{vy}/a animals representing five coat color classes used to classify phenotypes. The A^{vy} alleles of yellow mice are hypomethylated at the retrotransposon LTR, allowing maximal *agouti* expression. The A^{vy} alleles of brown mice are hypermethylated at the retrotransposon LTR, and *agouti* expression is under control of the hair cycle-specific promoter. (Reprinted with permission from Waterland, R.A., Jirtle, R.L. 2003. Transposable elements: targets for early nutritional effects on epigenetic gene regulation. *Molecular and Cellular Biology* 23:5293–5300 Copyright © 2003, American Society for Microbiology. Photograph courtesy of Randy Jirtle, Duke University Medical Center.)

21–26 nt by Dicer ribonucleases. Some of these siRNAs are incorporated into an RNA–induced silencing complex (RISC) and used to guide the degradation of target transcripts by sequence complementarity (see Fig. 9.12).

Alternatively, dsRNA and/or siRNAs derived from transcripts generated by transposable elements and repetitive DNA can also direct the methylation of DNA with complementary sequences. These RNAs

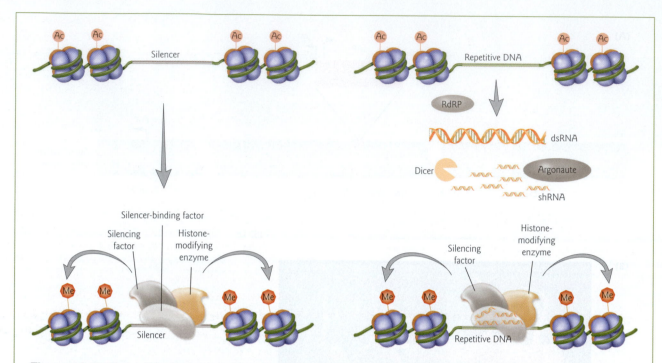

Figure 12.15 Targeting transposable elements for heterochromatin formation. Two mechanisms for heterochromatin formation are depicted. Heterochromatin can be initiated by specific *cis*-acting DNA sequences, called silencers, which are recognized by DNA-binding proteins (left). Alternatively, repetitive DNA elements such as transposons in the genome may serve as signals for heterochromatin formation (right). Transcripts generated by repetitive DNA are processed into shRNAs by a mechanism requiring components of the RNAi machinery such as RNA-dependent RNA polymerase (RdRP), Dicer, and Argonaute proteins. (Modified from Grewal, S.I.S., Moazed, D. 2003. Heterochromatin and epigenetic control of gene expression. *Science* 301:798–802).

are called "small heterochromatic" RNAs (shRNAs) because of their role in promoting heterochromatin formation (Fig. 12.15). In plants, RNA polymerase IV specializes in synthesizing the shRNA precursors. It is not yet clear how transposable elements, but not normal host genes, generate the shRNAs that trigger DNA methylation. Processing of shRNAs requires components of the RNAi machinery, including the RNA-dependent RNA polymerase, Dicer, and Argonaute proteins. RNA-directed DNA methylation typically involves heavy methylation of both CG and non-CG cytosines in the target region by specific DNA methyltransferases that either establish or maintain the pattern of methylation. The methylation process is highly sequence-specific and does not significantly spread beyond the boundary of the sequence complementarity with the shRNA.

12.7 Allelic exclusion

Monoallelic gene expression, where only one allele of a gene or gene cassette family is selected for expression, plays an important role in cell differentiation in a diversity of organisms. One major mechanism mediating allelic exclusion is programmed gene rearrangement. Programmed gene rearrangements are rearrangements of DNA that regulate the expression of some genes. In this section, three such systems will be compared: (i) mating-type switching in budding yeast where cells switch between expressing one of the two different mating-type cassettes; (ii) avoidance of the immune response in trypanosomes by switching which one of the many possible genes is active in making the major surface antigen; and (iii) the production of active immunoglobulin genes, which occurs by assembly of different DNA segments. In addition to recombination events, epigenetic mechanisms of gene regulation play an important role in maintaining allelic exclusion.

Yeast mating-type switching and silencing

Some strains of yeast have the remarkable ability to switch between two mating types called *a* and α. "Homothallic strains" with the dominant allele encoding the HO endonuclease can change mating type frequently, as often as once every generation (Fig. 12.16). Strains with the recessive allele *ho* have a stable mating type and change with a frequency of only one in 1,000,000. The study of mating-type (MAT) gene switching in two distantly related yeast species, *Saccharomyces cerevisiae* and *Schizosaccharomyces pombe*, has provided insights into two important aspects of gene regulation: gene silencing by epigenetic alterations in chromatin structure, and the basis of preferential recombination between a recipient locus and one of two possible donors of genetic information. Mating-type switching is accomplished by a highly programmed site-specific homologous recombination event. During recombination, the mating-type locus switches between two alternative alleles by copying information from one of the two silent donor loci. This section will focus on mating-type switching in the budding yeast, *S. cerevisiae*.

Under conditions of starvation, mating *a* and α haploids gives rise to heterozygous diploids that sporulate to generate haploid spores (Fig. 12.16). The yeasts are attracted to each other by pheromones (diffusible signaling molecules). *MATa* produces the *a* factor that binds to a receptor produced by *MATα*, and vice versa. Through the process of meiosis and sporulation, four haploid spores are produced. Under nutrient-rich conditions a haploid spore germinates and divides by budding. The larger mother cell can undergo mating-type switching, which always occurs in pairs. The smaller "buds" or daughter cells cannot undergo switching until they bud and divide at least once, becoming mother cells capable of switching. "Experienced" (mother) cells switch with high probability. The asymmetric pattern of switching is due to the selective expression of the gene encoding the HO endonuclease in the mother cell.

Yeast mating-type switching represents a simple form of cellular differentiation. The key molecular features of switching are:

1 Two mating types defined by the expression of one of two gene cassettes.
2 DNA rearrangement by homologous recombination (gene conversion).
3 Directionality of switching.
4 Silent cassettes are repressed through epigenetic mechanisms.

DNA rearrangement by homologous recombination (gene conversion)

Mating-type switching occurs when the active "cassette" (expressed locus) is replaced by information from a silent cassette (donor locus). The selective expression of only one gene cassette is achieved by the chromatin state of the three mating type loci, *HMRa*, *HMLα*, and *MAT*, on chromosome 3. The mating type is determined by the expressed *MAT* locus, which can contain either *a* or α information. The silent donor loci are located in opposite subtelomeric regions of chromosome 3 (Fig. 12.17). In the mother cell, mating-type switching is initiated by the HO endonuclease-dependent formation of a double-strand break at the *MAT* locus. HO is activated by SWI proteins that play a role in chromatin remodeling (see Section 11.6). The double-strand break induces DNA damage repair by a homologous recombination pathway (see Fig. 7.12). The mating-type donor cassette containing the opposite mating-type information is used for this reaction, where homology is provided at flanking homology domains present at all three cassettes, *MAT*, *HMLα*, and *HMRa*.

In the example shown in Fig. 12.17, prior to the switch, *MATα* is expressed in the active locus. The end result of gene conversion is that a copy of each *HMRa* gene is produced and the copied genes are inserted into the *MAT* locus. During the process of homologous recombination, the original *MATα* genes are "ejected" and degraded. However, the MATα information is not lost, because there is still a copy at the silent *HMRα* mating locus. The difference in cell identity is established by mating-type-specific gene products that ultimately cause the cell to produce the appropriate mating factor and receptor. Each of the mating-type cassettes contains two open reading frames. *MATα1* encodes a transcription factor that activates

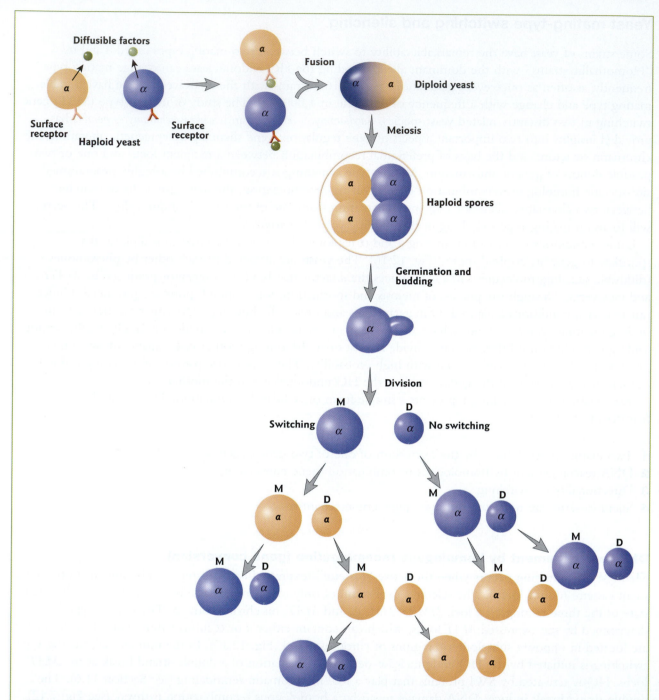

Figure 12.16 Homothallic life cycle of *Saccharomyces cerevisiae*. A haploid yeast cell of mating-type *a* that has divided can switch to the opposite mating type and the original cell and its switched partner can conjugate to form an *a*/α diploid cell. Meiosis and sporulation will regenerate haploid cells. Homothallic (mating-type) switching is confined to cells that have previously divided (mother cells, M). Daughter cells (D) must have budded and divided once before switching.

other α-specific genes, whereas *MATα2* encodes a repressor of *a*-specific genes. In haploid cells, *MATa1* and *MATa2* are not required for specifying cell type, because the default cell type is *a*. Instead, the *MATa1* gene is important in diploids, where its product interacts with the *MATα2* gene product and represses haploid–specific genes, including *HO* and *MATα1*.

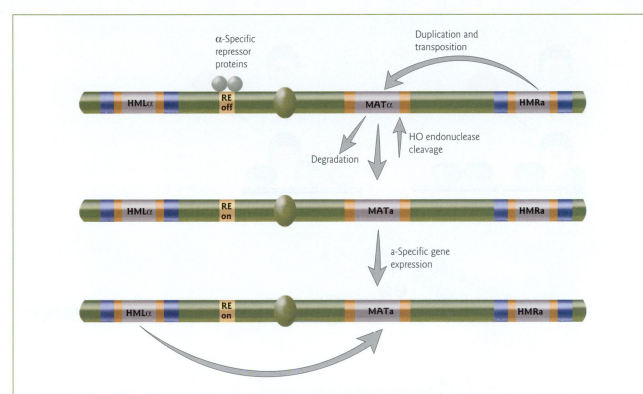

Figure 12.17 Mating-type loci on *Saccharomyces cerevisiae* chromosome 3. In addition to the expressed *MAT* locus, chromosome 3 has two unexpressed donor loci, *HMLα* and *HMRa*, near opposite ends of the chromosome (not drawn to scale). The centromere is indicated as a green oval. These donors are maintained in a silenced state mediated by two adjacent silencer sequences (blue boxes). When the HO endonuclease is expressed, *MAT* alleles can be switched by gene conversion. Regions of homology are indicated by red boxes. The homologous recombination enhancer (RE) (orange box) regulates the directionality of switching. In *MATα* cells, the RE is turned off by α-specific repressor proteins.

Directionality of switching

There is preferential switching to the opposite mating type. Switching from α to *a* (or *a* → α) occurs in 85–90% of switches. Switching from *a* to *a* (or α → α) with no change in mating type occurs rarely. A 244 bp DNA sequence element designated the recombination enhancer (RE), located between *HMLα* and *MAT*, controls the directionality of switching (Fig. 12.17). The recombination enhancer is turned "off" by binding of repressor proteins encoded by (or induced by) the α-mating-type cassette gene *MATα2*. In the "off" state, the recombination enhancer represses recombination between *MATα* and *HMLα*, and recombination preferentially occurs with *HMRa*. When *MATa* is expressed the recombination enhancer is "on" by default, since the repressor proteins are no longer produced. In this state, recombination preferentially occurs with *HMLα*.

Silent cassettes are repressed by epigenetic mechanisms

Expression of the silent mating-type donor sequences is repressed through epigenetic mechanisms. DNA regulatory elements called "silencers," designated E and I, flank the silent mating loci (Fig. 12.18). These *cis*-acting sequences act to recruit a variety of specific regulatory proteins, histone deacetylases, and heterochromatin assembly factors. Together these factors act to establish heterochromatin to repress transcription, but not recombination, at the donor loci.

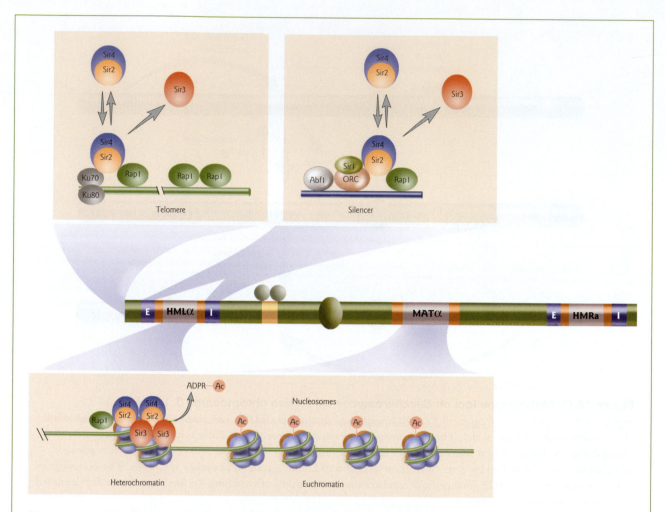

Figure 12.18 Model for assembly of silent chromatin domains in yeast. Assembly of the Sir complex at silencers of the silent mating loci and telomeres is proposed to occur in a stepwise fashion. First, the Sir2/4 heterodimer transiently binds to the silencer or the telomere via interactions with Rap1/Sir1 and Rap1/yKu70, respectively. This binding does not require the enzymatic activity of Sir2 and is independent of Sir3. Two other complexes are initially associated with the silencer, Abf1 (ARS-binding factor 1) and ORC (origin recognition complex). The association of Sir3 with the silencer can also occur independently of Sir2 but requires Sir4. Stable association of the complex with chromatin and spreading of the repressed domain of heterochromatin requires all three Sir proteins and the NAD+-dependent deacetylation activity of Sir2 (see Fig. 12.19). The acetyl group is transferred from the histone to ADP-ribose to generate a novel compound, 2′,3′-O-acetyl-ADP-ribose (ADPRAc). Chromatin at the expressed MAT locus is in an active conformation and the histone N-terminal tails are hyperacetylated. (Adapted from Hoppe, G.J., Tanny, J.C., Rudner, A.D., Gerber, S.A., Danaie, S., Gygi, S.P., Moazed, D. 2002. Steps in assembly of silent chromatin in yeast: Sir3-independent binding of a Sir2/Sir4 complex to silencers and role for Sir2-dependent deacetylation. *Molecular and Cellular Biology* 22:4167–4180.)

Silent information regulatory (Sir) proteins are recruited both to the silencers and the telomeres. The silencers act as nucleation sites, initially binding a Sir2/4 complex, and then recruiting Sir3. The assembled Sir complex then interacts with deacetylated lysines in the N-terminal tails of H3 and H4. Further deacetylation is catalyzed by Sir2, an NAD+-dependent histone deacetylase. Along with characterizing the steps in the assembly of silent chromatin, Danesh Moazed and colleagues demonstrated that the enzymatic

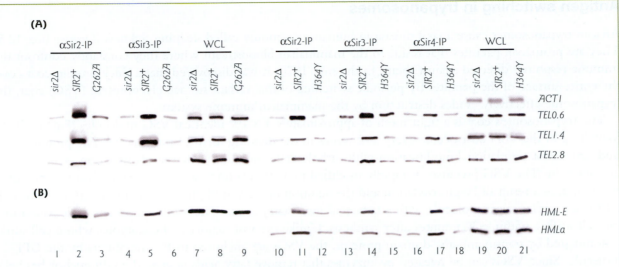

Figure 12.19 Enzymatic activity of Sir2 is required for association of Sir proteins with telomeric DNA regions and the *HML* mating-type locus. (A, B) Chromatin immunoprecipitation (ChIP) assays (see Fig. 9.15E for methods) were performed using a *sir2* deletion strain (*sir2Δ*), a *SIR2* wild-type strain (*SIR2+*), and strains containing *sir2* alleles that encode enzymatically inactive Sir2 proteins (*sir2-H364Y* and *sir2-G262A*) with anti-Sir2, anti-Sir3, and anti-Sir4 antibodies. Panels show phosphorimager data of PCR amplifications corresponding to input (WCL) and immunoprecipitated chromatin for the indicated regions of the telomere and the *HML* locus. The association of Sir2, Sir3, and Sir4 with DNA fragments 1.4 and 2.8 kb from the telomere (TEL1.4 and TEL2.8) and the HMLa locus were all greatly reduced in the mutant strains. There was a low level of Sir2 association with DNA fragments encompassing the HML-E silencer or at 0.6 kb from the telomere (TEL0.6) in the mutant strains, whereas association of Sir3 and Sir4 was greatly reduced. This suggests that the enzymatic activity of Sir2 is not absolutely required for initial binding to DNA. ACT1 is a positive control for the PCR. (Reprinted with permission from Hoppe, G.J., Tanny, J.C., Rudner, A.D., Gerber, S.A., Danaie, S., Gygi, S.P., Moazed, D. 2002. Steps in assembly of silent chromatin in yeast: Sir3-independent binding of a Sir2/Sir4 complex to silencers and role for Sir2-dependent deacetylation. *Molecular and Cellular Biology* 22:4167–4180. Copyright © 2002, American Society for Microbiology. Photograph courtesy of Danesh Moazed, Harvard Medical School.)

activity of Sir2 is required for the association of silencing proteins with regions distal from nucleation sites (Fig. 12.19). Interestingly, Sir2 has been shown to play an important role in aging. Lifespan is increased with higher doses of Sir2, suggesting that increased deacetylase activity increases the longevity of mother cells – aging in yeast is determined by the number of buds a mother cell produces. Usually yeast reaches senescence after 20–30 divisions.

At the telomeres, Sir2, Sir3, and Sir4 interact with the Rap1 protein (see Fig. 12.18). Rap1 binds to the simple repeated sequences at the telomeres and to the DNA repair/telomere-binding protein yKu70. The three-dimensional architecture of the nucleus provides an additional layer of epigenetic control. Fluorescence *in situ* hybridization (FISH) studies have shown that yeast telomeric and centromeric sequences are localized to the nuclear periphery. Since the silent mating-type loci are tightly linked to telomeres, they are also localized to the nuclear periphery. They appear to be tethered via the yKu70 protein to specific elements within the nuclear envelope that restrict their mobility. In addition, Rap1 interacts with the hypoacetylated N-terminal tails of histones H3 and H4. Following recruitment, spreading of a Sir2/3/4 complex leads to a repressed domain of "closed chromatin" that is more stable at silent mating-type loci than at telomeres. The DNA is largely inaccessible at the silent loci. In contrast, at the *MAT* locus, chromatin is in an active conformation. The histone N-terminal tails are hyperacetylated and there are DNase I hypersensitive sites in the promoter region.

Antigen switching in trypanosomes

African trypanosomes cause a fatal disease in humans commonly called sleeping sickness (Disease box 12.5). They are protozoan parasites that reside in the mammalian bloodstream where they constantly confront the immune responses directed against them. In the mid-1970s it was discovered that a thick protein coat covers the entire surface of the parasite. By periodic switching of this variant surface glycoprotein (VSG) coat, the trypanosome effectively evades destruction by the mammalian immune system.

The trypanosome's coat is a tight mesh of approximately 1×10^7 identical VSG molecules (Fig. 12.20). Switching of the VSG coat occurs every 1–2 weeks in the dividing form of the parasite. The protein undergoes post-translational modification with carbohydrate moieties, hence its classification as a glycoprotein. The VSG precursor is rapidly modified to a 59 kDa form within the endoplasmic reticulum and Golgi, as a result of N-glycosylation and the addition of a glycosyl phosphatidylinosital (GPI) molecule. GPI is a complex sugar molecule with a fatty acid (myristate) chain. The modified protein is delivered to the cell surface within ~30 minutes after synthesis. Unlike the vast majority of eukaryotes whose cell surface is dominated by transmembrane domain proteins, the VSGs are anchored to the membrane by the GPI molecule. Since VSGs can be released by enzymes that remove fatty acids *in vitro*, the GPI anchor has been proposed to provide a "quick release" mechanism *in vivo*. Each VSG has a region of homology and a variable region. Because of the three-dimensional shape of VSG, when attached to the cell membrane, the homology region is buried within the protein and only the variable region is exposed.

Characteristics of variant surface glycoprotein (*VSG*) genes

Central to the process of antigenic variation is monoallelic gene expression and switching of which one of the many *VSG* genes is expressed (Fig. 12.21). The trypanosome can change its coat by replacing the transcribed *VSG* gene in the expression site by a different *VSG* gene. Most potential donor *VSG* genes are clustered as tandem repeats in nontelomeric chromosome regions on any of the 11 pairs of chromosomes. These "chromosome-internal" genes can be transposed to an active expression site by a gene conversion event that displaces the resident gene (see below). Southern blots probed with *VSG* cDNAs under low hybridization stringency indicate that in the genome of *Trypanosoma brucei* there are over 1000 different *VSG* genes and pseudo-VSGs. This large repertoire of chromosome-internal *VSG* genes is further expanded by some 100 minichromosomes of 50–100 kb carrying a *VSG* gene at their termini. Two to 10% of the genome is devoted to antigenic variation, and VSG represents approximately 10% of the total protein produced.

A single telomeric expression site would be sufficient to express the entire repertoire of *VSG* genes, but in fact there are 20 different expression sites. A switch in *VSG* gene expression can either be due to a switch in the *VSG* gene in an active expression site, or a switch in which the expression site is active (see below). The latter occurs at a low frequency. Different expression sites do not serve primarily to provide an alternative way to switch coats. Instead, they are used to express different sets of expression site-associated genes (ESAGs). Within each expression site, there are 12 distinct ESAGs (see Fig. 12.20). The ESAGs are cotranscribed with the *VSG* gene as part of a long transcription unit. Most of the ESAGs encode surface proteins. For example, two of them encode the subunits of a transferrin receptor. Trypanosomes can make as many as 20 slightly different transferrin receptors with differing affinity for the transferrin of different mammals. Transferrin is an iron-transporting protein found in blood plasma.

VSG switching by homologous recombination

Antigen switching in trypanosomes appears to occur predominantly by DNA homologous recombination between short blocks of sequence homology adjacent to donor and target genes (Fig. 12.21). However, very little is known about the trypanosome genes that regulate antigenic variation, or that encode the enzymes that catalyze the switch from one VSG to another. To date, the DNA repair protein Rad51 (see Fig. 7.12)

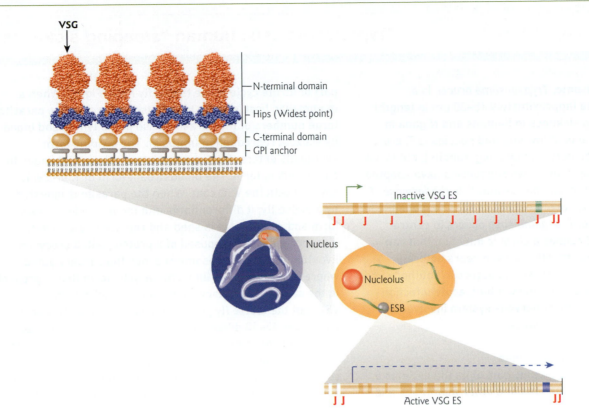

Figure 12.20 Variant surface glycoprotein (VSG) transcriptional control and the expression site body. The long, slender replicative form of *Trypanosoma brucei*, found in the mammalian bloodstream, is depicted. (Left inset) Space-filling model of the trypanosome surface showing the dense monolayer of VSG, based on X-ray crystal structures of the N-terminal. The remainder of the protein for which the three-dimensional structure is not available is depicted schematically. The GPI anchor is attached to the plasma membrane. (Reprinted with permission from Field, M.C., Carrington, M. 2004. Intracellular membrane transport systems in *Trypanosoma brucei*. *Traffic* 5:905–913. Copyright © 2004 Blackwell Munksgaard.) (Right inset) The nucleus is depicted, containing the nucleolus and the expression site body (ESB) in which *VSG* gene expression occurs. For simplicity, only four *T. brucei* chromosomes are depicted in the nucleus, and the inserts depict two telomeric sites of *VSG* transcription, termed expression sites (ES). The active *VSG* (blue or green box) is always found adjacent to the telomere (vertical line), and is cotranscribed with several ES-associated genes (orange boxes) and flanked upstream by an array of degenerate 70 bp repeats (brown lines). The inactive *VSG* gene (top) contains the modified base J (indicated in red) throughout its coding sequence, as well as in sequences at the telomere and upstream of the promoter (green arrow). The active ES is present within the ESB, and transcription (blue broken line) occurs over all the protein-coding sequences, where J is undetectable. (Adapted from McCulloch, R. 2004. Antigenic variation in African trypanosomes: monitoring progress. *Trends in Parasitology* 20:117–121. Copyright © 2004, with permission from Elsevier).

remains the only gene to be implicated in mediating VSG switching. But Rad51 mutants can still undergo antigenic variation by recombination.

Gene conversion, analogous to yeast mating-type switching, is the most frequent mode of antigen switching (Fig. 12.21A). The *VSG* in an active expression site is degraded and replaced with a duplicated copy of an unexpressed "donor" *VSG* by homologous recombination, either from an internal chromosomal site or another telomere. Further antigenic diversity can be created by point mutations generated during gene conversion (Fig. 12.21B). In some cases analyzed, up to 11 scattered point mutations were found in the duplicated copy. Switching can also occur by reciprocal recombination; however, this is a rare event

Trypanosomiasis: human "sleeping sickness"

The African trypanosome, *Trypanosoma brucei*, is a unicellular flagellate (approximately 15–30 µm in length) that causes sleeping sickness in humans and N'gana in livestock (Fig. 1). The best investigated species is *T. brucei*. It grows well in laboratory animals (e.g. rabbits), but is not infectious to the researcher. Two subspecies have adapted to the human host, *T. b. gambiense* and *T. b. rhodesiense*. *T. congolense* and *T. vivax* are the main species responsible for the livestock disease. *T. b. gambiense* occurs in central and West Africa and causes a chronic infection that can sometimes persist for months or even years before symptoms appear. *T. b. rhodesiense* occurs in southern and East Africa, and causes a highly virulent, acute infection whose symptoms of central nervous system involvement can emerge after only a few weeks.

Life cycle of African trypanosomes

The African trypanosome spends part of its life cycle as a parasite in the blood of humans and other mammals, and part of its life cycle in the tsetse fly host (Fig. 1). When a mammalian host is bitten by a tsetse fly carrying the parasite, large numbers of the parasite arise in the lymph and blood by asexual reproduction (mitosis). The dividing form eventually enters the nervous system, where it changes to a nondividing form. The bloodstream form of *T. brucei* is covered with the VSG coat. When the parasite is ingested by a tsetse fly, it differentiates into the procyclic or insect form and its VSG coat is shed and replaced by an invariant glycoprotein coat composed of a protein called procyclin. Eventually, the trypanosome reaches the salivary gland, where it differentiates into a metacyclic form that is present in the saliva. The metacyclic form is coated with only one VSG, but the tsetse fly population as a whole collectively expresses 15–20 different *VSG* genes. It is this form that is introduced into a new mammalian host by insect bite. After the trypanosome enters the bloodstream of the host, it continues to express the metacyclic VSG for as long as 7 days, and then switches to the expression of bloodstream VSG.

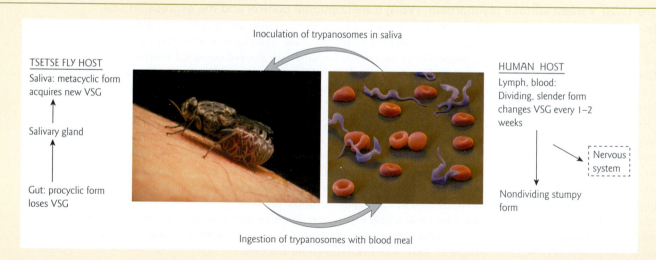

Inoculation of trypanosomes in saliva

TSETSE FLY HOST
Saliva: metacyclic form acquires new VSG

Salivary gland

Gut: procyclic form loses VSG

HUMAN HOST
Lymph, blood:
Dividing, slender form changes VSG every 1–2 weeks

Nervous system

Nondividing stumpy form

Ingestion of trypanosomes with blood meal

Figure 1 Life cycle of the African trypanosome. Scanning electron micrograph of *Trypanosoma brucei* protozoa and red blood cells is shown on the right. (Credit: Eye of Science / Photo Researchers, Inc.) The ribbon-like *T. brucei* are carried in the saliva of the blood-drinking tsetse fly, *Glossina morsitans*. The macrophotograph (left) shows a tsetse fly with a bloated abdomen towards the end of a meal on a human arm. (Credit: Martin Dohrn / Photo Researchers, Inc.) The trypanosomes enter a human host through the wound made by the fly when it feeds. They infect the blood, lymph, and spinal fluid, and begin to divide. In this form, the trypanosome undergoes periodic switching of its VSG coat. Damage to the nervous system by the infection eventually leads to lethargy, tremors, and mental and physical deterioration. The sufferer finally enters a comatose state and dies. A nondividing form is ingested by another tsetse fly during its blood meal. Within the fly host, the trypanosome changes form, losing the VSG coat in the procyclic form and acquiring a new VSG in the metacyclic form.

Trypanosomiasis: human "sleeping sickness"

Symptoms of trypanosomiasis

Human sleeping sickness has probably existed in Africa for many centuries, but the first clear descriptions of the disease came from European explorers and colonists in the 1800s and early 1900s. In 1910, physicians recorded a marked periodicity of patients' temperatures. It was determined that the cyclical waves of fever corresponded to spikes in the trypanosome population in the blood (Fig. 2). There are two stages in the disease. In the first stage, there is infection of blood vessels and lymph glands which leads to intermittent fever, rash, and swelling, and complete fatigue due to persistent infection. Proliferation of

trypanosomes in the blood induces an immune system response by the patient. The response involves the production of antibodies that are shaped to bind a particular VSG, which sets in action a cascade of events to destroy the parasites. The antibodies mediate killing of ~99% of the trypanosomes, but a few trypanosomes escape by spontaneous switching of their VSG coat. In the lab one in 10^2–10^7 trypanosomes switches its coat per doubling time of 5–10 hours. The VSG switch rate is much higher in trypanosomes recently isolated from the field than in laboratory-adapted strains. The trypanosome with the new VSG coat now escapes immune attack by the host and

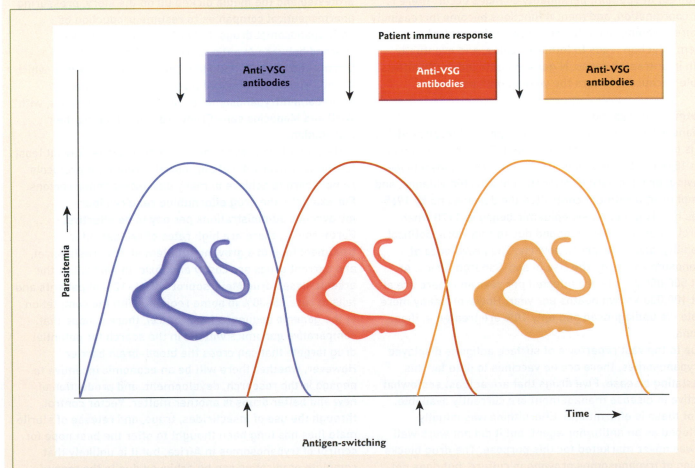

Figure 2 Symptoms of trypanosomiasis. The characteristic cyclical waves of fever (parasitemia) in patients with sleeping sickness (trypanosomiasis) is largely due to the expression by the trypanosomes of different VSGs (blue, red, orange) and the subsequent immune response over time.

proliferates. Because the patient has acquired immunity only to the first expressed VSG, the available antibodies do not recognize this new VSG coat as a foreign antigen. The cycle continues. By the time the immune system has made new antibodies, the trypanosome has shed its coat and replaced it with yet another.

The second stage of the disease is invasion of the central nervous system by the parasite, which causes inflammation of the brain outer membrane. As this stage progresses, headaches become severe and sleep disorders emerge. The name "sleeping sickness" was coined because of observations that patients are very sleepy in the daytime but have night-time insomnia, although overall sleep time is unaltered. In addition, personality changes occur, there is poor coordination, and mental functions become increasingly impaired leading to confusion. Progressive central nervous system infection leads to lethargy, coma, and eventually death in untreated cases. N'gana in cattle indirectly affects people because it lowers the yield of milk and meat.

Treatment of trypanosomiasis

Trypanosomiasis is endemic to large parts of tropical Africa and is generally fatal if left untreated. Epidemics occurred from 1896 until 1906 and in 1920. In the 1950s, insecticide spraying, brush clearing, relocation of affected villages, and prompt drug treatment controlled the disease and by 1965 it was nearly gone. A new epidemic began in 1970 when control measures were stopped due to civil wars, political instability, and widespread poverty. The prevalence of trypanosomiasis is on the rise in sub-Saharan Africa. At least 500,000 people are infected per year and there are at least 100,000 known deaths per year. However, many more people die undiagnosed and only 10% are treated in these regions.

Due to the vast repertoire of surface antigens displayed by trypanosomes, there are no vaccines to date for this devastating disease. Five drugs that are at least somewhat effective in disease management are currently available. One of these is eflornithine. Eflornithine was initially developed as an antitumor agent, but it did not work well and was never marketed for this purpose. The drug blocks division of trypanosomes growing in culture, but does not kill them. Thus, a functioning immune system is presumed

to be necessary to completely eliminate the parasite. At one point, the pharmaceutical industry stopped producing most antitrypanosomal drugs, including eflornithine, as their sale generated insufficient profit. Fortuitously, eflornithine came to the rescue. At the same time as production was stopped of eflornithine as an antitrypanosomal drug, eflornithine-containing cosmetics were introduced to the North American market. It had been discovered that the drug (sold as "Vaniqa") could be used in depilatory creams to slow the growth of unwanted facial hair in women. Eflornithine inhibits ornithine decarboxylase, a major enzyme of polyamine biosynthesis that is needed for hair to grow. Soon after Vaniqa came on the market, health organizations protested and the media picked up on the story, pressuring pharmaceutical companies to resume production of antitrypanosomal drugs. In 2001, an agreement was reached between Aventis, Bayer, the World Health Organization (WHO), and Médécins sans Frontières in which Aventis and Bayer ensured free production of the five essential antitrypanosomal drugs for the next 5 years, with WHO and Médécins sans Frontières coordinating their distribution.

Despite this success in ensuring drug availability (at least for the very near future) appropriate treatment protocols remain hard to achieve in many disease-endemic regions. For example, the drug eflornithine requires four intravenous administrations per day to be effective. Furthermore, there are high rates of relapse after treatment (due to a growing problem of drug resistance), and current drugs are costly and toxic. In particular, the arsenic-based drug Melarsoprol kills 4–12% of patients and fails to cure 25–30% in some regions. With the completion of the genome sequence for *T. brucei*, there is hope that comparative genomics will aid in the search for potential drug targets that can cross the blood–brain barrier. However, whether there will be an economic incentive to engage in the research, development, and production of new and better drugs is another matter. Vector control through the use of insecticides, traps, and release of sterile male flies has long been thought to offer the best hope for control of trypanosomes in Africa, but it is unlikely that complete eradication can be achieved due to the many technical and political obstacles.

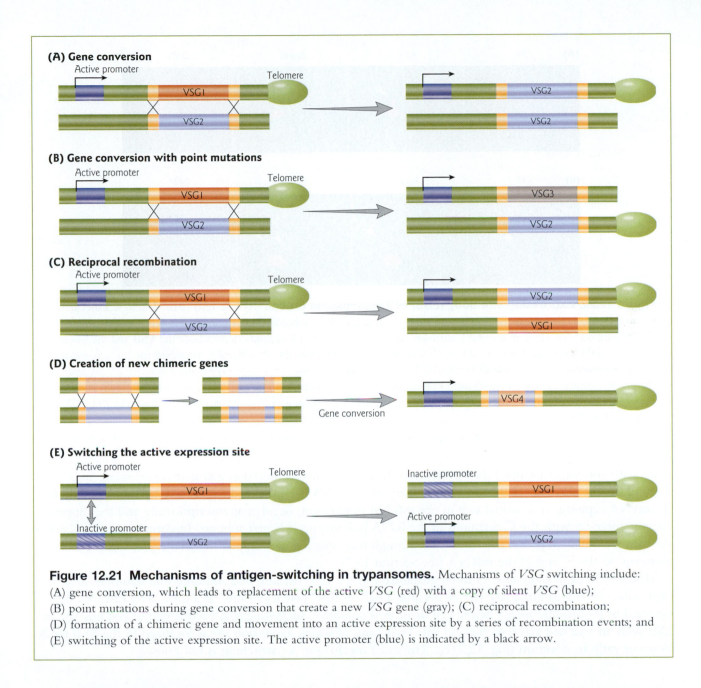

Figure 12.21 Mechanisms of antigen-switching in trypansomes. Mechanisms of *VSG* switching include:
(A) gene conversion, which leads to replacement of the active *VSG* (red) with a copy of silent *VSG* (blue);
(B) point mutations during gene conversion that create a new *VSG* gene (gray); (C) reciprocal recombination;
(D) formation of a chimeric gene and movement into an active expression site by a series of recombination events; and
(E) switching of the active expression site. The active promoter (blue) is indicated by a black arrow.

(Fig. 12.21C). In this mechanism, there is rearrangement of which *VSG* gene is in the active site without duplication or degradation of either *VSG* gene. Not only can trypanosomes change their coats, they also can create additions to their wardrobes. Homologous recombination events can construct new functional genes by combining together segments of two or more closely related *VSG* genes or segments of pseudogenes prior to movement to an expression site (Fig. 12.21D).

Epigenetic regulation of active expression site switching

Monoallelic *VSG* gene expression is maintained by an epigenetic control mechanism that silences all but one of the 20 possible expression sites. *VSG* gene transcription is mediated by RNA polymerase I, and transcription is localized in a discrete nuclear compartment called the "expression site body" (ESB). The ESB is also referred to as the "extranucleolar body" because of its proximity to the nucleolus. The ESB

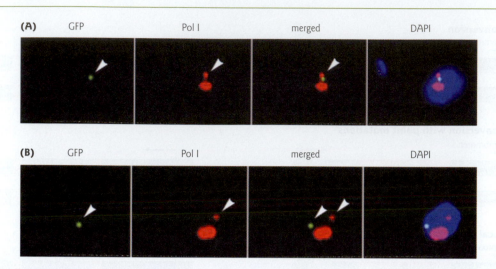

Figure 12.22 Expression site body (ESB) of *Trypanosoma brucei*. (A) The ESB (arrowheads) is associated with GFP-tagged active expression site sequences, as visualized by double-labeling using an anti-RNA polymerase I (Pol I) antibody (red) and an anti-GFP antibody (green). The anti-Pol I antibody detects both the ESB and the larger nucleolus. DNA is stained with DAPI (4′,6′-diamidino-2-phenylindole dihydrochloride), revealing the nucleus and the smaller kinetoplast (mitochondrial) DNA. (B) The ESB body is not associated with GFP-tagged inactive ES sequences. (Reprinted with permission from Nature Publishing Group and Macmillan Publishers Ltd: Navarro M., Gull, K. 2001. A pol I transcriptional body associated with VSG mono-allelic expression in *Trypanosoma brucei*. *Nature* 414:759–763. Copyright © 2001.)

stains with antibodies against RNA polymerase I, whereas there is no staining at inactive telomere sites (Fig. 12.22; see also Fig. 12.20). In other eukaryotes, mRNA is synthesized exclusively by RNA polymerase II. mRNA capping, an essential step in pre-mRNA maturation, occurs cotranscriptionally and the capping enzyme interacts with the C-terminus of the RNA polymerase II largest subunit. In contrast, for *VSG* genes the caps are obtained post-transcriptionally through *trans*-splicing of a capped "spliced leader" (SL) RNA transcript to the 5′ end of the mRNA (see Section 13.7).

Monoallelic expression appears to be guaranteed in the ESB because one and only one telomere-linked expression site is permitted to enter or become a part of this "privileged site" at a time. The ESB can be detected in postmitotic nuclei before cell division is complete, suggesting that each daughter cell inherits an ESB with an attached active expression site. Switching of which expression site is localized to the ESB tends to occur early in infection (Fig. 12.21E). The exact mechanism for switching is not known.

The modified base "J"

In trypanosomes, a fraction of thymine is replaced by the modified base β-D-glucosylhydroxymethyluracil, called "J." A bulky glucose moiety is attached to the thymine base such that it extends into the major groove of the DNA. J might fulfill the role played in mammals by methylcytosine, a modified base not present in *T. brucei*. J is present within telomeric repeats and the ~ 19 inactive telomeric *VSG* gene expression sites, but not in the active expression site (see Fig. 12.20). Exactly what role base J plays in antigenic variation remains to be determined.

V(D)J recombination and the adaptive immune response

There are two main branches of the immune system in vertebrates: the innate immune response and the adaptive immune response. Innate immunity is conferred by a variety of systems that destroy pathogens nonspecifically, including the action of natural killer cells and phagocytes. Adaptive immunity allows the host to resist specific pathogens and is mediated by humoral (blood-borne) and cell-based responses. The humoral response involves B cells and the cell-based response involves T cells. Foreign antigens (named for *anti*body *gene*rator), such as bacteria and viruses, are recognized by B and T cells via a vast repertoire of antigen-specific receptors. Similar to yeast mating-type switching and trypanosome antigen switching, programmed gene rearrangements that lead to differentiation take place during development. The diversity of the antigen receptor repertoire is created from a relatively small number of V, D, and J gene segments that are rearranged in somatic cells in a process called V(D)J recombination (the D in parentheses applies to the heavy chain only). Only one of the two alleles of the antigen receptor is expressed in B and T cells. Similar processes of allelic exclusion operate in B and T cells for immunoglobulin (antibody) and T cell receptor genes, respectively. Here, immunoglobulin genes will be highlighted as one example of such remarkable regulation. Interestingly, there is evidence that the system for rearranging these gene segments evolved from a captured DNA transposon (Focus box 12.1).

Immunoglobulin genes and antibody diversity

The immunoglobulin (antibody) molecule is composed of two identical heavy chains and two identical light chains, where each chain consists of a constant (C) region and a variable (V) region (Fig. 12.23; see also Tool box 9.4). These observations at the level of protein sequence and structure raised the question of how variable and constant regions were incorporated into a single gene. Evidence that the variable and constant portions of the light chain gene arose by novel DNA rearrangements came from independent work in the late 1970s by Susumu Tonegawa's group and Philip Leder's group. DNA fragments derived from mouse embryos and mature lymphocyte cell lines were hybridized to light chain RNA from the cell line. While the RNA hybridized to two specific fragments from embryo DNA, it hybridized to a novel fragment from the mature lymphocyte DNA. Results from both electron microscopy R-looping experiments and Southern blots suggested that some distance separated the variable and constant region DNA in the embryo and that these regions were joined in the mature lymphocyte (Fig. 12.24).

The development of molecular cloning and DNA sequencing techniques allowed for rapid progress in determining the major features of V(D)J recombination. In germline cells, immunoglobulin genes exist as linear arrays of V, diversity (D) (only in heavy chains), and joining (J) fragments upstream of the C region (Fig. 12.25). There are 100–1000 V segments, 12–16 D segments, and four J segments on the mouse chromosome that carry the heavy chain genes, and about 300 V and four J segments of the light chain. The heavy chain C region contains several segments corresponding to different classes of antibodies. In addition, there are two types of light chain, κ and λ, with a separate gene array for each type. Both types of light chain can interact with a heavy chain, but in a given cell only one type of light chain is expressed.

During development, millions of B cell types are produced. Each B cell has a unique antigen receptor generated from the V, D, and J gene segments by a series of site-specific recombination events. When the receptor binds a specific antigen, the B cells undergo division and secretion of antibodies. It is estimated that the human immune system can generate 10^{12} different antibody molecules. Antigens are recognized with remarkable specificity. For example, related proteins with only a single amino acid substitution can be distinguished by antibodies.

Mechanism for V(D)J recombination

The most important advance in the field after the original discovery of gene rearrangement was the discovery of the RAG recombinases. V(D)J recombination can be divided into two phases: DNA cleavage and joining. Cleavage is initiated by the RAG1/2 (recombination *activating gene*) recombinase complex. RAG1/2 introduces single-stranded DNA breaks in the target recombination signal sequence (RSS) that flanks each gene segment (Fig. 12.25). To fully cleave the DNA, the 3′-OH freed by the nicking attacks the phosphodiester backbone on the opposite strand in a direct transesterification reaction. After cleavage

FOCUS BOX 12.1

Did the V(D)J system evolve from a transposon?

Because of the parallels between V(D)J recombination and transposition, it has been postulated that the V(D)J system might have evolved from a transposon. First, both processes involve short specific sequences at the ends of the mobile segments that are recognized and acted upon by a recombinase. Second, the individual steps of V(D)J recombination are similar to steps in the "cut and paste" mechanism of transposition (compare Fig. 12.12A with Fig. 12.25). In both reactions, double-strand breaks are introduced that separate the recognition sequences from the flanking DNA. In "cut and paste" transposition, after excision of the transposon, the broken DNA ends at the donor site are joined by host DNA repair mechanisms. Similarly, ligation of the coding ends created by V(D)J recombination occurs via a nonhomologous end-joining pathway that requires cellular DNA double-strand break repair functions. A key difference between transposition and V(D)J recombination is that transposon ends are inserted

Figure 1 A speculative model for a transposon origin of the V(D)J system.
(1) A schematic representation of a hypothetical primordial gene for a cell surface protein (red). (2) The gene is interrupted by insertions of a DNA transposon (magenta; sites of transposition are shown by the dark magenta triangles). (3) Another insertion disrupts the gene further. (4) Duplication of the gene segments then builds up loci encoding the germline gene segments flanked by RSS signals (formerly transposon ends). The RAG proteins, encoded from a transposable element now lacking sites of transposon action, then mediate DNA rearrangements. (Modified from Bushman, F. 2002. *Lateral DNA Transfer. Mechanisms and Consequences*. Cold Spring Harbor Laboratory Press, Cold Spring Harbor, NY.)

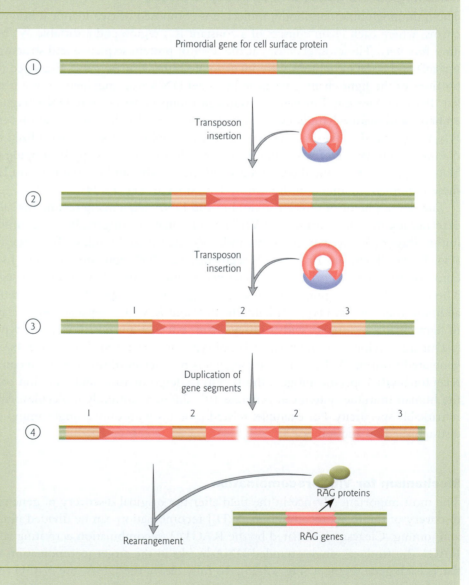

Did the V(D)J system evolve from a transposon?

into the target DNA molecule by the transposase, whereas RSS ends are joined by ligation using factors involved in the cellular double-strand break repair pathway.

The derivation of the V(D)J recombination system is hypothesized to have begun with lateral transfer of an ancient DNA transposon – related to the family of transposons that includes maize *Ac/Ds* – into a gene encoding a cell surface protein (Fig. 1). Amplification of the

DNA unit encoding the gene segment and the flanking transposon end sequences would create multiple gene segments each flanked by transposase-binding sites. The transposase is proposed to have been encoded elsewhere in the genome. The *RAG1* and *RAG2* genes are also postulated to have a transposon origin. In support of this model, they can catalyze DNA transposition of a transposon analog into a target DNA *in vitro*.

of both strands is complete, the segments are joined by the nonhomologous end-joining pathway (see Fig. 7.15). Nonhomologous end-joining brings together one V, one D (heavy chain only), and one J segment with a μ–type C segment. Later, during B cell differentiation, this μ–type C segment can be replaced by one of the other C segments (α, γ, δ, or ε) in a subsequent rearrangement event.

Epigenetic regulation of monoallelic recombination and expression

Each cell contains a paternal and maternal allele of the heavy chain locus and of each type of the light chain, κ and λ. For many years it was thought that rearrangement occurred randomly with respect to the two alleles. Further recombination was assumed to be inhibited once a functional receptor was expressed on the surface of the cell. More recently, evidence has been provided for epigenetic control of the initial selection of the allele to be rearranged, as well as the maintenance of allelic exclusion. All antigen receptor gene segments are flanked by the RSS; therefore, the accessibility of these sequences to the RAG recombinase must be highly regulated. *In vivo*, their accessibilty is mediated by DNA methylation, histone acetylation, and replication timing during the S phase of the cell cycle. The open chromatin state, as judged by histone acetylation and DNA demethylation, correlates with increased recombination. The correct order of rearrangements of the immunoglobulin gene loci is as follows: first, a functional heavy chain gene is assembled, and only then can the light chain locus undergo rearrangement (Fig. 12.26).

Assembly of a functional heavy chain gene   For the heavy chain, V(D)J recombination is a temporally ordered process. First, D-J recombination occurs on both alleles of the heavy chain genes. Subsequently, V-D-J recombination takes place only on one allele, which suggests that only the V region is subject to allelic exclusion (Fig. 12.26). Chromatin immunoprecipitation experiments have shown that histone acetylation correlates strikingly with V(D)J recombination, and the histones in the immunoglobulin heavy chain locus are hyperacetylated in a stepwise manner, domain by domain, such that the DJ region becomes accessible first while the V region still remains hypoacetylated.

Once the immunoglobulin heavy chain with a μ–type C segment (Igμ) is formed, it interacts with surrogate light chain and two other immunoglobulin molecules with different types of C segments (Igαβ and Igβ) to form a pre-B cell receptor. The pre-B cell receptor provides a feedback signal that is thought to inhibit further rearrangement of the second heavy chain allele. This feedback signal also directs the recombinase complex to the light chain locus.

Assembly of a functional light chain gene   At the light chain locus, the two alleles are differentially methylated (Fig. 12.26). The unmethylated light chain allele has a more open chromatin configuration (i.e.

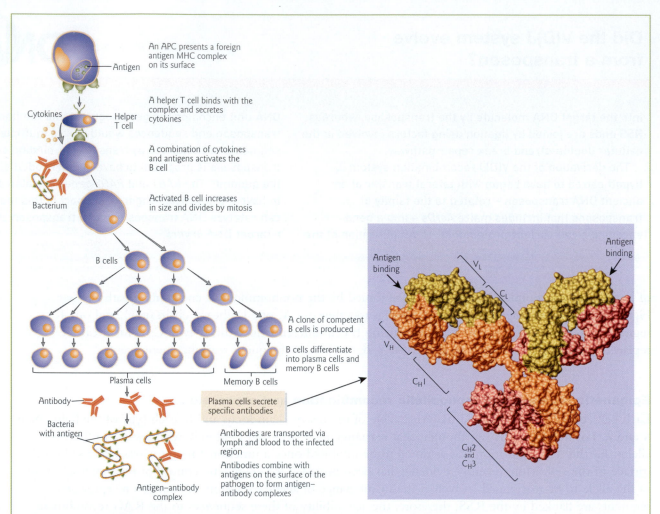

An APC presents a foreign antigen MHC complex on its surface

A helper T cell binds with the complex and secretes cytokines

A combination of cytokines and antigens activates the B cell

Activated B cell increases in size and divides by mitosis

A clone of competent B cells is produced

B cells differentiate into plasma cells and memory B cells

Plasma cells secrete specific antibodies

Antibodies are transported via lymph and blood to the infected region

Antibodies combine with antigens on the surface of the pathogen to form antigen–antibody complexes

Antigen

Cytokines — Helper T cell

Bacterium

B cells

Plasma cells

Memory B cells

Antibody

Bacteria with antigen

Antigen–antibody complex

Antigen binding

V_L

C_L

V_H

C_H1

C_H2 and C_H3

Antigen binding

Figure 12.23 Antibody-mediated immunity. In the example shown, an antigen-presenting cell (APC), such as a macrophage, ingests a bacterial cell and presents the foreign antigens on its cell surface as a foreign antigen–MHC complex (major histocompatibility complex class II proteins). A specific B cell becomes activated when it is exposed to cytokines produced by an activated helper T cell and when it binds with a specific antigen. A clone of cells is produced when the activated B cell divides. Most of the cells differentiate into plasma cells and secrete antibodies that are transported to the infection site by the lymph and blood. A small population of cells become memory B cells and provide acquired immunity. If the antigen reappears, a more rapid and stronger immune response is initiated. (Inset) Structure of an antibody molecule. The light chains are shown in gold, the heavy chains in orange and pink. The antigen-binding sites are as marked. (Protein Data Bank: 1IGT, 1IGY.)

as measured by sensitivity to DNase I) and is much more accessible to RAG-mediated cleavage as compared with the methylated allele. FISH experiments have shown that the initial event, very early in development, which discriminates the two parental alleles of the light chain, is asynchronous replication timing. One allele becomes early-replicating in the S phase (characteristic of active euchromatin) and the other late-replicating (characteristic of inactive heterochromatin). This asynchrony of replication is random with respect to the parental origin of the two alleles. Once established, the asynchrony is stable through many cell divisions, at least in cell culture. The early-replicating allele is usually the one that undergoes rearrangement, whereas the late-replicating allele is moved to the heterochromatic subdomains of the nucleus next to the centromeric heterochromatin.

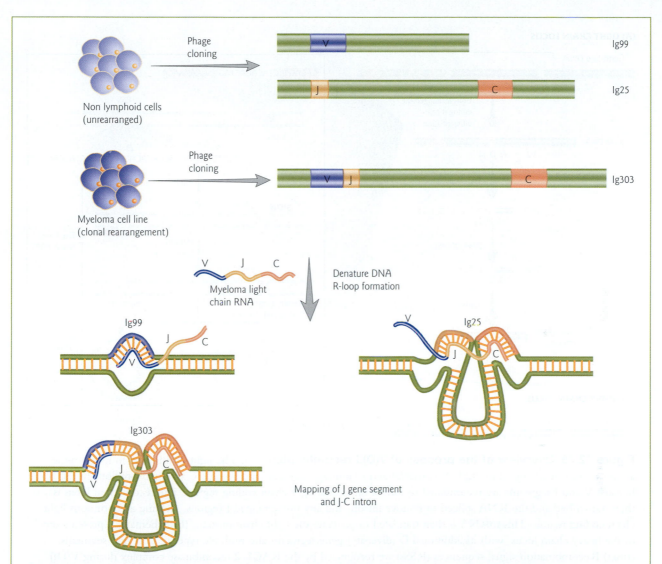

Figure 12.24 Mapping of cloned germline and rearranged immunoglobulin gene segments. Tonegawa and colleagues cloned DNA fragments from embryos (unrearranged nonlymphoid cells) and myeloma cells (rearranged lymphocytes). The embryonic clones Ig99 and Ig25 contained one copy each of the immunoglobulin light chain V or C region, and Ig25 also contained the J region. The myeloma DNA clone Ig303 contained the V, J, and C regions. The DNA was denatured and hybridized to myeloma light chain RNA and analyzed for "R loop" formation by electron microscopy. Mapping of the J region and the J-C intron demonstrated that the variable and constant regions had been joined in the mature lymphocytes. (Modified from Jung, D., Alt, F.W. 2004. Unraveling V(D)J recombination: insights into gene regulation. *Cell* 116:299–311.)

The expression of functional light chain and the formation of the B cell receptor lead to repression of RAG1/2 and of the surrogate light chain. This feedback mechanism is thought to be important for maintenance of light chain allelic exclusion. However, rearrangement can occur, with lower probability, on the "inactive" allele. In case of unsuccessful rearrangement of one allele, the second allele undergoes recombination until a functional B cell receptor is assembled.

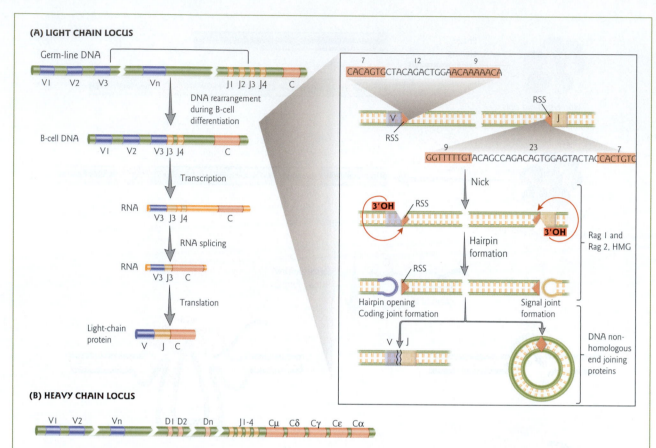

(A) LIGHT CHAIN LOCUS

Germ-line DNA

VI V2 V3   Vn   JI J2 J3 J4   C

DNA rearrangement during B-cell differentiation

B-cell DNA

VI V2 V3 J3 J4   C

Transcription

RNA

V3 J3 J4   C

RNA splicing

RNA

V3 J3   C

Translation

Light-chain protein

V   J   C

CACAGTGCTACAGACTGGAACAAAAACA

7   12   9

RSS

V   J   RSS

RSS

GGTTTTTGTACAGCCAGACAGTGGAGTACTACCACTGTC

9   23   7

Nick

3'OH RSS   3'OH

Rag I and Rag 2, HMG

Hairpin formation

RSS

Hairpin opening Coding joint formation   Signal joint formation

V   J

DNA non-homologous end joining proteins

(B) HEAVY CHAIN LOCUS

VI V2   Vn   DI D2   Dn   JI-4   Cμ Cδ   Cγ   Cε   Cα

Figure 12.25 Overview of the process of V(D)J recombination. (A) The light chain locus in its germline arrangement is made up of multiple V (variable) and J (joining) segments and a single C (constant) region. In the B cell lineage, V and J segments are recombined to form a functional light chain coding region. The recombined locus is then transcribed and the RNA spliced to remove introns and any unrearranged J regions, yielding a continuous light chain coding region. This mRNA is then translated to generate the light chain protein. (B) Schematic representation of the heavy chain locus, with its additional D (diversity) gene segments and multiple types of C region segments. (Inset) Recombination signal sequences (RSSs) are recognized by the RAG1/2 recombinase complex during V(D)J recombination. The RSSs are depicted as red triangles. The signals are in inverted orientation, allowing deletion of the intervening DNA to fuse the V (blue) and J (orange) segments. The RSS is composed of a heptamer (7 bp sequence) and a nonamer (9 bp sequence). Recombination takes place between one RSS with 12 bp between the heptamer and nonamer and a second with a spacing of 23 bp (the 12.23 rule). The RAG1/2 recombinase, in association with the HMG protein, catalyze single-strand cleavage at the ends of the RSSs, leaving a free 3'-OH. Each 3'-OH then initiates attack on the opposite strands to form a hairpin intermediate. The hairpin structures are subsequently hydrolyzed and then joined together by DNA nonhomologous end-joining, to form a coding joint between the V and J regions. This structure undergoes further recombination. The two ends carrying the RSSs are also joined to form a signal joint which is discarded.

Chapter summary

Some genes are transcribed preferentially from a single allele in each diploid cell. In most cases of monoallelic gene expression, selection of the active allele is random. An exception is genomic imprinting where selection of the active allele is nonrandom and based on the parent of origin. Epigenetic silencing mechanisms play a role in monoallelic expression, as do programmed gene rearrangements. Epigenetics is the study of mitotically and/or meiotically heritable changes in gene expression without changes in DNA

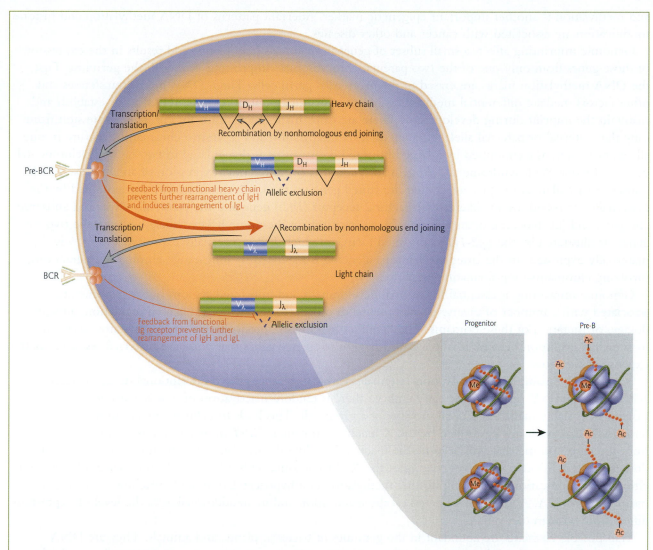

Figure 12.26 Epigenetic control of the ordered rearrangement of immunoglobulin genes. The heavy chain locus undergoes rearrangement first. Once a productive rearrangement has occurred with one allele, this allele is expressed (black arrows), forming together with surrogate light chain a pre-B cell receptor (pre-BCR). Pre-BCR signaling provides a feedback signal (red arrows and lines), leading to inactivation of the second heavy chain allele and redirecting the recombination machinery to the light chain locus. A functionally rearranged light chain becomes part of the B cell receptor (BCR), which, by a mechanism similar to that for the pre-BCR, provides a feedback inhibition of further rearrangements of both heavy and light chain loci. (Inset) The two light chain alleles are shown packaged into a nucleosome. (Left panel) In progenitor B cells, both alleles are methylated (Me) and the histones are deacetylated, but one allele replicates earlier (bottom) than the other (top). (Right panel) In pre-B cells, both alleles are packaged with hyperacetylated histones (*Ac*). The early allele (bottom) undergoes monoallelic demethylation, making it susceptible to rearrangement. (Adapted from Bergman, Y., Fisher, A., Cedar, H. 2003. Epigenetic mechanisms that regulate antigen receptor gene expression. *Current Opinion in Immunology* 15:176–181. Copyright © 2003, with permission from Elsevier.)

sequence. Cytosine DNA methylation at CG dinucleotides marks genes for silencing. Methylation is maintained through DNA replication by a semiconservative process. Small regions of unmethylated CG–rich DNA, called CpG islands, are found associated with gene promoters. Because of the lack of methylation, they are protected from C → T deamination. Stable maintenance of histone modifications such as acetylation and methylation is another important epigenetic marker. Aberrant patterns of DNA methylation and histone modification are associated with cancer and other diseases.

Genomic imprinting affects a small subset of genes in mammals (and plants) and results in the expression of those genes from only one of the two parental chromosomes. Imprinting is reset in the germline. First, the DNA methylation marks are erased in the primordial germ cells. Then, DNA methyltransferases and other factors mediate differential methylation of the imprinting control regions (ICRs) to re-establish and maintain the imprint during development. There are three main mechanisms for ensuring expression from only the maternal or paternal allele: (i) altered DNA methylation and repressive chromatin structure in one allele's promoter, as exemplified by a large imprinted domain that is deregulated in two neurodevelopmental diseases, Prader–Willi syndrome and Angelman syndrome; (ii) differential expression of an antisense RNA transcript that silences the protein-coding imprinted gene on the same chromosome, as exemplifed by the maternally expressed insulin-like growth factor 2 receptor (*Igf2r*) gene and the paternally expressed antisense *Air* gene; and (iii) blocking of an enhancer by an insulator on the unmethylated allele between the two genes, as illustrated by the *Igf2-H19* imprinted locus, in which *Igf2* is paternally expressed and *H19* is maternally expressed. In the latter mechanism, silencing is mediated by long-range chromatin interactions, involving chromatin-loop formation.

Genomic imprinting is essential for normal development and imprinted regions of the genome are associated with a number of diverse developmental disorders that result from impaired regulation, altered dosage, or mutation of these domains. The "conflict hypothesis" proposes that imprinting arose early in mammalian evolution because it is to a male's benefit to silence genes that conserve maternal resources at the expense of the fetus.

Random X chromosome inactivation in female mammals balances the transcriptional dosage between XX females and XY males. In the early embryo there is random inactivation of one X chromosome. The inactive X is stably maintained in all progeny of a given cell. This leads to cellular mosaicism in adult females. X inactivation is controlled by the X inactivation center. *XIST* transcript levels are upregulated from the chromosome that will become the inactive X, while the overlapping antisense *Tsix* transcript is downregulated. Silencing involves coating of the X chromosome with *XIST* RNA and a series of epigenetic chromatin modifications, including histone methylation and hypoacetylation, and enrichment of variant histone macroH2A. About 15% of genes escape inactivation and in an additional 10% the level of expression differs from woman to woman.

Transposable elements are abundant in the genomes of bacteria, plants, and animals. They are DNA sequences that have the ability to move and integrate into the genome at a new site within their cell of origin. Epigenetic mechanisms exist that suppress uncontrolled transposition of these elements, and the primary function of DNA methylation is proposed to be defense of the genome.

Transposable elements were first discovered by Barbara McClintock in the 1940s to 1950s. A high level of phenotypic variegation in maize was attributed to movement of the *Ac/Ds* mobile elements. DNA transposons have a wide host range and transpose through a DNA intermediate by a cut and paste mechanism. These elements consist of a transposase gene flanked by inverted terminal repeats that bind transposase and mediate movement. DNA transposons include insertion sequences in bacteria that often carry antibiotic resistance genes, P elements in *Drosophila*, and the ubiquitous *Tc1/mariner* superfamily. Retrotransposons are derived from ancestral retroviruses. They move by a copy and paste mechanism that duplicates the element through an RNA intermediate. They either contain long terminal repeats (LTRs) at both ends or lack the LTRs and have a poly(A) sequence at their 3′ end. Some autonomous LTR retrotransposons are active in the mammalian genome, including the human endogenous retrovirus (HERV) and intracisternal A particles (IAPs) in the mouse. Non-LTR retrotransposons include autonomous long interspersed nuclear elements (LINEs) and nonautonomous short interspersed nuclear elements (SINEs). LINEs are widespread in the human genome. Currently active members of the family are *L1* elements. The *L1* family has expanded to make up about 21% of human DNA. These elements act as insertional mutagens, regulatory control elements of neighboring genes, and as long-distance modifiers of chromatin. *L1* elements that are clustered around the X inactivation center may play a role in X inactivation. *Alu* elements are active

SINEs that make up some 11% of the human genome. Insertional mutagenesis by *Alu* elements and *L1* elements is associated with a number of human diseases.

Phenotype-altering mutations caused by the insertion of transposable elements are much less frequent than are classic point mutations. At least two control methods are known that silence transposition. The first is methylation of transposable elements. For example, diet can alter heritable phenotypic change in *agouti* mice by changing the DNA methylation pattern of the mouse genome and thereby activating a retrotransposon promoter upstream of the *agouti* gene. The second control method is heterochromatin formation mediated by RNAi and RNA-directed DNA methylation. Transposable elements are transcribed and processed into small heterochromatic RNAs (shRNAs) by the RNAi machinery. These shRNAs then associate with other factors to mediate RNA-directed DNA methylation of the transposable element.

Monoallelic gene expression plays an important role in cell differentiation in a diversity of organisms. One major mechanism mediating allelic exclusion is programmed gene rearrangement, involving recombination events. In mating-type switching in budding yeast, cells switch between expressing one of the two different mating-type cassettes. Information is copied from one of the two silent donor loci, *HMLa* or *HMLα*, into the *MAT* locus. DNA rearrangement occurs by site-specific homologous recombination (gene conversion), mediated by the HO endonuclease. There is preferential switching to the opposite mating type in cells that have divided at least once. Directionality of switching is controlled by a recombination enhancer. The recombination enhancer is turned off by the binding of repressor proteins encoded by the α-mating type. Expression of the silent mating-type donor sequences is repressed through epigenetic mechanisms. Silencer sequences flank the donor loci and recruit a variety of specific regulatory proteins (Sir1–Sir4 and Rap1), histone deacetylases, and heterochromatin assembly factors. Together these factors act to establish a domain of closed chromatin to repress transcription.

Avoidance of the immune response in African trypanosomes, the causative agent of human "sleeping sickness," occurs by switching which one of the many possible genes is active in making the major surface antigen. The trypanosome can change its variant surface glycoprotein (VSG) coat by replacing the transcribed *VSG* gene in the telomeric expression site by one of > 1000 different *VSG* genes, or by switching which expression site is active. There are 20 different expression sites, but only one is active at a time. Antigen switching mainly occurs by DNA homologous recombination, either by gene conversion or reciprocal recombination. Active expression site switching is regulated by epigenetic mechanisms that control which expression site is located in the expression site body (ESB) within the nucleus. The modified base "J" (thymine with a bulky glucose moiety) may play a similar role to methylcytosine in gene silencing.

The production of active immunoglobulin genes occurs by the assembly of different DNA segments. The immunoglobulin (antibody) protein is composed of two identical heavy chains and two identical light chains, where each chain consists of a constant (C) region and a variable (V) region. In germline cells, immunoglobulin genes exist as linear arrays of V, diversity (D) (only in heavy chains), and joining (J) regions upstream of the C region. A series of site-specific recombination events in B cells generate unique combinations of V(D)J sequences that encode unique antigen receptors. V(D)J recombination is mediated by the RAG1/2 recombinase complex. A target recombination signal sequence (RSS) flanks each gene segment. After cleavage of both strands is complete, the segments join by the nonhomologous end-joining pathway. Because of the parallels between V(D)J recombination and transposition, it has been postulated that the V(D)J system might have evolved from a transposon. There is epigenetic control of the initial selection of the allele to be rearranged, as well as maintenance of allelic exclusion. For the heavy chain, V(D)J recombination is a temporally ordered process. D-J recombination occurs on both alleles, but V-D-J recombination takes place on only one allele, mediated by progessive histone hyperacetylation. A feedback signal inhibits further rearrangement of the second heavy chain allele and directs the recombinase complex to the light chain locus. At the light chain locus, the two alleles are differentially methylated. One allele becomes early-replicating in the S phase and the other late-replicating. The early-replicating allele is usually the one that undergoes rearrangement, while the late-replicating allele is moved to heterochromatic subdomains of the nucleus.

Analytical questions

1 You find that the maternal allele of the mouse *MAMA* gene is active, while the paternal allele is repressed. The gene encodes a growth factor.

(a) Predict the effect on allelic expression of treating adult mice with 5-aza-deoxycytidine.

(b) During the course of the experiment you find that mouse pups are born with completely yellow fur instead of the expected brown fur. Provide an explanation.

2 The figure below shows the results of methylation analysis of DNA from a "normal" child, a child with Prader–Willi syndrome, and two children with Angelman syndrome, using a Southern blot of genomic DNA digested with *Hind*III and *Hpa*II, probed with PW71B, a chromosome 15-specific probe.

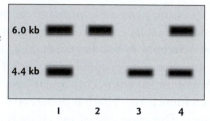

Lane 1: "Normal" child

Lane 2: Child with Prader-Willi Syndrome

Lane 3: Child with Angelman Syndrome

Lane 4: Child with Angelman Syndrome

Interpret the results for each lane.

3 On the inactive X chromosome many CpG islands are heavily methylated. Does this hold true for the CpG island upstream of the *XIST* gene? Why or why not?

4 Assume you have two cell-free transposition systems for a yeast *Ty* element and a vertebrate *Tc1/mariner* element. What effect would the following inhibitors have on these two systems, and why?

(a) Inhibitors of translation.

(b) Inhibitors of transcription.

(c) Inhibitors of double-stranded DNA replication.

(d) Inhibitors of reverse transcription.

5 Molecular biologists have demonstrated that there is preferential switching to the opposite mating type in budding yeast. Switching from *a* → *a* (or α → α) with no change in mating type occurs rarely. Since the DNA sequence in the *MATa* locus would be the same as the donor *HMRa*, how would you show whether *a* → *a* had occurred or not?

6 In the most common mode of trypanosome antigen switching, the *VSG* gene in the active expression site is degraded and replaced with the donor *VSG* gene. Why isn't there progressive loss of *VSG* genes over time?

7 Is the genomic DNA in a bone marrow stem cell identical to the genomic DNA in a mature B cell? Why or why not?

Suggestions for further reading

Abe, T., Toyota, M., Suzuki, H. et al. (2005) Upregulation of BNIP3 by 5-aza-2′–deoxycytidine sensitizes pancreatic cancer cells to hypoxia-mediated cell death. *Journal of Gastroenterology* 40:504–510.

Ash, C., Jasny, B.R., section eds (2005) The trypanosomatid genomes *Science* 309:399–442.

Bender, J. (2004) Chromatin-based silencing mechanisms. *Current Opinion in Plant Biology* 7:521–526.

Bergman, Y., Fisher, A., Cedar, H. (2003) Epigenetic mechanisms that regulate antigen receptor gene expression. *Current Opinion in Immunology* 15:176–181.

Bhattacharyya, M.K., Smith, A.M., Ellis, T.H., Hedley, C., Martin, C. (1990) The wrinkled-seed character of pea described by Mendel is caused by a transposon-like insertion in a gene encoding starch-branching enzyme. *Cell* 60:115–122.

Borst, P. (2002) Antigenic variation and allelic exclusion. *Cell* 109:5–8.

Brack, C., Hirama, M., Lenhard-Schuller, R., Tonegawa, S. (1978) A complete immunoglobulin gene is created by somatic recombination. *Cell* 15:1–14.

Buchholz, T., Smith, E. (1997) Testing for genomic imprinting. *Today's Life Science* 9:36–39.

Bushman, F. (2002) *Lateral DNA Transfer. Mechanisms and Consequences.* Cold Spring Laboratory Press, Cold Spring Harbor, NY.

Bushman, F. (2004) Selfish elements make a mark. *Nature* 429:253–255.

Carrel, L., Willard, H.F. (2005) X-inactivation profile reveals extensive variability in X-linked gene expression in females. *Nature* 434:400–404.

Clayton-Smith, J., Laan, L. (2003) Angelman syndrome: a review of the clinical and genetic aspects. *Journal of Medical Genetics* 40:87–95.

Comfort, N.C. (2001) *The Tangled Field. Barbara McClintock's Search for the Patterns of Genetic Control.* Harvard University Press, Cambridge, MA.

Dalgaard, J.Z., Vengrova, S. (2004) Selective gene expression in multigene families from yeast to mammals. *Science's STKE* 256:1–10 (www.stke.org/cgi/content/full/sigtrans;2004/256/re17).

Delaval, K., Feil, R. (2004) Epigenetic regulation of mammalian genomic imprinting. *Current Opinion in Genetics and Development* 14:188–195.

DiPaolo, C., Kieft, R., Cross, M., Sabatini, R. (2005) Regulation of trypanosome DNA glycosylation by a SWI2/SNF2-like protein. *Molecular Cell* 17:441–451.

Druker, R., Whitelaw, E. (2004) Retrotransposon-derived elements in the mammalian genome: a potential source of disease. *Journal of Inherited Metabolic Disorders* 27:319–330.

Emerson, J.J., Kaessmann, H., Betran, E., Long, M. (2004) Extensive gene traffic on the mammalian X chromosome. *Science* 303:537–540.

Ferguson-Smith, A.C., Sasaki, H., Cattanach, B.M., Surani, M.A. (1993) Parental-origin-specific epigenetic modification of the mouse *H19* gene. *Nature* 362:751–755.

Feschotte, C., Jiang, N., Wessler, S.R. (2002) Plant transposable elements: where genetics meets genomics. *Nature Reviews Genetics* 3:329–341.

Field, M.C., Carrington, M. (2004) Intracellular membrane transport systems in *Trypanosoma brucei*. *Traffic* 5:905–913.

Fraga, M.F., Ballestar, E., Villar-Garea, A. et al. (2005) Loss of acetylation at Lys16 and trimethylation at Lys20 of histone H4 is a common hallmark of human cancer. *Nature Genetics* 37:391–400.

Ferber, D. (2003) Triple-threat microbe gained powers from another bug. *Science* 302:1488.

Gartenberg, M.R., Neumann, F.R., Laroche, T., Blaszczyk, M., Gasser, S.M. (2004) Sir-mediated repression can occur independently of chromosomal and subnuclear contexts. *Cell* 119:955–967.

Goldmit, M., Bergman, Y. (2004) Monoallelic gene expression: a repertoire of recurrent themes. *Immunological Reviews* 200:197–214.

Grewal, S.I.S., Moazed, D. (2003) Heterochromatin and epigenetic control of gene expression. *Science* 301:798–802.

Haber, J.E. (1998) Mating-type gene switching in *Saccharomyces cerevisiae*. *Annual Reviews of Genetics* 32: 561–599.

Hajkova, P., Surani, M.A. (2004) Programming the X chromosome. *Science* 303:633–634.

Hekimi, S., Guarente, L. (2003) Genetics and the specificity of the aging process. *Science* 299:1351–1354.

Hoppe, G.J., Tanny, J.C., Rudner, A.D., Gerber, S.A., Danaie, S., Gygi, S.P., Moazed, D. (2002) Steps in assembly of silent chromatin in yeast: Sir3-independent binding of a Sir2/Sir4 complex to silencers and role for Sir2-dependent deacetylation. *Molecular and Cellular Biology* 22:4167–4180.

Horie, K., Kuroiwa, A., Ikawa, M., Okabe, M., Kondoh, G., Matsuda, Y., Takeda, J. (2001) Efficient chromosomal transposition of a *Tc1/mariner*-like transposon *Sleeping Beauty* in mice. *Proceedings of the National Academy of Sciences USA* 98:9191–9196.

Horn, D. (2004) The molecular control of antigenic variation in *Trypanosoma brucei*. *Current Molecular Medicine* 4:563–576.

Iida, S., Morita, Y., Choi, J.D., Park, K.I., Hoshino, A. (2004) Genetics and epigenetics in flower pigmentation associated with transposable elements in morning glories. *Advances in Biophysics* 39:141–159.

Jiang, Y., Bressler, J., Beaudet, A.L. (2004) Epigenetics and human disease. *Annual Review of Genomics and Human Genetics* 5:479–510.

Jones, J.M., Gellert, M. (2004) The taming of a transposon: V(D)J recombination and the immune system. *Immunological Reviews* 200:233–248.

Jung, D., Alt, F.W. (2004) Unraveling V(D)J recombination: insights into gene regulation. *Cell* 116:299–311.

Jurka, J. (2004) Evolutionary impact of human *Alu* repetitive elements. *Current Opinion in Genetics and Development* 14:603–608.

Kaneda, M., Okano, M., Hata, K., Sado, T., Tsujimoto, N., Li, E., Sasaki, H. (2004) Essential role for *de novo* DNA methyltransferase Dnmt3a in paternal and maternal imprinting. *Nature* 429:900–903.

Kato, Y., Sasaki, H. (2004) Imprinting and looping: epigenetic marks control interactions between regulatory elements. *BioEssays* 27:1–4.

Kazazian, H.H., Wong, C., Youssoufian, H., Scott, A.F., Phillips, D.G., Antonarakis, S.E. (1988) Haemophilia A resulting from de novo insertion of L1 sequences represents a novel mechanism for mutation in man. *Nature* 332:164–166.

Kazazian, H.H. (2004) Mobile elements: drivers of genome evolution. *Science* 303:1626–1632.

Kishino, T., Lalande, M., Wagstaff, J. (1997) UBE3A/E6-AP mutations cause Angelman syndrome. *Nature Genetics* 15:70–73.

Kobayshi, K., Nakahori, Y., Miyake, M. et al. (1998) An ancient retrotransposal insertion causes Fukuyama-type congenital muscular dystrophy. *Nature* 394:388–392.

Kobayashi, S., Goto-Yamamoto, N., Hirochika, H. (2004) Retrotransposon-induced mutations in grape skin color. *Science* 304:982.

Lin, T.H., Tsai, K.C., Lo, T.C. (2003) Homology modeling of the central catalytic domain of insertion sequence ISLC3 isolated from *Lactobacillus casei* ATCC 393. *Protein Engineering* 16:819–829.

Lund, A.H., van Lohuizen, M. (2004) Epigenetics and cancer. *Genes and Development* 18:2315–2335.

Makedonski, K., Abuhatzira, L., Kaufman, Y., Razin, A., Shemer, R. (2005) MeCP2 deficiency in Rett syndrome causes epigenetic aberrations at the PWS/AS imprinting center that affects UBE3A expression. *Human Molecular Genetics* 14:1049–1058.

Matsuura, T., Sutcliffe, J.S., Fang, P. et al. (1997) *De novo* truncating mutations in E6-AP ubiquitin-protein ligase gene (*UBE3A*) in Angelman syndrome. *Nature Genetics* 15:74–77.

McCulloch, R. (2004) Antigenic variation in African trypanosomes: monitoring progress. *Trends in Parasitology* 20:117–121.

Murphy, S.K., Jirtle, R.L. (2003) Imprinting evolution and the price of silence. *BioEssays* 25:577–588.

Niemitz, E.L., Feinberg, A.P. (2004) Epigenetics and assisted reproductive technology: a call for investigation. *American Journal of Human Genetics* 74:599–609.

Oldridge, M., Zackai, E.H., McDonald-McGinn, D.M. et al. (1999) De novo *Alu*-element insertions in *FGFR2* identify a distinct pathological basis for Apert syndrome. *American Journal of Human Genetics* 64:446–461.

Runt, M., Varon, R., Horn, D., Horsthemke, B., Buiting, K. (2005) Exclusion of the C/D box snoRNA gene cluster *HBII-52* from a major role in Prader–Willi syndrome. *Human Genetics* 116:228–230.

Russo, V.E.A., Martienssen, R.A., Riggs, A.D. (1996) *Epigenetic Mechanisms of Gene Regulation.* Cold Spring Harbor Laboratory Press, Cold Spring Harbor, NY.

Sakatani, T., Kaneda, A., Iacobuzio-Donahue, C.A. et al. (2005) Loss of imprinting of *Igf2* alters intestinal maturation and tumorigenesis in mice. *Science* 307:1976–1978.

Scott, R.J., Spielman, M. (2004) Epigenetics: imprinting in plants and mammals – the same but different? *Current Biology* 14:R201–R203.

Stoyanova, V., Oostra, B.A. (2004) The CGG repeat and the FMR1 gene. *Methods in Molecular Biology* 277:173–184.

Waterland, R.A., Jirtle, R.L. (2003) Transposable elements: targets for early nutritional effects on epigenetic gene regulation. *Molecular and Cellular Biology* 23:5293–5300.

Chapter 13

RNA processing and post-transcriptional gene regulation

The discovery of split genes along with some other findings of recent years, shows that the genetic apparatus of the cell is more complex, more variable, and more dynamic than any of us had suspected.

Pierre Chambon, *Scientific American* (1981) 244:60.

Outline

13.1 Introduction

Scientists who study RNA have been faced with more revolutionary and unexpected discoveries in the past several decades than in any other area of molecular biology. Investigations into the post-transcriptional processing of RNA led to the discovery of "split genes," catalytic RNA, and guide RNAs that alter the nucleotide sequence of a pre-messenger RNA (pre-mRNA) post-transcriptionally. Previously undetected small noncoding microRNAs (miRNAs) now have been found in such abundance they have been dubbed the "dark matter" of the cell. In Chapter 11 we saw that whether or not a gene is transcribed is a major way that gene expression is regulated. This chapter focuses on the emerging importance of post-transcriptional processes in eukaryotic gene regulation. Typically, RNA is first transcribed as heterogeneous nuclear RNA (hnRNA) and includes exons and introns. Characteristic features of the five major classes of introns are described: group I and group II self-splicing introns, archael and nuclear tRNA introns, and nuclear pre-mRNA introns. A recurring theme in processing events that involve RNA cofactors – whether splicing or base modification – is that the RNA is used to provide specificity by complementary base pairing. A mature mRNA ready for translation is formed by splicing out introns and joining exons. Alternative splicing provides a means of generating protein diversity from a small set of genes. Levels of post-transcriptional control of mRNA expression include the addition of a 5′ cap, polyadenylation, RNA processing, nuclear export, RNA quality control, and nonsense-mediated mRNA decay. All these processes are intimately coupled to transcription.

13.2 RNA splicing: historical perspective and overview

For many years after the genetic code was deciphered, it was assumed that the linear sequence of nucleotides in a discrete, contiguous stretch of DNA corresponded directly to a linear sequence of amino acids in a protein. In 1977 this principle of "colinearity" was proved wrong by Phillip Sharp, Richard Roberts, and colleagues. Comparison of the complementarity of sequences expressed as cytoplasmic mRNAs during the late stage of adenovirus infection with the viral DNA from which it was transcribed, resulted in the discovery of "split genes." Sharp and colleagues hypothesized that intervening sequences were removed or "spliced" from the primary RNA transcript, joining together four separate regions to make the mature adenovirus mRNA. Shortly after the discovery of split genes and RNA splicing in adenoviruses, a number of cellular genes were also shown to have intervening sequences. For example, the ovalbumin gene was shown by R loop mapping to be split into eight sets of sequences (Fig. 13.1).

In 1978, Walter Gilbert coined the term "intron" for these intervening sequences that split genes. Exons were defined as the functional, expressed sequences of the mature RNA molecule. Introns were found to range in size from less than 100 nt to hundreds of thousands of nucleotides. In comparison, exons are generally short, ranging in size from 50 to 300 nt. Until the early 1990s, the view was that introns are transcribed along with the exons, but are "junk" that is excised and degraded (Fig. 13.2). This passive role of introns was modified slightly upon finding that introns may contain transcriptional regulatory elements, such as enhancers or silencers (see Focus box 11.1). However, researchers still had a tendency to only look in detail at the exon sequence when establishing the function of a newly identified gene. Then, two important discoveries led to a major shift in focus. First, introns were discovered that code for small RNAs (e.g. small nucleolar RNAs and microRNAs). Subsequently, "inside-out" genes were discovered where the introns code for function and the exons are degraded (Focus box 13.1). Accordingly, introns are now defined as sequences that remain physically separated after excision. Exons are defined as sequences that are ligated together after excision. Conclusions as to which parts of a gene express function must be based on other analyses.

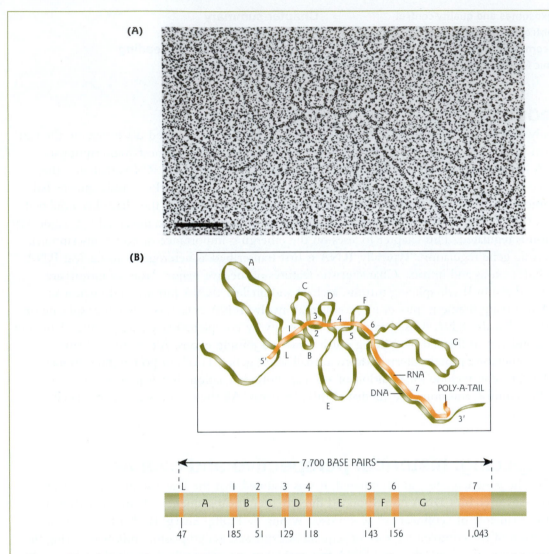

(A)

(B)

7,700 BASE PAIRS

47   185   51   129   118   143   156   1,043

Figure 13.1 Intervening sequences in the ovalbumin gene. The split gene structure of the DNA encoding the egg-white protein ovalbumin is demonstrated by the electron micrograph (A; magnification ×180,000) and its R loop map (B). A fragment of single-stranded DNA containing the ovalbumin gene was allowed to hybridize with purified ovalbumin mRNA. Segments of the DNA (green line on map) and RNA (orange) that are complementary to each other have hybridized, forming double-stranded regions. The DNA sequences in those regions are the eight exons (L, 1–7). Seven intervening sequences or introns (A–G) of DNA loop out from the hybrid, as they have no complementary sequences in the RNA. The two ends of the mRNA (5' and 3') are indicated, as is the short poly(A) tail at the 3' end. (C) The schematic representation of the gene shows the seven introns (green) and eight exons (orange) and the number of base pairs in each of the exons. The intron size ranges from 251 bp (B) to about 1,600 (G). (Reproduced from Chambon, P. 1981. Split genes. *Scientific American* 244:60–71. Image courtesy of Pierre Chambon, Institut de Genetique et de Biologie Moleculaire et Cellulaire, Strasbourg, France).

RNA splicing is the process by which introns are removed from a primary RNA transcript at precisely defined splice points, and the ends of the remaining RNA are rejoined to form a continuous mRNA, ribosomal RNA (rRNA), or transfer RNA (tRNA). The excision of introns and the joining of the exons is directed by special sequences at the intron–exon junctions called splice sites (Fig. 13.2). The 5' splice site marks the exon–intron junction at the 5' end of the intron. At the other end of the intron, the 3' splice site

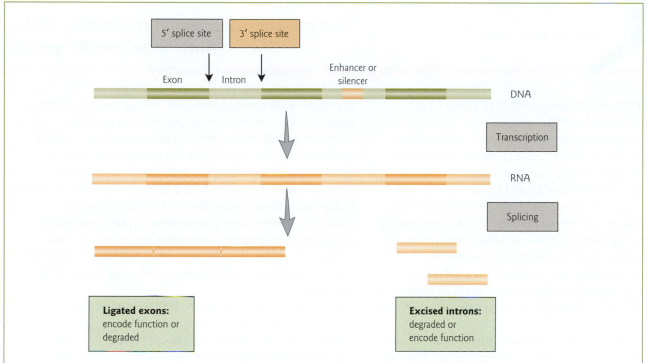

Figure 13.2 Conventional "split" genes are comprised of exons and introns. The DNA sequence
of a gene is transcribed into a primary transcript which is then processed by splicing. Introns (light orange) are the
sequences that remain physically separated after excision, while exons (dark orange) are ligated together. Introns may
contain important transcriptional regulatory elements, such as enhancers and silencers. Introns are usually nonfunctional
and are degraded; however, some introns encode functional RNA products, such as snoRNAs or miRNAs. Exons
typically encode function (e.g. mRNA, rRNA, tRNA), but in a few genes the exons are nonfunctional and are
degraded (see Focus box 13.1).

marks the junction with the next exon. Five major classes of introns are recognized based on their splicing
mechanism: autocatalytic group I and group II introns (Section 13.3), tRNA and/or archael introns
(Section 13.4), and spliceosomal introns in nuclear pre-mRNA (Section 13.5).

13.3 Group I and group II self-splicing introns

Group I and group II introns are large catalytic RNAs distinguished by their structure and mechanism of
splicing. As if self-splicing was not remarkable enough, both types of introns have mobile members that have
evolved mechanisms for inserting into intronless genes.

Group I introns require an external G cofactor for splicing

Group I introns were first characterized as self-splicing elements in the nuclear rRNA from *Tetrahymena*,
a ciliated protozoan (see Chapter 4 for details, Fig. 4.15). They are now known to be widely distributed
in the mitochondrial, chloroplast, and nuclear genomes of diverse eukaryotes, with over 1500 described to
date. Interestingly, nuclear group I introns are limited to rRNA, whereas in organelles they are found in
rRNA, tRNA, and mRNA. Ninety percent of all group I introns are found in fungi, plants, and red and
green algae. In animals, group I introns have only been found so far in the mitochondrial genomes of
one sea anenome (*Metridium senile*) and a coral (*Acropora tenuis*). No group I introns have been identified
to date in Archaea. They are rare in bacteria and occur infrequently in viruses and bacteriophages. Their

Intron-encoded small nucleolar RNAs and "inside-out" genes

In the early 1990s, it was discovered that genes for small nucleolar RNAs (snoRNA, pronounced "snow" RNA) are often encoded within introns, and are liberated from the pre-mRNA via a splicing-dependent reaction. After intron excision, an exonuclease trims the intron to form a mature snoRNA. Other snoRNAs, such as U3 and U8, are encoded by small nuclear RNA (snRNA) type RNA polymerase II genes. Intron-encoded snoRNAs are typically located in parent genes coding for proteins involved in nucleolar function, ribosome biogenesis, or mRNA translation. For example, the *Xenopus* U14 snoRNA is encoded in the intron of the ribosomal protein S13 gene. These findings led to closer scrutiny of intron sequences in other genes.

In 1994 the surprising discovery was made of an "inside-out" gene. The U22 snoRNA, which is required for 18S rRNA processing (see Section 13.9), was discovered within the introns of the U22 host gene. Partial sequence analysis of mouse and human U22 host genes revealed that, unlike in conventional genes, the introns were highly conserved and

the exons were poorly conserved. Within the introns there was extensive sequence complementarity to rRNA, which is a hallmark of snoRNAs. Researchers found that the U22 host gene encodes eight different snoRNAs in separate introns. Molecular biologists in the snoRNA field were greatly excited by this so-called "blizzard" of new snoRNAs. In contrast, the spliced exons of the U22 host gene transcript showed little potential for protein coding and were rapidly degraded in the cytoplasm. Researchers concluded that for the U22 host gene, the introns code for function and the spliced exons are left-over "junk" to be degraded.

It was soon realized that the U22 host gene structure was not unique. In 1998, the human U19 host gene was characterized as another "inside-out" gene, in which U19 snoRNA is processed from a long transcript with little potential for protein coding. Now scientists are finding microRNAs within the introns of such "host" genes (see Section 13.10). Undoubtedly, intron sequences will reveal many more surprises in the future.

1986 discovery in bacteriophage T4 marked the first time that RNA splicing had been found in a prokaryote.

Group I introns self-splice by a well-characterized and distinctive two-step pathway that relies on an external guanosine (G) nucleotide as a cofactor (Fig. 13.3A). Only seven nucleotides are conserved among the group I introns, but all of them adopt the same basic structure and carry out the same chemical reaction. The RNA forms a complex three-dimensional folding pattern of side by side stacks of helices, which brings the two exons close together. Successful catalysis is dependent on the correct folding of the intron. Self-splicing is initiated by the free 3'–OH of GTP, which binds to a pocket in the catalytic core of the intron called the G-binding site. The tertiary structure places the 3'–OH in a favorable position for attacking the phosphate group at the 5' splice site (typically a U). Hydroxyl (nucleophilic) attack initiates the first transesterification reaction; i.e. phosphate transfer or put simply, the "swapping" of phosphodiester bonds. In this reaction, the GTP is added to the 5' end of the intron, resulting in the release of the upstream exon. The newly added G then leaves the G-binding site. The last nucleotide of the intron, which is always a G, now binds in the G-binding site. The second transesterification step is initiated by an attack by the 3' end of the released exon on the 3' splice site. Splicing gives rise to two products, the mature RNA with the two exons ligated together, and the excised intron, which is degraded.

Although many group I introns can self-splice *in vitro*, proteins are required for their efficient splicing *in vivo*. These protein cofactors help to fold the intron RNA into the catalytically active structure. Some of the proteins required for splicing are encoded by the introns themselves. In some cases, splicing is protein-dependent both *in vivo* and *in vitro*. For example, several proteins are required for splicing cytochrome-b pre-mRNA in yeast mitochondria. The required proteins include a "maturase" encoded by an open reading frame within the intron itself.

Group II introns require an internal bulged A for splicing

Group II introns are less common than group I introns. They are present in the mitochondrial and chloroplast genomes of certain protists, fungi, algae, and plants, and in bacterial genomes. They self-splice through a pathway that is different from group I introns, but similar to the mechanism of pre-mRNA splicing by the spliceosome (see Section 13.5). The 5′ splice junction (GUGYG, where Y stands for any pyrimidine, C or U) and the 3′ splice junction (AY) of group II introns are conserved. In addition, the RNA folds into a conserved structure that forms an active site containing catalytically essential Mg^{2+} ions. The secondary structure resembles a "flower" with six double-helical domains radiating from a central wheel (Fig. 13.3B). Splicing occurs by two sequential transesterification reactions. The first nucleophilic attack is initiated by the 2′-OH of a bulged, internal adenosine (A) nucleotide within the intron. This results in cleavage of the 5′ splice site coupled with the formation of a lariat structure that can be visualized by electron microscopy. One end of the intron forms a novel 2′–5′-phosphodiester bond with the bulged A, which is near the 3′ end of the intron. This leaves a stub of RNA extending beyond the loop. The second nucleophilic attack is initiated at the 3′ splice site by the 3′-OH of the cleaved 5′ exon. This results in exon ligation and release of the intron lariat. Subsequently, the lariat is debranched and degraded.

Although some group II introns self-splice *in vitro*, this reaction generally requires nonphysiological conditions (e.g. high salt). For efficient splicing *in vivo* many, if not all, group II introns require proteins to fold the intron RNA into the catalytically active structure. As in the group I introns, the proteins are either encoded by the introns themselves or encoded by other genes of the host organism.

Mobile group I and II introns

In addition to being self-splicing, their unusual ability to move about the genome sets group I and II introns apart from other classes of introns. Some mobile members can spread efficiently into a homologous position in an allele that lacks the intron by a process termed "homing" (Fig. 13.4). Because they carry their own intrinsic splicing apparatus, they can potentially spread from organelle to nuclear genomes, or vice versa, or even to different organisms. Their insertion into a new locus would, in principle, have minimal effects on gene expression as long as they were still spliced efficiently.

The movement of mobile group I introns is mediated by highly site-specific homing endonucleases that are typically encoded by the self-splicing intron. The vast majority of group I introns in nature do not contain a homing endonuclease gene. In contrast, about a third of group II introns encode an open reading frame with homology to reverse transcriptases. The excised intron RNA reverse splices directly into a DNA target site and is then reverse transcribed by the intron-encoded protein (Fig. 13.4). Typically, they undergo homing. However, at a much lower frequency, they can move into nonhomologous sites by retrotransposition. In addition to reverse transcriptase activity, splicing requires a protein with "maturase" activity. Maturases are often intron-encoded and bind specifically to the intron RNA to stabilize the active structure. Group II introns that lack open reading frames may be assisted by maturase-related proteins encoded by nuclear genes. In yet another twist, these nuclear genes may themselves have originated from a mobile organelle-encoded group II intron. Other host-encoded splicing factors also function in conjunction with maturases, acting as RNA chaperones. The similarity of mobile group II introns to transposable elements has led to the proposal that these introns are the ancestors of nuclear non-LTR retrotransposons (see Section 12.5).

The first bacterial group II introns were discovered in the early 1990s. With the advent of genome sequencing projects, it has become clear that mobile group II introns are widely distributed in bacteria. About one-quarter of sequenced bacterial genomes contain a group II intron, with up to two dozen present in a given organism. Most bacterial group II introns are inserted between genes or within mobile DNA such as insertion elements or plasmids.

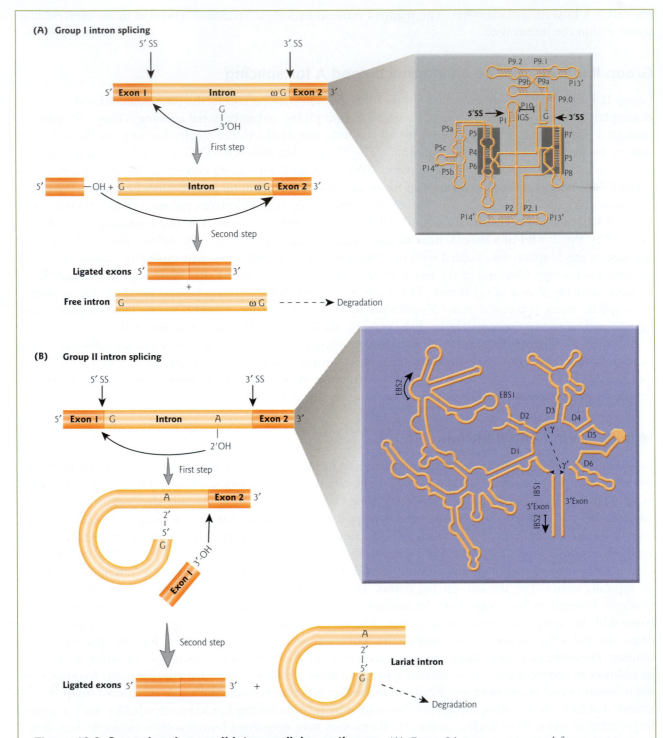

Figure 13.3 Group I and group II intron splicing pathways. (A) Group I introns are removed from precursor RNAs through a two-step splicing reaction. (1) The guanosine (G) cofactor attacks the 5′ splice site (5′ SS) and attaches to the intron resulting in the release of the upstream exon. The G cofactor leaves the G-binding site and is replaced by the last nucleotide of the intron, which is always a G (ωG). (2) Attack by the 3′ end of the released exon on the 3′ SS, which results in ligation of the exons and release of the intron RNA. (Inset) The typical secondary structure of a group I intron has approximately 10 base paired helical elements (P1–P10, plus the optional P11–P17), which are organized into three stacked domains at the tertiary structure. The intron recognizes the 5′ exon sequence by a 4–6 nt base pairing – the internal guide sequence (IGS) interaction. The central catalytic core of the intron RNA

13.4 Archael and nuclear transfer RNA introns

The group I, group II, and nuclear mRNA-type introns that are found in eukaryotes and/or bacteria have not yet been found in Archaea. Instead the latter carry introns in their tRNA, rRNA, and mRNA that are spliced by an apparently archael-specific mechanism. This pathway is completely different from that of spliceosomal introns, but has some similarity to the splicing mechanism for eukaryotic nuclear tRNA introns. Most archael tRNA introns and nuclear tRNA introns are located one nucleotide 3' to the anticodon sequence.

Archael introns are spliced by an endoribonuclease

All archael introns are enzymatically removed by a cut-and-rejoin mechanism that requires ATP, an endoribonuclease, and a ligase. Archael intron transcripts generate a "bulge-helix-bulge" motif at the exon–intron junction. This motif is recognized by the splicing endoribonuclease, which then cuts at symmetric positions within the three-nucleotide bulges (Fig. 13.5A). In the case of the large archael rRNA introns, the excised intron is circularized.

Some nuclear tRNA genes contain an intron

The presence or absence of an intron defines two classes of eukaryotic nuclear tRNA genes whose transcripts differ in a requirement for splicing. All known eukaryotic nuclear genomes code for at least some intron-containing tRNA genes. Intron-containing tRNA genes were first discovered in the yeast *Saccharomyces cerevisiae*. In contrast to introns from other RNA classes, they are comparatively small, ranging in size from 14 to 60 nt and from 12 to 25 nt in yeast and plant intron-containing pre-tRNAs, respectively. About one-fifth of the yeast tRNAs contain introns. In plants, only two kinds of nuclear tRNA genes coding for tRNA^{Tyr} and tRNA^{Met} are known to be interrupted by introns. In humans, tRNA^{Tyr} and tRNA^{Leu} contain introns.

The general mechanism of tRNA splicing involves two main reactions. First, the intron-containing pre-tRNAs are cleaved by a tRNA splicing endoribonuclease at the 5' and 3' boundaries of the intervening sequence. This liberates paired tRNA halves with 2',3'-cyclic phosphates, 5'-OH ends, and a linear intron (Fig. 13.5B). Second, the paired halves are joined by tRNA ligase in a GTP- and ATP-dependent reaction. This leaves a 2'-phosphate and 3'-OH at the splice junction. In plants and fungi, the 2'-phosphate is removed by an NAD$^+$-dependent phosphotransferase. The major ligation reaction mechanism in vertebrates differs from that in fungi or plants. In vertebrates, the 3' terminal phosphate of the 5'-tRNA half is incorporated into a normal 3',5'-phosphodiester linkage at the splice junction. So far, neither the vertebrate ligase protein nor the exact mechanism of action of this ligation process is known.

is approximately 100 nt (shaded region). (Redrawn from Adams, P.L., Stahley, M.R., Kosek, A.B., Wang, J., Strobel, S.A. 2004. Crystal structure of a self-splicing group I intron with both exons. *Nature* 430:45–50.) (B) Group II intron splicing also occurs by a two-step reaction. (1) Nucleophilic attack is initiated by the 2'-OH of a bulged adenosine (A) nucleotide within the intron. This results in cleavage of the 5' SS and the formation of a lariat structure with a 2',5'-phosphodiester bond. (2) Nucleophilic attack is initiated at the 3' SS by the 3'-OH of the 5' exon. This results in exon ligation and release of the intron lariat. Subsequently, the lariat is debranched and degraded. (Inset) Group II intron structure. Intron-binding sites (IBS1 and 2) located in the 5' exon are base paired to two intronic exon-binding sites (EBS1 and 2). The intron structure is characterized by six major domains (D1–D6) radiating from a central wheel. A tertiary interaction important for correct folding and splice site selection is marked by a broken line and labeled with Greek lettering. (Redrawn from Lehmann, K., Schmidt, U. 2003. Group II introns: structure and catalytic versatility of large natural ribozymes. *Critical Reviews in Biochemistry and Molecular Biology* 38:249–303).

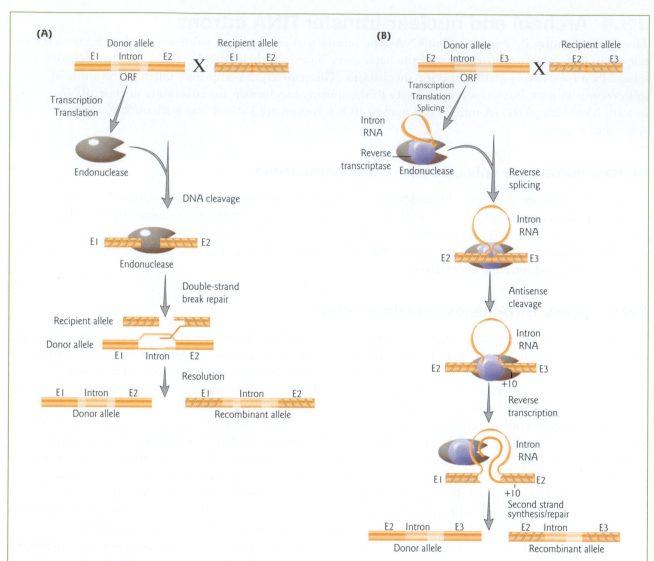

Figure 13.4 Mobile group I and II introns. (A) Group I intron "homing." The mobile intron in the donor allele has an open reading frame (ORF) that encodes a site-specific DNA endonuclease. The endonuclease recognizes the junction sequence between exon 1 (E1) and exon 2 (E2) in the recipient "intronless" allele, and makes a double-strand break at or near the intron insertion site. After DNA cleavage, the intron is copied from the donor allele into the recipient allele by a double-strand break repair mechanism involving cellular enzymes. Homing leads to insertion of the intron along with flanking 5′ and 3′ exon sequences. (B) Group II intron "retrohoming." The mobile intron has an ORF that encodes a DNA endonuclease and a reverse transcriptase. Retrohoming is initiated by an RNP complex containing the intron-encoded proteins and the excised intron lariat RNA. The endonuclease cleaves the recipient DNA in two steps. First, the intron RNA cuts the sense strand of the DNA at the intron insertion site by a reverse-splicing reaction. This leads to insertion of linear intron RNA between E1 and E2 of an intronless allele. Next, the endonuclease cuts the antisense DNA strand after position +10 of the 3′ exon. After the double-strand break, the 3′ end of the cleaved antisense strand is used as a primer for reverse transcription of the inserted intron RNA. The process of integration is completed by second strand synthesis and repair mechanisms involving cellular enzymes. (Redrawn from Lambowitz, A.M., Caprara, M.G., Zimmerly, S., Perlman, P.S. 1999. Group I and Group II ribozymes as RNPs: clues to the past and guides to the future. Chapter 18 in Gesteland, R.F., Cech, T.R., Atkins, J.F. eds., *The RNA World*, 2nd edn. Cold Spring Harbor Laboratory Press, Cold Spring Harbor, NY., pp. 451–485.)

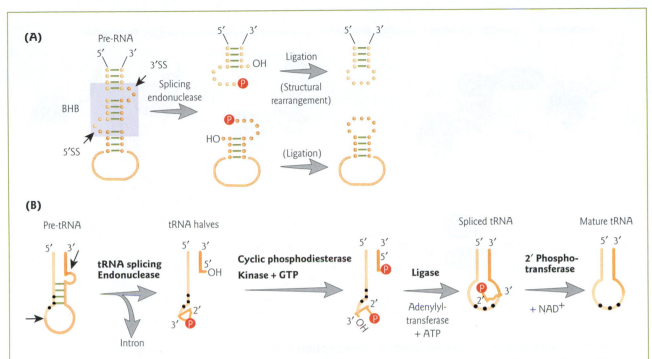

Figure 13.5 Splicing mechanisms of archael and tRNA introns. (A) Archael intron splicing pathway. The conserved secondary structure of an archael intron is shown. The bulge-helix-bulge motif is shaded in blue, and the intron is dark orange. The splicing endoribonuclease cleaves at the exon–intron junctions at the 3′ and 5′ splice site (3′ SS and 5′ SS, indicated by arrows) and generates 5′-OH and 2′,3′-cyclic phosphates. Subsequently, the exons are ligated by an unknown mechanism, and sometimes structural rearrangment occurs post-ligation. The large rRNA introns circularize. (Adapted from Kjems, J., Garrett, R.A. 1991. Ribosomal RNA introns in archaea and evidence for RNA conformational changes associated with splicing. *Proceedings of the National Academy of Sciences, USA* 88:439–443.) (B) The tRNA splicing pathway of plants and fungi. Intron-containing pre-tRNAs are first cleaved by a tRNA endonuclease at the 5′ and 3′ boundaries of the intervening sequence producing paired tRNA halves with 2′,3′-cyclic phosphate, 5′-OH ends, and a linear intron. Second, the halves are ligated by a complex reaction requiring GTP and ATP. Finally, the 2′-phosphate at the splice junction is removed by a NAD⁺-dependent phosphotransferase. (Redrawn from Englert, M., Beier, H. 2005. Plant tRNA ligases are multifunctional enzymes that have diverged in sequence and substrate specificity from RNA ligases of other phylogenetic origins. *Nucleic Acids Research* 33:388–399.)

13.5 Cotranscriptional processing of nuclear pre-mRNA

Eukaryotic pre-mRNA is covalently processed in three ways prior to export from the nucleus: (i) transcripts are capped at their 5′ end with a methylated guanosine nucleotide; (ii) introns are removed by splicing; and (iii) 3′ ends are cleaved and extended with a poly(A) tail. All three processing events are intimately linked with transcription and require the carboxy-terminal domain (CTD) of RNA polymerase II (Fig. 13.6). The CTD functions as a "landing pad" for RNA processing factors, localizing these factors close to their substrate RNAs. The binding of processing factors is regulated by phosphorylation of specific amino acids in the CTD, namely the serines at position 2 and 5 in the heptapeptide repeat (see Fig. 11.9D). During transcription initiation, the CTD is phosphorylated at serine 5 in each repeat. This triggers a conformational change in the CTD and allows binding of the 5′-capping enzyme for the first RNA processing event. CTD phosphorylation at serine 2 in each heptapeptide repeat is required for transcript elongation and for efficient 3′-RNA processing.

Deletion of the RNA polymerase II CTD has been shown to inhibit capping, splicing, and poly(A) site cleavage. Taking this observation further, Nova Fong and David L. Bentley demonstrated that different

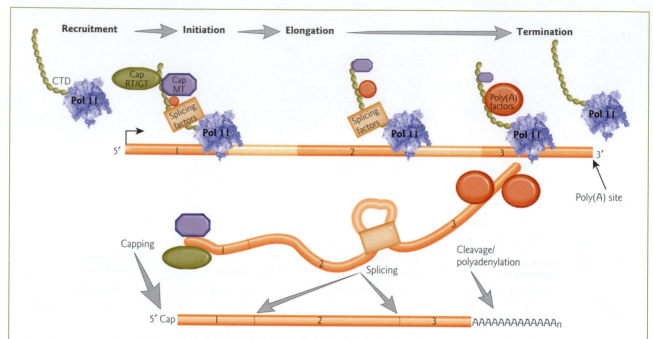

Figure 13.6 RNA processing is coupled to transcription. RNA processing factors interact with the RNA polymerase II (pol II) machinery via the carboxyl-terminal domain (CTD) of RNA pol II. Capping enzymes (GT, RNA guanylyltransferase; MT, RNA guanine 7-methyltransferase; RT, RNA triphosphatase) and 3′ end modifying factors (poly(A)) are recruited at the 5′ ends of genes. As RNA pol II moves through the gene, splicing factors associate with the transcription complex. The increased size of the symbols for processing factors corresponds to increased level of *in vivo* crosslinking, measured by chromatin immunoprecipitation (ChIP) assays (see Fig. 9.15E). Exon numbers are marked in dark orange boxes. Introns are shown in light orange boxes. (Redrawn from: Zorio, D.A.R., Bentley, D.L. 2004. The link between mRNA processing and transcription: communication works both ways. *Experimental Cell Research* 296:91–97.)

regions of the CTD serve distinct functions in pre-mRNA processing. The C-terminal heptapeptide repeats support capping, splicing, and 3′ processing, but the amino-terminal repeats only support capping (Fig. 13.7).

Addition of the 5′-7-methylguanosine cap

A unique feature of RNA polymerase II transcripts, including mRNAs and snRNAs, is the 5′ terminal cap. A distinguishing chemical feature of the cap is the $5' \rightarrow 5'$ linkage of 7-methylguanosine to the initial nucleotide of the mRNA (Fig. 13.8). The name of the cap is often abbreviated to m^7GpppN to reflect this linkage. Capping enzymes are brought to the right place at the right time by binding to the CTD of RNA polymerase II when it becomes phosphorylated during transcription initiation. Once transcription starts and the nascent RNA is about 22–40 bases long, the cap is added by a sequence of enzyme-mediated steps. RNA triphosphatase cleaves the γ-phosphate off the triphosphate at the 5′ end of the growing RNA, leaving a diphosphate. Next, RNA guanylyltransferase attaches GMP from GTP to the diphosphate at the end of the RNA, forming the $5' \rightarrow 5'$ triphosphate linkage. Metazoans have a single bifunctional polypeptide with RNA triphosphatase and RNA guanylyltransferase domains. Finally, RNA (guanine 7) methyltransferase transfers a methyl group from *S*-adenosylmethionine to position 7 in the carbon–nitrogen ring of the capping guanine. Another methyl transferase methylates the 2′-OH of the penultimate nucleotide.

The importance of the cap in mRNA processing and translation is well documented experimentally. The chemical nature of the cap makes the 5′ end of the mRNA inaccessible to 5′-exoribonucleases, and specific decapping activities are required for its degradation (see Section 13.11). In addition to protecting the mRNA

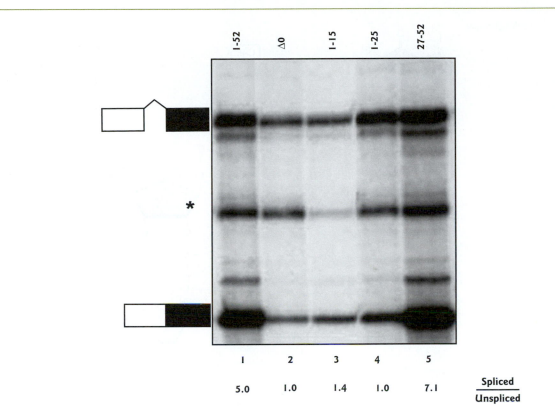

	1	2	3	4	5	
	5.0	1.0	1.4	1.0	7.1	Spliced / Unspliced

Figure 13.7 The carboxyl terminus of the RNA polymerase II (pol II) CTD is necessary and sufficient to enhance splicing. Mammalian cells were cotransfected with a β-globin gene expression vector and expression vectors for α-amanitin-resistant RNA pol II large subunits with full-length (1–52), deleted (Δ0), or truncated CTDs. α-Amanitin was added 12–16 hours after transfection to prevent further transcription by endogenous RNA pol II. mRNA made after this time is synthesized by RNA pol II that has incorporated the resistant large subunit. α-Amanitin is a highly toxic substance found in poisonous mushrooms of the genus *Amanita*. At very low concentrations, the toxin inhibits RNA pol II completely. RNA was harvested from cells after 36–48 hours and splicing levels were measured by RNase protection assay (see Fig. 9.8C) with a labeled antisense probe that spans the 3′ splice site of intron I. Ratios of spliced (shorter protected fragment) to unspliced (longer protected fragment) transcripts are given. Efficient splicing of β-globin intron I is supported by heptad repeats 27–52 but not by heptads 1–25. Similar experiments were performed to show that heptads 1–25 support efficient 5′ cap addition, but not splicing or 3′ processing (not shown). ★, irrelevant undigested probe. (Reprinted with permission from Fong, N., Bentley, D.L. 2001. Capping, splicing, and 3′ processing are independently stimulated by RNA polymerase II: different functions for different segments of the CTD. *Genes and Development* 15:1783–1795. Copyright © 2001 Cold Spring Harbor Laboratory Press.)

from degradation, the cap also serves other functions which are mediated by specific cap-binding proteins. The efficiency of pre-mRNA splicing, 3′ end formation, and U snRNA nuclear export are enhanced by the cap-binding complex (CBC). CBC is a heterodimer that consists of a small cap-binding protein (CBP20) subunit and a large (CBP80) subunit. In the cytoplasm, CBC is replaced by the cytoplasmic cap-binding factor, eukaryotic translation initiation factor 4E (eIF4E), which stimulates translation of the mRNA by the ribosome (see Section 14.4).

Termination and polyadenylation

A great deal is known about the mechanisms of transcription initiation and elongation (see Section 11.4 and 11.8), but the mechanism of termination is poorly understood. Initially it was presumed that there would be

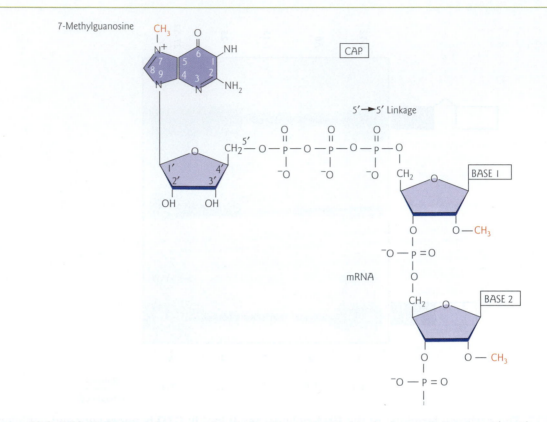

Figure 13.8 7-Methylguanosine cap structure. The cap is added in the "wrong" direction, 5′ to 5′, to the initial nucleotide of the mRNA. In vertebrates, methyl groups are added on the 2′-hydroxyl of the ribose of the first (base 1) and second (base 2) nucleotides, while only the first nucleotide is methylated in other animals and in cells of higher plants. Yeast lacks either of these methyl groups.

discrete "stop" sequences. However, termination of transcription on most genes occurs at various positions rather than at a single site, and no consensus termination sequence has been identified. Most metazoan mRNA 3′ ends are produced by cleavage of the pre-mRNA between conserved AAUAAA and G/U-rich sequence elements, which are separated by approximately 20 nt. The regions are recognized by cleavage and polyadenylation specificity factor (CPSF) and cleavage stimulation factor (CstF), respectively. Cleavage requires two additional multisubunit complexes, mammalian cleavage factor I and II (CFIm and CFIIm).

Recent research suggests that the mRNA is cleaved while it is still being synthesized. The released 5′ region forms the mature mRNA, and the remaining transcript is attacked by the Xrn2 exonuclease (Rat1 in yeast) while it is still associated with RNA polymerase II. Xrn2 "chases" after the polymerase and when it catches up with the polymerase, transcription is terminated (Fig. 13.9). Alexandre Teixeira and co-workers recently made the unexpected observation that a novel ribozyme mediates termination. They showed that newly synthesized β-globin mRNA is cut at two positions in human cells. One cut is at the well-characterized site of poly(A) addition. The other is within a "cotranscriptional cleavage site," which is self-cleaving.

The coupling of 3′ end formation with termination ensures that RNA polymerase II is only released from the template after it has completed synthesis of a full-length transcript. After cleavage and release of the mRNA, the 3′ end of almost all eukaryotic mRNAs is polyadenylated. This is accomplished by the enzyme poly(A) polymerase which adds 100–250 adenosine 5′-monophosphates (AMP) to the 3′ end. Without CPSF, poly(A) polymerase is barely active and has a low affinity for RNA. Metazoan mRNAs encoding histones are the only eukaryotic mRNAs known not to be polyadenylated. Instead, these mRNAs have a

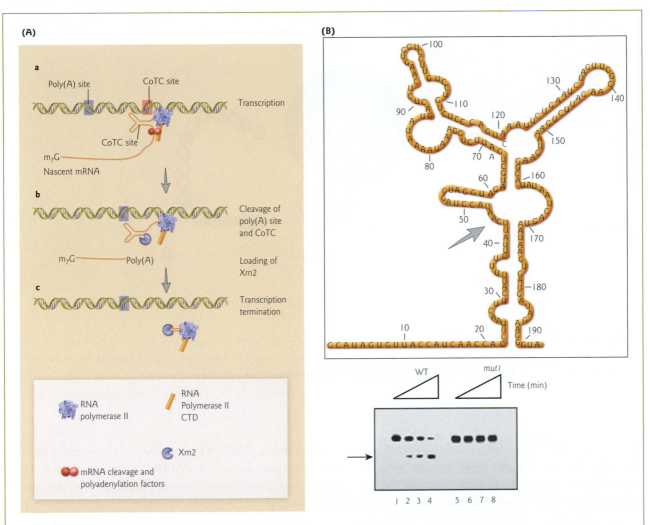

Figure 13.9 Self-cleavage of the human β-globin pre-mRNA promotes transcription termination. (A) Model for transcription termination. The nascent β-globin mRNA is cut at two positions: the site of poly(A) addition and within the cotranscriptional cleavage (CoTC) site. This allows entry of Xrn2, which progressively cleaves the mRNA until it catches up with RNA pol II. (Redrawn from West, S., Gromak, N., Proudfoot, N.J. 2004. Human 5′ → 3′ exonuclease Xrn2 promotes transcription termination at co-transcriptional cleavage sites. *Nature* 432:522–525.) (B) Secondary structure of the catalytic CoTC core. The arrow marks the position of the primary autocatalytic cleavage site at C44. (C) The minimal catalytic core is cleaved in a time course conducted under protein-free conditions (lanes 1–4 correspond to 0, 5, 15, and 120 minutes, respectively). Cleavage was performed with the 5′-^{32}P core RNA in the presence of 200 mM KCl. The arrow marks the primary cleavage site; *mut1* denotes a double mutant (C45C46) at the cleavage site that lacks autocatalytic activity. (Reprinted by permission from Nature Publishing Group and Macmillan Publishers Ltd: Teixeira, A., Tahiri Alaoui, A., West, S. et al. 2004. Autocatalytic RNA cleavage in the human β-globin pre-mRNA promotes transcription termination. *Nature* 432:526–530. Copyright © 2004.)

conserved stem-loop structure at their 3′ ends, which is bound by a specific protein. On a practical note, the poly(A) tail provides a convenient means of purifying mRNA from total cellular RNA (which is mostly rRNA) in the lab (see Tool box 8.2).

The poly(A) tails are coated with sequence-specific poly(A)–binding proteins, PABPN1 in the nucleus and PABPC in the cytoplasm (Fig. 13.10). PABPN1 is involved in the synthesis of poly(A) tails, increasing the processivity of poly(A) polymerase and defining the length of a newly synthesized tail. PABPC functions

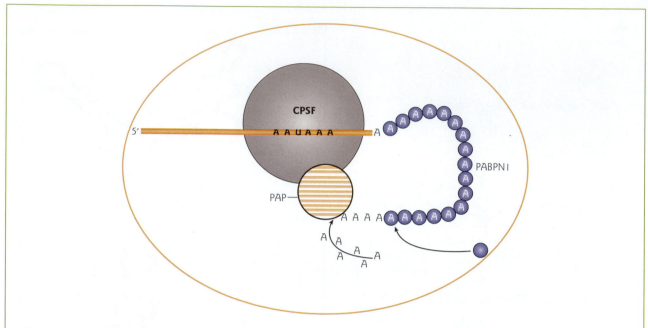

Figure 13.10 The polyadenylation complex. Cleavage and polyadenylation specificity factor (CPSF) binds the polyadenylation signal AAUAAA at the 3′ end of the mRNA and the poly(A)-binding protein PABPN1 binds the growing poly(A) tail. Together, they recruit and increase the processivity of poly(A) polymerase.

in the initiation of translation and the regulation of mRNA decay. PAPBN1 probably stays associated with the mRNA until it docks at the nuclear pore complex, and is exchanged for PAPBC in the cytoplasm. The critical role that poly(A)-binding proteins play in the control of gene expression was highlighted by the recent discovery of a link between a defect in a poly(A)-binding protein and a genetic disorder (Disease box 13.1).

Splicing

Spliceosomal introns are the most common intervening sequences found in the nuclear pre-mRNA transcripts of all multicellular eukaryotes. Although the basic mechanism of intron excision is identical to that of self-splicing group II introns, pre-mRNA introns do not themselves fold into conserved secondary or tertiary structures. The three-dimensional structure that is required for splicing is generated in a huge RNA–protein complex termed the spliceosome (Fig. 13.11).

Components of the spliceosome

The splicing apparatus contains five different RNAs and more than 200 additional proteins. This makes the spliceosome the largest and most complex molecular machine known at present. Improved purification of spliceosomes coupled with advances in mass spectrometry analysis of complex mixtures has led to the identification of more proteins associated with splicing than anticipated. The exact function of all the proteins and their interactions remain to be defined. These studies have also revealed the dynamic nature of spliceosomal complexes. The composition of the splicesome changes as splicing proceeds; components are removed and others are added as the particle is assembled and the reactions catalyzed.

The major components of the spliceosome are the five small RNA–protein complexes termed snRNPs (pronounced "snurps") for "small nuclear ribonucleoprotein particles." These snRNPs are designated U1, U2, U4, U5, and U6 snRNPs, numbered in order of the discovery of the uridine (U) rich snRNA

Oculopharyngeal muscular dystrophy: trinucleotide repeat expansion in a poly(A)-binding protein gene

DISEASE BOX 13.1

Oculopharyngeal muscular dystrophy (OPMD) is a dominantly inherited form of muscular dystrophy that begins in the muscles of the eyes and throat. It usually appears between the ages of 40 and 60, and progresses slowly. Initial symptoms include eyelid drooping and difficulty swallowing, and may lead to progressive limb muscle weakness. Although OPMD has a worldwide distribution, its prevalence is highest in French-Canadians who are descended from French immigrants, a man and wife, who emigrated to Canada in 1634.

OPMD was first described in 1915, but the genetic basis for the disease was not determined until 1998. OPMD is caused by stable trinucleotide repeat expansion in exon 1 of the *PABPN1* gene. The normal protein contains a series of 10 alanines in a row encoded by $(GCG)_6 (GCA)_3 GCG$. In most OPMD patients, $(GCG)_6$ is expanded by 2–7 repeats resulting in a total length of 12–17 alanines. Mutations introducing two or more additional alanines are autosomal dominant. Rare homozygotes for a $(GCG)_7$ variant have been reported.

In this case, the single additional alanine exhibits an autosomal recessive phenotype.

Mutations lead to the intranuclear aggregation of PABPN1 in the form of regular filaments. The reason for cell damage in OPMD patients is unclear. The mutant protein appears to be active, at least *in vitro*, and patients do not appear to have a severe polyadenylation defect. It is likely that protein aggregation leads to insoluble toxic inclusions and cell death. As noted in Section 7.2, the identification of trinucleotide repeat expansions in specific genes has become a common theme in inherited neurological diseases, including such diseases as Huntington's disease, fragile X syndrome, spinobulbar muscular atrophy or Kennedy's disease, and myotonic dystrophy. No medical treatment is presently available for these diseases. It is hoped that understanding of the molecular basis of OPMD, and other trinucleotide repeat diseases, will lead to treatment strategies in the future.

component. The "U3 snRNA" was later determined to be a snoRNA involved in rRNA processing (see Section 13.9). The U1, U2, U4, and U5 snRNAs are associated with a set of seven Sm common core proteins and several particle-specific proteins. U6 snRNA is associated with Sm-like core proteins and several particle-specific proteins. The Sm proteins are named after the patient whose autoimmune serum was first used to detect them. They form a doughnut-shaped complex that binds to a U-rich element in the snRNA (Fig. 13.12). The Sm class snRNPs are assembled in a highly orchestrated, stepwise process that takes place in multiple subcellular compartments. The process is mediated by the SMN (survival of motor neurons) protein and is defective in spinal muscular atrophy (Disease box 13.2).

Assembly of the splicing machinery

The splicing machinery is recruited to intron-containing transcripts cotranscriptionally and the cap-binding complex is important for this recruitment. The "splicing cycle" involves the stepwise assembly of the spliceosome, at least *in vitro*. Whether this is the case *in vivo* is still a subject of some debate. The "splicing cycle" pathway for spliceosomal introns was mostly determined from *in vitro* studies using an exogenous pre-mRNA and crude nuclear extract from yeast or mammalian cell nuclei. In such experiments, spliceosome assembly proceeds "stepwise" through a series of short-lived intermediate stages termed E, A, B, and C (Fig. 13.13). These subcomplexes can be distinguished by their different mobilities in native gels or density gradients, their snRNP composition, and the stage of processing of the pre-mRNA. As one illustration, Stephanie Ruby and John Abelson used an affinity purification method to show that U1 is the first snRNP to bind to a pre-mRNA in yeast nuclear extract (Fig. 13.14).

An alternative "penta-snRNP" model has been proposed, based on *in vivo* analyses. In this model, intron removal *in vivo* is mediated by pre-existing complexes, rather than by spliceosomes assembled in a stepwise

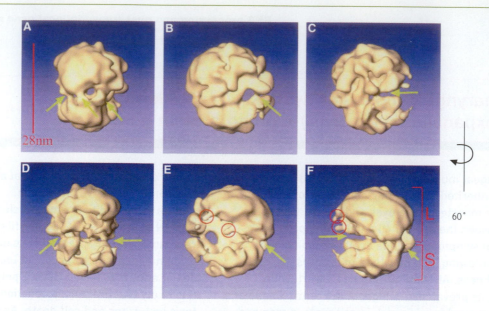

Figure 13.11 Spliceosome structure. Native spliceosomes were isolated from human (HeLa) cell nuclei and visualized by cryoelectron microscopy. At 20 Å (2.0 nm) resolution, the spliceosome appears as a globular particle composed of two distinct subunits. The two subunits are interconnected to each other with a tunnel in between. Each of the six three-dimensional reconstructions shown is separated by a rotation of 60° about the central axis. Green arrows indicate connecting points between the large (L) and small (S) subunits. Red circles indicate similar protruding bodies. (Reprinted from Azubel, M., Wolf, S.G., Sperling, J., Sperling, R. 2004. Three-dimensional structure of the native spliceosome by cryo-electron microscopy. *Molecular Cell* 15:833–839. Copyright © 2004, with permission from Elsevier.)

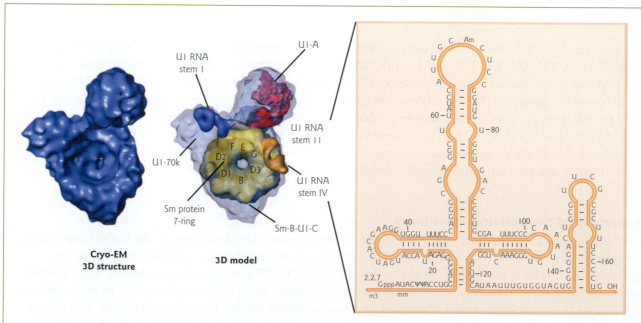

Figure 13.12 The structure of human U1 snRNP. On the left is a three-dimensional (3D) reconstruction based on cryoelectron microscopy images at 1.0 nm (10 Å) resolution. The 3D model on the right shows the positions of the Sm proteins, and the sites at which the proteins U1-70K and U1A bind to the U1 snRNA. (Reprinted by permission of Nature Publishing Group and Macmillan Publishers Ltd: Stark, H., Dube, P., Lührmann, R., Kastner, B. 2001. Arrangement of RNA and proteins in the spliceosomal U1 small nuclear ribonucleoprotein particle. *Nature* 409:539–542. Copyright © 2001.) (Inset) Primary and secondary structure of human U1 snRNA.

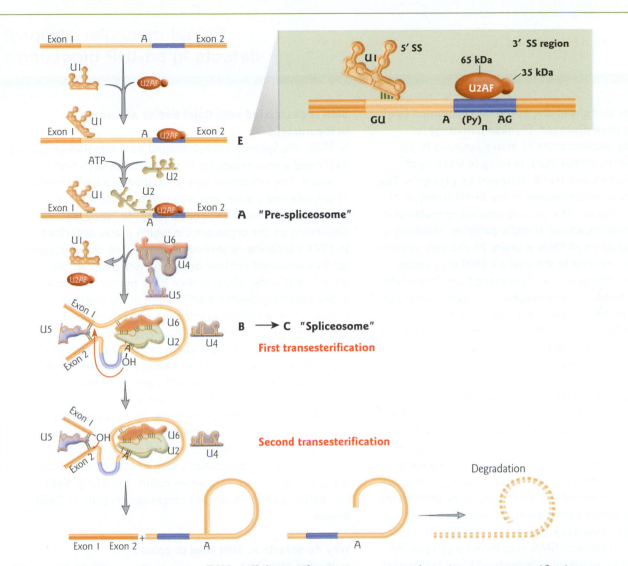

Figure 13.13 The nuclear pre-mRNA splicing pathway. Intron removal requires two transesterification steps (red arrows) at the 5′ and 3′ splice sites (5′ SS, 3′ SS), mediated by the spliceosome. Spliceosome assembly starts by formation of the E complex: U1 snRNP binds to the 5′ splice site, the 65 kDa subunit of U2 auxiliary factor (U2AF) binds to the polypyrimidine (Py) tract, and the 35 kDa subunit binds to AG at the 3′ splice site. Subsequently, U2 snRNP binds to the branch site region to form the A complex or "pre-spliceosome." As indicated, both snRNA components base pair with pre-mRNA sequences. This defines the reaction partners of the first transesterification event. The branch site adenosine (A) bulges from the RNA duplex formed by the U2 RNA. The A complex is joined by the U4–U6–U5 tri-snRNP to form the B complex. The active spliceosome (C complex) is formed by several rearrangements of RNA–RNA and RNA–protein interactions. The base pair interactions between U4 and U6 snRNP are resolved, and U6 base pairs with U2 snRNP as well as with the 5′ end of the intron, therefore replacing U1 snRNP. This positions the 2′-OH of the bulged adenosine for the nucleophilic attack at the 5′ end of the intron (red arrow). U5 snRNP helps in the correct positioning of the nucleophile (3′-OH of exon 1) for the second transesterification reaction (red arrow) by interacting with sequences of the 5′ and 3′ exon. Following exon ligation and the release of the mRNA and intron lariat, the spliceosome is disassembled and reassembled to carry out the next round of splicing. The lariat intron undergoes debranching and degradation.

Spinal muscular atrophy: defects in snRNP biogenesis

Spinal muscular atrophy is one of the most common fatal autosomal recessive diseases. Disease pathology is characterized by degeneration of motor neurons in the anterior horn of the spinal cord, leading to wasting of muscles in the limbs and trunk, followed by paralysis. The disorder results from mutations in the *SMN1* (survival of motor neurons 1) gene. The gene is deleted or mutated in over 98% of spinal muscular atrophy patients, resulting in diminished production of SMN protein. Phenotypic severity is inversely proportional to the overall SMN expression level. When the *SMN* gene was first cloned and "translated," the amino acid sequence showed no significant homology with any other protein sequence in the database. Determining its function was thus a challenge for molecular biologists. The 40 kDa SMN protein is now known to be required for snRNP biogenesis.

SMN-mediated snRNP assembly in the cytoplasm

Assembly of Sm class snRNPs is a stepwise process that takes place in multiple subcellular compartments. SMN plays a role in recognition of the snRNA nuclear export complex, assembly of the Sm core, hypermethylation of the snRNA 5′ cap, and trimming of its 3′ end. Along with other core components required for these functions, SMN forms large macromolecular complexes, collectively called "gemins." These complexes are distributed evenly throughout the cytoplasm. SMN also forms a cytoplasmic, pre-import RNA complex with importin-β1 and the import adaptor snurportin 1 (see Table 11.3). U snRNP import involves a bipartite nuclear localization signal (NLS) composed of the snRNA 5′-trimethylguanosine cap and the Sm protein core. SMN is proposed to link the Sm core with importin-β1 (Fig. 1). The assembled snRNPs enter the nucleus and are targeted to specific subnuclear compartments that appear as discrete foci. In adult tissues and most cultured cells these foci correspond to Cajal bodies. In contrast, in embryonic cells, SMN accumulates in separate nuclear structures called "gemini of coiled bodies" or "gems" (Fig. 1). In the Cajal bodies, the snRNAs undergo further modification, which includes 2′-O-methylation and pseudouridylation. Mature snRNPs eventually leave the Cajal bodies and travel to sites where splicing takes place, but the details of these steps are not known.

SMN is associated with Cajal bodies and gems in the nucleus

In 1903, the Spanish neurobiologist Santiago Ramon y Cajal described a new organelle in the nuclei of vertebrate neurons. This organelle was described in a wide variety of animals and plants, and was given a wide variety of names, ranging from "coiled body" to "spheres" depending on the organisms in which it was described. In 1999, researchers agreed on a common nomenclature for this universal nuclear organelle and adopted the term "Cajal body." Cajal bodies have numerous roles in the assembly and/or modification of the nuclear transcription and RNA-processing machinery. However, there is no evidence that splicing, rRNA processing, or transcription actually take place in the Cajal bodies. The characteristic protein of Cajal bodies, identified using patient autoimmune sera, is p80-coilin. In addition, the Cajal bodies contain an array of other proteins and RNAs, including SMN, snRNPs, snoRNPs, RNA polymerases, transcription factors, and the recently discovered small Cajal body-specific RNAs (scaRNAs) that are involved in snRNA modification. Studies using knockout mice lacking the coilin gene show that when coilin is missing, SMN and splicing snRNPs are not targeted correctly to Cajal bodies.

Why do defects in SMN lead to spinal muscular atrophy?

Nuclear functions for the SMN complex have not been described, but the loss of SMN nuclear foci correlates with the disease phenotype. In normal individuals, the SMN protein is expressed in all tissues, but at particularly high levels in motor neurons. This suggests that neuronal differentiation and maintenance have particularly stringent demands for splicing. In addition, there are nuclear import defects in a small subset of SMN mutants derived from spinal muscular atrophy patients, thus the link between SMN and snRNP import provides a potential target for SMN dysfunction. Major advances have been made in understanding the role of SMN in snRNP biogenesis and RNA splicing. However, the development of treatment strategies for this fatal disease awaits further insights.

Spinal muscular atrophy: defects in snRNP biogenesis

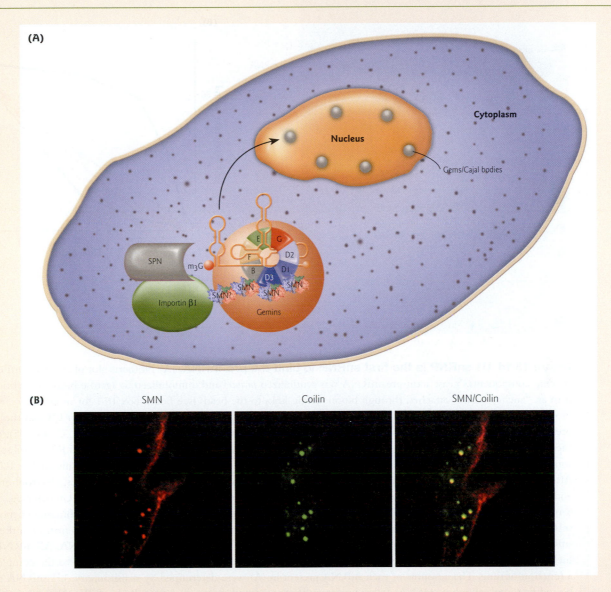

Figure 1 SMN-mediated snRNP assembly and localization. (A) A model for the U snRNP nuclear import complex is depicted. Snurportin 1 (SPN) binds the trimethylguanosine cap (m³G) of U1 snRNA and interacts with importin-β1. SMN plays a role in the assembly of the Sm core of the snRNP and remains bound to Sm-B, -D1, and -D3. Additional members of the SMN complex (gemins) are required for import. The connection to importin-β1 may be provided by SMN itself (SMN?), or by an unidentified adaptor. (Adapted from Narayanan, U., Achsel, T., Lührmann, R., Matera, A.G. 2004. Coupled *in vitro* import of U snRNPs and SMN, the spinal muscular atrophy protein. *Molecular Cell* 16:223–234.) (B) Colocalization of gems and Cajal bodies in the nucleus. Cells were double-immunostained with anti-SMN (red, left panel) and anti–coilin (green, middle panel) to locate the gems and Cajal bodies, respectively. Note the significant colocalization of the two types of nuclear bodies (yellow, right panel). (Photograph courtesy of C.-K. James Shen, National Taiwan University.)

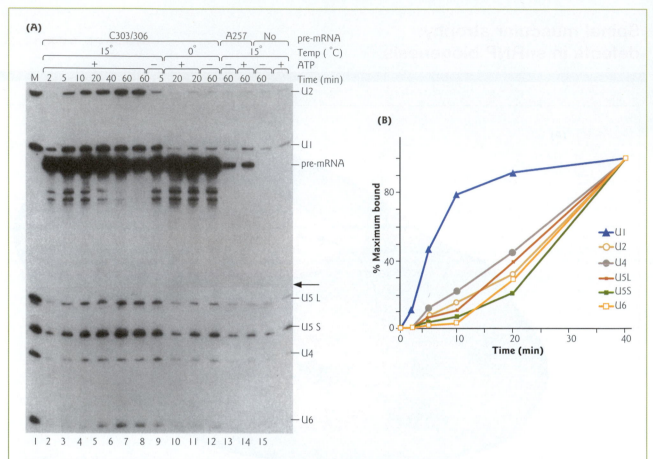

Figure 13.14 U1 snRNP is the first snRNP to bind the pre-mRNA. (A) Nothern blot of affinity purified splicing components. Yeast actin pre-mRNA was synthesized *in vitro* and immobilized to agarose beads by hybridizing it to an "anchor RNA" attached through biotin–avidin links to the beads (see Focus box 10.1 for more on biotin–avidin). The RNA bead complexes were incubated with yeast nuclear extract either at 15 or 0°C, in the presence of ATP, for 2–60 minutes as indicated. The pre-mRNA was mutated in the 3′ splice site (C303/305), or in the conserved branch point (A257). The former would assemble a spliceosome on the immobilized RNA and splice some of the RNA, but the latter would not. The lanes marked "No" contained no pre-mRNA, only anchor RNA. After incubation, unbound material was washed away, and RNA was extracted from the complexes, electrophoresed, and analyzed by Northern blot, using probes specific for actin pre-mRNA, U1, U2, U4, U5, and U6 snRNAs. Two forms of U5 (U5 L and U5 S) were detected. Lane 15, with no pre-mRNA, showed background binding of most snRNAs and served as a control for the other lanes. The time course shows that U1 snRNA binds first, then the other snRNAs bind. None of the snRNAs bound in significant amounts to the branch point mutant RNA. All snRNPs, including U1 and U4, remained bound after 60 minutes. (B) Graph depicting each snRNA bound to the complex as a function of time. The amounts of each snRNA present in (A) were determined by densitometry. U1 snRNA clearly bound first, with all other snRNAs following later. (Reprinted with permission from Ruby, S.W., Abelson, J. 1988. An early hierarchic role of U1 small nuclear ribonucleoprotein in spliceosome assembly. *Science* 242:1028–1035. Copyright © 1988 AAAS.)

fashion. There are a number of lines of evidence in support of this model. First, large RNP particles that are functional in pre-mRNA splicing and contain all five snRNPs plus more than 60 pre-mRNA splicing factors, have been isolated from yeast nuclei. This "penta-snRNP" complex was proposed to assemble prior to binding of the pre-mRNA substrate. Proponents of the penta-snRNP model argue that the "stepwise" *in vitro* assembly intermediates simply represent key structural transitions that become stabilized as the splicing machinery proceeds through splice site recognition and catalysis. Although such distinct complexes may represent intermediate states during *in vitro* assembly, they may not, in fact, occur *in vivo*. The debate is not over. Another recent study by Scott A. Lacardie and Michael Rosbash investigating spliceosome assembly in yeast came to the opposite conclusion. *In vivo* crosslinking and chromatin immunoprecipitation (ChIP) assays were used to determine the distribution of spliceosome components along intron-containing genes. This study detected patterns of progressive spliceosome assembly that were consistent with stepwise recruitment of individual snRNPs, rather than a preformed penta-snRNP. Further investigation will be required to determine which model is most consistent with the splicing pathway *in vivo* – or to propose an entirely new model.

Splicing pathway

For simplicity, only the key components of the major nuclear pre-mRNA splicing pathway are depicted in Fig. 13.13. The 5′ splice site of pre-mRNA includes a GU dinucleotide at the intron end within a larger, less conserved consensus sequence. At the other end of the intron, the 3′ splice site region has three conserved sequence elements: the branch point, followed by a polypyrimidine (Py) tract, and the terminal AG at the extreme 3′ end of the intron. The spliceosome assembles on these sequences and catalyzes the two chemical steps of the splicing reaction. Because of the shared splicing mechanism of group II introns and spliceosomal introns, the snRNA components of the spliceosome are proposed to be derived from a mobile group II intron (see Section 13.3).

During assembly of the spliceosome, the U1 snRNP binds to the 5′ splice site (GU) via base pairing between the splice site and the U1 snRNA (Fig. 13.13). This is an excellent illustration of the fundamental concept that processing events involving RNA cofactors use the RNA to provide specificity by complementary base pairing. The 65 kDa subunit of the U2 snRNP auxiliary factor (U2AF) binds to the polypyrimidine tract, and the 35 kDa subunit of U2AF binds to the AG at the 3′ splice site. The association of U1 and U2AF at the two intron ends forms the earliest defined complex in spliceosome assembly, called the E (early) or commitment complex. The E complex is joined by the U2 snRNP, whose snRNA base pairs at the branch point to form the A complex. Collectively, the components of the E and A complexes form the "pre-spliceosome." The A complex is joined by the U4/U6/U5 tri-snRNP to form the B complex. The B complex is rearranged to form the C complex. In the C complex, the U1 snRNP interaction at the 5′ splice site is replaced with U6 snRNP, and the U1 and U4 snRNPs are lost from the complex. It is the C complex that catalyzes the two transesterification steps of splicing.

In the first transesterification reaction, the 2′-OH of the bulged A residue at the branch point attacks the phosphate at the 5′ splice site (Fig. 13.13). This leads to cleavage of the 5′ exon from the intron and the ligation of the 5′ end of the intron to the branch point 2′-OH. This intron/3′-exon fragment is in a lariat configuation, containing a 2′–5′-phosphodiester bond as in the group II introns. In the second transesterification reaction, the 3′-OH of the detached exon attacks the phosphate at the 3′ end of the intron. This leads to ligation of the two exons and release of the intron, still in the form of a lariat. The spliceosome complex is subsequently released from the lariat intron and disassembled. This frees up the components for *de novo* synthesis of other spliceosomes. The lariat is debranched and degraded.

Protein factors that help mediate splicing

The dynamic remodeling of the spliceosomal RNP complex requires unwinding of many short helices in the RNA transcripts, and disruption of RNA–protein interactions. DEXH/D-box proteins have been implicated in these, and other, RNA rearrangements. The DEXH/D-box proteins are ATP-dependent RNA helicases that contain a DEXH/D motif (Asp-Glu-X-His/Asp, where X represents any amino acid).

Other important protein factors in splicing are SC35 and ASF/SF2. These factors are members of the highly conserved serine/arginine (SR) rich family of splicing factors that have been shown to have a dual role in stimulating constitutive and regulated splicing. SR proteins have a common modular domain structure of one or two N-terminal RNA recognition motifs (RRMs), followed by a C-terminal "RS

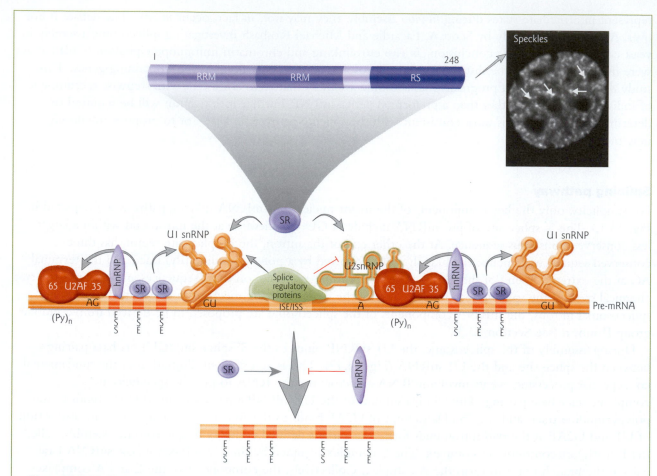

Figure 13.15 A schematic model of the regulation of splicing. The recruitment of splicing factors is controlled by activating and inhibitory splice regulatory proteins, which can bind to exonic splicing enhancers and silencers (ESEs and ESSs, respectively) (see Section 13.6), as well as to intronic splicing enhancers and silencers (ISEs and ISSs, respectively). SR proteins bind to splicing enhancers and stimulate binding of U1 snRNP to the 5′ splice site (GU). In addition, SR proteins stimulate binding of the U2 auxiliary factor (U2AF) subunits U2AF^{65} and U2AF^{35} to the polypyrimidine tract $(Py)_n$ and a dinucleotide (AG) at the 3′ splice site, respectively. U2AF guides the U2 snRNP to the branch point (A). Members of the hnRNP protein family bind to splicing silencers and inhibit RNA processing. (Redrawn from Schwerk, C., Schulze-Osthoff, K. 2005. Regulation of apoptosis by alternative pre-mRNA splicing. *Molecular Cell* 19:1–13.) (Insets) (Left) Domain structure of a respresentative SR protein, ASF/SF2, showing the two N-terminal RNA recognition motifs (RRM) and a C-terminal domain rich in repeating arginines and serines (RS). (Right) The distribution of the SR protein SC-35 in "speckles" (white arrows) in the nucleus. The image is a 3-D maximum point projection of a deconvolved data set, collected from a live mouse 10T1/2 embryonic fibroblast cell. The cell was transfected with an expression vector for SC-35 tagged with enhanced green fluorescent protein. Deconvolution excludes out of focus blur at a given point like confocal microscopy does, but the mechanism is mathematical processing by computer algorithms. (Photograph courtesy of Michael J. Hendzel, University of Manitoba, Canada).

domain." The RS domain contains repeated arginine-serine dipeptides that can be phosphorylated at multiple positions. The RRMs of SR proteins are sufficient for sequence-specific RNA binding, but the RS domain is required to enhance splicing. In general, SR proteins bind to exonic splicing enhancer elements to stimulate U2AF binding to the upstream 3′ splice site. They also bind to the branch point and 5′ splice site to stimulate U1 snRNP binding to the 5′ splice site. In the interphase nucleus, the bulk of SR proteins reside in subnuclear domains called "speckles" (Fig. 13.15).

Prp8 gene mutations cause retinitis pigmentosa   DISEASE BOX 13.3

Retinitis pigmentosa (RP) is a genetic disorder causing degeneration of the photoreceptors in the retina. The disease affects about one in 4000 people worldwide and can be inherited in X-linked, autosomal dominant or recessive modes. The hallmark of the disease is the presence of dark pigmented spots in the retina. Because the cells controlling night vision, called rods, are most likely to be affected, the disease leads initially to night blindness (decreased vision at night or in reduced light). This is followed by loss of peripheral vision, and eventually blindness in later stages, although typically not complete blindness. Twelve disease loci have been identified. Most defects are in retina-specific genes, such as the gene encoding the photoreceptor

molecule rhodopsin. However, a severe autosomal dominant form of the disease (RP13) was recently found to be associated with mutations in the gene encoding Prp8, an essential component of the spliceosome.

Prp8 is unique, having no obvious homology to other proteins. However, it does have a conserved RNA recognition motif (RRM) and a putative nuclear localization sequence (NLS). The association of defective Prp8 with a retina-specific disorder is puzzling, since Prp8 function is required in all cell types. It may be that the rod photoreceptor cells are particularly sensitive to alterations in the splicing machinery because of their rapid turnover.

The U12-type intron splicing pathway

Recently, some genes have been found to contain a minor class of rare introns called "U12-type introns." They have noncanonical consensus sequences and are excised by a distinct splicing machinery. The low-abundance spliceosome responsible for the excision of these introns includes four snRNPs (U11, U12, U4atac, and U6atac) that are different from but functionally analogous to the U1, U2, U4, and U6 snRNPs. The U5 snRNP is shared by both spliceosomes. The frequency of occurrence of U12-type introns is in the range of only 0.15–0.34% of the major class of introns.

Is the spliceosome a ribozyme?

A question that remains to be unequivocally answered is whether the spliceosome is a ribozyme. There are only a few components of the spliceosome known to interact directly with the pre-mRNA substrate. These included the protein Prp8 (Disease box 13.3), U2 snRNA, and U6 snRNA. Prp8 occupies a central position in the catalytic core of the spliceosome. It is a component of the U5 snRNP and can be photochemically crosslinked to the 5′ splice site, the branch point, and the 3′ splice site. The protein is hypothesized to function as a cofactor for RNA-mediated catalysis but not to have catalytic activity on its own. There is both indirect and direct evidence that U2 and U6 snRNAs contribute to the catalysis of pre-mRNA splicing. Since they directly base pair with the pre-mRNA intron, the U2 and U6 snRNAs are certain to reside near the active site of the spliceosome. In addition, many parallels exist between the spliceosome and group II self-splicing introns which are true ribozymes. Finally, there is strong experimental evidence for RNA-based catalysis. Saba Valadkhan and James L. Manley demonstrated that a protein-free human U2–U6 snRNA complex is capable of stimulating a slow, inefficient reaction *in vitro* that mimics the first step of splicing (Fig. 13.16). Higher resolution structural information will be required to determine whether the spliceosome has a catalytic RNA core, as has been shown for the ribosome (see Section 14.5).

13.6 Alternative splicing

A typical human or mouse gene contains 8–10 exons, which can be joined in different arrangements by alternative splicing (Fig. 13.17). Originally, it was thought that the majority of genes are transcribed into

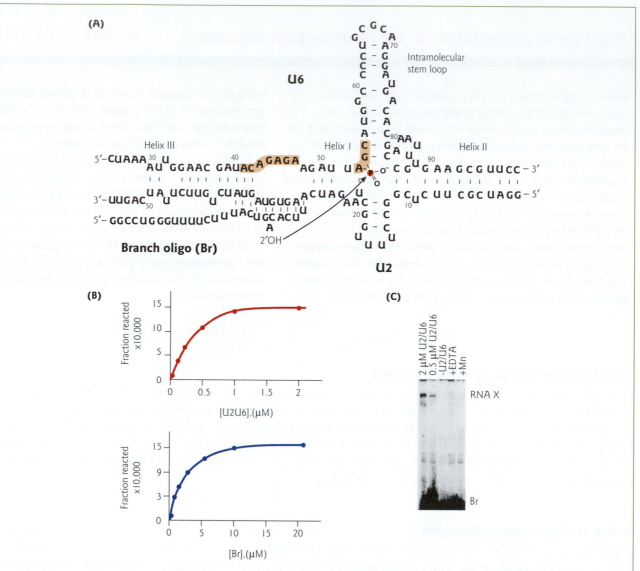

Figure 13.16 RNA X formation mediated by the catalytic activity of U2 and U6 snRNA. (A) Base pairing interactions in an *in vitro*-assembled complex of U2 snRNA, U6 snRNA, and the branch oligonucleotide (Br) which contains the branch site consensus sequence. Shaded boxes mark the invariant regions in U6. The three base–paired helices between U2 and U6 and the intramolecular stem loop of U6 are indicated. The long arrow indicates the reaction between U6 and Br that results in the formation of an X-shaped product (RNA X), in a reaction similar to the first step in splicing (see Fig. 13.13). The covalent bond formed in RNA X is a phosphotriester linkage between the 2′-OH of a bulged A in Br and the phosphate between the A and G in the AGC triad of U6. The numbers indicate nucleotide positions from the 5′ end. (B) The graphs plot the amount of RNA X formed in various concentrations of U2/U6 (top panel) or Br (bottom panel). RNA X formation shows characteristics of an equilibrium reaction. (C) RNA X formation by a U2/U6 chimera. The concentration of the U2/U6 chimera is noted above each lane. The reaction analyzed in the lane marked −U2/U6 contained only Br. The reaction in the lane marked +Mn contained Mn^{2+} instead of the Mg^{2+} present in the other reactions. The lane marked +EDTA contained this chelating agent, which binds Mg^{2+}. The location of RNA X and Br are shown to the right. RNA X formation reactions using the chimeric construct were identical to those using separate U2 and U6 RNAs (not shown) with respect to reaction kinetics and a requirement for Mg^{2+}. (Reprinted with permission from Cold Spring Harbor Laboratory Press: Valadkhan, S., Manley, J.L. 2003. Characterization of the catalytic activity of U2 and U6 snRNAs. *RNA* 9:892–904. Copyright © 2003 RNA Society).

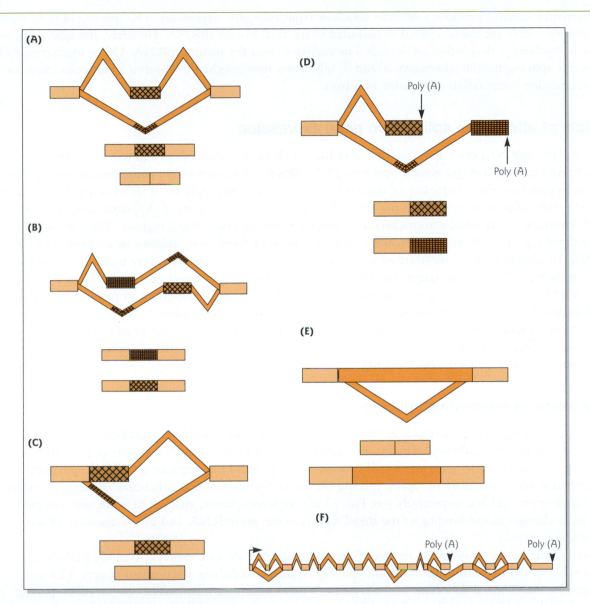

Figure 13.17 Patterns of alternative splicing. Exons that are present in all final mRNAs are shown as light orange. Alternative exons that may or may not be included in the mRNA are shown as hatched boxes. The darker orange lines indicate intron splicing patterns. (A) An exon can be either included in the mRNA or excluded. (B) Mutually exclusive exons occur when two or more adjacent exons are spliced such that only one exon in the group is included at a time. (C) Alternative 5′ and 3′ splice sites allow the lengthening or shortening of a particular exon. (D) Alternative poly(A) sites (black arrows) switch the 5′- or 3′ most exons of a transcript. (E) A retained intron (dark orange) can be excised or retained in the mRNA. (F) A single pre-mRNA can exhibit multiple sites of alternative splicing. These are often used in a combinatorial manner to produce many different mRNAs.

an RNA that gives rise to a single type of spliced mRNA. This is the case in budding yeast, which only contains a few known examples of alternative splicing. In contrast, at least 60% of human genes are alternatively spliced, and the percentage may actually be higher. Recent computational studies using quantitative microarray methods (see Chapter 16) have estimated that as many as 74% of human multiexon genes are alternatively spliced.

Alternative splicing provides a versatile means of regulating gene expression. The splicing of most exons is constitutive – they are always spliced or included in the final mature mRNA. However, the splicing of some exons is regulated – they either are included or excluded from the mature mRNA. Other mechanisms of alternative splicing include alternative 5′ and 3′ splice sites that lengthen or shorten a particular exon, or intron retention – the failure to remove an intron.

Effects of alternative splicing on gene expression

Particular pre-mRNAs often have multiple positions of alternative splicing, giving rise to a family of related proteins from a single gene (Focus box 13.2). This mechanism for generating protein diversity may explain in part how the complexity of mammals arises from a surprisingly small set of genes. Controlling the expression of protein isoforms is not the only effect of alternative splicing. Approximately 20% of mRNA variability that results from alternative splicing is within untranslated regions. These 5′ or 3′ untranslated regions commonly contain elements that regulate translation, stability, or localization of the mRNA. In addition, up to one-third of alternative splicing events insert premature termination codons in the transcript. This usually targets the mRNA for degradation by nonsense-mediated decay (see Section 13.11). Regulation of gene expression by alternative splicing is important in many cellular and developmental processes, including sex determination, apoptosis, axon guidance, cell excitation, and cell contraction, to name but a few. Consequently, errors in splicing regulation have been implicated in a number of different diseases (e.g. see Disease box 11.4). It is estimated that 15% of point mutations that cause human genetic diseases affect splicing.

Regulation of alternative splicing

A major goal of molecular biologists in the splicing field is to understand how alternative splicing is regulated. *Cis*-acting regulatory elements have been identified in exons and/or introns of pre-mRNAs. These RNA sequence elements are bound by *trans*-acting proteins that regulate splicing. Regulatory elements that act to stimulate or repress splicing are called "exonic splicing enhancers" (ESEs) or "exonic splicing silencers" (ESSs), respectively (see Fig. 13.15). In most systems, changes in splice site selection arise from changes in the binding of the initial factors to the pre-mRNA, and in the assembly of the spliceosome.

Some of the best understood examples of alternative splicing are in *Drosophila*. Genetic analysis of mutants led to the identification of specific splicing regulators and their downstream targets. This was followed by biochemical dissection of the regulatory mechanisms. In mammalian systems most splicing factors have been biochemically identified. To meet the criteria for a splicing regulator, a protein must be capable of influencing splice site choices both *in vitro* and in transfected cells. In general, the molecular mechanisms regulating tissue-specific and developmental stage-specific control of alternative splicing remain poorly defined. Three examples of alternative splicing are provided below. These examples illustrate the diversity of physiological processes regulated by alternative splicing, the complexity of patterns of splicing, and the role of splicing regulatory proteins in the process.

Alternative splicing of caspase-2 mRNA during apoptosis

The regulation of alternative splicing of caspase-2 mRNA illustrates the role of SR proteins as positive regulators and hnRNP proteins as negative regulators of splicing (Fig. 13.18A). Caspase-2 has two isoforms with strikingly divergent roles in mediating apoptosis (Focus box 13.3). Alternative splicing results in mRNA that either includes or lacks exon 9. Importantly, exon 9 contains a stop codon. hnRNP A1 facilitates exon 9 inclusion, possibly by blocking U2 snRNP binding to the branch point. This results in premature termination and the production of the cell survival-promoting (anti-apoptotic) short isoform of caspase-2, Casp2S. SC35 and ASF/SF2 promote skipping of exon 9. This results in the death–promoting

An extreme example of alternative splicing occurs in the *Drosophila* gene encoding the Downs syndrome cell adhesion molecule (Dscam). The *Dscam* gene has 95 variable exons out of a total of 115, and can potentially generate 38,016 different protein isoforms (Fig. 1). As the name suggests, the gene was first described in humans and maps to a Downs syndrome region of chromosome 21. The Dscam isoforms are a novel class of transmembrane neuronal cell adhesion molecules that are required for the formation of neuronal connections in *Drosophila* and humans. The extracellular domain of human DSCAM is highly homologous to *Drosophila* Dscam. Surprisingly, their intracellular domains share no obvious sequence homology. Despite these differences, human DSCAM and the *Drosophila* counterpart appear to have similar biological functions in axon guidance. In contrast to the *Drosophila* gene, the human *DSCAM* gene has 30 exons and only three different alternatively spliced transcripts have been identified.

In *Drosophila*, four variable domains are encoded by blocks of alternative exons (Fig. 1). Different isoforms are expressed in specific spatial and temporal patterns in neurons, and individual cells express around 50 different isoforms each. The array of isoforms expressed by each neuron is proposed to play an important role in the specificity of neural wiring in the fruitfly, by regulating interactions between neurons during axon guidance, targeting, or synapse formation. In addition, *Drosophila* immune-competent cells (fat body cells and hemocytes – insect blood cells) have the potential to express more than 19,008 isoforms of this immunoglobulin superfamily member with different extracellular domains. Although mechanistically entirely different from V(D)J somatic rearrangements that generate antibody diversity (see Fig. 12.25), this alternative splicing mechanism acts in a similar manner and may enhance the adaptive potential of fly populations to changing environmental and pathogenic threats.

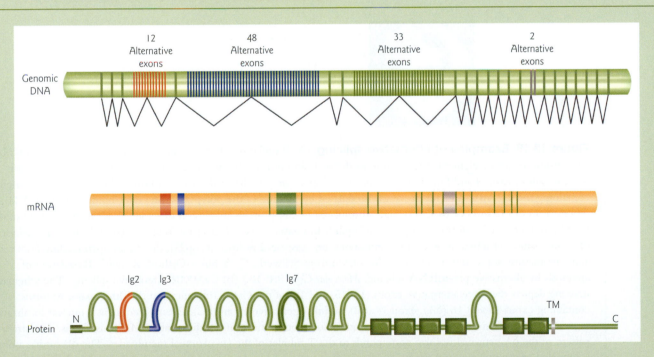

Figure 1 Extreme alternative splicing. Schematic representation of the *Dscam* gene, mRNA, and protein. The Dscam protein contains both constant and variable domains. The four variable domains are encoded by alternative exons (indicated by different colors). A transcript contains only one alternative exon from each block. The *Dscam* gene encodes 12 alternative exons for the N-terminal half of immunoglobulin 2 (Ig2, red), 48 alternative exons for the N-terminal half of Ig3 (blue), and 33 alternative exons for Ig7 (dark green). There are two alternative transmembrane domains (gray). (Reprinted from Wojtowicz, W.M., Flanagan, J.J., Millard, S.S., Zipursky, S.L., Clemens, J.C. 2004. Alternative splicing of *Drosophila* Dscam generates axon guidance receptors that exhibit isoform-specific homophilic binding. *Cell* 118:619–633. Copyright © 2004, with permission from Elsevier.)

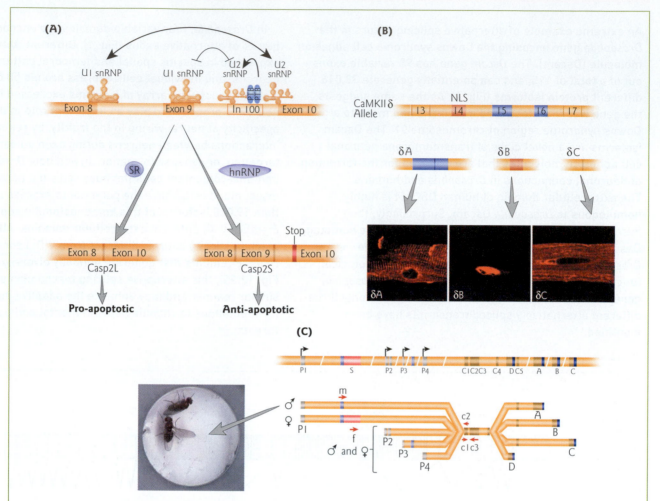

Figure 13.18 Examples of alternative splicing. (A) Regulation of exon 9 inclusion into the caspase-2 mRNA. An intronic sequence element (In100) inhibits the inclusion of exon 9 into the caspase-2 mRNA. In100 is located downstream of exon 9 and functions as a "decoy" 3′ splice site via the formation of a nonproductive splicing complex containing the U2 snRNP. This process is modulated by binding of the polypyrimidine tract-binding protein (PTB) downstream of U2 snRNP to In100, and is stimulated by SR proteins. Exclusion of exon 9 leads to production of the long "pro-apoptotic" isoform of caspase-2 (Casp2L). In contrast, exon 9 inclusion is facilitated by hnRNP1. Inclusion of exon 9 causes production of a short "anti-apoptotic" caspase-2 isoform (Casp2S), due to an open reading frame shift and a premature stop codon in exon 10. (Redrawn from Schwerk, C., Schulze-Osthoff, K. 2005. Regulation of apoptosis by alternative pre-mRNA splicing. *Molecular Cell* 19:1–13.) (B) CaMKIIδ alternative splicing. The schematic drawing depicts the three alternative exons (14, 15, and 16) of the CaMKIIδ gene, and the three major isoforms containing exons 15 and 16 (δA), exon 14 (δB), and no alternative exons (δC). Exon 14 contains a nuclear localization sequence (NLS). Immunostaining shows the intracellular localizations of the different CaMKIIδ isoforms. The striated pattern of the δA neuronal isoform corresponds to the T tubules of the sarcolemmal membranes. The δB cardiac isoform is localized to the nucleus, and the δC cardiac isoform shows a largely diffuse cytoplasmic localization. Inappropriate expression of the neuronal isoform leads to defects in heart development. (Reprinted from Xu, X., Yang, D., Ding, J.H. et al. 2005. ASF/SF2-regulated CaMKIIδ alternative splicing temporally reprograms excitation–contraction coupling in cardiac muscle. *Cell* 120:59–72. Copyright © 2005, with permission from Elsevier.) (C) Organization of the alternatively spliced *Drosophila fru* gene (top) and its transcripts (bottom). P1–P4 indicate alternative promoters; S, the sex-specifically spliced exon found only in P1 transcripts; C1–C5, common exons; and A–D, alternative 3′ exons. The single sexually dimorphic splicing event regulates male courtship behavior. (Inset) A male fly has one wing extended to "sing" to the female fly. (Reprinted from Demir, E., Dickson, B.J. 2005. *fruitless* splicing specifies male courtship behavior in *Drosophila*. *Cell* 121:785–794. Copyright © 2005, with permission from Elsevier.).

(pro-apoptotic) long isoform, Casp2L. Accordingly, SC35 overexpression triggers apoptosis, whereas hnRNP A1 overexpression is anti-apoptotic.

Alternative splicing in mammalian heart development

Alternative splicing of Ca^{2+}/calmodulin–dependent kinase II δ (CaMKIIδ) is regulated by the SR protein ASF/SF2, and has been shown to play a critical role in mammalian heart development. CaMKIIδ is a major intracellular mediator of Ca^{2+} action in cells. Within the heart, it functions as a regulator of excitation–contraction coupling, the process by which electrical stimulation results in physical contraction of the heart muscle. Heart contraction is initiated by Ca^{2+} entry into the cell and its release from stores in the sarcoplasmic reticulum. During muscle relaxation, a small fraction of cytoplasmic Ca^{2+} is pumped out of the cell, but the majority of cytoplasmic Ca^{2+} is recycled back to the sarcoplasmic reticulum through a Ca^{2+} pump. The phosphorylation of key Ca^{2+}-handling proteins is regulated, in part, by CaMKIIδ.

In the developing fetal heart, the two cardiac isoforms (CaMKIIδB and CaMKIIδC) and one neuronal isoform (CaMIIδA) are expressed. Between 1 and 2 months of age, CaMKIIδ undergoes an alternative splicing transition, switching to the expression of the cardiac isoforms only. The different isoforms exhibit similar catalytic and regulatory activity. However, they differ completely in their intracellular localization (Fig. 13.18B). The neuronal isoform (δA) is targeted to the transverse (T) tubules of the sarcolemmal membranes that surround individual muscle fibers. In contrast, one cardiac isoform (δC) is located in the cytoplasm of muscle cells, and the other (δB) is targeted to the nucleus. The nuclear localization sequence is inserted via the alternative splicing event. The different intracellular localization of the three isoforms has an important functional consequence – the isoforms have different targets for phosphorylation. Consequently, inappropriate expression of the neuronal isoform leads to defects in heart development and function.

Alternative splicing in *Drosophila* courtship behavior and sex discrimination

In the fruitfly, a single splicing event regulates the function of a neural circuit dedicated to male courtship behavior. The *Drosophila fruitless* (*fru*) mutant affects every step of the stereotyped courtship behavior of male flies. In an elaborate ritual, a wild-type fly begins by tapping a female with his forelegs. He then sings a species-specific courtship song by extending and vibrating one wing (Fig. 13.18C). In contrast, a male *fru* mutant will mate indiscriminately with both males and females. The behavioral defect of *fru* mutants is often vividly represented by *fru* male "mating chains," where mutant males are both trying to mate with and are being mated by another fly. *fru* females show the same behavioral phenotypes as wild-type females.

The *fru* pre-mRNA is transcribed from multiple promoters and displays a vast range of differential splicing events. Sex-specific transcripts are initiated from a 5′-located promoter (P1), whereas transcripts initiating from other promoters are not sex-specifically spliced. A single sexually dimorphic splicing event generates the male (Fru^M) or the female (Fru^F) specific protein, both of which are putative zinc finger transcription factors. Fru^M is expressed in a subset of olfactory, auditory, and taste sensory neurons that are known to play a role in male courtship.

13.7 *Trans*-splicing

In the vast majority of pre-mRNAs, the exons are joined only within an individual pre-mRNA. However, in rare cases, an exon from one pre-mRNA can join to an exon from another pre-mRNA in a process called *trans*-splicing. Except for its intermolecular nature, *trans*-splicing exactly parallels the two steps of *cis*-splicing in group II introns and pre-mRNA spliceosomal introns. *Trans*-splicing is rare, but is now known to occur in some special situations in organisms as diverse as flatworms, the protist *Euglena gracilis*, plant organelles, nematodes, and *Drosophila*. *Trans*-splicing occurs in certain highly abundant viral pre-mRNAs and cellular pre-mRNAs, but at present there is no evidence that the low level of *trans*-splicing observed in

Apoptosis (from the Greek words *apo* for "from" and *ptosis* for "falling," pronounced ap-a-tow-sis) is a tightly regulated form of cell death, also called programmed cell death (Fig. 1). Apoptosis is characterized by chromatin condensation and cell shrinkage in the early stage. In later stages, the nucleus and cytoplasm fragment, forming membrane-bound apoptotic bodies, which can be engulfed by phagocytes. In contrast to necrosis – a form of cell death that results from acute tissue injury – apoptosis is an important process during normal development. For example, the differentiation of human fingers in a developing embyro requires the cells between the fingers to

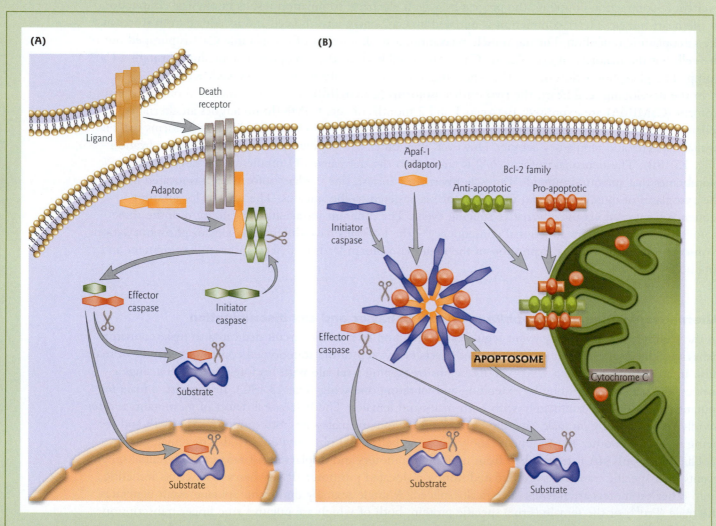

Figure 1 Schematic representation of the two major pathways to apoptosis in mammalian cells. (A) The extrinsic (death receptor) pathway. Binding of ligands to the tumor necrosis factor (TNF) family leads to oligomerization of death receptors, followed by recruitment and activation of initiator caspases (caspase-2, -8, -9, -10) via adaptor proteins. The activated initiator caspases target effector caspases (caspase-3, -6, -7) for proteolysis and activation. (B) The intrinsic (mitochondrial) pathway. In response to an apoptotic stimulus, Bcl-2 proteins induce permeabilization of the outer mitochondrial membrane and the release of cytochrome *c*. Subsequently, cytochrome *c* binds to apoptotic protease–activating factor 1 (Apaf-1), an adaptor protein. Together they mediate recruitment of an intitiator caspase and formation of the caspase–activating apoptosome complex. The activated initiator caspase in turn activates effector caspases. In both pathways, the activated caspases cleave selected nuclear and cytoplasmic target proteins, leading to cell death. (Reprinted from Schwerk, C., Schulze-Osthoff, K. 2005. Regulation of apoptosis by alternative pre-mRNA splicing. *Molecular Cell* 19:1–13. Copyright © 2005, with permission from Elsevier.)

Apoptosis

FOCUS BOX 13.3

initiate apoptosis so that the fingers can separate. Defective apoptotic processes have been implicated in a wide variety of diseases. For example, failure of apoptosis is a main contributor to tumor development, and inappropriate apoptosis occurs in neurological disorders such as Alzheimer's disease.

Apoptosis can be triggered by a multitude of different stimuli but occurs by either one of two major pathways (Fig. 1). The extrinsic "death receptor" pathway involves the transmission of extracellular signals to the intracellular death machinery. The intrinsic pathway requires the release of pro-apoptotic factors from the mitochondria. In both pathways, the activation of enzymes called caspases (for cysteine *aspartic* acid prote*ases*) leads to cleavage of nuclear and cytoplasmic target proteins, resulting in the apoptotic phenotype.

---

mammalian cells leads to the production of proteins with essential functions. There are three major types of *trans*-splicing: discontinuous group II, spliced leader, and tRNA *trans*-splicing.

Discontinuous group II *trans*-splicing

Discontinuous group II *trans*-splicing was first discovered when two plant chloroplast genomes were being sequenced. Coding sequences for the amino-terminal part of a ribosomal protein were found some 30–60 kb away from the rest of the gene and transcribed from opposite DNA strands. Unlike the "spliced leader" category (see below), this "discontinuous group II" form of *trans*-splicing is not confined to the 5′ termini of mRNAs and may occur more than once within the same gene. Complex splicing processes are required to create meaningful open reading frames for some plant mitochondrial proteins. For example, there is extensive *trans*-splicing of the NADH dehydrogenase 1 (*nad1*) pre-mRNA transcript. NADH dehydrogenase is an enzyme involved in cellular respiration (Fig. 13.19A). In *Oenothera* (evening primrose) and *Arabidopsis* there are three separate transcripts involved, with five exons that need to be joined together. Two *cis*-splicing and two *trans*-splicing events are required to produce the functional mRNA. In petunia and wheat, there are four separate transcripts involved. One *cis*-splicing and three *trans*-splicing events are required to produce the NADH dehydrogenase mRNA. The scattering of exons probably results from recombination events that frequently rearrange plant mitochondrial genomes. Having bits of coding sequence scattered around the genome adds to the challenge of cloning "the gene" for a particular protein.

Spliced leader *trans*-splicing

Several years after the discovery of introns in eukaryotic genes, it was recognized that mRNAs from African trypanosomes (see Section 12.7) were not strictly colinear with their corresponding gene. Each mRNA has an identical short noncoding sequence at its 5′ end. These "mini-exons" were found to be encoded elsewhere in the trypanosome genome. The mini-exons were designated "spliced leader" RNA (SL RNA). The SL RNA is transcribed by RNA polymerase II from intronless genes that reside on short, tandemly repeated DNA sequences. The only known function of these genes is to donate the short (~15–50 nt) leader sequence. Although they have no significant sequence identity, SL RNAs have a striking overall similarity to Sm class U-rich snRNAs. Shared features include a small size, a modified 5′ cap structure, and a U-rich Sm protein-binding site. The SL RNA is transferred to pre-mRNA splice acceptor sites, and becomes the 5′ end of the mature mRNA (Fig. 13.19B). The splicing process is mediated by the spliceosome.

The *trans*-splicing process provides 5′ noncoding exons to anywhere from a minority to 100% of the pre-mRNAs in many invertebrate organisms and in some chordates, such as the ascidian tunicate *Ciona*

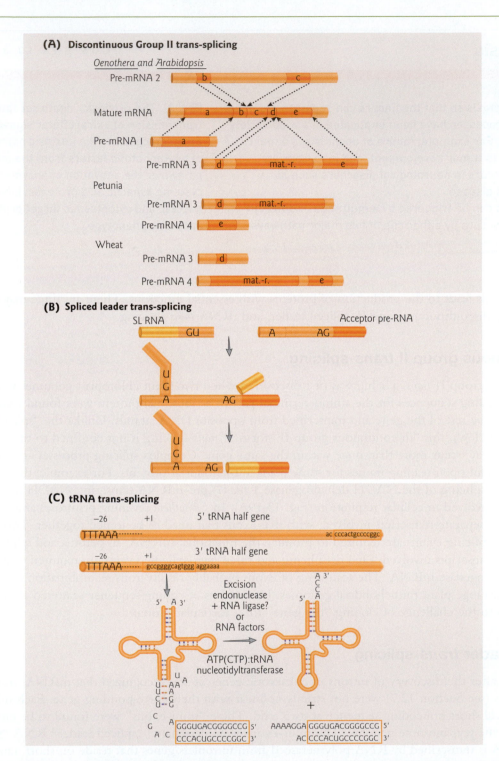

Figure 13.19 Three major types of *trans*-splicing. (A) Discontinuous group II *trans*-splicing in plant mitochondria. Exons of the *nad1* genes in *Oenothera*, *Arabidopsis*, *Petunia*, and wheat require several *trans*-splicing events for assembly of the open reading frame. The two *trans*-splicing reactions between exons a/b and c/d are conserved in all species. The mature *nad1* mRNA is assembled from three separate pre-mRNA transcripts in *Oenothera* and *Arabidopsis*. The *Oenothera* and *Arabidopsis cis*-splicing intron between exons d and e is interrupted in wheat and *Petunia* in different positions, resulting in an additional *trans*-splicing event in these plants. In *Petunia* and wheat the mature *nad1* mRNA is

intestinalis and the larvacean *Oikopleura dioica*. Unlike alternative splicing, the addition of the spliced leader probably does not lead to protein diversification. Instead, spliced leader *trans*-splicing has several distinct roles in mRNA function:

1 To provide a 5′ cap structure. Addition of the cap structure is essential for the protein-coding RNAs that are transcribed by RNA polymerase I in trypanosomes (see Fig. 12.22).
2 To resolve polycistronic RNA polymerase II transcripts into individual, capped mRNAs. SL RNAs are *trans*-spliced to acceptor sites that are located upstream of each open reading frame. This type of *trans*-splicing occurs in trypanosomes, nematodes, and flatworms.
3 To enhance mRNA translational efficiency. In trypanosomes and nematodes the addition of a hypermodified cap structure and/or leader sequence by *trans*-splicing can increase the efficiency of mRNA recruitment to the ribosome.

tRNA *trans*-splicing

Nanoarchaeum equitans is currently the only characterized member of the kingdom Nanoarchaeota. Already, the genome sequence of this small archael parasite is revealing surprises. When the genome was sequenced, it appeared to be missing genes encoding the glutamate, histidine, tryptophan, and initiator methionine tRNAs. A search of the genome using a computational approach revealed that nine genes encode separate tRNA halves. Together, they account for the "missing" tRNA genes. The tRNA sequences are split one nucleotide after the anticodon – the normal location of tRNA introns (see Fig. 13.5B). The two halves are proposed to join by a *trans*-splicing mechanism aided by intervening reverse complementary sequences (Fig. 13.19C).

13.8 RNA editing

It once seemed that the nucleotide sequence of every mRNA would be a simple copy of the sequence of its DNA template. The 1977 discovery that introns interrupted many genes dramatically altered this viewpoint. A second major change to this simple model occurred 11 years later with the finding of RNA editing – the post-transcriptional modification of the base sequence of mRNA. For the molecular biologist trying to predict protein sequences from starting genes, this represents yet another challenge. For the molecular biologist studying gene regulation, it represents a novel and intriguing level of control. RNA editing was first discovered in trypanosomes, and later found to occur in viruses, plants, slime molds, mammals (including humans and marsupials), squid, and dinoflagellates. It is now recognized as a widespread mechanism for changing gene-specified codons and thus protein structure and function. Two examples of very different mechanisms for editing will be highlighted, one in trypanosomes, the other in mammals.

thus assembled from four separate pre-mRNA transcripts. (Binder, S., Marchfelder, A., Brennicke, A., Wissinger, B. 1992. RNA editing in trans-splicing intron sequences of nad2 mRNAs in *Oenothera* mitochondria. *Journal of Biological Chemistry* 267:7615–7623.) (B) Spliced leader *trans*-splicing. Intermolecular *trans*-splicing occurs in a two-step process, as in *cis*-splicing (compare with Figs 13.3 and 13.13). In the first transesterification step, cleavage at the spliced leader (SL) donor site yields free 5′ SL exon. The remainder of the SL RNA is joined to the branch point adenosine in the acceptor pre-mRNA by a 2′,5′-phosphodiester bond to form a Y intermediate. In the second step, cleavage at the 3′ splice site and exon ligation yield the spliced exons and an excised Y intron. In trypanosomes, this reaction results in a common 39 nt sequence at the 5′ end of all mRNAs. (C) tRNA *trans*-splicing. Schematic representation of a 5′ tRNA half gene (tRNA^{Glu}) and the corresponding 3′ tRNA half gene found in *Nanoarchaeum equitans*. The archael RNA polymerase II promoter consensus sequence (TTTAAA), the tRNA half genes, and the intervening reverse complementary sequences that may facilitate joining of the halves (orange boxes) are indicated. Possible mechanisms for *trans*-splicing are shown. (Redrawn from Randau, L., Münch, R., Hohn, M.J., Jahn, D., Söll, D. 2005. *Nanoarchaeum equitans* creates functional tRNAs from separate genes for their 5′- and 3′-halves. *Nature* 433:537–541.)

RNA editing in trypanosomes

In the early 1980s, researchers began a routine study of the mitochondrial genome in African trypanosomes, looking at gene content and organization of the mtDNA, with no idea of finding anything unusual. Trypanosomes are parasitic protozoans that, at the time, were already well known for their remarkable ability to outwit the mammalian immune system by their repertoire of coat changes (see Disease box 12.5). However, the initial sequence analysis of mitochondrial maxicircles (see below) from *Trypanosoma brucei* and two related species, *Leishmania tarentolae* and *Crithidia fasciculata*, were puzzling to say the least. There were internal frameshifts in protein-coding regions conserved among the three species, the absence of ATG codons for translation initiation for many of the genes, and the apparent absence of some highly conserved electron transport proteins that exist in the mitochondria of all other organisms that had been studied. Upon closer scrutiny, major RNA transcripts encoding important electron transport chain enzymes were found in mitochondria that contained nucleotides not encoded in the DNA. The insertion of four uridines in the cytochrome oxidase subunit II transcript was first reported in 1986 by Rob Benne and colleagues. In 1987 Jean Feagin and colleagues described the addition of 34 uridines at several different sites within the 5′ end of the cytochrome-b transcript. These editing events were shown to correct internal frameshifts, create AUG initiation and UAG/UAA stop codons, and even to create open reading frames (Fig. 13.20). The cytochrome-b mRNA was found to be edited in procyclic-form parasites in the nutrient-poor environment of the tsetse fly host, and primarily unedited bloodstream forms within the mammalian hosts.

These initial ground-breaking reports were rapidly followed by reports of "pan-editing" events involving hundreds of U insertions in mitochondrial transcripts. In some mRNAs analyzed there were so many U insertions that the RNA was doubled in length. It is now known that 12 of the 17 mtRNAs in trypanosomes undergo RNA editing in which precursor mRNA (pre-mRNA) sequences are changed, often extensively, by the insertion of Us and less frequently by the deletion of Us. The editing is so extensive in most cases that it is not possible to deduce the protein-coding sequence from the "encrypted" gene sequence.

For a time, RNA editing in trypanosomes seemed to shake "the central dogma." The finding that nearly half of the sequence information in an mRNA was not present in the gene encoding it was hard to rationalize. If "DNA codes for RNA codes for protein," what was coding for these added uridines? In 1990 the mechanism for RNA editing was discovered, revealing an unprecedented mode of controlling gene expression: a guide RNA is used to alter the nucleotide sequence of the pre-mRNA.

Guide RNAs

Guide RNAs (gRNAs) are 50–70 nt long RNAs that specify the edited sequence. Multiple gRNAs are required to edit each pre-mRNA transcript. gRNAs are precise complementary versions of the mature mRNAs in the edited region (allowing for GU base pairs). Again, the principle of RNA complementarity is used to establish the specificity of function. The mitochondrial genome of *T. brucei* is termed "kinetoplast DNA." It consists of two molecular species – approximately 50 identical 22 kb "maxicircles" and ~10,000 heterogeneous 1 kb "minicircles" – that form a disk-shaped structure of catenated DNA circles *in situ* (Fig. 13.21). The pre-edited mRNAs are encoded by the larger maxicircles, whereas the smaller minicircles each encode three to four gRNAs that specifiy editing. The maxicircles of different trypanosome species encode the same mRNAs (and rRNAs) but differ in which RNAs are edited and to what extent.

Mechanism of editing

The general mechanism of editing has been determined, aided by the development of *in vitro* editing assays. The first well-characterized systems were mitochondrial extracts from procyclic-form trypanosomes that are present in the tsetse fly host. More recently, Kari Halbig and co-workers developed an *in vitro* system from mammalian bloodstream-form trypanosomes that recreates complete cycles of both insertion and deletion (Fig. 13.22). Editing is catalyzed by a multiprotein complex called the "editosome" that has not yet been fully defined or characterized. A complex that sediments at 20S on glycerol gradients contains the four key

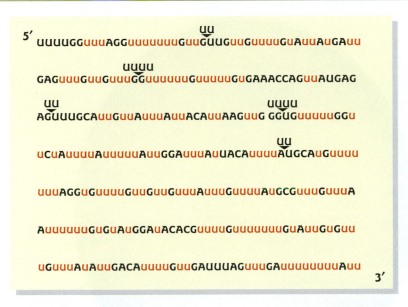

Figure 13.20 Extensive post-transcriptional editing of the cytochrome oxidase subunit III transcript in *Trypanosoma brucei.* Part of the edited sequence of the cytochrome oxidase subunit III mRNA is shown. The Us deleted in the mRNA (present as Ts in the gene) are shown in black above the sequence. (Redrawn from Feagin, J.E., Abraham, J.M., Stuart, K. 1988. Extensive editing of the cytochrome c oxidase III transcript in *Trypanosoma brucei. Cell* 53:413–422.)

enzyme activities and catalyzes *in vitro* editing. Depending on the purification scheme, editing complexes contain as few as seven to as many as 20 major proteins. RNA editing is restricted to the mitochondrion, but the protein components of the editing machinery are encoded by the nuclear DNA (Fig. 13.21).

The overall directionality of editing is from 3′ to 5′ along the mRNA. The process of U insertion and deletion occurs by a series of enzymatic reactions through the following steps (Fig. 13.23):

1 Anchoring: formation of an "anchor" duplex between the pre-mRNA and the gRNA by complementary base pairing near the editing site. The 5′ regions of the gRNAs recognize the pre-mRNA by a complementary region of 4–14 nt adjacent to the region to be edited.

2 Cleavage: at the first point of mismatch between the pre-mRNA and gRNA upstream of the anchor duplex, an endoribonuclease cleaves the pre-mRNA. gRNAs also have a 3′ oligo (U) tail of unknown function that is added post-transcriptionally. The U tail is dispensable *in vitro* but other reports imply that it is essential *in vivo*. The tail is proposed to stabilize the interaction of the 3′ region of the gRNA with the 5′ cleavage product of mRNA.

3 Uridine insertion or deletion: at the site of cleavage, a U is inserted by a 3′ terminal uridylyl transferase (TUTase), or removed by a U-specific 3′ → 5′ exoribonuclease (ExoUase).

4 Ligation: the RNA ends at the site of U addition or deletion are ligated by an RNA ligase.

5 Repeat of editing cycle: the process is repeated at the next mismatch until the RNA is fully edited. The central information regions of the gRNAs specify the editing sites and the numbers of Us to be added or removed. Complete editing results in continuous base pairing between the gRNA and the pre-mRNA.

This extreme form of editing is apparently limited to the mitochondria of trypanosomes. However, its discovery made molecular biologists consider the possibility of editing events occurring in other systems. Within 5 years of the first report of editing in trypanosomes, RNA editing events, involving very different mechanisms, had been described in a number of organisms, including mammals.

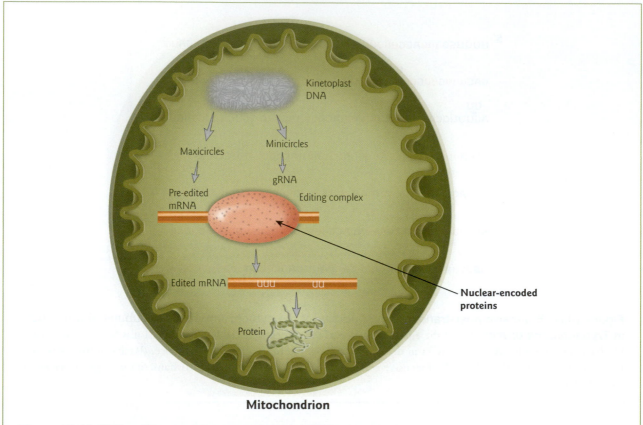

Mitochondrion

Figure 13.21 RNA editing requires proteins and RNAs encoded by mitochondrial and nuclear genomes within the trypanosome. The catenated network of maxicircles and minicircles that make up the kinetoplast DNA represents the mitochondrial genome. Pre-edited mRNAs are transcribed from the maxicircles while the gRNAs are transcribed from the minicircles. Editing complex proteins are nuclear encoded and imported into the mitochondria (long arrow). The pre-edited mRNA is edited by uridine insertion/deletion to generate a mature, translatable RNA.

RNA editing in mammals

In mammals, there are two main classes of editing enzymes, both of which deaminate encoded nucleotides. One generates inosine (I) from adenosine (A), and the other generates uridine (U) from cytidine (C) (Fig. 13.24). Editing plays important roles in regulating a diversity of processes, including aspects of neurotransmission and lipid metabolism.

Adenosine to inosine (A → I) editing

A → I editing occurs frequently, affecting > 1600 genes. It is catalyzed by members of the double-stranded RNA-specific ADAR (*adenosine deaminase acting on RNA*) family (Fig. 13.24). The recent crystal structure of ADAR revealed an unusual feature. The enzyme requires inositol hexakisphosphate as a cofactor. Inositol hexakisphosphate is abundant in mammalian cells, but only recently has been implicated as a versatile molecule with important roles in controlling diverse cellular activities.

In a few examples, A → I editing changes an amino acid in the translated protein, resulting in a change in its function. However, most reported editing events occur within introns and untranslated regions. The high frequency of RNA editing in humans may be explained by the prevalence of *Alu* elements (see

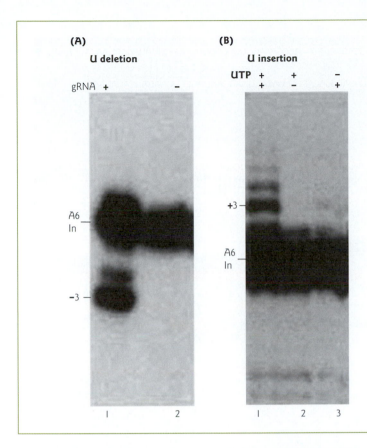

(A)
U deletion

gRNA +   –

A6
In

–3

1   2

(B)
U insertion

UTP +   +   –
      +   –   +

+3

A6
In

1   2   3

Figure 13.22 Complete cycles of trypanosome RNA editing *in vitro*.
Uridine deletion and insertion cycles in crude mitochondrial extracts from bloodstream-form trypanosomes were performed using 3′ end-labeled ATPase 6 pre-mRNA (A6 input) and gRNAs. (A) Deletion reaction with and without gRNA. (B) Insertion reaction with gRNA and UTP or without either gRNA or UTP, as indicated. Accurate deletion and insertion products are indicated as −3 and +3, respectively. Partially edited products are also detected. Editing depends on exogenous gRNA, and U insertion also depends on exogenous UTP. (Reprinted by permission from Cold Spring Harbor Laboratory Press: Halbig, K., De Nova-Ocampo, M., Cruz-Reyes, J. 2004. Complete cycles of bloodstream trypanosome RNA editing in vitro. *RNA* 10:914–920. Copyright © 2004 RNA Society.)

Section 12.5), since most of the editing sites documented reside in *Alu* elements within untranslated regions. Whether there is functional significance to the editing that occurs within these untranslated regions has yet to be determined. These editing events could affect splicing, RNA localization, RNA stability, and translation. A → I RNA editing is particularly abundant in brain tissues, occurring primarily in receptors and ion channels. Not surprisingly, editing defects occur in a number of neurological disorders. Altered editing patterns are associated with inflammation, epilepsy, depression, malignant gliomas (brain tumors), and amyotrophic lateral sclerosis (ALS) (Disease box 13.4).

Cytidine to uridine (C → U) editing

In contrast to widespread A → I editing, C → U editing has only been identified in the apolipoprotein B (ApoB) (see below) and neurofibromatosis type 1 (NF1) transcripts. Neurofibromatosis type 1 is a dominantly inherited disease that predisposes affected individuals to various forms of cancer. An editing event that introduces a C → U modification results in a premature stop codon. The truncated form of the NF1 (neurofibromin) protein lacks tumor suppressor function.

Apolipoprotein B editing in humans   Apolipoprotein B (ApoB) is a plasma protein that plays a key role in the assembly, transport, and metabolism of plasma lipoproteins. There are two forms of ApoB which circulate in the plasma. The long form, ApoB-100, is synthesized by the liver. It is the main component of very low-density and low-density lipoproteins (LDLs), and is a ligand for the LDL receptor. The LDL receptor removes these lipoprotein particles from the bloodstream. The small intestine produces a short form, ApoB-48, which is required for the absorption and transport of dietary lipid. ApoB-48 lacks the receptor-binding domain for the LDL receptor and is removed from the blood circulation by a different receptor pathway. The two different proteins are products of a single gene generated by RNA editing.

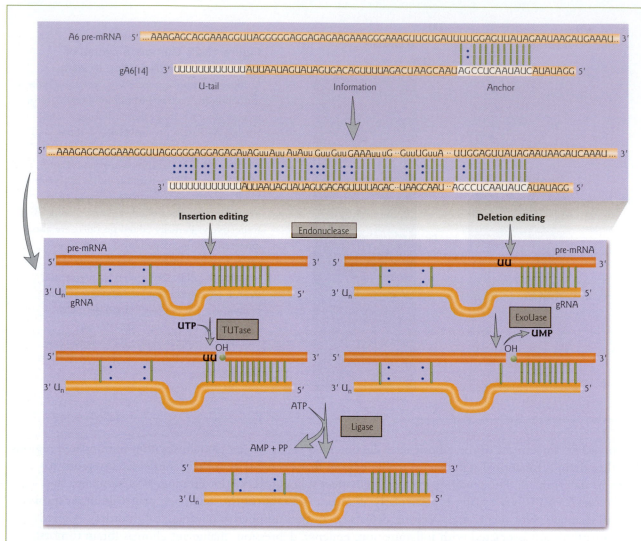

Figure 13.23 General mechanisms of insertion and deletion RNA editing. Pre-mRNAs (dark orange strands) are edited progressively from 3′ to 5′ with each gRNA (light orange strands) specifying the editing of several sites. Interaction between the RNAs by Watson-Crick base pairs (unbroken green lines) and GU base pairs (blue dots) determines the sites of cleavage and the number of U nucleotides that are added or removed. The gRNAs have a 3′ oligo(U) tail that is added post-transcriptionally. Editing occurs by a series of catalytic steps. Cleavage of the pre-mRNA by an endoribonuclease occurs upstream of the anchor duplex between the pre-mRNA and its gRNA (arrow). Us are either added to the 5′ cleavage fragment by a 3′ terminal uridylyl transferase (TUTase) or removed by a U-specific exoribonuclease (ExoUase), as specified by the sequence of the gRNA. The 5′ and 3′ mRNA fragments are then ligated by an RNA ligase. The process is repeated until all of the sites specified by a gRNA are edited, resulting in complementarity (GU, AU, and GC base pairing) between the edited mRNA and the gRNA, except at the gRNA terminals. Editing by each gRNA creates a sequence that is complementary to the anchor region of the next gRNA to be used. This allows for the sequential use of the multiple gRNAs that are required to edit the mRNAs in full. (Inset) Editing of the first block of *Trypanosoma brucei* ATPase 6 pre-mRNA. The 3′ part of the pre-mRNA is shown with its cognate guide RNA (gA6[14]). The gRNA specifies the insertion of 19 Us and the deletion of four Us. Inserted Us are shown in lowercase letters and the positions of Us that have been deleted from the precursor RNA are indicated by black dots. (Inset redrawn from Kable, M.L., Seiwert, S.D., Heidmann, S., Stuart K. 1996. RNA editing: a mechanism for gRNA-specified uridylate insertion into precursor mRNA. *Science* 273:1189–1195.)

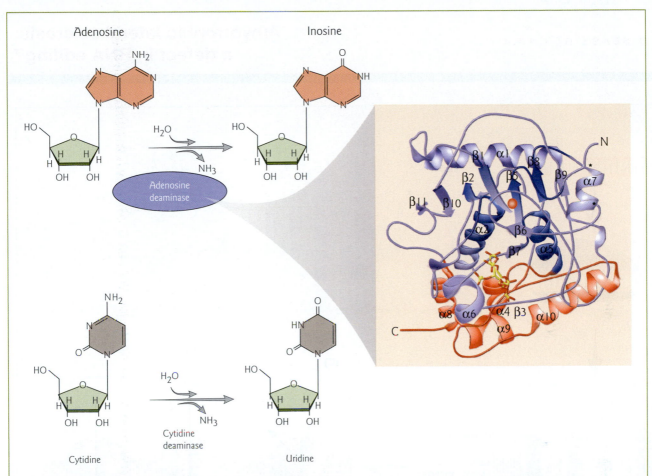

Figure 13.24 The two main classes of RNA editing enzymes in mammals. Adenosine deaminase (e.g. ADAR) generates inosine from adenosine, and cytidine deaminase generates uridine from cytidine (e.g. APOBEC1). (Inset) Ribbon model of the catalytic domain of human ADAR2. The active-site zinc atom is represented by a magenta sphere. The N-terminal domain is colored cyan; the deamination motif region is dark blue; and the C-terminal helical domain which, with contributions from the deamination motif, makes the major contacts to inositol hexakisphosphate (IP_6, ball and stick) is colored red. (Protein Data Bank, PDB:1ZY7. Reprinted with permission from Macbeth, M.R., Schubert, H.L., VanDemark, A.P., Lingam, A.T., Hill, C.P., Bass, B.R. 2005. Inositol hexakisphosphate is bound in the ADAR2 core and required for RNA editing. *Science* 309:1534–1539. Copyright © 2005 AAAS.)

Editing occurs post-transcriptionally within the nucleus of cells in the small intestine (Fig. 13.25). The *apoB* gene is 43 kb in length and has 29 exons and 28 introns. Within the nucleus, the pre-mRNA transcript undergoes splicing, polyadenylation, and editing. The C → U transition converts a glutamine codon (CAA) at position 6666 to an in-frame premature stop codon (UAA). In the unedited RNA present in liver cells, a second stop codon is used to produce ApoB-100.

Significant advances in understanding the molecular mechanism for C → U editing were made through the development of a simple, quantitative *in vitro* RNA editing assay (Fig. 13.25). Using this assay, researchers were able to fractionate cell extracts to determine which fraction contained the editing activity. Active fractions were further purified by column chromatography. A two-subunit editing enzyme complex was shown to carry out the separate functions of recognition and catalysis. The catalytic subunit is an RNA-specific cytidine deaminase designated APOBEC1 for "ApoB-editing catalytic subunit 1." APOBEC1 was the first identified mammalian cytidine deaminase. APOBEC1 has lax sequence specificity and will edit any

Amyotrophic lateral sclerosis: a defect in RNA editing?

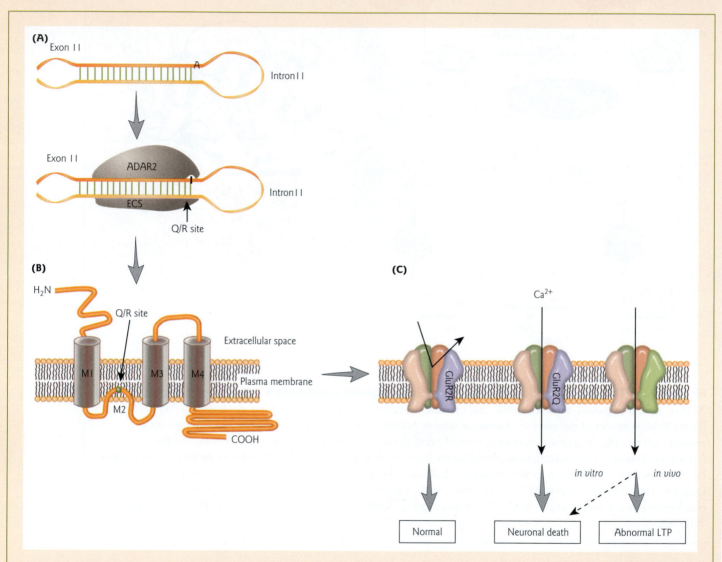

Figure 1 RNA editing and AMPA receptor Ca²⁺ permeability. (A) Editing mechanism. The adenosine deaminase ADAR2 generates inosine (I) from the adenosine residue (A) at the Q/R site in exon 11 of the GluR2 pre-mRNA. Recognition of this specific adenosine is via the structure of the duplex that is formed between the Q/R editing site and its editing site complementary sequence (ECS) located in intron 11. Editing results in substitution of a glutamine (Q) codon for an arginine (R) codon. (B) Structure of the GluR2 protein: the Q/R site is located in the putative second membrane domain (M2). (C) Ca²⁺ permeability of AMPA receptors and the GluR2 subunit. Functional AMPA receptors are homo- or heterotetramers with the four subunits GluR1, GluR2, GluR3, and GluR4 in various combinations. The Ca²⁺ conductance of AMPA receptors differs markedly depending on whether they contain the GluR2 subunit. AMPA receptors that contain at least one edited GluR2R subunit have low Ca²⁺ conductance (left), whereas those lacking a GluR2 subunit (right) and those containing unedited GluR2Q (middle) have high Ca²⁺ conductance. Increased Ca²⁺ permeability leads to neuronal death. LTP, long-term potentiation. (Adapted from Kwak, S., Kawahara, Y. 2005. Deficient RNA editing of GluR2 and neuronal death in amyotropic lateral sclerosis. *Journal of Molecular Medicine* 83:110–120.)

Amyotrophic lateral sclerosis: a defect in RNA editing?

The most common motor neuron disease, amyotrophic lateral sclerosis (ALS) (also known commonly as Lou Gehrig disease) is characterized by a selective loss of upper and lower motor neurons. The disease typically initiates in mid-life and leads to muscle wasting and progressive paralysis. First identified by J.-M. Charcot in 1874, ALS has a worldwide prevalence of 0.8–7.3 cases per 100,000 individuals, with the risk of disease increasing in an age-dependent manner after age 60. Only approximately 5–10% of all ALS cases are familial. Familial cases are linked to defects in genes encoding superoxide dismutase 1 (SOD1), ALS2, and senataxin. However, the mechanism underlying motor neuron death has not been determined.

Sporadic (nonhereditary) ALS accounts for most cases of the disease. The cause remains unknown; however, one hypothesis for selective neuronal death in sporadic ALS is excitotoxicity mediated by defective α-AMPA (amino-3-hydroxy-5-methyl-4-isoxazolepropionate) receptors. AMPA receptors are a subtype of glutamate receptors. Excitotoxicity was first described in the 1970s and involves the activation of glutamate receptors in the central nervous system. High concentrations of glutamate can cause cell death through excessive activation of these receptors. Normally, nerve cells prevent glutamate build-up through glutamate transporters. Motor neurons express abundant glutamate receptors and are sensitive to exaggerated receptor activation by glutamate. Recently, a defect in RNA editing of AMPA receptors has been proposed to play a role in ALS.

RNA editing of the AMPA receptor

Functional AMPA receptors are composed of four subunits, GluR1, GluR2, GluR3, and GluR4, in various combinations. Almost all GluR2 mRNA undergoes an RNA editing event in which a glutamine (Q) codon is substituted by an arginine (R) codon at the "Q/R" site in the second transmembrane domain (Fig. 1). The Ca^{2+} conductance of AMPA receptors varies depending on whether the edited GluR2 subunit is a component of the receptor. AMPA receptors lacking GluR2 or containing unedited GluR2 (Q remains at the Q/R site) are Ca^{2+} permeable, and those with at least one edited GluR2 subunit (R is found at the Q/R site) have low Ca^{2+} conductance. An increased influx of Ca^{2+} through activated AMPA receptor-coupled channels appears to play a key role in neuronal death.

In ALS spinal motor neurons, there is a reduction in GluR2 RNA editing. The editing efficiency varies greatly, from 0 to 100%, among the motor neurons from individuals with ALS. In one study by Shim Kwak and Yukio Kawahara (2005), editing was incomplete in 56% of ALS motor neurons, compared with 100% editing in normal controls. This marked reduction in editing results in the number of GluR2Q-containing, Ca^{2+}-permeable AMPA receptors increasing to a deleterious amount, and leads to selective motor neuron death. The mechanism behind the deficiency in RNA editing in ALS patients appears to be due to a reduction in ADAR2 deaminase activity. Expression levels of ADAR2 mRNA are reduced in ALS motor neurons. Restoration of this enzyme activity in ALS motor neurons may provide a specific therapeutic strategy for ALS.

C in its general vicinity. The other subunit in the editing complex is the APOBEC1 complementation factor (ACF) which mediates sequence specificity. ACF is an RNA-binding protein that recognizes an 11 nt "mooring sequence" that is located four nucleotides downstream of the edited C in the *apoB* mRNA. Binding of ACF to the RNA transcript thus positions APOBEC1 over the correct cytidine. There are 375 CAA triplets in the *apoB* gene, of which 100 are in-frame glutamine codons. Even so, the editing complex is able to confer almost absolute specificity at this particular CAA.

13.9 Base modification guided by small nucleolar RNA molecules

Small nucleolar RNAs (snoRNAs) primarily function as RNA chaperones in the processing of ribosomal RNA (rRNA) in the nucleolus. The two major classes of snoRNA base pair with specific sites in the

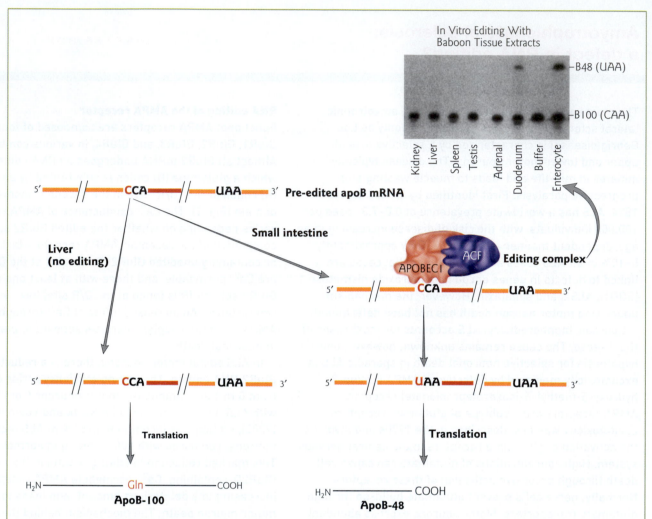

Figure 13.25 Apolipoprotein B (ApoB) editing occurs post-transcriptionally within the nucleus of small intestine cells. The ApoB mRNA is depicted as an orange line; slash marks indicate a long section of transcript not shown (not to scale). The two-subunit editing complex converts a glutamine (Gln) codon (CAA) to a stop codon (UAA), by deamination of the C at position 6666 (red). APOBEC1 (red) is the catalytic RNA-specific cytidine deaminase. ACF (blue) is an RNA-binding protein that recognizes an 11 nt "mooring sequence" (red line). The premature stop codon is used during translation to produce the shorter ApoB-48. In the unedited ApoB mRNA present in liver cells, a second stop codon is used during translation to produce the longer ApoB-100. (Inset) Synthetic pre-edited apoB mRNA was incubated with whole-cell extract prepared from various baboon tissues in an *in vitro* editing assay. This assay involves a ^{32}P end-labeled 35 nt primer complementary to a sequence located just downstream of the target C, and primer extension with reverse transcriptase in the presence of high concentrations of dideoxy-GTP. Extension products were separated by polyacrylamide gel electrophoresis. The extension of apoB RNAs that are unedited stops at the first upstream C (the editing site), are designated as B100 (CAA) on the autoradiogram. Transcripts that have undergone a C → U editing event terminate at the second C in the substrate, generating a product 10 nt longer, and are designated as B48 (UAA). Editing only occurs in cell extracts of small intestine origin (duodenum and enterocytes). (Republished with permission from Driscoll, D.M., Casanova, E. 1990. Characterization of the apolipoprotein B mRNA editing activity in enterocyte extracts. *The Journal of Biological Chemistry* 265:21401–21403. Copyright © 1990 by the American Society for Biochemistry and Molecular Biology, Inc; permission conveyed through Copyright Clearance Center, Inc.)

rRNA, and guide their modification. Again, the principle of RNA complementary is used to establish specificity and, in this case, the essential modification enzymes are recruited by RNA–protein interactions. Box C/D snoRNAs guide 2′-O-methylation, whereas H/ACA snoRNAs direct pseudouridine formation. The two classes are categorized based on the presence of short consensus sequence motifs (Fig. 13.26). Each snoRNA can direct one, or at most two, rRNA modifications through a single or a pair of appropriate antisense elements. Mature box C/D snoRNAs associate with four common core proteins to form snoRNPs (pronounced "snorps"). One of these core proteins is the methyltransferase fibrillarin. Mature H/ACA snoRNPs are also associated with four core proteins, including dyskerin which is a pseudouridine synthase.

Recent studies indicate that the role of snoRNAs is not limited to ribosome synthesis. Novel members target spliceosomal snRNAs in vertebrates, tRNAs in Archaea, and possibly even mRNAs for modification. An increasing number of "orphan" guides without known RNA targets have been identified, some of which are expressed in a tissue-specific fashion and subjected to genomic imprinting (see Section 12.3 and Fig. 12.5A).

13.10 Post-transcriptional gene regulation by microRNA

The recent discovery of hundreds of genes that encode small regulatory RNA molecules represents yet another landmark discovery in molecular biology. Because the focus of many scientists had been on the characterization of protein–coding genes, a whole level of eukaryotic gene regulation was overlooked for

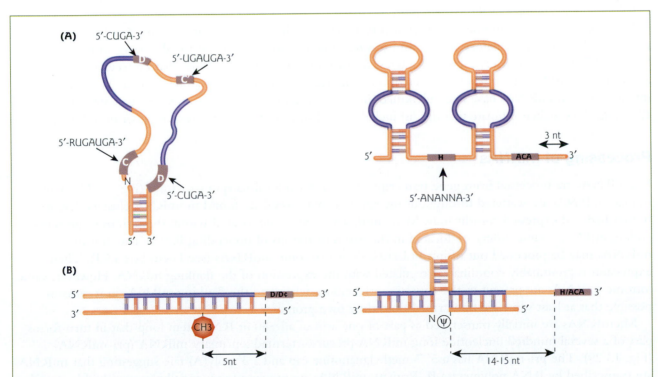

Figure 13.26 Structural features of the two major classes of snoRNAs. (A) Schematic secondary structures of the box C/D and H/ACA classes of eukaryotic snoRNAs. Their conserved box motifs (gray boxes), and sequence tracts complementary to the rRNA target – also termed antisense elements (thick blue lines) – are indicated. For box C/D snoRNAs (left), the 5′,3′-terminal stem allowing the formation of the box C/D structural motif is represented. (B) Canonical structure of each type of guide RNA duplex, indicating the site of methylation or pseudouridylation in the cognate rRNA. (Redrawn from Bachellerie, J.-P., Cavaillé, J., Hüttenhofer, A. 2002. The expanding snoRNA world. *Biochimie* 84:775–790.)

decades. Small RNAs fall into two major classes: microRNAs (miRNAs) and small interfering RNAs (siRNAs), which were introduced in Section 9.7. miRNA sequences are nearly always conserved in related organisms, whereas endogenous siRNA sequences are rarely conserved. miRNAs specify "heterosilencing" – they are derived from unique genes that specify the silencing of very different genes. In contrast, siRNAs typically specify "autosilencing" – they guide silencing of the same genetic locus or a very similar locus from which they originate. siRNAs may be derived from viruses, transposable elements, heterochromatin, or exogenous genes inserted into a cell by the bench scientist. A significant function of siRNAs is in defense of the genome from viruses and mobile DNA elements (see Figs 9.12 and 12.15).

Historical perspective: the discovery of miRNA in *Caenorhabditis elegans*

miRNAs were first discovered in the early 1990s in the nematode worm *C. elegans*. This unexpected discovery was made by Victor Ambros and colleagues while they were studying mutations that changed the timing of developmental events in the worm. A mutation that caused an increase in the translation of mRNA to protein was found not in the gene encoding the protein, but in a second gene encoding a very small RNA molecule. The wild-type small RNA (*lin-4*) was shown to bind through complementary base pairing to the 3′ untranslated region of the target mRNA, thereby controlling a "heterochronic gene hierarchy" (Fig. 13.27). The original interpretation was that the binding of *lin-4* RNA to the mRNA directly blocked mRNA translation, without destabilization of the mRNA. More recent results, however, suggest that *lin-4* miRNA also promotes target mRNA degradation (Fig. 13.28). Bagga and co-workers demonstrated that there was a significant decrease in *lin-14* and *lin-28* target mRNA levels at the point in development when *lin-4* miRNA first appears.

When *let-7* (*lethal-7*), another small regulatory RNA, was discovered 7 years after *lin-4*, it became clear that this type of regulatory mechanism was not unique. However, it was the discovery of RNAi (see Section 9.7) that dramatically increased interest in small regulatory RNAs. Now that molecular biologists are looking, miRNAs have been discovered in all multicellular eukaryotes examined so far. More than 200 different human miRNAs have been identified, including a number of human miRNAs that are conserved in *C. elegans*. Researchers estimate that this list includes less than half of the actual total number of miRNA genes.

Processing of miRNAs

MicroRNAs are processed from gene transcripts that can form local hairpin structures (Fig. 13.29). Most human miRNAs are scattered throughout the genome, but several are found in miRNA clusters that are transcribed and expressed coordinately. Many miRNAS (~56%) are located within the introns of protein-coding mRNAs, while others are located in the exons or introns of noncoding RNA. The intronic miRNAs may be processed out of intron lariats, similar to some snoRNAs (see Focus box 13.1). Their expression is presumably coordinately regulated with the expression of the flanking mRNA. However, some intronic miRNAs are present in the antisense orientation, relative to the flanking mRNA, so it remains possible that at least some are transcribed from their own promoter.

MicroRNAs are initially transcribed as part of one arm of an ~80 nt RNA stem loop that in turn forms part of a several hundred nucleotide long miRNA precursor termed a primary miRNA (pri-miRNA) (Fig. 13.29). The pri-miRNA has a 5′ 7-methylguanosine cap and a 3′ poly(A) tail, suggesting that miRNAs are transcribed by RNA polymerase II. Primary miRNAs are processed sequentially by two double-stranded RNA (dsRNA) specific ribonuclease III (RNase III) enzymes, Drosha and Dicer. The stem loop is cleaved by Drosha to liberate a shorter ~60–70 nt hairpin, termed a pre-miRNA, that contains a two nucleotide 3′ overhang. Exportin 5, a member of the karyopherin family of nucleocytoplasmic transport factors (see Table 11.3), mediates nuclear export of the pre-miRNAs. In the cytoplasm, Dicer binds to the 3′ overhang present at the base of the pre-miRNA hairpin, and liberates a ~22 bp RNA duplex with 3′ overhangs of two nucleotides. A single miRNA:miRNA duplex is generated from each hairpin precursor. In contrast, a multitude of siRNA duplexes are generated from each siRNA precursor molecule (see Fig. 9.12). The

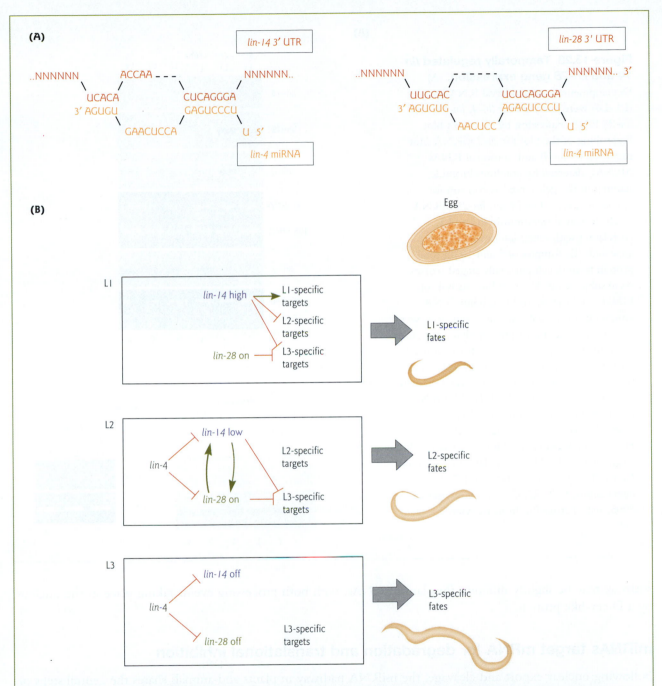

Figure 13.27 *Lin-4* miRNA regulates the timing of larval development in *Caenorhabditis elegans*. (A) The *lin-4* mRNA is shown with its complementary sites in the 3′ untranslated region (3′-UTR) of *lin-14* (left panel) and *lin-28* mRNAs (right panel). (Redrawn from Ambros, V. 2004. The functions of animal microRNAs. *Nature* 431:350–355.) (B) The heterochronic gene hierarchy. During larval development, *lin-4* mediates the downregulation of *lin-14* and *lin-28*, which in turn regulates the expression of stage-specific developmental events in the first three larval stages (L1–L3). A high level of *lin-14* is necessary for L1-specific fates and negative regulation of L2-specific targets. When *lin-4* is on, both *lin-14* and *lin-28* are downregulated, but each must retain a certain level of activity and mutual positive interaction to allow L2-specific fates to occur. L3-specific fates occur later when both *lin-14* and *lin-28* activities become fully reduced. (Redrawn from Moss, E.G., Lee, R.C., Ambros, V. 1997. The cold shock domain protein LIN-28 controls developmental timing in *C. elegans* and is regulated by the *lin-4* RNA. *Cell* 88:637–646.)

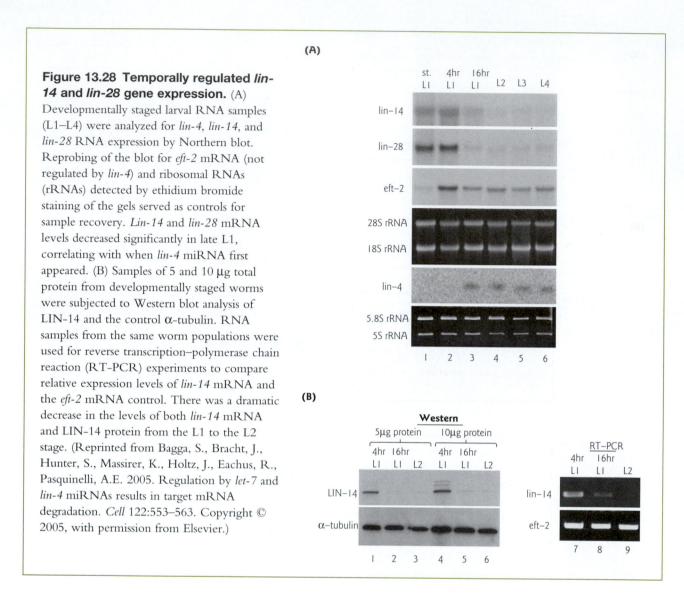

Figure 13.28 Temporally regulated *lin-14* and *lin-28* gene expression. (A) Developmentally staged larval RNA samples (L1–L4) were analyzed for *lin-4*, *lin-14*, and *lin-28* RNA expression by Northern blot. Reprobing of the blot for *eft-2* mRNA (not regulated by *lin-4*) and ribosomal RNAs (rRNAs) detected by ethidium bromide staining of the gels served as controls for sample recovery. *Lin-14* and *lin-28* mRNA levels decreased significantly in late L1, correlating with when *lin-4* miRNA first appeared. (B) Samples of 5 and 10 μg total protein from developmentally staged worms were subjected to Western blot analysis of LIN-14 and the control α-tubulin. RNA samples from the same worm populations were used for reverse transcription–polymerase chain reaction (RT-PCR) experiments to compare relative expression levels of *lin-14* mRNA and the *eft-2* mRNA control. There was a dramatic decrease in the levels of both *lin-14* mRNA and LIN-14 protein from the L1 to the L2 stage. (Reprinted from Bagga, S., Bracht, J., Hunter, S., Massirer, K., Holtz, J., Eachus, R., Pasquinelli, A.E. 2005. Regulation by *let-7* and *lin-4* miRNAs results in target mRNA degradation. *Cell* 122:553–563. Copyright © 2005, with permission from Elsevier.)

pathway may be slightly different for plant miRNAs, with both processing events taking place in the nucleus by a Dicer-like protein.

miRNAs target mRNA for degradation and translational inhibition

Following nuclear export and cleavage, the miRNA pathway in plants and animals shares the central steps of RNA silencing pathways known as post-transcriptional gene silencing (PTGS) in plants, quelling in fungi, and RNAi in animals (see Fig. 9.12). The short miRNA duplex is first unwound by an unidentified RNA helicase activity in the RNA-induced silencing complex (RISC) loading complex (Fig. 13.29). One strand of the duplex is incorporated into activated RISC. RISC can target protein-coding mRNAs either for inhibition, by blocking their translation into protein, or destruction by cleavage. Complementary base pairing between the miRNA and its target mRNA gives the process its specificity. Whether translational inhibition or mRNA cleavage occurs is thought to be regulated by the degree of mismatch between the miRNA and its target mRNA. Degradation by cleavage is the outcome for the most complementary targets. After cleavage of the target mRNA, the miRNA remains intact and can guide the recognition and destruction of additional mRNAs. Efficient translational repression may require the cooperative action of multiple RISCs. Whether certain miRNAs target chromatin for heterochromatin formation, like some small

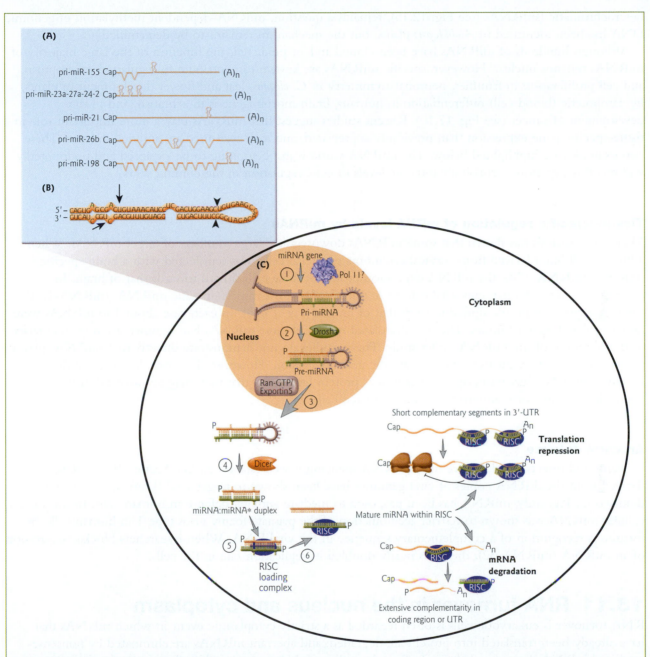

Figure 13.29 Biogenesis of miRNA. (A) Schematic diagram of the structure of five human pri-miRNAs. The miRNA stems are shown in lavender, noncoding sequences in light orange, introns in light orange V-shapes, and coding exons in dark orange (not drawn to scale). (B) Structure of the human pri-miRNA-30a RNA hairpin. Drosha cleavage sites are shown by arrows and Dicer cleavage sites by arrowheads. (Adapted from Cullen, B.R. 2004. Transcription and processing of human microRNA precursors. *Molecular Cell* 16:861–865. Copyright © 2004, with permission from Elsevier.) (C) The pathway of miRNA biogenesis. (1) The miRNA gene is transcribed, possibly by RNA polymerase II (Pol II?). (2) The pri-miRNA hairpin is processed by Drosha in the nucleus and (3) the pre-miRNA exits the nucleus by an exportin 5-mediated pathway. (4) In the cytoplasm pre-miRNA is cleaved into duplex miRNA by Dicer. (5) The duplex is unwound by a helicase in the RISC loading complex and the mature miRNA associates with activated RISC. (6) RISC targets mRNAs for either degradation or translational repression.

heterochromatic (sh)RNAs (see Fig. 12.15), remains a question. miRNA-dependent methylation of genomic DNA has been identified in *Arabidopsis* plants, but the mechanism remains to be determined.

Although hundreds of miRNAs have been cloned and/or predicted, the function of the large majority of miRNAs remains unclear. However, specific miRNAs are known to contribute to regulation of apoptosis and cell proliferation in fruitflies, neuronal asymmetry in *C. elegans*, leaf and flower development in plants, hematopoietic (blood cell) differentiation in humans, brain morphogenesis in zebrafish, and in the development of cancer (see Fig. 17.10). Recent studies suggest that miRNAs play a more important role in tissue-specific gene expression than previously appreciated, and may even mediate antiviral defense. These two examples are highlighted below. The miRNA world is just beginning to be explored and undoubtedly will reveal many more surprisingly intricate levels of gene regulation in the coming years.

Tissue-specific regulation of mRNA levels by miRNAs

Microarray analysis has shown that some miRNAs downregulate large numbers of target mRNAs. When a human cell line (derived from cervical carcinoma epithelial cells) was transfected with a brain-specific miRNA (miRNA-124), the mRNA expression profile of these cells shifted towards that of brain. In contrast, when the human epithelial cells were transfected with a muscle-specific miRNA (miRNA-1), the mRNA expression profile shifted towards that of muscle (Fig. 13.30). In each case about 100 mRNAs were downregulated after 12 hours. The 3′ untranslated region of these mRNAs had significant complementarity to the 5′ region of the miRNA under study. These studies do not demonstrate directly that miRNAs play a primary role in the regulation of tissue-specific gene expression. However, the work does suggest that human miRNAs affect transcript level as well as protein levels, and that they may be involved in the determination and/or maintenance of differentiated cell fates.

Antiviral defense

In plants and insects the defensive role for RNA silencing is well established (see Section 9.7). siRNAs derived from the dsRNA copies of viral genomes have been shown to target viral RNA transcripts for destruction. Recently, miRNA has been proposed to mediate antiviral defense in human cells. In this case, a cellular miRNA was shown to restrict accumulation of the primate foamy virus type 1 in human cells, by fortuitous recognition of a complementary sequence in the viral RNA. When researchers blocked expression of an miRNA (miRNA-32), the virus nearly doubled its replication rate in the cells.

13.11 RNA turnover in the nucleus and cytoplasm

RNA turnover is eukaryotic cells is often regarded as a strictly cytoplasmic event in which mRNAs that have already been translated into protein are degraded, and aberrant mRNAs are eliminated by nonsense-mediated mRNA decay (see below). In fact, the bulk of RNA turnover takes place in the nucleus. Pre-mRNA introns that are removed by splicing, and the short-lived processing intermediates of rRNAs, snoRNAs, and snRNAs, are degraded by the nuclear exosome. In addition, all pre-mRNAs are subject to a process called "quality control." There are specific nuclear decay pathways that destroy defective pre-mRNAs, or those failing to be exported from the nucleus.

Nuclear exosomes and quality control

A central mediator of pre-mRNA quality control is the nuclear exosome, a multisubunit complex which contains several exoribonucleases. Nuclear decay of pre-mRNA primarily occurs in the 3′ → 5′ direction. In yeast, RNA degradation by the nuclear exosome is promoted by a complex that contains an ATP-dependent RNA helicase (Mtr4p), a zinc knuckle protein (Air2p) and a poly(A) polymerase (Trf4p). The complex is called TRAMP for *Trf4p–Air2p–Mtr4p* polyadenylation complex. TRAMP interaction with RNAs or RNP

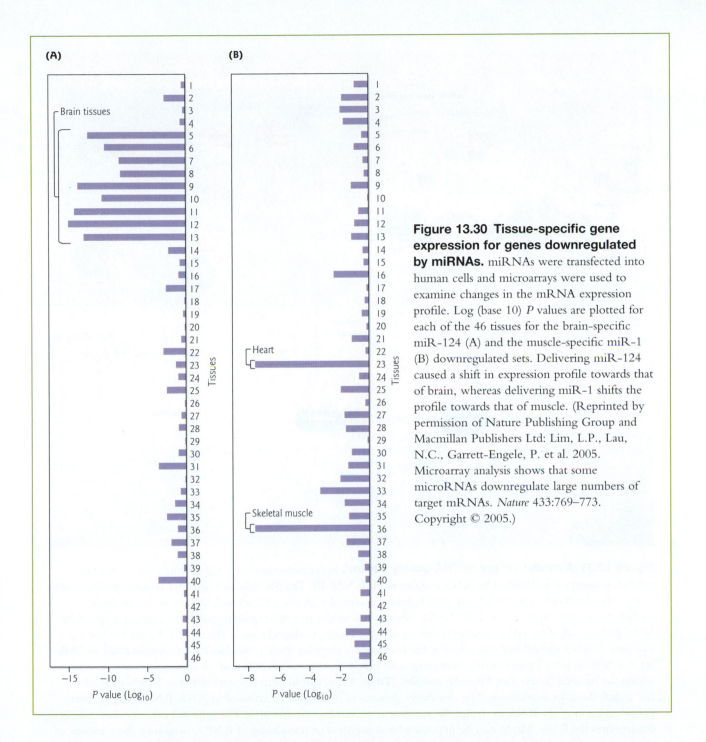

Figure 13.30 Tissue-specific gene expression for genes downregulated by miRNAs. miRNAs were transfected into human cells and microarrays were used to examine changes in the mRNA expression profile. Log (base 10) *P* values are plotted for each of the 46 tissues for the brain-specific miR-124 (A) and the muscle-specific miR-1 (B) downregulated sets. Delivering miR-124 caused a shift in expression profile towards that of brain, whereas delivering miR-1 shifts the profile towards that of muscle. (Reprinted by permission of Nature Publishing Group and Macmillan Publishers Ltd: Lim, L.P., Lau, N.C., Garrett-Engele, P. et al. 2005. Microarray analysis shows that some microRNAs downregulate large numbers of target mRNAs. *Nature* 433:769–773. Copyright © 2005.)

is proposed to target them for degradation. A characteristic feature of this process is the addition of a short poly(A) tail to the target molecules before the recruitment and activation of the exosome (Fig. 13.31). Homologs of TRAMP are present in plant, fruitfly, and human genomes, suggesting that the quality control process is conserved in eukaryotes.

Quality control and the formation of nuclear export-competent RNPs

Prior to being available for translation, the mRNA must be transported out of the nucleus to the cytoplasm. Nuclear export is a highly regulated process mediated by the nuclear pore complexes. The nuclear pore

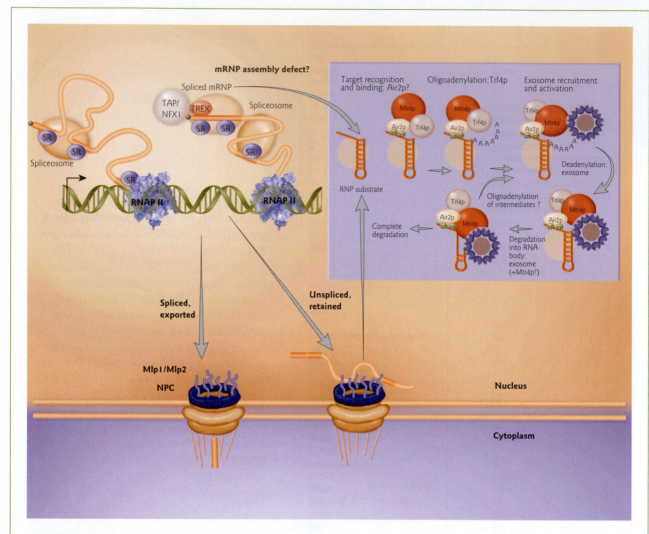

Figure 13.31 A model for pre-mRNA quality control. Spliceosome assembly and splicing occur as the pre-mRNA transcript is synthesized by RNA polymerase II (RNAP II). The SR splicing regulatory proteins remain bound to the spliced mRNA. The TREX complex becomes associated with pre-mRNA and, along with SR proteins, recruits the nuclear export factor TAP/NFX1. Aberrant messenger ribonucleoprotein (mRNPs) particles at any point in the pathway are detected by quality control surveillance leading to degradation of the mRNA by the nuclear exosome. mRNA surveillance also occurs at the nuclear pore complex (NPC). Nuclear basket proteins Mlp1 and Mlp2 act as a "sorting filter," preferentially interacting with export-competent mRNPs, and retaining unspliced transcripts within the nucleus. (Inset) The TRAMP complex (Trf4p, Air2p, Mtr4p) interacts with defective RNAs or RNPs and targets them for degradation. The zinc finger domains of Air2p may be involved in RNA/RNP binding. After polyadenylation of the RNA by Trf4p, the exosome is recruited to the complex. The activated exosome then rapidly deadenylates the RNA. Mtr4p may be important for dissociation or remodeling of RNP structures to allow passage of the exosome. Complete degradation occurs via the exoribonuclease activity of the exosome. (Adapted from LaCava, J., Houseley, J., Saveanu, C., Petfalski, E., Thompson, E., Jacquier, A., Tollervey, D. 2005. RNA degradation by the exosome is promoted by a nuclear polyadenylation complex. *Cell* 121:713–724.)

complexes are dynamic channels that perforate the nuclear membrane (see Section 11.9 and Focus box 11.6). mRNAs can escape nuclear destruction by forming export-competent messenger RNPs (mRNPs). Nuclear production of export-competent mRNPs is a complicated process, and any aberrant mRNPs are rapidly attacked by quality control surveillance leading to degradation of the mRNA by the nuclear exosome. Most likely there are multiple, possibly overlapping, pathways that direct mRNP export. Two candidate key players are TREX (*t*ranscription/*ex*port) and members of the SR family of splicing regulators.

TREX and SR proteins provide "adapter" function by recruiting the general nuclear export factors (NXFs), such as TAP/NXF1, to mRNAs (Fig. 13.31). The TREX adaptor complex is conserved from yeast to humans. The two main components in humans are the multisubunit THO complex and the mRNA export adaptor protein Aly/REF and its partner, the ATPase/RNA helicase UAP56. The TREX complex is recruited to pre-mRNA during a late step in splicing in mammals, whereas in yeast TREX is recruited by the transcription machinery during elongation. TAP/NXF1 interacts preferentially with shuttling SR proteins that are hypophosphorylated. SR proteins are initially recruited to pre-mRNAs for splicing in their hyperphosphorylated forms and are dephosphorylated postsplicing. The phosphorylation state of SR proteins may thus contribute to the ability of the export machinery to discriminate between spliced and unspliced mRNPs.

mRNP surveillance also appears to occur at the nuclear periphery. Two filamentous proteins that are anchored at the nuclear basket of the nuclear pore complex in yeast (Mlp1p and Mlp2p) are proposed to function as a "sorting filter," preferentially interacting with properly assembled mRNP particles. They may serve to retain unspliced transcripts within the nucleus, possibly via recognition of a component associated with the 5′ splice site (Fig. 13.31).

Cytoplasmic RNA turnover

Once an mRNP successfully navigates the quality control machinery in the nucleus and is translocated through the nuclear pore complex into the cytoplasm, there are several possible fates (Fig. 13.32). The mRNA may be held in a translationally silent state, it may be translated and then degraded, or, if it contains a premature termination codon, the mRNA is rapidly degraded by a process called nonsense-mediated mRNA decay.

Storage of translationally silent mRNA

Ribosomes may or may not translate an mRNA immediately after its export into the cytoplasm. For example, in multicellular animals the oocyte (immature egg) accumulates all the mRNAs required for early development, since no new transcription occurs until after several embryonic cell divisions. In frog oocytes a number of these maternally stored mRNAs are translationally silenced until the proper time in development. Silencing occurs through a mechanism involving shortening of their poly(A) tails from their initial length of 200–250 adenosines to 20–40. This shortening is mediated by CPEB, a protein that binds the cytoplasmic polyadenylation element (CPE) in the 3′ untranslated region of the mRNA (Fig. 13.32A). CPEB also interacts with Maskin, a protein the competes with eukaryotic translation initiation factor 4G (eIF4G) for binding to eIF4E, and blocks the binding of cytoplasmic poly(A)-binding protein (PABPC). The poly(A) tail is not essential for translation, but when RNAs lacking a poly(A) tail compete with polyadenylated RNAs for limiting translational machinery, the polyadenylated RNA is translated more efficiently. When oocytes are induced to complete meiosis upon fertilization, CPEB becomes phosphorylated and stimulates the readdition of the poly(A) tail by cytoplasmic poly(A) polymerases. The new poly(A) tail binds PABPC, which then recruits eIF4G to initiate translation (see Section 14.4).

General mRNA decay pathways

In Section 13.10, we discussed how miRNAs can target specific mRNAs for degradation via the RNA-induced silencing complex (RISC), but there are also general mRNA decay pathways. Typically, mRNAs are subject to relatively rapid turnover to allow a continuous adjustment of gene expression to physiological needs. In most cells, the core degradation machinery attacks mRNA from its ends. The 3′ poly(A) tail is removed by deadenylases and the 5′ cap is removed by decapping enzymes (Fig. 13.32B). Whether a particular mRNA is degraded primarily from 3′ to 5′ or 5′ to 3′ depends on which set of enzymes is most active in a particular cell type and which set is recruited most efficiently to that mRNA. Decay occurs in

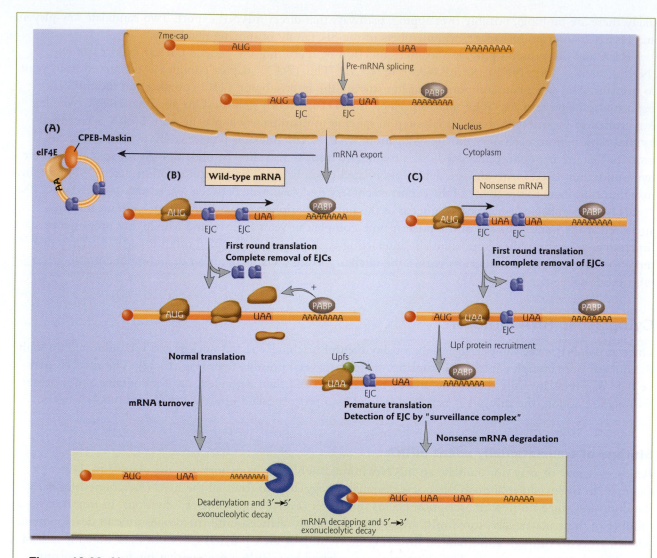

Figure 13.32 Alternate mRNA fates in the cytoplasm. Nuclear pre-mRNA processing events form mRNPs that are exported to the cytoplasm for translation. (A) Some mRNPs are stored in the cytoplasm in a translationally silent state. (B) Translation of wild-type mRNA. During the pioneer round of translation, ribosomes displace exon–exon junction complexes (EJCs) from wild-type mRNAs. The absence of EJCs and positive signals from remaining RNPs, such as the poly(A)-binding protein (PABP), ensure continued translation. Post-translational mRNA turn-over involves general mRNA decay pathways. A common pathway is deadenylation-independent decapping followed by 5′ → 3′ exonucleolytic digestion by Xrn1; an alternative pathway involves deadenylation and exosome-mediated 3′ → 5′ decay. (C) Model for nonsense-mediated decay of mRNA. Failure of the ribosome to remove an EJC due to a premature termination codon (UAA) in a nonsense mRNA leads to recruitment of Upf proteins and 5′ → 3′ "surveillance complex" scanning of the mRNA. Interactions between a nonsense RNA and the surveillance complex result in translational repression and rapid degradation of the RNA involving multiple decay pathways.

specialized cytoplasmic processing bodies (P bodies) that are enriched in the decay machinery. One common degradation pathway is deadenylation–independent decapping. Because the mRNA is normally protected from degradation by the 5′ cap, its removal by decapping enzymes (DCP1 and DCP2) causes rapid degradation of the mRNA by a 5′ → 3′ exonuclease (XRN1). In an alternate mRNA decay pathway, deadenylation is followed by 3′ → 5′ degradation. This requires the exosome (a multimeric assembly of 3′ → 5′ exonucleases) and the Ski complex (a trimeric protein complex that regulates exosome activity).

In the deadenylation-dependent decapping pathway, after the poly(A) tail has been reduced to approximately 10 nt, the 7-methylguanosine cap structure is hydrolyzed. Subsequently, the rest of the mRNA is degraded by the combined action of $5' \rightarrow 3'$ and $3' \rightarrow 5'$ exonucleases.

Nonsense-mediated mRNA decay

Premature termination codons arise due to gene mutation or errors in transcription or splicing. Nonsense-mediated mRNA decay reduces the levels of these nonsense codon-bearing mRNAs, which may encode truncated proteins that are detrimental to a cell (Fig. 13.32C). In mammals, recognition of an mRNA with a premature termination codon involves the assembly of protein complexes within the open reading frame of the mRNA. These exon junction complexes (EJCs) are assembled approximately 20–24 nt upstream of each exon–exon boundary after mRNA splicing.

When the first ribosome begins translating an mRNA, the EJCs are normally displaced as the mRNA enters the decoding center of the ribosome. If a premature termination codon is present in the mRNA, then the surveillance machinery is activated. UPF1, UPF2, and UPF3 proteins are recruited to the EJC to form a "surveillance complex." The RNA helicase UPF1 associates with both translation release factors (i.e. eRF1 and eRF3) and with UPF2 and UPF3, providing a link between the surveillance complex and the translation machinery. The surveillance complexes interact with the prematurely terminating ribosome, and promote mRNA degradation by the general mRNA decay pathways described above.

Chapter summary

Post-transcriptional processing of RNA is an important level of gene regulation in eukaryotes. RNA splicing is the process by which introns are removed from a primary RNA transcript at precisely defined splice points, and the ends of the remaining RNA are rejoined to form a continuous mRNA, rRNA, or tRNA. The excision of introns and joining of exons is directed by 5' and 3' splice sites at the intron–exon junctions. Introns are defined as sequences that remain physically separated after excision. Exons are defined as sequences that are ligated together after excision. Some introns encode snoRNA and miRNA, and some exons do not encode a functional product.

Group I introns, such as the one in the *Tetrahymena* pre-rRNA, fold into a conserved tertiary structure and self-splice *in vitro* in the absence of protein cofactors. The reaction begins with an attack by an external guanine nucleotide on the 5' splice site, adding the G to the 5' end of the intron, and releasing the first exon. In the second transesterification step, the first exon attacks the 3' splice site, ligating the two exons together, and releasing the linear intron. RNAs containing group II introns fold into a conserved tertiary structure and self-splice via an A-branched lariat intermediate with a novel 2'–5'-phosphodiester bond. The first of two transesterification steps is initiated by the 2'-OH of the internal bulged A. Some group I and group II introns are mobile and can move into an allele that lacks the intron. Movement is mediated by highly site-specific homing endonucleases that are typically encoded by the self-splicing intron.

Archael introns are removed by a cut-and-rejoin mechanism mediated by an endoribonuclease that recognizes a bulge-helix-bulge motif at the exon–intron junction. If a tRNA precursor has an intron, it is always in the same place, one nucleotide to the 3' side of the anticodon. The general mechanism of tRNA splicing involves two main reactions. First, the intron containing pre-tRNAs is cleaved by an endoribonuclease to generate paired tRNA halves. The paired halves are then joined by tRNA ligase.

Eukaryotic pre-mRNA is covalently processed in three ways prior to export from the nucleus: transcripts are capped at their 5' ends with a methylated guanosine nucleotide, introns are removed by splicing, and 3' ends are cleaved and extended with a poly(A) tail. All three processing events are coupled with transcription and mediated by the RNA polymerase II CTD. Capping, polyadenylation, and splicing proteins all associate with the CTD during transcription.

The 5'-7-methylguanosine cap is made in steps: First, an RNA triphosphate removes the terminal phosphate from pre-mRNA; next a guanylyl transferase adds the capping GMP from GTP. Next, two methyl transferases methylate the N-7 of the capping guanosine and the 2′-O-methyl group of the penultimate nucleotide. The cap protects the mRNA from degradation and enhances the efficiency of splicing, nuclear export, and translation.

Most eukaryotic mRNAs have a chain of A residues about 250 nt long at their 3′ ends. This poly(A) tail is added post-transcriptionally by poly(A) polymerase, and functions to enhance mRNA stability and translation efficiency. An efficient mammalian polyadenylation signal consists of an AAUAAA motif about 20 nt upstream of a polyadenylation site in pre-mRNA, follwed by a GU-rich motif. Transcription of eukaryotic genes extends beyond the polyadenylation site. The transcript is cleaved at the site of poly(A) addition, and is polyadenylated at the 3′ end created by the cleavage. Cleavage and polyadenylation in mammals require several proteins: CPSF, CstF, CFIm, CFIIm, poly(A) polymerase, and the RNA polymerase II CTD. In addition, cleavage also occurs at a cotranscriptional cleavage site which, at least in the case of β-globin mRNA, is self-cleaving. The remaining transcript still associated with the polymerase is attacked by an exonuclease that chases after the polymerase; when it catches up, transcription is terminated.

Nuclear pre-mRNA splicing takes place on a dynamic RNP particle called the spliceosome. The spliceosome contains five different snRNAs and more than 200 additional proteins. snRNAs exist in the cell associated with proteins in complexes called snRNPs, designated U1, U2, U4, U5, and U6. The five snRNPs all contain a common set of seven Sm or Sm-like proteins and several other proteins specific to each snRNP. The Sm proteins form a doughnut-shaped structure that binds to a U-rich element in the snRNA.

The spliceosome cycle includes the cotranscriptional assembly, splicing activity, and disassembly of the spliceosome. Experimental evidence suggests that spliceosome assembly proceeds "stepwise" through a series of short-lived intermediate stages, although an alternative penta-snRNP model proposes that intron removal is mediated by pre-existing complexes. Assembly begins with the binding of U1 snRNP to the conserved 5′ GU splice site of the pre-mRNA to form a commitment complex. U2 then binds to the branch point, the polypyrimidine tract, and the 3′ AG. The U1–U2 complex is joined by the U4–U6–U5 tri-snRNP. When U6 dissociates from U4, it displaces U1 at the 5′ splice site. This activates the spliceosome and allows U1 and U4 to be released. Splicing proceeds by two transesterification reactions via a lariat-shaped intermediate as in group II intron splicing. There is strong evidence that U2 and U6 snRNAs contribute to the catalysis of pre-mRNA splicing. Dynamic remodeling of the spliceosomal RNA complex requires DEXH/D box RNA helicases. SC35 and ASF/SF2 are important splicing factors that have a dual role in stimulating constitutive and regulated splicing. A minor class of introns, with variant but highly conserved 5′ splice sites and branch points, can be spliced with the help of variant classes of snRNAs, including U11, U12, U4atac, and U6atac.

Alternative splicing provides a versatile means of regulating gene expression. The splicing of some exons is regulated; they are either included or excluded from the mature mRNA. Particular mRNAs often have multiple positions of alternative splicing, giving rise to a family of related proteins with different functions from a single gene. Alternative splicing can introduce premature termination codons, or modular domains such as a nuclear localization sequence. mRNA variability also arises from alternative splicing within untranslated regions that contain elements that regulate translation, stability, or localization of the mRNA. Alternative splicing is regulated by *cis*-acting exonic splicing enhancers and exonic splicing silencers. Changes in splice site selection arise from changes in the binding of the intitial factors to the pre-mRNA, and in the assembly of the spliceosome.

In most pre-mRNAs, the exons are joined only within an individual pre-mRNA. However, in rare cases, an exon from one pre-mRNA can join to an exon from another pre-mRNA by *trans*-splicing. Discontinuous group II *trans*-splicing joins coding sequences from separate transcripts to form a complete open reading frame. In spliced leader *trans*-splicing, a short sequence is added to the 5′ termini of mRNAs. This leader exon functions to provide a 5′ cap structure, to resolve polycistronic transcripts into individual capped RNAs, or to enhance mRNA translational efficiency. In *Nanoarchaeum equitans*, nine genes encode separate tRNA halves which are then joined by *trans*-splicing.

RNA editing is the post-transcriptional modification of the base sequence of mRNA. Trypanosome mitochondrial transcripts undergo "pan-editing" events involving hundreds of U insertions or deletions that create open reading frames. Editing is mediated by multiple short guide RNAs (gRNAs) that specify the edited sequence, and is catalyzed by a multiprotein complex called the editosome. An anchor duplex is formed between the pre-mRNA and the gRNA by complementary base pairing near the mismatch. An endoribonuclease cleaves the pre-mRNA. At the site of cleavage, a U is inserted by a 3′ terminal uridylyl transferase, or removed by a U-specific exoribonuclease. The RNA ends are joined by RNA ligase and the cycle is repeated until the RNA is fully edited.

In mammals there are two main classes of editing enzymes, adenosine deaminases and cytidine deaminases, that catalyze A → I and C → U editing, respectively. A → I editing occurs frequently and altered editing patterns are associated with a number of neurological disorders, including amyotrophic lateral sclerosis. C → U editing has only been identified in apolipoprotein B and neurofibromatosis type 1 transcripts. Editing generates an mRNA encoding a short isoform of the apolipoprotein B protein in the small intestine by a C → U event that converts a glutamine codon (CAA) to a stop codon (UAA). Site-specific deamination is mediated by an editing complex with a catalytic subunit, APOBEC1, and an RNA-binding recognition subunit, ACF.

Small nucleolar RNAs mediate base modification of ribosomal RNA and possibly other RNAs. Box C/D snoRNAs guide 2′-O-methylation and H/ACA snoRNAs direct pseudouridine formation. By complementary base pairing with the rRNA, the snoRNAs recruit the modification enzymes to their target sites.

Small noncoding microRNAs target mRNA for degradation or translational inhibition. They are derived from unique genes that specify the silencing of very different genes. In contrast, siRNAs (see Section 9.7) are derived from viruses, transposable elements, repetitive DNA, or exogenous genes and guide silencing of the same (or a very similar) genetic locus from which they originate. miRNAs are processed from gene transcripts that form local hairpin structures. The primary miRNA is cleaved by the endoribonuclease Drosha to form shorter hairpins. This pre-miRNA is exported to the cytoplasm where it is cleaved by Dicer into a ~22 bp duplex miRNA. The short miRNA duplex is unwound by a helicase and one strand is incorporated into the RNA-induced silencing complex (RISC). RISC targets protein-coding mRNAs for degradation or blocks their translation. Degradation by cleavage is the outcome for targets that have the most complementarity. Since the discovery of *lin-4* in *C. elegans*, hundreds of miRNAs have been found. The function of the large majority of miRNAs remains unclear, but specific miRNAs contribute to the regulation of apoptosis, development timing, tissue-specific gene expression, and cell cycle control, and may mediate antiviral defense.

The short-lived processing intermediates of rRNAs, snoRNAs, and snRNAs are degraded by the nuclear exosome, a multisubunit complex that contains several exoribonucleases. In addition, there are specific nuclear decay pathways that destroy defective pre-mRNAs or those failing to form export-competent mRNPs. The TRAMP complex adds a short poly(A) tail to target molecules before the recruitment and activation of the exosome. TREX and SR proteins recruit general nuclear export factors, and contribute to the ability of the export machinery to discriminate between spliced and unspliced mRNPs. Further quality control surveillance and retention of unspliced transcripts occurs at the nuclear pore complex.

Once in the cytoplasm, there are several possible fates for an mRNP. Some mRNAs are stored in a translationally silent state, often mediated by deadenylation, until a particular stage of development. Most mRNAs are immediately translated and later degraded by general decay pathways. The 5′ cap is removed by decapping enzymes followed by exonuclease digestion. Alternatively, the 3′ poly(A) tail is removed by deadenylases and the rest of the mRNA is degraded by the exosome. Finally, mRNAs that contain a premature termination codon are rapidly degraded by a process called nonsense-mediated mRNA decay. Exon junction complexes (EJCs) are assembled near exon–exon boundaries after mRNA splicing. When the first ribosome begins translating an mRNA, the EJCs are normally displaced. If a premature termination codon is present, this activates a surveillance complex. The surveillance complex interacts with the prematurely terminating ribosome and targets the mRNA for degradation.

Analytical questions

1 You are investigating a gene with two large introns (introns 1 and 2) and three short exons (exons A, B, and C). Show the results of R-looping experiments performed with:

   (a) mRNA with exons A, B, and C, and single-stranded DNA.

   (b) pre-mRNA and single-stranded DNA.

   (c) mRNA with exons A and C only, and single-stranded DNA.

2 Sequence analysis of a novel cloned gene reveals that the exons are poorly conserved and the introns are highly conserved. What features would you look for in the RNA that might help in determining the possible function of the intronic RNA?

3 You have developed an *in vitro* assay for both splicing and polyadenylation. You generate *in vitro* radioactively labeled pre-mRNA substrates that either include a 5′ cap or lack a 5′ cap. You incubate these labeled pre-mRNA substrates with mammalian cell nuclear extract and run out the products on a sequencing gel. You then distinguish the products based on their relative size in the gel. You get the following results where the number of pluses is related to the relative amount of radioactivity found in that band on the gel:

	Pre-mRNA	Spliced RNA	Poly(A)
RNA uncapped	+++	+	+
RNA capped	+	+++	+++

Propose a hypothesis that explains these results.

4 You have discovered a minor class of rare introns that use a distinct splicing machinery involving snRNPs U110, U120, and U140. An experiment was performed to determine the order of assembly. A pre-mRNA was attached to agarose beads and incubated with purified splicing components. At time points from 0 to 20 minutes, RNA was extracted from the beads and analyzed by Northern blot with labeled probes specific for each RNA. In what order are the snRNPs assembled?

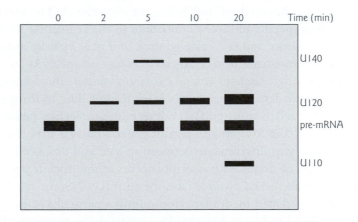

5 You determine that a single gene encodes a longer protein in embryos and a shorter protein in the adult organism. Describe experiments to determine whether this results from alternative splicing or RNA editing. Provide sample positive results.

6 You have cloned what you thought was the complete sequence of a gene. However, closer scrutiny reveals that the sequence lacks an entire exon normally found in other organisms. Propose a hypothesis that explains these results.

7 The results of the experiment shown in Fig. 13.30 suggest that miRNAs play a role in the tissue-specific suppression of certain genes during development. Propose an experiment to follow up on these studies. For example, when does suppression by miR-1 and miR-124 occur naturally?

8 What is the likely fate of an RNA transcript if:

   (a) Alternative splicing produces a premature termination codon?

   (b) There is a mutation in a splice site that results in unspliced RNA?

Suggestions for further reading

Ambros, V. (2004) The functions of animal microRNAs. *Nature* 431:350–355.

Azubel, M., Wolf, S.G., Sperling, J., Sperling, R. (2004) Three-dimensional structure of the native spliceosome by cryo-electron microscopy. *Molecular Cell* 15:833–839.

Bachellerie, J.-P., Cavaillé, J., Hüttenhofer, A. (2002) The expanding snoRNA world. *Biochimie* 84:775–790.

Baehr, W., Chen, C.K. (2001) RP11 and RP13: unexpected gene loci. *Trends in Molecular Medicine* 7:484–486.

Bagga, S., Bracht, J., Hunter, S., Massirer, K., Holtz, J., Eachus, R., Pasquinelli, A.E. (2005) Regulation by *let-7* and *lin-4* miRNAs results in target mRNA degradation. *Cell* 122:553–563.

Baker, K.E., Parker, R. (2004) Nonsense-mediated mRNA decay: terminating erroneous gene expression. *Current Opinion in Cell Biology* 16:293–299.

Bartel, D.P. (2004) MicroRNAs: genomics, biogenesis, mechanism, and function. *Cell* 116:281–297.

Black, D.L. (2003) Mechanisms of alternative pre-messenger RNA splicing. *Annual Reviews in Biochemistry* 72:291–336.

Bortolin, M., Kiss, T. (1998) Human U19 intron-encoded snoRNA is processed from a long primary transcript that possesses little potential for protein coding. *RNA* 4:445–454.

Brecht, M., Niemann, M., Schlüter, E., Müller, U.F., Stuart, K., Göringer, H.U. (2005) TbMP42, a protein component of the RNA editing complex in African trypanosomes, has endo-exoribonuclease activity. *Molecular Cell* 17:621–630.

Butcher, S.E., Brow, D.A. (2005) Towards understanding the catalytic core structure of the spliceosome. *Biochemical Society Transactions* 33:447–449.

Chambon, P. (1981) Split genes. *Scientific American* 244:60–71.

Conti, E., Izaurralde, E. (2005) Nonsense-mediated mRNA decay. Molecular insights and mechanistic variations across species. *Current Opinion in Cell Biology* 17:316–325.

Cullen, B.R. (2004) Transcription and processing of human microRNA precursors. *Molecular Cell* 16:861–865.

Davidson, N.O. (2002) The challenge of target sequence specificity in C → U RNA editing. *Journal of Clinical Investigation* 109:291–294.

Davies, M.S., Wallis, S.C., Driscoll, D.M., Wynne, J.K., Williams, G.W., Powell, L.M., Scott, J. (1989). Sequence requirements for apolipoprotein B RNA editing in transfected rat hepatoma cells. *Journal of Biological Chemistry* 264:13395–13398.

Demir, E., Dickson, B.J. (2005) *fruitless* splicing specifies male courtship behavior in *Drosophila*. *Cell* 121:785–794.

Driscoll, D.M., Casanova, E. (1990) Characterization of the apolipoprotein B mRNA editing activity in enterocyte extracts. *Journal of Biological Chemistry* 265:21401–21403.

Eisenberg, E., Nemzer, S., Kinar, Y., Sorek, R., Rechavi, G., Levanon, E.Y. (2005) Is abundant A-to-I editing primate-specific? *Trends in Genetics* 21:77–81.

Englert, M., Beier, H. (2005) Plant tRNA ligases are multifunctional enzymes that have diverged in sequence and substrate specificity from RNA ligases of other phylogenetic origins. *Nucleic Acids Research* 33:388–399.

Ernst, N.L., Panicucci, B., Igo, R.P., Jr., Panigrahi, A.K., Salavati, R., Stuart, K. (2003) TbMP57 is a 3′ terminal uridylyl transferase (TUTase) of the *Trypanosoma brucei* editosome. *Molecular Cell* 11:1525–1536.

Feagin, J.E., Abraham, J.M., Stuart, K. (1988) Extensive editing of the cytochrome c oxidase III transcript in *Trypanosoma brucei*. *Cell* 53:413–422.

Fong, N., Bentley, D.L. (2001) Capping, splicing, and 3′ processing are independently stimulated by RNA polymerase II: different functions for different segments of the CTD. *Genes and Development* 15:1783–1795.

Gall, J.G. (2003) The centennial of the Cajal body. *Nature Reviews Molecular and Cellular Biology* 4:975–980.

Gesteland, R.F., Cech, T.R., Atkins, J.F. (1999) *The RNA World*, 2nd edn. Cold Spring Harbor Laboratory Press, Cold Spring Harbor, NY.

Giraldez, A.J., Cinalli, R.M., Glasner, M.E. et al. (2005) MicroRNAs regulate brain morphogenesis in zebrafish. *Science* 308:833–838.

Görnemann, J., Kotovic, K.M., Hujer, K., Neugebauer, K.M. (2005) Cotranscriptional spliceosome assembly occurs in a stepwise fashion and requires the cap binding complex. *Molecular Cell* 19:53–63.

Grainger, R.J., Beggs, J.D. (2005) Prp8 protein: at the heart of the spliceosome. *RNA* 11:533–557.

Halbig, K., De Nova-Ocampo, M., Cruz-Reyes, J. (2004) Complete cycles of bloodstream trypanosome RNA editing in vitro. *RNA* 10:914–920.

Hastings, K.E.M. (2005) SL trans-splicing: easy come or easy go? *Trends in Genetics* 21:240–246.

Haugen, P., Simon, D.M., Bhattacharya, D. (2005) The natural history of group I introns. *Trends in Genetics* 21:111–119.

Huang, Y., Steitz, J.A. (2005) SRprises along a messenger's journey. *Molecular Cell* 17:613–615.

Johnson, J.M., Castle, J., Garrett-Engele, P. et al. (2003) Genome-wide survey of human alternative pre-mRNA splicing with exon junction microarrays. *Science* 302:2141–2144.

Koslowsky, D.J. (2004) A historical perspective on RNA editing. How the peculiar and bizarre became mainstream. *Methods in Molecular Biology* 265:161–197.

Kwak, S., Kawahara, Y. (2005) Deficient RNA editing of GluR2 and neuronal death in amyotropic lateral sclerosis. *Journal of Molecular Medicine* 83:110–120.

Lacadie, S.A., Rosbash, M. (2005) Cotranscriptional spliceosome assembly dynamics and the role of U1 snRNA:5′ss base pairing in yeast. *Molecular Cell* 19:65–75.

LaCava, J., Houseley, J., Saveanu, C., Petfalski, E., Thompson, E., Jacquier, A., Tollervey, D. (2005) RNA degradation by the exosome is promoted by a nuclear polyadenylation complex. *Cell* 121:713–724.

Lambowitz, A.M., Zimmerly, S. (2004) Mobile group II introns. *Annual Review of Genetics* 38:1–35.

Lecellier, C.H., Dunoyer, P., Arar, K. et al. (2005) A cellular microRNA mediates antiviral defense in human cells. *Science* 308:557–560.

Lehmann, K., Schmidt, U. (2003) Group II introns: structure and catalytic versatility of large natural ribozymes. *Critical Reviews in Biochemistry and Molecular Biology* 38:249–303.

Li, W., Guan, K.L. (2004) The Down syndrome cell adhesion molecule (DSCAM) interacts with and activates Pak. *Journal of Biological Chemistry* 279:32824–32831.

Lim, L.P., Lau, N.C., Garrett-Engele, P. et al. (2005) Microarray analysis shows that some microRNAs downregulate large numbers of target mRNAs. *Nature* 433:769–773.

Lykke-Andersen, J., Aagaard, C., Semionenkov, M., Garrett, R.A. (1997) Archael introns: splicing, intercellular mobility and evolution. *Trends in Biochemical Sciences* 22:326–331.

Macbeth, M.R., Schubert, H.L., VanDemark, A.P., Lingam, A.T., Hill, C.P., Bass, B.R. (2005) Inositol hexakisphosphate is bound in the ADAR2 core and required for RNA editing. *Science* 309:1534–1539.

Masuda, S., Das, R., Cheng, H., Hurt, E., Dorman, N., Reed, R. (2005) Recruitment of the human TREX complex to mRNA during splicing. *Genes and Development* 19:1512–1517.

Moore, M. (2005) From birth to death: the complex lives of eukaryotic mRNAs. *Science* 309:1514–1518.

Moss, E.G., Lee, R.C., Ambros, V. (1997) The cold shock domain protein LIN-28 controls developmental timing in *C. elegans* and is regulated by the *lin-4* RNA. *Cell* 88:637–646.

Narayanan, U., Achsel, T., Lührmann, R., Matera, A.G. (2004) Coupled *in vitro* import of U snRNPs and SMN, the spinal muscular atrophy protein. *Molecular Cell* 16:223–234.

Pan, Q., Shai, O., Misquitta, C. et al. (2004) Revealing global regulatory features of mammalian alternative splicing using a quantitative microarray platform. *Molecular Cell* 16:929–941.

Patel, A.A., Steitz, J.A. (2003) Splicing double: insights from the second spliceosome. *Nature Reviews Molecular Cell Biology* 4:960–970.

Randau, L., Münch, R., Hohn, M.J., Jahn, D., Söll, D. (2005) *Nanoarchaeum equitans* creates functional tRNAs from separate genes for their 5′- and 3′-halves. *Nature* 433:537–541.

Reed, R., Cheng, H. (2005) TREX, SR proteins and export of mRNA. *Current Opinion in Cell Biology* 17:269–273.

Ruby, S.W., Abelson, J. (1988) An early hierarchic role of U1 small nuclear ribonucleoprotein in spliceosome assembly. *Science* 242:1028–1035.

Rusché, L.N., Cruz-Reyes, J., Piller, K.J., Sollner-Webb, B. (1997) Purification of a functional enzymatic editing complex from *Trypanosoma brucei* mitochondria. *EMBO Journal* 16:4069–4081.

Saguez, C., Olesen, J.R., Jensen, T.H. (2005) Formation of export-competent mRNP: escaping nuclear destruction. *Current Opinion in Cell Biology* 17:287–293.

Schwerk, C., Schulze-Osthoff, K. (2005) Regulation of apoptosis by alternative pre-mRNA splicing. *Molecular Cell* 19:1–13.

Sharp, P.A. (1994) Split genes and RNA splicing (Nobel lecture). *Cell* 77:805–815.

Shen, H., Green, M.R. (2004) A pathway of sequential arginine-serine-rich domain splicing signal interactions during mammalian spliceosome assembly. *Molecular Cell* 16:363–373.

Sontheimer, E.J., Carthew, R.W. (2005) Silence from within: endogenous siRNAs and miRNAs. *Cell* 122:9–12.

Stevens, S.W., Ryan, D.E., Ge, H.Y., Moore, R.E., Young, M.K., Lee, T.D., Abelson, J. (2002) Composition and functional characterization of the yeast spliceosomal penta-snRNP. *Molecular Cell* 9:31–44.

Stuart, K.D., Schnaufer, A., Ernst, N.L., Panigrahi, A.K. (2004) Complex management: RNA editing in trypanosomes. *Trends in Biochemical Sciences* 30:96–106.

Tarn, W.-Y., Steitz, J.A. (1996) A novel spliceosome containing U11, U12, and U5 snRNPs excises a minor class (AT-AC) intron *in vitro*. *Cell* 84:801–811.

Teixeira, A., Tahiri-Alaoui, A., West, S. et al. (2004) Autocatalytic RNA cleavage in the human β-globin pre-mRNA promotes transcription termination. *Nature* 432:526–530.

Tollervey, D. 2004. Termination by torpedo. *Nature* 432:456–457.

Tycowski, K.T., Shu, M.-D., Steitz., J.A. (1996) A mammalian gene with introns instead of exons generating stable RNA products. *Nature* 379:464–466.

Valadkhan, S., Manley, J.L. (2003) Characterization of the catalytic activity of U2 and U6 snRNAs. *RNA* 9:892–904.

Vinciguerra, P., Stutz, F. (2004) mRNA export: an assembly line from genes to nuclear pores. *Current Opinion in Cell Biology* 16:285–292.

Watanabe, Y., Yokobori, S., Inaba, T. et al. (2002) Introns in protein-coding genes in Archaea. *FEBS Letters* 510:27–30.

Watson, F.L., Püttmann-Holgado, R., Thomas, F. et al. (2005) Extensive diversity of Ig-superfamily proteins in the immune system of insects. *Science* 309:1874–1878.

Wissinger, B., Brennicke, A., Schuster, W. (1992) Regenerating good sense: RNA editing and trans-splicing in plant mitochondria. *Trends in Genetics* 8:322–328.

Wojtowicz, W.M., Flanagan, J.J., Millard, S.S., Zipursky, S.L., Clemens, J.C. (2004) Alternative splicing of *Drosophila* Dscam generates axon guidance receptors that exhibit isoform-specific homophilic binding. *Cell* 118:619–633.

Wyers, F., Rougemaille, M., Badis, G. et al. (2005) Cryptic pol II transcripts are degraded by a nuclear quality control pathway involving a new poly(A) polymerase. *Cell* 121:725–737.

Xu, X., Yang, D., Ding, J.H. et al. (2005) ASF/SF2-regulated CaMKIIδ alternative splicing temporally reprograms excitation–contraction coupling in cardiac muscle. *Cell* 120:59–72.

Yong, J., Pellizzoni, L., Dreyfuss, G. (2002) Sequence-specific interaction of U1 snRNA with the SMN complex. *EMBO Journal* 21:1188–1196.

Zorio, D.A.R., Bentley, D.L. (2004) The link between mRNA processing and transcription: communication works both ways. *Experimental Cell Research* 296:91–97.

Chapter 14

The mechanism of translation

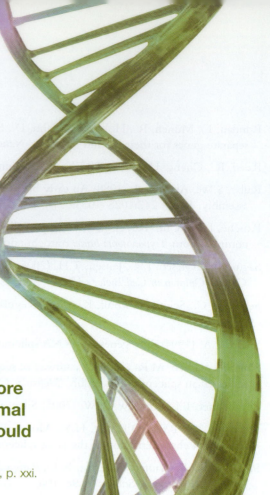

We can do experiments on rRNAs today that are far more powerful than anything we ever attempted with ribosomal proteins in the past. So today we work on rRNA; we would be crazy not to.

Peter B. Moore, *The Ribosome. Structure, Function and Evolution* (1990), p. xxi.

Outline

14.1 Introduction

This chapter covers the mechanics of protein synthesis. The focus is on eukaryotic translation, but references are made to translation in prokaryotes for comparison and where details are better understood. Full appreciation of the process of translation requires an understanding of the intricate machinery involved in deciphering the genetic code. Before reaching the key step where a peptide bond forms between two amino acids, many events must take place. Ribosomes must be assembled in the nucleolus of the cell from a diversity of different gene products. Transfer RNAs (tRNAs) must be "charged" with their appropriate amino acid, and all the players must join together in the cytoplasm. Translational control provides yet another level of gene regulation in eukaryotes.

14.2 Ribosome structure and assembly

Major advances were made in the ribosome field following the development of cell fractionation techniques, such as ultracentrifugation, and electron microsopy. By the early 1940s researchers had demonstrated an RNA-rich, particulate fraction from cytoplasmic extracts called "microsomes." Physical studies in the mid-1950s characterized these microsomes as ribonucleoprotein (RNP) particles that are roughly half protein and half RNA. They were shown to be composed of two nonequivalent subunits that form a more or less spherical complex, with diameters of approximately 250 Å (25 nm) and molecular weights of several million (Fig. 14.1). The word "ribosome" was proposed by Dick Roberts in 1958, to replace the many different phrases used to designate these large RNP particles at the time. In his introduction to the volume of a symposium on microsomal particles and protein synthesis he wrote: "This seems a satisfactory name, and it has a pleasant sound." By around 1966, the role of the ribosome in protein synthesis was well established, and much of the biochemistry of translation had been worked out. In essence, the ribosome is an enzyme, a polypeptide polymerase. As we shall see, however, important new observations continue to be made and the field remains as vibrant as ever.

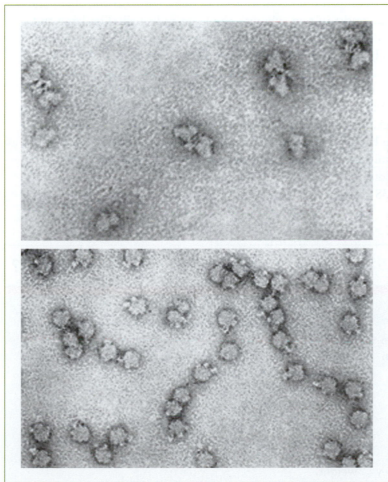

Figure 14.1 Electron micrographs of *E. coli* small ribosomal subunits (top) and large ribosomal subunits (bottom). Small subunits are connected by antibodies against small subunit ribosomal protein S6. These are attached to the platform of the small subunit. Between 1974 and 1990, the locations of 18 (of 21) small subunit proteins, five regions of 16S rRNAs, five large subunit proteins, and the nascent chain exit site were mapped using immuno- and DNA hybridization-electron microscopy, and subsequently confirmed by the high resolution X-ray structures. (Electron micrographs courtesy of James Lake, University of California, Los Angeles.)

Table 14.1 Components of ribosomes.

Ribosomes	Subunits	Ribosomal RNAs	Ribosomal proteins
Bacterial (*E. coli*)			
70S **Mass 2.5 × 10⁶ Da** **66% RNA**	50S (large)	23S (2904 nt) 5S (120 nt)	32 (L1, L2, L3, etc.)
	30S (small)	16S (1542 nt)	21 (S1, S2, S3, etc.)
Mammalian			
80S **Mass 4.2 × 10⁶ Da** **60% RNA**	60S (large)	28S (4718 nt) 5.8S (160 nt) 5S (120 nt)	49 (L1, L2, L3, etc.)
	40S (small)	18S (1874 nt)	33 (S1, S2, S3, etc.)

Structure of ribosomes

The structure of ribosomes has been determined through a combination of X-ray crystallography and cryoelection microscopy analyses, along with biochemical and genetic data. These studies have revealed intricate folding patterns of ribosomal RNAs (rRNAs) and protein that join together to form a remarkable molecular machine. An overview of the basic structure of the ribosome is provided here, with structural details highlighted in Section 14.5. Table 14.1 compares the composition of eukaryotic (mammalian) and prokaryotic (bacterial) ribosomes. Ribosomes consist of two subunits, the large and small. The large and small subunits of the bacterial ribosomes are known as the 50S and 30S subunits, respectively, from their rates of sedimentation in a centrifuge (Focus box 14.1). The intact ribosome is referred to as the 70S ribosome. The 70S ribosome is less than the sum of the 50S and 30S subunits because sedimentation velocity is not strictly a measure of mass; it is determined by both shape and size. The large and small subunits are composed of rRNA and many ribosomal proteins (Table 14.1). In bacteria, the 50S subunit is assembled from a 23S rRNA and a 5S rRNA, whereas the 30S subunit contains a single 16S rRNA. Eukaryotic ribosomes are slightly larger than bacterial ribosomes overall; the intact ribosome is referred to as the 80S ribosome. The large and small subunits of mammalian ribosomes are known as 60S and 40S, respectively. The 60S subunit includes an additional 5.8S rRNA. The sequence of the 5.8S rRNA corresponds to the 3′ end of the bacterial 23S rRNA.

FOCUS BOX 14.1   **What is "S"?**

Ultracentrifugation was developed in the 1920s by the physical chemist Theodor Svedberg. The term "S" refers to a Svedberg unit. One Svedberg unit equals a sedimentation coefficient of 10^{-13} seconds; i.e. the rate of sedimentation during density-gradient centrifugation in sucrose. The technique gives information on the size and shape of a molecule, while isolating it in native form. The gradient is preformed by layering of decreasingly concentrated sucrose solutions in a tube. When sedimenting particles are subjected to centrifugal force in an ultracentrifuge, they will differ in sedimentation velocity depending on particle mass and shape. Centrifugation is stopped after a fixed length of time. Samples are analyzed by punching a hole in the bottom of the tube and collecting drops in sequentially numbered tubes. The first fraction will have the heaviest molecular weight particles (highest S values). S increases with particle mass, but the increase is not strictly linear or simply due to the frictional coefficient, which also depends on shape. So, S is only a rough estimate of molecular weight.

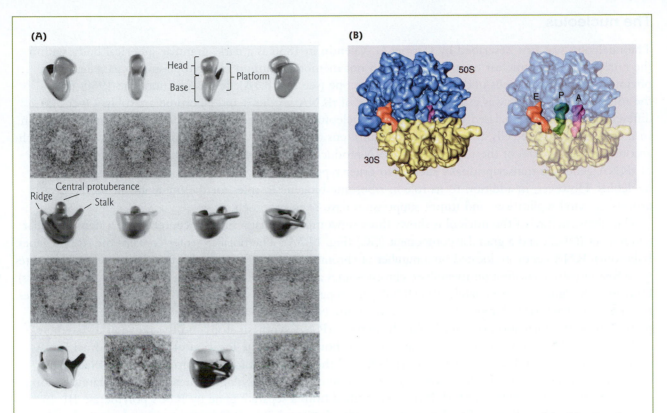

Figure 14.2 Three-dimensional models of the ribosome. (A) Classic three dimensional model of the *E. coli* ribosome showing the asymmetric shape of the two ribosomal subunits. Electron micrographs are compared with the model. Small subunits are shown in the top two rows, large subunits in the next two rows, and monomeric ribosomes in the last row. The small subunit includes a head, base, and platform. The larger subunit includes a central protuberance, ridge, and stalk. The subunit structures, their relative orientations in the complete ribosome, and their enantiomorphs (mirror images), at about 35 Å resolution, were all subsequently confirmed by high resolution X-ray structures. (Images courtesy of James Lake, University of California, Los Angeles. Adapted with permission of the author from Lake, J.A. 1976. Ribosome structure determined by electron microscopy of *Escherichia coli* small subunits, large subunits, and monomeric ribosomes. *Journal of Molecular Biology* 105:131–159. (B) Representation of the 70S *E. coli* ribosome from cryoelectron microscopy studies. On the left, the map is shown in solid. On the right is a semitransparent representation of the large (50S) and small (30S) subunits showing the location of tRNAs in the E (exit) site, P (peptidyl) site, and A (acceptor) site. (Reprinted from Valle, M., Zavialov, A., Sengupta, J., Rawat, U., Ehrenberg, M., Frank, J. 2003. Locking and unlocking ribosomal motions. *Cell* 114:123–134. Copyright © 2003, with permission from Elsevier.)

The ribosome acts in concert with a variety of smaller factors that help to orchestrate the process of protein synthesis. The two main functions, however, are carried out by the ribosome itself: decoding the genetic code in the messenger RNA (mRNA) and catalyzing the formation of peptide bonds between amino acids resulting in a polypeptide. The peptidyl transferase center, the catalytic site, is in the large subunit. The small subunit serves as the assembly guide for all the factors needed in protein synthesis. Decoding the mRNA occurs on the small subunit. There are three tRNA-binding sites on the ribosome, the A (acceptor), P (peptidyl), and E (exit) sites, each of which is occupied in succession by a particular tRNA during the protein synthesis cycle (Fig. 14.2). The tRNAs bridge the large and small subunits, with the anticodon arm of the tRNA pointing towards the small subunit for decoding and the acceptor arm of the tRNA pointing into the large subunit for peptidyl transferase.

Ribosomes function in the cytoplasm, but their assembly occurs in the nucleus, within a special subcompartment called the nucleolus. The structure of the nucleolus and the orchestration of ribosome biogenesis are described in the following sections.

The nucleolus

The nucleolus was first described by Gabriel G. Valentin in 1836 as a dark, granular spherical body within the nucleus. This subnuclear organelle, although not membrane-bound like cytoplasmic organelles, is an obvious component of cells under the light microscope (see Fig. 3.1A). It was not until the 1960s that the role of the nucleolus was established as the site of rRNA synthesis and ribosome assembly. Over 500 different proteins are associated with the human nucleolus; 30% of these are currently of unknown function. In metabolically active cells, the nucleolus contains tens to hundreds of active ribosomal RNA genes, which account for about 50% of the total cellular RNA production. In a typical somatic cell a significant fraction of rRNA genes are transcriptionally silent. No other types of active genes have been identified in the nucleolus. However, in addition to its role in ribosome biogenesis, roles for the nucleolus in RNA processing, viral replication, and tumor suppression have been proposed.

The ultrastructure of the nucleolus shows three subcompartments: a fibrillar center (FC), a dense fibrillar component (DFC), and a granular component (GC) (Fig. 14.3A). The fibrillar center contains the rRNA genes. Ribosomal RNA genes are located on a number of chromosomes and form arrays of tandem repeats. These sites were first visually identified on metaphase chromosomes and were called nucleolus organizer regions (NORs). Human cells contain approximately 400 rRNA gene repeats within five pairs of NOR-bearing chromosomes. During interphase, rRNA repeats from more than one NOR-bearing chromosome cluster together and participate in the formation of a nucleolus. In mitosis, rRNA synthesis stops and the nucleoli disassemble.

The dense fibrillar component of the nucleolus is thought to be the location of active ribosomal RNA genes. It contains a high concentration of rRNA, and the first steps of rRNA processing take place here. With the exception of 5S rRNA, rRNA genes are transcribed by RNA polymerase I to generate one long precursor rRNA (pre-rRNA). 5S rRNA is transcribed from a separate gene by RNA polymerase III. In mammalian cells the 45S pre-rRNA contains the 18S, 5.8S, and 28S rRNA sequences and internal and external transcribed spacers which are cleaved during processing of pre-rRNA (Fig. 14.3B). An active rRNA gene has a "Christmas tree" structure and was first shown in spreads of nuclear contents from amphibian oocytes. The fibrillar center and dense fibrillar component are embedded in a mass of 15 nm granules comprising the granular component, which represents, at least in part, the preribosomal particles and is the site of later steps of pre-rRNA processing.

Ribosome biogenesis

Eukaryotic large and small ribosomal subunits are assembled in the nucleolus before export to the cytoplasm. The pathway of ribosome biogenesis in humans has not yet been defined, but its orchestration in the budding yeast *Saccharomyces cerevisiae* is well characterized. Ribosome assembly is a highly coordinated, dynamic process that requires synthesis, processing, and modification of pre-rRNAs, assembly with ribosomal proteins, and transient interaction of more than 150 nonribosomal factors with the maturing preribosomal particles. These include the small nucleolar RNPs (snoRNPs) (see Section 13.9) and numerous nonribosomal proteins that process, modify, and fold the pre-rRNAs (e.g. nucleases, pseudouridine synthases, methyltransferases, RNA helicases, and GTPases). Eukaryotic ribosomes contain more than 70 ribosomal proteins (r-proteins) and four different rRNA species (25S/28S, 18S, 5.8S, and 5S), and the coordinated assembly of all of these components depends on the action of all three RNA polymerases. Despite this amazing diversity of participants, the assembly is highly efficient. In a human cell, as many as 14,000 ribosomal subunits are assembled and leave the nucleoli per minute.

The earliest precursor particle in ribosome biogenesis is the 90S precursor. This complex was first identified in the early 1970s by the laboratories of John Warner and Rudi Planta. The 90S precursor includes a partially processed 35S rRNA synthesized by RNA polymerase I, 5S rRNA made by RNA polymerase III, and snoRNAs, ribosomal proteins, and other nonribosomal proteins made by RNA polymerase II (Fig. 14.4). Cleavage of the 35S pre-rRNA splits the 90S precursor into the 40S and 60S preribosomal subunits. After export into the cytoplasm via the nuclear pore complexes, the remaining nonribosomal factors dissociate from the mature 60S and 40S ribosomal subunits.

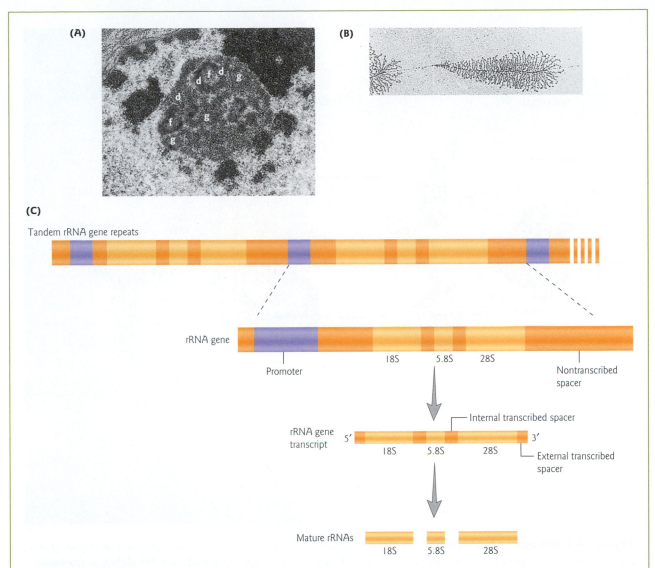

Figure 14.3 Ribosomal RNA gene transcription. (A) Electron micrograph of a nucleolus from a pancreatic acinar cell showing the fibrillar center (f), dense fibrillar component (d) and granular component (g). (Electron micrograph courtesy of Joe Scott, College of William and Mary.) (B) Spread Christmas tree structure of an active ribosomal RNA gene. (Electron micrograph courtesy of Aimee Hayes Bakken, University of Washington). (C) Schematic representation of human ribosomal RNA genes and transcripts.

In addition to the ribosome, other key players in the process of protein synthesis are the mRNA carrying the genetic information in the form of the genetic (triplet) code (see Section 5.3) and tRNA. The next section discusses the charging of tRNA with the appropriate amino acid.

14.3 Aminoacyl-tRNA synthetases

The overall fidelity of translation is dependent on the accuracy of two processes: codon–anticodon recognition and aminoacyl-tRNA synthesis. Aminoacyl-tRNAs are synthesized by the 3′-esterification of tRNAs with the appropriate amino acids. An "uncharged" tRNA is aminoacylated to generate a "charged" tRNA, which then interacts with a translation elongation factor. This high-fidelity reaction is catalyzed by a family of enzymes known as the aminoacyl-tRNA synthetases.

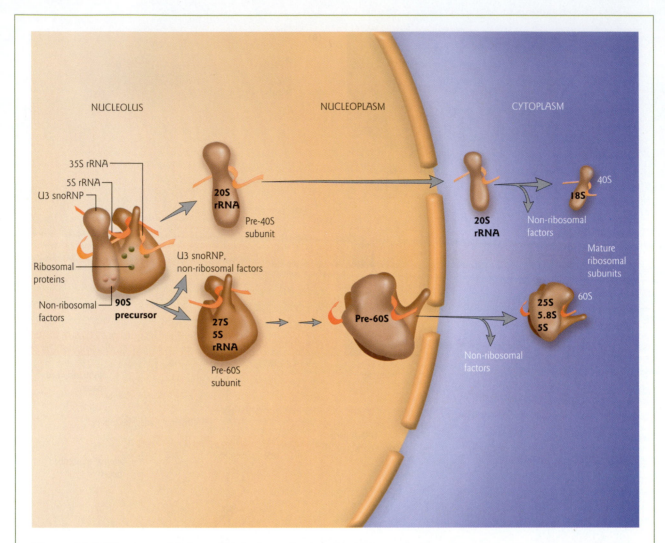

NUCLEOLUS   NUCLEOPLASM   CYTOPLASM

35S rRNA
5S rRNA
U3 snoRNP

20S
rRNA

Pre-40S
subunit

U3 snoRNP,
non-ribosomal factors

Ribosomal
proteins

Non-ribosomal
factors   90S
                 precursor

27S
5S
rRNA

Pre-60S
subunit

Pre-60S

40S
18S

20S   Non-ribosomal
rRNA   factors

Mature
ribosomal
subunits

25S   60S
5.8S
5S

Non-ribosomal
factors

Figure 14.4 Ribosome biogenesis. Pathway for assembly in the nucleolus, maturation, and nuclear export of 40S and 60S ribosomal subunits in yeast. In yeast the largest rRNA is 25S rRNA, compared with 28S rRNA in mammals. Multiple arrows indicate the presence of additional intermediates not shown for simplicity.

Aminoacyl-tRNA charging

All aminoacyl–tRNA synthetases attach an amino acid to a tRNA in two enzymatic steps (Fig. 14.5). In the first step, the amino acid reacts with ATP to become adenylylated (addition of AMP) and pyrophosphate is released. The amino acid is attached by a high-energy ester bond between the carbonyl group of the amino acid and the phosphoryl group of AMP. In the second step, AMP is released and the amino acid is transferred to the 3′ end of tRNA via the 2′-OH for class I enzymes and via the 3′-OH for class II enzymes.

Each member of the family is able to precisely match a particular amino acid with the tRNA containing the correct corresponding anticodon. Specific aminoacyl–tRNA synthetases are denoted by their three-letter amino acid designation, e.g. MetRS for methionyl–tRNA synthetase. Methionine tRNA or tRNA^{Met} indicates the uncharged tRNA specific for methionine, while methionyl–tRNA or Met–tRNA denotes tRNA aminoacylated with methionine. A special tRNA called tRNA$_{CUA}$, and a novel archael aminoacyl tRNA synthetase, PylS, are required for incorporation of pyrrolysine. PylS has been shown to attach pyrrolysine to tRNA$_{CUA}$ *in vitro* in a reaction specific for pyrrolysine (Fig. 14.6).

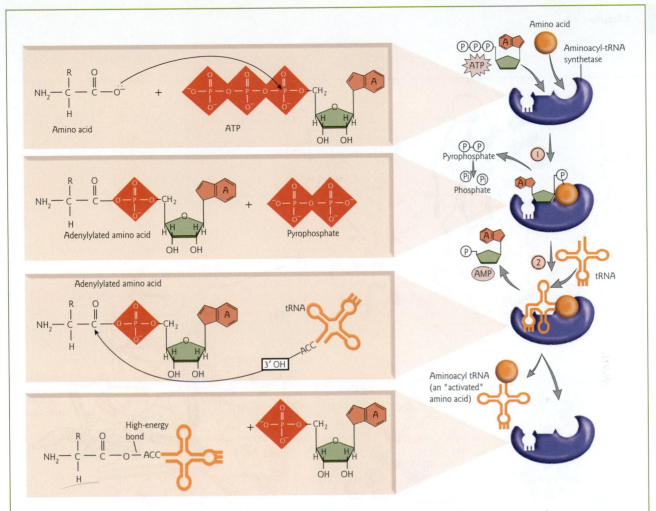

Figure 14.5 The two steps of aminoacyl-tRNA charging. (1) The amino acid and a molecule of ATP enter the active site of the enzyme. Simultaneously, ATP loses pyrophosphate, and the resulting AMP bonds covalently to the amino acid. The pyrophosphate is hydrolyzed to two P_is. (2) The tRNA forms a high-energy bond with the adenylylated amino acid, displacing the AMP. The aminoacyl-tRNA is then released from the enzyme. The process is shown for a class II aminoacyl-tRNA synthetase.

Figure 14.6 Charging of tRNA$_{CUA}$ with pyrrolysine. Aminoacylation of tRNA$_{CUA}$ in cellular tRNA pools was analyzed by Northern blotting with a labeled tRNA$_{CUA}$-specific probe. Both tRNA$_{CUA}$ and aminoacyl–tRNA$_{CUA}$ were present in cellular tRNA, so alkaline hydrolysis was used to deacylate the cellular charged species. Aminoacylation of tRNA$_{CUA}$ by recombinant His–tagged PylS (PylS-His6) was assayed. Lane 1 contains the alkali deacylated cellular tRNA pool with only uncharged tRNA$_{CUA}$; lane 2 contains cellular tRNA as isolated and shows the charged (upper band) and uncharged (lower band) tRNA$_{CUA}$. Each lane represents reactions with the following: 3, no amino acid; 4, pyrrolysine; 5, pyrrolysine; 6, lysine; 7, pyrrolysine plus lysine; 8, a mixture of the 20 canonical amino acids; 9, pyrrolysine plus a mixture of the 20 canonical amino acids; and 10, pyrrolysine but lacking PylS-His6. Sample groups 1–4 and 5–10 are from two different experiments. Only reactions that contained pyrrolysine and PylS–His6 resulted in the conversion of 50% of deacylated tRNA$_{CUA}$ to a species that migrated with the same electrophoretic mobility as the aminoacyl-tRNA$_{CUA}$ present in the extracted cellular tRNA pool. (Reprinted by permission from Nature Publishing Group and Macmillan Publishers Ltd: Blight, S.K., Larue, R.C., Mahapatra, A. et al. 2004. Direct charging of tRNA$_{CUA}$ with pyrrolysine *in vitro* and *in vivo*. *Nature* 431:333–335.)

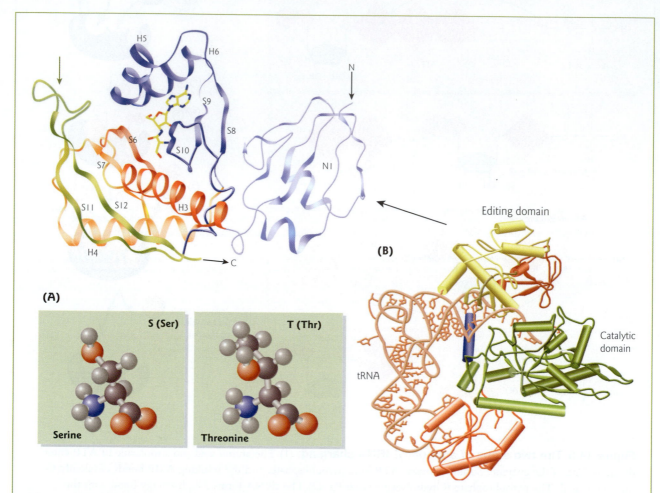

Figure 14.7 Editing domain of ThrRS. (A) Comparison of the spatial arrangement of atoms in the amino acids serine and threonine. (B) Overall view of one monomer of ThrRS interacting with tRNA^{Thr}, showing the catalytic and editing domains. (Protein Data Bank, PDB: 1QF6. Adapted from Sankaranarayanan, R., Dock-Bregeon, A.C., Romby, P. et al. 1999. The structure of threonyl-tRNA synthetase-tRNA (Thr) complex enlightens its repressor activity and reveals an essential zinc ion in the active site. *Cell* 97:371–381. Copyright © 1999, with permission from Elsevier.) (Inset) Three-dimensional structure of the N-terminal domains of *E. coli* ThrRS. The secondary structure elements of the editing domain N2 are labeled. The editing cleft, colored in red and blue is shown with the ligand, SerA76, a nonhydrolyzable analog of serine. The structure responsible for tRNA binding (β-strands S11 and S12) is in green. The green arrow indicates the hairpin interacting with the two first base pairs of the tRNA. (Protein Data Bank, PDB: 1TKG. Adapted from Dock-Bregeon, A.C., Rees, B., Torres-Larios, A., Bey, G., Caillet, J., Moras, D. 2004. Achieving error-free translation: the mechanism of proofreading of threonyl-tRNA synthetase at atomic resolution. *Molecular Cell* 16:375–386. Copyright © 2004, with permission from Elsevier.)

Proofreading activity of aminoacyl-tRNA synthetases

The aminoacyl–tRNA synthetases display an overall error rate of about one in 10,000. This very low frequency of errors is achieved by two means. First, the aminoacyl–tRNA synthetases make an intricate series of contacts with both their amino acid and tRNA in the enzyme active site, which ensures for the most part that only the correct substrates are selected from the large cellular pool of similar candidates (Fig. 14.7). Second, the enzymes possess a variety of proofreading (editing) activities that serve to hydrolyze the mismatched amino acid either before or after transfer to tRNA. Misactivation becomes a significant issue when the side chains of amino acids are very similar, as for threonine and serine (Fig. 14.7; see also

Fig. 5.3). Crystal structures of the editing domain of threonyl–tRNA synthetase have recently been solved in complexes with serine and with a nonhydrolyzable serine analog. These studies have provided a structural basis for proofreading. The proofreading mechanism involves water-mediated hydrolysis of the mischarged tRNA. The correct product, Thr-tRNA^{Thr}, is not hydrolyzed due to steric exclusion. The deep editing pocket where serine is bound and removed by hydrolysis is too narrow to accommodate threonine, whose methyl group would clash with the side chains of the amino acids at the bottom of the pocket.

14.4 Initiation of translation

The process of protein synthesis is fundamentally the same throughout bacteria and eukaryotes. The differences lie in the details of some of the steps and in the components used to accomplish each step (Table 14.2). Although the focus of this chapter is on eukaryotic protein synthesis, reference to bacterial protein synthesis is made for comparison and where details are better understood. Translation can be divided into three main stages: initiation, elongation, and termination. Each stage of protein synthesis involves multiple accessory factors and energy from GTP hydrolysis. Initiation is the most complex and the most tightly controlled of the steps in protein synthesis. In this slow, rate-limiting step, the ribosome is assembled at the initiation codon in the mRNA with a methionyl initiator tRNA bound in its peptidyl (P) site. The next step, elongation, occurs rapidly. Aminoacyl–tRNAs enter the acceptor (A) site where decoding takes place. If they are the correct (cognate) tRNAs, the ribosome catalyzes the formation of a peptide bond between the incoming amino acid and the growing polypeptide chain. After the tRNAs and mRNA are translocated such that the next codon is moved to the A site, the process is repeated. Termination takes place when a stop codon is encountered and the completed polypeptide chain is released from the ribosome. Finally, the ribosomal subunits are dissociated, releasing the mRNA and deacylated tRNA, leaving them ready for another round of initiation.

The initiation stage can be further subdivided into four steps: ternary complex formation and loading onto the 40S subunit, loading of the mRNA, scanning and start codon recognition, and joining of the 40S and 60S subunits to form the functional 80S ribosome.

Table 14.2 Some comparisons between prokaryotic and eukaryotic translation.

	Prokaryotes	Eukaryotes
Ribosome binding site on mRNA	Shine–Dalgarno sequence UAAGGAGG (near AUG, base pairs with 16S rRNA)	Kozak consensus sequence ACCAUGG, sequence context around AUG
Initiation codon (downstream of binding site)	AUG, GUG, UUG	AUG
Initiation amino acid	*N*-formyl methionine	Methionine
Initiation tRNA	tRNA$_i^{f\text{-Met}}$	tRNA$_i^{Met}$
Initiation factors	IF1, IF2, IF3	~13 eIF factors: eIF1, eIF1A, eIF2, eIF2B, eIF3, eIF4A, eIF4B, eIF4E, eIF4F, eIF4G, eIF4H, eIF5, eIF5B
Elongation factors	EF-Tu, EF-Ts, SelB	eEF1A, eEF1B, mSelB
Translocation factor	EF-G	eEF2
Release factors **Class I**	RF1 (UAA, UAG) RF2 (UGA, UAA)	eRF1 (UAG, UAA, UGA)
Class II	RF3 (GTPase)	eRF3 (GTPase)

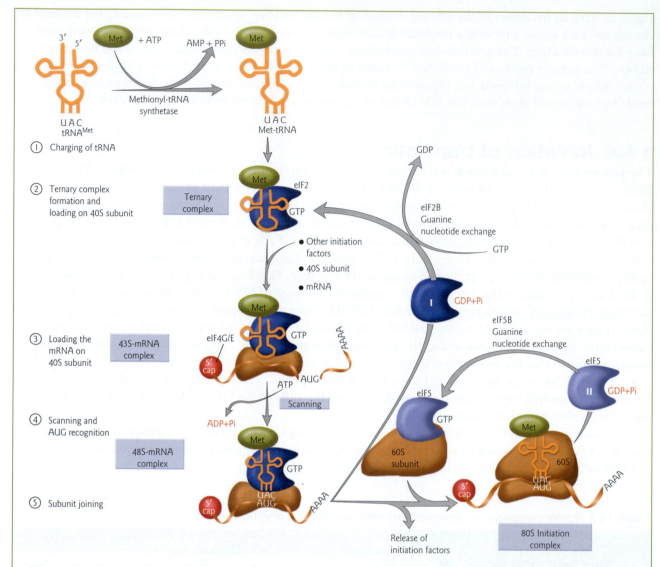

Figure 14.8 Translation initiation. An uncharged tRNA^Met is first charged with methionine by methionyl–tRNA synthetase. The aminoacyl–tRNA interacts with the initiation factor eIF2 to form a ternary complex. This complex is loaded on the 40S ribosomal subunit to form a 43S complex. In association with other initiation factors, including eIF4G/E, the mRNA is loaded on the 43S complex. The 40S subunit scans for the AUG start codon, and initiation factors are released in a process that requires the first GTP hydrolysis step (I). The 60S subunit then joins with the 40S subunit to form the 80S initiation complex, in a process that requires a second GTP hydrolysis step (II).

Ternary complex formation and loading onto the 40S ribosomal subunit

The first step in the initiation pathway is the assembly of a ternary complex comprised of eukaryotic initiation factor 2 (eIF2), GTP, and the amino acid–charged initiator tRNA (Met–tRNA) (Fig. 14.8). This complex binds to the 40S ribosomal subunit, in association with other initiation factors, to form a 43S complex. The exact order of subsequent steps – when particular initiation factors join and dissociate from the complex – is not completely worked out.

Loading the mRNA on the 40S ribosomal subunit

In bacteria, where transcription and translation are coupled processes, as soon as the Shine–Dalgarno sequence (named after its discoverers, see Table 14.2) emerges from the transcriptional apparatus it can be

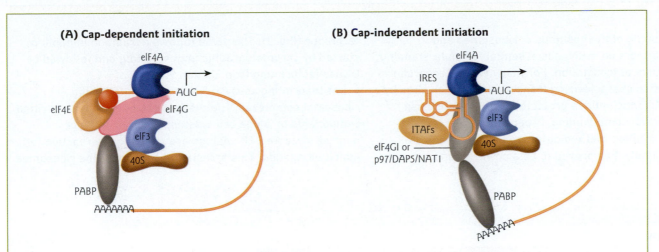

Figure 14.9 The closed-loop model of translation. (A) Cap-dependent initiation. The initation factor eIF4E (orange) binds the 5′ cap of the mRNA (red). The capped end of the mRNA is bridged to the 40S ribosomal subunit (brown) by an adapter molecule eIF4G (pink), which binds to eIF3 (light blue). eIF4A (dark blue) is an RNA-dependent ATPase and RNA helicase. Poly(A)-binding protein (PABP, dark gray) circularizes the mRNA by binding the poly(A) tail and eIF4G. (B) Cap-independent initiation. Internal ribosome entry site (IRES) *trans*-acting factors (ITAFs, orange) and proteolytic fragments of eIF4GI or p97/DAP5/NAT1, a distant homolog of eIF4G (light gray), stimulate IRES translation. Only the key components of the translation machinery are depicted. (Adapted from Holcik, M., Sonenberg, N., Korneluk, R.G. 2000. Internal ribosome initiation of translation and the control of cell death. *Trends in Genetics* 16:469–473. Copyright © 2000, with permission of Elsevier.)

bound by the small ribosomal subunit. In contrast, in eukaryotes, the mRNA to be translated is a fully processed mRNA, most likely folded into secondary and tertiary structures and coated with protective RNA-binding proteins. Two features of the eukaryotic mRNA become important at this point: the 5′ cap and the poly(A) tail. The m^7GpppN cap structure identifies the 5′ end of the message. Specific initiation factors with RNA helicase activity (eIF4G and eIF4E) associate with the 5′ cap of the mRNA, unwind any secondary and tertiary structures, and remove any RNA-binding proteins. Other initiation factors associate with the poly(A)-binding protein (PABP) bound to the 3′-poly(A) tail. The poly(A) tail helps to recruit the 43S complex to the mRNA and when bound with PABP may signal that the mRNA has not been degraded and is fit for translation. Together, the 5′ and 3′ initiation factor complexes work to load the mRNA onto the 43S complex.

One widely accepted model suggests that the 5′ and 3′ ends join to form a closed loop with eIF4G serving as a bridge between them (Fig. 14.9). It has been shown experimentally that the rate of translation of an RNA carrying both a cap and a poly(A) tail is much higher than the sum of the translation rates of RNAs carrying either modification itself, suggesting that the poly(A) tail acts synergistically with the m^7GpppN cap at the 5′ end of the RNA. However, as much as 3–5% of cellular mRNAs are translated by a 5′ cap-independent mechanism. In this case, ribosomes are directly recruited by an internal ribosome entry site (IRES). This bypasses the requirement for the mRNA 5′ cap structure and eIF4E. IRESs were first discovered in picornaviruses, where they initiate the translation of uncapped viral RNAs. Efficient IRES-dependent translation requires additional cellular proteins known as IRES *trans*-acting factors (ITAFs). It is not clear how scanning to find the AUG start codon works in this case.

Scanning and AUG recognition

Once the mRNA is loaded, the 43S complex then scans along the message from 5′ to 3′ looking for the AUG start codon (see Fig. 14.8). One model suggests that the ribosome moves by a ratchet mechanism in which an ATP-dependent RNA helicase located on the 3′ side of the 40S subunit unwinds any secondary

Precise mapping of the positions of ribosomes and associated factors on mRNAs is essential for understanding the mechanism of translation. For many years, researchers were limited to characterizing translation initiation sites by analyzing oligonucleotides protected from ribonuclease digestion by ribosome binding. "Toeprinting" – a primer extension inhibition method developed by Gould and co-workers in 1988 – has proved to be a powerful and versatile technique (Fig. 1). The name follows a tradition in naming started by chromatographic fingerprinting and followed by DNase footprinting (see Section 9.8).

In a toeprinting assay, mRNA is translated using ribosomal complexes. The method can be used with purified components or crude cell lysates. Cycloheximide or nonhydrolyzable GTP analogs are added to the reaction to inhibit elongation. This arrests the position of the ribosomes

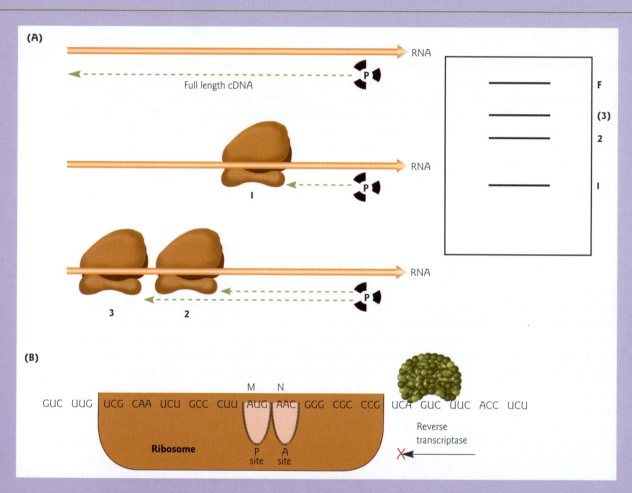

Figure 1 Translation toeprint assay. (A) Principle of the assay. A radiolabeled oligonucleotide primer (star) anneals to the mRNA template (solid arrow) and is extended by reverse transcriptase to form a cDNA (dotted arrow). On naked mRNA, extension proceeds to the 5′ end of the template. When ribosomes and their associated factors are bound at discrete positions on the RNA they inhibit primer extension, generating shorter cDNA fragments (e.g. fragments 1, 2, and 3). cDNA products are analyzed on denaturing sequencing gels (right) on which the full-length cDNA (F) migrates most slowly. (B) Relationship between toeprint sites and the A and P sites of the ribosome. A ribosome with the AUG initiation codon in its P site blocks the movement of reverse transcriptase on the RNA template, causing reverse transcription to stop (arrow) 16 nt downstream of the A of the AUG initiator. (Adapted from Sachs, M.S., Wang, Z., Gaba, A. et al. 2002. Toeprint analysis of the positioning of translation apparatus components at initiation and termination codons of fungal mRNAs. *Methods* 26:105–114.)

Translation toeprinting assays

on the transcript. The mRNA complex is then copied into complementary DNA (cDNA) by reverse transcriptase using a complementary labeled primer. Traditionally, a radiolabeled primer is used, but recently researchers have adapted the assay for use with a fluorescently labeled primer. In either case, where the reverse transcriptase meets the ribosome bound to the mRNA, polymerization is halted and a "toeprint" cDNA fragment is generated.

Typically, the position of the P site of the stalled ribosome is 15–17 nt upstream of the toeprint. If the position of the ribosomes along the transcript is nonrandom, the corresponding cDNA will be in abundance. This is seen on an autoradiogram of a gel as a band, or as a peak in the electropherogram when fluorescently labeled cDNAs are run on an automated DNA sequencing machine.

structures in the mRNA, allowing the ribosome to slide past via diffusion. When the ribosome randomly slides in the 3′ direction over the unwound structures, they reform behind it, preventing backsliding. The 43S complex eventually encounters an AUG codon (usually the first AUG) that is embedded in a favorable sequence context called the Kozak consensus sequence. This sequence is named after its discoverer Marilyn Kozak (see Table 14.2). The ribosomal complex arrests at the initiation codon, forming a stable 48S complex. Upon arrival at the start codon, codon–anticodon base pairing takes place between the AUG and the initiator tRNA in the ternary complex. The codon–anticodon interaction occurs by complementary base pairing with an antiparallel orientation between the tRNA and mRNA (see tRNA structure, Fig. 4.5).

Ribosomes sometimes bypass the first AUG triplet in an mRNA by "leaky scanning" and initiate translation at a second AUG further downstream. This can happen if the context of the first AUG deviates from the consensus sequence, or if the first AUG is located very close to the 5′ cap. Researchers have investigated which factors are responsible for the discrimination of initiation codon context using toeprinting assays (Tool box 14.1). In one such experiment an mRNA substrate was designed with either a "good" (ACC**AUG**A) or "bad" (CAA**AUG**C) context upstream of the initiation codon in the β-glucuronidase (GUS) mRNA. These RNA constructs were tested for their ability to assemble 48S complexes in the presence or absence of various initiation factors. Toeprinting analysis showed the importance of eIF1, eIF4A, eIF4B, and eIF4F in codon–anticodon interaction (Fig. 14.10). Additional experiments demonstrated that 43S complexes retained some capacity to scan along mRNA in the absence of eIF1, but they could no longer discriminate between initiation codons in "good" and "bad" contexts.

Joining of the 40S and 60S ribosomal subunits

Initiation involves two GTP hydrolysis events, one catalyzed by eIF2 upon initiation codon recognition and another at the end of the pathway, after 80S complex formation. GTP hydrolysis by eIF2 releases the initiator tRNA into the P site of the 40S subunit and then dissociates eIF2 from the complex (see Fig. 14.8). In order to recycle into its active form and bind another initiator Met-tRNA, eIF2·GDP must be converted to eIF2·GTP through a nucleotide exchange reaction. eIF2 has a 400-fold greater affinity for GDP than for GTP. The exchange of tightly bound GDP for GTP requires the guanine nucleotide exchange factor eIF2B, which is present in limiting concentrations at 15–25% of the amount of eIF2. The essential nature of eIF2B is illustrated by the devastating effects of loss of function on the central nervous system (Disease box 14.1).

After binding of the Met-tRNA in the P site, eIF5-GTP binds to the growing complex. Once the other initiation factors are released, eIF5 helps to join the large (60S) ribosomal subunit to the 40S/Met-tRNA/mRNA complex. Exactly how eIF5 facilitates joining is not understood at a molecular level. What is understood is that this event triggers GTP hydrolysis by eIF5 and it dissociates from the ribosome. In order

Figure 14.10 Factor dependence of initiation codon selection. (A) Sequences of the 5'-UTRs of (CAA)n-AUG_{good}-GUS and (CAA)n-AUG_{bad}-GUS mRNAs, showing the AUG initiation codons in bold. Context residues for each initiation codon are underlined; the A of the AUG codon is designated as +1. (B, C) Toeprinting analysis of 48S complex formation on (CAA)n-AUG_{good}-GUS (B) and (CAA)n-AUG_{bad}-GUS mRNA (C). The reaction mixtures contained 40S subunits and translation initiation factors as indicated (see Tool box 14.1 for methods). Full-length cDNA is labeled E. The label "48S (GUS)" indicates the position of toeprints caused by 48S complexes assembled on the GUS initiation codon; the labels "48S 'good'" and "48S 'bad'" indicate 48S complexes assembled at upstream initiation codons in these mRNAs. The position of these initiation codons are shown to the left of reference lanes (C, T, A, G), which show cDNA sequences derived using the same primer as for toeprinting. The negative control lane shows that in the absence of 40S subunits no 48S complex is assembled on the mRNA and a full-length cDNA is generated by primer extension. Approximately 60% of scanning ribosomes stopped at the "good" context AUG and the remainder scanned to the initiation codon. In contrast, approximately 90% of ribosomes scanned past the "bad" context AUG to the GUS initiation codon. Leaving out eIF1A alone, or simultaneous omission of eIF4A, -4B, and -4F or eIF1A, -4A, -4B, and -4F, from reactions significantly reduced 48S complex formation on the GUS initiation codon of the (CAA)n-AUG_{good}-GUS mRNA. These same treatments did not alter the pattern of 48S complex formation on the (CAA)n-AUG_{bad}-GUS mRNA. (Reprinted with permission from Pestova, T.V., Kolupaeva, V.G. 2002. The roles of individual eukaryotic translation initiation factors in ribosomal scanning and initiation codon selection. *Genes & Development* 16:2906–2922. Copyright © 2002 Cold Spring Harbor Laboratory Press.)

(A)

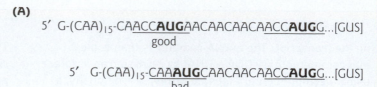

5' G-(CAA)$_{15}$-CA<u>ACC</u>**AUG**AACAACAACAACCAUGG...[GUS]
good

5' G-(CAA)$_{15}$-<u>CAA</u>**AUG**CAACAACAACC**AUG**G...[GUS]
bad

(B)

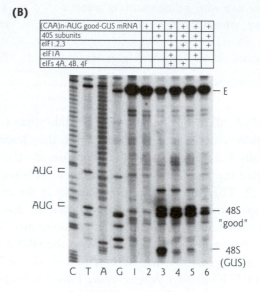

(C)

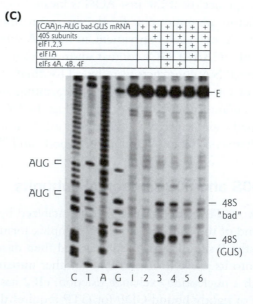

Eukaryotic initiation factor 2B and vanishing white matter

For many years the cause of a rare childhood disease called leukoencephalopathy with vanishing white matter (also known as childhood ataxia with central hypomyelination) remained a mystery. This recessively inherited, fatal disease leads to progressive loss of movement and speech usually before age 5, and sometimes seizures or coma. There are episodes of rapid and major deterioration following fever or minor head trauma. Over time, white matter vanishes to become replaced by cerebrospinal fluid. White matter is the region of the brain and spinal cord that consists chiefly of myelinated axons, which are white in appearance. In some cases the decline takes years. In others, it is rapid.

A child may have a fever one day and 4 days later be in a coma.

In 2001 mutations in the genes encoding two subunits of eukaryotic translation initiation factor 2B (eIF2B) were found to be the cause of the disease in some patients. eIF2B is the guanine nucleotide exchange factor for eIF2. It is composed of five subunits (α, β, δ, ϵ, γ) encoded by five separate genes. Mutations in genes encoding all five subunits have been found in patients with "vanishing white matter" syndrome. Most of these are missense mutations. The severity of the disease depends on where the mutation is located in the gene. Why the cells that make myelin are particularly sensitive to defects in eIF2B remains unknown.

to recycle into its active form and bind another 60S subunit, eIF5·GDP must be converted to eIF5·GTP through a nucleotide exchange reaction mediated by eIF5B. The poly(A) tail and poly(A)-binding proteins may also influence subunit joining by direct interaction of the poly(A)-binding proteins with the 60S ribosomal subunit.

14.5 Elongation

In contrast to the initiation and termination stages of protein synthesis, the machinery used during the elongation stage has been highly conserved in bacteria and eukaryotes. Peptide chain elongation begins with a peptidyl-tRNA in the ribosomal P site next to a vacant A site (Fig. 14.11). An aminoacyl-tRNA is carried to the A site as part of a ternary complex with GTP and elongation factor 1A (eEF1A). The amino group of the first amino acid is aimed into the A site, ready to link to the next amino acid to be delivered. The critical decoding event occurs as the charged tRNA binds to the A site (see below). Cognate (correct) codon–anticodon base pairing causes three bases in the 18S rRNA to swing out and interact with the resulting mRNA–tRNA duplex. This in turn appears to activate the GTPase activity of eEF1A. eEF1A· GDP releases the aminoacyl-tRNA into the A site in a form that can continue with peptide bond formation. In order to recycle into its active form and bind another aminoacyl-tRNA, eEF1A·GDP must be converted to eEF1·GTP through a nucleotide exchange reaction mediated by eEF1B.

In prokaryotes, the incorporation of selenocysteine at a specific UGA site depends upon tRNA^{Sel}, a specialized translation elongation factor, SelB, and a secondary structural element of the RNA, called a selenocysteine insertion sequence (SECIS), located immediately after the UGA (Fig. 14.12A). In eukaryotes, the SECIS elements are located in the 3′ untranslated regions (UTR) of the mRNA (Fig. 14.12B). Pyrrolysine insertion elements (PYLIS) in archaebacterial mRNA signal for pyrrolysine insertion at the UAG codon, in a manner similar to SECIS.

Decoding

There is increasing recognition from studies in bacteria that tRNA plays an active role in translation, as opposed to being a static "adaptor." First, conformational changes in tRNA are the basis for induced fit, which is

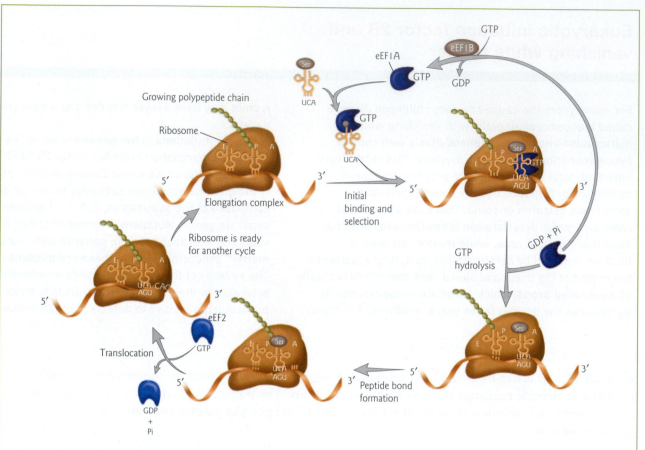

Figure 14.11 Translation elongation in eukaryotes. An aminoacyl-tRNA (in this example, Ser-tRNA) is carried to the A site as part of a complex with GTP-bound eEF1A. The Ser-tRNA enters the acceptor (A) site where decoding takes place. If it is the correct tRNA, GTP hydrolysis releases it into the A site and the ribosome catalyzes peptide bond formation. After the tRNAs and mRNA are translocated such that the next codon is moved to the A site, the process is repeated. Translocation is mediated by eEF2 and requires GTP hydrolysis.

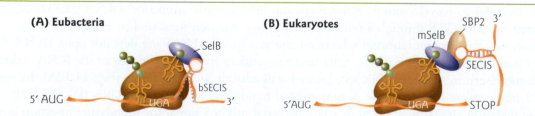

Figure 14.12 Incorporation of selenocysteine at a stop codon. Insertion of selenocysteine at UGA codons requires a selenocysteine insertion sequence (SECIS). (A) In eubacteria, elongation factor SelB binds both the SECIS located immediately 3′ to the UGA and tRNA^{Sel}. (B) In eukaryotes, the elongation factor requires an adaptor, SBP2, and SECIS is located in the 3′ untranslated region. (Modified from Namy, O., Rousset, J.P., Napthine, S., Brierley, I. 2004. Reprogrammed genetic decoding in cellular gene expression. *Molecular Cell* 13:157–168.)

essential for high-fidelity tRNA selection. Second, tRNA itself acts as a direct functional link between the decoding center of the small ribosomal subunit and remote regions of the large ribosomal subunit that promote GTP hydrolysis and peptide bond formation. The ribosome selects aminoacyl-tRNA matching the mRNA codon from the bulk of nonmatching aminoacyl-tRNAs in two consecutive selection steps: initial selection and proofreading (Fig. 14.13A). The two steps are separated by irreversible hydrolysis of GTP. tRNA selection at the initial selection step is entirely kinetically controlled and is due to much faster (650-fold) GTP hydrolysis when the ternary complex contains the correct tRNA. The fidelity of discrimination is very high, with only one incorrect amino acid per 1000–10,000 correct amino acids incorporated into the protein. Induced fit further increases fidelity by selectively accelerating the forward rates of two steps in the selection process – activation of EF-Tu for GTP hydrolysis and accommodation of aminoacyl-tRNA into the A site. This induced fit results from correct codon–anticodon interactions somehow accelerating the rate-limiting conformational changes that are required for GTP hydrolysis and peptidyl transfer.

Experimental evidence for the importance of tRNA conformation in decoding comes from studies of a mutant tRNA^{Trp} called the "Hirsh suppressor" (after its discoverer). This tRNA variant has a single G24A substitution in its D arm that changes the U11–G24 base pairs to U11–A24 (Fig. 14.13B). Kinetic studies demonstrated that the mutant accelerates the forward rate constants independent of codon–anticodon pairing in the decoding center. The mutant tRNA recognizes both UGG and UGA stop codons even though the mutation does not alter the anticodon (Fig. 14.13C).

Peptide bond formation and translocation

The ribosomal peptidyl transferase center catalyzes the formation of a peptide bond between the incoming amino acid and the peptidyl-tRNA. The resulting deacylated tRNA is moved into the exit (E) site of the large ribosomal subunit, the peptidyl-tRNA is moved into the P site, and the mRNA is moved by three nucleotides to place the next codon of the mRNA into the A site (see Fig. 14.11). Translocation is mediated by eEF2 and requires GTP hydrolysis. GDP has a lower affinity for eEF2 than does GTP and is rapidly released. The unbound eEF2 binds a new GTP molecule. The tRNA in the E site is released as a new tRNA enters the A site. This cycle is repeated until a stop codon is reached and the process of termination begins.

Peptidyl transferase activity

The chemical reaction catalyzed by the ribosome is a simple one – the joining of amino acids through peptide bonds. However, as we have seen in the preceding sections, the orchestration of protein synthesis is complex and highly regulated. The ribosome accelerates the rate of peptide bond formation by at least 10^7-fold. A central question for many years was whether the "peptidyl transferase activity" that catalyzes peptide bond formation is the result of a protein or an RNA enzyme. Approaches used to answer this question have relied on studies of bacterial ribosomes.

Since the discovery that ribosomes are the sites of cellular protein synthesis almost 50 years ago, theories about ribosomal proteins have changed drastically. It had been known since the 1960s that the peptidyl transferase region is located at the base of the central protuberance in the 50S subunit of the bacterial ribosome (Fig. 14.14). Early hypotheses ascribed all of the important functions of the ribosomes to the ribosomal proteins. The ribosomal proteins were thought to be held together and properly positioned by an otherwise inert rRNA framework. Proteins of the 50S subunit were proposed to catalyze peptide bond formation, while the rRNA was presumed to hold things in place by base pairing with the tRNA and mRNA. These models were based on the conventional protein-centric view of biochemists at the time. Only proteins were thought to act as biological catalysts.

A reversal of these roles took shape in the 1970s and 1980s, however, when the remarkable capacities of RNA molecules were recognized. The dogma that nucleic acids are the repository of information and that proteins catalyze chemical reactions was overturned in 1981 by the demonstration of catalytic RNAs, called ribozymes (see Section 4.6). Support increased for the "RNA world hypothesis" which sees RNA-based life

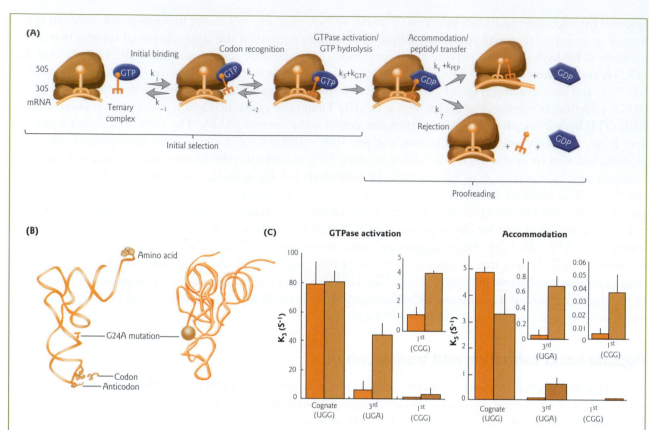

Figure 14.13 Kinetics of tRNA selection on the ribosome. (A) tRNA selection on the ribosome involves two main stages: initial selection and proofreading. Rate constants are shown for kinetically resolved steps. The two forward steps are GTP hydrolysis and accommodation. A new aminoacyl-tRNA is brought into the A site of the bacterial ribosome in complex with elongation factor EF-Tu. Upon GTP hydrolysis, the aminoacyl-tRNA is released by EF-Tu, and swings into the peptidyl transferase center of the 50S subunit. The tRNA is distorted when bound with EF-Tu on the ribosome. EF-Tu is shown in different conformations in the GTP- and GDP-bound form (Adapted from Cochella, L., Green, R. 2005. An active role for tRNA in decoding beyond codon:anticodon pairing. *Science* 308:1178–1179.) (B) Location of the G24A mutation in the Hirsh suppressor tRNA. The mutation is far from the anticodon, but close to the region of distortion in the tRNA when it first binds to the ribosome. (Adapted from Daviter, T., Murphy, F.V. IV, Ramkrishnan, V. 2005. A renewed focus on transfer RNA. *Science* 308:1123–1124.) (C) Calculated rate constants are shown for wild-type (orange) and G24A (brown) tRNA^{Trp} on UGG (cognate codon), UGA, or CGG programmed ribosomes. Each bar represents the average of two to four experiments and the error bars represent their standard deviations. Rates of GTP hydrolysis were measured by mixing programmed bacterial ribosomes with purified ternary complexes composed of EF-Tu, [γ-^{32}P]GTP, and either wild-type or variant Trp-tRNA^{Trp}. tRNA accommodation was measured by assaying peptide bond formation. Programmed ribosomes containing f-[^{35}S]Met-tRNA^{fMet} in the P site were mixed with the wild-type and mutant Trp-tRNA^{Trp} ternary complexes. The amount of [^{35}S]fMet-Trp dipeptide formed over time was quantitated. The experiment demonstrated that the G24A tRNA^{Trp} variant accelerates forward rates of GTPase activation and accommodation, independent of correct codon–anticodon pairing. The mutation allows movement of tRNA into the peptidyl transferase center even on the stop codon UGA. (Reprinted with permission from Cochella, L., Green, R. 2005. An active role for tRNA in decoding beyond codon:anticodon pairing. *Science* 308:1178–1179. Copyright © 2005 AAAS.)

forms as the progenitors of modern cells (see Focus box 4.1). However, most biologists still did not seriously consider the possibility that ribosomal RNA could be playing more than a minor role in reactions such as protein synthesis. Beginning in the early 1990s, new lines of biochemical and structural evidence begin to point towards the primacy of rRNA in ribosome function.

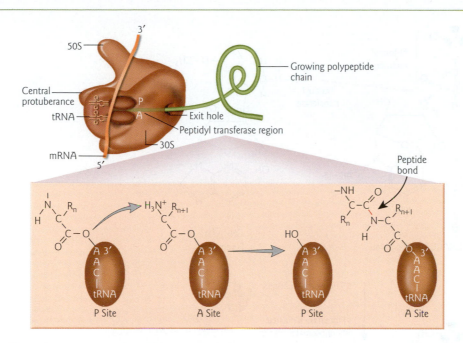

Figure 14.14 Peptidyl transferase region in the ribosome. The site of peptide bond formation is located at the base of the central protuberance in the 50S large ribosomal subunit. Peptidyl transferase activity transfers a growing polypeptide chain from peptidyl tRNA in the P site to an amino acid esterified with another tRNA in the A site.

Biochemical evidence that 23S rRNA is a ribozyme

A landmark experiment was performed in Harry Noller's lab at the University of California–Santa Cruz in 1992. Noller's group used the "fragment reaction," which is a simple *in vitro* assay for peptidyl transferase activity, to test whether ribosomal RNA alone could catalyze the formation of a model peptide bond. In this assay, the P-site tRNA is replaced by a short oligonucleotide (CAACCA) attached to the initiator amino acid, *N*-formyl methionine (f–Met). The A-site tRNA is replaced by the antibiotic puromycin. Puromycin is an aminoacylated tRNA analog. The assay measures transfer of ^{35}S-labeled f–Met to the amino group of puromycin to form a model peptide bond (Fig. 14.15A). Previously, peptidyl transferase activity had been shown to reside in the 50S ribosomal subunit. There is no requirement for the small ribosomal subunit, mRNA, other protein factors, or GTP to be included in the fragment reaction.

Noller's group treated 50S ribosomes from the bacterium *Thermus aquaticus* with proteinase K and sodium dodecyl sulfate (SDS). Proteinase K is a protease that degrades proteins and SDS is a strong ionic detergent commonly used to remove proteins from nucleic acids. They showed that after protein extraction, the remaining 23S rRNAs still had peptidyl transferase activity (Fig. 14.15B). Importantly, they also showed that this catalytic activity was inhibited by treatment with ribonucleases and by known peptidyl transferase inhibitors. Their findings strongly suggested that 23S rRNA alone can catalyze peptide bond formation. Noller's findings were confirmed by a Japanese research group in 1998. Further evidence was needed, however, to convince the scientific community that 23S rRNA was a ribozyme, since it was considered possible that residual ribosomal proteins were still present in these assays.

Structural evidence that rRNA forms the active site of the ribosome

The year 2000 marked another major advance, with the publication of atomic resolution views of the large ribosomal subunit from the archaeon *Haloarcula marismortui*. Prior X-ray crystallographic images of the ribosome only resolved the structure to 20 Å resolution, while the new images were at 2.4 Å resolution

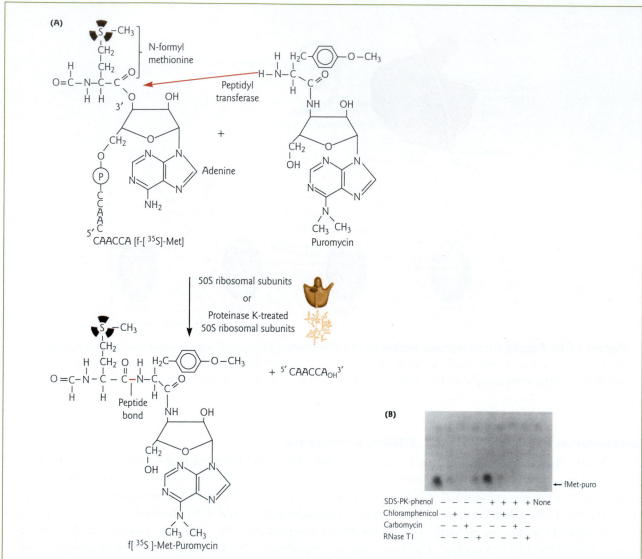

Figure 14.15 Sensitivity of *Thermus aquaticus* fragment reaction to peptidyl transferase inhibitors and RNase. (A) The fragment reaction measures the transfer of ^{35}S-labeled *N*-formyl methionine from a short oligonucleotide (peptidyl–tRNA analog) to the amino group of puromycin (aminoacyl–tRNA analog) to form a model peptide bond. (B) *T. aquaticus* 50S ribosomal subunits were treated with protein denaturants (SDS–proteinase K–phenol), peptidyl transferase inhibitors (chloramphenicol, carbomycin), or RNase T1 as indicated, then assayed for peptidyl transferase by the fragment reaction shown in (A). Fragment reactions carried out by either intact or protein denaturant–treated *T. aquaticus* 50S subunits were inhibited by carbomycin, chloramphenicol, and RNase as expected if rRNA is a key factor in peptidyl transferase. (Reprinted with permission from Noller, H.F., Hoffarth, V., Zimniak, L. 1992. Unusual resistance of peptidyl transferase to protein extraction procedures. *Science* 256:1416–1419. Copyright © 1992 AAAS.)

(Fig. 14.16). These images revealed that about two-thirds of the ribosome's mass is composed of RNA. The structures showed that rRNA both creates a structural framework for the ribosome and, at the same time, forms the main features of its functional sites. The rRNA forms most of the intersubunit interface, the peptidyl transferase center, the decoding site, and the A and P sites. In striking contrast, the ribosomal proteins are abundant on the exterior of the ribosome, but not in the active site. The ribosomal proteins follow the contours of the RNA. No protein is located within 15 Å of the site of catalysis.

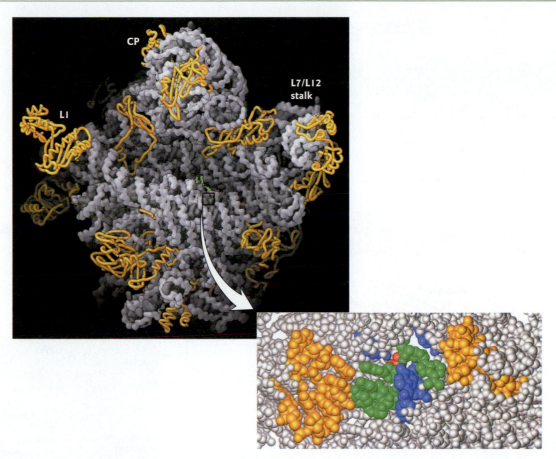

Figure 14.16 Atomic resolution structure of the ribosome. The large ribosomal subunit of the archaeon *Haloarcula marismortui* is shown in the rotated crown view. The L7/L12 stalk is to the right, the L1 stalk is to the left, and the central protuberance (CP) is at the top. The surface of the large subunit that interacts with the small subunit faces the reader. RNA is shown in gray in a pseudo-space-filling rendering. Protein backbones are shown in gold. The Yarus transition state analog (CCdA–phosphate–puromycin, green) marks the peptidyl transferase site. The large subunit is approximately 250 Å across. (Reprinted with permission from Ban, N., Nissen, P., Hansen, J., Moore, P.B., Steitz, T.A. 2000. The complete atomic structure of the large ribosomal subunit at 2.4 Å resolution. *Science* 289:905–920. Copyright © 2000 AAAS.) (Inset) Two shells of nucleotides in the active site of the ribosome. The innermost active site nucleotides critical for peptide release are shown in green (A2451, U2506, U2585, A2602) (see Fig. 14.17). The more remote A and P loops critical for orienting the CCA ends of the tRNA substrates are shown in gold with the A loop on the right. The transition state analog is shown in blue with its tetrahedral phosphate at the center of the active site (red). (Reprinted from Youngman, E.M., Brunelle, J.L., Kochaniak, A.B., Green, R. 2004. The active site of the ribosome is composed of two layers of conserved nucleotides with distinct roles in peptide bond formation and peptide release. *Cell* 117:589–599. Copyright © 2004, with permission from Elsevier.)

   Biochemical and kinetic experiments, mutagenesis studies, and structural analyses all lead to the widely accepted conclusion that the ribosome is a ribozyme. Exactly how it mediates catalysis remains a subject of debate. The approximately 10^7-fold rate of enhancement of peptide bond formation by the ribosome appears to be mainly due to substrate positioning by the 23S rRNA within the active site, rather than to chemical catalysis. The 23S rRNA is organized in six major domains composed of connecting loops and helices. These irregular helices are packed against one another to form the final globular structure, through coaxial stacking of helices and formation of noncanonical structures. The peptidyl transferase center is

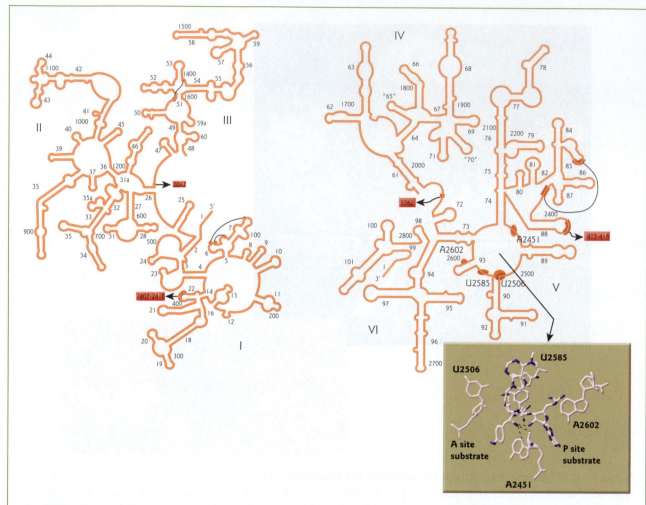

Figure 14.17 Secondary structure model of *E. coli* 23S rRNA. The peptidyl transferase region is located in domain V and positions the substrates for catalysis. (Adapted from Noller, H.F., Kop, J., Wheaton, V., Brosius, J., Gutell, R.R., Kopylov, A.M., Dohme, F., Herr, W., Stahl, D.A., Gupta, R., Waese, C.R. 1981. Secondary structure model for 23S ribosomal RNA. *Nucleic Acids Research* 9:6167–6189; and Gutell, R.R., Lee, J.C., Cannone, J.J. 2002. The accuracy of ribosomal RNA comparative structure models. *Current Opinion in Structural Biology* 12:301–310.) (Inset) The "inner shell" of conserved nucleotides plays an important role in peptide release during termination of protein synthesis (see Fig. 14.16). (Protein Data Bank, PDB: 1M90. Adapted from Hansen J.L., Schmeing, T.M., Moore, P.B., Steitz, T.A. 2002. Structural insights into peptide bond formation. *Proceedings of the National Academy of Sciences USA.*)

located in domain V of the 23S rRNA (Fig. 14.17). The 3′ terminal adenosines of both the A- and P-site tRNAs are positioned in the peptidyl transferase site by A-minor interactions with the 23S rRNA (see Fig. 4.8). Many mutagenesis studies have been performed on the universally conserved nucleotides in the so-called "inner shell" of the peptidyl transferase center. Evidence first pointed to a role in peptide bond formation. However, more recent studies indicate that they are not directly involved in the catalysis of peptide bond formation but are critical instead for the catalysis of peptide release during the termination step of protein synthesis.

Another question that remains to be answered is what are the functions of ribosomal proteins in the ribosomes? Ribosomal RNA folds correctly only by assembling with ribosomal proteins, which orchestrate the order of folding to avoid kinetic misfolding traps (see Fig. 4.12). In addition to protecting the rRNA from nuclease attack and helping the rRNA to fold into a more effective conformation, the ribosomal proteins appear to make contributions to efficient and accurate protein synthesis. For example, L27 contributes to peptide bond formation by facilitating proper placement of the tRNA at the peptidyl transferase center, rather than by contributing active groups for catalysis. In addition, in the *Escherichia coli* 50S subunit, the extended internal loops of proteins L4 and L22 line part of the tunnel through which newly synthesized polypeptide chains exit and may help regulate their movement.

Events in the ribosome tunnel

One of the key features of the ribosome is the long "tunnel" that conducts the nascent protein through the large subunit from the site of peptidyl transferase activity to the peptide exit hole (see Fig. 14.2). Three-dimensional structures of translating ribosomes determined by cryoelectron microscopy show that the tunnel may not be a narrow passage as originally thought. It is a dynamic structure that can expand to allow polypeptides to begin to fold into their rudimentary globular conformation.

The components of the ribosomal tunnel somehow sense the requirements of nascent polypeptides and connect them to downstream processes. One example is translocation of secreted and integral membrane proteins to the endoplasmic reticulum (ER). The signal recognition particle (SRP) is a ribonucleoprotein complex that binds to ribosomes translating polypeptides that bear a signal sequence for ER targeting. SRPs arrest translation and target the ribosome to the ER membrane. When translation resumes after release of the SRP, the nascent polypeptide is either integrated into the ER membrane or translocated through a channel called the translocon into the ER lumen for secretion. The SRP is a rod-shaped complex that can stretch from the tunnel exit to the interface of the two ribosomal subunits (Fig. 14.18). It is composed of two domains, the *Alu* domain (which contains the 7S RNA component) and the S domain. Binding of the S domain to the signal sequence in the nascent polypeptide allows the *Alu* domain to contact the elongation factor-binding site, which causes translation arrest. While in the tunnel, the transmembrane domain of the nascent polypeptide adopts an α-helical conformation. This may provide the initial signal for membrane integration.

The SRP is just one of a number of macromolecular factors that bind to ribosome-nascent chain complexes. The *E. coli* trigger factor is an example of a ribosome-associated molecular chaperone that helps the emerging nascent polypeptide chain to adopt its correct three-dimensional fold. The trigger factor "hunches over the exit like a crouching dragon" and creates a protected folding space where nascent polypeptides are shielded from proteases and aggregation as they emerge from the ribosome tunnel (Fig. 14.19).

14.6 Termination

The ribosomal peptidyl transferase center is responsible for two fundamental reactions: peptide bond formation and nascent peptide release, during the elongation and termination phases of protein synthesis, respectively. The termination of translation occurs in response to the presence of a stop codon (UAG, UAA, or UGA) in the ribosomal A site. The end result of this process is the release of the completed polypeptide following hydrolysis of the ester bond linking the polypeptide to the P-site tRNA (Fig. 14.20).

The termination of protein synthesis by the ribosome requires two release factor (RF) classes. Class 1 factors decode stop codons, while class 2 factors are GTPases that stimulate the activity of class 1 release factors. When any of the stop codons are present in the ribosome A site, they are recognized by eukaryotic release factor 1 (eRF1). Eukaryotes possess a single class 2 release factor, eRF3. A role for eRF3 in triggering the release of eRF1 from the ribosome following peptidyl-tRNA hydrolysis has yet to be experimentally shown. In prokaryotes, it is known that GDP-bound RF3 binds to a class 1 release factor, followed by exchange of GDP for GTP and release of the class 1 release factor. GTP hydrolysis triggers the release of RF3.

The final steps in protein synthesis involve the release of the completed polypeptide, expulsion of the tRNA from the E site, dissociation of the ribosome from the mRNA, and dissociation of the 40S and 60S subunits. The events involved in the process of recycling of the ribosomal subunits so that they can be used in another round of initiation are not well understood in eukaryotes.

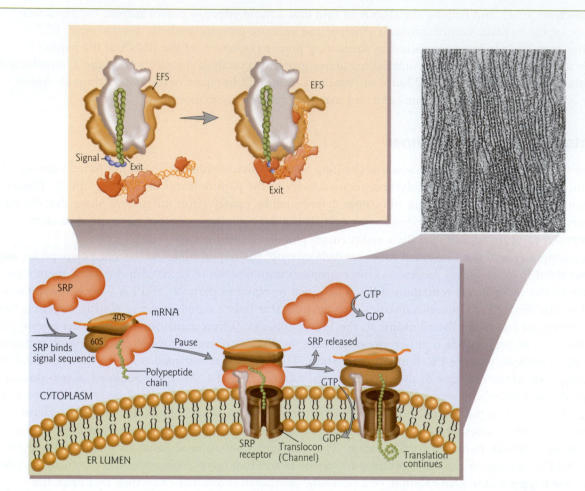

Figure 14.18 Cotranslation translocation pathway from the ribosome to the endoplasmic reticulum (ER) lumen. The signal recognition particle (SRP) and SRP receptor use a cycle of recruitment and hydrolysis of GTP to control delivery of the ribosome–mRNA complex to the ER translocon. (Upper left inset) The SRP interacts with the ribosome, stretching from the peptide tunnel exit to the elongation factor-binding site (EFS) in the intersubunit space, where it causes elongation arrest (pause) by competition with elongation factors. The nascent polypeptide ER signal sequence is shown in blue. (Adapted from Halic, M., Becker, T., Pool, M.R., Spahn, C.M.T., Grassucci, R.A., Frank, J., Beckmann, R. 2004. Structure of the signal recognition particle interacting with the elongation-arrested ribosome. *Nature* 427:808–814.) (Upper right inset) Electron micrograph of rough ER, which is especially abundant in pancreatic acinar cells due to their predominant role in protein synthesis. On the surface of the ER membrane are numerous ribosomes (tiny spheres). They give the ER its granular appearance and give rise to the name "rough" ER (Photograph courtesy of Evguenia Orlova, College of William and Mary.)

14.7 Translational and post-translational control

Investigations into mechanisms that control gene expression in eukaryotes have focused largely on transcription initiation (see Chapter 11), and to a lesser degree on post-transcriptional events related to the fate of specific mRNAs (see Chapter 13). Increasing evidence points to translational control as another important level of gene regulation. Translational control by eIF2 was the first example of regulation of eukaryotic gene expression at the level of protein synthesis, and remains the subject of intensive investigation. The fundamental mechanism for translational control by eIF2 is conserved from budding yeast to mammals. The mechanism centers on protein phosphorylation and thus also provides an excellent illustration of post-translational control by protein modification (see Section 5.5 for further discussion of post-translational modification).

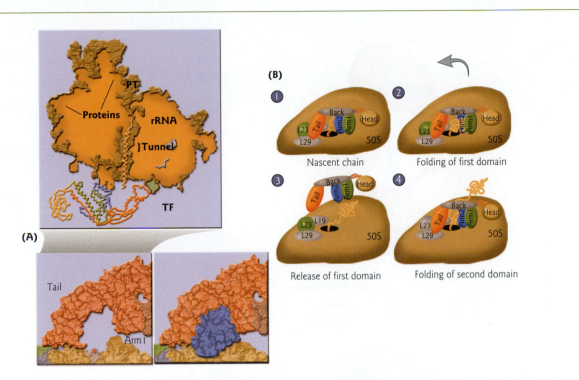

Figure 14.19 The *E. coli* trigger factor. (A) (Left) view through the arch–like structure delineated by the tail and arms of trigger factor (red) covering the peptide exit tunnel, indicated by an asterisk. (Right) Identical view to the left hand figure – the size of the arch can accommodate an entire molecule of lysozyme (blue), manually fitted into this region for illustration. (Inset) Structure of the trigger factor (TF) bound to a slice of the 50S ribosomal subunit along the peptide exit tunnel. A modeled nascent polypeptide chain in magenta is shown extending from the peptidyl transferase centre (PT). (B) Schematic representation of the proposed mechanism of trigger factor action. (1) The trigger factor is bound to an unfolded nascent polypeptide chain. (2) Upon folding of this domain, contacts between the trigger factor and the newly synthesized polypeptide are weakened. (3) Trigger factor dissociates from the ribosome. (4) Trigger factor reassociates with the ribosome when the next stretch of newly synthesized, unfolded polypeptide becomes exposed. (Adapted with permission from Nature Publishing Group and Macmillan Publishers Ltd: from Ferbitz, L., Maier, T., Patzelt, H., Bukau, B., Deuerling, E., Ban, N. 2004. Trigger factor in complex with the ribosome forms a molecular cradle for nascent proteins. *Nature* 431:590–596. Copyright © 2004.)

Phosphorylation of eIF2α blocks ternary complex formation

Many different types of stress, such as hypoxia, viral infection, amino acid starvation, and heat shock repress translation by triggering the phosphorylation of the α-subunit of eIF2 at residue Ser51. This inhibits the exchange of GDP for GTP on the eIF2 complex, which is catalyzed by eIF2B, and thereby prevents formation of the ternary complex (Fig. 14.21; see also Fig. 14.8). eIF2α is present in excess over eIF2B, so even small changes in the phosphorylation of eIF2α have a notable effect on the formation of the ternary complex and translation. eIF2α phosphorylation represses translation of most mRNAs. There are a few exceptions where phosphorylation enhances the translation of selected mRNAs which encode proteins that function in adaptation to stress and the recovery of translation. For example, the yeast transcriptional activator GCN4 and mammalian activating transcription factor 4 (ATF4) both show increased protein synthesis during stress. The 5′ regions of their mRNAs include multiple upstream open reading frames (uORFs), with four in GCN4 and two in ATF4. Each uORF has its own AUG and stop codon. The upstream AUGs are used by unphosphorylated eIF2α and result in termination, while the AUG closest to the gene-coding region is initiated at more frequently by phosphorylated eIF2α. Similarly, translation of

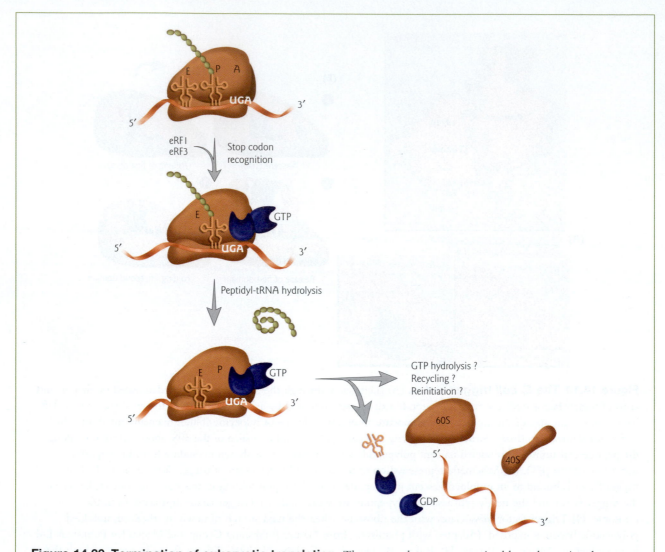

Figure 14.20 Termination of eukaryotic translation. The stop codons are recognized by eukaryotic release factor 1 (eRF1) in association with eRF3. The completed polypeptide is cleaved from the peptidyl-tRNA, followed by expulsion of the tRNA, dissociation of the ribosome from the mRNA, and dissociation of the 40S and 60S subunits. GTP hydrolysis may trigger the release of eRF1 and eRF3.

GADD34 mRNA is enhanced by phosphorylated eIF2α. GADD34 binds the catalytic subunit of protein phosphatase 1 (PP1) and promotes dephosphorylation of eIF2α during recovery from stress.

eIF2α phosphorylation is mediated by four distinct protein kinases

The phosphorylation of eIF2α is mediated by four distinct protein kinases in mammals: heme–regulated inhibitor kinase (HRI), protein kinase RNA (PKR), PKR-like endoplasmic reticulum kinase (PERK), and the mammalian ortholog of yeast general control non–derepressible 2 (GCN2). These protein kinases share extensive homology in their kinase domain, and integrate diverse stress signals into a common translational control pathway (Fig. 14.21). The actions of PKR and HRI are highlighted to illustrate the diversity of signaling pathways involved in translational control

PKR is activated by viral double-stranded RNA (dsRNA). This leads to dimer formation, autophosphorylation, and functional activation of the kinase. In its activated conformation, PKR binds

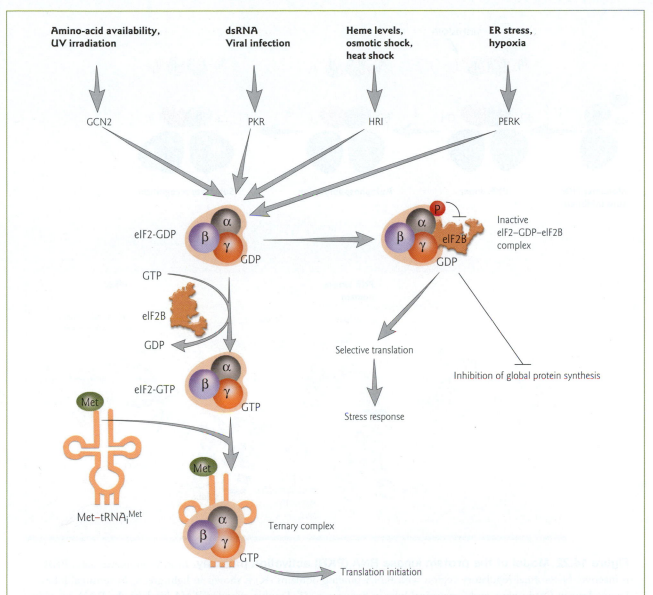

Figure 14.21 Translational control by phosphorylation of eukaryotic initiation factor 2α. Four distinct protein kinases are activated in response to different stress conditions: GCN2, PKR, HRI, and PERK. Phosphorylation of eIF2α by these kinases inhibits GDP–GTP exchange by reducing the dissociation rate of the nucleotide exchange factor eIF2B. This leads to inhibition of translation by blocking ternary complex formation. Selective translation of a subset of mRNAs continues, which allows cells to adapt to stress conditions. ER, endoplasmic reticulum.

eIF2α and phosphorylates the initiation factor at Ser51 (Fig. 14.22). X-ray crystal structures show that the catalytic domain of PKR adopts a bilobed structure typical of protein kinases (see Fig. 5.15). A small N-terminal lobe and a larger C-terminal lobe are connected by a short hinge. eIF2α binds to the C-terminal catalytic lobe of PKR while catalytic-domain dimerization is mediated by the N-terminal lobe. Binding to PKR induces a local unfolding of the Ser51 residue in eIF2α and positions the phosphorylation site within the catalytic cleft.

Biochemical studies have implicated translational control in regulating gene expression. Researchers are now turning to *in vivo* studies to validate the physiological significance of *in vitro* assays. Direct evidence for a physiological role of translational control comes from targeted disruption of the gene encoding HRI in mice

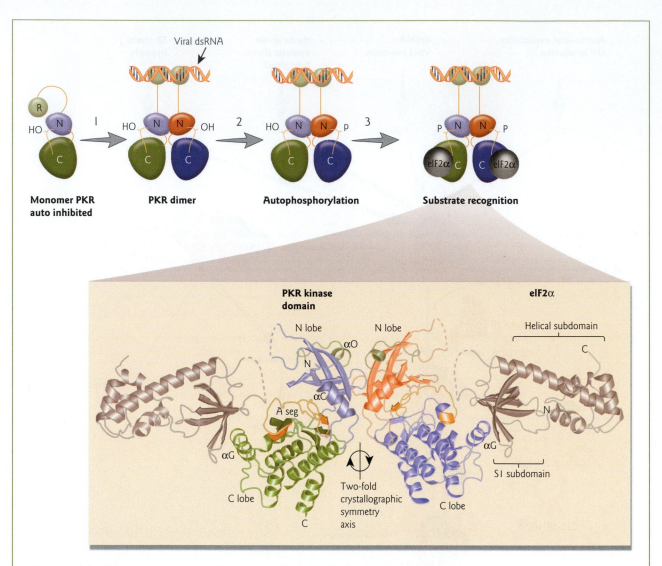

Figure 14.22 Model of the protein kinase RNA (PKR) activation pathway. In its monomeric state, PKR is inactive. N-terminal regulatory region with RNA-binding domains (R) is shown in light green; N-terminal lobe kinase domain (N) in blue; and C-terminal lobe in dark green. (1) Double-stranded RNA binds to the RNA-binding domains and promotes catalytic-domain dimerization. (2) Catalytic-domain dimerization promotes autophosphorylation on activation-segment (orange) residue Thr446 (P). (3) Autophosphorylation of PKR enhances its catalytic activity and is required for specific recognition of eIF2α (gray). (Adapted from Dey, M. et al. (2005) *Cell* 122:901–913.) (Inset) Ribbon representation of the PKR–eIF2α complex. Regions not modeled due to disorder are shown as dashed lines. (Adapted from Dar, A.C., Dever, T.E., Sicheri, F. 2005. Higher-order substrate recognition of eIF2α by the RNA-dependent protein kinase PKR. *Cell* 122:887–900. Copyright © 2005, with permission from Elsevier.)

(see Fig. 15.4). HRI is expressed predominantly in erythroid cells (red blood cell precursors) and is activated by heme, the prosthetic group of hemoglobin, through two heme-binding domains. Analysis of HRI$^{-/-}$ knockout mice demonstrated the importance of this kinase in regulating globin gene expression and cell survival in erythroid cell lineages. In normal mice, when the intracellular concentration of heme declines, HRI is activated and phosphorylates eIF2α. This prevents the synthesis of globin in excess of heme. In iron-deficient HR1$^{-/-}$ mice, much of eIF2α remains unphosphorylated (Fig. 14.23). Synthesis of both α- and β-globin continued, resulting in aggregation of globin in red blood cells, anemia, and accelerated apoptosis in bone marrow and spleen.

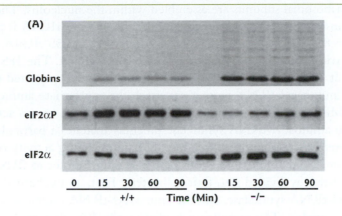

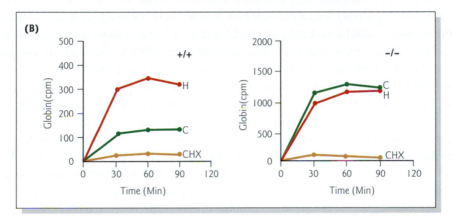

Figure 14.23 Protein synthesis and eIF2α phosphorylation in reticulocytes from HRI⁻/⁻ knockout mice. (A) Reticulocytes (red blood cell precursors) were isolated from HRI$^{+/+}$ or HRI$^{-/-}$ mice and metabolically labeled with [^{35}S]methionine. Samples were taken every 15 minutes to monitor the rate of globin protein synthesis (top, autoradiogram). Phosphorylation of eIF2α was analyzed by Western blot (middle) using an antibody specific to the phosphorylated protein. The level of phosphorylated eIF2α (eIF2αP) was significantly decreased in HRI$^{-/-}$ reticulocytes compared with HRI$^{+/+}$ reticulocytes. The lower levels of eIF2αP result from phosphorylation by other kinases, such as PKR. Total eIF2α levels were not altered significantly (bottom, Western blot). (B) HRI$^{+/+}$ or HRI$^{-/-}$ reticulocytes were incubated in [^{35}S]methionine in the presence of hemin (H) or cycloheximide (CHX). Hemin is the Fe^{3+} oxidation product of heme. Untreated reticulocytes were used as the controls (C). Samples were taken every 30 minutes and analyzed for globin synthesis. Proteins were separated by SDS-PAGE, transferred to nitrocellulose membranes and visualized by autoradiography. Nitrocellulose strips containing globin were quantified by liquid scintillation counting. There was a significant loss of heme-dependent globin synthesis in HRI$^{-/-}$ reticulocytes; protein synthesis was not increased by the addition of hemin. In contrast, hemin significantly increased globin synthesis in +/+ reticulocytes, by 208% compared to controls. Protein synthesis of both +/+ and −/− reticulocytes was inhibited equally well by cycloheximide, which inhibits translation elongation and is independent of HRI. (Reprinted by permission from Nature Publishing Group and Macmillan Publishers Ltd: Han, A.P., Yu, C., Lu, L. et al. 2001. Heme-regulated eIF2α kinase (HRI) is required for translational regulation and survival of erythroid precursors in iron deficiency. *EMBO Journal* 20:6909–6918. Copyright © 2001.)

Chapter summary

Protein synthesis takes place on the ribosome. Ribosomes consist of a large and small subunit composed of ribosomal RNA and many ribosomal proteins. The peptidyl transferase center is in the large subunit. Decoding the mRNA occurs on the small subunit. There are three tRNA-binding sites on the ribosome that bridge the large and small subunits, the A (acceptor), P (peptidyl), and E (exit) sites.

Eukaryotic large and small ribosomal subunits are assembled within the nucleolus. The nucleolus is the site of 45S preribosomal RNA transcription by RNA polymerase I. The 45S pre-rRNA is processed to form the 28S rRNA, 18S rRNA, and 5.8S rRNA. The 28S rRNA, 5.8S rRNA, and 5S rRNA (transcribed by RNA polymerase III) are assembled with ribosomal proteins to form the large subunit. The 18S rRNA and associated proteins form the small subunit. After assembly, the 40S and 60S subunits are exported to the cytoplasm.

In a high-fidelity reaction, aminoacyl-tRNA synthetases attach the appropriate amino acid to a tRNA in two enzymatic steps. First, AMP is bound to the amino acid, and second, the amino acid is transferred to the 3′ end of the tRNA. Each aminoacyl-tRNA synthetase precisely matches a particular amino acid with the tRNA containing the correct corresponding anticodon. The proofreading activity of aminoacyl-tRNA synthetases hydrolyzes the mismatched amino acid either before or after transfer to tRNA.

The initiation stage of translation is divided into four main steps. First, an uncharged $tRNA^{Met}$ is charged with methionine by methionyl-tRNA synthetase, and the aminoacyl-tRNA interacts with the initiation factor eIF2 to form a ternary complex. This complex is loaded on the 40S ribosomal subunit to form a 43S complex. The 5′ cap and 3′ poly(A) tail of the mRNA join to form a closed loop with eIF4G serving as the bridge between them. Some cellular mRNAs are translated by a 5′ cap-independent mechanism in which ribosomes are directly recruited by an internal ribosome entry site (IRES). Second, in association with other initiation factors, including eIF4G/E, the mRNA is loaded on the 43S complex. Toeprinting (primer extension inhibition) assays have been used to characterize the initiation factors required for AUG recognition. Third, the 40S subunit scans for the AUG start codon, and initiation factors are released in a process that requires GTP hydrolysis. Fourth, the 60S subunit joins with the 40S subunit to form the 80S ribosome initiation complex, in a process that requires a second GTP hydrolysis step.

During the elongation phase of protein synthesis, an aminoacyl-tRNA is carried to the A site as part of a complex with GTP-bound eEF1A. The aminoacyl-tRNA enters the acceptor (A) site where decoding takes place. tRNA selection involves two steps: initial selection and proofreading. The two steps are separated by irreversible hydrolysis of GTP. When the ternary complex contains the correct tRNA, the initial selection step occurs more rapidly and GTP hydrolysis releases the tRNA into the A site. The site of peptide bond formation is located at the base of the central protuberance in the 50S large ribosomal subunit. Peptidyl transferase activity transfers a growing polypeptide chain from peptidyl-tRNA in the P site to an amino acid esterified with another tRNA in the A site. After the tRNAs and mRNA are translocated such that the next codon is moved to the A site, the process is repeated. Translocation is mediated by eEF2 and requires GTP hydrolysis.

Structural studies and biochemical analyses have demonstrated that the ribosome is a ribozyme. The 23S rRNA alone can catalyze peptide bond formation and the main features of the ribosome functional sites are formed of rRNA. Proteins are located on the periphery of the active site. The rate enhancement of peptide bond formation by the ribosome is mainly due to substrate positioning by the 23S rRNA within the active site, rather than to chemical catalysis. Universally conserved nucleotides in the peptidyl transferase center are critical for the catalysis of peptide release during termination.

Proteins emerging from the ribosomal tunnel often associate with other factors. Signal recognition particles (SRPs) bind to ribosomes translating polypeptides that bear a signal sequence for targeting to the endoplasmic reticulum. The SRP and SRP receptor use a cycle of recruitment and hydrolysis of GTP to control delivery of the ribosome–mRNA complex to the ER translocon. The bacterial trigger factor creates a protected folding space where nascent polypeptides are shielded from proteases and aggregation as they emerge from the peptide exit hole.

During termination of translation, the stop codons are recognized by release factor eRF1 in association with eRF3. The completed polypeptide is cleaved from the peptidyl-tRNA, followed by expulsion of the tRNA, dissociation of the ribosome from the mRNA, and dissociation of the 40S and 60S subunits. GTP hydrolysis may trigger the release of eRF1 and eRF3.

Translational control provides another level of gene regulation in eukaryotes. A classic example of both translational and post-translational control is the phosphorylation of eIF2. Four distinct protein kinases are activated in response to different stress conditions: GCN2, PKR, HRI, and PERK. For example, double-stranded RNA binds to the RNA-binding domains of PKR and promotes catalytic-domain dimerization. Catalytic-domain dimerization promotes autophosphorylation, and autophosphorylation of PKR enhances its catalytic activity and is required for specific recognition of eIF2α. HRI is activated by low heme levels. Activated HRI phosphorylates eIF2α and prevents the synthesis of globin in excess of heme. Phosphorylation of eIF2α by these kinases inhibits GDP–GTP exchange by reducing the dissociation rate of the nucleotide exchange factor eIF2B. This leads to inhibition of translation by blocking ternary complex formation. Selective translation of a subset of mRNAs continues, which allows cells to adapt to stress conditions.

Analytical questions

1 Describe a toeprint assay involving mammalian ribosomal subunits and an mRNA in a cell-free extract that contains all the factors necessary for translation.
   (a) What results would you expect to see with 40S ribosomal subunits alone?
   (b) With 60S subunits alone?
   (c) With both subunits and all amino acids except serine, which is required in the 15th position of the polypeptide?
2 What would be the effect on the function of eIF2 of inhibiting its guanine exchange factor? Would the outcome be different if eIF2 were phosphorylated?
3 You are investigating a tRNA whose charging specificity appears to be affected by a U11–G24 to U11–A24 base pair in the D stem. Design an experiment using an *in vitro* reaction to show that changing this base pair changes the charging specificity of the tRNA.
4 A skeptic thinks that only proteins can be enzymes. Provide X-ray crystallographic and biochemical evidence in support of the view that the ribosome is a ribozyme.

Suggestions for further reading

Abbot, C.M., Proud, C.G. (2004) Translation factors: in sickness and in health. *Trends in Biochemical Sciences* 29:25–31.

Ban, N., Nissen, P., Hansen, J., Moore, P.B., Steitz, T.A. (2000) The complete atomic structure of the large ribosomal subunit at 2.4 Å resolution. *Science* 289:905–920.

Blight, S.K., Larue, R.C., Mahapatra, A. et al. (2004) Direct charging of tRNA_CUA with pyrrolysine *in vitro* and *in vivo*. *Nature* 431:333–335.

Cochella, L., Green, R. (2005) An active role for tRNA in decoding beyond codon:anticodon pairing. *Science* 308:1178–1180.

Dar, A.C., Dever, T.E., Sicheri, F. (2005) Higher-order substrate recognition of eIF2α by the RNA-dependent protein kinase PKR. *Cell* 122:887–900.

Daviter, T., Murphy, F.V. IV, Ramkrishnan, V. (2005) A renewed focus on transfer RNA. *Science* 308:1123–1124.

Dey, M., Cao, C., Dar, A.C., Tamura, T., Ozato, K., Sicheri, F., Dever, T.E. (2005) Mechanistic link between PKR dimerization, autophosphorylation, and eIF2α substrate recognition. *Cell* 122:901–913.

Dock-Bregeon, A.C., Rees, B., Torres-Larios, A., Bey, G., Caillet, J., Moras, D. (2004) Achieving error-free translation: the mechanism of proofreading of threonyl-tRNA synthetase at atomic resolution. *Molecular Cell* 16:375–386.

Egea, P.F., Stroud, R.M., Walter, P. (2005) Targeting proteins to membranes: structure of the signal recognition particle. *Current Opinion in Structural Biology* 15:213–220.

Ferbitz, L., Maier, T., Patzelt, H., Bukau, B., Deuerling, E., Ban, N. (2004) Trigger factor in complex with the ribosome forms a molecular cradle for nascent proteins. *Nature* 431:590–596.

Francklyn, C., Perona, J.J., Puetz, J., Hou, Y.M. (2002) Aminoacyl-tRNA synthetases: versatile players in the changing theater of translation. *RNA* 8:1363–1372.

Gilbert, R.J.C., Fucini, P., Connell, S. et al. (2004) Three-dimensional structures of translating ribosomes by cryo-EM. *Molecular Cell* 14:57–66.

Gould, P.S., Bird, H., Easton, A.J. (2005) Translation toeprinting assays using fluorescently labeled primers and capillary electrophoresis. *BioTechniques* 38:397–400.

Gromadski, K.B., Rodnina, M.V. (2004) Kinetic determinants of high fidelity tRNA discrimination on the ribosome. *Molecular Cell* 13:191–200.

Halic, M., Becker, T., Pool, M.R., Spahn, C.M.T., Grassucci, R.A., Frank, J., Beckmann, R. (2004) Structure of the signal recognition particle interacting with the elongation-arrested ribosome. *Nature* 427:808–814.

Han, A.P., Yu, C., Lu, L. et al. (2001) Heme-regulated eIF2α kinase (HRI) is required for translational regulation and survival of erythroid precursors in iron deficiency. *EMBO Journal* 20:6909–6918.

Hill, W.E., Dahlberg, A., Garrett, R.A., Moore, P.B., Schlessinger, D., Warner, J.R., eds (1990) *The Ribosome. Structure, Function, and Evolution.* American Society for Microbiology, Washington, DC.

Holcik, M., Sonenberg, N. (2005) Translational control in stress and apoptosis. *Nature Reviews Molecular Cell Biology* 6:318–327.

Ibba, M., Söll, D. (2000) Aminoacyl-tRNA synthesis. *Annual Review of Biochemistry* 69:617–650.

Kapp, L.D., Lorsch, J.R. (2004) The molecular mechanics of eukaryotic translation. *Annual Review of Biochemistry* 73:657–704.

Klaholz, B.P., Myasnikov, A.G., Van Heel, M. (2004) Visualization of release factor 3 on the ribosome during termination of protein synthesis. *Nature* 427:862–865.

Krishnamoorthy, T., Pavitt, G.D., Zhang, F., Dever, T.E., Hinnebusch, A.G. (2001) Tight binding of the phosphorylated α subunit of initiation factor 2 (eIF2α) to the regulatory subunits of guanine nucleotide exchange factor eIF2B is required for inhibition of translation initiation. *Molecular and Cellular Biology* 21:5018–5030.

Leung, A.K., Andersen, J.S., Mann, M., Lamond, A.I. (2003) Bioinformatic analysis of the nucleolus. *Biochemical Journal* 376:553–569.

Namy, O., Rousset, J.P., Napthine, S., Brierley, I. (2004) Reprogrammed genetic decoding in cellular gene expression. *Molecular Cell* 13:157–168.

Nissen, P., Hansen, J., Ban, N., Moore, P.B., Steitz, T.A. (2000) The structural basis of ribosome activity in peptide bond synthesis. *Science* 289:920–930.

Nitta, I., Kamada, Y., Noda, H., Ueda, T., Watanabe, K. (1998) Reconstitution of peptide bond formation with *Escherichia coli* 23S ribosomal RNA domains. *Science* 281:666–669.

Noller, H.F., Hoffarth, V., Zimniak, L. (1992) Unusual resistance of peptidyl transferase to protein extraction procedures. *Science* 256:1416–1419.

Pestova, T.V., Kolupaeva, V.G. (2002) The roles of individual eukaryotic translation initiation factors in ribosomal scanning and initiation codon selection. *Genes and Development* 16:2906–2922.

Proud, C.G. (2005) eIF2 and the control of cell physiology. *Seminars in Cell and Developmental Biology* 16:3–12.

Raška, I. (2003) Oldies but goldies: searching for Christmas trees within the nucleolar architecture. *Trends in Cell Biology* 13:517–525.

Roberts, R.B., ed. (1958) *Microsomal Particles and Protein Synthesis.* Pergamon Press, New York.

Rodnina, M.V., Beringer, M., Bieling, P. (2005). Ten remarks on peptide bond formation on the ribosome. *Biochemical Society Transactions* 33:493–498.

Sachs, M.S., Wang, Z., Gaba, A. et al. (2002) Toeprint analysis of the positioning of translation apparatus components at initiation and termination codons of fungal mRNAs. *Methods* 26:105–114.

Sankaranarayanan, R., Dock-Bregeon, A.C., Romby, P. et al. (1999) The structure of threonyl-tRNA synthetase-tRNA (Thr) complex enlightens its repressor activity and reveals an essential zinc ion in the active site. *Cell* 97:371–381.

Tschochner, H., Hurt, E. (2003) Pre-ribosomes on the road from the nucleolus to the cytoplasm. *Trends in Cell Biology* 13:255–263.

Valle, M., Zavialov, A., Sengupta, J., Rawat, U., Ehrenberg, M., Frank, J. (2003) Locking and unlocking ribosomal motions. *Cell* 114:123–134.

Youngman, E.M., Brunelle, J.L., Kochaniak, A.B., Green, R. (2004) The active site of the ribosome is composed of two layers of conserved nucleotides with distinct roles in peptide bond formation and peptide release. *Cell* 117:589–599.

Chapter 15

Genetically modified organisms: use in basic and applied research

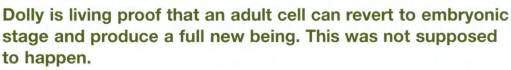

Dolly is living proof that an adult cell can revert to embryonic stage and produce a full new being. This was not supposed to happen.

Charles Krauthammer, *Time* (1997) 149:60.

Outline

15.1 Introduction

A favorite theme of science fiction writers is that of an evil or misguided scientist working in a secret lab on a remote island whose genetically altered monsters escape and wreak havoc. A classic example is the beast-people in H.G. Wells' 1896 novel *The Island of Doctor Moreau*. Another gem is a 1959 movie called *The Killer Shrews* – giant, flesh-eating, *and* poisonous! Sharing of genes between species and genetic manipulation of organisms once seemed outrageous notions. The creation of clones from the nuclear DNA of another

Table 15.1 Comparison of transgenic technology, gene targeting, and cloning by nuclear transfer.

	Method of DNA introduction	Recipient cell	Result	Common use in analysis of gene expression
Transgenic technology	Pronuclear microinjection	Fertilized egg	Random integration of transgene	Gain of function analysis
Gene targeting	Homologous recombination in embryo-derived stem (ES) cells	Blastocyst stage embryo	Disruption or mutation of targeted gene	Loss of function analysis
Cloning by nuclear transfer	Transfer of entire nucleus	Enucleated unfertilized egg	Genetically identical individual to donor nucleus	Analysis of genome reprogramming

individual also has the ring of science fiction, particularly if it takes place in a jungle laboratory as in the 1978 movie *The Boys from Brazil*. In fact, transgenic organisms have been a reality since 1980, strategies to alter an organism's genome at the level of specific nucleotide changes by gene targeting were developed in the late 1980s, and it is now possible to produce clones by nuclear transfer.

A transgenic organism is one that carries *trans*ferred *gene*tic material (i.e. the transgene) that has been inserted into its genome at a random site. Gene targeting – the replacement or mutation of a particular gene – provides the means for creating strains of "knockout" organisms with mutations in virtually any gene. Cloning is the production of genetically identical animals by nuclear transfer from adult somatic (body) cells to unfertilized eggs. Table 15.1 compares and contrast these three distinct methods for genetic manipulation of organisms.

In this chapter various methods of genetic manipulation in the mouse are described that have led to key advances in many fields, including the study of neurobiology, human genetic disease, immunology, cancer, and development. The ability to introduce foreign or altered genes into the mouse provides an unparalleled resource for the study of gene regulation. Unlike similar investigations performed *in vitro* or in cultured cells, the transgene can be studied in the context of the whole organism. Transgenic and gene targeting strategies can be used to overexpress, modify, or inactivate genes in the mouse. Such genetic manipulation can be directed to all tissues of the body, specific cell types, or specific stages in development. This allows for the study of tissue-specific and developmental gene expression, and the analysis of loss and gain of gene function effects.

Although the focus of this chapter is on basic research using the mouse as a model organism, other applications of transgenic technology are also addressed. This technology can potentially be used to increase the performance of commercially important animals and plants by adding new traits or improving on existing ones. Another important potential application of transgenic technology is the overexpression of foreign proteins for therapeutic use.

Finally, cloning by nuclear transfer is described in the context of its potential for providing insight into mechanisms of development, cell differentiation, nuclear reprogramming, genomic imprinting, and aging. There are many practical constraints with the current cloning methodology. However, the potential value for preservation of prize animal stocks, wildlife conservation, and therapeutic uses is discussed.

15.2 Transgenic mice

In 1980 the first transgenic mouse was produced by microinjection of foreign DNA into fertilized eggs. A recombinant viral DNA construct was successfully integrated into the mouse genome, but it was rearranged and did not express. The first visible phenotypic change in transgenic mice was described in 1982 by

Figure 15.1 "Super" mouse. A transgenic mouse (left) expressing the rat growth hormone gene under control of the mouse metallothionein gene promoter grew to twice the size of a normal sibling (right). (Photograph courtesy of Ralph Brinster, University of Pennsylvania.).

Richard Palmiter and Ralph Brinster. They engineered mice that successfully integrated and expressed the rat growth hormone gene coding sequence. The unexpected and dramatic result was that some mice grew to be twice the size of normal siblings. Images of these "super" mice captured the attention of both the general public and scientists alike (Fig. 15.1). Since then, transgenics has been a rapidly growing field, with many technological advances and refinements (Focus box 15.1).

How to make a transgenic mouse

The standard procedure for making transgenic mice by microinjection of foreign DNA into fertilized eggs is shown in Fig. 15.2. Microinjection results in the introduction of the transgene into the chromosomes of a fertilized mouse egg. If the transgene is integrated into one of the embryonic chromosomes, the mouse will be born with a copy of this new information in every cell. There are three main stages in the process: (i) microinjection of DNA into the pronucleus of a fertilized mouse egg; (ii) implantation of the microinjected embryo into a foster mother; and (iii) analysis of mouse pups and subsequent generations for the stable integration and expression of the transgene.

Oncomouse patent

FOCUS BOX 15.1

Is a transgenic mouse an invention? In 1988, the United States Patent Office decided just that when Philip Leder and Timothy Stewart of Harvard University received the first ever patent to be issued on any mammal. The "OncoMouse" patent was for all transgenic nonhuman mammals whose germ cells and somatic cells contain an activated oncogene sequence (originally the *myc* oncogene). DuPont subsequently bought the rights to the patent. Two patents followed in 1992 and 1999, for a method for establishing a cell line from a transgenic nonhuman mammal and a testing method using transgenic mice expressing an oncogene. DuPont packaged the three patents (US patents 4,736,866; 5,087,571; and 5,925,803) as the OncoMouse

Portfolio and now sells licensing rights to private companies that want to use transgenic lab mice in medical research. Nonprofit researchers funded by the National Institutes of Health are allowed to use the mice without paying royalties, following guidelines for use agreed upon between DuPont and the Public Health Service of the US Department of Health and Human Services.

The patent remains controversial worldwide. Under new restrictions issued by the European Patent Office in 2004, in Europe the OncoMouse patent for producing transgenic animals applies only to mice. In Canada, the Supreme Court ruled that the OncoMouse does not qualify as an invention and therefore cannot be patented at all.

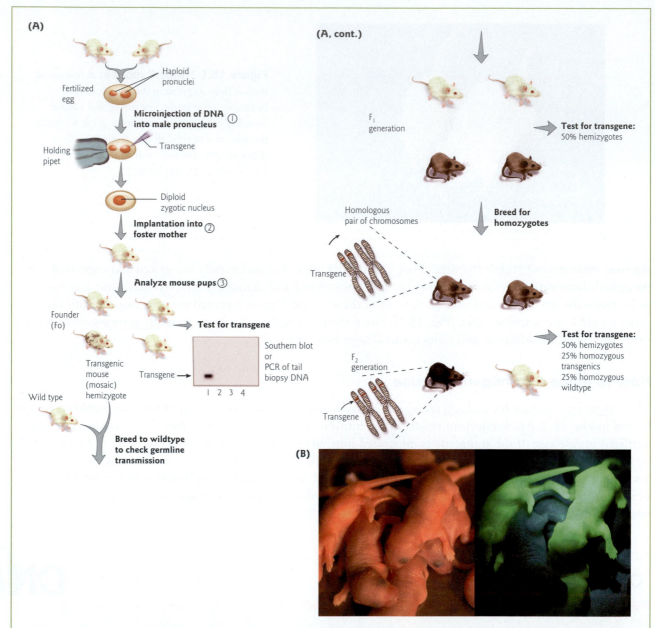

Figure 15.2 How to make a transgenic mouse. (A) The three main stages in the standard procedure for making transgenic mice by microinjection of foreign DNA into fertilized eggs are shown: (1) microinjection of DNA into the pronucleus of a fertilized mouse egg; (2) implantation of the microinjected embryo into a foster mother; and (3) analysis of the mouse pups and subsequent generations for the stable integration of the transgene (see text for details). (B) GFP-expressing transgenic mice. The photo in the left panel was taken in normal light, the photo in the right panel in UV light. A nontransgenic sibling is not fluorescent. (Reprinted by permission of Federation of the European Biochemical Societies from: Okabe, M., Ikawa, M., Kominami, K., Nakanishi, T., Nishimune, Y. 1997. "Green mice" as a source of ubiquitous green cells. *FEBS Letters* 407:313–319. Copyright © 1997. Photograph courtesy of Masaru Okabe, Osaka University, Japan).

Pronuclear microinjection

The first step in making a transgenic mouse is to surgically remove eggs from a female mouse and fertilize them with mouse sperm. The second step is microinjection of the foreign DNA into the fertilized egg. A transgenic mouse is usually designed to overexpress a gene of interest. It represents a "gain of function" mouse model. The transgenic construct at a minimum contains a promoter element, complementary DNA (cDNA) for the gene of interest, and a polyadenylation signal. The promoter sequence is required as a binding site for RNA polymerase and transcription factors (see Section 11.3). By choosing either a tissue-specific or inducible promoter, the expression of the foreign gene can be regulated spatially and temporally. For example, in the case of the "super" mice described above, fertilized eggs were injected with a gene construct (recombinant plasmid vector) containing the rat growth hormone cDNA under control of the mouse metallothionein (MT) gene promoter. The metallothionein promoter was chosen because it is inducible (turned on) by the heavy metals zinc and cadmium. The metallothionein gene codes for a heavy-metal-binding protein that is active in the liver. More recently developed strategies for inducible expression are described at the end of this section.

The timing of microinjection is critical. The foreign DNA must integrate into the mouse genome prior to the doubling of the genetic material that occurs before the first cleavage. If the foreign DNA integrates into the mouse genome after the first cleavage, a mosaic mouse may be produced in which many cells do not possess the new gene. For this reason, the transgene is introduced into the fertilized egg at the earliest possible stage; i.e., the pronuclear period immediately following fertilization.

For several hours following the entry of the sperm into the egg, the sperm nucleus and the egg nucleus – called the male and female pronuclei – are microscopically visible as individual structures. Injections must be done before the haploid sperm and egg pronuclei have fused to form a diploid zygotic nucleus. Several hundred copies of the transgene in solution are loaded into a glass microneedle, and the DNA is injected directly into one of the pronuclei (Fig. 15.2). Usually the male pronucleus is microinjected since it is larger in size and closer to the egg surface. The egg itself is only 50 μm in diameter, so this procedure requires great skill and patience. The injections are very tedious and even the experienced technician may only be able to complete 5–10 successful injections in a day.

Implantation into foster mother

To be able to become live-born transgenic mice, the manipulated embryos must be transferred into the reproductive tract of a female mouse (Fig. 15.2). Female mouse recipients for embryo transfer are prepared by mating with vasectomized males. Anywhere from two to 15 successfully injected embryos are surgically transferred to the uterus of the recipient "pseudopregnant" mouse. Pregnancy is visible about 2 weeks after embryo transfer and the litter is delivered about 1 week later. Mouse pups are weaned and analyzed at 3–4 weeks of age.

Analysis of mouse pups

There are two important questions to be answered once mouse pups are ready for analysis. First, is there stable integration of the transgene into the mouse chromosome? In other words, are the mice, in fact, transgenic? Second, if the transgene is shown to be present in the mouse genome, is it expressed appropriately? In this context, "appropriate" means whether the gene is expressed at the correct time during development, in the correct tissue(s), and in the correct amount.

Analysis of stable integration Tail biopsies are taken from mouse pups to obtain DNA for analysis. The DNA is tested for the presence of the transgene by Southern blot hybridization or polymerase chain reaction (PCR) (Fig. 15.2). A mouse in which there was successful integration of the transgene DNA is referred to as the founder (F_0) of a new transgenic lineage. The transgene will be heritable and passed on from generation to generation as part of the mouse genome. The success rate of stable integration is around 2.5–6% in mice.

The integration event usually occurs by nonhomologous recombination, which results in a random site of integration of foreign DNA into the embryonic genome. The mechanism is not clear, but nonhomologous recombination may be provoked by the action of DNA repair enzymes (see Fig. 7.15). These enzymes are thought to be induced by free ends of exogenous DNA (the transgene) and create double-stranded breaks in the chromosomal DNA. The probability of identical integration events in two embryos receiving the same transgene is highly unlikely. In addition, it is impossible to regulate exactly how many copies of the transgene will be introduced into the embryo and how many will join together to integrate (usually at a single site) as a single linear array, called a concatamer (anywhere from one to 150 copies). The number of copies of the transgene that integrate into the founder's genome is referred to as the copy number. The copy number rarely appears to be correlated with the degree of transgene expression in the mouse.

The first generation of transgenic mice is referred to as "mosaic" or "chimeric." In general, they have foreign DNA in some, but not all, germline and somatic cells. The transgenic founder mice are inbred to produce a second generation, called the F_1 generation. This generation of mice is also analyzed for stable integration of the transgene. If the germ cells of the founder (mosaic or not) transmit the transgene stably, then all descendants of this mouse are members of this unique transgenic lineage. The genotype of the founder is described as hemizygous for the transgene rather than heterozygous, since the new transgenic locus is present in only one member of a particular chromosome pair. A homozygous genotype, in which transgene alleles are present on each chromosome in a pair, may be produced by the mating of hemizygous F_1 siblings (see Fig. 15.2). Since the site of integration may be different, mating mice with identical transgenes but from different founder lineages will not result in a true homozygote in which independent segregation of the loci is predictable.

Analysis of transgene expression   If the transgene is shown to be present in the mouse genome, the next question is whether the transgene is regulated well enough to function in its new environment. To answer this question, RNA and protein expression levels are analyzed. A variety of techniques can be used to assess transcriptional activity of the transgene, including Northern blots, reverse transcription–PCR (RT-PCR), and *in situ* hybridization. For analysis of transgene expression at the level of translation, Western blots and immunohistochemistry are often used (see Figs 9.8 and 9.9).

Many studies have found dramatic differences in the expression of a specific transgene within individual sibling embryos simply due to different sites of random integration. As noted above, the first mice to express a transgene (the rat growth hormone gene) were over twice the size of untreated normal siblings. In addition, females had impaired fertility and both males and females died prematurely. The conclusion of investigators was that there was an unregulated, excessive amount of growth hormone produced by the introduced cDNA under control of the inducible metallothionein gene promoter. Additional gene-specific regulatory regions were most likely needed to achieve appropriately regulated expression (see Section 11.3 and Focus box 11.1).

In addition to problems with expression of the transgene itself, frequently the transgene randomly inserts into an endogenous mouse gene and interferes with its expression. Such mutations may be either inconsequential or lethal, but they can also result in a viable mouse with a distinct mutant phenotype. The identification of the site of transgene insertion is of great value in such cases of "insertional mutagenesis" because it maps the location of an important endogenous gene. These mutations can be distinguished from the true transgenic phenotype because only a single lineage exhibits the defect.

Inducible transgenic mice

The situation often arises where the introduction of a transgene results in death or such reduced viability that it is difficult or impossible to maintain the transgenic mouse line by breeding. For example, certain transgene sequences may be activated *in utero* and may affect embryo survival. In this case, it is helpful to design a transgene that can be activated after establishment of the transgenic line. The development of inducible expression systems has allowed researchers to overcome some of the problems associated with transgenic studies. For example, "Tet–off" is one commonly used expression system in which transgene expression is dependent on the activity of an inducible transcriptional activator. Expression of the transcriptional activator can be regulated both reversibly and quantitatively by exposing the transgenic animals to varying concentrations of the antibiotic tetracycline, or tetracycline derivatives such as doxycycline (Fig. 15.3).

15.3 Gene-targeted mouse models

Amino acid sequence prediction for a protein produced by a cloned gene and subsequent protein modeling can often provide information on function, especially if the protein of interest shares sequence homology with a well-characterized protein. But for an unknown protein, or for a gene with multiple effects, the best method for determining what a gene does in an organism is to analyze its function in a model organism. In Section 15.2, the production of transgenic mice by pronuclear microinjection was described. Transgenic mice engineered in this way are generally "gain of function" mutants since the transgene is designed either to express a novel gene product or to misexpress a normal gene. Another means of testing for gene function *in vivo* is by genetic "knockout" in which a particular gene is deliberately disrupted. Transgenic mice engineered by this approach are called "loss of function" mutants (Focus box 15.2).

The ability to create a mouse of any desired genotype was moved from science fiction to reality by the vision and perseverance of Mario Capecchi. In 1980, Capecchi requested funding from the National Institutes of Health (NIH) to test the feasibility of what he called "gene targeting." He proposed to precisely alter endogenous genes in mouse embryonic stem (ES) cells by homologous recombination with a foreign DNA sequence. His proposal was rejected because reviewers thought that getting homologous recombination to occur was next to impossible. Capecchi used funds from another project to carry out his experiments and acquired data showing that gene targeting was, in fact, possible. In 1984, he submitted a second request to the NIH for funding that was successful. According to Capecchi in a *Scientific American* article he wrote in 1994, "The critique of this proposal opened with the statement, 'We are glad that you didn't follow our advice'."

Generation of the first "knockout" mouse using this gene targeting method was reported by Capecchi in 1990. Mice homozygous for the loss of the *int-1* proto-oncogene showed severe abnormalities in the development of the midbrain and cerebellum. Since then, many variations of this technique have been developed. For example, "knockin" and "knockdown" mice have been engineered where expression of an endogenous gene is altered, but not necessarily inactivated. The targeted changes are introduced into the

A mouse for every need

FOCUS BOX 15.2

A US-based consortium is being organized to systematically knock out mouse genes one by one in embryonic stem (ES) cells. The goal is to have knockout ES cells available "off the shelf" for all the predicted 30,000 genes (see Section 16.5 and Table 16.4). Investigators could then order the desired gene-targeted cell line from a catalog, and use them to establish a lineage of knockout mice. Making the targeted ES cells is an achievable goal. However, there is no guarantee that the targeted cells will yield useful information about gene function in mice for two main reasons. First, up to 15% of genes are thought to be needed for embryonic development. Any ES cell missing one of these important genes will not allow development of a lineage of knockout mice. Second, some knockout mice find ways to compensate for the lost function of the knocked-out gene and so appear normal. To address these potential problems in making "off the shelf" mouse models, a European-based consortium is adopting a different approach. They are focusing on engineering knockout ES cells containing genes that the investigator will be able to switch on or off at any stage of development in the mutant mouse.

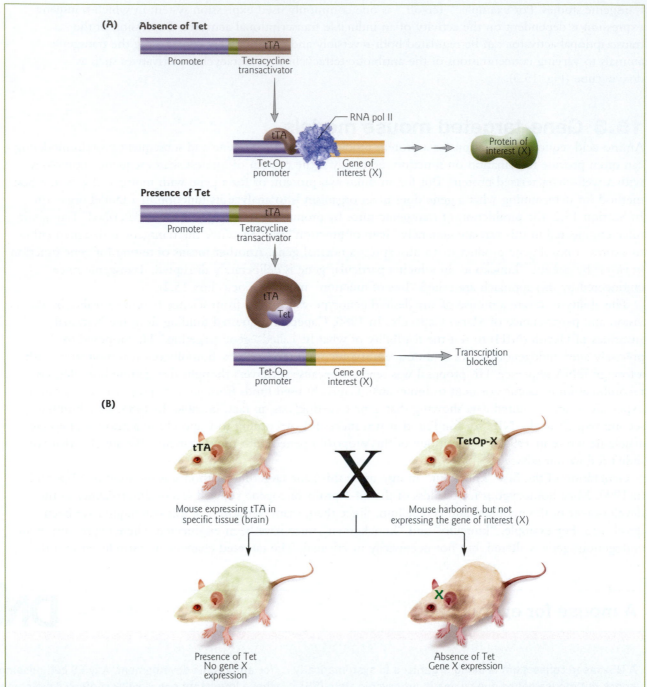

Figure 15.3 Inducible transgene expression. (A) In the tetracycline (Tet)-off system, transcription of a gene of interest (X) occurs when the tetracycline–controlled transactivator protein (tTA) binds to a tetracycline-responsive DNA regulatory element called the TetOp promoter. In the absence of the drug tetracycline, tTA binds to the TetOp promoter and activates transcription of the target transgene. In the presence of tetracycline, tTA cannot bind to the TetOp promoter and expression of the target transgene remains inactive. (B) The Tet-off system is introduced into mice by creating two independent transgenic mouse lines. The first transgenic mouse line expresses tTA in a specific tissue (e.g. brain). The second transgenic mouse line carries gene X under control of the TetOp promoter (TetOp-X). Gene X is not expressed in this mouse line. Mice from the two lines are then crossed to generate a Tet-inducible line of transgenic mice. In these mice, in the absence of tetracycline, those tissues expressing tTA (e.g. in the brain) will also express gene X. When Tet is present, transcription of gene X is blocked.

sequence of the coding region or upstream regulatory elements of the gene. In addition, inducible and conditional expression systems have been designed for gene-targeted mouse models.

Knockout mice

Making knockout mice requires exact knowledge of the complete DNA sequence of the gene of interest, or at least a significant portion of it. This means it is not a screening technique for finding genes involved in processes but rather a means of assigning function to known genes with unknown or poorly understood roles.

The method developed by Capecchi for making knockout mice involves five main stages: (i) construction of the targeting vector; (ii) gene targeting in embryo-derived stem (ES) cells; (iii) selection of gene-targeted cells; (iv) introduction of targeted ES cells into mouse embryos and implantation into a foster mother; and (v) analysis of chimeric mice and inbreeding (Fig. 15.4). The focus of this section is on transgenic mice, but knockout technology is not limited to this model organism. Genetic knockouts of zebrafish, fruitflies, and a host of other eukaryotic systems have been created using similar methods. Most recently, gene-targeted human embryonic stem cells were engineered, although in this case the experiments ended with the stem cells.

Construction of the targeting vector

A recombinant DNA molecule called the "targeting vector" is constructed in the laboratory by investigators (Fig. 15.4). The cloned gene of interest is typically disrupted by insertion of the *neo*^{r} gene, which encodes the enzyme neomycin phosphotransferase. It serves as a marker to indicate that the vector DNA was integrated in a mouse chromosome. The *neo*^{r} insert is flanked by DNA sequences from the two ends of the gene of interest. These gene-specific regions provide the region of homology for recombination with the corresponding endogenous mouse gene. The targeting vector is also engineered to carry a second marker at one end: the herpesvirus thymidine kinase (*tk*) gene. These markers are standard, but others could be used instead.

Gene targeting in ES cells

The next step in creating a knockout mouse is to obtain ES cells from an early mouse embryo. ES cells are used for two main reasons. First, they can be cultured in the lab indefinitely. Second, they are pluripotent; that is, they are capable of giving rise to all cells types because they are not yet specialized. Investigators find it convenient to use the coat color of future newborn mouse pups as a visual guide to whether the ES cells have survived in the embryo. In the example shown in Fig. 15.4, ES cells are obtained from a mouse with a brown coat and implanted into an embryo from a black strain of mice. The ES cells carry two copies of the *agouti* gene. This gene produces brown coloring by directing yellow pigment to be deposited next to black pigment in the hair shaft. Production of the yellow and black pigments is under the control of other genes. The *agouti* gene is dominant and generates a brown coat even when present in cells as a single copy. Alternatively, ES cells can be obtained from a mouse with a black coat (dominant) and implanted into an embryo from an albino (recessive) strain of mice.

The ES cells are transfected with the targeting vector, either by chemical transfection or electroporation. Three events can take place: no recombination, homologous recombination, or nonhomologous recombination. Homologous recombination is a very rare event in mammalian cells, occurring at best in only one in 1000 cells. Nonhomologous recombination is a more common event and involves random integration.

Selection of gene-targeted ES cells

The third stage in creating a knockout mouse is to select for the rare ES cell in which homologous recombination has taken place. The targeted cell(s) is selected for by a lengthy screening and enrichment procedure (Fig. 15.4). This is a very time-consuming step – lots of cells and lots of trials are required to obtain a pure culture of targeted ES cells.

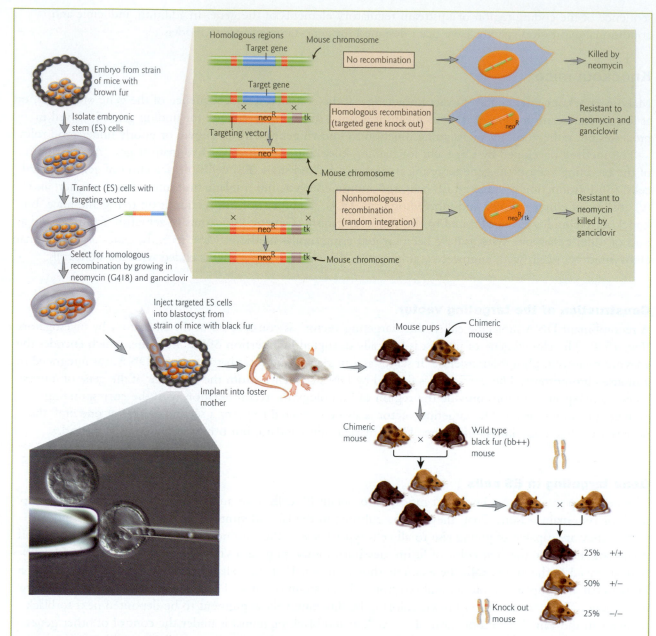

Figure 15.4 How to make a knockout mouse. The five main stages involved in making a knockout mouse are shown: (1) Construction of the targeting vector (inset). The targeting vector may contain any DNA sequence of interest, e.g. different alleles (both functional and nonfunctional), different genes, or reporter genes. Regardless of the insert, some flanking DNA that is identical in sequence to the targeted gene locus must be included. In addition to the positive selection marker (in this case, neomycin resistance, *neo*ʳ), often a negative selection marker (e.g. thymidine kinase, *tk*) is added to the targeting vector. The negative selection marker is outside the region of sequence similarity between the vector and targeted locus on the mouse chromosome. (2) Gene targeting in embryo-derived stem (ES) cells. Embryonic stem cells are isolated from a strain of mice with brown fur and grown in culture. The cells are transfected with the targeting vector and grown in neomycin (or the neomycin analog G418) and ganciclovir to select for homologous recombination events. The original targeting vector cannot replicate in a nucleus and it is lost quickly in dividing cells. (3) Selection of gene-targeted cells (inset). Three outcomes are possible. (i) No recombination takes place between the targeting vector and the mouse chromosomes. These cells are killed by neomycin. (ii) Homologous recombination takes place. By mechanisms that are poorly understood, the targeting vector finds the targeted gene and recombination takes place within the flanking homologous sequences. The end result is that the *neo*ʳ gene (positive

Introduction of ES cells into a mouse embryo and implantation into a foster mother

Mouse embryos at the blastocyst stage (4.5 days old) have not yet become attached to the mother's uterus and can be manipulated in a petri dish. When a pure population of targeted ES cells is obtained, the cells are injected into the blastocoel cavity of a blastocyst from a black strain of mice (Fig. 15.4). These recipient blastocysts lack the *agouti* gene which is present in the targeted ES cells. In the absence of the ES cells, the embryo would acquire a totally black coat. The altered embryo is implanted into the uterus of a foster mother by surgical procedures and allowed to grow to term.

Analysis of chimeric mice and inbreeding

When mouse pups are born several weeks later, investigators examine the coats of the newborns. Solid black coloring indicates that the ES cells did not survive the procedure. If the pups have a patchy brown and black coat, this indicates that the ES cells have survived and proliferated, contributing to the formation of most mouse tissues (Fig. 15.4). These mice are called "chimeras" because they contain cells derived from two different strains of mice. Some of the germ cells will be derived from the targeted ES cells and others will be derived from the recipient blastocyst. Getting the targeted ES cells into the germline is strictly a matter of chance. Thus, a lot of chimeric mice may have to be created and bred before researchers are successful in passing the targeted mutation on to the next generation.

To obtain a pure lineage, chimeric males are mated to black (non-*agouti*) females (Fig. 15.4). Any black offspring are discarded, since they represent mice arising from sperm without the gene knockout. The presence of a brown coat indicates that the mice are heterozygotes for the targeted gene knockout: one chromosome has the gene of interest disrupted, while the other chromosome in the pair has an intact gene. Males and females carrying the targeted gene knockout are then mated to each other to eventually produce homozygous mice, in which both chromosomes in the pair have the gene of interest disrupted. These mice are identified by direct analysis of their DNA, either by Southern blot or PCR (see Tool boxes 8.3 and 8.7). Next, investigators confirm that the mice really do lack the targeted gene product in all tissues of the body. RNA and protein expression levels are analyzed by such techniques as Northern blotting or RT-PCR for RNA, and Western blotting for proteins (see Figs 9.8 and 9.9). Finally, the mice are examined carefully for any sign of morphological or behavioral abnormalities. The phenotype of the mouse displays the impact of the particular gene on mouse development and physiology (Fig. 15.5).

selection marker) is inserted into the chromosome in place of the targeted mouse locus. The negative selection marker is not incorporated into the chromosome by homologous recombination. These cells are resistant to both neomycin and ganciclovir. (iii) Nonhomologous recombination takes place. If the targeting vector aligns in a nonhomologous region of the mouse genome, then recombination is random and both the positive (neo^r) and negative (*tk*) selection markers are integrated into the genome. The products of nonhomologous recombination are cells that are resistant to neomycin but that are killed by ganciclovir. (4) Introduction of targeted ES cells into mouse embryos and implantation into a foster mother. A pure population of gene-targeted ES cells, in which homologous recombination has occurred, are isolated and injected into a new blastocyst from a strain of mice with black fur. Several chimeric blastocysts are implanted into a pseudopregnant mouse. (Inset) Microinjection of a Day 4 blastocyst. (Photograph courtesy of the laboratory of Dr. Manfred Baetscher, Oregon Health & Science University Transgenic Core, Portland, Oregon.) (5) Analysis of chimeric mice and inbreeding. Some mouse pups will have normal black fur mice (indicating that the ES cells did not survive) but others will be chimeric mice with patches of brown fur. Chimeric mice are mated with wild-type black-fur mice. If the gonads of the chimeric mice were derived from gene-targeted ES cells, all the offspring will have brown fur. Every cell in brown–fur mice is heterozygous for the homologous recombination event. These heterozygous brown mice (+/−) are mated to produce a strain of brown mice whose cells carry the chosen mutation in both copies (alleles) of the target gene and thus lack a functional gene. Homozygotes for the gene knockout are identified by direct analyses of their DNA. This pure breeding mouse strain is a "knockout mouse."

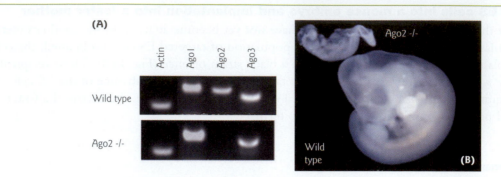

Figure 15.5 Argonaute2 knockout mouse. The *Argonaute2* (*ago2*) gene, which encodes the "Slicer" activity for RNAi (see Fig. 9.12), was disrupted by gene targeting methods similar to those described in Fig. 15.4. (A) Total RNA from wild-type or mutant embryos (Ago2–/–) was tested for expression of *ago2*, and two other Argonaute family members, Ago1 and Ago3 by RT-PCR. Actin was also examined as a control. The knockout embryos showed no expression of *ago2* mRNA, whereas *ago1* and *ago2* mRNA was present in normal levels. (B) At embryonic development day E10.5, Ago2–/– embryos show severe developmental delay as compared with heterozygous and wild-type littermates, indicating that *ago2* is essential for mouse development. (Reprinted with permission from Liu, J., Carmell, M.A., Rivas, F.V. et al. 2004. Argonaute2 is the catalytic engine of mammalian RNAi. *Science* 305:1437–1441. Copyright © 2004 AAAS.)

Knockin mice

Depending on the research question, an experimenter may be interested in expressing transgenes from known regulatory elements, or to modify rather than merely inactivate the targeted gene. Such expression of introduced DNA sequences can be achieved by generating a knockin mouse. The method is essentially the same as for generating knockout mice – a targeting vector with a foreign DNA insert is introduced into a specific locus by homologous recombination in ES cells. The knockin approach is often used for *in vivo* site-directed mutagenesis (see Fig. 9.7). Single base substitutions or deletions can be created to study the correlation of structure and function of the gene product in the whole animal. The knockin allele replaces the coding region of the endogenous allele, while keeping the endogenous upstream regulatory elements intact. The knockin allele is generally expressed precisely as the endogenous gene would have been expressed.

Knockdown mice

Gene targeting strategies can also be applied to inactivate or modify expression levels of a gene, by modifying the *cis*-regulatory elements (e.g. the promoter region or other upstream regulatory sequences; see Section 11.3). Making a "knockdown" mouse is analogous to "promoter bashing" (see Section 9.4 and Fig. 9.7) but with the advantage of analyzing effects on endogenous gene expression in the whole animal. Single base substitutions or deletions can be created to study the regulatory elements required, for example, to drive tissue-specific or developmental stage-specific gene expression. The knockdown targeting sequence disrupts endogenous upstream regulatory elements, while keeping the endogenous coding region intact. The effect of the targeted mutation on gene transcription is then analyzed in the knockdown mice.

Conditional knockout and knockin mice

Gene knockouts often result in embryonic lethality. To make it possible to study a gene's role later in development or in the adult mouse, it is useful to design a conditional mutation of the target gene. Genetic switches can be engineered to target expression or disruption of any gene (for which basic sequence information is available) to any tissue at any defined time. One of these genetic switches is the Cre/*lox* system. This system is based on the use of site-specific DNA recombination (Fig. 15.6).

Cre/*lox* system for site-specific recombination

Site-specific DNA recombinases have been known since the early days of molecular biology. The first member of the "Int" family of site-specific recombinases was discovered through study of integration-defective (int-) mutants of bacteriophage lambda (λ). These mutants were defective in the integration and excision of phage λ DNA into and out of the chromosome of the host cell (see Fig. 8.1). Since the 1960s, the Int family has grown to more than 100 members. One of the most widely known members of the family is the Cre (*cyclization recombination*) recombinase of bacteriophage P1. Having a detailed understanding of the properties and function of Cre allowed scientists to develop an elegant and precise method for controlling gene expression in mice.

Cre specifically recognizes a 34 bp site on the bacteriophage P1 genome called *lox* (locus of X-over) and catalyzes reciprocal recombination between pairs of *lox* sites (Fig. 15.6). Unlike many recombinases of the Int family, no accessory host factors are required for Cre-mediated recombination. This characteristic makes Cre very useful for genetic manipulation in eukaryotic cells. Figure 15.6 shows an example of activation of transgene expression by site-specific recombination. For conditional mutants of knockout mice, it is possible to modify the target gene by homologous recombination in ES cells so that it is flanked by *lox* sites. Mice containing such a modified gene are then crossed with mice expressing Cre in the desired target tissue. Cre-mediated excision results in tissue-specific gene knockout (Fig. 15.7).

15.4 Other applications of transgenic animal technology

The focus of this chapter so far has been primarily on basic research using transgenic or gene-targeted mice. Transgenic technology has been applied to many other animals with varying success, including farm animals, monkeys, and zebrafish. Transgenic animals have been explored as tools for applied purposes, ranging from artwork (Focus box 15.3) to pharmaceuticals.

Transgenic primates

Mice do not always provide an accurate model of human physiology and disease pathology. For example, a mouse model for Lesch–Nyhan syndrome does not mimic classic symptoms of the disease. Lesch–Nyhan is a rare X-linked disease in humans due to loss of activity of HPRT (hypoxanthine-guanine phosphoribosyltransferase), an enzyme that plays an important role in purine metabolism. This deficiency results in profound behavioral, motor, and intellectual abnormalities in children, including self-mutilating behaviors such as lip and finger biting and/or head banging. Mice have alternative pathways for purine metabolism that are not present in humans, so although the enzyme deficiency can be reproduced by knockout of the *HPRT* gene, knockout mice only show minor neurochemical or functional changes.

Due to such problems in developing mouse models, there has been some interest in extending transgenic and gene-targeting studies to nonhuman primates. Since primate development is much closer to human development than that of mice, they may offer opportunities for better understanding and treatment of human genetic diseases. However, the potential benefits from the use of primates in experimental medicine have to be weighed against the ethical and financial costs.

In January 2001, the first transgenic primate – a rhesus monkey – was engineered using an alternative approach to achieve gene transfer. The standard method of gene transfer, pronuclear microinjection, which is widely used in transgenic mouse production, has only had limited success in producing transgenic animals of larger species. Instead, researchers used a genetically modified retrovirus vector (see Focus box 17.3) to introduce the *GFP* gene into unfertilized rhesus monkey eggs. When successful, retrovirus vector-mediated

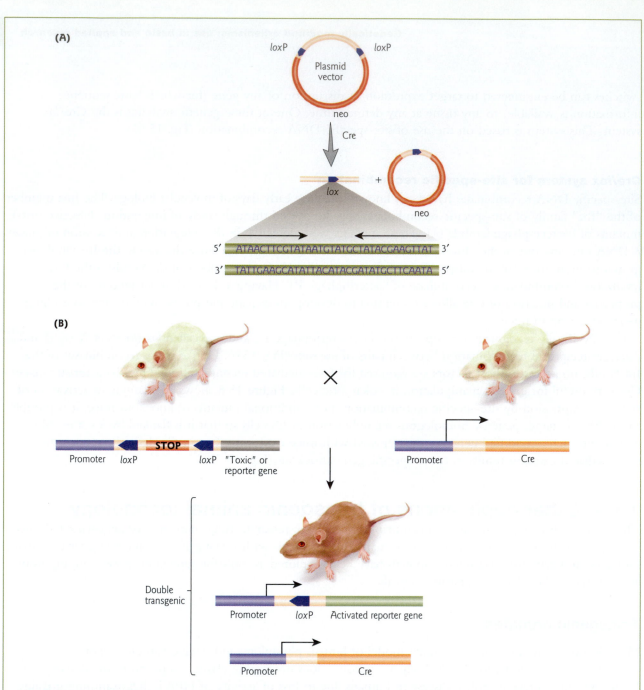

(A)

loxP   loxP

Plasmid
vector

neo

Cre

lox   +

neo

5′ ATAACTTCGTATAATGTATGCTATACGAAGTTAT 3′
3′ TATTGAAGCATATTACATACGATATGCTTCAATA 5′

(B)

×

Promoter loxP   STOP   loxP "Toxic" or
                                  reporter gene

Promoter   Cre

Double
transgenic

Promoter loxP   Activated reporter gene

Promoter   Cre

Figure 15.6 Inducible gene expression in mice using the Cre/*lox* system. (A) Cre-mediated
recombination. A plasmid vector is depicted that carries the neomycin resistance gene (*neo*) between two *lox* sites
(indicated by blue arrows). The *lox* site consists of two 13 bp inverted repeats separated by a 6 bp core region called the
"overlap" or "crossover" region. Each of the 13 bp inverted repeats of the *lox* site binds a single Cre monomer. In cells
carrying an integrated copy of this plasmid, Cre-mediated recombination results in excision of the DNA between them
as a covalently close circle. This leads to loss of the *neo* gene. The inverted repeats of the *lox* sequence are indicated
by thin arrows. (B) Cre-mediated activation of gene expression in transgenic mice. A *lox* "STOP" cassette is placed
between the promoter of interest and the "toxic" gene (e.g. a gene that results in death or such reduced viability that it
is difficult or impossible to maintain the transgenic mouse line by breeding) or reporter gene of interest. The "STOP"
signal blocks gene transcription. Mice carrying this inactive transgene are mated with mice transgenic for the *cre* gene.
In the doubly transgenic mouse, Cre-mediated recombination results in excision of the STOP sequence, which
activates the "toxic" or reporter gene. Tissue-specific expression of the transgene can be achieved by choosing the
appropriate promoters for both the reporter and *cre* transgenes. The blue arrows indicate the *lox* site (*loxP*); black
arrows indicate transcription. (Adapted from Sauer, B. 1998. Inducible gene targeting in mice using the Cre/lox
system. *Methods: A Companion to Methods in Enzymology* 14:381–392.)

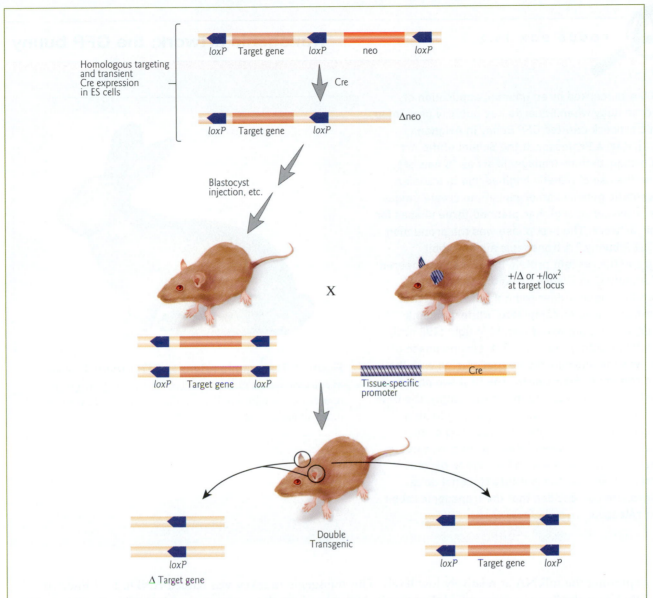

Figure 15.7 Conditional gene knockout by Cre-mediated recombination. The mouse gene of interest is modified by gene targeting in embryonic stem (ES) cells so that it is flanked by two *lox* (*loxP*) sites. Only one allele is shown for simplicity. A third *lox* site is included so that the neomycin resistance (*neo*) gene can be removed by transient expression of Cre in ES cells. After identification of ES cells lacking *neo* (Δ*neo*) but harboring the *loxP*-flanked target gene, mice are generated from ES cells by standard procedures (see Fig. 15.4). The *lox*-modified mouse is then mated with a mouse transgenic for the *cre* gene under control of a tissue-specific promoter. In double transgenic offspring, the *lox*-modified gene will be deleted in those tissues in which the *cre* transgene is expressed. In the example shown, Cre is only expressed in the ears (blue stripes) of the *cre* transgenic. In the double transgenic, deletion of the *lox*-modified target gene is restricted solely to the ears (gray). (Adapted from Sauer, B. 1998. Inducible gene targeting in mice using the Cre/lox system. *Methods: A Companion to Methods in Enzymology* 14:381–392.)

gene transfer results in random integration of the foreign gene into the chromosomes of the egg. The eggs were fertilized by intracytoplasmic injection of sperm a few hours later – a method developed previously in cattle. Fertilized embryos were then implanted into a female rhesus monkey. From multiple (224) attempts, one such fertilization resulted in the live birth of an infant carrying the *GFP* transgene and

Transgenic artwork: the GFP bunny

The year 2000 was marked by an unusual application of transgenic technology when Eduardo Kac publicly presented his transgenic artwork entitled *GFP Bunny* in Avignon, France (Fig. 1). Kac, a Professor at the School of the Art Institute of Chicago, defines transgenic art as "a new art form based on the use of genetic engineering to transfer natural or synthetic genes to an organism, to create unique living beings" (www.ekac.org). Kac planned three phases for his transgenic artwork. The first phase was the production of Alba, the "GFP bunny." A transgenic albino rabbit expressing green fluorescent protein (GFP) was engineered by a team of scientists in France using the pronuclear microinjection technique. Under ordinary environmental conditions, she appeared as completely white with pink eyes. However, when illuminated with blue light (488 nm), she glowed with a bright green light. The second phase of Kac's project was the ongoing debate that started with the public announcement of Alba's birth. The first two phases were successful; Alba glowed and, not surprisingly, the GFP bunny was an instant media sensation, sparking heated debate amongst scientists, artists, ethicists, and animal rights activists. The third phase of Kac's project was not to be. He planned to take Alba home to Chicago to become part of his family; however, the Institute National de la Recherche Agronomique decided that the transgenic rabbit should not be released.

Figure 1 Transgenic artwork: *GFP bunny*. Drawing of Alba the transgenic albino rabbit that carried the green fluorescent protein (*GFP*) gene and glowed bright green under ultraviolet light.

expressing the mRNA at relatively low levels. This transgenic monkey was named ANDi, for "inserted DNA" (in the "reverse transcribed direction"). However, for unknown reasons, ANDi did not glow green in any accessible tissue. Whether transgenic primates will prove useful in experimental medicine remains to be determined.

Transgenic livestock

Transgenic technology has been of interest in agriculture because of its potential to be used to increase the performance of commercially important animals by adding new traits (e.g. faster growth rate, disease resistance, decreased body fat) or improving on existing ones (e.g. yield or quality of meat and milk). Attempts to use pronuclear microinjection to produce transgenic livestock such as pigs, goats, sheep, and cattle have been made with only limited success. Less than 1% of offspring are transgenic.

A recently developed technique greatly improves the production efficiency of large transgenic animals. The method is called linker-based sperm-mediated gene transfer (LB–SMGT). The linker protein, a monoclonal antibody (mAbC) is a basic protein that binds DNA through an ionic interaction allowing exogenous DNA to be linked specifically to sperm. mAbC is reactive to a surface antigen on sperm of all tested species, including pig, mouse, chicken, cow, goat, sheep, and human. In one study using pigs, the transgene was successfully integrated into the genome with germline transfer to the F_1 generation at a highly efficient rate of 37.5% (Fig. 15.8).

Gene pharming

Another application of transgenic animal technology is "gene pharming" – turning animals into pharmaceutical bioreactors for protein-based human therapeutics. For example, a flock of transgenic sheep was engineered by a pharmaceutical company to produce α_1-antitrypsin (AAT). This enzyme is used in the treatment of hereditary emphysema and cystic fibrosis. However, due to questions about the purity of AAT produced in sheep milk, large-scale human trials and production have been put on hold.

An alternative strategy being explored is using transgenic chicken as bioreactors. In this approach, the human gene of interest is linked to the promoter for a chicken egg white protein. After establishing a line of transgenic chickens, the chickens would then be used to make commercial quantities of pure recombinant human proteins in the whites of the eggs they lay. Eggs, although not transgenic ones, are already used as bioreactors for the production of vaccines (including those against influenza). The difficulty to date has been in creating the transgenic chicken in the first place. Using the standard pronuclear injection method is particularly challenging because of the size of the egg, the presence of a shell, the large quantity of viscous yolk, and the difficulty in harvesting an egg before it has begun growing into a chick. So far, chickens have been engineered that produce the reporter protein β-galactosidase in their eggs. Other researchers have used rooster sperm gene transfer techniques to produce transgenic chickens expressing GFP (green fluorescent protein) linked to the lysozyme gene promoter in egg white. In this method, a monoclonal antibody that binds specifically to the surface of sperm (mAbC; see preceding section) was used. This allows DNA linked to the antibody to enter the sperm cell and incorporate itself into the sperm genome.

15.5 Cloning by nuclear transfer

Clones are genetically identical individuals. Identical twins are naturally occurring clones, but it is also possible to produce artificial clones by nuclear transfer. In this method, the nucleus is removed from a somatic (body) cell and placed in an egg whose own nucleus has been removed. The first animal cloning experiments were conducted in the 1950s when developmental biologists Robert Briggs and Thomas King developed a method for nuclear transplantation in the leopard frog, *Rana pipiens*. Their reason for cloning was not for the purpose of producing lots of frogs. Instead, they were interested in directly testing the question of the genetic equivalence of somatic cell nuclei in development and cell differentiation. In other words, does cell differentiation depend on changes in gene expression or changes in the content of the genome? This had been a long-standing question in developmental biology.

Genetic equivalence of somatic cell nuclei: frog cloning experiments

In 1893, August Weismann proposed that as cells differentiate, genes no longer required for other specialized cell types are lost or permanently inactivated. Hans Spemann performed experiments in 1914 that seemed to refute this proposal. His results suggested that the complete genome is replicated during cell division, at least during early cleavage. However, the definitive experiments were performed by Briggs and King nearly four decades later. They showed that a normal hatched tadpole of *R. pipiens* can be obtained by transplanting the nucleus of a blastula cell (very early embryo) into an enucleated egg. As cells differentiated into older embryos, the ability to develop into a frog was dramatically reduced (Fig. 15.9A). In the 1960s, John Gurdon and co-workers carried out similar nuclear transplantation experiments using *Xenopus laevis*, the South African clawed frog. Nuclei from intestinal epithelial cells of tadpoles were used and approximately 1% of the embryos developed into normal, fertile adult frogs (Fig. 15.9B).

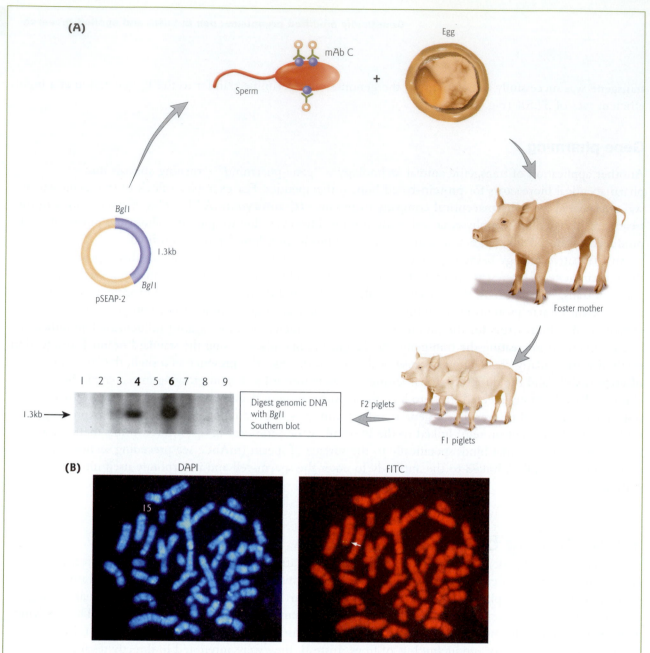

Figure 15.8 Transgenic pigs generated by linker-based sperm-mediated gene transfer. (A) A basic monoclonal antibody (mAbC) that binds DNA and is specific for a sperm surface antigen was used for linker-based sperm-mediated gene transfer (LB-SMGT) into pig eggs. The introduced plasmid DNA encodes secreted human alkaline phosphatase under control of the simian virus 40 (SV40) promoter and enhancer (pSEAP-2). The pSEAP-2 plasmid has two internal sites for the restriction endonuclease *Bgl*I, that generate a 1.3 kb fragment. Genomic DNA from ear biopsies of F_2 offspring from a line of transgenic pigs was digested with *Bgl*I and analyzed by Southern blotting. The ^{32}P-labeled DNA probe corresponded to the vector region of pSEAP-2, to avoid any cross hybridization with endogenous alkaline phosphatase gene(s). An autoradiogram of a Southern blot is shown. The numbers in bold (lanes 4 and 6) indicate positive detection of the transgene on the blot. (B) Fluorescent *in situ* hybridization (FISH) analysis for transgene localization. A metaphase chromosome spread was prepared from a blood sample from a transgenic pig (F_1 generation). The pSEAP-2 transgene was located to chromosome 15, region q25–q28 (right panel: arrow, two yellowish–green spots). A FITC (fluorescein isothiocyanate) labeled 3.1 kb DNA fragment from the plasmid vector was used as a probe (pseudocolored red in the right panel). DAPI (4′,6′-diamidino-2-phenylindole dihydrochloride) is a DNA-specific blue fluorescent dye (left panel). (Reprinted with permission from: Chang, K., Qian, J., Jiang, M. et al. 2002. Effective generation of transgenic pigs and mice by linker-based sperm-mediated gene transfer. *BMC Biotechnology* 2:5–17. http://www.biomedcentral.com/1472-6750/2/5. Copyright © 2002 Chang et al., licensee BioMed Central Ltd.).

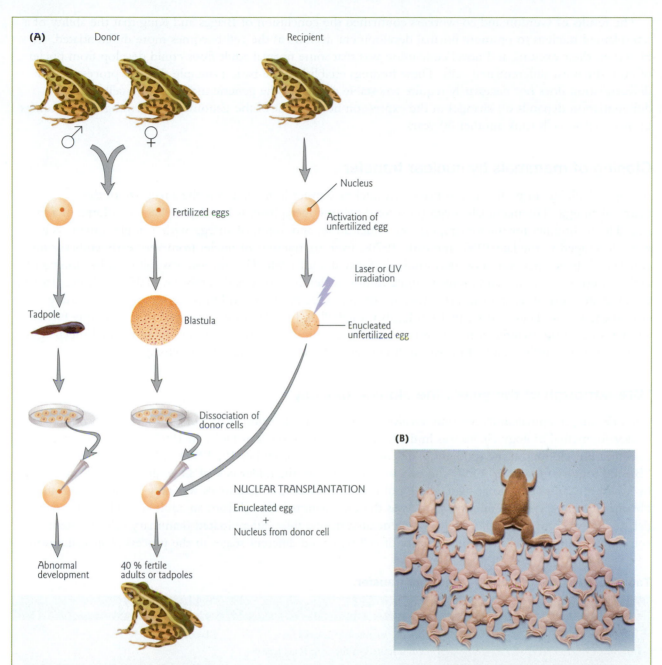

Figure 15.9 Amphibian cloning. (A) Cloning of leopard frogs. Adult leopard frogs (*Rana pipiens*) are mated and the fertilized eggs are permitted to develop to the blastula stage, with approximately 3000 cells, or to later stages, such as the tadpole stage. The embryos are dissociated to yield single cells that will be used as nuclear donors. Another female provides a recipient egg which is artificially activated by pricking with a sharp glass needle. The activated egg is enucleated by laser or UV irradiation. A dissociated donor cell is inserted into the enucleated egg with a pipet (nuclear transplantation). When cells from a blastula were used, up to 40% of the embryos developed into a normal frog. By the tadpole stage, nuclei give only abnormal development. (B) Cloning of South African clawed frogs (*Xenopus laevis*). The photo shows clones of *X. laevis* that were produced by transfer of nuclei from cells of an albino female tailbud-stage tadpole into the enucleated egg from a wild-type (pigmented) female frog. The resulting frogs were all female and albino. (Photograph courtesy of John Gurdon, Wellcome Trust / Cancer Research UK Gurdon Institute, Cambridge, UK).

The results of Gurdon and co-workers confirmed the conclusion of Briggs and King that the ability of a transplanted nucleus to promote normal development declines as the cell becomes more differentiated. However, their exciting and novel conclusion was that some normal adult frogs could develop from nuclei of even the most differentiated cells. These findings established the basic principle that the process of cell differentiation does not necessarily require any stable change to the genetic make-up of a cell. Cell differentiation depends on changes in the expression not content of the genome. Confirming these findings in mammalian cells took another 30 years.

Cloning of mammals by nuclear transfer

A major challenge in performing somatic cell nuclear transfer in mammals is the small size of the mammalian egg. The mammalian egg (in second meiotic metaphase) is less than 0.1% the volume of a frog egg. The techniques for micromanipulation, enucleation, and fusion of an egg with a single somatic cell were developed in the late 1960s and early 1970s. Even so, transfers of nuclei from very early embryos into enucleated sheep eggs were not successfully performed until 1986. This success was followed by cloning of rabbits, pigs, mice, cows, and monkeys using donor nuclei from very early embryos (Table 15.2, Fig. 15.10). Cloning attempts of nonhuman primates have proved difficult; Neti and Ditto are the only successfully cloned rhesus monkeys to date. In 1996 Keith Campbell and Ian Wilmut developed this technique further and performed the transfer of nuclei from an established embryonic stem cell line to enucleated sheep eggs. These nuclear transfers resulted in two healthy cloned sheep named Megan and Morag.

"Breakthough of the year": the cloning of Dolly

For a decade, mammalian clones only received slightly more attention than cloned frogs (which outside of developmental biology circles was limited). Then, with the birth of Dolly in 1997, a flurry of media attention put cloning in the spotlight. The cover of *Science* (December 19, 1997) called Dolly the "Breakthrough of the year." Dolly the ewe was cloned from the udder cell of an adult sheep in Scotland by Wilmut, Campbell, and co-workers. Until this time, cloning in mammals was only possible using nuclei obtained from very early embryos. Dolly was the first mammal cloned from an adult cell. There was some debate, however, over whether Dolly was generated from a fully differentiated mammary cell or a stem cell, since mammary cells contain a variety of cell types and different stages in the process of specialization.

Table 15.2 History of cloning by nuclear transfer.

Year	Animal cloned	Source of donor nucleus
1952	Cloning by nuclear transfer first demonstrated in the frog *Rana pipiens*	Blastula cells
1966	Cloning in the frog *Xenopus laevis* achieved by nuclear transfer	Intestinal epithelial cells of feeding tadpoles
1986–1997	Cloning of mammals (sheep, rabbits, pigs, mice, cows, monkeys)	Very early embryos
1996	Cloning of the sheep Megan and Morag	Embryonic stem cell line (from day 9 embryo)
1997	Cloning of the sheep Dolly	Adult sheep cell
1997	Cloning of the sheep Polly	Stably transfected (transgenic) fetal sheep cells
1998–2005	Cloning of mice, cows, pigs, goats, rabbits, cats, horses, mules, and dogs	Adult cells

Figure 15.10 A photo gallery of clones. (A) Dolly (left) was cloned using the nucleus of a mammary gland cell from a Finn Dorset ewe. Next to her is the Scottish Blackface ewe which was the recipient (right) (Reprinted by permission from Nature Publishing Group and Macmillan Publishers Ltd: Wilmut, I., Schnieke, A.E., McWhir, J., Kind, A.J., Campbell, K.H.S. 1997. Viable offspring derived from fetal and adult mammalian cells. *Nature* 385:810–813. Copyright © 1997). (B) Hooper the mouse shown with his "father." He was cloned by using the nucleus from an E14 embryonic stem cell. (Reprinted with permission from Wakayama, T., Rodriguez, I., Perry, A.C., Yanagimachi, R., Mombaerts, P. 1999. Mice cloned from embryonic stem cells. *Proceedings of the National Academy of Sciences USA* 96:14984–14989. Copyright © 1999 National Academy of Sciences, USA). (C) Three cloned mules together as yearlings. The mules were cloned using nuclei from a fibroblast cell line established from a 45-day-old fetus. (Credit: University of Idaho / Kelly Weaver).

Not surprisingly, the use of adult somatic cells stimulated the general public, scientists, novelists, journalists, filmmakers, politicians, theologians, and ethicists to think about the application of cloning to humans.

On a more fundamental note, the cloning of Dolly confirmed the two key principles of genetic equivalence. First, differentiated animal cells on their own are unable to develop into complete animals but the nuclei of most differentiated cells retain all the necessary genetic information. Second, transfer of a nucleus from a differentiated cell to the environment of the enucleated egg reprograms the nucleus (makes it forget its history) and allows the full development of a viable animal that is genetically identical to the donor of the somatic cell.

Since Dolly, many other mammals have been successfully cloned from adult donor cell nuclei. These include cows, pigs, mice, domestic cats, horses, and even a mule (Table 15.2, Fig. 15.10). The latter clone was particularly novel because a mule is a sterile hybrid that results from breeding a male donkey with a female horse. The efficiency of mammalian nuclear transfer experiments is very similar to that obtained from frogs. Less than 1% of all nuclear transfers from adult differentiated cells result in normal-appearing offspring.

Method for cloning by nuclear transfer

The basic technique for cloning by nuclear transfer or "reproductive cloning" is shown in Fig. 15.11, using sheep as an example. Cloning involves four main steps: (i) preparation of donor cells; (ii) enucleation of unfertilized eggs; (iii) nuclear transfer by cell fusion; and (iv) implantation of the embryo into a surrogate mother and analysis of clones.

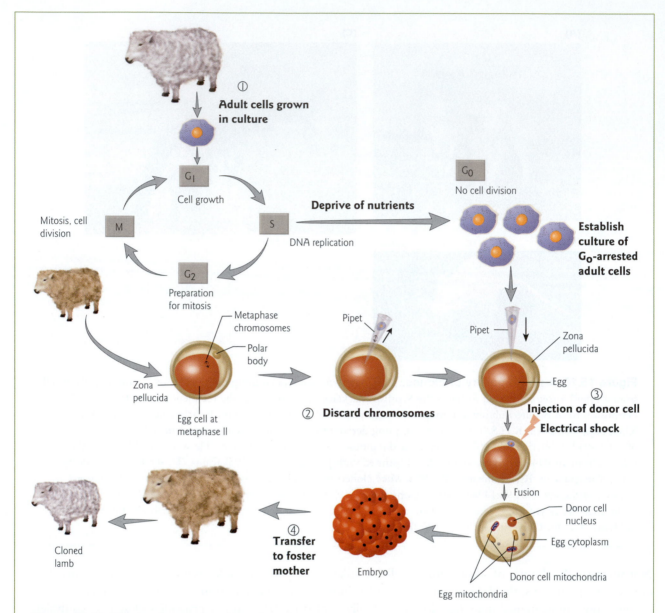

Figure 15.11 Cloning by nuclear transfer. Sheep are used as an example to illustrate the four main steps in cloning by nuclear transfer or "reproductive cloning." (1) Preparation of donor cells. Adult cells are arrested in the G0 phase of the cell cycle by nutrient deprivation. (2) Enucleation of unfertilized eggs. The chromosomes, spindle, and first polar body are withdrawn using a pipet that is inserted through the zona pellucida (noncellular envelope surrounding the egg). (3) Nuclear transfer by cell fusion. After injection of the donor cell into the space between the egg membrane and the zona pellucida, the cells are fused by an electric shock and activated to divide. The reconstructed embryo will contain egg cytoplasm and the donor nucleus with its accompanying cytoplasm. Thus, the embryo will contain mitochondria (with mitochondrial DNA) from both sources. (4) Implantation of the embryo into a surrogate mother. After about 7 days, the embryo is transferred to the uterus of a foster mother. The resulting cloned offspring will be genetically identical to the original adult cell nucleus.

Preparation of donor cells

The source of donor cells for nuclear transfer may be either totipotent embryonic stem cells or differentiated somatic cells. A key aspect of success of nuclear transfer is having the donor nucleus and the egg at the same phase of the cell cycle. Before fertilization, the egg's nucleus is quite inactive. The nucleus of the donor cell must also be made inactive, otherwise it will not be reprogrammed and development may fail. To achieve inactivation, the donor cells are cultured and then starved for certain essential nutrients. This puts them into a nondividing state in the cell cycle in which there is no gene expression (G0 phase) (Fig. 15.11). This quiescent state is compatible with nuclear transfer.

Enucleation of unfertilized eggs

The chromosomes from an unfertilized egg, also referred to as an oocyte (egg cell arrested at metaphase II of meiosis) are removed using a micropipet (Fig. 15.11). Recently, some investigators have begun using a gentler procedure. Withdrawing the nucleus using a micropipet may damage the protein machinery that controls cell division. A newer method involves nicking a small hole in the egg's membrane and gently squeezing out the nucleus.

Nuclear transfer

After enucleation, a micropipet is used to inject an intact, donor cell into the cavity between the egg's plasma membrane and the noncellular membrane surrounding the egg (Fig. 15.11). The two cells are usually fused by electrofusion. The egg and donor cell are given an electric pulse that causes breaks in the cell membranes of the two cells, permitting the contents of each to mingle before membrane healing. The nuclear transfer step provides the unfertilized egg with a diploid set of chromosomes so there is no need for sperm. The electric discharge can activate some eggs to undergo cleavage. In most cases, however, additional treatment is required (such as an additional electric pulse). Alternatively, in some mouse experiments, isolated nuclei may be injected into eggs that are then activated.

Implantation and analysis of clones

After about 7 days in a petri dish, the embryo at the blastocyst stage is implanted into the uterus of a surrogate mother. If the pregnancy is successful, the resulting cloned offspring are genetically identical to the original donor nucleus (Fig. 15.11). To confirm this, investigators use DNA typing techniques to compare species-specific short tandem repeat (STR) loci between the offspring and the original donor animal (see Section 16.2 for methods).

Source of mtDNA in clones

When the cell fusion method is used, the reconstructed embryo will contain egg cytoplasm and the donor nucleus with its accompanying cytoplasm. Thus, the embryo will contain mitochondria from both sources; i.e. it will be heteroplasmic for mitochondrial DNA (mtDNA) (Fig. 15.11). Mitochondria inherited from the recipient egg could have a major influence over functions such as muscle development and physiology that depend on mitochondrial gene expression. Thus, a clone is not really genetically identical if the mtDNA is taken into account. The exception would be if a clone is made using a donor nucleus from the somatic cell of an adult female and one of her eggs is used as the recipient.

Why is cloning by nuclear transfer inefficient?

As of yet the cloning procedure has not been proven to be without risk. Currently, the technology has very low efficiency, less than 1–10% of cloned mammalian embryos result in offspring. For example, to create Dolly, it took 277 trials; i.e. 276 embryos were created that did not survive the procedure. Further, the chance of abnormal offspring is very high. The successes and failures of cloning have generated heated debate in many countries, particularly as some scientists have suggested cloning humans. Many argue that human reproductive cloning should be made illegal, if for no other reason because of the many defects observed after birth in cloned mammals.

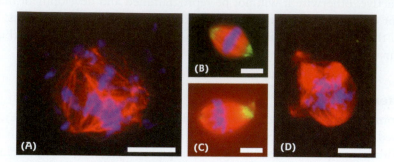

Figure 15.12 Primate nuclear transfer failures. In rhesus monkey embryos generated by nuclear transfer, the proteins that help organize the mitotic spindle are missing and the chromosomes do not divide properly. This leads to aneuploid embryos (having more or fewer that an exact multiple of the haploid number of chromosomes) after nuclear transfer. No pregnancies resulted from 33 embryo transfers during this study. (A) Defective mitotic spindle with misaligned chromosomes in embryos after nuclear transfer. (B–D) NuMA (nuclear mitotic apparatus) is a matrix protein responsible for spindle pole assembly that normally concentrates at the centrosomes in dividing cells. Centrosomal NuMA is present during meiosis (B) and mitosis (C) in normal primate cells but not in mitotic spindles after nuclear transfer (D). Blue, DNA; red, β-tubulin; green, NuMA. Scale bar, 10 μm. (Reprinted with permission from Simerly, C., Dominko, T., Navara, C. et al. 2003. Molecular correlates of primate nuclear transfer failures. *Science* 300:297. Copyright © 2003 AAAS.)

Multiple gene defects can be found in cloned mice. When 10,000 genes were screened, 4% were shown to be functioning incorrectly. Cloned animals suffer from developmental abnormalities including extended gestation, large birth weight, inadequate formation of the placenta, and histological defects in most organs, including the kidney, brain, cardiovascular system, and muscle. There is a tendency towards obesity, liver failure, pneumonia, and premature death. In some species tested so far (mice and pigs), these abnormal phenotypes were not transferred to subsequent offspring.

These effects, at least on first generation clones, may be due to inefficient reprogramming; i.e. the nucleus is not able to forget its "history." Nuclear reprogramming is the reversal of the gene expression pattern of a differentiated cell to one that is totipotent and capable of directing normal development. In addition, there may be effects of cellular aging, due to the age of the donor nucleus. Finally, genetic defects can arise from improper segregation of chromosomes during embryonic cell divisions. Rhesus monkey embryos generated by nuclear transfer have been shown to be missing important components of the mitotic spindle (Fig. 15.12).

Reprogramming the genome

Very early on in development of an embryo the cells are "totipotent" – that is, they are capable of forming any cell type and if separated from the embryo can develop into complete individuals. Later embryonic cells become "pluripotent" – that is they are capable of developing into several different cell types. As development proceeds, the cells become differentiated (specialized) towards specific functions. This restriction in developmental potential is associated with differential gene expression. Tissue-specific genes are activated only in particular cell types. In contrast, some genes – called "housekeeping" genes – are active in most cell types because they are required for basic functions such as protein synthesis or maintenance of the cytoskeleton. In a differentiated cell, a relatively small number of genes – estimated at less than 10% of all the genes in the genome – are expressed at high levels, and a relatively large fraction of genes are silenced.

An important question is whether and how these cells are genetically reprogrammed to adopt a different fate during cloning. Successful cloning by somatic cell nuclear transfer requires reprogramming of the donor nuclei from differentiated cells to an undifferentiated state, so that the cells are capable of supporting the

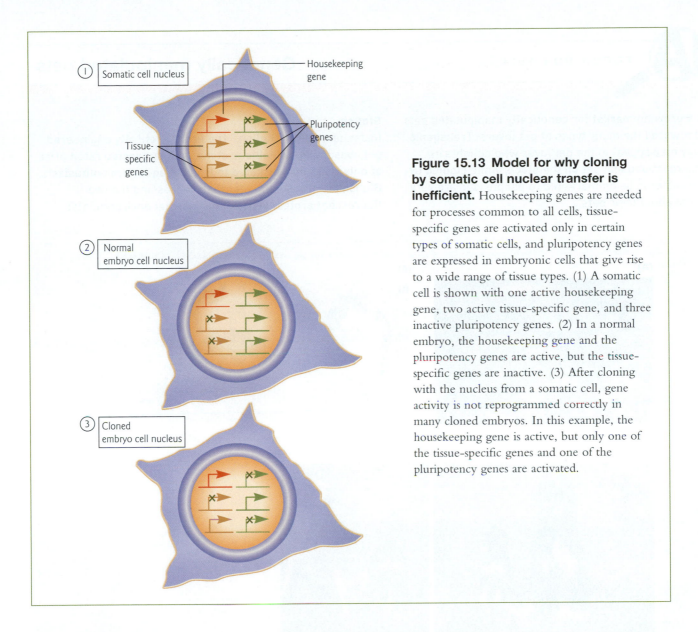

Figure 15.13 Model for why cloning by somatic cell nuclear transfer is inefficient. Housekeeping genes are needed for processes common to all cells, tissue-specific genes are activated only in certain types of somatic cells, and pluripotency genes are expressed in embryonic cells that give rise to a wide range of tissue types. (1) A somatic cell is shown with one active housekeeping gene, two active tissue-specific gene, and three inactive pluripotency genes. (2) In a normal embryo, the housekeeping gene and the pluripotency genes are active, but the tissue-specific genes are inactive. (3) After cloning with the nucleus from a somatic cell, gene activity is not reprogrammed correctly in many cloned embryos. In this example, the housekeeping gene is active, but only one of the tissue-specific genes and one of the pluripotency genes are activated.

development of all the different cell types needed to build an animal. Cloning is inefficient, in part, because gene silencing is difficult to reverse in cloned embryos. Once cells have differentiated into specific types, the silencing of unwanted gene expression is very tightly controlled. For example, in cloned mouse embryos, the reactivation of genes that are silent in most adult tissues but are needed for early development – the so-called "pluripotency genes" – appears to be defective (Fig. 15.13). Some forms of gene silencing appear to be more easily reversed than others. X chromosome inactivation is efficiently reversed and randomized in the embryo, as illustrated by the orange and black coat color of a cloned calico cat (Focus box 15.4). In contrast, the outcome is less clear for DNA methylation and gene imprinting. Modification of DNA with methyl groups (methylation) is commonly associated with gene silencing (see Section 12.2), as is the methylation of the histone proteins that compact the DNA into chromatin (see Section 11.6). Imprinting is a process that occurs naturally during gametogenesis (the differentiation of sperm and eggs) and early development, and controls whether certain genes are expressed from the maternal or paternal chromosomes (see Section 12.3).

Further exploration of the mechanisms involved in reprogramming should provide insights into how cell differentiation is maintained. This is of interest both in basic research and in understanding cancer biology, and might open the way to the cloning of stem cells for therapeutic purposes (see below).

FOCUS BOX 15.4

Genetically manipulated pets

There is a growing market for genetically manipulated pets targeted towards the many types of pet lovers. Transgenic technology may appeal to the pet lover who delights in having the most unusual pet on the block, while cloning by nuclear transfer is being marketed to people who are very sentimental about their pets.

Glofish

In January 2004, "Glofish," the first genetically engineered pet, was made available in stores at a suggested retail price of only US$5.00. Marketed to tropical aquarium enthusiasts, Glofish are transgenic zebrafish expressing the red fluorescent protein gene. Under normal environmental

Figure 1 Genetically modified pets. (A) Fluorescent zebrafish for sale. Glofish are transgenic zebrafish expressing the red fluorescent protein gene. Under normal environmental conditions, the zebrafish appear silver with black stripes. Under ultraviolet light the fish fluoresce a reddish color. (Credit: www.glofish.com). (B) Copy Cat – the first cloned cat. The first cloned domestic cat was derived from nuclei from the cumulus cells surrounding the ova of a calico research cat named Rainbow (i). Copy Cat (ii) was not an exact copy of her calico donor, however, because coat markings result partly from random events during development. Copy Cat is shown with her surrogate mother. (Reprinted by permission from Nature Publishing Group and Macmillan Publishers Ltd: Shin, T., Kraemer, D., Pryor, J. et al. 2002. A cat cloned by nuclear transplantation. *Nature* 415:859. Copyright © 2002.) (C) (i) Snuppy, the first cloned dog, at 67 days after birth (right), with the 3-year-old male Afghan hound (left) whose somatic skin cells were used to clone him. Snuppy is genetically identical to the donor Afghan hound. (ii) Snuppy (left) was implanted as an early embryo into a surrogate mother, the yellow Labrador retriever on the right, and raised by her. (Reprinted by permission from Nature Publishing Group and Macmillan Publishers Ltd: Lee, B.C., Kim, M.K., Jang, G. et al. 2005. Dogs cloned from adult somatic cells. *Nature* 436:641. Copyright © 2005.)

Genetically manipulated pets

FOCUS BOX 15.4

conditions, the zebrafish appear silver with black stripes. Under ultraviolet light the fish fluoresce a reddish color (Fig. 1A). The Food and Drug Administration (FDA) declared that no formal review for environmental safety was required for sale of these fish, since they are thought to represent no public health risk. The original objective in creating fluorescent zebrafish was for use in the analysis of gene expression during development. Another interest is their use as a biomonitoring organism for surveillance of environmental pollutants, such as heavy metals. This would require engineering a regulatory switch that would cause the zebrafish to only fluoresce in the presence of environmental toxins.

Cloned cats

The first cloned domestic cat (*Felis domesticus*) was derived from nuclei from the cumulus cells surrounding the ova of a calico research cat named Rainbow. The lively, normal-looking kitten born on December 22, 2001, was named CC, short for Copy Cat (also Carbon Copy) (Fig. 1B). CC was not an exact copy of her calico donor, however, because these coat markings result partly from random events during development. Two alleles affecting coat color are present on the X chromosome. One allele specifies orange coat color, the other black coat color. A female can be heterozygous for the orange and black alleles, resulting in a "calico" or mosaic coat with orange and black patches. These patches result from the natural process of X chromosome inactivation that occurs randomly in developing embryos (see Section 12.4). In some cell lineages the X chromosome with the orange allele is inactivated and the X chromosome with the black allele is active. The result is black fur. In cell lineages in which the X chromosome with the black allele is inactivated, the orange allele on the active X chromosome specifies orange fur. The white patches are due to an autosomal gene *S* for white spotting which prevents pigment formation in the cell lineages in which it is expressed. It is not clear why the *S* gene is expressed in some cell lineages and not others. Of great interest to molecular biologists is the fact that both X chromosomes must have been "reprogrammed" in the cloned kitten. If X inactivation had not been reversed, the cloned cat would have been only black with white spots, or only orange with white spots, depending on which X chromosome was inactivated in the donor nucleus.

Fundamental knowledge aside, with the birth of CC, the possibilities for the cloning of beloved family pets captured the interest of some wealthy pet owners. In December 2005, San Francisco-based "Genetic Savings and Clones" cloned a kitten called Little Nicky. Little Nicky was sold to a Dallas resident who paid US$50,000 for the cloned kitten! The company hopes to reduce the cost of pet cloning substantially in the future.

Cloned dogs

The first cloned dog was born in South Korea in June 2005. Snuppy (for Seoul National University puppy) was cloned from somatic skin cells from an adult Afghan hound (Fig. 1C). It is unlikely that cloning pet dogs will become of commercial interest in the near future, since the process was exceptionally inefficient. There were only two live births and one survivor from a total of 1095 embryos implanted in 123 surrogate mothers.

Effects of cellular aging

Another question that has arisen from cloning using adult cells, is whether there are side effects from the age of the donor cell. For example, is telomerase reactivated to restore the length of telomeres (see Section 6.9)? Recent research suggests that animals produced by cloning from adult cells may age prematurely, but further investigation is necessary. When Dolly the cloned sheep was only 2 years old she started to show signs of wear more typical of an older animal, and she developed arthritis at the relatively young age of 5.5 years. Dolly was euthanized on February 14, 2003 at age 6, because of complications from a virally induced lung cancer commonly found in older sheep kept indoors. Dolly's cells showed a telomere loss of 20%. The age of the donor nucleus (the udder cell was from a 6-year-old ewe) and proliferation in culture prior to transfer are thought to have contributed to telomere loss.

The age of the donor nucleus may not be a consideration in some species. As an example, telomere length is rebuilt in cloned cattle. When donor nuclei from either cultured adult cells or fetal cells were used, in both cases telomerase activity was reprogrammed at the blastocyst stage, resulting in telomeres similar in length to age-matched controls.

Applications of cloning by nuclear transfer

The impetus behind cloning, like other developments in basic science, was to seek new fundamental knowledge, not to make legions of clones. However, there are a number of commercial and medical interests in cloning by nuclear transfer. These include the cloning of transgenic animals, cloning prize farm animals or race horses, cloning pets (Focus box 15.4) and endangered species, and cloning for stem cells.

Cloning transgenic animals

One interest in cloning by nuclear transfer is an extension of "gene pharming." Cloning could be used to rapidly produce herds of transgenic or gene-targeted farm animals that produce valuable human proteins. In July 1997, "Polly" the sheep was cloned by nuclear transfer. The difference between "Dolly" and "Polly" was that the donor nucleus used to create Polly came from stably transfected (transgenic) sheep fibroblast cells. The transgene carried in the donor nucleus was designed to express human clotting factor IX in sheep milk. The disease hemophilia B is caused by low levels or a complete absence of this blood protein, which is essential for clotting. Expression constructs were composed of the coding sequences for neomycin resistance (neo^r) and human clotting factor IX placed under control of the promoter sequence for ovine β-lactoglobin. Neo^r was included for the selection of cells with the transgene stably integrated in their genome. The promoter sequence directs tissue-specific expression in the mammary gland. As noted in Section 15.4, questions about the purity of human proteins produced in sheep milk have halted human trials and production.

Other researchers are interested in the cloning of agriculturally important animals that are transgenic for a particular trait of interest. A noteworthy example is a group of researchers in South Korea who have the goal of creating a herd of cloned, genetically modified cows that are resistant to bovine spongiform encephalopathy ("mad cow disease," see Disease box 5.1). Another application is xenotransplantation – the use of genetically modified pigs as a source of organs suitable for transfer to humans.

Cloning of prize animals

Animal cloning has the potential to overcome the limitations of the normal breeding cycle. Animal breeders have an interest in maintaining and rapidly propagating elite herds of animals with superior traits, such as cows with high milk production or a champion racehorse. A prize animal is not generally identified until it matures, so differentiated adult cells must be used for cloning. Many male racehorses are castrated (geldings) "to keep their mind on the game and not think about the opposite sex." Cloning would enable these gelding racing champions to contribute their genotype to future generations. Currently, the Thoroughbreds' Jockey Club and the American Quarter Horse Association will not register clones, so there may not be a practical application of cloning. In terms of basic research, however, cloning racehorses would provide the opportunity to test for the reproducibility of traits such as character and sporting performance. However, the heteroplasmic nature of the mtDNA in the clone would need to be taken into account. As noted in an earlier section, mitochondria inherited from the recipient egg could have a major influence over critical aspects of sporting performance, such as muscle development.

Wildlife conservation

Cloning is one of several ways of increasing the number of individuals within a population when the population is in decline. Populations with low numbers of individuals have lost some genetic diversity.

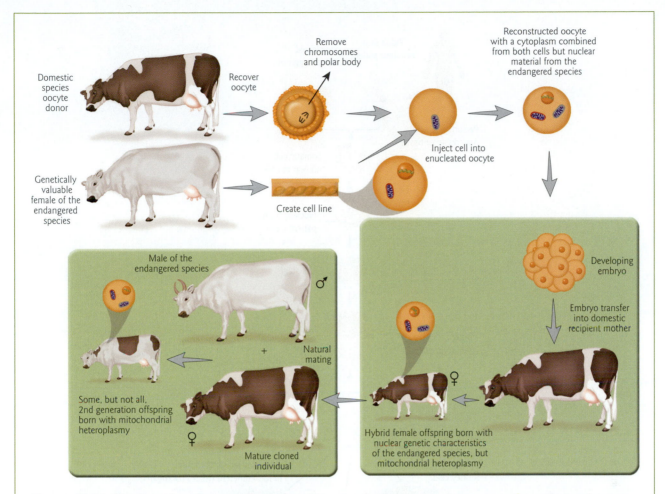

Figure 15.14 Cloning of endangered species. The example illustrates the importance of considering the source of mitochondria (and hence the mitochondrial DNA) when carrying out trans-species cloning by somatic cell nuclear transfer. In this example, a domestic cow provides the oocyte (egg) cytoplasm, while the donor nucleus, and its accompanying cytoplasm, comes from a female of the endangered species (e.g. a banteng). The reconstructed embryo would be a hybrid containing mitochondria from both species (mitochondrial heteroplasmy). If the cloned female bred naturally with a male of the endangered species, the offspring would still have mitochondria that descended from the original donor domestic cow. (Adapted from Holt, W.V., Pickard, A.R., Prather, R.S. 2004. Wildlife conservation and reproductive cloning. *Reproduction* 127:317–324.)

If individuals are cloned, in theory, this would improve the long-term fitness of the species through the retention of genetic diversity. Since current success rates with cloning by nuclear transfer are very low, hundreds of nuclear transfers might need to be performed to obtain just one offspring. This poses a problem for endangered species where it might not be possible to obtain eggs in such high numbers. For example, the giant panda only ovulates one or two oocytes per year.

Some endangered species have been cloned in an attempt to preserve the species; however, since these used "trans-species" cloning they are really genetic hybrids (Fig. 15.14). Animals resulting from trans-species cloning are valuable scientifically, but are not directly useful for supporting endangered populations. One of the first examples of trans-species cloning was a gaur (*Bos gaurus*) produced by scientists at Advanced Cell Technology in Worcester, Massachusetts. The gaur is a wild ox native to South East Asia. Unfortunately, the clone, created using a skin cell of an adult male gaur, died of dysentery 2 days after birth. In an another effort, scientists cloned a banteng (*Bos javanicus*), a wild, hoofed mammal from Java, using skin cell nuclei from frozen tissue from a banteng that died in 1980. The banteng calf appeared healthy. In both these cases,

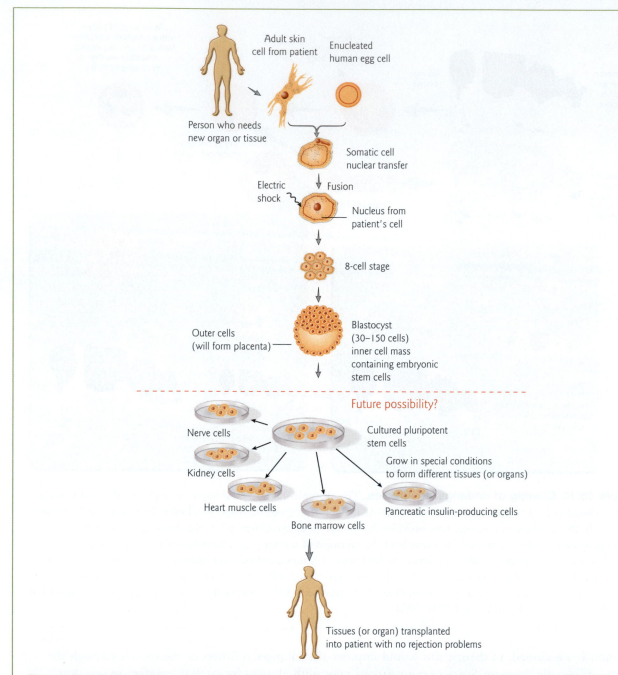

Figure 15.15 Cloning for stem cells. The diagram shows a hypothetical strategy for the therapeutic use of cloning for stem cells. Actual therapeutic use is in the distant future. An embryo is created by transferring the nucleus from a patient's cell into a human egg cell stripped of its own nucleus. Cloned cells are allowed to develop into the blastocyst stage (30–150 cells). Cells are then removed from the inner cell mass and cultured to produce stem cells. The stems cells could then be cultured in special conditions that promote the formation of different tissues (or organs). These tissues could then be used to treat such conditions as Parkinson's disease, diabetes, or spinal cord injury.

the egg cytoplasm being used to create the embryos was derived from a common domesticated cow (*Bos taurus*). Thus, the embryo contained mitochondria of both species, forming an unusual type of hybrid. If such a cloned female were to breed naturally with a male that shared its nuclear genome, the offspring would still possess mitochondria that derived form the original donor and the cow egg, since mtDNA is inherited via the maternal cytoplasm. However, male clones of breeding age would not transfer the hybrid mitochondria to the next generation, so the nuclear genome of a male genetic line of particular value could be restored.

Cloning for stem cells

Since the birth of Dolly the cloned sheep in 1997 and the isolation of human ES cells in 1998, scientists have hoped to combine the two techniques. Their aim is to create ES cells with genes that match those of a patient, an approach called alternatively "therapeutic cloning," "cloning for stem cells," or "cell replacement cloning." Cloning for stem cells is the most accurate term at this point, since actual therapeutic use is in the distant future (if at all). All these terms are designed to make a clear distinction between this approach, in which there is no attempt to produce another live human, and so-called "reproductive cloning" in which the end goal is to produce another living organism.

Cloning for stem cells would involve creating an embryo by transferring the nucleus from a patient's cell into a human egg cell stripped of its own nucleus. Cloned cells would be allowed to develop to the blastocyst stage (30–150 cells). Cells then would be removed from the inner cell mass and cultured to produce stem cells (Fig. 15.15). The hope is to develop techniques of growing the ES cells into specific cell types to treat such conditions as Parkinson's disease, diabetes, or spinal cord injury. There would, in theory, be no problem with immune rejection in using grafts derived from these cells to repair diseased or damaged tissues, since they originated from the patient's own cells.

There is heated debate over how such research should be regulated; opinion varies around the world. Currently, this type of stem cell research is not supported by federal funds in the USA. The primary ethical objection to cloning for stem cells is the creation of a 30–150-cell human embryo that is subsequently dispersed into a cell culture. Some people view the blastocyst as a potential human being that should not be commercialized and used for "spare parts." Others view the embryo as simply a collection of cells that has no possibility of becoming a human being in the absence of implantation. Another constraint to cloning for stem cells may be obtaining a sufficient supply of human recipient eggs.

One underlying objective of researchers in the field of nuclear transfer is to identify the molecular components that mediate the reprogramming activity of the egg. When this remarkable process is better understood, it might one day be possible to reprogram adult human somatic cells *in vitro* and grow them in culture. If these differentiated cells could be turned back to an embryonic "totipotent" state without the cloning step, many ethical objections to cell replacement therapy would be resolved.

15.6 Transgenic plants

Transgenic technology can potentially be used to increase the performance of commercially important plants by adding new traits (e.g. herbicide and pest resistance) or improving on existing ones (e.g. yield or quality of fruits and grains). However, the introduction of genetically modified (GM) crops into the food chain is a source of controversy and public concern, particularly in Europe. Some human health concerns with GM crops are their potential to express novel antigenic proteins causing an allergic reaction in people eating the GM food, the potential for other unintended or toxic effects, and the potential for detrimental changes in nutrient composition. At least so far, scientific evidence suggests that GM foods are as safe and nutritious as conventional counterparts (Focus box 15.5).

The route to making transgenic dicotyledonous plants is relatively simple for a number of reasons. First, the naturally occurring and highly efficient Ti plasmid-based gene delivery system can be used (see below). Second, in contrast to animal cells, differentiated plant cells are totipotent, i.e. they still have the ability to produce all cell types. Third, at least in some species, differentiated plant cells will regenerate into whole adult plants under appropriate conditions. Dicotyledons (or dicots) are a class of flowering plants having an embryo with two cotyledons (seed leaves). They include familiar plants such as tomatoes, potatoes, beans, and peas, and the model plant *Arabidopsis* (see Section 9.11). Since a similar natural gene delivery system is

FOCUS BOX 15.5

Genetically modified crops: are you eating genetically engineered tomatoes?

Most people are familiar with the flavorless tomatoes sold in supermarkets. Their lack of flavor results from the practice of shipping green tomatoes and chemically ripening them later. Vine-ripened tomatoes cannot be shipped very far because they spoil too quickly. In the early 1990s, scientists used transgenic technology to try to counteract this problem. They genetically engineered tomatoes to alter an aspect of fruit ripening called "softening." The process of fruit softening is caused in part by the breakdown of pectins (compounds which give support to the cell walls of fruit). The tomatoes were engineered to have reduced levels of polygalacturonase, an enzyme that breaks down pectin. In addition, for selection purposes, the tomatoes carried resistance to the antibiotic kanamycin. Tomato plants were transformed with an antisense polygalacturonase gene sequence, which encodes an RNA transcript that is complementary to polygalacturonase mRNA. The two RNAs (sense and antisense) bind to one another so that they are degraded and translation of the polygalacturonase protein is prevented. The result was the "Flavr-Savr" tomato which spoiled less quickly after harvesting and thus could be left to ripen on the vines longer, developing more flavor and allowing later shipment to stores.

In late 1991, Calgene had a variety of Flavr-Savr tomato ready for marketing. They requested the opinion of the Food and Drug Administration (FDA), since this was the first genetically modified food to reach this point in product development. Public concerns about food safety also prompted Calgene to request a ruling from the FDA regarding the safety of antibiotic resistance genes in food. Calgene received FDA approval for its tomatoes in mid-1994 and started selling tomatoes in markets in the Chicago area. The tomatoes were clearly labeled as "genetically modified" and supplied with information pamphlets. Despite growing protests by activist groups, the tomatoes were well received. However, it was difficult to ship the delicate tomatoes without damage, the tomatoes did not grow well in Florida production fields the first year, and high development costs were followed by several years of low tomato prices. By March of 1997, the Flavr-Savr tomato had been taken off the market, and Calgene sold its interests in the tomato to Monsanto.

So, is that slice of tomato on your sandwich genetically modified? The answer is "No." Several companies are currently developing new varieties of GM tomatoes, but there are currently no GM tomatoes present in US markets either as whole tomatoes or in processed tomato foods.

---

not available for monocotyledonous plants, alternative techniques have been developed. Monocotyledons (monocots) are a class of flowering plants having an embryo with one cotyledon, such as daffodils, lilies, cereals, and grasses.

T-DNA-mediated gene delivery

Living plants and plants cells in culture can be transformed by transferred DNA (T-DNA) excised from the Ti (tumor-inducing) plasmid of *Agrobacterium tumefaciens*. *A. tumefaciens* is a Gram-negative soil bacterium that causes crown gall disease in many dicotyledonous plants. Thus, recombinant T-DNA is a highly efficient gene transfer vector for dicotyledonous plants. The T-DNA region normally carries genes encoding plant hormones and opine (modified amino acids) synthesis enzymes. Crown gall disease results from transformation of the plant genome with this part of the Ti plasmid in a process analogous to bacterial conjugation. This tumor-inducing portion of the plasmid can be replaced with foreign genes and is still transferred into the host genome.

The general strategy is to cut leaf disks (causing cell injury) and then incubate with *Agrobacterium* carrying recombinant disarmed Ti vectors (Fig. 15.16). The disks can then be transferred to shoot-inducing medium

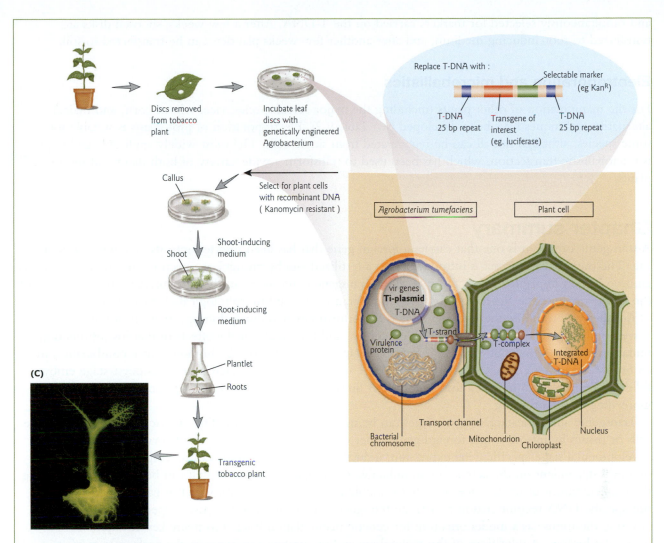

Figure 15.16 How to make a transgenic plant. A common method for making transgenic plants is outlined using tobacco as a model. A disk is cut out of a leaf from a tobacco plant and incubated in culture with *Agrobacterium tumefaciens* carrying a recombinant Ti plasmid with both a selectable marker and the transgene of interest. (A) *A. tumefaciens* contains a tumor-inducing (Ti) plasmid, which includes virulence (*vir*) genes and a transferred DNA (T-DNA) region. Genes of interest can be inserted into the T-DNA region. In this example, the T-DNA sequence is modified by substituting a selectable marker (kanamycin-resistance gene, *Kan*R) and the transgene of interest (luciferase) between the 25 bp T-DNA repeats. (B) The wounded cells at the edge of the leaf disk release phenolic defense compounds that attract *A. tumefaciens* and trigger the expression of the *vir* genes. The encoded virulence proteins excise the T-DNA region from the Ti plasmid, producing a T strand. After *A. tumefaciens* attaches to the plant cell, the T strand and several types of virulence proteins are transferred into the plant cell through a transport channel. Inside the plant cell, the virulence proteins interact with the T strand to form a T complex. This complex is targeted to the nucleus where the T-DNA integrates into the plant genome and expresses the encoded genes. Only those plant cells that take up the plasmid DNA and express the selectable marker gene will survive and grow to form a callus (a mass of undifferentiated cells). Growth factors applied to the callus induce it to form shoots and then roots. The rooted plantlet is then grown into an adult plant that carries the transgene. (C) Image of a transgenic tobacco plant expressing the gene for the firefly enzyme luciferase under control of a tissue-specific promoter. When watered with a solution of the substrate luciferin, luciferase produces a characteristic yellowish green fluorescence. Note that the veins but not the leaves of the plant fluoresce. (Reprinted with permission from Ow, D.W., Wood, K.V., DeLuca, M., de Wet, J.R., Helinski, D.R., Howell, S.H. 1986. Transient and stable expression of the firefly luciferase gene in plant cells and transgenic plants. *Science* 234:856–859. Copyright © 1986 AAAS.)

and simultaneously selected for markers carried on the T-DNA. After a few weeks, shooted disks are transferred to root-inducing medium, and after another few weeks plantlets can be transferred to soil.

Electroporation and microballistics

For the manipulation of other plants (including the major cereal species, rice, wheat, corn, and maize), alternative techniques have been developed (see Table 9.2). Electroporation of protoplasts is suitable for some species, although not all can be regenerated from single cells. The most widely applicable technique is microballistic transfection, which has been used to transform a wide variety of both dicot and monocot plants.

Chapter summary

A transgenic organism is one that carries a foreign gene that has been inserted into its genome at a random site. The transgene construct is introduced into a fertilized egg by pronuclear injection and the microinjected embryo is then implanted into a foster mother. Transgene expression can then be analyzed in the context of the whole organism, at the levels of transcription, translation, and morphological characteristics.

Another way to probe the role of a gene is to perform targeted disruption or mutation of the corresponding gene in a mouse (or other organism), and then look for the effects of that loss of function mutation on the "knockout" mouse. Targeted disruption is carried out by homologous recombination in embryo-derived stem cells. Gene-targeted cells are selected for and injected into a blastocyst-stage embryro, followed by implantation into a foster mother. A lineage of homozygous knockout mice can be created by inbreeding of the offspring.

Transgenic and gene-targeting strategies can be used to overexpress, modify, or inactivate genes in selected cells or tissues by introducing a foreign gene under control of tissue-specific regulatory elements. To study developmental stage-specific gene expression, inducible expression systems can be employed. One such system is dependent on the activity of an inducible transcriptional activator that can be regulated by varying the concentrations of tetracycline. Another example is the Cre/*lox* system, which is based on the used of site-specific DNA recombination to activate transgene expression or induce tissue-specific gene knockouts.

Using the mouse as a model organism for genetic manipulation has led to many key advances in molecular biology. Applications of this technology include strategies to increase the performance of commercially important animals by adding new traits or improving on existing ones, or for the overexpression of foreign proteins for therapeutic use (gene pharming).

Cloning by nuclear transfer involves transfer of an entire nucleus from a donor cell into an enucleated unfertilized egg. The cells are fused and the embryo is implanted into a surrogate mother. The resulting clone is genetically identical to the donor nucleus; however, the mtDNA derives from both the donor cell and recipient egg. Cloning by somatic cell nuclear transfer was first carried out in amphibians in the 1950s. However, cloning of mammals by adult somatic cell nuclear transfer was not thought to be possible until the cloning of Dolly the sheep in 1997. The technique is still highly inefficient, particularly in primates, and first generation cloned animals often suffer from developmental abnormalities. Cloning by nuclear transfer is used to provide insights into mechanisms of development, cell differentiation, nuclear reprogramming of differentiated cells to an undifferentiated (totipotent) state, imprinting, and aging. Reproductive cloning by nuclear transfer also has potential value for cloning of transgenic animals, preservation of prize animal stocks, and wildlife conservation. Therapeutic use of cloning for stem cells is an area of much interest and controversy.

Transgenic technology has been applied to plants in an attempt to increase the performance of commercially important plants by adding new traits or improving on existing ones. However, genetically modified crops have not always received a warm reception from the public, in part because of human health concerns. Transfer of cloned genes to plant leaf disks is performed using a plant vector such as the Ti plasmid carried by *Agrobacterium*, or by electroporation or microballistic transfection. The leaf disks are then transferred to shoot- and root-inducing media to form plantlets.

Analytical questions

1 Design a Southern blot experiment to check a transgenic mouse's DNA for integration of the *GFP* cDNA under control of a constitutive promoter. You may assume any array of restriction sites you wish in the target mouse chromosome and in the *GFP* transgene.

(a) Show sample results for a successful and an unsuccessful integration.

(b) Will mice from different founder lineages have the same site of integration?

(c) Describe techniques you would use to analyze the transgenic mice for transcription of the transgene and translation of the transgene.

2 What would be the outcome if the procedure used to make gene-targeted (knockout) mice was altered in the following ways (consider each situation separately)?

(a) The embryo-derived stem (ES) cells did not possess a gene responsible for visible phenotypic trait (such as *agouti* coat color) that differed from that of the cell in the recipient blastocyst.

(b) The inserted "disrupting" gene (e.g. *neo*ʳ) did not insert into the gene of interest.

(c) The chemical transfection or electroporation procedure failed.

(d) The gene knockout was embryonic lethal.

3 Consider the following hypothetical scenario:

> **A young scientific genius, Dolly, is terminally ill with hereditary nonpolyposis colon cancer (HNPCC). When she dies, her parents feel that one of the most remarkable minds in science will die with her, and they feel they owe it to the world to not let this happen. The family travels to a secret lab on a small offshore island that allegedly performs "reproductive cloning" (cloning by somatic cell nuclear transfer). Dolly's parents hope to clone her from one of her skin cells.**

(a) If the lab is successful in cloning Dolly, can they guarantee that her clone will be as academically gifted as Dolly? Why or why not?

(b) Would Dolly and her clone have identical DNA "fingerprints?" Why or why not?

(c) Would Dolly and her clone share the same mitochondrial DNA? Why or why not?

(d) Would you expect there to be a difference in the length of telomeres in Dolly's skin cells versus those in the cells of her embryonic clone? Why or why not?

5 A team of molecular biologists seeks your advice regarding whether it would be more important to include regulatory elements such as insulators and matrix attachment regions in a gene construct for generating transgenic mice by gene targeting in embryo-derived stem cells, or for generating transgenic mice by microinjection of foreign DNA into pronuclei of fertilized eggs? Provide advice and explain your rationale.

Suggestions for further reading

Alper, J. (2003) Hatching the golden egg: a new way to make drugs. *Science* 300:729–730.

Betts, D.H., Bordignon, V., Hill, J.R., Winger, Q., Westhusin, M.E., Smith, L.C., King, W.A. (2001) Reprogramming of telomerase activity and rebuilding of telomere length in cloned cattle. *Proceedings of the National Academy of Sciences USA* 98:1077–1082.

Campbell, K.H.S., McWhir, J., Ritchie, W.A., Wilmut, I. (1996) Sheep cloned by nuclear transfer from a cultured cell line. *Nature* 380:64–66.

Capecchi, M.R. (1994) Targeted gene replacement. *Scientific American* 270:34–41.

Chan, A.W.S., Chong, K.Y., Martinovich, C., Simerly, C., Schatten, G. (2001) Transgenic monkeys produced by retroviral gene transfer into mature oocytes. *Science* 291:309–312.

Chang, K., Qian, J., Jiang, M. et al. (2002). Effective generation of transgenic pigs and mice by linker-based sperm-mediated gene transfer. *BMC Biotechnology* 2:5–17 (http://www.biomedcentral.com/1472-6750/2/5).

Galli, C., Lagutina, I., Crotti, G. et al. (2003) A cloned horse born to its dam twin. *Nature* 424:635.

Gong, Z., Ju, B., Wan, H. (2001) Green fluorescent protein (GFP) transgenic fish and their applications. *Genetica* 111:213–225.

Gurdon, J.B., Byrne, J.A. (2003) The first half-century of nuclear transplantation. *Proceedings of the National Academy of Sciences USA* 100:8048–8052.

Holt, W.V., Pickard, A.R., Prather, R.S. (2004) Wildlife conservation and reproductive cloning. *Reproduction* 127:317–324.

Lee, B.C., Kim, M.K., Jang, G. et al. (2005) Dogs cloned from adult somatic cells. *Nature* 436:641.

Liu, J., Carmell, M.A., Rivas, F.V. et al. (2004) Argonaute2 is the catalytic engine of mammalian RNAi. *Science* 305:1437–1441.

Malarkey, T. (2003) Human health concerns with GM crops. *Mutation Research* 544:217–221.

Mikkola, H.K.A., Orkin, S.H. (2004) Gene targeting and transgenic strategies for the analysis of hematopoietic development in the mouse. In: *Methods in Molecular Medicine*, Vol. 105. *Developmental Hematopoiesis: Methods and Protocols* (ed. M.H. Baron), pp. 3–22. Humana Press, Inc., Totowa, NJ.

Meng, L., Ely, J.J., Stouffer, R.L., Wolf, D.P. (1997) Rhesus monkeys produced by nuclear transfer. *Biology of Reproduction* 57:454–459.

Nash, S. (2004) Glofish gives new shine to GM debate. *The Scientist* 18:46–47.

Ow, D.W., Wood, K.V., DeLuca, M., de Wet, J.R., Helinski, D.R., Howell, S.H. (1986) Transient and stable expression of the firefly luciferase gene in plant cells and transgenic plants. *Science* 234:856–859.

Palmiter, R.D., Brinster, R.L., Hammer, R.E., Trumbauer, M.E., Rosenfeld, M.G., Birnberg, N.C., Evans, R.M. (1982) Dramatic growth of mice that develop from eggs microinjected with metallothionein-growth hormone fusion genes. *Nature* 300:611–615.

Picciotto, M.R., Wickman, K. (1998) Using knockout and transgenic mice to study neurophysiology and behavior. *Physiological Reviews* 78:1131–1164.

Reik, W., Dean, W. (2003) Silent clones speak up. *Nature* 423:390–391.

Sauer, B. (1998) Inducible gene targeting in mice using the Cre/lox system. *Methods: a Companion to Methods in Enzymology* 14:381–392.

Schnieke, A.E., Kind, A.J., Ritchie, W.A. et al. (1997) Human factor IX transgenic sheep produced by transfer of nuclei from transfected fetal fibroblasts. *Science* 278:2130–2133.

Shin, T., Kraemer, D., Pryor, J. et al. (2002) A cat cloned by nuclear transplantation. *Nature* 415:859.

Simerly, C., Dominko, T., Navara, C. et al. (2003) Molecular correlates of primate nuclear transfer failures. *Science* 300:297.

Surani, M.A. (2001) Reprogramming of genome function through epigenetic inheritance. *Nature* 414:122–128.

Thomas, K.R., Capecchi, M.R. (1990) Targeted disruption of the murine int-1 proto-oncogene resulting in severe abnormalities in midbrain and cerebellar development. *Nature* 346:847–850.

Vogel, G. (1999) Mice cloned from cultured stem cells. *Science* 286:2437.

Vogel, G. (2001) Cloned gaur a short-lived success. *Science* 291:409.

Weismann, A. (1893) *The Germ-Plasm: a Theory of Heredity.* Walter Scott Ltd, London.

Woods, G.L., White, K.L., Vanderwall, D.K. et al. (2003) A mule clone from fetal cells by nuclear transfer. *Science* 301:1063.

Chapter 16

Genome analysis: DNA typing, genomics, and beyond

Some scientists said there was no reason to do it [The Human Genome Project] over 15 years. Why not do it over 25? One important reason is that if you did it over 25 years, most of the experienced scientists involved in it might be dead, at least mentally, by the time it was finished . . . Most people like to do things where they can see the results.

James Watson, *Genetics and Society* (1993), p. 18.

Outline

16.1 Introduction

The development of recombinant DNA techniques greatly facilitated studies of gene organization and expression. Advances in technology and understanding of genome structure have allowed molecular biologists to compare the genomes of individuals within species and across species with a degree of precision not possible even a decade ago. Levels of analysis range from personal identification based on analysis of variability in regions of repetitive DNA between individuals, to comparative analysis of entire genomes. The principles underlying DNA typing and its applications, to forensic casework in particular, are discussed. In addition overviews of bioinformatics tools and the rapidly expanding field of genomics and proteomics are included. It is now possible to apply molecular approaches to determine the nucleotide sequence of entire genomes, to carry out large-scale genome comparisons of both the coding and noncoding regions of various organisms, and to analyze the full complement of expressed RNAs and proteins in a cell.

16.2 DNA typing

DNA typing, as it is now known, is one of the most reliable and conclusive methods available for the identification of an individual. First called "DNA fingerprinting," this technique was developed in 1985 by Alec Jeffreys and co-workers. The name has since been changed to distinguish it from traditional skin fingerprinting. DNA typing can be used in many different contexts, including: (i) to establish paternity and other family relationships (Fig. 16.1); (ii) to match organ donors with recipients in transplant programs; (iii) to identify catastrophe victims (e.g. victims of the 2004 tsunami in South East Asia); (iv) to detect bacteria and other organisms that may pollute air, water, soil, and food; (v) to determine whether a clone is genetically identical to the donor nucleus (see Chapter 15); (vi) to trace the source of different marijuana plants (Focus box 16.1); and (vii) to identify endangered and protected species as an aid to wildlife officials (Focus box 16.2). The most well-known use of DNA typing is to identify potential suspects whose DNA may match evidence left at crime scenes, or to exonerate persons wrongly accused of crimes.

Frequently, the forensic scientist is asked to characterize biological remains from the scene of a crime and match specimens with suspects, e.g. blood stains, semen, hair, saliva, and other body tissues. Originally, forensic scientists had to rely on the limited information available from analyzing blood types and polymorphic proteins and enzymes. Since 1987, forensic DNA analysis has been used in the USA as evidence in criminal trials and paternity cases. Genetic evidence is introduced in less than 1% of prosecutions, yet these cases usually involve murder or rape and the tests often have a dramatic effect on verdicts. When DNA profiles were first presented in court, skeptics wondered whether they could uniquely identify a person. Now, judges and juries readily accept the data, though mistakes in the collection and testing of samples do occur.

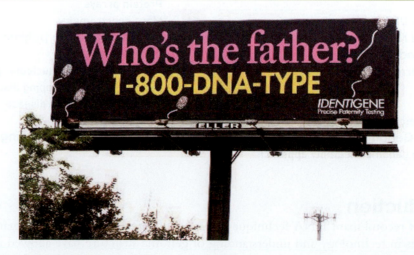

Figure 16.1 A billboard advertising paternity testing (Photograph courtesy of Identigene.)

Marijuana (*Cannabis sativa*) is a naturally occurring weed species that was bred and cultivated into a plant noted for its hallucinogenic, euphoria-inducing properties (Fig. 1A). Although not legal in the USA, cannabis is effective in the treatment of a variety of ailments, including glaucoma, chronic pain, and nausea (particularly in cancer patients undergoing chemotherapy). The active component in marijuana is Δ9-tetrahydrocannabinol (THC). Forensic scientists at the Connecticut State Forensic Science Laboratory have applied DNA typing methods to the marijuana plant with the aim of creating a database of DNA profiles of different marijuana plants which will help them to trace the source of any sample. Showing that two samples have matching DNA profiles does not by itself prove they came from the same grower or the same plants. Although seed-generated marijuana plants would be expected to have unique DNA profiles analogous to a human population, identical clones of plants are easily created by taking cuttings. This is a method growers often use to propagate particularly potent (high THC content) strains of marijuana. However, growers tend to not give away cuttings of their best plants, so samples may still provide a potential link from the user, to the distributor, to the grower.

The forensic scientists are using a technique called amplified fragment length polymorphism (AFLP, pronounced "a-flip"). AFLP involves the PCR amplification of restriction fragments to which adaptor oligomer sequences have been attached. The PCR primers recognize the adaptor oligomers and bind to amplify different sized DNA fragments. The DNA fragments generated are detected with a DNA sequencer. The sequencer has a laser that excites the fluorescent dye that was incorporated into the DNA fragments during PCR. Fluorescing DNA fragments are detected by a CCD camera as they pass by the laser and the band patterns are recorded by a computer and data are analyzed. AFLP profiles from unrelated marijuana plants are easily distinguishable (Fig. 1B).

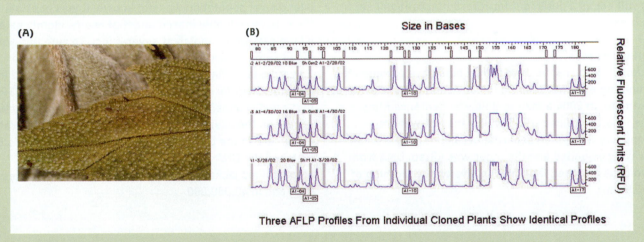

Three AFLP Profiles From Individual Cloned Plants Show Identical Profiles

Figure 1 DNA profiles of marijuana by amplified fragment length polymorphism (AFLP). (A) Photograph of a *Cannabis sativa* (marijuana) leaf shows serration and resin droplets. The resin droplets may capture foreign pollen that could be used to link a sample back to a source country if appropriate reference pollen databases are available. Pollen databases can be used to correlate plant species, blooming seasons and geographic locations to help identify a season and location where marijuana samples were originally grown and harvested. (B) AFLP profiles are generated by (1) restriction enzyme digestion of a DNA plant sample (2) polymerase chain reaction (PCR) amplification of DNA fragments and (3) visualization and separation of band patterns of DNA profiles on a DNA sequencer. AFLP is a technology useful for potentially individualizing plant samples, requires no prior DNA sequence knowledge and has a sufficient number of markers to distinguish between even highly inbred plant varieties (e.g. plants with common ancestry). The three AFLP profiles shown here are from three separate plants derived from one individual source ("mother") plant. (Photographs courtesy of Heather Miller Coyle, Timothy M. Palmbach, and Henry C. Lee, Forensic Science Program, University of New Haven, West Haven, Connecticut).

The analysis of nonhuman biological evidence has been an occasional requirement of forensic scientists and other investigators. For example, DNA typing can be used to identify endangered and protected species as an aid to wildlife officials, and nonhuman DNA evidence based predominantly on STR polymorphisms has found its way into some criminal courts.

DNA of protected whales found at Japanese markets

DNA typing is part of an increased effort to abolish illegal whaling practices. In July 2004, DNA typing was used to show that whale meat from the Japanese retail market matched the profile of the sei (pronounced "say") whale's southern hemisphere stock. The sei whale (*Balaenoptera borealis*) is the third largest baleen whale after the blue and finback whale. It is an endangered species and is supposedly protected from hunting under the International Whaling Commission and the Endangered Species Act. Wildlife officials hope that such cases of successful DNA typing will act as a deterrent for future marketing of sei whale meat.

Incriminating pets

In a number of cases, beloved family pets have provided the incriminating evidence leading to conviction of their owners. For example, in 2004 a man in Ohio was sentenced to 6 months in jail and fined US$5000 for his Rottweilers attacking two women on separate occasions. One woman was gravely wounded and the other was killed. DNA from the two dogs matched DNA from saliva on the clothing of both victims.

In another well-publicized case, a white American shorthaired cat called Snowball provided evidence that led to his owner being convicted of second-degree murder. In the fall of 1994, a woman disappeared from her home on Prince Edward Island. A man's leather jacket stained with the victim's blood was found about 8 km from the victim's home. Several white cat hairs were found in the coat lining. It was known that the victim's boyfriend owned a white cat named Snowball. When the victim's body was finally found in a shallow grave the following spring, the boyfriend was charged with murder. Coincidentally, about that time molecular biologists Marilyn Menotti-Raymond, Stephen J. O'Brien, and co-workers had just published a paper characterizing the location of STR markers in the cat genome. Menotti-Raymond and O'Brien agreed to perform DNA typing on blood from Snowball and DNA extracted from crime-scene cat hairs. Enough DNA was extracted from one cat hair to do the identity test. The two samples matched at 10 STRs. However, the genetic diversity of the island's cat population was not known. It was possible that the cats were highly inbred and that two randomly selected island cats would match at several loci. To rule this out, blood samples were analyzed from 19 cats on Prince Edward Island that were thought to be unrelated and from nine cats in different parts of the USA. By studying STRs in these two groups, it was shown that the genetic structure of cats on the island was about as diverse as that of cats elsewhere in the USA. The probability that Snowball's DNA matched the DNA from the white hairs found on the bloodstained jacket by chance was calculated to be less than one in 40,000,000.

DNA polymorphisms: the basis of DNA typing

Only about 0.1% of the human genome (about 3 million bases) differs from one person to another. With the exception of the human leukocyte antigen (HLA) region, genetic variation is relatively limited in coding DNA. This results from expressed genes being subjected to selection pressure during evolution to maintain their function. However, less than 40% of the human genome is comprised of genes and gene-related sequences (e.g. introns, pseuodogenes) (Fig. 16.2). The majority of the human genome represents intergenic DNA. Intergenic regions of the genome are not generally controlled by selection pressure and mutations in these regions are usually maintained and transmitted to the next generation. Intergenic DNA consists of unique or low copy number sequences and moderately to highly repetitive sequences. The repetitive sequences are divided into two major classes: (i) interspersed elements; and (ii) tandem repetitive sequences.

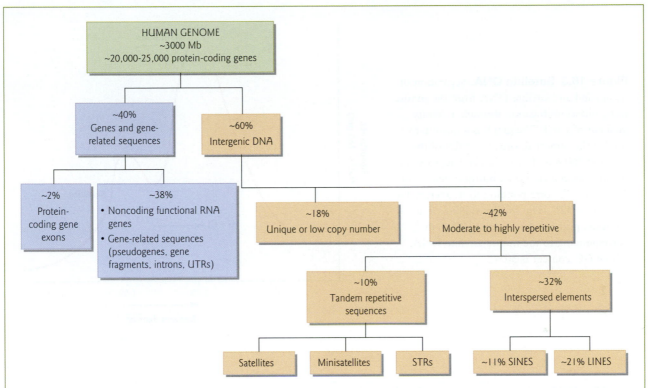

Figure 16.2 Organization of the human genome. The human genome consists of many different types of DNA sequence, the majority of which do not encode proteins. The flow chart shows the distribution of each of the various types of sequence. LINEs, long interspersed nuclear elements; SINEs, short interspersed nuclear elements; STRs, short tandem repeats; UTRs, untranslated regions.

Interspersed elements

Interspersed elements are genome-wide repeats that are primarily degenerate copies of transposable elements (unstable DNA elements that can move to different parts of the genome; see Section 12.5). The repeats are not clustered, but are dispersed at numerous locations within the genome, and make up about 32% of the genome. Interspersed repetitive DNA elements are further subdivided into two categories based on their length. Sequences of fewer than 500 bp are called SINEs (short interspersed nuclear elements) and sequences of 500 bp or more are called LINEs (long interspersed nuclear elements).

Tandem repetitive sequences

The majority of DNA typing systems used in forensic casework are based on genetic loci with tandem repetitive DNA sequences. These tandem repeats make up approximately 10% of the genome and are classified into three subdivisions based on their length. Satellites are very highly repetitive DNA with repeat lengths of one to several thousand base pairs whose buoyant density during density gradient centrifugation differs from that of the bulk of the DNA (Fig. 16.3). These sequences typically are organized as large clusters in the heterochromatic regions of chromosomes, near centromeres and telomeres. They are also found abundantly on the Y chromosome. The regions of interest for forensics are much shorter than the bulk satellite DNA. Depending on the length of the repeat, these shorter tandem repetitive sequences are classified as either minisatellite or short tandem repeats (STRs) (Table 16.1).

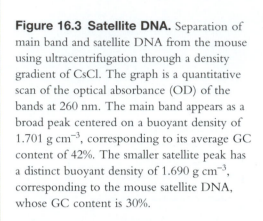

Figure 16.3 Satellite DNA. Separation of main band and satellite DNA from the mouse using ultracentrifugation through a density gradient of CsCl. The graph is a quantitative scan of the optical absorbance (OD) of the bands at 260 nm. The main band appears as a broad peak centered on a buoyant density of 1.701 g cm^{-3}, corresponding to its average GC content of 42%. The smaller satellite peak has a distinct buoyant density of 1.690 g cm^{-3}, corresponding to the mouse satellite DNA, whose GC content is 30%.

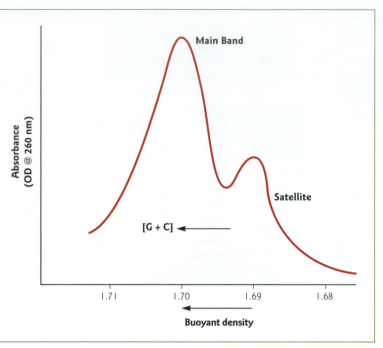

Table 16.1 DNA polymorphisms used in forensic genetics.

Preferred name	Alternate name	Length of repeat	Total length of tandem array	Location in genome	Current use
Minisatellite	Variable number tandem repeat (VNTR) loci	15–50 bp	500 bp to 20 kb	More common in subtelomeric regions	Used in some paternity testing labs
Short tandem repeat (STR)	Microsatellite	2–6 bp	50–500 bp	Widely distributed throughout entire genome	Method of choice in forensic DNA typing labs

Minisatellites, also known as variable number tandem repeat (VNTR) loci, are composed of sequence motifs ranging from around 15 to 50 bp in length. The total length of tandem repeats is between 50 bp and 20 kb. Short tandem repeats (also referred to as microsatellites) are much shorter. The repeat unit ranges from 2 to 6 bp for a total length of between 50 and 500 bp. The most common STR sequences are dinucleotide repeats. The genetic variation between individuals in these minisatellites and STRs is mainly based on the number of tandemly arranged repetitive elements, but there may also be slight differences in the sequence. These variable regions are particularly useful for forensic genetics because they can be used to generate a DNA profile of an individual, but do not give information about phenotypic traits of that individual.

The power of DNA evidence lies in statistics. Forensic technicians typically analyze multiple variable regions, called polymorphic markers. Their aim is to calculate a probability that only one person in, for example, a quadrillion (10^{15}) could have the same profile of markers. A variety of DNA technologies are used in forensic investigations, including minisatellite analysis, polymerase chain reaction (PCR) based analysis, STR analysis, mitochondrial DNA analysis, Y chromosome analysis, and random amplified polymorphic DNA (RAPD) analysis.

Minisatellite analysis

Minisatellite analysis is one of the original applications of DNA analysis to forensic investigations. Until the introduction of STR analysis, minisatellite (VNTR) analysis was very popular in forensic labs. Two major drawbacks of minisatellite analysis are that it requires relatively large amounts of DNA, and it does not work well with samples degraded by environmental factors, such as dirt or mold.

As described in Chapter 8 (Section 8.10), restriction fragment length polymorphism (RFLP) is a technique for analyzing the variable lengths of DNA fragments that result from digesting a DNA sample with a restriction endonuclease. Traditional RFLPs result from base pair changes in restriction sites. In contrast, minisatellites are a special class of RFLP in which the variable lengths of the DNA fragments result from a change in the number, not base sequence, of minisatellite repeats. There is wide variation in the size of repeat clusters between individuals; i.e. minisatellite loci are highly polymorphic with a very large numbers of alleles. This variability arises from misalignment of the repeats during chromosome pairing and unequal crossing-over. Minisatellite repeats or VNTRs are dispersed throughout the genome, but tend to be clustered near telomeres. Minisatellites number in the thousands but each locus shows a distinctive repeat unit. However, some of these loci share a shorter common core sequence (e.g. a 10–15 bp sequence that is GC rich), which makes it possible to analyze multiple loci at one time.

Classic "DNA fingerprinting": minisatellite analysis with a multilocus probe

It was the "hypervariable" nature of minisatellites that led Alec Jeffreys and his colleagues to develop a technique in 1985 for genetic mapping in humans. A simplified example of a classic DNA profile, or "DNA fingerprint," obtained by minisatellite analysis is shown in Fig. 16.4. In this example, three different individuals possess different numbers of copies of a specific tandemly repeated base sequence (minisatellite). For a pair of homologous chromosomes within an individual, the number of copies of a specific tandemly repeated base sequence may also vary. DNA from each individual is cut with a restriction enzyme that does not cut the minisatellite array itself. The best restriction endonucleases to use for minisatellite analysis are those with 4 bp recognition sites such as *Hae*III, *Hin*fI or *Sau*3A. With these enzymes it is unlikely that they will cut within the relatively short minisatellite unit sequence, but they are highly likely to cut within several hundred base pairs of flanking sequence on both sides. *Hin*fI is usually used in Europe and *Hae*III in the United States.

The DNA fragments resulting from the restriction endonuclease digestion are separated via agarose gel electrophoresis, and Southern blotting is carried out (see Tool box 8.7). The blot is probed with a multilocus probe that hybridizes with the basic repeated units (the common core sequence) and detects multiple minisatellites at different locations (although only one locus is shown in this example for simplicity). Probes can be labeled using isotopic or nonisotopic methods. Nonisotopic chemiluminescent labeling and detection methods have gained in popularity in recent years. Different patterns of bands are observed in each individual on an autoradiogram. Actual DNA fingerprint patterns, as shown in Fig. 16.5, are much more complex than the schematic representation in Fig. 16.4. The simultaneous detection of 10–40 unlinked and highly polymorphic loci provides a whole genome "fingerprint" pattern which is very likely to show differences between any two unrelated individuals.

Because of the unique pattern of 30 or more bands produced by minisatellite analysis with multilocus probes, it is very unlikely that any two people would by chance share the same pattern. The DNA profile is a unique biological identifier for each individual, except for identical twins. The profile is essentially constant for an individual, irrespective of the source of DNA (whether from blood, semen, saliva, etc.). A match between band sets gives a positive result. If the sample profiles do not match, the person, for example, did not contribute DNA at the crime scene (Fig. 16.5B). Importantly, there is a simple Mendelian pattern of inheritance. For a child, 50% of the bands are derived from the mother and 50% are derived from the father. Thus, in a paternity case, the bands that come from the mother can be identified; the remaining bands must come from the father, and can be matched with DNA from possible fathers (Fig. 16.5A). Occasionally a new mutant band appears, but with very low frequency.

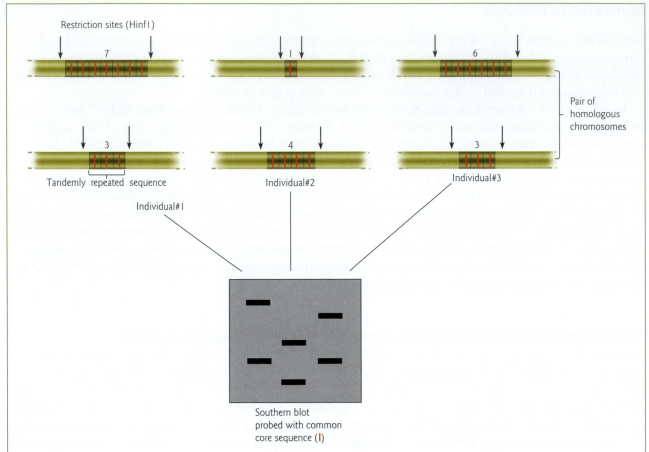

Figure 16.4 Simplified representation of minisatellite repeat analysis. For a given minisatellite locus, the number of repeats is highly variable among individuals and heterozygosity is high. Three different individuals (1, 2, and 3) have different numbers of copies of a specific tandemly repeated base sequence at a particular locus. The number of repeats is indicated above the array. DNA from each individual is cut with a restriction enzyme and Southern blotting is carried out using a probe that hybridizes with a common core sequence (red) in the tandem repeats. Different patterns of bands are observed in each individual. Analyzing the number of repeats at one or more such loci provides a highly sensitive measure of individual identity.

Multilocus probes were never very successful in forensic caseworks. Early on it became clear that there were at least two main drawbacks to their use. First, there were statistical problems with evaluation of the evidence in cases of a match. Second, there were also problems with standardization of the technique. Forensic investigators who still analyze minisatellite loci have mainly switched to the use of single-locus probes.

Minisatellite analysis with single-locus probes

A single-locus probe allows the detection of a single minisatellite DNA locus on one chromosome. At most, two bands are observed on a Southern blot, one for each of the two alternate alleles present at the locus on each member of the chromosome pair (Fig. 16.5A). Only one band is observed if the alleles are the same size. To increase the sensitivity of discrimination, 3–5 single-locus probes are mixed in a

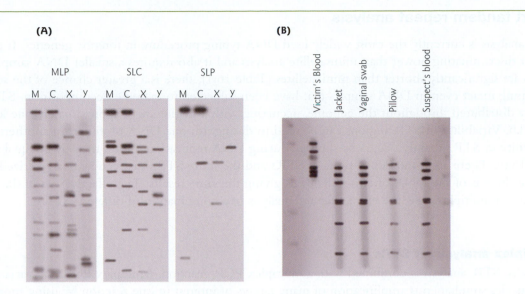

Figure 16.5 Comparison of minisatellite repeat analysis using multilocus versus single-locus probes in paternity and forensic testing. (A) Three examples of Southern blots for a paternity test (M, mother's DNA; C, child's DNA; X, true father's DNA; Y, excluded alleged father's DNA). (i) Southern blot probed with a multilocus probe (MLP) that hybridizes to a common core sequence. (ii) Blot probed with a single-locus cocktail (SLC) containing four different single-locus probes. (iii) Blot probed with a single-locus probe (SLP) to a single minisatellite locus on one chromosome. (B) An example of a Southern blot from a criminal case. The blot was probed with a single-locus cocktail containing four different single-locus probes. The DNA band patterns from specimens retrieved at the crime scene (from a jacket, vaginal swab, and a pillow) and a blood sample from the suspect were compared. The banding patterns match indicating the suspect as the possible source of evidence DNA. The band patterns from the samples are not likely to be identical solely by chance.

single-locus "cocktail." This mixture of probes is used to examine different loci on a Southern blot simultaneously. Minisatellite analysis with single-locus probes is still used by a few labs working in paternity testing.

Polymerase chain reaction-based analysis

The PCR is a powerful technique for directly amplifying short segments of the genome (see Tool box 8.3). Millions of exact copies of DNA from a biological sample can be made. PCR requires that the sequence on either side of the target region is known, and then allows the region between two defined sites to be amplified. Use of this technique in forensics extends DNA typing to the level of a single cell. For example, sufficient DNA can be collected from saliva on a postage stamp, or a cigarette butt, or bones from skeletons. Even pieces of highly degraded DNA can still be amplified, as long as the target sequence is intact. Of course, scientists must be very careful to avoid contamination with their own DNA, or other biological materials, during the identifying, collecting, and preserving of the sample.

Once PCR has been used to generate large copies of a DNA segment of interest, different approaches may be taken to detect genetic variation within the segment amplified. Some years ago, simultaneous amplification of six specific genetic loci was popular in forensics labs. These loci included the highly variable HLA class II gene. Now, the analysis of STRs by PCR (see next section) is the method of choice for forensic indentification.

Short tandem repeat analysis

STR analysis is currently the most widely used DNA typing procedure in forensic genetics. It provides higher discriminating power than minisatellite analysis and it also requires a smaller DNA sample size. Since STRs are significantly shorter than minisatellites (Table 16.1), there is a greater chance of the sequence remaining intact even in DNA samples that have been subjected to extreme decomposition. STRs are widely distributed throughout the genome, occurring with a frequency of approximately one locus every 6–10 kb. Variability in STR regions can be used to distinguish one DNA profile from another. The variability in STRs mainly occurs by slippage during DNA replication, rather than by unequal crossing-over (Fig. 16.6). There are hundreds of types of STRs and the more STRs that can be characterized, the less chance there is of the DNA of two individuals giving the same results. The effectiveness of this technique has lead to multiplexing – the extraction and analysis of a combination of different STRs.

Multiplex analysis of STRs

Typically, STR analysis is combined with a multiplex PCR reaction. Multiplex PCR is a variant of PCR that enables simultaneous amplification of many targets of interest in one reaction by using more than one pair of primers. The introduction of fluorescent-based technology and the use of automated DNA sequencers (see Fig. 8.16) have allowed the typing of large multiplexes, as well as automation of the typing procedure (Fig. 16.7). Although gel electrophoresis is an adequate method for separation of STR loci, multichannel capillary electrophoresis is the preferred separation technique. Data interpretation and designation of alleles have become predominantly computer-based. In this process, three different colored fluorescent dyes are incorporated into the DNA fragments during PCR. Different STR loci and alleles are separated by size (migration rate in gel) and detected by color after laser-induced excitation of the fluorescent dyes. One commercial multiplex system also amplifies the amelogenin locus to allow gender identification; a 106 bp X-specific band and a 112 bp Y-specific band are generated (Fig. 16.7). Amelogenin is the major protein component of tooth enamel. It is expressed from two nonallelic genes on the sex chromosomes.

The Federal Bureau of Investigation (FBI) in the USA uses a standard set of 13 specific STR regions for CODIS (Combined DNA Index System). CODIS is a software program that operates local, state, and national databases of DNA profiles from convicted offenders, unsolved crime scene evidence, and missing persons. The international growth of DNA databases has been rapid. The odds that two individuals will have the same 13-loci DNA profile is about one in a billion. Extremely discriminative 15-plexes are becoming more and more popular in forensic labs. The probabilities of two unrelated individuals matching by chance are less than one in 10^{17} for some of these large multiplexes (Table 16.2).

Mitochondrial DNA analysis

Sometimes there is not enough nuclear DNA for analysis, or it is highly degraded. Because every cell has hundreds of mitochondria with several hundred mtDNA molecules and mtDNA degrades less rapidly than nuclear DNA, older biological samples (e.g. strands of hair, solid bone, or teeth) often lack usable nuclear DNA but have abundant mtDNA. mtDNA has even been successfully isolated from the fossil bones of a 60,000-year-old anatomically modern man from Lake Mungo in Australia. This analysis of ancient mtDNA revealed the intriguing finding that a sequence was present in the mtDNA of the Lake Mungo man that differs from other fossils and from modern people. In modern people, this sequence now exists as an insert on chromosome 11 in the nuclear genome.

Mitochondrial DNA analysis is used to examine DNA from forensic samples when experts presume that nuclear DNA, which can provide a more precise match, is not present in sufficient quantities to warrant STR (or minisatellite) analysis. Analysis typically involves PCR amplification and direct sequencing of two highly variable regions in the D loop region of the mtDNA called hypervariable (HV) regions 1 and 2

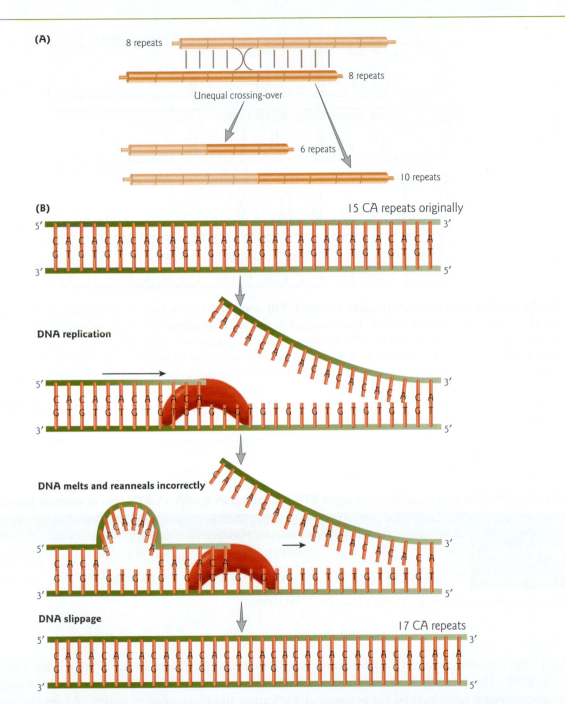

Figure 16.6 Length variations can be generated at tandem repeat loci by unequal crossing-over and slippage during DNA replication. (A) Unequal crossing-over (nonreciprocal recombination) occurs when one copy of a tandem repeat such as a minisatellite in one chromosome misaligns for recombination during meiosis with a different copy of the repeat in the homologous chromosome, instead of with the corresponding copy. Recombination increases the number of repeats in one chromosome (duplication) and decreases it in the other (deletion). Individual repeat units are represented by boxes. In the example shown, crossing-over generates chromosomes with six and 10 repeats, instead of the eight repeats of each parent. (B) The variable number of short tandem repeats present in the genome primarily results from DNA slippage. In the example shown, there are 15 CA dinucleotide repeats in the original template DNA. During replication, the DNA melts and then reanneals incorrectly in the repeated region (slippage), resulting in re-replication of an additional two repeats. The newly synthesized DNA contains 17 CA repeats.

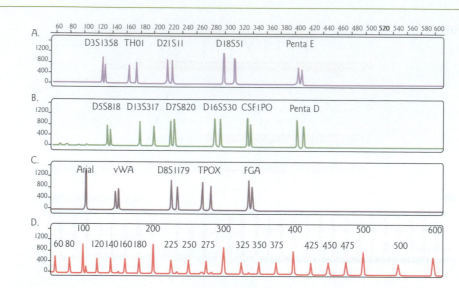

Figure 16.7 Multiplex short tandem repeat (STR) analysis. Fifteen different STRs and a gender-specific marker were amplified simultaneously from a single DNA template by polymerase chain reaction (PCR), using a multiplex primer pair mix (PowerPlex™ 16 System, Promega). One primer in each pair is labeled with a fluorescent tag for four-color detection. The PCR amplification products were mixed with labeled internal lane standards and detected using an automated sequencer. Computer software was used to generate electropherograms showing the peaks of the five fluorescein-labeled STR loci (blue, panel A), the six JOE-labeled loci (green, panel B), the four TMR-labeled loci and a gender-specific marker, Amelogenin (black, panel C), and the sizes of fragments of the internal lane standard (red, panel D). (Reprinted with permission from Promega Corporation).

Table 16.2 Matching probabilities of eight STR loci compared with 15 STR loci in various populations.[*]

STR system	African-American	Caucasian-American	Hispanic-American	Asian-American
Eight STR loci	1 in 2.77×10^8	1 in 1.15×10^8	1 in 1.45×10^8	1 in 1.32×10^8
Fifteen STR loci	1 in 1.41×10^{18}	1 in 1.83×10^{17}	1 in 2.93×10^{17}	1 in 3.74×10^{17}

[*] Adapted from Promega Corporation Technical Manual No. DO12 (10/02).

(Fig. 16.8). The first time mtDNA evidence was allowed in a courtroom in the USA was for a murder case in 1996. The case involved analysis of mtDNA from a hair found on the victim. mtDNA analysis is now considered a valid method for personal identification and is accepted in courts all over the world. Statistical interpretation in cases of match, however, is more complicated than for nuclear DNA and appropriate corrections must take into account population substructure and sampling error. Mutation rate, heteroplasmy (see Focus box 3.1), and the statistical approach used sometimes make interpretations difficult. For this reason, nuclear DNA is preferred for forensic analysis, because clear similarities and differences are easier to establish.

In the investigation of cases that have gone unsolved for many years, mtDNA has proved to be extremely valuable. In addition, comparing the mtDNA profile of unidentified remains with the profile of a potential maternal relative can be an important technique in missing persons investigations. If there is a close match, it can be concluded that individuals are related. A limitation of mtDNA analysis, however, is that it can only identify a person's maternal lineage since it is only passed on to the next generation by the egg.

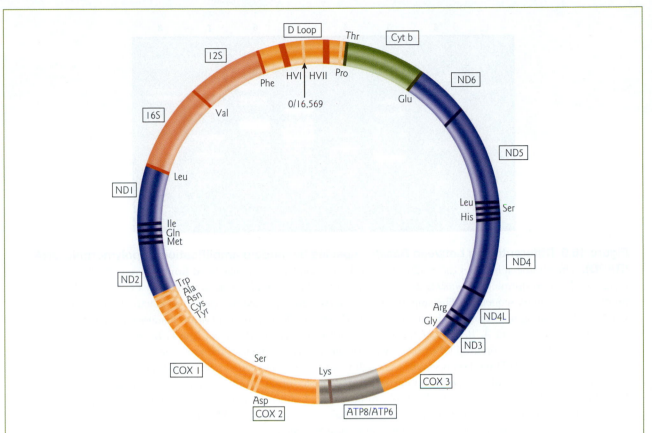

Figure 16.8 Map of human mitochondrial DNA (mtDNA). The D loop (or "displacement" loop) includes the region where replication of the heavy strand of mtDNA begins (0/16,569). Within the D loop, hypervariable region I (HVI) spans nucleotides 57–372, and hypervariable region II (HVII) spans nucleotides 16,024–16,383 of the 16,569 kb mtDNA genome. Cyt b, cytochrome *b* gene; ND1–4, NADH dehydrogenase genes; COX1–3, cytochrome oxidase genes; 12S and 16S, ribosomal RNA genes; ATP8/ATP6, ATP synthase genes; amino acid abbreviations, tRNA genes.

Recently, it has been shown that nuclear DNA can, in fact, be used successfully to examine skeletal remains. In Bosnia, the success rate for identifying victims in the Balkan war zone using nuclear DNA was greater than 90%, and multiplex analysis of STRs is currently being used on teeth and bone samples from victims of the tsunami of 2004 in South East Asia. However, forensic technology still has its limitations. DNA typing methods have been unsuccessful in the identification of 42% of the victims of the September 11, 2001 terrorist attacks on the World Trade Center.

Y chromosome analysis

The Y chromosome is passed directly from father to son, so the analysis of genetic markers on the Y chromosome is especially useful for tracing relationships among males. Although the variation in the Y chromosome is low overall, there are some polymorphisms, including STRs.

Analysis of Y chromosome-specific STRs is useful in paternity testing when a male offspring is in question, or for analyzing biological evidence in criminal casework involving multiple male contributors. Mixed male–female stains often occur as a result of sexual assault on a female victim by a male offender(s). Selective targeting of the male DNA removes the possibility of amplifying female DNA at the same time during PCR. This allows unambiguous determination of the male profile.

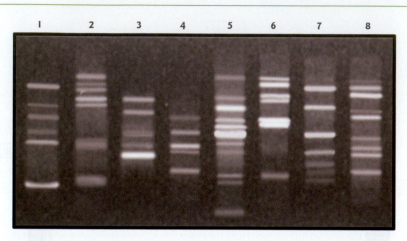

Figure 16.9 Differentiation between *Bacillus* species by random amplification of polymorphic DNA (RAPD). The photograph shows the results of a RAPD reaction. DNA was extracted from different *Bacillus* strains, including the spore-forming *B. anthracis* that is the cause of anthrax. Samples were amplified by PCR using short primers with arbitrary sequences. The amplification products were separated by agarose gel electrophoresis, and visualized by staining with ethidium bromide. Lane 1, *B. anthracis*; lane 2, *B. thuringiensis thuringiensis*; lane 3, *B. thuringiensis israeliensis*; lane 4, *B. cereus* 569; lane 5, *B. licheniformis*; lane 6, *B. subtilis*; lane 7, *B. cereus / thuringiensis*; lane 8, *B. mycoides*. Very few common DNA bands were obtained even in strains of the same species (e.g. lanes 2 and 3) indicating that the RAPD method is sensitive enough to distinguish between closely related species, and to identify *B. anthracis* in case of hostile dissemination during an act of bioterrorism. (Reproduced with permission of Blackwell Publishing from Levy, H., Fisher, M., Ariel, N., Altboum, Z., Kobiler, D. 2005. Identification of strain specific markers in *Bacillus anthracis* by random amplification of polymorphic DNA. *FEMS Microbiology Letters* 244:199–205. Copyright © 2005 Federation of European Microbiological Societies.)

Randomly amplified polymorphic DNA (RAPD) analysis

Another PCR-based technique for DNA typing is called RAPD (pronounced "rapid") for "randomly amplified polymorphic DNA." RAPD markers are generated in a single standard PCR reaction where the PCR primers consist of random sequences (typically 10–15 bp in length). Wherever the PCR primers have sequence homology with the DNA template, they will bind and PCR products of variable size will be generated (Fig. 16.9). No knowledge of an organism's DNA sequence is required to perform RAPD analysis; however, the PCR conditions must be held constant to obtain reproducible banding patterns. This technique is useful when little is known about the genome of the organism being analyzed. RAPD analyses are commonly used for typing strains of bacteria, and for analyzing genome variation in higher plants.

The RAPD method has been used and accepted in court for both criminal and civil cases. One well-documented case involved the use of DNA profiles from Palo Verde tree (*Cercidium* spp.) seed pods to link a suspect's vehicle back to a homicide crime scene. Each Palo Verde tree has a unique RAPD profile. The seed pods found in the suspect's vehicle matched seed pods from a Palo Verde tree with a broken branch at the scene of the crime.

16.3 Genomics and beyond

Walk into a molecular biology lab even a decade ago and you might have seen one lonely computer used primarily for word processing. Walk into a molecular biology lab now and you will probably see multiple computers, many dedicated to particular pieces of equipment, and in use not only for word processing and image analysis, but for DNA and protein sequence analysis. Computer sequence analysis is now considered a routine lab tool for the new generation of molecular biologists.

Whereas gene discovery once drove DNA sequencing, now the sequencing of entire genomes drives gene discovery. The aim of the Human Genome Project was to complete a genetic and physical map and sequence of the genome. The availability of this sequence information is hoped to provide many benefits in medicine, drug development, and understanding of the basis of inherited disorders. There are some fears that information could be misused. For example, there could be discrimination in the workplace or by insurance companies against an individual with a particular genotype. This section will provide an overview of bioinformatics tools used in the scaling up of biological research programs in the age of genomics and beyond.

What is bioinformatics?

Bioinformatics is a discipline combining information technology with biotechnology that has arisen in the last few years due to the enormous volume of data available on the internet. It is an area of computer science devoted to collecting, organizing, and analyzing DNA and protein sequences, and the wealth of data being generated by genomics and proteomics laboratories. As such, it is a rapidly growing and changing field.

The tools of bioinformatics allow researchers to analyze data in many different ways, including the following: (i) to locate sequences in public databases by accession number or keyword searching; (ii) to align two sequences or multiple sequences; (iii) to find sequences in a database similar to the sequence of interest; (iv) to assemble short overlapping fragments of sequence into consensus sequences; (v) to analyze physical and chemical properties of proteins; (vi) to analyze sequence patterns to locate restriction sites, promoters, DNA-binding domains, and protein motifs; and (vii) to perform phylogenetic (evolutionary) analysis. The most commonly used genome tool is BLAST (basic local alignment search tool). BLAST programs find regions of similarity between different protein-coding genes. A BLAST search often involves searching a genome, or the genomes of many different organisms, for all of the predicted protein sequences that are related to a so-called "query sequence" (the sequence of interest). Statistical methods are used by BLAST programs to determine the likelihood that the "hits" (the genes or encoded proteins identified by the query sequence) have a similar function. In the example shown in Fig. 16.10, genes that encode regulatory proteins containing a specialized form of the zinc finger DNA-binding motif (see Fig. 11.16) are identified when a similar motif from the thyroid hormone receptor is used as a query.

The first generation of bioinformatics began with the Human Genome Project. It ended in 2001 with the sequencing of the human genome by Celera Genomics and the Human Genome Project International Consortium. In this second generation of bioinformatics, there is a shift away from emphasis on gene sequence database services to discovery of therapeutics and diagnostics. The web browser is the primary means of access to bioinformatics data and programs for most molecular biologists. There are over 400 key databases, including the NCBI (National Center for Biotechnology Information) Map Viewer (genome browser), UCSC (University of California, Santa Clara) Genome Bioinformatics, and the Swiss Protein Knowledgebase. Besides basic sequence information, these databases provide information on gene expression, intermolecular interactions, metabolic pathways, pathology, biological function, and more (Table 16.3).

An important trend affecting bioinformatics is the scaling up of biological research programs for "high-throughput" screening methods. This scaling up has led to more automated approaches to projects such as DNA sequencing and analysis of gene expression. These types of large-scale genomic and proteomic research programs require a much greater degree of computational support.

Genomics

Genomics is the comprehensive study of whole sets of genes and their interactions rather than single genes or proteins. It is the comparative analysis of genomes based on the availability of complete genome

>☐ gi│55932│emb│CAA31237.1│ [G] c-erb-A thyroid hormone receptor [Rattus norvegicus]

   Length = 398

   Score = 73.6 bits (179),   Expect = 2e-12
   Identities = 30/30 (100%),   Positives = 30/30 (100%), Gaps = 0/30 (0%)

   Query   1   CVVCGDKATGYHYRCITCEGCKGFFRRTIQ   30
                 CVVCGDKATGYHYRCITCEGCKGFFRRTIQ
   Sbjct   41   CVVCGDKATGYHYRCITCEGCKGFFRRTIQ   70

>☐ gi│47216914│emb│CAG02086.1│ unnamed protein product [Tetraodon nigroviridis]

   Length = 385

   Score = 73.6 bits (179),   Expect = 2e-12
   Identities = 30/30 (100%),   Positives = 30/30 (100%), Gaps = 0/30 (0%)

   Query   1   CVVCGDKATGYHYRCITCEGCKGFFRRTIQ   30
                 CVVCGDKATGYHYRCITCEGCKGFFRRTIQ
   Sbjct   28   CVVCGDKATGYHYRCITCEGCKGFFRRTIQ   57

>☐ gi│209662│ gb │ AAA42393.1│ polyprotein gag-p75-erbA [Avian erythroblastosis virus]

   Length = 455

   Score = 71.2 bits (173),   Expect = 9e-12
   Identities = 29/30 (96%),   Positives = 29/30 (96%), Gaps = 0/30 (0%)

   Query   1   CVVCGDKATGYHYRCITCEGCKGFFRRTIQ   30
                 CVVCGDKATGYHYRCITCEGCK FFRRTIQ
   Sbjct   94   CVVCGDKATGYHYRCITCEGCKSFFRRTIQ   123

>☐ gi│2208725│ gb │ AAM90896.1│ [G] farnesoid X receptor [Gallus gallus]

   gi│45383880│ ref │ NP_989444.1│ [G] nuclear receptor subfamily 1, group H, member 4

      Length = 473

   Score = 62.4 bits (150),   Expect = 4e-09
   Identities = 24/29 (82%),   Positives = 27/29 (93%), Gaps = 0/29 (0%)

   Query   1   CVVCGDKATGYHYRCITCEGCKGFFRRTI   29
                 CVVCGDKA+GYHY +TCEGCKGFFRR+I
   Sbjct   128 CVVCGDKASGYHYNALTCEGCKGFFRRSI   156

Figure 16.10 Example of a BLAST search. A sequence of amino acid residues from the specialized zinc finger DNA-binding domain of thyroid hormone receptor α from the brown Norway rat (*Rattus norvegicus*) was used to "query" the protein sequence database via the publicly available BLAST website (www.ncbi.nlm.nih.gov/BLAST). The thyroid hormone receptor α is a ligand-activated transcription factor. The results of a search are usually available within less than a minute. Examples of some of the "hits" with the highest scores are shown. The first hit is the rat thyroid hormone receptor α (c-ErbA) itself. The second hit shown is an unnamed protein product from the pufferfish (*Tetraodon nigroviridis*). A score of 179 is assigned to the match between the rat thyroid hormone receptor α and the pufferfish protein. A total of 30 out of 30 amino acid residues are identical between the two (100% identity), suggesting the pufferfish protein may also function as a ligand-activated nuclear receptor. A score of 173 was obtained for the polyprotein Gag-p75-ErbA (v-erbA), which is the oncogenic homolog of the thyroid hormone receptor α. In this case there are 29 out of 30 exact matches (96% identity). The oncoprotein v-ErbA is discussed in Chapter 17 (Fig. 17.13). The fourth hit shown is for the farnesoid X receptor from chicken (*Gallus gallus*). A score of 150 was assigned to the match. A total of 24 out of 29 amino acid residues are identical between the two (82%), and 27 of the residues are either identical or similar (that is, they represent conservative amino acid substitutions). The farnesoid X receptor has recently been identified as a bile acid–activated nuclear receptor that controls bile acid synthesis, conjugation, and transport, as well as lipid metabolism.

Table 16.3 Selected bioinformatics databases.

Name	URL
Major genome browsers	
Ensembl	www.ensembl.org
NCBI Map Viewer	www.ncbi.nlm.nih.gov/mapview/
UCSC Genome Bioinformatics	genome.ucsc.edu
Major public DNA sequence databases	
DNA Data Bank of Japan	www.ddbj.nig.ac.jp
EMBL Nucleotide Sequence Database	www.ebi.ac.uk/embl/index.html
NCBI GenBank	www.ncbi.nlm.nih.gov
Expressed sequence tag clustering databases	
NCBI Unigene	www.ncbi.nlm.nih.gov/UniGene
South African STACK Project	www.sanbi.ac.za/Dbases.html
TIGR Gene Indices	www.tigr.org/tdb/tgi.shtml
Gene identification and physical maps	
Eukaryotic Promoter Database	www.epd.isb-sib.ch
SNP Consortium	snp.cshl.org
HapMap Consortium	www.hapmap.org
Genetic and physical maps	
GeneLoc	www.genecards.org/
Metabolic pathways and cellular regulation	
EcoCyc	www.ecocyc.org
Kyoto Encyclopedia of Genes and Genomes	www.genome.ad.jp/kegg
Model organism database	
AceDB (*Caenorhabditis elegans*)	www.acedb.org
Arabidopsis Information Resource	www.arabidopsis.org
Arabidopsis Genome Initiative	mips.gsf.de/proj/thal/db
Berkeley *Drosophila* Genome Project	www.fruitfly.org
EcoGene (*Escherichia coli*)	bmb.med.miami.edu/EcoGene/EcoWeb
FlyBase (*Drosophila*)	www.flybase.org
Mouse Genome Informatics	www.informatics.jax.org
Rat Genome Database	rgd.mcw.edu
Saccharomyces Genome Database	www.yeastgenome.org
Schizosaccharomyces pombe Genome Project	www.sanger.ac.uk/Projects/S_pombe
WormBase (*C. elegans*)	www.wormbase.org
Zebrafish Information Network	www.zfin.org
Proteins	
Swiss Protein Knowledgebase	www.expasy.ch/sprot
The RCSB Protein Data Bank	www.rcsb.org/pdb

sequences. Genomics is based on automated, high-throughput methods of generating experimental data, ranging from gene sequences to gene expression levels (DNA microarrays) to analysis of single nucleotide polymorphisms (SNPs). The massive data sets produced by genomics are analyzed and interpreted using the tools of bioinformatics. In essence, genomics is based on the experimental aspects of molecular biology, while bioinformatics is based on computer science, mathematics, and theoretical approaches to the data.

Proteomics

Proteomics is the comprehensive study of the full set of proteins encoded by a genome. Proteomics in an overall sense means protein biochemistry on a high-throughput scale. Instead of having to examine one protein at a time, researchers are now able to study proteins *en masse*. Studies include functional and comparative analysis using the tools of bioinformatics. The term proteome was first coined to describe the set of proteins encoded by the genome. The study of the proteome, proteomics, now encompasses not only all the proteins in any given cells, but also the set of all protein isoforms and modifications, the interactions between them, and the structural description of proteins and their higher order complexes – basically almost everything "post-genomic." Of course, proteomics would not be possible without genomics, which provides the blueprint of possible gene products that are the focal point of proteomics studies, and the computer algorithms of bioinformatics.

The age of "omics"

The coining of the terms genomics and proteomics has spawned a whole set of related terms to describe the comparative analysis of databases. These include transcriptomics (study of all the transcriptional units, coding and noncoding, in the genome), metabolomics (metabolic pathways), kinomics (kinases), glycomics (carbohydrates), lipidomics (lipids), and interactomics (study of macromolecular machines, mapping protein–protein interactions throughout a cell). For example, the interactome of *Caenorhabditis elegans* links 2898 proteins by 5460 interactions. The interactome emphasizes one of the fundamental principles in molecular biology: much of a cell's work is done not by individual proteins but by large macromolecular complexes. The "localizome," also referred to as the "organeller proteome" identifies where each protein resides within a cell.

16.4 The Human Genome Project

As described in Chapter 8 (Section 8.11), a major milestone in genome sequencing was the development of automated DNA sequencing in 1986. In 1987, the US Department of Energy (DOE) officially started the Human Genome Project. A year later the National Institutes of Health (NIH) took over the project and in 1989 established the National Human Genome Research Institute (NHGRI) in Maryland. Over the next few years, other nations joined the effort. The publicly funded international sequencing project run by the NHGRI and the Sanger Centre in Cambridge, UK started sequencing human and model organism genomes in 1990. At the same time, the BLAST algorithm was developed for similarity searches. In addition, bacterial artificial chromosomes (BACs) became available in 1992 (see Section 8.4). BACs and the BLAST algorithm were crucial to the clone by clone genome assembly approach used by the public consortium.

Clone by clone genome assembly approach

The clone by clone approach to sequencing is the traditional method for sequencing genomic DNA (Fig. 16.11A). This method works as follows. Many copies of the genome are cut into fragments of about 150,000 bp by partial digestion with restriction endonucleases. Partial digestion means that the reaction is not carried out for long enough to allow every possible cleavage to occur. The large DNA fragments are cloned into BACs and amplified in bacterial hosts. The clones are purified for further analysis. Each clone is completely digested with a restriction endonuclease, chosen to produce a characteristic pattern of small fragments for each clone. Comparison of the patterns reveals overlap between the clones, allowing them to be lined up in order, while the sites in the genome at which the restriction endonucleases cut are mapped. This results in what is called a "physical map". Individual BAC clones are then sheared into smaller fragments and cloned. Enough of the resulting subclones are sequenced to ensure that each part of the original clone is analyzed several times. The sequences of individual BAC clones are assembled using the

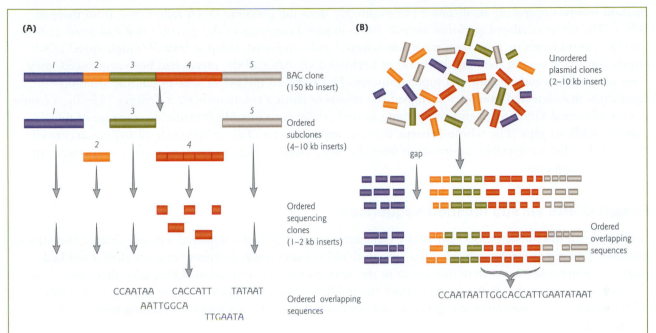

Figure 16.11 Genome sequencing methods. (A) The clone by clone approach. Restriction fragments of ~150 kb are cloned into bacterial artificial chromosomes (BACs). Each BAC clone is first physically mapped by restriction digests (i.e. it is necessary to know that subclone 1 comes before subclone 2) before it is broken up into smaller fragments and subcloned. Once the BACs have been ordered, each subclone is sequenced and, using overlapping sequences of neighboring subclones, the sequence of the whole genome is put together. For simplicity, the process for just one BAC clone is depicted. (B) Shotgun sequencing. Plasmid clones with 2–10 kb inserts are prepared directly from fragmented genomic DNA. From the mixture of clones, a fragment is randomly selected for sequencing. To ensure that the whole source clone has been sequenced, this stretch of DNA must be sequenced numerous times to produce an overlapping sequence. Gaps in this process will occur where a clone is not fully sequenced.

physical map as a guide. Finally, the whole genome sequence is assembled. Information from lower resolution genetic and cytogenetic maps can also be used to help order and orient the sequences. A "genetic map" (also called a linkage map) is a diagram of the order of genes on a chromosome in which the distance between adjacent genes is proportional to the frequency of recombination between them during meiosis. The "cytogenetic map" is based on variation in the staining properties of different regions of the genome, as viewed by light microscopy.

The clone by clone method is time-consuming but precise. This approach generally produces an accurate representation of all regions of a genome, including both euchromatin ("gene rich") and heterochromatin ("gene poor"). However, initially, gaps in the continuity of the sequence may exist because there is not always a BAC clone to cover every part of the genome, and overlaps between clones can be missed by sequencing errors or the presence of large-scale repeats in the genome. Using this procedure, the public consortium was able to determine the order of the bases in a series of approximately 25,000 overlapping human DNA BAC clones. The source of human genomic DNA was bits and pieces from numerous anonymous donors.

Whole-genome shotgun approach

A privately funded sequencing project carried out by J. Craig Venter's company Celera Genomics Corporation used an alternate, quicker method called the whole-genome shotgun approach (Fig. 16.11B). They did not even get started until 1998, but finished at the same time as the public consortium. This method involves preparing small-insert clones directly from the genomic DNA rather than from mapped BACs. The set of overlapping clones were 2–10 kb in size. Fragments of the genome that had been cloned into the plasmid vectors were then randomly selected and sequenced, using at least 300 high-speed DNA sequencers (each cost in the neighborhood of US$300,000). After all the pieces had been sequenced, they were reassembled with the aid of a supercomputer. This involved the use of whole-genome assembly algorithms that detected overlap and aligned the inserts to form a less-than-perfect "contig." Ideally, a contig is a set of cloned DNA fragments overlapping in such a way as to provide continuous coverage of the genome without gaps. The whole-genome shotgun approach often fails to represent all regions accurately; however, by also using public information from the BAC clones Celera researchers were able to come up with a draft sequence.

Rough drafts versus finished sequences

In 2001, both the public consortium and Celera announced they had completed a rough draft of the human genome. The sequence was considered a "rough draft" because it was of lower accuracy than a finished sequence – some segments were missing or in the wrong order or orientation. Originally, there were thousands of gaps in the draft sequence. Both the public and private sequences were missing about 10% of the euchromatin portion of the genome and 30% of the genome as a whole (including regions of heterochromatin).

On October 21, 2004, the public group announced the "finished" sequence. In this final version, bases were identified to an accuracy of less than one error per 100,000 DNA bases and were placed in the right order and orientation along a chromosome with almost no gaps (there are only 341 gaps remaining). Scientists estimate that the "finished" sequence now covers 99% of the euchromatin portion of the genome.

The sequencing phase of the Human Genome Project provided a massive data set of ordered bases, but did not show where genes begin or end or what they do. With the sequencing phase essentially complete, each of the 24 human chromosomes is now being annotated. Annotation of the genome is an in-depth analysis of all functional elements of the genome. Much of the emphasis is on the gene content, with the aim to characterize all of the genes and their functions. Results from genome sequencing are coupled with high-throughput studies of gene expression, for example, by use of DNA and protein microarrays (see Section 16.6).

16.5 Other sequenced genomes

Genome sequences of a number of bacteria were completed well before the first eukaryotic genome because of their small size. The first genome sequence of a free-living organism was published in 1995 for *Haemophilus influenzae*, an agent of bacterial meningitis. A sampling of sequenced eukaryotic genomes is provided in Table 16.4. This table is by no means all-inclusive, since some new animal, plant, or other organism is sequenced almost on a weekly to monthly basis. In 1996, *Saccharomyces cerevisiae* (the budding yeast) became the first eukaryote to have its genome sequenced. Two years later, *Caenorhabditis elegans* (a roundworm) gained status as the first multicellular eukaryote to have its genome sequenced.

Comparing the complete sequences of many diverse organisms has revealed many intriguing and puzzling similarities and differences. Prior to the draft sequence of the human genome, scientists estimated that the human genome contained at least 100,000 genes. However, when the draft sequence was analyzed, the number of genes predicted was in the range of 30,000–40,000. What came as a surprise was that humans have less than twice the number of genes of a worm (Table 16.4). Current estimates from the finished sequence suggest 20,000–25,000 genes, which places us even closer to the worm! However, it is still not clear how many genes there really are in the human genome (see below).

Another interesting finding was that the pufferfish, which has a small genome compared with most vertebrates, has a similar number of genes compared with humans. This may be due, in part, to the loss of

Table 16.4 Sampling of sequenced eukaryotic genomes.

Organism	Year	Millions of bases sequenced	Predicted number of protein-coding genes
Saccharomyces cerevisiae (budding yeast)	1996	12	6000
Caenorhabditis elegans (roundworm)	1998	97	19,099
Drosophila melanogaster (fruitfly)	2000	116	14,000
Arabidopsis thaliana (wild mustard weed)	2000	115	25,498
Human: rough draft/finished (public sequence)	2001–2004	2693/2900	20,000–25,000
Human: rough draft (Celera sequence)	2001	2654	39,114
Takifugu rubripes /Tetraodon nigroviridis (pufferfish)	2001/2004	365/342	20,000–25,000
Oikopleura dioica (marine chordate, larvacean)	2001	~65	15,000
Mus musculus (mouse)	2002	2600	30,000
Anopheles gambiae (malarial mosquito)	2002	278	14,000
Ciona intestinalis (sea squirt)	2002	153	16,000
Rattus norvegicus (brown Norway rat)	2004	2750	30,000
Gallus gallus (red jungle fowl, ancestor of domestic chicken)	2004	1000	20,000–23,000
Oryza sativa (rice)	2005	389	37,544
Pan troglodytes (common chimpanzee)	2005 (draft)	2843	20,000–25,000

noncoding regions. Analysis of the pufferfish genome has yielded much information on genomic evolution (Focus box 16.3). Additional insights into vertebrate origins come from analysis of the marine larvacean *Oikopleura dioica*, which has the smallest genome ever in a chordate. The chordates include the vertebrates, the cephalochordates (e.g. amphioxus) and the urochordates (e.g. sea squirts and larvaceans).

Other genomes that have been sequenced or are in the process include many bacterial genomes, chicken (*Gallus gallus*) (Focus box 16.3), rice (*Oryza sativa*), zebrafish (*Danio rerio*), dog (*Canis familiaris*), cow (*Bos taurus*), South American gray short-tailed opossum (*Monodelphis domesticus*), Rhesus macaque (*Macaca mulatta*), pig (*Sus scrofa*), African savannah elephant (*Loxodonta africana*), domestic cat (*Felis catus*), orangutan (*Pongo pygmaeus*), rabbit (*Oryctolagus cuniculus*), European common shrew (*Sorex araneus*), slimemold (*Dictyostelium discoideum*), a single-celled ciliate (*Oxytricha trifallax*), a flatworm (*Schmidtea mediterranea*), honeybee (*Apis mellifera*), and the sea lamprey (*Petromyzon marinus*). All mammalian genome sequences play an important role in helping researchers to interpret the human genome. Sequence and comparative analyses of nonmammalian genomes help to provide unique perspectives on the evolution of anatomy, physiology, development, and behavior across many diverse species. Our knowledge of the human genome has been enhanced by comparing the human and chimpanzee (*Pan troglodytes*) sequences. The answer to the question of what makes us uniquely human lies somewhere within the 35 million single-nucleotide substitutions, 5 million small insertions and deletions, local rearrangements, and a chromosomal fusion that distinguish the chimp genome from the human genome.

Comparative analysis of genomes: insights from pufferfish and chickens

Comparative genome analysis allows researchers to assess changes in gene structure and sequence that have occurred during evolution. These comparisons help to identify protein coding and regulatory regions within a genome. Two sequences are considered homologous if they share a common evolutionary ancestry. Genes that are homologous may be either orthologous (*ortho*, "exact") or paralogous (*para*, "in parallel"). Orthologs are genes in different species that are homologous because they are derived from a common ancestral gene (e.g. α-globin genes in humans and mice). Paralogs are two genes in a genome that are similar because they arose from a gene duplication (e.g. human α- and β-globin genes, or mouse α- and β-globin genes). Comparative genome analysis has also revealed a striking degree of synteny – conservation in genetic linkage – between the genes of distantly related animals. This means that genetic loci lie in the same order on the same chromosome in different species, suggesting that this conserved order may be of importance for gene regulation.

Insights from the pufferfish genome

The genomes of two species of pufferfish have now been sequenced using the whole-genome shotgun approach. The first genome to be completed was that of the Japanese pufferfish, *Takifugu rubripes*, a poisonous marine fish best known to connoisseurs of sushi. The second genome to be sequenced was that of the spotted green pufferfish, *Tetraodon nigroviridis*, a popular freshwater aquarium fish. The pufferfish is also known as the blowfish, swellfish, or globefish. These various common names vividly describe how the fish puffs up to about twice its normal size by gulping water when it is threatened. Many parts of the Japanese pufferfish (including the liver, muscles, skin, and ovaries) contain an extremely potent, paralyzing poison called tetrodotoxin. The poison selectively blocks the sodium conductance channels in neurons and muscle fibers and thus blocks the generation of nerve impulses and muscle contraction. There is no known antidote for this poison, which is about a thousand times deadlier than cyanide. Pufferfish (called fugu on the menu) is eaten in Japan, but is only cooked by specially trained chefs who can minimize the amount of poison in each bite. Despite these precautions, many Japanese have died from eating this delicacy.

Genome sequence analysis suggests that the Japanese pufferfish and the spotted green pufferfish diverged from a common ancestor between 18 and 30 million years ago and from a common ancester with mammals – a primitive bony fish – about 450 million years ago (Fig. 1). Both species possess about eight or nine times less DNA than humans, in part due to a lack of interspersed repeats (transposable elements are very rare in the *Tetraodon* genome). The estimated number of protein-coding genes is 20,000–25,000, which is comparable to the estimate for humans. By matching up the genes on each pufferfish chromosome to the related genes on human chromosomes, it was deduced that the long-extinct ancestor of actinopterygians (ray-finned fish, including pufferfish) and sarcopterygians (lobe-finned fish, the lineage that gave rise to humans) had 12 pairs of chromosomes. In contrast, the modern pufferfish genome has 21 pairs chromosomes. Further, sequence analysis suggests that a duplication of the whole genome occurred sometime within the ray-finned fish lineage, possibly close to the origin of the teleost fish 230 million years ago.

Insights from the chicken genome

The genome of the red jungle fowl (*Gallus gallus*), the wild ancestor of domestic chickens, was sequenced by a combination of approaches: the whole-genome shotgun data were revised and correctly ordered according to a physical map of 180,000 BAC clones. Interestingly, the draft chicken genome (approximately 98% complete) is one-third the size of sequenced mammalian genomes (e.g. humans, mice, and rats). This is apparently due to a smaller amount of interspersed and tandem repetitive elements, fewer duplicated copies of genes overall, and a reduction in the number of pseudogenes.

The widespread interest in the chicken genome stems from the potential for using comparative sequence analysis to map regulatory elements present in the human genome. Enough differences have accumulated between the human and chicken sequences over the last 310 million years, to highlight any base pairs that have been conserved. By

Comparative analysis of genomes: insights from pufferfish and chickens

comparison, the mouse, which split from humans only 75 million years ago, is too similar at the base pair level, leading to difficulties in identifying functional elements. Preliminary analysis describes 70 million base pairs of

sequence that are highly conserved between humans and chickens. This includes base pairs in gene-coding regions, but also base pairs that are between coding regions and therefore may relate to potential regulatory elements.

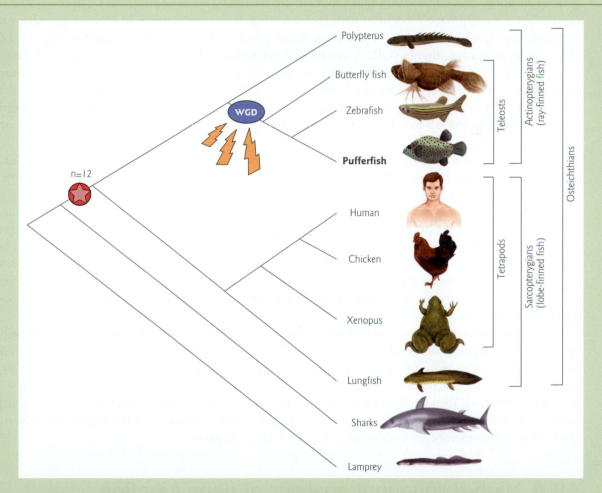

Figure 1 Evolutionary relationship between humans, pufferfish, and other vertebrates. By comparing the genome sequence of the spotted green pufferfish, *Tetraodon nigroviridis*, with that of humans, researchers have deduced that the extinct ancestor of actinopterygians (ray-finned fish, including pufferfish) and sarcopterygians (lobe-finned fish, the lineage that gave rise to humans) had 12 pairs of chromosomes (*n* = 12). In addition, a whole-genome duplication (WGD) occurred during the evolution of ray-finned fish. (Adapted by permission of Nature Publishing Group and Macmillan Publishers Ltd: Mulley, J., Holland, P. 2004. Small genome, big insights. *Nature* 431:916–917. Copyright © 2004.)

What is a gene and how many are there in the human genome?

The term "gene" was first used in the early 1900s as an abstract concept to explain the hereditary basis of traits. In the 1930s, George Beadle and Edward Tatum suggested that each gene codes for just one protein (see Section 1.2). We now know that one gene can code for multiple proteins or for a functional RNA product. To take into account this new understanding, a gene is now defined as a complete chromosomal segment responsible for making a functional product. This definition highlights three essential features of a gene: the expression of a product, the requirement that it be functional, and the inclusion of both coding and regulatory regions. Often complementary DNAs (cDNAs) are referred to loosely as "genes" despite only containing the coding regions. This is an oversimplified and misleading use of the term, given all that is now known about the importance of regulatory elements for appropriate gene expression (see Section 11.3).

Gene hunting (prediction) computer programs are becoming more sophisticated. Most use two main approaches. One approach recognizes genes by detecting distinctive patterns in DNA sequences, such as start and stop signals for translation, promoter regions, and exon–intron splice junctions (Fig. 16.12). However, open reading frames (ORFs) can be difficult to find when genes are short, or when they undergo an appreciable amount of RNA splicing with small exons in between large introns. Another approach uncovers new genes based on their similiarity to known genes. This approach would miss genes that have no obvious similarity to other genes. Eventually, of course, the investigator must leave the computer terminal, return to the lab bench, and test a predicted gene for its function using the many tools of molecular biology.

The best estimate of the number of genes in humans is that there are approximately 20,000–25,000 protein-coding genes, of which the exons take up only 2% of the total DNA (see Fig. 16.2). However, some of these "genes" could be pseudogenes. Pseudogenes are similar in sequence to normal genes, but they usually contain obvious disruptions such as frameshifts or stop codons in the middle of coding domains. This prevents them from producing a functional product or having a detectable effect on the organism's phenotype. Pseudogenes occur in a wide variety of animals, fungi, plants, and bacteria. They can be quite prevalent; for example, there are 80 ribosomal protein genes in the human genome versus > 2000 associated pseudogenes. Furthermore, this estimate does not even begin to take into account genes coding for functional RNA molecules, or the tremendous variability introduced by alternative splicing and post-translational modification.

To provide a better estimate of the number of genes, the *Encyclopedia of DNA Elements* (ENCODE) Project aims to identify all functional elements in the human genome sequence – this would include protein-coding genes, nonprotein-coding genes, transcriptional regulatory elements, and sequences that mediate chromosome structure and dynamics. The ENCODE Consortium plans to develop high-throughput methods for identifying functional elements. The pilot phase is focused on 30 Mb distributed across many chromosomes, including the α- and β-globin gene clusters, and the region encompassing the cystic fibrosis transmembrane conductance regulator (*CFTR*) gene.

16.6 High-throughput analysis of gene function

With the sequencing of genomes comes the challenge of annotating the genome and determining the function of predicted gene products. Methods for genome and proteome analysis are called "high-throughput" analysis because the activities of thousands of genes and their products are studied at the same time. Such methods include DNA microarrays, protein arrays, and MALDI-TOF and tandem mass spectrometry.

DNA microarrays

Over the past few years, DNA microarrays, also known as DNA chips, have transformed molecular biology. The very first DNA microarray results were reported in 1995 by Mark Schena, Ron Davis, Dari Shalon,

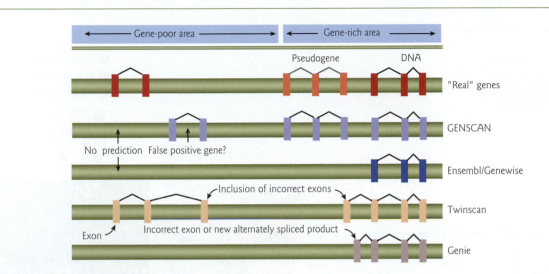

Figure 16.12 Gene prediction programs. Although gene prediction programs have become much more sophisticated, no program predicts all genes correctly. Some see protein-coding genes (shown here as coding regions, or exons, connected by bent lines) where there are none, some miss a gene altogether; and some do not put all the gene's parts in the right places. In this example, the "real genes" are shown in red, and the accuracy of their prediction by four different gene prediction programs (GENSCAN, Ensembl/Genewise, Twinscan, and Genie) is shown. (Source: Ewan Birney / Sanger Institute / EBI, from Pennisi, E. 2003. Bioinformatics: Gene counters struggle to get the right answer. *Science* 301:1040–1041. Reprinted with permission from AAAS.)

and Patrick Brown at Stanford University Medical Center. They measured the differential expression of 45 *Arabidopsis* genes by means of simultaneous, two-color fluorescence hybridization. Further refinement of DNA microarray technology now allows researchers to study the transcriptional activity of thousands of genes simultaneously. The complete transcription program of an organism during specific physiological responses and developmental processes or diseased states can be analyzed. In the example shown in Fig. 16.13, the aim of the experiment is to identify genes that become more or less active in a developing tumor, by comparing the pattern of gene expression in healthy tissues with those that are cancerous. The development of a classification system may allow for the prediction of clinical outcome (response to therapy) of different types of cancer. DNA microarrays have been used for this type of prediction for breast cancer and medulloblastoma (a childhood brain tumor). The hope is to be able to design a patient–tailored therapy strategy, based on the classification of tumor gene expression.

Protein arrays

Protein array formats allow rapid analysis of protein activity on a proteomic scale. These arrays may be based on recombinant proteins or reagents that interact specifically with proteins, including antibodies, peptides, and small molecules. Data from protein-based arrays can derive from protein interactions, protein modifications, or enzymatic activities. Two common types of arrays, analytical and functional, are depicted in Fig. 16.14. Analytical protein arrays are used for monitoring protein expression levels and for clinical diagnostics. Commercially prepared functional protein arrays contain collections of full-length ORFs expressed as glutathione-*S*-transferase (GST) fusion proteins. For example, a yeast protein array is available that contains more than 4000 unique proteins, which is nearly the full protein complement. These chips are used to analyze enzymatic activities, protein–protein interactions, and post-translational modifications (such as phosphorylation). With the proper detection method, functional protein arrays can be used to identify the substrates of enzymes of interest. Consequently, this class of chips is particularly useful in drug and drug-target identification and in mapping biological pathways.

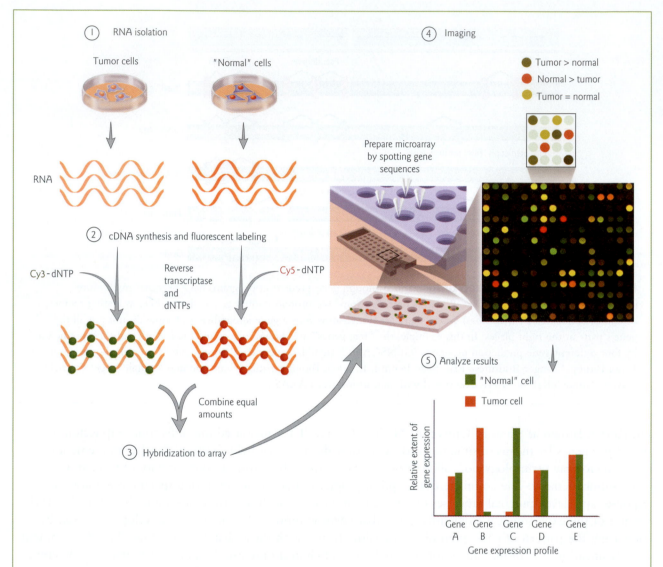

Figure 16.13 DNA microarray. To prepare the DNA microarray or "chip," individual gene sequences (partial cDNAs or oligonucleotides) are applied to a glass microscope slide by a robotic device and amplified by PCR. Commercial arrays (e.g. Affymetrix) are made by synthesis of oligonucleotides on the surface of the slide, not by deposition of PCR products. The gene sequences are chemically or heat-treated to attach the DNA to the glass slide and denature it. The array represents a grid of thousands of different gene sequences bound to closely spaced regions on the surface of the glass slide. The method involves five main steps. (1) Total RNA is extracted, for example, from tumor cells and "normal cells." (2) Labeled cDNA is synthesized from the purified RNA using the enzyme reverse transcriptase and deoxynucleoside triphosphates (dNTPs). The tumor cell cDNA is labeled with a green fluorescent tag by including Cy3-tagged dNTPs in the reaction. The "normal" cell cDNA is labeled with a red fluorescent tag by including Cy5-tagged dNTPs in the reaction. (3) The two types of labeled cDNA are mixed in equal amounts and hybridized to the prepared array. (4) The array is scanned by fluorescence microscopy. The result is a grid of fluorescent spots that represent hybridization of complementary sequences to the array, therefore indicating that a particular gene was expressed in this cell type. In the example, a bright green spot indicates greater expression in cancer cells, a bright red spot indicates greater expression in normal cells, and a bright yellow spot indicates expression in both cell types (Photograph courtesy of Katie Hoadley and Charles Perou, The University of North Carolina at Chapel Hill). (5) The results are analyzed by computer digital imaging. Quantitative data analysis will reveal the relative extent of gene expression based on the intensity of fluorescence. A graph of the gene expression profile indicates genes that may be of importance in cancer, by the relative enhancement or repression (often by only 1.5–2-fold) of their expression in cancer cells.

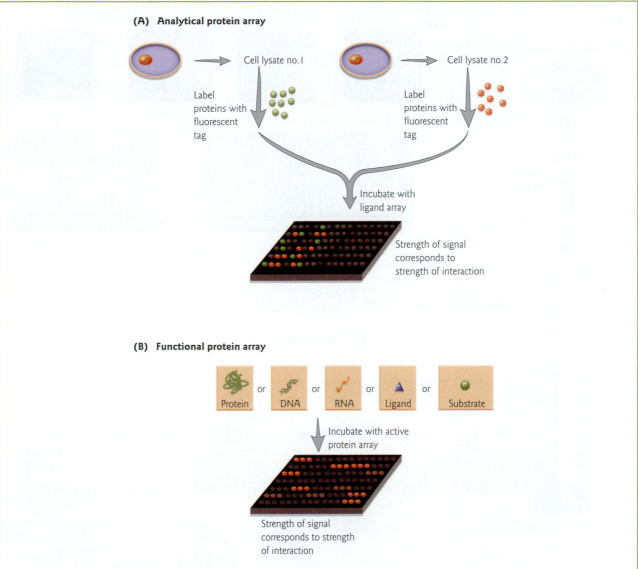

(A) Analytical protein array

Cell lysate no.1

Label proteins with fluorescent tag

Cell lysate no.2

Label proteins with fluorescent tag

Incubate with ligand array

Strength of signal corresponds to strength of interaction

(B) Functional protein array

Protein   or   DNA   or   RNA   or   Ligand   or   Substrate

Incubate with active protein array

Strength of signal corresponds to strength of interaction

Figure 16.14 Protein arrays. (A) In an analytical array, different types of ligands – including antibodies, antigens, DNA or RNA, carbohydrates, or small molecules – with high affinity and specificity are spotted on a specially treated surface. Similar to the procedure in DNA microarray experiments, protein samples from two biological states to be compared are conjugated to red or green fluorescent dyes, mixed, and incubated with the chips. Fluorescent spots in red or green color identify an excess of proteins from one state over the other that interact with a specific ligand. (B) In a functional array, native proteins or peptides are individually purified or synthesized using high-throughput approaches and thousands of proteins are spotted in duplicate on nitrocellulose or modified-glass microarray surfaces. Labeled proteins, DNA, RNA, other ligands, or enzyme substrates are then incubated with the chips.

Mass spectrometry

Mass spectrometry is a powerful technique for measuring the mass of molecules like proteins and peptides. As such, it has become a fundamental tool in proteomics research. Two popular strategies for mass spectrometry-based protein identification are: (i) peptide mass fingerprinting of a single isolated protein, using matrix-assisted laser desorption/ionization–time of flight (MALDI-TOF) mass spectrometry; and (ii) shotgun proteomics of an entire "proteome," using tandem mass spectrometry (MS/MS).

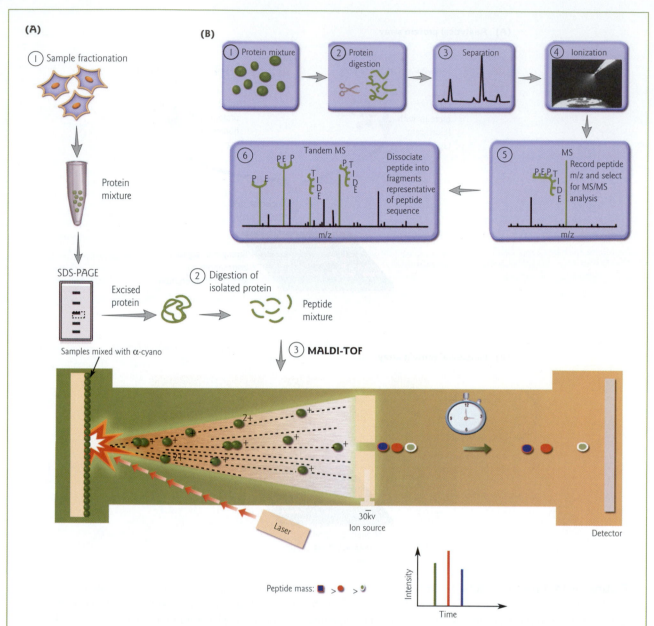

Figure 16.15 Mass spectrometry. (A) Peptide mass fingerprinting using MALDI-TOF. The process involves three main steps: (1) fractionation of cells or tissues to separate and isolate a protein of interest; (2) enzymatic digestion of the protein with trypsin; and (3) analysis of the resulting mixture of peptides by MALDI-TOF mass spectrometry. The MALDI-TOF mass spectrometer consists of an ion source, a mass analyzer that measures the mass to charge ratio (m/z) of the ionized peptides, and a detector that registers the number of ions at each m/z value. The peptides are embedded in a crystalline matrix (e.g. α-cyano-4-hydroxy-cinnamic acid) and arrayed on a metal target. Ultraviolet laser pulses vaporize the peptides into the gas phase, where they become ionized. The ions are injected into a tube with the assistance of an electrical field and they "fly" down the tube toward a detector. Their time of flight is inversely proportional to their mass and directly proportional to the charge on the protein. Molecules of smaller mass are accelerated to higher velocities and hence reach the detector earlier compared to those with greater mass. By calibration of the instrument with molecules of known masses, the flight time can be converted to mass accordingly. (B) Shotgun proteomics using tandem mass spectrometry (MS/MS). The process involves six main steps. (1) A mixture of proteins is extracted from any number of sources. (2) The proteins are enzymatically digested resulting in a complex mixture of peptides (approximately 40 peptides per protein). (3) Digested peptides are separated by chromatographic

Peptide mass fingerprinting using MALDI-TOF

The process of peptide mass fingerprinting typically begins either with a protein spot extracted from a one- or two-dimensional protein gel, or with liquid chromatography. The isolated protein is then enzymatically digested, typically with trypsin. Trypsin specifically cleaves proteins on the carboxyl side of the amino acids arginine and lysine (except where proline is C-terminal). The molecular weights of the resulting collection of peptides are then measured by MALDI-TOF mass spectrometry (Fig. 16.15A). The final step in the process is the application of computer algorithms that match the MALDI-TOF spectra against a database of proteins that have been digested *in silico* to identify the protein from which the peptides originated. *In silico* literally means "in silicon" and refers to manipulations performed on a computer or via computer simulation. With the increasing availability of genome sequences, this approach has almost eliminated the need to chemically sequence a polypeptide.

Shotgun proteomics using MS/MS

In the method called shotgun proteomics, an entire proteome is enzymatically digested, separated by chromatography, and "interrogated" (analyzed) using MS/MS. During MS/MS, peptide ions are bombarded with rare gas atoms that provide the energy required to cleave amide bonds of the peptide backbone. The process produces a collection of peptide ion fragments that differ in mass by a single amino acid. Measurement of mass to charge (m/z) ratios of the fragments allows the amino acid sequence to be read (Fig. 16.15B). Through such methods, there has been partial analysis of the proteomes of yeast and some bacteria, and various organellar proteomes (Focus box 16.4).

16.7 Single nucleotide polymorphisms

All individual humans share genome sequences that are approximately 99.9% the same. The remaining variable 0.1% is responsible for the genetic diversity between individuals. Single nucleotide polymorphisms (SNPs) (pronounced "snips"), where two or more possible nucleotides occur at a specific mapped location in the genome, are the most common type of genomic sequence variation among individuals. For example, a SNP might change the DNA sequence ATGCCTA to ATGCTTA. Individuals may be homozygotes (e.g. T/T or C/C), or heterozygotes with different bases (e.g. T/C) at polymorphic sites.

For a variation to be considered a SNP it must occur in at least 1% of the population. It has been estimated that there are ~7 million common SNPs with a population frequency of at least 5% across the entire human population. An additional 4 million SNPs exist with an allele frequency of between 1 and 5%. The least variability occurs on the sex chromosomes (X and Y). Two of every three SNPs characterized so far involve the replacement of cytosine (C) with thymine (T) (see Section 7.2 for a discussion of single base changes). SNPs can occur in both coding and noncoding regions of the genome. Resequencing certain parts of the genome using DNA from different individuals generates a map of SNPs. This map is used by researchers to scan the human genome for haplotypes associated with common diseases (Disease box 16.1). Haplotypes are patterns of sequence variation, i.e. stretches of continuous DNA containing a distinctive set

methods, e.g. liquid chromatography or 2D-PAGE. (4) Peptides are individually ionized via electrospray ionization and introduced into the mass spectrometer. (5) The ionized peptides are separated based on their m/z value. (6) Abundant peptide cations are selected for fragmentation via MS/MS. Following peptide ion dissociation into fragments representative of the peptide sequence, the newly generated fragments are m/z analyzed to produce the MS/MS spectrum. In the example, the MS/MS spectrum contains the consecutive peptide fragment ions PE, PEP, TIDE, etc. (Reprinted with permission from Coon, J.J., Syka, J.E.P., Shabanowitz, J., Hunt, D.F. 2005. Tandem mass spectrometry for peptide and protein sequence analysis. *BioTechniques* 38:519-523. Copyright © 2005.)

The nucleolar proteome

Recently mass spectrometry was applied to characterize the protein components of the human nucleolus (Fig. 1), a subnuclear organelle best known for its role in ribosome biogenesis. Nucleoli were purified from a cultured human cell line by sucrose density centrifugation and their proteins were extracted. The protein mixtures were resolved by two-dimensional polyacrylamide gel electrophoresis (2D-PAGE).

Gel spots were cut into slices, digested with trypsin, and analyzed by mass spectrometry. The resulting spectra revealed the mass of the peptides in the mixture. The amino acid sequence was obtained by using a second round of fragmentation and mass analysis (tandem mass spectrometry, see Fig. 16.15B). Computer algorithms were applied to match the spectra against protein sequence

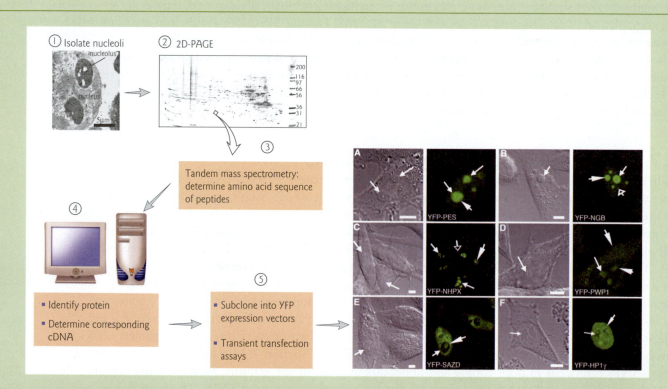

Figure 1 The nucleolar proteome. The steps in analysis of the proteome of the nucleolus are illustrated. (1) Isolation of nucleoli. Transmission electron microscopy (TEM) image of thin sections of a HeLa (human cell line) cell nucleus with two nucleoli (arrow) shown prior to isolation. (2) Fractionation of proteins by 2D-PAGE. The photo shows a silver-stained two-dimensional gel of nucleolar protein samples. Molecular weight size markers are shown on the right (in kDa). (3) Tandem mass spectrometry. (4) Identification of proteins and corresponding cDNAs by database searches. (5) Confirmation of nucleolar localization of candidate proteins. Candidate proteins were expressed as YFP fusion proteins in HeLa cells. For the six proteins illustrated (A–F), each panel shows a confocal fluorescence micrograph on the right with the corresponding light microscope image on the left showing cell morphology. The fusion proteins accumulate within the nucleoli, with varying patterns. Small arrows indicate nucleoli, large arrowheads indicate the localization of the fusion proteins within the nucleoli, and open arrowheads highlight small nuclear bodies also labeled by the fusion proteins (in the case of YFP-NGB and YFP-NHPX). Scale bars, 5 μm. (Adapted from Andersen, J.S., Lyon, C.E., Fox, A.H., Leung, A.K., Lam, Y.W., Steen, H., Mann, M., Lamond, A.I. 2002. Directed proteomic analysis of the human nucleolus. *Current Biology* 12:1–11. Copyright © 2002, with permission from Elsevier.)

The nucleolar proteome

databases to identify the proteins from which the peptide originated. Subsequently, a select group of candidate nucleolar proteins identified by mass spectrometry, which were either novel or not known to be nucleolar, were chosen for further analysis. The corresponding cDNAs were identified, and these were subcloned into YFP expression vectors. The tagged proteins were shown by confocal microscopy to accumulate within the nucleoli following transient transfection in cultured cells. Additional studies suggest the dynamic nature of the nucleolus – there is probably no unique, complete proteome for organelles, but rather an overlapping set of proteomes that are relevant to different cell states or conditions.

of alleles. SNPs do not necessarily cause disease, but they can help to determine the likelihood that someone will develop a particular disease.

Chapter summary

Modern DNA typing uses a panel of DNA probes to detect variable sites in individual organisms, including humans. As a forensic tool, DNA typing can be used to test parentage, to identify criminals, or to remove innocent people from suspicion. Only about 0.1% of the human genome differs from one person to another. Most of this variability lies in the intergenic regions of DNA that make up > 60% of the genome. Intergenic DNA consists of unique or low copy number sequences and moderately to highly repetitive sequences. The repetitive sequences include interspersed elements that are degenerate copies of transposable elements, and tandem repetitive sequences.

The majority of DNA typing systems in forensic casework are based on RFLP analysis of genetic loci with a variable number of tandem repetitive DNA sequences, either minisatellite or short tandem repeats (STRs). Currently, the most widely used DNA typing strategy is multiplex PCR analysis of the variation in number of STRs at 13–15 different loci. PCR amplification and direct sequencing of two highly variable regions in the D loop region of mtDNA is used when there is not enough nuclear DNA for analysis. Analysis of Y chromosome-specific STRs is used in paternity testing or for analyzing biological evidence in criminal casework involving multiple male contributors. For organisms where little sequence information is available, randomly amplified polymorphic DNA (RAPD) or amplified fragment length polymorphism (AFLP) analysis is often used for DNA typing.

Bioinformatics is the collection, organization, and manipulation of biological databases. The most complex of these databases contain the DNA sequences of genomes. The tools of bioinformatics are essential for mining the massive amount of genomic data to gain knowledge about gene structure and expression. Public databases contain a vast store of biological information, including genomic and proteomic data. The most commonly used genome tool is BLAST (basic local alignment search tool). Using this tool, a researcher can start with a DNA sequence or amino acid sequence and discover the gene/protein it belongs to, and then compare this sequence with that of similar genes and their products.

Genomics is the comprehensive study of whole sets of genes (genomes) and their interactions. Similarly, the sum of all proteins produced by an organism is its proteome, and the study of these proteins, even smaller sets of them, is called proteomics. Many related terms have been coined to describe the comparative analysis of other databases. For example, the sum of all RNAs produced by an organism is its transcriptome, and the study of these transcripts is transcriptomics.

Mapping disease-associated SNPs: Alzheimer's disease

Identifying a polymorphic sequence does not necessarily mean the discovery of a marker for a disease. At the moment, mapping disease-associated SNPs has proved to be a slow and painstaking process. So far, there are only a few examples with clear links to disease. Two well-characterized examples of SNPs include the missense mutation that causes sickle cell anemia (see Fig. 8.14), and the apolipoprotein E4 allele implicated in susceptibility to late-onset Alzheimer's disease.

Alzheimer's disease is the most common form of dementia among older people. The term "dementia" describes symptoms resulting from changes in parts of the brain that control thought, memory, and language. Symptoms may include becoming lost in familiar places and not recognizing friends and family, being unable to follow directions, asking the same questions repeatedly, and neglecting personal safety and hygiene. The disease is named after Dr Alois Alzheimer, a German doctor who, in 1906, described abnormal clumps (now called amyloid plaques) and tangled bundles of fibers (now called neurofibrillary tangles) in the brain tissue of a woman who had died of an unusual mental illness. These plaques and tangles in the brain are considered hallmarks of Alzheimer's disease (Fig. 1A).

Up to 4 million people in the USA suffer from this disease. Scientists do not yet fully understand the causes of Alzheimer's disease. What is understood is that there is probably not one single cause, but several factors that affect each person differently. The polygenic nature of disorders such as Alzheimer's disease makes genetic testing complicated. Age is the most important known risk factor. The number of people with the disease doubles every 5 years beyond age 65. Family history is another risk factor. For example, familial Alzheimer's disease is a rare form that can be inherited and usually occurs between the ages of 30 and 60. However, in the most common form of the disease, called late-onset Alzheimer's disease, no obvious family pattern is seen. Symptoms usually begin after age 60, and the risk goes up with age. The disease progresses slowly, starting with mild memory problems and ending with severe brain damage. On average, Alzheimer's patients live from 8 to 10 years after they are diagnosed. One risk factor of late-onset Alzheimer's is the apolipoprotein E4 allele.

Apolipoprotein E4 allele: a risk factor for late-onset Alzheimer's disease

Apolipoprotein E (ApoE) is a protein involved in lipid metabolism. The apolipoprotein E (apoE) gene contains two SNPs that result in three possible alleles for this gene: ε2, ε3, and ε4. Each allele differs by one DNA base, and the variable protein product (isoform) of each allele differs by one amino acid. These single amino acid changes have major effects on protein structure and function. Each individual inherits a maternal and a paternal copy of the apoE gene. The apoE ε3 allele is the most common allele in all populations studied so far. In this allele, TGC encodes the cysteine at position 112 in the protein, and CGC encodes the arginine at position 158. In the ε2 allele, another TGC codon results in a cysteine at position 158 instead. In the ε4 allele a CGC codon gives rise to an arginine at position 112. The three apoE alleles determine six genotypes: three homozygotes (ε4/4, ε3/3, and ε2/2) and three heterozygotes (ε3/4, ε2/3, and ε2/4). Figure 1B shows an example of apoE genotyping by restriction endonuclease digestion of a PCR-amplified fragment. The restriction endonucleases AflIII and HaeII recognize the allele-specific nucleotide substitutions at codons 112 and 158, respectively. The 218 bp amplified product generates 145, 168, and 195 bp fragments that are specific for the apoE ε3, ε2, and ε4 alleles, respectively.

Researchers have shown that an individual who inherits at least one apoE ε4 allele will have a greater chance of getting Alzheimer's disease. Possession of the ε4 allele may account for 50% of Alzheimer's disease in the USA. The ε4 allele appears to increase the rate and extent of amyloid plaque and neurofibrillary tangle formation. In addition, individuals with the ε4 allele show increased total serum cholesterol and have a greater risk of coronary heart disease. In contrast, the ε2 allele is associated with longevity, and individuals with this allele are less likely to develop Alzheimer's. However, it is important to note that SNPs are not absolute indicators of disease development. Someone who has inherited two ε4 alleles may never develop Alzheimer's, while another who has inherited two ε2 alleles may. The apoE gene is just one gene of several that have been linked to Alzheimer's disease.

Mapping disease-associated SNPs: Alzheimer's disease

DISEASE BOX 16.1

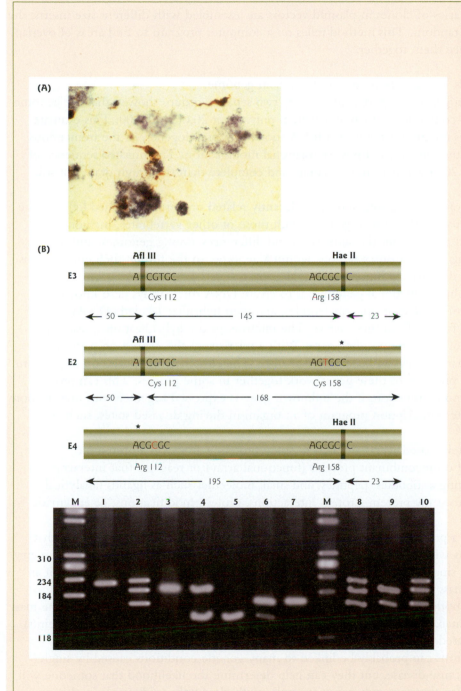

Figure 1 Single nucleotide polymorphism (SNP) in the apolipoprotein E gene is linked to Alzheimer's disease. (A) Light micrograph of entorhinal cortex brain tissue affected by Alzheimer's disease. Two characteristic features are seen here: neurofibrillary tangles (brown) and neuritic plaques (gray). Neurofibrillary tangles are formed of abnormal filaments of tau protein, an internal structural component of healthy nerve cells. The pathogenic abnormal form has no structural role, and damages the cells. The neuritic plaques are formed mainly of aggregations of the protein amyloid. Tissues sections are immunostained with anti-tau serum. (Credit: Dr. M. Goedert / Photo Researchers, Inc.) (B) *apoE* genotyping by simultaneous *Afl*III and *Hae*II digestion of a 218 bp amplified fragment. (Upper panels) The recognition sequences of *Afl*III and *Hae*II are depicted, and the vertical lines represent their cleavage sites. The red letters T and C indicate the formation of cysteine or arginine codons, respectively. The sizes (base pairs) of the digested fragments are shown between the arrows. Asterisks designate losses of restriction sites in the *apoE* E2 and E4 alleles. (Lower panel) Electrophoretic separation of all *apoE* genotypes on an agarose gel. Lane 1 depicts the 218 bp amplified segment, and lanes 2–7 show the E2/E4, E4/E4, E3/E4, E3/E3, E2/E3, and E2/E2 genotypes, respectively. Lanes 8–10 illustrate that the uncut 218 bp fragment is an E2/E4 heteroduplex. The marker (M) *Hae*III-digested X174 was used as a standard. (Reprinted with permission from Zivelin, A., Rosenberg, N., Peretz, H., Amit, Y., Kornbrot, N., Seligsohn, U. 1997. Improved method for genotyping apolipoprotein E polymorphisms by a PCR-based assay simultaneously utilizing two distinct restriction enzymes. *Clinical Chemistry* 43:1657–1659. Copyright © 1997 American Association for Clinical Chemistry, Inc.)

The DNA base sequences of organisms ranging from bacteria to animals and plants have been obtained. Massive sequencing projects can take two forms:

1 In the clone by clone genome assembly approach, a physical map of the genome is produced, then the clones (mostly BACs used in the mapping) are sequenced. This places the sequences in order so they can be pieced together.

2 In the shotgun approach, libraries of clones in plasmid vectors are assembled with different size inserts; the inserts are then sequenced at random. This method relies on a computer program to find areas of overlap among the sequences and piece them together.

A combination of these methods was used to sequence the human genome.

The rough draft of the human genome reported in 2001 by two separate groups showed that the genome probably contains fewer protein-coding genes than anticipated: only about 20,000–25,000. This estimate does not take into account genes coding for functional RNA molecules, pseudogenes, or the tremendous variability introduced by alternative splicing and post-translational modification. The finished sequence of the human genome reported in 2004 is much more accurate and complete than the rough draft, but still contains some gaps.

Comparison of the genomes of closely related and more distantly related organisms can shed light on the evolution of these species. Comparing the human genome with those of other vertebrates, including the chimpanzee, has already revealed much about the similarities and differences among genomes, and has helped to identify protein-coding and regulatory regions within a genome. In the future, such comparisons will help find the genes that are defective in human genetic diseases.

One approach to high-throughput functional genomics is to create DNA microarrays (also known as DNA chips) holding thousands of cDNAs or oligonucleotides, and then hybridize labeled cDNAs corresponding to RNA isolated from cells to these arrays. The intensity of the hybridization to each spot reveals the extent of expression of the corresponding gene. With a microarray, the expression patterns (both temporal and spatial) of many genes can be surveyed at once. The clustering of expression of genes in time and space suggests that the products of these genes work together in some process. This can provide information about genes of unknown function, if the unknown gene is expressed together with one or more well studied genes. The complete transcription program of an organism during diseased states, such as cancer, can be analyzed.

Analytical and functional protein-based arrays are used to study protein activity on a proteomic scale. These arrays may be based on recombinant proteins (functional array) or reagents that interact specifically with proteins, including antibodies, peptides, and small molecules such as ligands (analytical array). Such arrays can be used to study protein–protein interactions, protein modifications, or enzymatic activities.

After they are separated from a proteome, proteins must be identified. The best method for doing that involves digestion of the proteins with proteases, and identifying the resulting peptides by mass spectrometry. Peptide mass fingerprinting of a single isolated protein using MALDI-TOF, and shotgun proteomics of an entire proteome using tandem mass spectrometry (MS/MS), are two popular strategies for protein identification. The final step in both strategies is the application of computer algorithms that match the mass spectrometry spectra against a database of proteins that have been digested *in silico* to identify the protein(s) from which the peptides originated, and subsequently the genes.

Single nucleotide polymorphisms can probably be linked to many genetic conditions caused by single genes. SNPs do not necessarily cause disease, but they can help determine the likelihood that someone will develop a particular disease. So far, there are only a few examples with clear links to disease, e.g. the missense mutation that causes sickle cell anemia, and the apolipoprotein E4 allele implicated in susceptibility to late-onset Alzheimer's disease. Sorting out the important SNPs from those with no effect is proving to be very challenging.

Analytical questions

1 A cigarette butt found at the scene of a violent crime is found to have a sufficient number of epithelial cells stuck to the paper for the DNA to be extracted and typed. Shown below are the results of typing for three probes (locus 1, locus 2, and locus 3) of the evidence (X) and four suspects (A through D). Which of the suspects can be excluded? Which cannot be excluded? Can you identify the criminal? Explain your reasoning.

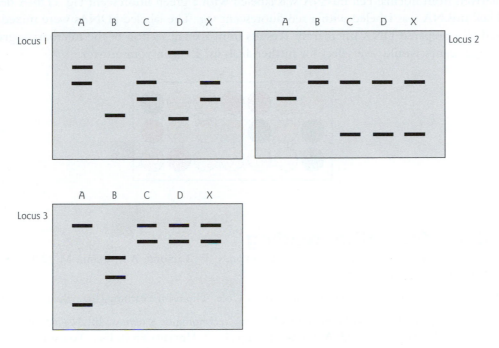

2 Based on the DNA fingerprinting (minisatellite analysis with a multilocus probe) results shown in the figure below, is Mr X or Mr Y the child's father? Explain your answer.

Lane 1: mother.

Lane 2: mother's child.

Lane 3: Mr Y.

Lane 4: Mr X.

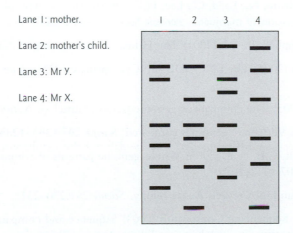

3 You separate a mixture of proteins by 2D-PAGE. From one protein spot, you obtain the following peptide sequence by mass spectrometry:

EQAGGDATENFEDVGHSTDAR

Use the BLAST tool to identify the unknown protein.

4 After sequencing the entire genome of a novel organism by the whole-genome shotgun approach, you use a computer program to search for start/stop signals for translation and homology searches based on similarity to cDNAs in the database. You predict approximately 20,000 genes. Is this an accurate representation of the total number of genes in the organism? Why or why not?

5 After SNP analysis of a region of the *apoE* gene, you determine that a patient carries the *apoE* ε4 allele. This allele is linked with Alzheimer's disease. Counsel the patient regarding their prognosis.

6 cDNA derived from normal cell mRNA was labeled with a green fluorescent tag. cDNA derived from diseased cell mRNA was labeled with a red fluorescent tag. The labeled cDNAs were mixed and hybridized to a prepared DNA microarray. Results from a small section of the microarray grid are shown below. Which genes would you select for further analysis? Explain your answer.

Suggestions for further reading

Andersen, J.S., Lam, Y.W., Leung, A.K.L., Ong, W.E., Lyon, C.E., Lamond, A.I., Mann, M. (2005) Nucleolar proteome dynamics. *Nature* 433:77–83.

Barnum, S.R. (2005) *Biotechnology. An Introduction.* Brooks/Cole: Thomson Learning, Inc., Belmont, CA.

Carracedo, A., Sánchez-Diz, P. (2004) Forensic DNA-typing technologies: a review. In: *Methods in Molecular Biology 297: Forensic DNA Typing Protocols* (ed. A. Carracedo), pp. 1–12. Humana Press, Inc., Totowa, NJ.

Coon, J.J., Syka, J.E.P., Shabanowitz, J., Hunt, D.F. (2005) Tandem mass spectrometry for peptide and protein sequence analysis. *BioTechniques* 38:519–523.

Coyle, H.M., Palmbach, T., Juliano, N., Ladd, C., Lee, H.C. (2003) An overview of DNA methods for the identification and individualization of marijuana. *Forensic Sciences* 44:315–321.

Dennis, C., Gallagher, R., Campbell, P., eds (2001) The Human Genome Issue of *Nature* 409:813–958.

ENCODE Project Consortium (2004) The ENCODE (ENCyclopedia Of DNA Elements) Project. *Science* 306:636–640.

Gunter, D., Dhand, R., eds (2005) The chimpanzee genome [series of articles]. *Nature* 437:47–108.

Hedges, S.B., Kumar, S. (2002) Vertebrate genomes compared. *Science* 297:1283–1285.

Hinds, D.A., Stuve, L.L., Nilsen, G.B. et al. (2005) Whole-genome patterns of common DNA variation in three human populations. *Science* 307:1072–1079.

Holden, C. (2001) Oldest human DNA reveals Aussie oddity. *Science* 291:230–231.

International Chicken Genome Sequencing Consortium (2004) Sequence and comparative analysis of the chicken genome provide unique perspectives on vertebrate evolution. *Nature* 432:695–716.

International Human Genome Sequencing Consortium (2004) Finishing the euchromatic sequence of the human genome. *Nature* 431:931–945.

International Rice Genome Sequencing Project (2005) The map-based sequence of the rice genome. *Nature* 436:793–800.

Jaillon, O., Aury, J.M., Brunet, F. et al. (2004) Genome duplication in the telost fish *Tetraodon nigroviridis* reveals the early vertebrate proto-karyotype. *Nature* 431:946–957.

Jasny, B.R., Kennedy, D., eds (2001) The Human Genome Issue of *Science* 291:1145–1434.

Jeffreys, A.J., Wilson, V., Thein, S.L. (1985) Hypervariable minisatellite regions in human DNA. *Nature* 314:67–73.

Marte, B., ed. (2003) Proteomics: nature insight [series of articles]. *Nature* 422:191–237.

Mills, J.C., Roth, K.A., Cagan, R.L., Gordon, J.I. (2001) DNA microarrays and beyond: completing the journey from tissue to cell. *Nature Cell Biology* 3:E175–E178.

Mulley, J., Holland, P. (2004) Small genome, big insights. *Nature* 431:916–917.

Pevsner, J. (2003) *Bioinformatics and Functional Genomics*. John Wiley & Sons, Hoboken, NJ.

Raber, J., Huang, Y., Ashford, J.W. (2004) ApoE genotype accounts for the vast majority of AD risk and AD pathology. *Neurobiology of Aging* 25:641–650.

Reilly, P.R. (2000) *Abraham Lincoln's DNA and Other Adventures in Genetics*. Cold Spring Harbor Laboratory Press, Cold Spring Harbor, NY.

Schena, M., Shalon, D., Davis, R.W., Brown, P.O. (1995) Quantitative monitoring of gene expression patterns with a complementary DNA microarray. *Science* 270:467–470.

Snyder, J., Gerstein, M. (2003) Defining genes in the genomics era. *Science* 300:258–260.

Van de Vijver, M.J., He, Y.D., van't Veer, L.J. et al. (2002) A gene-expression signature as a predictor of survival in breast cancer. *New England Journal of Medicine* 15:1999–2009.

Walsh, S.J. (2004) Recent advances in forensic genetics. *Expert Reviews in Molecular Diagnostics* 4:31–40.

Weiner, M.P., Hudson, T.J. (2002) Introduction to SNPs: discovery of markers for disease. *BioTechniques* (Suppl.) June:4–13.

Zivelin, A., Rosenberg, N., Peretz, H., Amit, Y., Kornbrot, N., Seligsohn, U. (1997) Improved method for genotyping apolipoprotein E polymorphisms by a PCR-based assay simultaneously utilizing two distinct restriction enzymes. *Clinical Chemistry* 43:1657–1659.

Chapter 17
Medical molecular biology

As to diseases, make a habit of two things – to help, or at least to do no harm.

Epidemics, in *Hippocrates*, translated by W.H.S. Jones (1923), Vol. I, 165.

Outline

17.1 Introduction

Discoveries in molecular biology have led to many advances in medicine, both in understanding of the nature of human disease and in developing treatment strategies. This chapter begins with an overview of the molecular mechanisms underlying the development of cancer. The second section focuses on progress in gene therapy. While once considered the realm of fantasy, somatic cell gene therapy is now being used with variable success to treat some inherited and acquired diseases. The chapter ends with a discussion of the role of genes versus the environment in determining human behavior, with a focus on aberrant behavior and psychiatric disorders.

17.2 Molecular biology of cancer

The oldest description of cancer is found in the Edwin Smith Papyrus – an ancient Egyptian medical manuscript that dates back to approximately 1600 BC. The writing states: "If thou examinest a man having tumors on his breast . . . There is no treatment." Today, cancer remains the second leading cause of death in the United States. Half of all men and one-third of all women in the USA will develop cancer during their lifetimes.

Normal cells have a defined lifespan – they grow, divide, and die in an orderly fashion. A critical balance is maintained between cell growth, proliferation, differentiation, and apoptosis. In the presence of agents that promote tumor formation, such as an inherited genetic defect, a chemical carcinogen, viral infection, or irradiation, this critical balance is disrupted. When cells in a part of the body begin to grow out of control, cancer develops. There are three major changes that occur when a cell becomes cancerous (Fig. 17.1):

1 **Immortalization.** The cancer cell acquires the ability to grow and divide indefinitely.
2 **Transformation.** The cancer cell fails to observe the normal constraints of growth; i.e. growth occurs independently of cell growth factors. Transformed cells may form a sold tumor. In order for the tumor to grow it must develop its own blood supply through the process of angiogenesis – the growth of new blood vessels.
3 **Metastasis.** Cancer cells often travel from the tissue of origin to other parts of the body where they begin to grow and replace normal tissue. This process, called metastasis, occurs as cancer cells gain the ability to invade the bloodstream or lymph vessels and other tissues. When cells from a cancer like colon cancer spread to another organ like the lung, the cancer is still called colon cancer, not lung cancer. The trajectory is characteristic of a tumor type; e.g. prostate cancer often spreads to bone, and breast cancer to lung. Hippocrates introduced the Greek "carcinos" for crab, to describe this grasping growth of a malignant tumor.

Our understanding of what causes cancer is continually progressing and increasing in complexity, as more interconnecting pathways are revealed. When molecular biologists first learned about oncogenes (see below), the focus was on looking for the "cancer gene." It is now clear that cancer is not just one disease but a group of genetically diverse disorders, such that each individual tumor can have its own "genetic signature." The gene mutations acquired by a tumor cell become a heritable trait of all cells of subsequent generations. Cancer is a multistep disease. The emergence of a tumor cell requires the accumulation of many – estimated to be between four and eight – genetic changes over the course of years.

Gene mutations that increase the risk for developing cancer can be inherited (Table 17.1) or acquired. Acquired genetic changes result from spontaneous or environmentally inflicted errors during DNA replication. The genetic changes associated with tumorigenesis can be divided into two major categories: those that result from a gain of function and those that result from a loss of function. Gain of function involves inappropriate activation of oncogenes. These genes are stimulatory for growth and can cause cancer when hyperactive. Loss of function involves inactivation of tumor suppressor genes. These are genes that inhibit cell growth and which cause cancer when they are not expressed.

Activation of oncogenes

Oncogenes are defined as genes whose products have the ability to cause malignant transformation of eukaryotic cells. They were originally identified as the "transforming genes" carried by some DNA and RNA tumor viruses (see section on Viruses and cancer, below). The term "proto-oncogene" refers to cellular genes with the potential to give rise to oncogenes. They are the normal counterparts in the eukaryotic genome to the oncogenes carried by retroviruses.

Proto-oncogenes are highly conserved from yeast to humans. Proto-oncogenes that were originally identified as resident in transforming retroviruses are designated "c-," indicative of their cellular origin, as opposed to "v-" to signify original identification in retroviruses. Oncogenes are referred to by a three letter

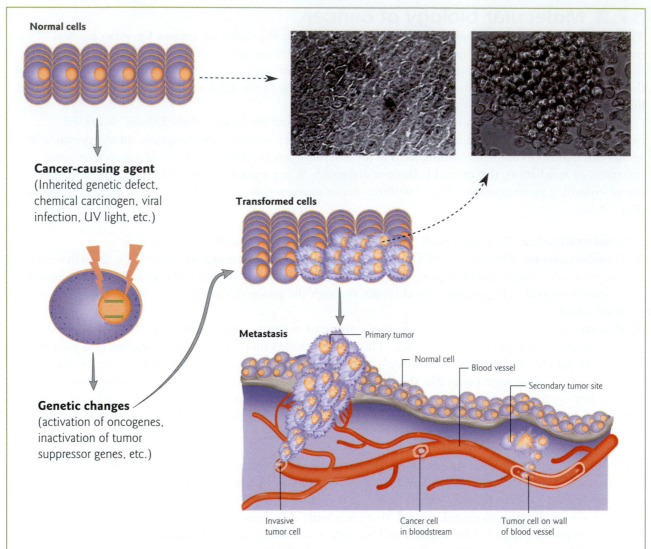

Normal cells

Cancer-causing agent
(Inherited genetic defect,
chemical carcinogen, viral
infection, UV light, etc.)

Genetic changes
(activation of oncogenes,
inactivation of tumor
suppressor genes, etc.)

Transformed cells

Metastasis   ─ Primary tumor

─ Normal cell   ─ Blood vessel

─ Secondary tumor site

Invasive   Cancer cell   Tumor cell on wall
tumor cell   in bloodstream   of blood vessel

Figure 17.1 The progression of cancer. (1) The growth of normal cells is restricted by properties of individual tissue. (2) Cancer-causing agents induce gene mutations and cells begin to proliferate out of control. (3–5) Metastasis occurs when cancer cells travel via the bloodstream to start more tumors elsewhere. (Inset) Transformed cells in culture. A normal rat fibroblast culture is shown in the left panel. Transformed cells are shown in the right panel. Cells are rounding up and detaching from the culture plate (Reprinted by permission from Nature Publishing Group and Macmillan Publishers Ltd: Yeatman, T.J. 2004. A renaissance for *src Nature Reviews Cancer* 4:470–480. Copyright © 2004.)

code (e.g. *src*, *myc*), generally reflecting the virus from which they were first isolated. Some viruses carry more than one oncogene (e.g. *erbA*, *erbB*) (Table 17.2). By convention, the gene name is given in lower case and italics, whereas the protein name has the first letter capitalized and is in plain type. Typically, all capital letters are used for human genes and proteins.

Inappropriate activation of proto-oncogenes may be due to qualitative or quantitative changes. Qualitative changes result from mutations in the coding sequence such as point mutations or insertion mutations, whereas quantitative changes are due to inappropriate regulation of expression. This may result from retroviral promoter/enhancer integration, chromosomal translocation, or amplification and overexpression of the gene (Fig. 17.2). Whether the oncogene is introduced by a virus (e.g. DNA tumor virus or retrovirus), or results from a mutation of a proto-oncogene in the genome, it is dominant over its allelic proto-oncogene.

Table 17.1 Selective list of cancer predisposition syndromes.

Cancer syndrome	Primary tumor	Defective gene(s)	Normal function of gene product	Cross reference
Hereditary nonpolyposis colon cancer	Colorectal cancer	*MSH2, MLH1, PMS1, PMS2, MSH6*	DNA mismatch repair	Chapter 7
Xeroderma pigmentosum	Skin cancer	*XPA, XPB, XPC, XPD, XPF, XPG*	Nucleotide excision repair	Chapter 7
Familial breast cancer	Breast and ovarian cancer	*BRCA1, BRCA2*	Repair of DNA double-strand breaks	Chapter 7
Beckwith–Wiedemann syndrome	Wilms' tumor (kidney), adrenocortical carcinoma, hepatoblastoma (liver cancer)	Loss of imprinting in region of chromosome 11 including: *IGF2* *p57^{KIP2}*	Growth promoting Growth suppressing	Chapter 12
Neurofibromatosis type 1	Tumors that grow on a nerve or nerve tissue	*Neurofibromin (NF1)*	Tumor suppressor	Chapter 13
Li–Fraumeni syndrome	Sarcomas, breast cancer, leukemia, brain tumors	*p53*	Transcription factor: cell cycle regulation, apoptosis	Chapter 17
Familial retinoblastoma	Tumor of retina, osteogenic sarcoma	*RB1*	Cell cycle regulator	Chapter 17

Proto-oncogenes can be classified into many different groups based on their normal function within cells or based upon sequence homology to other known proteins (Fig. 17.3). As predicted, proto-oncogenes have been identified at all levels of various signal transduction cascades that control cell growth and proliferation, differentiation, and apoptosis. They include genes encoding secreted growth factors, cytoplasmic proteins such as serine kinases and tyrosine kinases, surface and membrane-associated proteins involved in signal transduction, and transcription factors. A complete discussion of all oncogenes identified to date (> 100) is

Table 17.2 Selective list of viral oncogenes.

Retrovirus	Oncogene	Function of proto-oncogene-encoded proteins
Rous sarcoma virus	v-*src*	Tyrosine kinase
Simian sarcoma virus	v-*sis*	Platelet-derived growth factor (PDGF) B chain
Avian erythroblastosis virus	v-*erbA* v-*erbB*	Thyroid hormone receptor Epidermal growth factor receptor
Kirsten strain of murine sarcoma virus Harvey strain of murine sarcoma virus	v-K-*ras* v-Ha-*ras*	Membrane-associated GTP-binding protein Membrane-associated GTP-binding protein
MC29 avian myelocytoma virus	v-*myc*	Transcription factor
Abelson murine leukemia virus	v-*abl*	Tyrosine kinase
Feline osteosarcoma virus	v-*fos*	Transcription factor (often interacts with a second proto-oncogene, c-*jun* to form the AP-1 transcription complex) (see Figs 11.17 and 11.18)

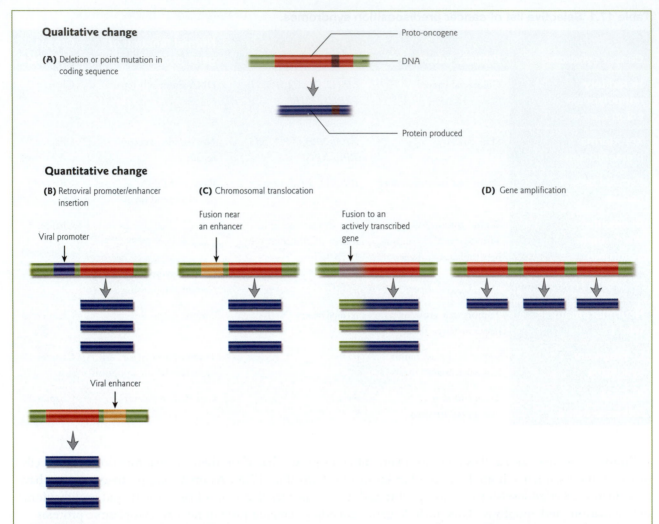

Figure 17.2 Inappropriate activation of proto-oncogenes may be due to qualitative or quantitative changes. (A) Deletion of point mutation in coding sequence: constitutively active mutant protein is produced in normal amounts. (B) Retroviral promoter/enhancer insertion: placement of a strong promoter upstream by retroviral integration causes overproduction of normal protein. Placement of a strong enhancer nearby (upstream or downstream) also causes overproduction of normal protein. (C) Chromosome rearrangements: placement of a strong enhancer nearby causes overproduction of normal protein, or fusion to another actively transcribed gene results in either increased levels of the fusion product or the fusion protein being hyperactive. (D) Gene amplification: normal protein is produced in much higher amounts.

beyond the scope of this textbook. Instead, only a selective survey of some genes that have been highly characterized is included in this section. Examples of a tyrosine kinase and a transcription factor are used to illustrate the complex pathways impacted by oncogene activation.

v-*src* tyrosine kinase

Tyrosine protein kinases are enzymes that transfer phosphate from adenosine triphosphate (ATP) to specific tyrosines on a target protein. Phosphorylation of target proteins leads to the activation of signal-transduction pathways which have a critical role in a variety of processes, including cell growth, differentiation, and death. Many oncogenes encode tyrosine kinases that are deregulated and overexpressed in human cancers.

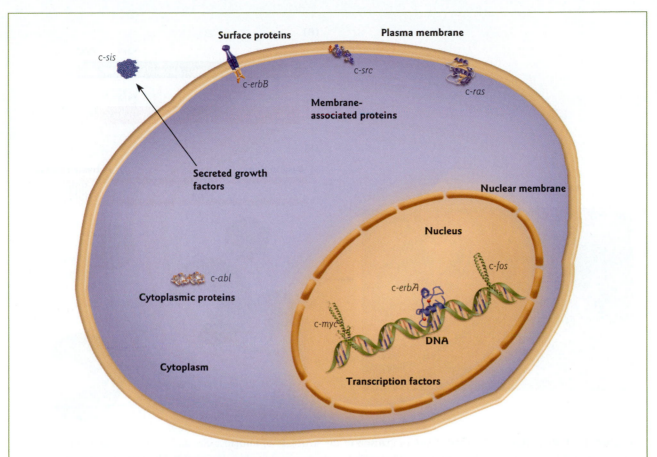

Figure 17.3 Cellular localization and function of proto-oncogene-encoded proteins. Proto-oncogenes encode secreted growth factors, cytoplasmic proteins (e.g. serine and tyrosine kinases), surface and membrane-associated proteins involved in signal transduction (e.g. tyrosine kinases, growth factor receptors, GTP-binding proteins), and transcription factors (see Table 17.2 for more detail).

One of these, v-*src*, was the first oncogene discovered. It was identified in the 1970s as an "extra" gene not needed for replication of the Rous sarcoma virus. The c-*src* gene encodes a membrane-associated, nonreceptor tyrosine kinase. Inactivation of human c-Src occurs when a tyrosine in its C-terminal domain is phosphorylated by the c-Src tyrosine kinase (CSK) (Fig. 17.4). When phosphorylated, the C-terminus changes conformation and interacts with SH (SRC homology) domains in the protein. When the C-terminus is dephosphorylated by phosphatases such as protein tyrosine phosphatase 1B (PTP1B), c-Src is activated. In the active conformation, c-Src phosphorylates target proteins. v-Src differs from c-Src by substitution of sequences at the C-terminus. This results in the loss of amino acids that normally interact with the SH domains of the enzyme to stabilize its "closed" or inactive conformation. v-Src thus remains in the "open" or active conformation (Fig. 17.4).

Overexpression of c-Src and an increase in its enzymatic activity has been observed in numerous cancer types including colorectal, hepatocellular, pancreatic, gastric, esophageal, breast, ovarian, and lung cancers. Despite the fact c-*src* is one of the most investigated proto-oncogenes, its role in cancer is not completely understood. However, many of the pathways that intersect with c-Src are beginning to be dissected. In addition to stimulating cellular proliferation, a primary role of c-Src in cancer is to regulate cell adhesion, invasion, and motility (Focus box 17.1). All of these basic cellular processes are deregulated during tumor progression and metastasis.

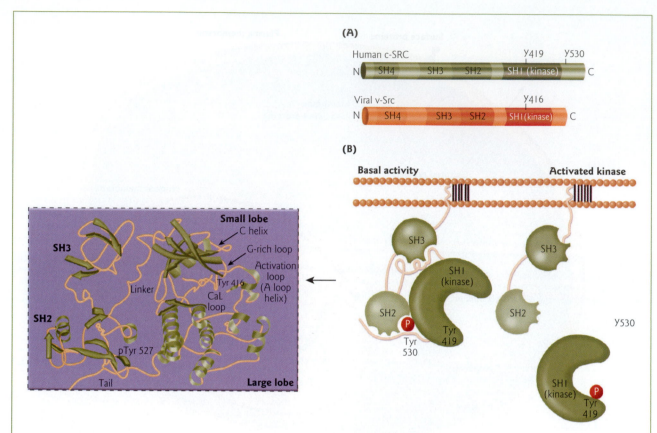

Figure 17.4 Activation of c-Src. (A) Comparison of the structure of human c-Src and viral v-Src. Both proteins contain four SRC homology (SH) domains. The SH1 domain contains the kinase activity. The viral oncoprotein v-Src lacks the carboxy (C) terminal negative regulatory domain and contains numerous amino acid substitutions throughout the protein. (B) Inactivation of human c-Src occurs when its C-terminal Tyr530 (Y530) is phosphorylated and it binds back to the SH2 domain. This interaction, and an interaction between the SH3 domain and the kinase domain, results in a "closed" structure that inhibits access of substrates to the kinase domain. c-Src activation occurs when the C-terminal phosphotyrosine is removed. This results in an "open" structure that permits access of substrates to the kinase domain. Full activation involves autophosphorylation at Tyr419 (Y419) in c-Src or Tyr416 (Y416) in v-Src. P, phosphorylation. (Adapted by permission from Nature Publishing Group and Macmillan Publishers Ltd: Martin, G.S. 2001. The hunting of the Src. *Nature Reviews Molecular Cell Biology* 2:467–475. Copyright © 2001.) (Inset) Ribbon diagram illustrating the structure of human c-Src. AMP-PNP (red) is bound in the active site. The A-loop helix occurs between the small and large lobes of the kinase and sequesters Tyr416. (Reprinted from Xu, W., Doshi, A., Lei, M., Eck, M.J., Harrison, S.C. 1999. Crystal structures of c-Src reveal features of its autoinhibitory mechanism. *Molecular Cell* 3:629–638. Copyright © 1999, with permission from Elsevier.)

c-*myc* transcription factor

c-*myc* is a central "oncogenic switch" that regulates a diversity of cellular functions through altering gene expression. The v-*myc* gene was originally identified in the avian myelocytoma virus (Table 17.2). c-*myc* encodes a helix-loop-helix transcription factor (see Chapter 11) that dimerizes with Max, its helix-loop-helix partner. In normal cells, c-*myc* is only expressed when cells actively divide and it inhibits terminal differentiation of most cell types. The Myc–4Max heterodimer protein regulates the expression of many genes involved in cell cycle regulation, apoptosis, or cell adhesion (Fig. 17.5). c-*myc* overexpression is often correlated with highly aggressive tumors, including breast, colon, cervical, prostate, skin, hepatocellular (liver), and small-cell lung carcinomas, and with some types of leukemia (Table 17.3). It is estimated that approximately 100,000 US cancer deaths per year are associated with changes in the c-*myc* gene or its expression.

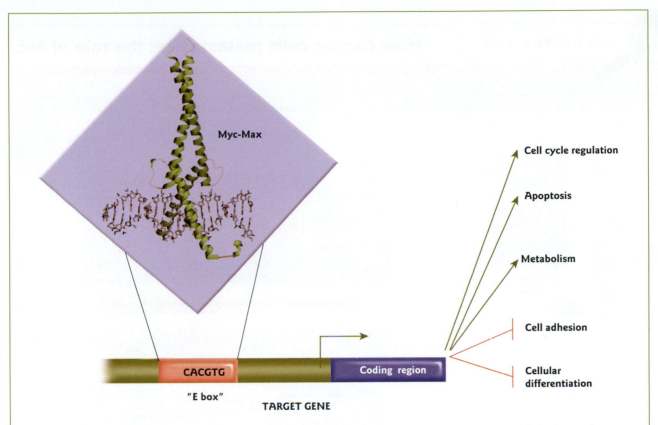

Figure 17.5 Function of the c-Myc transcription factor. The Myc–Max basic helix-loop-helix heterodimer binds a target DNA site, termed the E box. c-Myc activates (green arrow) or inhibits (red line) the expression of target genes involved in cell cycle regulation, apoptosis, cell metabolism, cell adhesion, and cellular differentiation. (Protein Data Bank, PDB: 1NKP).

Inactivation of tumor suppressor genes

The discovery of cellular proto-oncogenes opened the way to figuring out mechanisms by which cancers may be caused. In addition, their discovery led to the identification of another class of cellular genes, the tumor suppressor genes. Tumor suppressor gene products are required for normal cell function and, like oncogene products, play a central role in regulating cell growth and division. Cancer arises when there are two independent mutations or "hits" that lead to loss of function of both tumor suppressor alleles at a locus. Loss can be either sporadic or heritable. If loss of one allele is inherited through the germline, an individual is said to have a "genetic predisposition" to cancer. Sporadic mutation of the second allele in a somatic cell leads to tumorigenesis. Loss of function of a tumor suppressor gene can also occur at a much lower frequency by two sporadic mutations in the same cell. Two of the best-characterized tumor suppressor genes encode the retinoblastoma (pRB) and p53 proteins.

Retinoblastoma protein: the cell cycle master switch

The retinoblastoma (*RB*) protein gene was the first tumor suppressor gene to be cloned. How it acts in the childhood hereditary cancer syndrome for which it was named remains unclear (Disease box 17.1). The retinoblastoma protein (pRB) is a member of the "pocket" protein family. The family is named for a shared viral oncoprotein-binding domain. Like other proteins once thought to have one job, pRB is now known to have many. Because of this, pRB is often called the "cell cycle master switch." The tumor suppressor

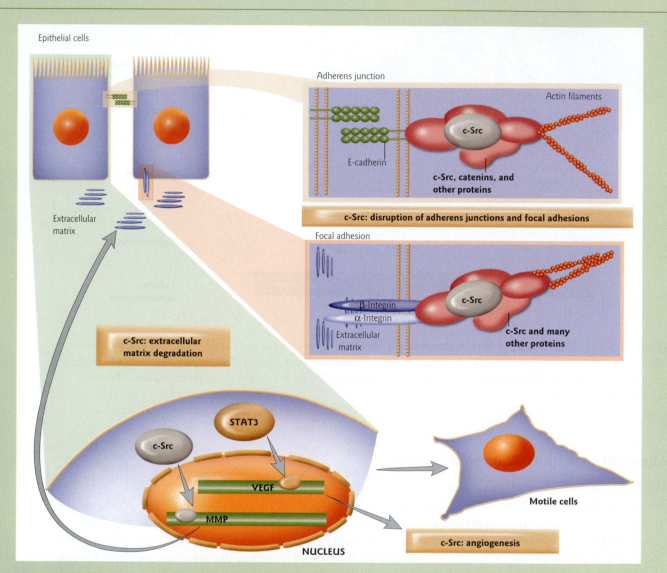

Figure 1 Role of c-Src in cancer cell metastasis. c-Src plays a role in regulating the disruption of the adherens junctions and focal adhesions, extracellular matrix degradation, and angiogenesis. For simplicity, only some of the key players in various pathways are shown. Adherens junctions facilitate cell–cell adhesion through binding between E cadherin molecules on adjacent cells. A cytoplasmic complex consisting of various catenins links E cadherin homodimers to the actin cytoskeleton. c-Src associates with this complex and, when activated, is able to promote the disruption of the adherens junctions. At focal adhesions, heterodimers of α- and β-integrin subunits bind the extracellular matrix through their extracellular domains. Their cytoplasmic domains bind to a complex consisting of a range of proteins, including c-Src, which can promote the turnover of the focal adhesion when activated, to promote cellular motility. c-Src regulates matrix-metalloproteases (MMPs) that degrade the extracellular matrix, promoting invasion of surrounding tissues. Finally, activation of STAT3 (signal transducer and activator of transcription 3) leads to increased expression of VEGF (vascular endothelial growth factor), a signaling molecule that promotes tumor blood vessel formation (angiogenesis). (Adapted from Yeatman, T.J. 2004. A renaissance for *src*. *Nature Reviews Cancer* 4:470–480.)

How cancer cells metastasize: the role of Src   FOCUS BOX 17.1

Metastasis – the spread of cancer from its tissue of origin and its subsequent growth in other organs – is the most life-threatening aspect of this disease (see Fig. 17.1). For cells to spread, they must lose their normal cell–cell connections and become motile. Inappropriate expression of anywhere from 20 to 95 different genes or gene families has been implicated in cancer cell invasion. Increasing evidence suggest that the *src* oncogene plays a pivotal role in mediating metastasis (Fig. 1).

Cell motility is a highly regulated, multistep event. One of the first steps is formation of cellular protrusions such as lamellipodia and filopodia. Ha- and K-*ras* (see Table 17.2) are small GTPases that promote lamellipod extension and transcription of other invasion genes. These proto-oncogenes are often inappropriately expressed in cancer cells. Cells are thought to move directionally by forming and extending protrusions, forming stable attachments to the extracellular matrix near the leading edge of these protrusions, and propulsion forward. This is followed by release of adhesions from other cells (adherens junctions) and from the basement membrane matrix (focal adhesions) and retraction of the rear end. Adherens junctions facilitate

cell–cell adhesions through binding between E cadherin glycoproteins on adjacent cells (Fig. 1). A cytoplasmic complex composed of various catenins and c-Src links the E cadherin homodimer to the actin cytoskeleton. At focal adhesions, heterodimers of α- and β-integrin subunits bind the extracellular matrix through their extracellular domains. Their cytoplasmic domains bind to a complex of proteins, including c-Src, which connect the integrins to the actin cytoskeleton. Disruption of adherens junctions and focal adhesions is regulated by c-Src.

c-Src may also affect cancer cell motility by regulating matrix-metalloproteases and tissue inhibitors of them. Matrix-metalloproteases (MMPs) degrade the extracellular matrix which would otherwise hinder movement and invasion of surrounding tissues. In addition, activation by c-Src of STAT3 (signal transducer and activator of transcription 3) leads to increased expression of VEGF (vascular endothelial growth factor), a signaling molecule that promotes tumor angiogenesis (Fig. 1). These diverse functions of c-Src all contribute to tumor progression and metastasis.

protein has been implicated in the regulation of DNA replication, cell differentiation, DNA repair, cell cycle checkpoints, and apoptosis. In fact, pRB may interact with more than 100 different cellular proteins. One of the normal cellular functions of pRB is to prevent cells from entering S phase, the phase in which DNA is replicated (Fig. 17.6). Hypophosphorylated pRB binds the E2F transcription factor complex and prevents E2F from binding to regulatory elements in target genes. The E2F complex is required for activation of genes essential for DNA replication and cell proliferation. During mid to late G1 phase of the cell cycle, pRB is phosphorylated by cyclin–CDK (cyclin-dependent kinase) complexes. When pRB is phosphorylated,

Table 17.3 Some chromosomal translocations associated with cancer.

Disease	Translocation	Result
Burkitt's lymphoma	8 to 14	c-*myc* proto-oncogene inserted close to immunoglobulin heavy chain promoter
Chronic myelogenous leukemia	9 to 22 (Philadelphia chromosome named after the city of its discovery in 1960)	Fuses gene *bcr* and proto-oncogene c-*abl*, encoding a BCR–ABL fusion protein that has unregulated tyrosine kinase activity
Acute promyelocytic leukemia	15 to 17	PML (or PLZF) RAR fusion protein that inhibits p53 activity by promoting its deacetylation and degradation

Knudson's two-hit hypothesis and retinoblastoma

For many years a puzzle in cancer biology was how some forms of cancer could be inherited and others be sporadic. The "two-hit" hypothesis proposed by Dr Alfred G. Knudson in 1971 provided a unifying model for understanding cancer that occurs in individuals who carry a "susceptibility gene" and cancers that develop because of randomly induced mutations in otherwise normal genes (Fig. 1).

Knudson was studying children with retinoblastoma, a tumor of the retina. He noted differences among patients whose tumor was inherited and those who appeared to have no "susceptibility" to the disease. Since some children who inherit a defective allele do not get the tumor, it was apparent that inheriting the mutation was not sufficient for tumor formation. What Knudson proposed was that the inherited mutant allele is the "first hit," but another hit after conception produces a tumor. Loss of the second allele is the "rate limiting" step and occurs by sporadic mutation in cells of the retina. The same gene, known as *RB*, is involved in children with the nonhereditary form, but these children sustain "two hits" after conception (Fig. 1). The "hits" can occur in many ways – from an environmental toxin, dietary factors, radiation, or the kind of random mutation that sometimes occurs during normal cell division. For example, "loss of heterozygosity" can arise by gene conversion, nondisjunction, or recombination during mitosis.

In 1976 Knudson and others showed that some patients with retinoblastoma are missing a segment of chromosome 13. In 1986, other scientists cloned the *RB* gene from this locus, and its function as a tumor suppressor was studied in detail. Loss of *RB* also occurs in diverse sporadic cancers including osteosarcomas and small-cell lung cancer. Additional mutations beyond *RB* disruption must occur for retinoblastoma or other cancers to develop. Mouse models suggest that these mutations occur in genes that regulate cell differentiation. Why retinal cells are especially susceptible to transformation following disruption of *RB* remains to be explained.

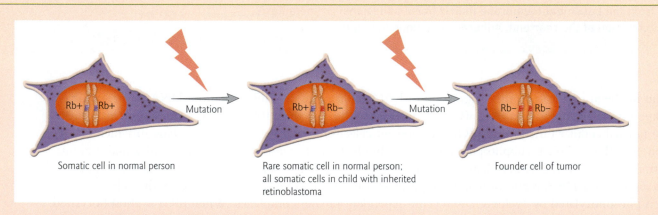

Somatic cell in normal person

Rare somatic cell in normal person; all somatic cells in child with inherited retinoblastoma

Founder cell of tumor

Figure 1 Knudson's two-hit hypothesis and the development of retinoblastoma. Sporadic retinoblastoma – a tumor of the retina – requires two hits (mutations) while the inherited form requires only one hit and is quite highly penetrant.

the E2F complex is released and can activate its target genes. If pRB is absent, the E2F complex continues to stimulate S phase-specific genes, resulting in unrestrained cell growth.

In acute lymphoblastic leukemia, the most common childhood cancer, pRB is effectively "absent" because CDK2 catalytic activity in the acute lymphoblastic leukemia cells keeps pRB in a phosphorylated state (Fig. 17.7). Although the proliferation rate of these cells is not high, they have lost their ability to differentiate or undergo apoptosis, causing an accumulation of malignant cells. The central importance of pRB is further highlighted by knockout mouse studies. pRB$^{-/-}$ mice die *in utero* around embryonic day

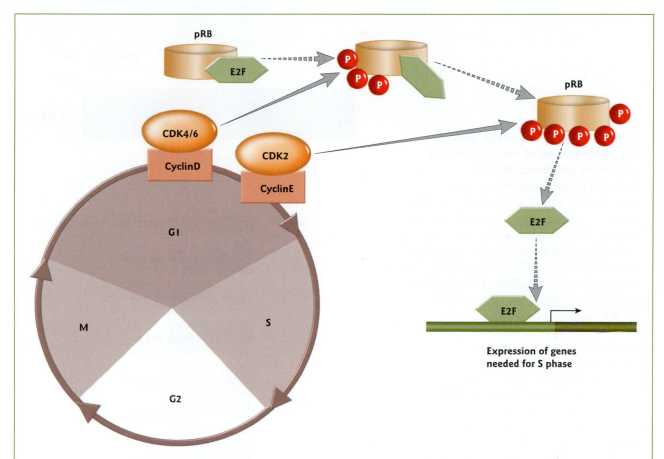

Figure 17.6 Retinoblastoma tumor suppressor protein (pRb): the cell cycle master switch. The cell cycle is driven by the coordinated activation of the cyclin-dependent kinases (CDKs). The CDKs are expressed throughout the cell cycle, but their activating subunits – the cyclins – oscillate between rapid synthesis and degradation. CyclinD- and cyclinE-dependent kinases phosphorylate (P) and inactivate pRb, which is a principal checkpoint controlling the progression from G1 to S phase. Inactivated Rb releases E2F transcription factors, which stimulate the expression of downstream cyclins and other genes that are required for DNA synthesis during S phase.

13.5, with severe defects in the development of many tissues, including the lens of the eye, muscle, blood, bone, and nerve cells.

p53: the "guardian of the genome"

The tumor suppressor protein p53 (named for its molecular size of 53 kDa) is often referred to as the "guardian of the genome." It regulates multiple components of the DNA damage control system, and is one of the genes most commonly induced in response to cellular stress signals. As a transcription factor, p53 can activate or repress many target genes. Over 100 genes regulated by p53 have been identified through microarray analysis. Bioinformatic studies predict that the number may be much greater – greater than 4000 human genes contain putative p53-binding sites in their promoter region.

p53 regulation is impressively complex. In normal cells there are low levels of p53, and these are constantly turning over. The expression levels of p53 can be regulated at both transcriptional and post-translational stages. However, regulation mainly occurs at the protein level, by increasing the stability of p53. Under normal circumstances, p53 is targeted for proteasomal degradation by the binding of MDM2 (mouse double minute 2), a ubiquitin E3 ligase. This process is enhanced by the transcription factor YY1 (Yin Yang 1).

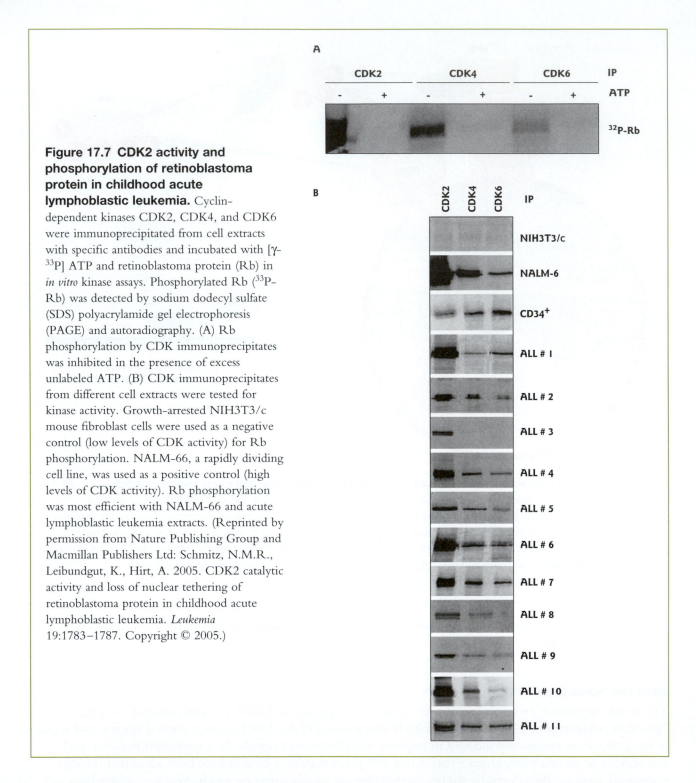

Figure 17.7 CDK2 activity and phosphorylation of retinoblastoma protein in childhood acute lymphoblastic leukemia. Cyclin-dependent kinases CDK2, CDK4, and CDK6 were immunoprecipitated from cell extracts with specific antibodies and incubated with [γ-^{33}P] ATP and retinoblastoma protein (Rb) in *in vitro* kinase assays. Phosphorylated Rb (^{33}P-Rb) was detected by sodium dodecyl sulfate (SDS) polyacrylamide gel electrophoresis (PAGE) and autoradiography. (A) Rb phosphorylation by CDK immunoprecipitates was inhibited in the presence of excess unlabeled ATP. (B) CDK immunoprecipitates from different cell extracts were tested for kinase activity. Growth-arrested NIH3T3/c mouse fibroblast cells were used as a negative control (low levels of CDK activity) for Rb phosphorylation. NALM-66, a rapidly dividing cell line, was used as a positive control (high levels of CDK activity). Rb phosphorylation was most efficient with NALM-66 and acute lymphoblastic leukemia extracts. (Reprinted by permission from Nature Publishing Group and Macmillan Publishers Ltd: Schmitz, N.M.R., Leibundgut, K., Hirt, A. 2005. CDK2 catalytic activity and loss of nuclear tethering of retinoblastoma protein in childhood acute lymphoblastic leukemia. *Leukemia* 19:1783–1787. Copyright © 2005.)

Various inhibitory proteins prevent p53 degradation by a range of mechanisms. For example, the protein p14^{Arf} binds MDM2 and inhibit its E3 ligase activity, while PML (promyelocytic leukemia) protein binds MDM2 and sequesters it to the nucleolus (Fig. 17.8A).

Cellular stress signals induce p53 expression. For example, DNA damage from ultraviolet irradiation activates p53 in sunburned skin cells. Dependent on the stage in the cell cycle, DNA damage leads to two possible outcomes: cell cycle arrest and DNA repair before replication, or apoptosis (e.g. peeling of skin

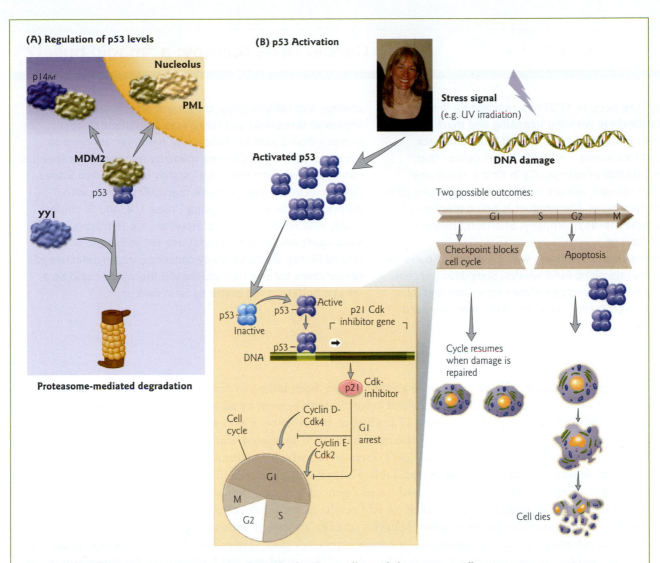

Figure 17.8 Tumor suppressor protein p53: the "guardian of the genome." (A) Mechanisms of p53 degradation. The E3 ubiquitin ligase MDM2 binds p53 and targets it for proteasomal degradation, enhanced by the transcription factor YY1. Various inhibitory proteins prevent this degradation by a range of mechanisms. In the example shown, p14^Arf inhibits MDM2 action, and PML sequesters it in the nucleolus. (B) Activation of p53. p53 is maintained at a low level by degradation. Stress signals such as UV irradiation induce an increase in p53 levels and transcriptional activity. If DNA damage occurs early in G1, p53 regulates the expression of genes such as the CDK inhibitor p21 and mediates cell cycle arrest, followed by DNA repair. If DNA damage occurs later in the cell cycle, then p53 promotes apoptosis. (Inset) DNA damage from UV irradiation activates p53 in sunburned skin cells (photo of the author in evening wear after a day out on the ice in Antarctica).

after bad sunburn). If the DNA damage occurs early in G1, a checkpoint is triggered by p53 that blocks the cell cycle. p53 activates the *p21* gene which encodes a CDK inhibitor. The inhibitor blocks the action of cyclin D–CDK4 and cyclin E–CDK2 and causes G1 arrest (Fig. 17.8B). As a consequence, pRB remains hypophosphorylated, leading to sequestration of the E2F transcription complex required for activation of S phase-specific genes (see Fig. 17.6). The cycle resumes when the damage is repaired. The ability of p53 to promote apoptosis is less well understood, although it is known to play a role in both mitochondrial-mediated and death receptor–mediated apoptotic pathways (see below).

Cancer gene therapy: a "magic bullet?"

Gene therapy (see Section 17.3) has reached its first commercial milestone with the licensing and marketing in China of Gendicine for the treatment of head and neck squamous cell carcinoma – a highly lethal cancer that strikes some 300,000 people yearly in China. Gendicine contains a recombinant adenovirus vector carrying the *p53* tumor suppressor gene. The product is being marketed by China's first gene therapy company, Shenzen SiBiono Gene Technologies Co. Ltd.

In the Chinese study, Gendicine was given to 120 patients in single weekly injections for 8 weeks. Sixty-four percent of patient's tumors showed complete regression and 32% showed partial regression. In combination with standard chemo- and radiotherapy, treatment effectiveness was improved three-fold and there was no patient relapse in more than 3 years of follow-up. A similar adenovirus vector containing a *p53* gene made by Introgen Therapeutics has been given "orphan drug" status in the United States. Advexin is being used in more than 500 patients with different cancers in 20 ongoing Phase I, II, and III clinical trials, with notable success. However, no therapies have been approved for commercial use yet in Europe or the United States. Because many cancers contain defective *p53*, researchers believe that *p53*-based therapies could be a "magic bullet" in the battle against cancer.

Role of p53 in cancer   Alterations in the *p53* gene have been linked to many cancers including cervix, breast, bladder, prostate, liver, lung, skin, and colon. Eighty percent of all human cancers show either deletion of both alleles on chromosome 17 leading to the absence of p53 protein (similar to pRB), or a missense point mutation in one allele. Mutations in *p53* are associated with the hereditary cancer syndrome, Li–Fraumeni syndrome (named after its 1969 discoverers), which predisposes patients to brain tumors, sarcomas, leukemia, and breast cancer (see Table 17.1). Because *p53* is defective in so many different types of cancers, many treatment strategies based on *p53* gene therapy are being tested (Disease box 17.2).

A missense point mutation in one allele of *p53* results in the production of a dominant negative mutant protein. This type of dominant negative behavior, usually associated with oncogenes, led to confusion early on about whether *p53* was an oncogene or tumor suppressor gene (Focus box 17.2). Dominant negative *p53* not only has lost tumor suppressor function but also inhibits wild-type p53 activity by the formation of inactive heterotetramers (Fig. 17.9). A homotetramer of wild-type p53 retains tumor suppressor function, whereas a heterotetramer with only one mutant *p53* subunit leads to unrestrained cell growth. Very little is known about the regulation of mutant *p53* function, but it is known that that it can have a different set of binding partners from wild-type p53. Over 80% of *p53* mutations found in human cancer are located in the central region of the protein. This region mediates DNA binding and interaction with the ASPP (apoptosis-stimulating protein of p53) family of proteins and other proteins such as Bcl-2 (B cell chronic lymphocytic leukemia (CLL)/lymphoma 2) and BAK (Bcl-2-antagonist/killer 1) that regulate the apoptotic function of p53 (Fig. 17.9). Consequently, the majority of point mutations prevent p53 from binding DNA and from promoting apoptosis.

Inappropriate expression of microRNAs in cancer

A number of lines of evidence suggest that inappropriate expression of microRNAs (miRNAs) plays a role in cancer. Although miRNAs currently only represent 1% of the mammalian genome, more than 50% of miRNA genes are located within regions associated with amplification, deletion, and translocation in cancer. First, recent studies show that there is altered expression of miRNA genes in several human malignancies, including chronic lymphocytic leukemia, Burkitt's lymphoma, gastric cancer, and lung cancer. Second, the pattern of miRNA expression has been shown to vary dramatically across tumor types. Interestingly, the

The discovery of p53

p53 was originally identified in 1979 as a major nuclear antigen in transformed cells. It was found associated with the simian virus 40 (SV40) large T antigen (see section on DNA tumor viruses). Early on, a mutant form of the *p53* gene rather than a wild-type gene was inadvertently used in experiments. The mutant *p53* gene was cloned from a mouse liver library which was apparently contaminated by some clones from a transformed cell library. These complementary DNA (cDNA) clones were shown to have transforming activity on cells in culture and thus it was concluded that *p53* was an oncogene (Fig. 1). However, at the same time, a *p53* cDNA clone derived from another library (mouse F9 teratocarcinoma cells) was not shown to have transforming activity. Finally, researchers realized that the cDNA clone from the "normal" liver cells in fact encoded a mutant form of p53, while the *p53* cDNA from the teratocarcinoma cells encoded wild-type p53. With the demonstration that wild-type p53 could block the ability of other oncogenes to transform cells in culture, it became clear that the *p53* gene was, in fact, an "anti-oncogene" or tumor suppressor gene.

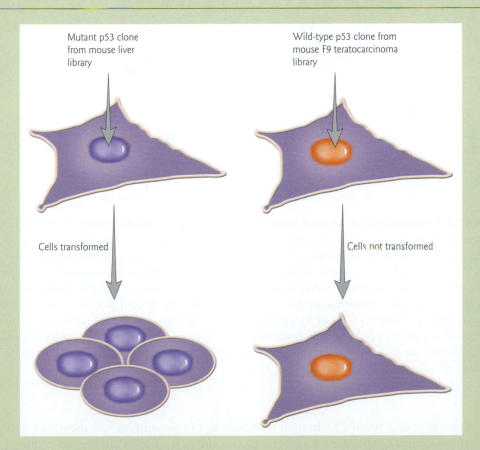

Figure 1 Wild-type p53 is a tumor suppressor protein. A p53 cDNA clone derived from "normal" liver cells turned out to be a mutant form of p53 that had transforming activity on cells in culture. A cDNA clone for wild-type p53 derived from mouse F9 teratocarcinoma cells did not have transforming activity.

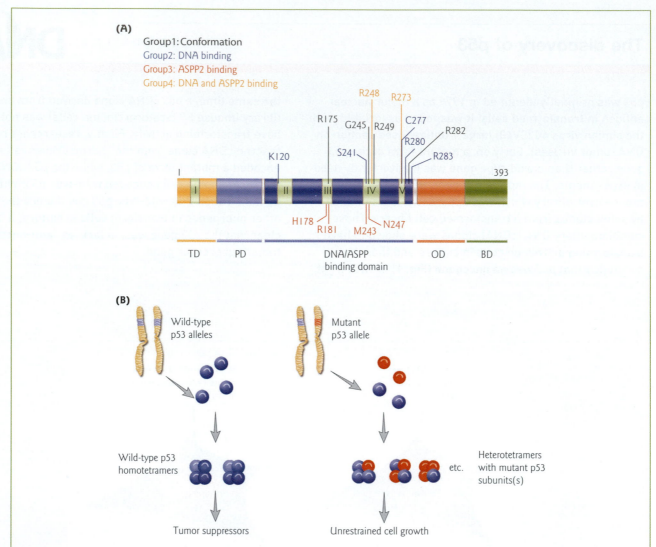

Figure 17.9 Mutant p53 acts as a dominant negative protein in cancer. (A) The functional domains of p53 and the location of common mutations found in human tumors are shown. All tumor-derived p53 mutants are defective in inducing apoptosis. However, some of them are as competent as wild-type p53 in inducing cell cycle arrest. The conserved regions of p53 are illustrated as boxes I–V. ASPP, apoptosis–stimulating protein of p53 family of proteins; BD, nonspecific DNA-binding domain; OD, oligomerization domain; PD, proline–rich region; TD, transcriptional domain. (Adapted from: Lu, X. 2005. p53: a heavily dictated dictator of life and death. *Current Opinions in Genetics and Development* 15:27–33. Copyright © 2005, with permission from Elsevier.) (B) A missense point mutation in one allele of p53 results in production of a dominant negative mutant protein. A homotetramer of wild-type p53 retains tumor suppressor function, whereas a heterotetramer with only one mutant p53 subunit promotes unrestrained cell growth and proliferation.

expression pattern of a set of 217 human miRNAs in cancer samples was shown to define the cancer type better than microarray expression data from 16,000 mRNAs. Finally, in certain human lymphomas (cancers derived from immune cells), extra copies of a chromosome 13 fragment accumulate in cancer cells. The region of amplification contains a "host" gene (*c13orf25*) which encodes the precursors of seven miRNAs (*mir-17-19b* polycistron), which are overexpressed in B cell lymphoma samples (Fig. 17.10). It is unlikely that the *c13orf25* transcript encodes a protein, as predicted open reading frames encode only short peptides (< 70 amino acids) which are not conserved.

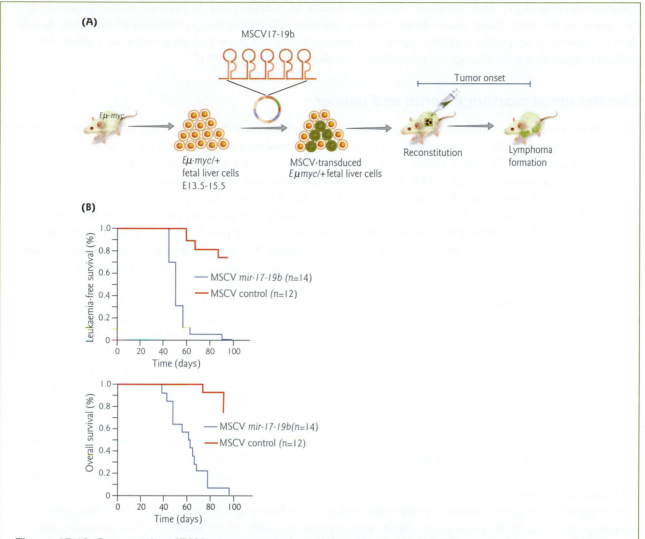

Figure 17.10 Oncogenic miRNA: overexpression of the *mir-17-19b* miRNA gene cluster accelerates c-*myc*-induced B cell lymphoma. (A) Schematic representation of the experimental strategy. Fetal liver-derived hematopoietic stem cells were isolated from a transgenic mouse carrying the c-*myc* oncogene under control of the immunoglobulin heavy-chain enhancer (Eμ-myc). Lethally irradiated Eμ-myc mice were then reconstituted with hematopoietic stem cells expressing *mir-17-19b* in an MSCV retroviral vector (MSCV *mir-17-19b*), or infected with control MSCV virus. The viral vector also contained a GFP transgene, allowing the altered stem cells to be tracked *in vitro* and during lymphoma formation *in vivo*. (B) The percentage of leukemia-free survival or overall survival of the mice was analyzed starting 5 weeks after transplantation. (Reprinted by permission from Nature Publishing Group and Macmillan Publishers Ltd: He, L., Thomson, J.M., Hemann, M.T. et al. 2005. A microRNA polycistron as a potential human oncogene. *Nature* 435:828–833. Copyright © 2005.)

In 2005, Lin He, J. Michael Thomson, and co-workers used a transgenic mouse model of B cell lymphoma to investigate the biological consequences of the overexpression of this miRNA gene cluster (Fig. 17.10). These transgenic mice carry the proto-oncogene c-*myc* under control of the immunoglobulin heavy-chain enhancer. The mice develop B cell lymphomas by 4–6 months of age. The researchers infected hematopoietic (blood-forming) stem cells from these mice with a retrovirus vector carrying a portion of the miRNA cluster, and then transplanted the cells back into recipient lymphoma-prone mice. The retrovirus vector also contained a green fluorescent protein (*GFP*) transgene, allowing the altered stem cells to be followed *in vitro* and *in vivo*. Compared with control mice, there was a striking increase in how rapidly leukemia developed (51 days compared with 3–6 months in controls) and an increase in the frequency of the cancer in the mice (from about 30 to 100% of animals). The intriguing conclusion of this study is that altered expression of specific miRNA genes contributes to the initiation and progression of cancer. The authors suggest that such oncogenic microRNAs be designated "oncomiRs."

Chromosomal rearrangements and cancer

The combination of technical advances in molecular cytogenetics and the sequencing of the human genome has allowed for major advances in cancer biology. Chromosomal mapping allows the precise localization of the site of a gene on a particular chromosome. It is now apparent that many cancers are associated with alterations in chromosomes, particularly translocations (the breakage of a chromosome so that the two parts associate with two parts of another chromosome). Many breakpoints in tumor cells are very close to a known proto-oncogene (see Table 17.3). These chromosomal rearrangements can bring a cellular proto-oncogene under the control of the wrong promoter and/or enhancer (see Fig. 17.2). For example, in Burkitt's lymphoma, the c-*myc* proto-oncogene on chromosome 8 is translocated to a site on chromosome 14, close to the gene for immunoglobulin heavy chains. The proto-oncogene is thus brought under control of the immunoglobulin promoter which is very active in B lymphocytes.

In acute promyelocytic leukemia chromosome rearrangements create a fusion protein with oncogenic properties. The *PML* (promyelocytic leukemia) gene is found in a reciprocal translocation with the retinoic acid receptor (RAR) gene. The resulting sequence encodes the fusion protein PML-RAR. PML-RAR inhibits the transcription of retinoic acid-responsive target genes, thus blocking myeloid cell differentiation. The fusion protein also inhibits p53 function allowing enhanced survival of promyelocytic cells. Both the differentiation block and enhanced survival are mediated by recruitment of a histone deacetylase (HDAC) by the fusion protein (Fig. 17.11A). Pharmacological (nonphysiological) concentrations of retinoic acid cause dissociation of the PML-RAR/HDAC complex and degradation of the fusion protein, restoring the normal differentiation pathway. Retinoic acid treatment has been used successfully to cure some patients. Fusion proteins are probably not sufficient to induce leukemia. A "two-hit" model hypothesizes that they induce a "pre-leukemic" state, which then favors the occurrence of the secondary gene mutations required for the full leukemic phenotype.

A single event triggers chronic myelogenous leukemia: a chromosomal translocation fuses the genes encoding ABL and BCR (breakpoint cluster region protein) (Table 17.3). BCR is a GTPase-activating protein essential for activation of the ABL tyrosine kinase. c-*abl* is the cellular homolog of v-*abl* (Abelson murine leukemia viral oncogene homolog 1). The resulting fusion protein BCR-ABL has unregulated tyrosine kinase activity (Fig. 17.11B). Precisely how this kinase activity leads to leukemia is still under investigation. Recently, highly specific protein tyrosine kinase inhibitors have been developed that work to slow tumor progression. A small organic molecule called imatinib mesylate (Gleevac) has been shown to control chronic myelogenous leukemia by inhibiting BCR-ABL activity in early-stage patients. When BCR-ABL binds a molecule of ATP in the "kinase pocket", the substrate is activated by phosphorylation of one its tyrosine residues. It can then activate other downstream effector molecules. When imantinib occupies the kinase pocket, the action of BCR-ABL is inhibited, preventing phosphorylation of its substrate.

Viruses and cancer

Tumor cells can arise by nongenetic means through the actions of specific tumor viruses. Tumor viruses promote cancer because they either carry a copy of an oncogene or tumor suppressor gene or alter expression of the cell's copy of one of these genes. If a virus takes up residence in a cell and alters the properties of that cell, the cell is said to be "transformed." Transformation by a virus is defined as changes in the functions of a cell that result from regulation of the cell by viral genes. These changes often (but not always) result from integration of the viral genome into the host cell genome. Transformation of cells in culture is a visible event, leading to morphological changes. The cells round up, disaggregate, and begin to

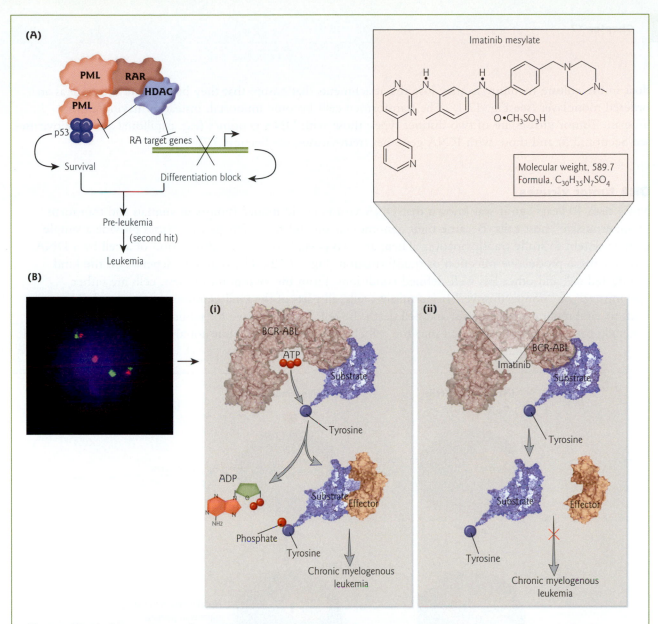

Figure 17.11 Chromosomal rearrangements can lead to cancer. (A) In acute promyelocytic leukemia, a chromosomal translocation t(15;17) brings together the *PML* and *RAR* genes. This leads to the synthesis of a fusion protein PML-RAR, which acts as a histone deacetylase (HDAC) dependent inhibitor of pathways regulated by wild-type PML (promyelocytic leukemia) and the retinoic acid receptor (RAR). PML-RAR inhibits the transcription of retinoic acid (RA) responsive target genes in the presence of ligand, thus blocking myeloid cell differentiation. PML-RAR also inhibits p53 function (in a manner requiring wild-type PML), allowing enhanced survival of leukemia cells. Both differentiation block and enhanced survival are mediated by recruitment of HDAC by the fusion protein. (Modified from Insinga, A., Monestiroli, S., Ronzoni, S. et al. 2004. Impairment of p53 acetylation, stability, and function by an oncogenic transcription factor. *EMBO Journal* 23:1144–1154.) (B) Mechanism of action of BCR–ABL and of its inhibition by the drug imatinib. (i) The BCR-ABL oncoprotein with a molecule of ATP in the kinase pocket. The substrate is activated by the phosphorylation of one of its tyrosine kinase residues. It can then activate other downstream effector molecules, leading to chronic myelogenous leukemia. (ii) When imatinib occupies the kinase pocket, the action of BCR-ABL is inhibited. (Modified from Savage, D.G., Antman, K.H. 2002. Imatinib mesylate – a new oral targeted therapy *New England Journal of Medicine* 346:683–693). (Left inset) *BCR-ABL* rearrangement in an interphase cell. The reciprocal t(9;22) generates two fusion signals. A normal *ABL* (red) and *BCR* (green) are present. (Reprinted from Debernardi, S., Lillington, D., Young, B.D. 2004. Understanding cancer at the chromosomal level: 40 years of progress. *European Journal of Cancer* 40:1960–1967. Copyright © 2004, with permission from Elsevier.) (Upper inset) The structure of imatinib mesylate, a tyrosine kinase inhibitor.

float in the culture medium, as they lose the attachments that ensure that they bind to the substrate as an ordered monolayer (see Fig. 17.1). The transformed cells become immortal, instead of having a finite lifespan. Tumor viruses are of two distinct types: those with DNA genomes (e.g. papilloma and adenoviruses, see Section 3.5) and those with RNA genomes (retroviruses, see Section 3.7).

DNA tumor viruses

More than 40 years ago it was known that DNA viruses could induce tumors in animals and transform "nonpermissive" host cells. Because their genomes encode relatively few genes, viruses provide a simple genetic system to study transformation. There are two possible outcomes of infection of a cell by a DNA tumor virus, a productive infection or transformation (Fig. 17.12). The outcome depends on the kind of infected cell and other less well-defined conditions. From the viral point of view, cells are either "permissive" or "nonpermissive." In permissive cells, all parts of the viral genome are expressed. This leads to viral replication, cell lysis, and cell death. In cells nonpermissive for replication, the viral DNA generally becomes stably integrated into the cell chromosomes (usually but not always) at random sites, and

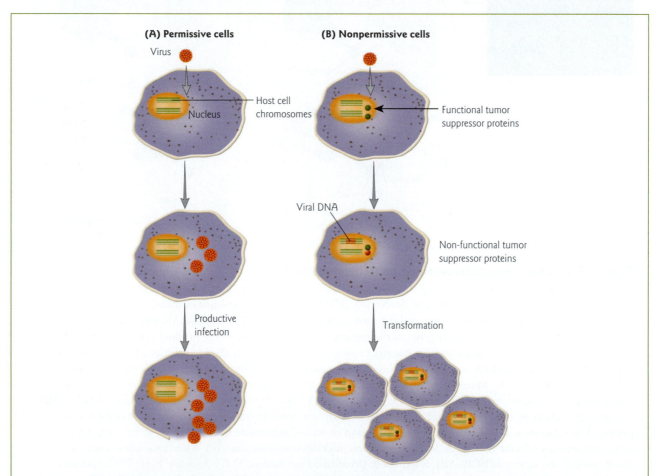

Figure 17.12 Possible outcomes of DNA tumor virus infection. (A) When a virus (red symbol) enters a permissive cell, virus components are produced and viral particles assemble. A productive infection results in lysis, cell death, and release of progeny viruses. (B) When a virus enters a nonpermissive cell, the viral DNA (red bar) may become stably integrated in the host cell genome (green bar). The cells undergo transformation and begin to proliferate. Cellular transformation is generally the result of interaction between viral-encoded proteins (red circles) and the host cell proteins (green circles) that inhibit their normal tumor suppressor function.

only part of the viral genome is expressed. Because viral structural proteins are not expressed, no progeny virus is released and the infection is abortive. For example, simian virus 40 (SV40) replicates and causes cell lysis in permissive African green monkey cells, whereas it transforms nonpermissive rodent cells.

In most cases, cellular transformation by DNA tumor viruses has been shown to be the result of protein–protein interactions. Proteins are encoded by the DNA tumor viruses that have no cellular counterpart, but are essential for the initiation and maintenance of the transformed state and tumorigenesis. The viral gene-encoded proteins are seen as foreign by animals harboring these tumors, and the animals produce antibodies directed against these viral antigens. For this reason, these viral antigens were termed large (T) and small (t) tumor antigens. These tumor antigens interact with cellular proteins, primarily tumor suppressor proteins. This interaction effectively sequesters the tumor suppressor proteins away from their normal functional locations within the cell. It is the loss of their normal suppressor functions that results in cellular transformation.

There are a number of DNA tumor viruses for which there is strong evidence of a role in human cancers. Hepatocellular (liver) carcinoma, which is one of the world's most common cancers, is linked to the hepatitis B virus. Human papilloma viruses are found associated with penile, uterine, and cervical carcinomas (Disease box 17.3). Epstein–Barr virus, a herpesvirus, is most commonly known for being the agent of infectious mononucleosis or "mono;" however, it is also strongly linked with certain types of cancer. Unlike some DNA tumor viruses, herpesviruses do not integrate into the host cell genome, but they can still induce chromosomal breaks. Epstein–Barr virus is causally associated with Burkitt's lymphoma (see Table 17.3) in Africa, nasopharyngeal cancer in China and South East Asia, B cell lymphomas in immune-suppressed individuals (such as in organ transplantation or AIDS), and Hodgkin's lymphoma. Why this virus causes a benign disease in some populations but malignant disease in others is unknown.

RNA tumor viruses (retroviruses)

In 1911 Peyton Rous described a sarcoma (connective tissue tumor) in chickens caused by an RNA tumor virus that later became known as the Rous sarcoma virus (RSV). He was awarded the Nobel Prize for this work in 1966. RNA tumor viruses are common in chickens, mice, and cats but rare in humans. Study of the oncogenes carried by retroviruses has led to major advances in understanding of the molecular basis of cancer. However, most human cancers are probably not the result of a retroviral infection. Human immunodeficiency virus 1 (HIV-1), the causative agent of AIDS, is associated with an increased risk of developing several cancers, especially Kaposi's sarcoma and non-Hodgkin's lymphoma. Human T cell leukemia virus 1 (HTLV-1) is sexually transmitted and linked to a type of adult T cell leukemia that is found in some Japanese islands, the Caribbean, Latin America, and Africa. Retroviruses can transform cells they infect by either of two main mechanisms: introduction of an oncogene or promoter/enhancer insertion.

Retroviral introduction of an oncogene   When a retrovirus infects a cell its RNA genome is converted into DNA by the viral-encoded reverse transcriptase. The DNA then integrates into the genome of the host cell where it is duplicated during cell division. At the ends of the retroviral genome there are highly efficient promoter sequences termed long terminal repeats (LTRs). The LTRs promote transcription of the viral DNA leading to the production of new virus particles. With some frequency, the integration process leads to rearrangement of the viral genome and the incorporation of a portion of the host genome into the viral genome (Fig. 17.13A). Occasionally, the retrovirus acquires a gene from the host that is normally involved in cellular growth control. After the gene is acquired by the virus, it is subject to mutation. Such changes include amino acid substitutions or deletions that give different translation products. As a result, the cellular homologs of viral oncogenes are not identical to their corresponding viral oncogene. Because of the mutation of the cellular gene during the process, as well as the gene being overexpressed due to its association with the retroviral LTRs, the oncogene confers a growth advantage to the infected cells. The end result of this process is unrestricted cellular proliferation leading to tumorigenesis.

Many retroviruses lose part of their genome during such rearrangements. This has at least two potential consequences: (i) the protein encoded by the oncogene is often part of a fusion protein with other virally

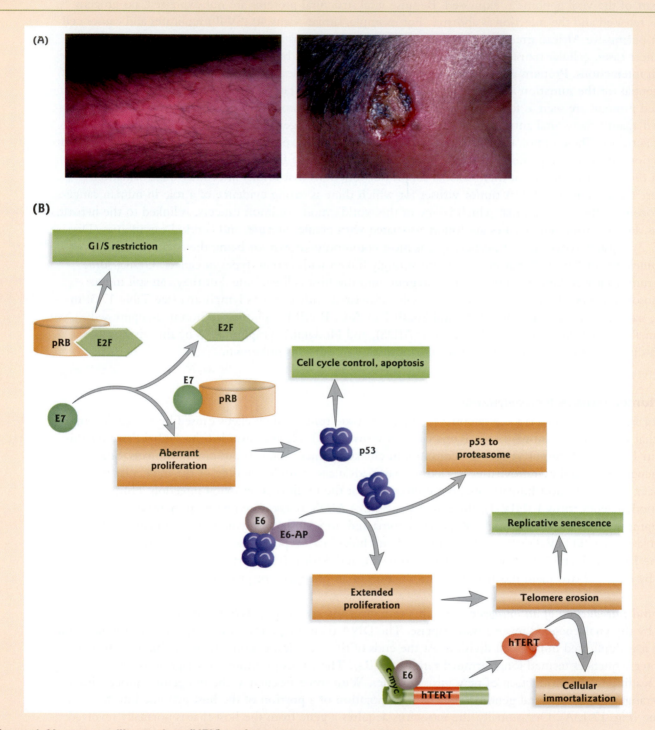

Figure 1 Human papilloma virus (HPV) and cancer. (A) The development of epidermodysplasia verruciformis. (Left) Numerous flat warts on the arm of a patient. (Right) squamous cell carcinoma of the face developed from epidermodysplasia verruciformis lesions. (Photographs courtesy of Daniel Wallach, M.D., Hôpital Tarnier-Cochin, Paris, France.) (B) Schematic outline of HPV-induced carcinogenesis. Inactivation of the pRB and p53 tumor suppressor pathways and expression of the catalytic telomerase subunit hTERT are a subset of the steps that have been shown to be necessary for the generation of fully transformed human epithelial cells *in vitro*. See text for details. (Modified from Münger, K., Baldwin, A., Edwards, K.M. et al. 2004. Mechanisms of human papillomavirus-induced oncogenesis. *Journal of Virology* 78:11451–11460.)

Human papilloma virus (HPV) and cervical cancer

DISEASE BOX 17.3

Papilloma viruses are small nonenveloped viruses that contain double-stranded DNA genomes of approximately 8000 bp. They are widely distributed throughout the animal kingdom. Papilloma viruses are wart-causing viruses that infect basal epithelial cells – the only cell layer in an epithelium that is actively dividing. Warts are usually benign but they can convert to malignant carcinomas (tumors of epithelial tissue). Papillomas may be the cause of approximately 10% of all cancers worldwide and 16% of female cancers. Epidemiological data strongly suggest that the viruses are the cause of the cancers. However, the fact that a virus is usually found in association with a tumor does not prove that the transformation of the cells is the result of the presence of the virus. The vital experiment, done in many animal systems, would be to inject the virus purified from a tumor into a human and see if the tumor redevelops. For obvious ethical reasons, that critical experiment has not been done.

Approximately 200 different HPVs have now been characterized. Individual viruses are designated as high risk or low risk according to their tendency to promote cancer. Most HPVs are low risk and produce localized benign warts that do not become cancerous even if left untreated. HPV-5 and HPV-8 are classified as high risk as they are associated with the development of epidermodysplasia verruciformis. This is a very rare skin condition that can progress from flat wart-like cutaneous lesions in early childhood to skin cancers later in life (Fig. 1A). Low-risk HPVs such as HPV-6 and HPV-11 cause genital warts (condyloma accuminata), whereas high-risk HPVs can cause lesions that progress to invasive squamous cell carcinoma. The vast majority of human cervical cancers are associated with high-risk HPV

infections. HPV-16 is by far the most prevalent, followed by HPV-18, HPV-31, and others.

HPV-mediated cervical cancer

One of the key events of HPV-induced cervical cancer is integration of the high-risk HPV genome into a host chromosome. Integration leads to loss of expression of some *HPV* genes; however, there is continued expression of the viral *E6* and *E7* genes (Fig. 1B). The E6 and E7 viral proteins inactivate the p53 and pRB tumor suppressors, respectively. E7 interacts directly with pRB, blocking it from carrying out its function in cell cycle control. E6 forms a complex with the cellular E6-AP protein, and induces ubiquitinylation and rapid proteasomal degradation of p53. In addition, viral E6 can activate the transcription of the gene encoding the reverse transcriptase subunit of telomerase (hTERT). This may occur by interaction with c-*myc* and binding to c-*myc* responsive elements in the *hTERT* gene promoter. Thus, the combined expression of high-risk HPV E6 and E7 proteins in cervical cancers causes inactivation of the pRB and p53 tumor suppressor pathways and induces telomerase activity. Other studies have shown that expression of SV40 large T antigen, SV40 small t antigen, hTERT, and the Ha-*ras* oncogene are the minimum requirements for generating fully transformed human epithelial cell lines *in vitro*. SV40 large T antigen inactivates p53 and pRB, whereas SV40 small t antigen interacts with and inhibits protein phosphatase 2A. E6 and E7 thus provide a subset of the minimum requirements for transformation. Additional oncogenic events are necessary in E6/E7-expressing cells to yield full transformation *in vivo* and *in vitro*.

encoded amino acids attached; and (ii) the virus may require a helper virus to replicate and bud from the host cell because it cannot make all of its own products. For example, the avian erythroblastosis virus (AEV) became replication-defective as a result of rearrangement of the viral genome. v-*erbA*, one of the oncogenes carried by AEV, encodes a fusion protein of part of the viral Gag polyprotein and a highly mutated homolog of the thyroid hormone receptor (c-*erbA*). The fusion protein acts as a dominant negative repressor of the thyroid hormone receptor, a ligand-dependent transcription factor (Fig. 17.13A). v-ErbA promotes cancer by competition with the thyroid hormone receptor for DNA-binding sites and/or auxiliary factors involved in the transcriptional regulation of thyroid hormone-responsive genes, and by mislocalization of the normal receptor to the cytoplasm.

Retroviral promoter/enhancer insertion   Some retroviruses do not carry oncogenes but can still promote cancer by the mechanisms of promoter or enhancer insertion. With some frequency, the integration of the

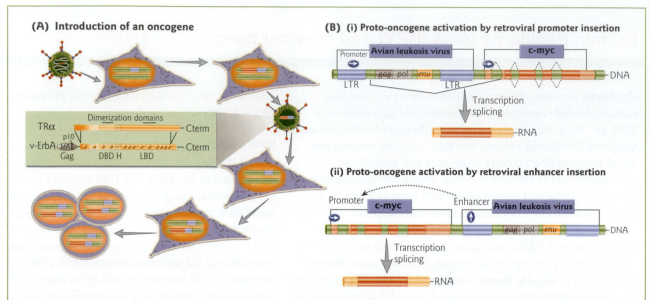

Figure 17.13 Transformation of cells by retrovirus infection. (A) An example of the introduction of an oncogene. The avian erythroblastosis virus (AEV) genome (red bar) acquired a portion of the coding region for the thyroid hormone receptor α (TRα) (blue bar), which was then subject to mutation over time. The resulting oncogene, v-*erbA*, encodes a fusion protein that acts as a dominant negative repressor of TRα. v-*erbA* promotes tumor formation in conjunction with v-*erbB*, a second oncogene acquired by AEV (see Table 17.2). During this process, AEV becomes replication defective and requires a helper virus for a productive infection. (Inset) Schematic diagram (not to scale) of the oncoprotein v-ErbA illustrating its homology with TRα. v-ErbA differs from TRα by fusion at its C-terminus (Cterm) with a retroviral Gag sequence and by several amino acid substitutions (circles), along with deletions at both the C- and N-termini. DBD, DNA-binding domain; LBD, ligand-binding domain; MA, matrix association domain; p10, domain containing a CRM1–dependent nuclear export sequence (see Section 11.9). (B) Example of promoter/enhancer insertion. The avian leukosis virus promotes tumor formation when it integrates either upstream or downstream of the c-*myc* proto-oncogene. The c-*myc* gene is inappropriately activated by the strong retroviral promoter (i) or enhancer (ii) in the long terminal repeat (LTR) region.

retrovirus genome leads to the placement of the LTRs close to a host gene that is normally involved in cellular growth control. If the gene is expressed at an abnormally elevated level under control of the LTR, this can result in cellular transformation. For example, the avian leukosis virus causes tumor formation in birds when it integrates either upstream or downstream of the c-*myc* proto-oncogene (see section on the c-*myc* transcription factor, above). The c-*myc* gene then comes under the influence of the strong viral promoter or enhancer, respectively, which leads to its overexpression (Fig. 17.13B).

Chemical carcinogenesis

Cancer was first experimentally produced by the application of coal tar to the ears of rabbits in 1915. Today, about 200 different chemical compounds and mixtures are classified as human carcinogens. These include such compounds as vinyl chloride in some plastics, asbestos, and polycyclic aromatic hydrocarbons (PAHs) such as benzo(*a*)pyrene present in cigarette smoke. Environmental exposures make a substantial contribution to sporadic human cancers. Eighty percent of cancer deaths in Western industrial countries can be attributed to factors such as tobacco, alcohol, diet, infections, and occupational exposure, with diet (35%) and tobacco (30%) as the major contributors.

When chemical carcinogens are taken up by cells they are often metabolized, and the resulting metabolic products are either excreted or retained by the cell. Inside the cell, carcinogens or their metabolic products can either directly or indirectly affect gene expression (Fig. 17.14). Some carcinogens act by "genotoxic"

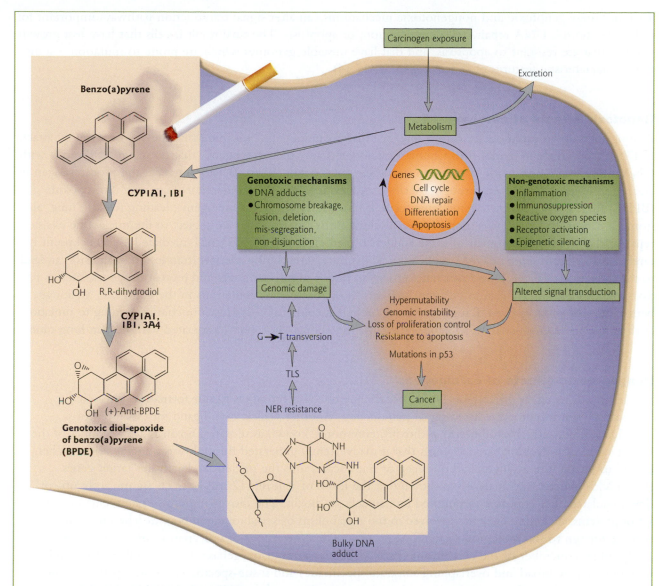

Figure 17.14 Overview of genotoxic and nongenotoxic effects of carcinogens. After chemical carcinogens are taken up by cells, metabolic products that are retained by the cell can either directly or indirectly affect gene expression. Benzo(a)pyrene, a carcinogen present in cigarette smoke, is converted to a genotoxic diol-epoxide through a series of enzymatic steps mediated by members of the cytochrome P450-dependent mono-oxygenase (CYP) family (CYP1A1, CYP1B1, and CYP3A4). The genotoxic diol-epoxide of benzo(a)pyrene forms a bulky DNA adduct that is resistant to nucleotide excision repair (NER). During DNA replication, translesion synthesis (TLS) across the adduct causes G → T transversions, leading to mutations in tumor suppressor genes, such as *p53*. Together, genotoxic and nongenotoxic mechanisms can alter signal transduction pathways, leading to cancer. (Adapted by permission from Luch, A. 2005. Nature and nurture – lessons from chemical carcinogenesis. *Nature Reviews Cancer* 5:113–125. Copyright © 2005.)

mechanisms. For example, benzo(a)pyrene can induce formation of DNA adducts that interfere with replication and transcription. The carcinogen can also induce chromosomal changes such as breakage, fusion, deletion, mis–segregation, and nondisjunction during cell division. Other carcinogens act by nongenotoxic mechanisms. Nongenotoxic mechanisms include induction of inflammation, immunosuppression, formation of reactive oxygen species, activation of receptors for arylhydrocarbon or estrogen, and epigenetic silencing.

Together these genotoxic and nongenotoxic mechanisms can alter signal transduction pathways important for cell cycle control, DNA repair, cell differentiation, or apoptosis. The final result is cells that have lost growth control, that are resistant to apoptosis, and that have unstable genomes which are prone to mutation – some of the characteristic features of cancer cells (Fig. 17.14).

Genotoxic effects of carcinogens

Among the known carcinogens, only a few are "direct carcinogens," meaning that they can covalently attach to DNA (e.g. ethylene oxide). Most human carcinogens are "procarcinogens" that do not react directly with DNA, and require enzymatic conversion into their carcinogenic form. Benzo(*a*)pyrene, the procarcinogen in cigarette smoke, is converted through a series of reactions to the genotoxic benzo(*a*)pyrene diol-epoxide (BPDE) (Fig. 17.14). In this carcinogenic form, it binds to DNA and is mutagenic, leading to skin, lung, and stomach cancer. Bulky adducts such as those resulting from diol-epoxides are subject to the nucleotide excision repair pathway. However, repair of these adducts is inefficient and the p53/p21-mediated G1 checkpoint is rarely activated. Subsequent DNA replication across unrepaired bulky adducts (translesion synthesis) can lead to the induction of G → T base pair transversions in tumor suppressor genes and proto-oncogenes, such as *p53*. Sixty percent of lung cancers in cigarette smokers show inactivating mutations in the *p53* gene. Strikingly, benzo(*a*)pyrene induces G → T transversions in bronchial epithelial cells grown in culture, leading to mutations at three specific codons of the *p53* gene – the same mutational "hot spots" associated with human lung cancer.

Nongenotoxic effects of carcinogens

In addition to its effect directly on DNA, benzo(*a*)pyrene also promotes tumor formation through arylhydrocarbon receptor-mediated signal transduction. Another well-known ligand for this receptor is TCDD (2,3,7,8,-tetrachlorodibenzo-*p*-dioxin), commonly known as dioxin. Dioxin is a by-product of the manufacture of polychlorinated phenols that is inadvertently generated through waste incineration. When bound to ligand, the arylhydrocarbon receptor translocates into the nucleus and activates target genes (Fig. 17.15). Microarray analysis has shown that the levels of at least 112 mRNAs are significantly up- or downregulated after exposure of cultured human cells to dioxin. The dioxin-responsive genes encode a diverse set of proteins, including enzymes involved in the metabolism of xenobiotic compounds (chemical compounds that are foreign to the biological system). Most enzymatic conversions of procarcinogens to carcinogens are catalyzed by a member of the cytochrome P450-dependent mono-oxygenase (CYP) family. CYP family members display broad and overlapping substrate specificities and tissue-specific expression patterns. For example, CYP1A1 is the most important form in human lung, while CYP1A2, CYP2A6, CYP2E1, or CYP3A4 are mainly expressed in liver, and CYP1B1 is expressed in almost all organs except liver and lung.

17.3 Gene therapy

During the past decade, the idea of using gene therapy to attempt to treat human diseases has moved from the lab bench to the clinic. The principle of somatic cell therapy is that a malfunctioning gene is replaced or compensated for by a properly functioning gene in somatic cells of a patient. Because only somatic cells are genetically altered, the treatment is not heritable and only affects the individual patient. Germline gene therapy would entail genetic modification of gametes or embryos, such that induced changes are passed on to the next generation. There are many ethical considerations and such therapy has not been attempted.

The first transfer of a marker gene for neomycin resistance (*neo*r) to a patient was carried out in 1989. This trial was followed a year later by the first USA-approved gene therapy protocol for an inherited immune system deficiency. Now, over 900 clinical protocols are in progress around the world, with about two-thirds for the treatment of various forms and stages of cancer. A wide variety of other trials are ongoing as well, most of which fall under the headings of cardiovascular disease, infectious disease (nearly all HIV-related), and inherited autosomal recessive disorders caused by defects in a single gene, such as cystic fibrosis.

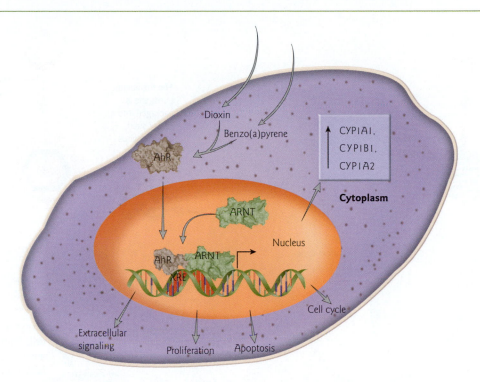

Figure 17.15 Tumor promotion through arylhydrocarbon receptor (AhR)-mediated signal transduction. Chemical compounds such as dioxin or benzo(*a*)pyrene result in tumor promotion through arylhydrocarbon receptor (AhR) mediated signal transduction. Binding of dioxin or benzo(*a*)pyrene to AhR leads to activation and translocation of the complex into the nucleus. After heterodimerization with the AhR nuclear translocator (ARNT), the complex binds to xenobiotic-responsive elements (XREs) and induces the expression of a variety of different genes involved in carcinogen metabolism, including cytochrome P450-dependent mono-oxygenase (CYP) forms 1A1, 1B1, and 1A2. It also changes the expression pattern of genes involved in extracellular signaling, proliferation, apoptosis, and cell cycle control. (Adapted by permission from Luch, A. 2005. Nature and nurture – lessons from chemical carcinogenesis. *Nature Reviews Cancer* 5:113–125. Copyright © 2005.)

Vectors for somatic cell gene therapy

There are two main strategies for somatic cell gene therapy: *in vivo* and *ex vivo*. For *in vivo* techniques the target gene is delivered to the desired cell type within the patient. For *ex vivo* strategies, cells are manipulated out of the body and then reimplanted into the patient (Fig. 17.16). Target cells need to be removable, returnable, long-lived, hardy, and accessible (e.g. by aerosol delivery, injection into the bloodstream, or surgical removal and manipulation). Often stem cells are used because they are actively dividing. However, they tend to be rare and not very tolerant of manipulation.

Gene therapy has proved disappointing time and time again in a large number of clinical studies where it has done absolutely nothing, or failed to outperform standard therapies, despite near miraculous cures achieved in animal models. Most researchers blame the current lack of success on difficulties with vectors. Overall, there is a lack of efficiency of gene delivery and sustained expression, or a potential host immune reaction. Most vectors are not targeted to a particular cell type, although this is an area of much research interest. The majority of gene therapy trials use viral vectors, but use of liposomes and naked DNA has been explored (Table 17.4). The three main viral vectors are retroviruses, adenoviruses, and adeno-associated viruses. Other viral vectors include the lentiviruses and poxviruses.

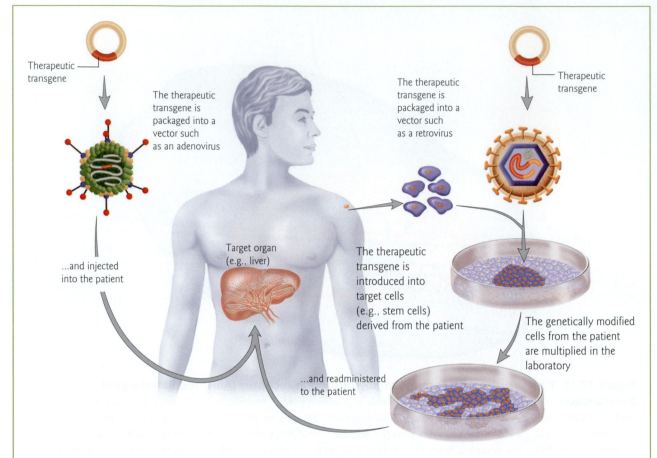

Figure 17.16 Two main strategies for somatic cell gene therapy. For *in vivo* techniques the target gene is delivered directly to the desired cell type within the patient. For *ex vivo* techniques, cells are removed from the body for delivery of the target gene, and then reimplanted into the patient.

Liposome vectors

The advantages of liposomes are that they are nontoxic and nonimmunogenic; however, gene transfer is highly inefficient and expression is transient (Table 17.4). Liposomes are tiny lipid spheres in which the phospholipid bilayer entraps DNA. The liposome bilayer can be constructed to enhance binding to certain cell types, e.g. through antibodies, peptides, or carbohydrate moieties. The plasmid–liposome complexes fuse with the plasma membrane and enter the cell. Most of the introduced material is not directed to the nucleus, and instead is shunted to the endosomes – cytoplasmic organelles where the material is degraded. Some make it to the nucleus, but the DNA is not integrated into the host chromosome (Fig. 17.17).

Retrovirus vectors

Some 34% of current gene therapy clinical trials involve retroviruses (Focus box 17.3). Retrovirus vectors integrate stably into host chromosomes, but they have a limited insert size which is not able to accommodate all the regulatory regions within a gene. Although retroviral vectors were long thought to have entirely random integration, in fact, recent studies show they are not entirely "random." There is a tendency to insert into genes more often than other regions of DNA. Furthermore, many somatic cells are differentiated and not actively dividing, and thus are not amenable to retrovirus infection (see Table 17.4).

Table 17.4 Gene therapy vectors.

Vector	Advantages	Disadvantages
Retrovirus	Enters cells efficiently Few immunity problems Integrates stably into host chromosomes	Hard to produce Limited insert size (~7 kb) Integration may cause insertional mutations; preferentially inserts near active genes Infects only dividing cells
Lentivirus	Same as retrovirus, but can infect both dividing and nondividing cells	Same as retrovirus
Adenovirus	Enters cells efficiently Larger capacity for foreign genes (> 30 kb) Produces high expression of therapeutic genes Infects both dividing and nondividing cells	Does not provide long-term gene expression due to lack of integration (extrachromosomal) Very immunogenic, can be toxic
Adeno-associated virus	Causes no known human diseases Does not produce an immune response	Small capacity for foreign gene sequences Does not provide long-term gene expression due to lack of integration (extrachromosomal) Rarely integrates into chromosome; but preferentially inserts near active genes
Herpesvirus	High expression of therapeutic genes Targets nondividing nerve cells	Hard to produce Possible immune response
Liposomes	Nonpathogenic, no immunity problems No limit to size of foreign gene	Much less efficient at transferring genes to cells Low rate of stable integration and long-term gene expression
Biolistics (gene gun and naked DNA)	Same as liposome-mediated transfer Promising as a vaccination method	Limited to dermal tissue Low rate of stable integration Less efficient

Adenovirus vectors

Adenoviruses account for nearly 27% of current gene therapy trials; however, these vectors bear the stigma of causing the first fatality directly attributed to gene therapy (Focus box 17.4). Adenoviridae are a family of large nonenveloped double-stranded DNA viruses that infect vertebrates and cause generally mild diseases of the respiratory tract, conjunctiva and cornea, gastrointestinal tract, and genitourinary tract. Adenoviruses were first isolated from adenoids of infected patients, hence the name. Due to the ability of adenoviruses to infect a wide variety of dividing and nondividing cells, they are one of the most efficient vectors for gene transfer into mammalian cells (see Table 17.4). In addition, they have a larger capacity for foreign genes. The major disadvantage is that because the viral DNA does not integrate, repeated introduction of the vector may be required. In turn, repeated introduction of viral proteins can lead to a potentially fatal patient immune response.

Adeno-associated virus vectors

Adeno-associated virus (AAV) is a parvovirus with a single-stranded DNA genome. The single-stranded virion DNA is converted to a double-strand DNA template for transcription and replication. Two important

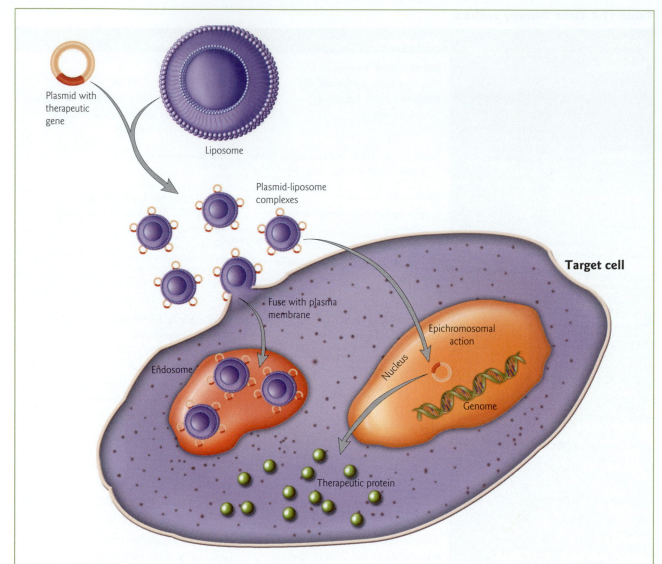

Figure 17.17 Liposome-mediated gene transfer. Complexes are formed between liposomes and a plasmid carrying the therapeutic gene. The complex fuses with the plasma membrane and enters the cell. Most of the plasmid DNA is degraded in endosomes. Some plasmid DNA enters the nucleus and is expressed epichromosomally (not integrated into the host cell genome).

features that initially attracted the attention of gene therapy researchers were: (i) AAV does not appear to cause any disease in humans; and (ii) it tends to remain inactive in the absence of a helper virus, most commonly adenovirus (see Table 17.4). One of the viral replication proteins (Rep) mediates integration of the natural virus into a specific site on human chromosome 19. Most AAV-based vectors do not include the *rep* gene because of the potential for a patient immune response to the viral protein. Recombinant AAV-based vectors (rAAV) lacking *rep* mainly function episomally. However, they can still integrate into host cell chromosomes, albeit inefficiently. At first it was thought that integration in the absence of *rep* was entirely random. A 2003 study by Hiroyuki Nakai, Mark A. Kay and colleagues showed this is not the case. Liver cells were extracted from mice whose DNA contained the rAAV vector and then the DNA around the vector was sequenced. The sequences were compared with the mouse genome sequence database to see whether they matched a known gene. What was found was that when rAAV does integrate into a host cell

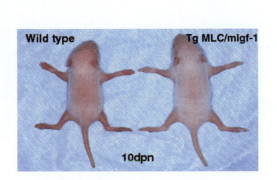

Figure 17.18 "Schwarzenegger mice". Transgenic mice were generated with a rat insulin-like growth factor 1 (mIgf-1) cDNA driven by skeletal muscle-specific regulatory elements from the rat myosin light chain (MLC) locus. The transgenic mice had twice as much muscle as a normal mice and built muscle without exercising. The photograph shows normal (Wild type) and transgenic (Tg MLC/mIgf-1) mouse pups at 10 days after birth (10 dpn). (Reprinted by permission from Nature Publishing Group and Macmillan Publishers Ltd: Musarò, A., McCullagh, K., Paul, A. et al. 2001. Localized Igf-1 transgene expression sustains hypertrophy and regeneration in senescent skeletal muscle. *Nature Genetics* 27:195–200. Copyright © 2001.)

chromosome, ~72% of the time it integrates into active genes. These findings will need to be considered in risk/benefit considerations until the consequences of vector integration are more fully understood.

Enhancement genetic engineering

In theory, the same techniques used to treat genetic diseases could be used for "enhancement" genetic engineering in which a gene is inserted to "improve" or alter a characteristic or a complex trait that depends on many genes plus extensive interactions with the environment. Along these lines, transgenic mice were engineered in 2001 by H. Lee Sweeney and colleagues at the University of Pennsylvania with an extra copy of the insulin-like growth factor 1 (*IGF-1*) gene. The transgenic mice had twice as much muscle as normal mice, lived longer, and recovered more quickly from injuries. They built muscle without exercising and seemed to defy the aging process. Dubbed the "Schwarzenegger mice" they immediately became a media sensation (Fig. 17.18). There was widespread speculation about athletes cheating in the future through use of "gene therapy" techniques instead of doping with performance-enhancing drugs such as testosterone and anabolic steroids. A step closer to this, Sweeney and colleagues used AAV to insert the *IGF-1* gene directly into muscle in rats, producing animals with 20–30% more muscle mass than controls. Associating genes and their variants with athletic ability has proven difficult in humans. However, particular variants of two genes, α-actinin and angiotensin-converting enzyme (ACE), have been found to be associated with elite athletic performance. A variant of α-actinin is found in sprinters, whereas the ACE variant is more commonly found in endurance runners.

Gene therapy for inherited immunodeficiency syndromes

The first ever clinical trial for treatment of an inherited disorder was initiated in September 1990. The gene therapy protocol was designed to treat a 4-year-old girl with adenosine deaminase (ADA) deficient severe combined immunodeficiency (SCID). ADA is an essential enzyme involved in metabolism of adenine and guanine. T lymphocytes are destroyed by a lack of ADA, resulting in a predisposition to recurrent, persistent infection. Gene therapy has also been used to treat an X chromosome-linked form of severe combined immundeficiency (SCID-X1). This is a lethal condition resulting from mutations in the gene encoding the γc-cytokine receptor, a receptor that responds to interleukins. The disorder leads to a deficiency in T lymphocytes and natural killer cells, due to a block in differentiation. Since 1999, gene therapy has helped to

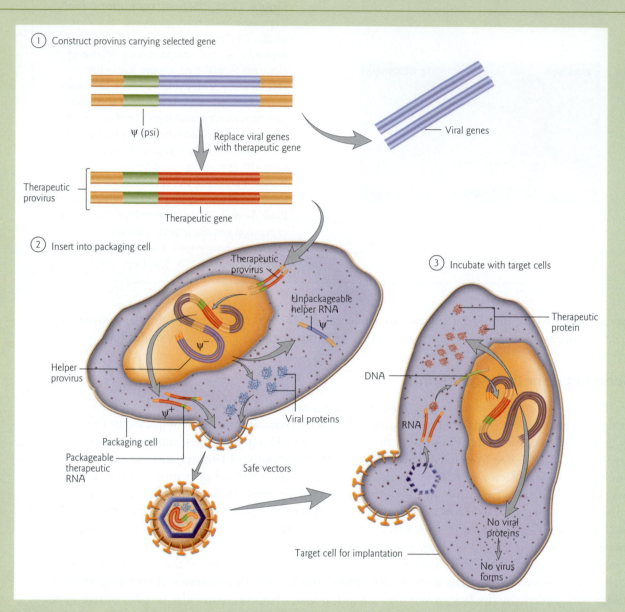

Figure 1 Retroviral-mediated gene transfer. (1) A provirus DNA is constructed in which the viral genes are replaced with a therapeutic gene. (2) The therapeutic provirus is then inserted into a packaging cell to assemble a "safe vector." The packaging cell line contains a helper provirus that codes for viral proteins but lacks the psi sequence ($\psi-$) required for inclusion of RNA in a viral particle. Upon entry into the packaging cell, the therapeutic provirus integrates into the cell genome and is transcribed. Because the therapeutic RNA is $\psi+$, the therapeutic RNA can be assembled into a viral particle. The "safe vectors" that are assembled in the packaging cells include the reverse transcriptase, the therapeutic RNA, and other viral proteins, but do not contain any genomic viral RNA. (3) After purification, the safe vectors are incubated with target cells from the patient *ex vivo*. The viral particles enter the cells, the RNA genome is reverse transcribed into DNA and is integrated into the genome, and therapeutic proteins are expressed. Since no viral proteins are encoded by the therapeutic provirus, the virus cannot reproduce or form infectious particles. The target cells are grown for a few days under selective conditions, and then reimplanted into the patient.

Retroviral-mediated gene transfer: how to make a "safe vector"

FOCUS BOX 17.3

The original method for gene transfer in gene therapy was developed in the early 1980s. The strategy was to capitalize on the native ability of retroviruses to enter cells highly efficiently. In this method, a construct of provirus DNA is engineered in which the viral genes for the virus core, envelope, and reverse transcriptase are replaced with a therapeutic gene (Fig. 1). Typically a selective marker gene encoding antibiotic resistance is also included in the construct. After construction in the lab, the provirus DNA is then electroporated into a special cell line for packaging into a "safe vector." The packaging cell line (often derived from mouse fibroblasts) contains a helper provirus that codes for viral proteins but lacks the psi sequence required for inclusion of RNA in a viral particle. Upon entry into the packaging cell, the therapeutic provirus integrates into the cell genome and is transcribed. Because the therapeutic RNA transcript has the psi sequence, the therapeutic RNA can be assembled into a viral particle. The "safe vectors" that are assembled in the packaging cells include the reverse transcriptase, therapeutic RNA, and other viral proteins, but do not contain any genomic viral RNA.

After purification, the safe vectors are incubated with target cells from the patient *ex vivo*. The viral particles enter the cells and the RNA genome is reverse transcribed into DNA. Because the viral DNA can only reach the host chromosomes when the nuclear membrane breaks down during cell division, integration of the provirus requires actively dividing cells. After integration into the genome of the patient's cell and transcription of the therapeutic mRNA, therapeutic proteins are produced. Since no viral proteins are encoded by the therapeutic provirus, the virus cannot reproduce or form infectious particles. The target cells are grown for a few days under conditions of antibiotic selection for neomycin resistance, and then re-implanted into the patient (Fig. 1).

Similar techniques are used to prepare "safe" adenovirus and adeno-associated virus (AAV) vectors. Prior to clinical trials, protocols are tested in cell culture and in animal models, for example mice and primates. Researchers check for pathology (diseased tissue), immune response, and malignancy (tumor formation). One concern is that integration of the gene (in the case of retroviral and AAV vectors) could alter the function of neighboring genes, or activate an oncogene. Another question addressed is whether the introduced "gene" is expressed appropriately. The retroviral provirus can only hold an insert size of < 7 kb, so typically a cDNA clone of the gene of interest is used under control of a viral promoter, instead of a genomic clone that would contain all the *cis*-regulatory elements. Other potential concerns are whether the "safe" viral vector could recombine with endogenous host or viral sequences to produce replication-competent, infectious virus, or whether exposure to viral proteins (in the case of adenovirus and AAV) will cause a toxic immune response.

restore the immune systems of at least 22 children with either ADA-SCID or SCID-X1. In the wake of Jesse Gelsinger's death (Focus box 17.4), for a brief moment, these successes made gene therapy seem full of promise.

ADA-SCID

The traditional treatment for ADA-SCID is bone marrow transplant and weekly injections of ADA. Gene therapy was explored as a treatment option because transplants often fail and enzyme therapy is ineffectual in many patients. One trial was started in 1990, followed by another in 1991. Since no side effects were observed, an additional 11 children were enrolled in the gene therapy trial in 1996. An *ex vivo* approach was used in which the ADA cDNA was introduced into T lymphocytes by retroviral-mediated gene transfer. Gene-corrected T lymphocytes were then infused into the patients.

At age 17, the first girl treated was continuing to lead a normal lifestyle and had a much stronger immune system. She was able to attend public school, had no more than the average number of infections, no side effects, and an increased number of T cells. T cells do not have the longevity of stem cells so it was thought that repeat treatments would be required. This turned out not to be the case: ADA gene expression in T cells persisted for at least 13 years. Although she was still receiving ADA injections, she was markedly better

The first gene therapy fatality

Prior to September 17, 1999 the general consensus was that somatic cell gene therapy for the purpose of treating a serious disease was an ethical therapeutic option. In the case of life-threatening or terminal disease it was felt that the potential benefits outweighed the costs and risks. When Jesse Gelsinger died from a hyperimmune response against the large quantity of vector needed in untargeted therapy, this general consensus was called into question.

Jesse Gelsinger was a relatively healthy 18-year-old with a mild form of partial ornithine transcarbamylase (OTC) deficiency that could be controlled through diet and drugs, although the regimen was very strict. The hallmark of this disease is the inability to break down dietary protein (nitrogen) correctly, resulting in a toxic build up of ammonia soon after birth. A gene therapy trial was begun on September 13, 1999 in an attempt to correct this disorder. Gelsinger was administered a high dose of recombinant adenovirus vector particles carrying the normal *OTC* gene via the portal vein, a blood vessel that feeds the liver. He was administered 38 trillion viral particles, the highest dose

ever given in a clinical trial. Even so, only 1% of transferred genes reached the liver and there was no significant gene expression in the liver. The viral particles invaded all other organs in his body examined, including the spleen, lung, thyroid, heart, kidney, testicles, brain, pancreas, lymph node, bone marrow, and bladder. Four days later, Gelsinger died from a systemic inflammatory response to the viral proteins that resulted in multiple organ failure.

In January 21, 2000 the US Food and Drug Administration (FDA) shut down five gene therapy trials at the University of Pennsylvania after finding "serious deficiencies" in the supervision and monitoring of this trial in 18 areas. For example, it was alleged that patients were not told of the deaths of two rhesus monkeys who received the experimental treatment, and that informed consent forms were not completed properly. Five years after Jesse's death, the US Department of Justice has reached a settlement with the researchers and their institutions, including substantial fines and clinical research restrictions.

than with ADA enzyme treatment alone. For the majority of other patients enrolled in the trial, it was not clear whether they were significantly helped by gene therapy. The continued administration of ADA during gene therapy was required to avoid the ethical dilemma of withholding or stopping a life-saving therapy to test an unknown treatment.

SCID-X1

In April 2000 two infants in France with SCID-X1, initially aged 8 and 11 months, were treated by gene therapy. Both infants were very ill, suffering from pneumonia, diarrhea, and skin lesions. Bone marrow samples were obtained and hematopoietic stem cells were treated *ex vivo* with a retroviral vector carrying the γc-receptor cDNA. Gene-corrected cells were then transfused back into the patients. No other treatment was provided or required. Initially the results were very encouraging, so the clinical trials were expanded. Then, two patients developed leukemia. One of these children has since died. In both cases, the retroviral vector had inserted near the *LMO2* proto-oncogene. *LMO2* encodes a transcription factor required for hematopoiesis. When overexpressed, it promotes leukemia (Fig. 17.19). Integration of the retroviral vector may have activated *LMO2*, cooperating with the γc-cytokine receptor to stimulate cell growth and proliferation.

Because of this disturbing development, in January 2003, 27 clinical trials in the United States were put on hold. In February 2004, it was decided that gene therapy for SCID-X1 could continue, but only for children who have had failed bone marrow transplants. In 2005, a third child developed leukemia, resulting in the French trials being halted again and the US Food and Drug Administration (FDA) suspending three SCID-X1 trials. The third leukemia case appears to involve multiple retroviral vector insertions, affecting *LMO2* and three other proto-oncogenes. So far the risk of cancer seems unique to SCID-X1. As of March 2005, gene therapy trials for ADA–SCID were being allowed to proceed.

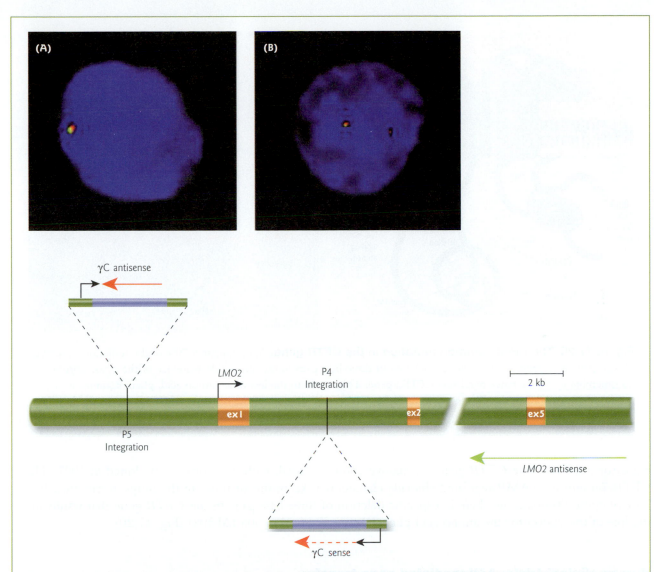

Figure 17.19 Insertional mutagenesis of the *LMO2* proto-oncogene in SCID-X1 gene therapy trials.
RNA fluorescence *in situ* hybdrization (FISH) analysis of the activated *LMO2* allele with single-stranded DNA probes. (A) An antisense LMO2 probe (green arrow, map) and sense γc-cytokine receptor probe (red dashed arrow, map) were hybridized to T cells from a patient (P4) who developed leukemia after gene therapy. Both probes detected transcription that originated at the *LMO2* promoter. (B) An antisense LMO2 probe and an antisense γc probe (red solid arrow, map) were hybridized to T cells from another patient (P5) who also developed leukemia. The antisense γc probe also detected transcription from the endogenous γc gene. DAPI (4′-6-diamidino-2-phenylindole) staining of DNA is shown in blue. ex, exon. (Reprinted with permission from Hacein-Bey-Abina, S., Von Kalle, C., Schmidt, M. et al. 2003. LMO2-associated clonal T cell proliferation in two patients after gene therapy for SCID-X1. *Science* 302:415–419. Copyright © 2003 AAAS.)

Cystic fibrosis gene therapy

Cystic fibrosis is a recessive genetic disease that affects one in 2000 Caucasians, but is less common in other ethnic groups. Individuals with cystic fibrosis have excessive salt loss in their sweat, thick sticky mucus in the air tubes, recurrent lung infections, irreversible lung damage, and often obstruction of the ducts of the pancreas. The disorder is due to a defect in the gene encoding the cystic fibrosis transmembrane conductance

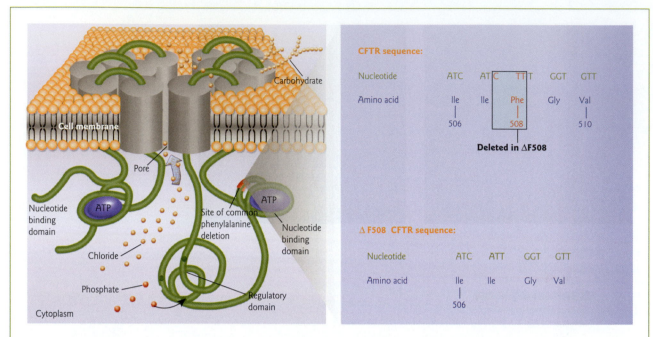

Figure 17.20 The most common mutation in the *CFTR* gene. Approximately 70% of the mutations in cystic fibrosis patients correspond to a specific deletion of three base pairs in the nucleotide sequence of the cystic fibrosis transmembrane conductance regulator (CFTR) gene. This results in the loss of an amino acid, phenylalanine, at position 508 (ΔF508) in the CFTR protein. (Inset) A schematic diagram of the structure of CFTR shows the location of the phenylalanine.

regulator (CFTR). The *CFTR* gene – a massive 250 kb in total, with 27 exons – was cloned in 1991. The CFTR protein is a cAMP-regulated chloride channel that keeps the air tubes in the lungs mucus-free. In 70% of cystic fibrosis cases, there is a specific deletion of three base pairs in the *CFTR* gene that results in the loss of the codon for the amino acid phenylalanine at position 508 (ΔF508) (Fig. 17.20).

Human clinical trials: AAV-mediated gene transfer

Although adenovirus- and liposome-mediated gene therapy were both effective in "curing" CFTR⁻/CFTR⁻ knockout mice, similar techniques proved ineffectual in cystic fibrosis patients. Adenovirus or liposome vectors carrying the *CFTR* cDNA were administered to the nasal airway epithelium or upper respiratory tract of patients by a nasal spray. Results were considered disappointing in both cases. Because of the immune response associated with adenovirus-mediated gene transfer, researchers have now turned to use of adeno-associated virus (AAV).

Recombinant adeno-associated virus serotype 2 (rAAV2) based gene therapy for cystic fibrosis has progressed through a series of preclinical studies and Phase I (vector safety) and Phase II (safety and efficacy) clinical trials. This agent has shown an encouraging safety profile, consistent levels of DNA transfer, and positive evidence of short-term clinical improvement in lung function in a placebo-controlled trial of aerosol administration. The use of the rAAV2-CFTR vector led to improved lung function for 30 days after treatment and demonstrated an excellent safety profile in the Phase II trial. Nasal cells harvested from gene therapy patients showed *CFTR* mRNA expression, and cAMP-activated chloride channel function.

The results of a study showing that the rAAV2 vector integrates itself more often into genes than it does into regions of DNA that do not contain genes has raised some concerns. This finding suggests that the vector could potentially cause inappropriate activation of an oncogene, similar to events that led to leukemia in SCID-X1 patients. Researchers are hoping that concerns are unwarranted, since AAV is different to retrovirus-based vectors. Retroviruses must insert themselves into DNA to work, but AAV integration is inefficient and occurs much less often. The persistence of AAV gene delivery vectors in tissues is largely attributable to episomal function.

HIV-1 gene therapy

HIV-1 infection and its associated disease, acquired immunodeficiency syndrome (AIDS), continue to be an expanding epidemic. Initially a disease primarily of male homosexuals, women now account for nearly half of the 40 million people living with HIV-1. For example, in sub-Saharan Africa, females account for 60% of those infected with HIV-1 and 75% of infected individuals are between the ages of 15 and 24. In the United States, the number of AIDS cases increased 15% among women and only 1% among men from 1999 to 2003. The majority of infections were due to heterosexual transmission (80%) or to injected drug use (19%). These same risk factors, especially injected drug use, have led to a 50% increase in HIV-1 infections in women in Asia and Central Europe over the past 2 years.

The development of drugs designed to delay progression of infection or to block replication of HIV-1 in infected individuals has been the subject of intense research efforts. So far, no drug treatments are totally effective, they have significant toxicity, and drug-resistant HIV-1 is increasingly frequent. The virus has extraordinary genetic diversity, making the creation of an HIV vaccine particularly challenging. Because of the inaccuracy of the replication machinery of the virus, new mutations are introduced into virtually every virion generated. It is estimated that as many as a billion new and unique viral particles can be created each day in an infected individual. Geographically, distinct HIV subtypes are clustered. The AIDS epidemic in the western hemisphere and Western Europe is caused by one subtype, while the epidemic in sub-Saharan Africa and the Indian subcontinent is due to another subtype. Investigation into additional therapeutic approaches, such as anti-HIV-1 gene therapy is thus of great interest. Designing such therapies requires in-depth understanding of the HIV-1 life cycle (Focus box 17.5) to determine potential targets.

HIV-1 life cycle

FOCUS BOX 17.5

HIV-1 is a lentivirus belonging to the retrovirus family. The virus contains two plus-stranded RNA copies of its 9 kb genome. The genome contains the *gag*, *pol*, and *env* genes and at least five supplementary genes termed *tat*, *rev*, *nef*, *vif*, *vpr*, and *vpu* (Fig. 1). The *tat* and *rev* genes encode regulatory proteins absolutely required for viral replication; *nef*, *vif*, *vpr*, and *vpu* encode for small "auxiliary" proteins. Their expression is usually dispensable *in vitro*, but essential for viral replication and pathogenesis *in vivo*. The life cycle of HIV-1 involves a series of steps necessary for the successful infection of human target cells, including virus–receptor interactions, virus entry, reverse transcription, proviral integration, transcription, splicing and nuclear export, translation, and assembly, release, and maturation (Fig. 1). HIV-1 primarily infects T lymphocytes, monocytes, and macrophages that display the CD4 cell surface receptor (CD4+ cells). However, some cells of the central nervous system are also targets (e.g. astrocytes and brain microglial cells). The infection spreads to the lymphatic tissue that contains follicular dendritic cells. These specialized cells of the immune system may act as a storage place for latent viruses. Over time, virus replication leads to a slow and progressive destruction of the immune system.

Virus–receptor interaction and viral entry
Receptor-mediated fusion at the plasma membrane is mediated by the HIV-1 envelope (Env) protein. Env is cleaved post-translationally by a cellular protease to yield two polypeptides, gp41 and gp120, which are noncovalently joined together as a trimeric structure. gp41 forms the transmembrane region and gp120 is the surface subunit that is responsible for specifically binding the CD4 cell surface receptor. Binding to CD4 causes a conformational

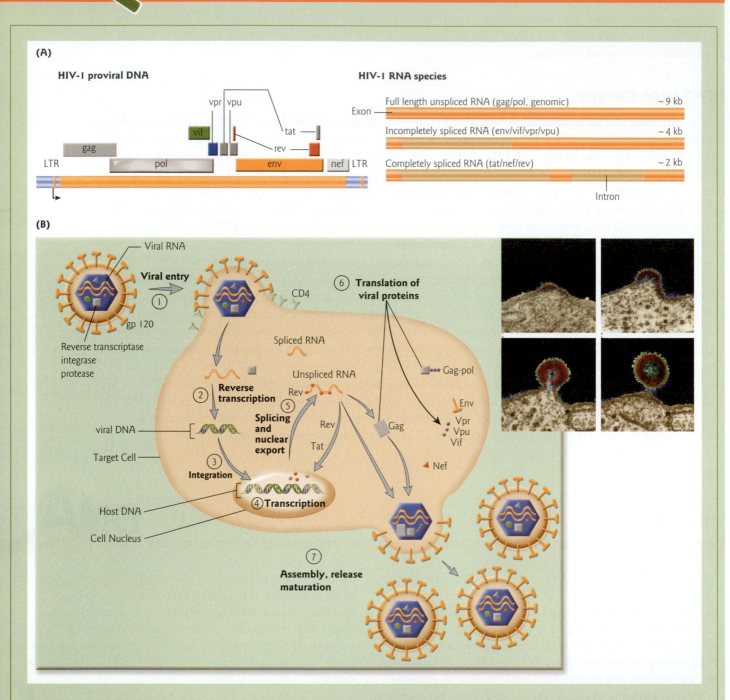

Figure 1 HIV-1 life cycle. (A) Schematic representation of the HIV-1 provirus and the different RNA species produced by splicing (the introns that are removed during splicing are shown as light orange regions in between darker orange exons): *gag*; group specific antigen; *gag-pol*, group-specific antigen polymerase; *env*, envelope; *tat*, transactivator of transcription; *rev*, regulator of expression of virion proteins; *nef*, negative effector; *vif*, virion infectivity factor; *vpr*, viral protein r; *vpu*, viral protein u. LTR, long terminal repeat. (Modified from Nielsen, M.H., Pederson, F.K., Kjems, J. 2005. Molecular strategies to inhibit HIV-1 replication. *Retrovirology* 2:10.) (B) Schematic of the seven major steps in the HIV-1 life cycle (see text for details): (1) virus–receptor interactions and virus entry; (2) reverse transcription; (3) proviral integration; (4) transcription; (5) splicing and nuclear export; (6) translation of viral proteins; and (7) assembly, release, and maturation of viral particles. (Inset) Colored transmission electron micrograph (TEM) of an AIDS virus (red/green) budding from the surface of a T lymphocyte white blood cell. (Credit: Eye of Science / Photo-Researchers, Inc.)

HIV-1 life cycle

change in gp120, which then exposes chemokine coreceptor-binding sites. Binding of the coreceptors triggers the insertion of the gp41 fusion peptide into the target cell membrane. This results in membrane fusion, thus allowing the viral core to enter the cell.

Reverse transcription

In the cytoplasm of the host cell, the RNA genome is converted to DNA form through the process of reverse transcription. The viral reverse transcriptase enzyme possesses three essential activities: reverse transcriptase, RNase H activity (cleaves the genomic RNA in RNA–DNA hybrids), and DNA polymerase activity (synthesis of the second strand of the proviral DNA). A cellular $tRNA_3^{lys}$ acts as a primer and initiates minus-strand DNA synthesis. Because reverse transcriptase is essential for viral replication it has been one of the most popular targets for antiviral therapies.

Proviral integration

After synthesis of the double-stranded proviral DNA, the next step is translocation of the DNA-containing capsid into the nucleus. This process is mediated by independent pathways involving either Vpr, the matrix protein, or the viral integrase. Vpr contains a nuclear localization signal and is thought to mediate the nuclear import of the preintegration complex through the nuclear pore complex in nondividing cells. The viral integrase cleaves the host cell DNA and inserts the provirus into the host genome.

Transcription of viral RNA

HIV-1 gene expression is controlled by interactions between a number of host cell transcription factors, including AP-1, NF-κB, CREB, and Sp1, and the viral Tat protein. The Tat (transactivator of transcription) protein recognizes a 5′ stem-loop structure in the primary viral transcript, called the transactivation responsive element region (TAR). Because the Tat–TAR interaction is essential for activation of HIV-1 gene transcription, it is a popular target for antiviral strategies.

Splicing, nuclear export, and translation

Viral gene expression can be divided into early and late phases, which are Rev-independent and Rev-dependent, respectively. In the early phase the RNA transcript is spliced by the cellular splicing machinery into multiple RNAs. Translation of the completely spliced RNA results in the Tat, Rev (regulator of expression of virion proteins), and Nef (negative effector) proteins. After translation in the cytoplasm, the Rev protein is imported to the nucleus via interaction with importin-β1. When Rev has accumulated in the nucleus to a critical level, mRNA production shifts from multiply spliced to singly spliced and unspliced transcripts, which are characteristic of late-phase gene expression.

Rev contains an RNA-binding motif that directly interacts with a stem-loop structure within the RNA termed the Rev response element (RRE). Rev–RRE complexes stimulate the export of unspliced and singly spliced RNA to the cytoplasm where translation can proceed. Nuclear export of Rev is mediated by the export factor CRM1. Translation of nonspliced RNA and single-spliced RNA results in a number of different proteins: Gag (group-specific antigen) and Gag-Pol (group-specific antigen-polymerase), Vpu (viral protein u), Vif (virion infectivity factor), Vpr (viral protein r), and Env.

Virus particle assembly

After synthesis, viral components assemble at the plasma membrane to form new virions that are able to propagate. A packaging signal sequence (psi) in the viral RNA genome is required for assembly. The highly conserved psi sequence contains four stem loops recognized by the nucleocapsid protein domain of the Gag protein. In addition to the viral genome, several cellular transfer RNAs (tRNAs) are packaged in the viral particle. Assembly is regulated, in part, by the Vpu and Vif proteins.

Release and viral maturation

After virus budding from the cell surface, the Gag and Gag-Pol polyproteins are proteolytically cleaved by the protease domain of Gag-Pol precursors. Cleavage of Gag results in the matrix (MA), capsid (CA), nucleocapsid (NC), and core envelope link (CEL) viral proteins. The MA proteins form a matrix under the viral envelope, and the CA proteins condense to form a conical core surrounding the NC-coated RNA genome. In addition to mediating packaging of Vpr, the CEL protein is thought to associate with the Env protein. Cleavage of the Gag-Pol polyproteins liberates the integrase (IN), reverse transcriptase (RT), and the protease (Pro). After maturation the virus is ready for another round of infection.

Current anti-HIV-1 gene therapy is aimed at reducing plasma viral load and improving patient quality of life. Wiping out every single cell infected with HIV-1 would be virtually impossible. The hope is to inhibit the virus enough to have some additive effect with another type of therapy. There are a number of different strategies being investigated including both protein-based (e.g. toxins and antibodies) and RNA-based. RNA approaches include antisense, ribozymes, RNA aptamers, and RNA interference. Currently, there are Phase I clinical trials under way testing the safety of such techniques as *ex vivo* retroviral gene delivery.

Anti-HIV-1 ribozymes

Ribozymes have been designed to specifically target different HIV-1 genes. One strategy to overcome the great variability of the virus is to combine the use of different ribozymes with different specificities. The HIV LTR region plays an important role in the viral life cycle (Focus box 17.5). Catalytic antisense RNAs have been developed that combine two inhibitory features: a stem-loop antisense motif that binds efficiently to the substrate RNA (antisense effect) and a hairpin or hammerhead motif that catalyzes its specific cleavage. The inhibitory RNAs consist of a TAR (transactivation responsive element region) complementary domain linked to a hairpin or hammerhead ribozyme specifically designed to cleave the LTR region (Fig. 17.21). These catalytic antisense RNAs achieve up to 90% inhibition of HIV-1 replication in eukaryotic cells as measured by reductions in p24 levels (capsid protein). Although results such as these are

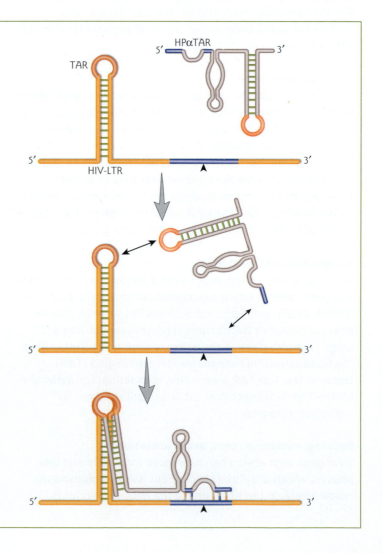

Figure 17.21 Anti-HIV-1 LTR catalytic antisense RNAs. Schematic representation of the hypothesized action of the hairpin-derived catalytic antisense RNA (HPαTAR) against the HIV LTR. The antisense domain is the TAR complementary stem-loop domain (αTAR). Note that the molecules are not drawn to scale. The substrate-binding domain of the catalytic domain is encircled. The arrowhead indicates the site of cleavage, and the blue box represents target sequences within the substrate RNA. (Modified from Puerta-Fernández, E., Barroso-del Jesus, A., Romero-López, C., Tapia, N., Martínez, M.A., Berzal-Herranz, A. 2005. Inhibition of HIV-1 replication by RNA targeted against the LTR region. *AIDS* 19:863–870.).

promising, much remains to be done before any type of anti-HIV-1 gene therapy reaches routine clinical practice.

17.4 Genes and human behavior

While it is relatively straightforward to determine the effect of a single gene defect, such as a mutation in the *CFTR* gene leading to cystic fibrosis, it is not so straightforward to relate gene polymorphisms to human behavior. Since Francis Galton first posed the question in 1874, the nature–nurture controversy has remained a subject of debate. Many advances have been made in answering the question of how genes produce behavior in model systems such as the study of foraging behavior in honey bees or *Drosophila* mating habits. However, studying complex human behaviors has proved much more difficult. The term "human behavior" encompasses everything from personality, temperament, and cognitive style to psychiatric disorders. The simplest model predicts a direct linear relationship between individual genes and behaviors. In reality the relationship is much more complex. Behavioral traits are polygenic, i.e. they are governed by more than one gene, and each gene may be polymorphic. Networks of polymorphic genes and multiple environmental factors affect brain development and function. This in turn will influence behavior (Fig. 17.22). Heredity definitely plays some role in behavior, but the problem is society's tendency to translate this into determinism or "DNA as destiny." In general, one should be wary of announcements in the popular media about scientists finding "the gene" for an aspect of human behavior.

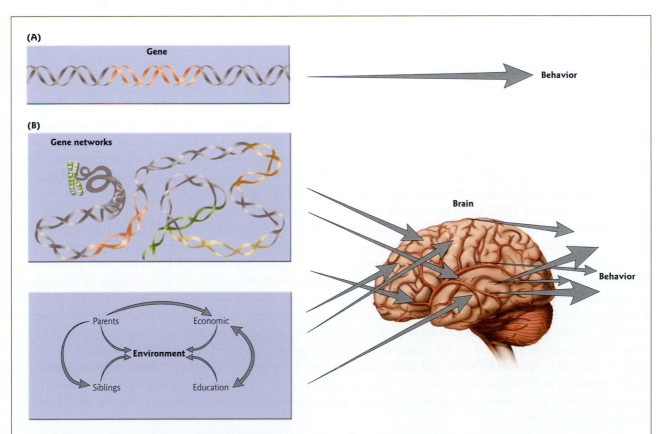

Figure 17.22 Two views of human behavior. (A) In an oversimplified model of human behavior, there is a direct linear relationship between individual genes and behaviors. (B) A more accurate model depicts complex gene networks and multiple environmental factors affecting brain development and function, which in turn will influence behavior. (Reprinted with permission from Hamer, D. 2002. Rethinking behavior genetics. *Science* 298:71–72. Illustration: Redrawn from original illustration by Katharine Sutliff. Copyright © 2002 AAAS.)

Many factors make studies of human behavior particularly challenging. Behavioral traits tend to be inexactly defined because they are hard to measure and quantify. For example, a wrong diagnosis might be made, or a trait may show up after a study is complete. Studies often have an inadequate sample size, show sample bias, or do not use statistical methods appropriately. It is often unclear what the prevalence of a polymorphism is in the general population and what the variability is in ethnic groups. Research teams around the world have been studying many aspects of human behavior for generations, including the search for genes linked to intelligence, novelty seeking, addiction, and depression. In this section, the efforts of researchers to find genes linked to aggressive behavior and schizophrenia are discussed. These are two examples where some progress has been made in understanding the interaction between genes and the environment.

Aggressive, impulsive, and violent behavior

Family, twin, and adoption studies have suggested a heritability of 0% to more than 50% for predisposition to violent behavior. Heritability represents the degree to which a trait stems from genetic factors. In the 1960s researchers reported an association between an extra Y chromosome and violent crime in males. Follow-up studies found that, in fact, there was no connection whatsoever. The original study used a biased sample – inmates in a prison. It was later realized that the extra Y was frequent in the general population as well. A much stronger case has been made for a link between aggressive behavior and a gene on the X chromosome that codes for the enzyme monoamine oxidase A (MAOA). MAOA metabolizes several neurotransmitters in the brain, such as dopamine and serotonin, thus preventing excess neurotransmitters from interfering with communication among neurons.

MAOA functional length polymorphism

In 1993 a chain termination mutation in exon 8 of the *MAOA* gene in a Dutch family was linked to abnormal criminal behavior. However, in this study there was no information on environmental effects influencing behavior. A functional length polymorphism in the transcriptional control region of the *MAOA* gene has now been characterized that affects transcriptional activity. A repeat sequence of 30 bp is present in 3, 3.5, 4, or 5 copies. Alleles with 3.5 or 4 copies of the repeat sequence are transcribed more efficiently and produce more MAOA enzyme (high-activity alleles), whereas those with 3 or 5 copies of the repeat produce less MAOA enzyme and fewer neurotransmitters are metabolized (low-activity alleles) (Fig. 17.23). A 2002 study in New Zealand that followed 1037 males from birth to adulthood provides convincing evidence that the repeat polymorphism influences aggressive behavior, but this depends on early environment. The authors showed that maltreated children with high-activity alleles were less likely to develop antisocial behavior. In contrast, maltreated children with low-activity alleles were more likely to develop antisocial behavior. Men who carried a low-activity allele, and so presumably produced a limited amount of enzyme, were four times more likely to be aggressive, impulsive, and commit violent crimes if they were abused as children or drank alcohol. Women also inherit the two alleles, but the effects are easier to study in men, who have only one X chromosome.

Similar results to those in humans have been obtained from studies of primate aggressive behavior. The same 30 bp repeat or a shorter 18 bp repeat in the *MAOA* gene, among other forms, was found in 600 primates sampled. Apes and Old World (Asian and African) monkeys carried these alleles, whereas New World (South American) monkeys do not. A recent study has shown that repeat polymorphism in the rhesus macaque *MAOA* gene promoter region influences aggressive behavior in male subjects, but also is dependent on early environment.

Schizophrenia susceptibility loci

Schizophrenia is a severe psychiatric disorder that affects about 1% of the general population. The disorder is characterized by psychotic symptoms, such as paranoid delusions and hallucinations, reduced interest and drive, altered emotional reactivity, and disorganized behavior. These characteristic features generally have

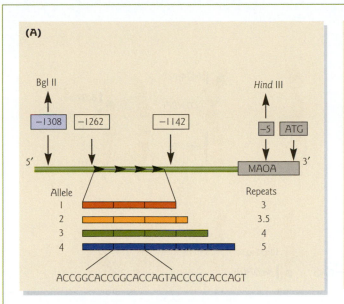

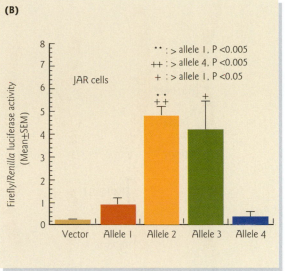

Figure 17.23 Functional polymorphism in the *monoamine oxidase A* (*MAOA*) gene is linked to aggressive, impulsive, and violent behavior. (A) Map of the promoter region of the human *MAOA* gene. The polymorphic region consists of a 30 bp repetitive sequence with length variation of 3, 3.5, 4, and 5 repeats. The numbers in boxes indicate the positions of the *MAOA* variable number tandem repeats and of the sites that were used for cloning into the promoter fusion vector. (B) Promoter fusion assays. JAR (human placental choriocarcinoma cell line) cells were transfected with plasmid constructs in which the coding sequence of firefly luciferase was fused to the *MAOA* promoter carrying alleles 1, 2, 3, or 4 or with the plasmid vector alone. The cells were cotransfected with a plasmid containing the coding sequence for *Renilla* luciferase under control of a constitutive viral promoter (*Herpes simplex* virus thymidine kinase promoter region). After 48 hours' incubation, cell extracts were prepared and sequentially assayed for firefly luciferase activity and *Renilla* luciferase activity. *Renilla* luciferase activity was unaffected by treatments and served as an internal control. The ratios of firefly to *Renilla* luciferase (mean ± SEM) in arbitrary luminescence units are shown for replicate experiments. Similar results were obtained using other cell lines. (Reprinted from Sabol, S.Z., Hu, S., Hamer, D. 1998. A functional polymorphism in the monoamine oxidase A promoter. *Human Genetics* 103:273–279. Copyright © 1998 Springer-Verlag, with kind permission of Springer Science and Business Media.)

their onset in the late teens and early twenties. Although outcomes are variable, even with treatment, the typical course is one of relapses followed by only partial remission. People suffering from schizophrenia find it extremely difficult to function in the workplace or social settings. Twin and adoption studies show conclusively that the risk of developing schizophrenia is increased among the relatives of affected individuals. The complex mode of transmission is compatible with a multilocus model. However, despite intensive efforts by many research teams, the number of susceptibility loci, the disease risk conferred by each locus, the extent of genetic variability, and the degree of interaction among loci all remain unknown.

Over the years, potential susceptibility loci have been described on chromosomes 1, 5, 6, 8, 10, 12, 13, 17, and 22. Most studies have not achieved stringent levels of significance, and many have been unable to be replicated. The putative susceptibility genes for which the most promising follow-up data are available are those encoding dysbindin (DTNBP1, also known as dystrobrevin-binding protein 1), neuregulin 1 (NRG1), disrupted in schizophrenia 1 (DISC1), D-amino acid oxidase (DAO), D-amino acid oxidase activator (DAOA), and regulator of G protein signaling 4 (RGS4) (Fig. 17.24). Of these, the evidence is thought to be very strong for a role for the genes encoding DTNBP1 and NRG1. Several other genes have also been implicated, but there is as yet insufficient follow-up data to draw conclusions. Although its function is largely unknown, variation in mRNA expression of dysbindin may confer risk by presynaptic effects on glutamate trafficking or release, particularly in the hippocampus. Neuregulin is thought to encode ~15 proteins with a

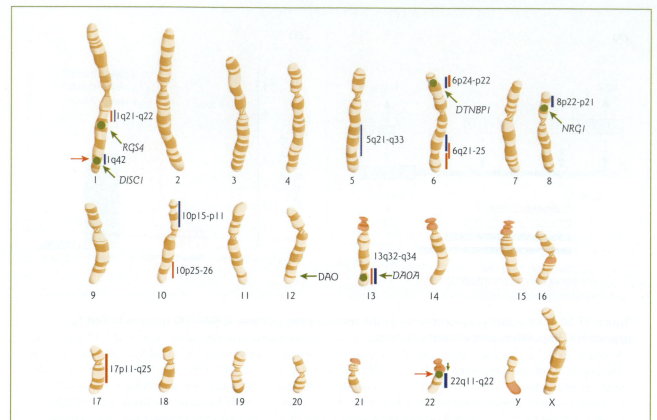

Figure 17.24 Locations of putative schizophrenia susceptibility genes. Linkages that reached genome-wide significance on their own (red) or those that have received strong support from more than one sample (blue) are shown. The red arrows refer to the location of chromosomal abnormalities associated with schizophrenia. The green arrows and circles show the locations of the genes discussed in the text. (Adapted from Owen, M.J., Craddock, N., O'Donovan, M.C. 2005. Schizophrenia: genes at last? *Trends in Genetics* 21:518–525. Copyright © 2005, with permission from Elsevier.)

diverse range of functions in the brain, including cell–cell signaling, axon guidance, synaptogenesis, glial differentiation, myelination, and neurotransmission. Altered expression or mRNA splicing of neuregulin may confer risk. The mechanisms remain unclear, but mediation of glutamatergic mechanisms might be involved.

Chapter summary

Discoveries in molecular biology have led to many advances in understanding the nature of cancer, and in developing treatment strategies. Cancer is a group of genetically diverse disorders in which the critical balance between cell growth, proliferation, differentiation, and apoptosis has been disrupted. Normal cells have a defined lifespan, whereas cancer cells are immortal. The emergence of a tumor cell requires the accumulation of at least 4–8 genetic changes over the course of years. Transformed cells often undergo metastasis and travel from the tissue of origin to other parts of the body. Agents that promote tumor formation include inherited genetic defects, chemical carcinogens, viral infection, or irradiation.

Oncogenes are genes whose products have the ability to cause malignant transformation of cells. They were originally identified as genes carried by some tumor viruses. Cellular genes with the potential to give rise to oncogenes are called proto-oncogenes. The gain of function or inappropriate activation of proto-oncogenes occurs due to qualitative changes (point or insertion mutations) or quantitative changes (inappropriate regulation of expression). Proto-oncogenes encode cellular products involved in the control of cell growth and proliferation, differentiation, and apoptosis. Examples include the tyrosine kinase c–Src and the transcription factor c–Myc. Qualitative or quantitative changes in the genes encoding these proteins is linked to many forms of cancer. The v-*src* oncogene carried by the Rous sarcoma virus encodes a tyrosine kinase that has lost amino acids from the C-terminus and remains in an open conformation in which it constitutively phosphorylates target proteins. c–Myc normally blocks cell differentiation; its inappropriate overexpression is correlated with highly aggressive tumors.

Tumor suppressor gene products are required for normal cell function. Cancer arises when there are two independent mutations or "hits" that lead to the loss of function of both tumor suppressor alleles at a locus. If the loss of one allele is inherited through the germline, an individual is said to have a genetic predisposition to cancer. Sporadic mutation of the second allele in a somatic cell leads to tumorigenesis. The retinoblastoma protein (pRB) and the p53 protein are classic examples of tumor suppressor genes. Deletion of both *RB* alleles leads to retinoblastoma, a childhood hereditary cancer syndrome. Normally, hypophosphorylated pRB prevents cells from entering S phase by binding the E2F transcription factor complex and preventing it from activating genes essential for DNA replication. Phosphorylation by cyclin–CDK complexes dissociates pRB from the E2F complex. If pRB is absent, E2F is active, resulting in unrestrained cell growth. Cellular stress signals such as UV irradiation induce p53 expression. DNA damage leads to two possible outcomes: cell cycle arrest and DNA repair before replication, or apoptosis. A missense point mutation in one allele of p53 results in the production of a dominant negative mutant protein that associates with wild-type p53 to form inactive heterotetramers. Cell proliferation thus occurs without repair of damaged DNA and apoptosis is blocked. Further mutations accumulate, eventually leading to tumor formation.

Upregulation or downregulation of some clusters of miRNAs is associated with a number of types of cancer, suggesting that the altered expression of miRNA contributes to the initiation and progression of cancer. Other genetic changes associated with cancer are chromosome rearrangements that bring a cellular proto–oncogene under the control of the wrong promoter and/or enhancer, or that create a fusion protein with oncogenic properties. Such rearrangements are common in leukemia.

Tumor cells can arise by nongenetic means through the actions of specific DNA or RNA tumor viruses which either carry a copy of an oncogene or tumor suppressor gene, or alter expression of the cellular copy by introduction of a promoter or enhancer. Transformed cells are cells that are nonpermissive for viral replication but ones in which a virus has taken up residence in the cell and alters its growth properties. For example, the vast majority of cervical cancers are associated with high-risk human papilloma viruses. After integration of the viral genome into the host cell, the E6 and E7 viral proteins are expressed. The viral proteins inactivate p53 and pRB, and activate expression of the reverse transcriptase subunit of telomerase, leading to unrestrained cellular proliferation and immortality. Retroviruses can acquire a proto–oncogene from the host during integration into the host genome. After the gene is acquired by the virus, it is subject to mutation and is often part of a fusion protein with other virally encoded amino acids. These genetic changes lead to oncogenic properties.

Chemical carcinogens or their metabolic products can either directly or indirectly affect gene expression. Some carcinogens, such as benzo(*a*)pyrene in cigarette smoke, act by genotoxic mechanisms and induce formation of DNA adducts that interfere with replication and transcription, or induce chromosomal rearrangements. Other carcinogens act by nongenotoxic mechanisms. For example, benzo(*a*)pyrene also promotes tumor formation by altering arylhydrocarbon receptor-mediated signal transduction pathways important for cell cycle control. Upregulated target genes include members of the cytochrome P450-dependent mono-oxygenase (CYP) family of enzymes that convert procarcinogens to carcinogens.

Over 900 clinical gene therapy trials are in progress around the world. About two-thirds are for the treatment of cancer. The remainder target inherited autosomal recessive disorders caused by defects in a single gene, or for infectious diseases (nearly all HIV-related). In somatic cell gene therapy, a malfunctioning gene is replaced or compensated for by a properly functioning gene in somatic cells of a patient. There are two main strategies for somatic cell gene therapy: *in vivo* and *ex vivo*. The majority of gene therapy trials use viral vectors, such as modified retroviruses, adenoviruses, and adeno-associated viruses, but some use liposomes and naked DNA. Most gene therapy trials have had disappointing results, mainly due to a lack of efficiency of gene delivery and sustained expression. The first fatality directly attributed to gene therapy was due to an immune reaction to an adenovirus vector.

In theory, the same techniques used to treat genetic diseases could be used to improve or alter characteristics through "enhancement" genetic engineering. Along these lines, rats with greater muscle mass than controls were engineered through "gene therapy" by viral–mediated introduction of the insulin–like growth factor 1 gene.

The first ever clinical trial for gene therapy of an inherited disorder was performed in 1990 to treat a young girl with adenosine deaminase (ADA) deficient severe combined immunodeficiency (SCID), in combination with traditional enzyme replacement therapy. Gene therapy has also been used to treat an X chromosome–linked form of SCID (SCID-X1). Since 1999, ADA-SCID gene therapy has helped to restore the immune systems of at least 22 children with either ADA-SCID or SCID-X1. However, three patients in the SCID-X1 trials developed leukemia due to the retroviral vector inserting near the *LMO2* proto-oncogene, leading to its inappropriate activation. Two of the children have since died. Although trials for ADA-SCID are being allowed, SCID-X1 trials have been suspended.

Gene therapy for the recessive genetic disease cystic fibrosis has so far been ineffectual. The use of an adeno–associated virus vector appears promising; however the finding that this vector integrates more often into genes than into other regions of DNA raises some concern.

HIV-1 gene therapy strategies are aimed at reducing viral load and improving patient quality of life. Although it is not possible to completely get rid of the virus, the hope is to inhibit it enough to have some additive effect with traditional drug therapy. Antisense ribozymes have been designed to specifically target HIV-1 RNA transcripts at different points in the viral life cycle. Experiments in cell culture are promising, but much remains to be done before anti-HIV-1 gene therapy reaches routine clinical practice.

Human behavioral traits are governed by more than one gene (polygenic) and are influenced by multiple environmental factors. Heredity plays some role in behavior, but DNA is not "destiny." A variable number or tandem repeats in the monoamine oxidase A (MAOA) gene promoter that leads to altered levels of gene transcription is linked to aggressive, impulsive, and violent behavior. Men who carried a low–activity allele were shown to be more likely to develop antisocial behavior, but only if they were abused as children or drank alcohol. A number of potential susceptibility loci have been described for schizophrenia. The loci for which there is the most evidence for linkage encode various proteins involved in neurotransmission, axon guidance, and cell–cell signaling in the brain. However, the disease risk conferred by each locus, the extent of genetic variability, and the degree of interaction among loci and the environment remain unknown.

Analytical questions

1 You suspect that a gene you have cloned is a proto-oncogene. Describe how you would test your hypothesis experimentally. Predict the results.

2 DNA microarray analysis reveals no expression of the pRB tumor suppressor gene in a bone cancer cell line. Would you expect this to be the only difference in expression pattern between normal and bone cancer cells? Why or why not?

3 Analysis of another cell line reveals normal expression levels of pRB and yet the cells exhibit unrestrained cell growth. Provide an explanation and suggest how to test your hypothesis.

4 A friend of yours tells you (as they light up) that they have heard that benzo(*a*)pyrene in cigarette smoke is not really a carcinogen. Set them straight.

5 You are designing a somatic cell gene therapy protocol for spinal muscular atrophy, a genetic disorder that is caused by a defect in the *SMN* (survival of motor neurons) gene. You have cloned the *SMN* cDNA.

(a) What is the minimal *cis*-acting DNA regulatory element you would need to ligate to the *SMN* cDNA to get transcription of the cDNA in cells?

(b) Assume that you decide to use retroviral-mediated gene transfer to introduce the *SMN* cDNA into the patient's target cells *ex vivo*. What are the advantages and disadvantages of this choice of vector for gene delivery? Is there another vector that would work better?

6 In a study that involved 100 Caucasian males, you find a polymorphism in a gene for a neurotransmitter receptor. The "long allele" is linked with reckless behavior. What conclusions, if any, can you draw about the results of this study? Should you hold a press conference to announce the discovery of the "reckless" gene?

Suggestions for further reading

Aiuti, A., Slavin, S., Aker, M. et al. (2002) Correction of ADA-SCID by stem cell gene therapy combined with nonmyeloablative conditioning. *Science* 296:2410–2413.

Blume-Jensen, P., Hunter, T. (2001) Oncogenic kinase signalling. *Nature* 411:355–365.

Boehm, J.S., Hahn, W.C. (2005) Understanding transformation: progress and gaps. *Current Opinion in Genetics and Development* 15:13–17.

Bonamy, G.M., Guiochon-Mantel, A., Allison, L.A. (2005) Cancer promoted by the oncoprotein v-ErbA may be due to subcellular mislocalization of nuclear receptors. *Molecular Endocrinology* 19:1213–1230.

Brunner, H.G., Nelen, M., Breakefield, X.O., Ropers, H.H., van Oost, B.A. (1993) Abnormal behavior associated with a point mutation in the structural gene for monoamine oxidase A. *Science* 262:578–580.

Caspi, A., McClay, J., Moffitt, T.E. et al. (2002) Role of genotype in the cycle of violence in maltreated children. *Science* 297:851–854.

Cavazzana-Calvo, M., Hacein-Bey, S., de Saint Basile, G. et al. (2000) Gene therapy of human severe combined immunodeficiency (SCID)-X1 disease. *Science* 288:669–672.

Cavazzana-Calvo, M., Thrasher, A., Mavilio, F. (2004) The future of gene therapy. *Nature* 427:779–781.

Debernardi, S., Lillington, D., Young, B.D. (2004) Understanding cancer at the chromosomal level: 40 years of progress. *European Journal of Cancer* 40:1960–1967.

DeFrancesco, L. (2004) The faking of champions. *Nature Biotechnology* 22:1069–1071.

Dyer, M.A., Bremner, R. (2005) The search for the retinoblastoma cell of origin. *Nature Reviews Cancer* 5:91–101.

Eccles, S.A. (2005) Targeting key steps in metastatic tumour progression. *Current Opinion in Genetics and Development* 15:77–86.

Flotte, T.R., Schwiebert, E.M., Zeitlin, P.L., Carter, B.J., Guggino, W.B. (2005) Correlation between DNA transfer and cystic fibrosis airway epithelial cell correction after recombinant adeno-associated virus sertotype 2 gene therapy. *Human Gene Therapy* 16:921–928.

Frame, M.C. (2004) Newest findings on the oldest oncogene; how activated *src* does it. *Journal of Cell Science* 117:989–998.

Guzman, E., Langowski, J.L., Owen-Schaub, L. (2003) Mad dogs, Englishmen, and apoptosis: The role of cell death in UV-induced skin cancer. *Apoptosis* 8:315–325.

Hacein-Bey-Abina, S., Von Kalle, C., Schmidt, M. et al. (2003) LMO2-associated clonal T cell proliferation in two patients after gene therapy for SCID-X1. *Science* 302:415–419.

Hamer, D. (2002) Rethinking behavior genetics. *Science* 298:71–72.

He, L., Thomson, J.M., Hemann, M.T. et al. (2005) A microRNA polycistron as a potential human oncogene. *Nature* 435:828–833.

Hyde, S.C., Gill, D.R., Higgins, C.F. et al. (1993) Correction of the ion transport defect in cystic fibrosis transgenic mice by gene therapy. *Nature* 362:250–255.

Insinga, A., Monestiroli, S., Ronzoni, S. et al. (2004) Impairment of p53 acetylation, stability, and function by an oncogenic transcription factor. *EMBO Journal* 23:1144–1154.

Ishikawa, T., Zhang, S.S., Qin, X., Takahashi, Y., Oda, H., Nakatsuru, Y., Ide, F. (2004) DNA repair and cancer: lessons from mutant mouse models. *Cancer Science* 95:112–117.

Kaiser, J. (2003) Seeking the cause of induced leukemias in X-SCID trial. *Science* 299:495.

Lee, S., Barton, E.R., Sweeney, H.L., Farrar, R.P. (2004) Viral expression of insulin-like growth factor-I enhances muscle hypertrophy in resistance trained rats. *Journal of Applied Physiology* 96:1097–1104.

Letvin, N.L. (2005) Progress toward an HIV vaccine. *Annual Reviews of Medicine* 56:213–223.

Levine, A.J., Finlay, C.A., Hinds, P.W. (2004) p53 is a tumor suppressor gene. *Cell* S116:S67–S69.

Lu, J., Getz, G., Miska, E.A. et al. (2005) MicroRNA expression profiles classify human cancers. *Nature* 435:834–838.

Lu, X. (2005) p53: a heavily dictated dictator of life and death. *Current Opinion in Genetics and Development* 15:27–33.

Luch, A. (2005) Nature and nurture – lessons from chemical carcinogenesis. *Nature Reviews Cancer* 5:113–125.

Marte, B., ed. (2004) Nature Insight. Cell division and cancer [series of articles]. *Nature* 432:293–341.

McCarty, D.M., Young, S.M., Jr., Samulski, R.J. (2004) Integration of adeno-associated virus (AAV) and recombinant AAV vectors. *Annual Review of Genetics* 38:819–845.

Minn, A.J., Gupta, G.P., Siegel, P.M. et al. (2005) Genes that mediate breast cancer metastasis to lung. *Nature* 436:518–524.

Münger, K., Baldwin, A., Edwards, K.M. et al. (2004) Mechanisms of human papillomavirus-induced oncogenesis. *Journal of Virology* 78:11451–11460.

Musarò, A., McCullagh, K., Paul, A. et al. (2001) Localized Igf-1 transgene expression sustains hypertrophy and regeneration in senescent skeletal muscle. *Nature Genetics* 27:195–200.

Nakai, H., Montini, E., Fuess, S., Storm, T.A., Grompe, M., Kay, M.A. (2003) AAV serotype 2 vectors preferentially integrate into active genes in mice. *Nature Genetics* 34:297–302.

Newman, T.K., Syagailo, Y.V., Barr, C.S. et al. (2005) Monoamine oxidase A gene promoter variation and rearing experience influences aggressive behavior in rhesus monkeys. *Biological Psychiatry* 57:167–172.

Nielsen, M.H., Pederson, F.K., Kjems, J. (2005) Molecular strategies to inhibit HIV-1 replication. *Retrovirology* 2:10 (http://www.retrovirology.com/content/2/1/10).

Owen, M.J., Craddock, N., O'Donovan, M.C. (2005) Schizophrenia: genes at last? *Trends in Genetics* 21:518–525.

Pelengaris, S., Khan, M., Evan, G. (2002) c-Myc: more than just a matter of life and death. *Nature Reviews Cancer* 2:764–776.

Puerta-Fernández, E., Barroso-del Jesus, A., Romero-López, C., Tapia, N., Martínez, M.A., Berzal-Herranz, A. (2005) Inhibition of HIV-1 replication by RNA targeted against the LTR region. *AIDS* 19:863–870.

Quinn, T.C., Overbaugh, J. (2005) HIV/AIDS in women: an expanding epidemic. *Science* 308:1582–1583.

Richards, L. (2004) Cystic fibrosis: pipeline of promise. *Modern Drug Discovery* 7:27–29.

Robinson, G.E. (2004) Beyond nature and nurture. *Science* 304:397–399.

Roskoski, R., Jr. (2004) Src protein-tyrosine kinase stucture and regulation. *Biochemical and Biophysical Research Communications* 324:1155–1164.

Sabol, S.Z., Hu, S., Hamer, D. (1998) A functional polymorphism in the monoamine oxidase A promoter. *Human Genetics* 103:273–279.

Sahai, E. (2005) Mechanisms of cancer cell invasion. *Current Opinion in Genetics and Development* 15:87–96.

Savage, D.G., Antman, K.H. (2002) Imatinib mesylate – a new oral targeted therapy. *New England Journal of Medicine* 346:683–693.

Schmitz, N.M.R., Leibundgut, K., Hirt, A. (2005) CDK2 catalytic activity and loss of nuclear tethering of retinoblastoma protein in childhood acute lymphoblastic leukemia. *Leukemia* 19:1783–1787.

Strayer, D.S., Akkina, R., Bunnell, B.A. et al. (2005) Current status of gene therapy strategies to treat HIV/AIDS. *Molecular Therapy* 11:823–842.

Varley, J.M. (2003) Germline TP53 mutations and Li–Fraumeni syndrome. *Human Mutation* 21:3113–320.

Verma, I.M. (1990) Gene therapy. *Scientific American* 263:34–41.

Vogel, G. (2004) A race to the starting line. *Science* 305:632–635.

Williams, D.A., Baum, C. (2003) Gene therapy – new challenges ahead. *Science* 302:400–401.

Yeatman, T.J. (2004) A renaissance for *src*. *Nature Reviews Cancer* 4:470–480.

Glossary

A

A site The ribosomal site to which new aminoacyl-tRNAs bind, with the exception of the initiator tRNA.

AAUAAA An element of the eukaryotic polyadenylation signal that dictates cleavage and polyadenylation about 20 nucleotides downstream.

abasic site A deoxyribose in a DNA strand that has lost its base through DNA damage.

Ac *Activator*, a maize transposon that encodes a transposase required for movement and can activate transposition of an inactive transposon like *Ds*.

acceptor stem The region of a tRNA molecule formed by base-pairing between the 5′ and 3′ ends of the molecule that becomes charged with an amino acid.

acentric chromosome A chromosome with no centromere.

acetylation The post-translational modification of a protein by the addition of an acetyl group(s) (e.g. histone acetylation).

acid A substance that releases an H⁺ ion or proton in solution.

acid blob A transactivation domain in a transcription factor that is rich in acidic amino acids.

acrocentric chromosome A chromosome with the centromere near one end.

activator A protein that binds to an enhancer or other DNA regulatory element and activates transcription from a nearby promoter.

active site The part of an enzyme at which substrate molecules bind and are converted into their reaction products.

acylated tRNA A tRNA molecule to which an amino acid is linked.

adenine (A) A nitrogenous purine base found in DNA and RNA. Adenine always pairs with thymine in DNA and uracil in RNA.

adeno-associated virus A parvovirus that is being tested for use as a gene therapy vector.

adenosine A nucleoside containing the base adenine.

adenosine deaminase acting on RNA (ADAR) An RNA editing enzyme that deaminates certain adenosines in RNAs, converting them to inosines.

adenosine deaminase deficiency (ADA) A severe immunodeficiency disease that results from a lack of the enzyme adenosine deaminase. It usually leads to death within the first few months of life.

adenovirus A group of DNA-containing viruses which cause respiratory disease, including one form of the common cold. Genetically modified adenovirus vectors have been used in gene therapy trials to treat cystic fibrosis, cancer, and other diseases.

A-DNA An alternate form of the right-handed DNA double helix found at low relative humidity, with 11 base pairs per helical turn. It is unlikely that A-DNA is present in any length sections in cells. This is the form adopted in solution by an RNA–DNA hybrid.

ADP-ribosylation The enzymatic transfer of an ADP-ribose residue from NAD⁺ to a specific amino acid of the acceptor protein by an ADP-ribosyltransferase.

affinity chromatography Chromatography that relies on affinity between the substance of interest and another substance immobilized on a resin.

AFLP Amplified fragment length polymorphism. The PCR amplification of restriction fragments to which adaptor oligomer sequences have been attached. Pronounced "a-flip."

agarose A polysaccharide extracted from seaweed; used as a gelling agent in electrophoresis of nucleic acids; its value is that few molecules bind to it, so it does not interfere with electrophoretic movement.

allele One of a number of alternative forms of a given gene (or DNA sequence) that can occupy a given genetic locus on a chromosome. Different alleles produce variation in inherited characteristics such as hair color or blood type. In an individual, one form of the allele (the dominant one) may be expressed more than another form (the recessive one).

allelic exclusion See monoallelic gene expression.

allolactose A rearranged version of lactose with a β–1,6-galactosidic bond; the real inducer of the *lac* operon.

allosteric protein Any protein whose activity is altered by a change of shape induced by binding a small molecule.

allosteric regulation Ligand-induced conformational (shape) changes that alter the activity of a protein.

alpha (α)–helix A fundamental unit of protein folding in which successive amino acids form a right-handed helical structure held together by hydrogen bonding between the amino and carboxyl components of the peptide bonds in successive loops of the helix.

alternative splicing Splicing the same pre-mRNA in two or more ways to yield two or more different mRNAs that produce two or more different protein products.

Alu **element** A human nonautonomous retrotransposon that contains the AGCT sequence recognized by the restriction enzyme *Alu*I. Also known as a short interspersed nuclear element (SINE).

amino acids A group of 20 common and two rare kinds of small molecules that link together in long chains to form proteins. Often referred to as the "building blocks" of proteins.

amino terminus The end of a polypeptide with a free amino group; i.e. the end at which protein synthesis began.

aminoacyl–tRNA An activated amino acid linked through a high-energy anhydride bond to the phosphate group of AMP. Created by aminoacyl-tRNA synthetase in the first step in tRNA charging.

aminoacyl–tRNA synthetase The enzyme that links a tRNA to its cognate amino acid.

A–minor motif A folding motif in RNA in which single-stranded adenosines make tertiary contacts with the minor grooves of RNA double helices by hydrogen bonding and van der Waals contacts.

ampicillin A derivative of penicillin that blocks synthesis of the peptidoglycan layer that lies between the inner and outer membranes of *E. coli*, and kills dividing cells.

amyloid fibrils Insoluble toxic deposits of aggregated proteins that accumulate in cells in certain types of protein misfolding diseases, such as Alzheimer's, Parkinson's, Huntington's, type II diabetes, and Creutzfeldt–Jakob disease.

anaphase A short stage of cell division in which the chromosomes move to the poles.

Angelman syndrome A neurodevelopmental disorder caused by defects in genomic imprinting in a region of chromosome 15, or by classic mutations in the UBE3A gene which is involved in the ubiquitin-mediated protein degradation pathway.

animal model See model organism.

annealing of DNA The process of bringing back together the two separate strands of denatured DNA to re-form a double helix. This is also referred to as renaturation.

annotated genes Genes or gene-like sequences from a genomic sequencing project that are at least partially characterized.

anode Positive (red) electrode.

antibody A blood protein that is produced in response to and counteracts an antigen with great specificity. Antibodies are produced in response to disease and help the body fight against the particular disease and develop immunity.

anticodon A 3-base sequence in tRNA that base pairs with a specific codon in mRNA.

anticodon loop The loop drawn by convention at the bottom of the tRNA molecule which contains the anticodon.

antigen A substance recognized and bound by an antibody.

antiparallel The relative polarities of the two strands in a DNA double helix (or in any double-stranded nucleic acid) ; if one strand runs 5′ → 3′, top to bottom, the other runs 3′ → 5′.

antisense DNA strand The noncoding strand in double-stranded DNA. The antisense strand serves as the template for mRNA synthesis.

antisense oligonucleotides Short oligonucleotides designed to selectively bind to a specific mRNA and induce antisense-mediated inhibition of gene expression. Either the mRNA strand in the hybrid duplex is cleaved by RNase H or translation is blocked.

antisense RNA An RNA complementary to an mRNA.

antiserum Blood serum containing an antibody or antibodies directed against a particular antigen.

AP endonuclease An enzyme that cuts a strand of DNA on the 5′ side of an apurinic or apyrimidinic site (AP site).

AP1 A transcription activator that is a heterodimer composed of one molecule each of Fos and Jun, or of a Jun–Jun homodimer.

APE1 (AP endonuclease 1) A mammalian enzyme that uses 3′ → 5′ endonuclease activity to edit the errors made by DNA polymerase β during base excision repair.

apoenzyme Inactive protein part of an enzyme remaining after removal of the prosthetic group or other necessary accessory protein subunit (see holoenzyme).

apolipoproteins Proteins involved in the metabolism of cholesterol.

apoptosis Programmed cell death, an organism's normal method of disposing of damaged, unwanted, or unneeded cells.

aptamer The domain of a riboswitch that selectively binds to the target metabolite.

apurinic site (AP site) A deoxyribose sugar in a DNA strand that has lost its purine base.

apyrimidinic site (AP site) A deoxyribose sugar in a DNA strand that has lost its pyrimidine base.

AraC The negative regulator of the arabinose operon.

Archaea A third domain of life. Although prokaryotic, they are as phylogenetically distinct from bacteria as they are from eukaryotes. The Archaea typically live in extreme environments, e.g. hot springs and salt lakes. Some are methane producers.

Argonaute One of the components of the RISC complex that has "Slicer" activity and cleaves the target mRNA during RNAi.

Artemis An endonuclease involved in end-processing of double-strand break sites in DNA during repair by nonhomologous end-joining.

A–T rich DNA sequences that have many adenine-thymine bases that promote DNA melting; e.g. at origins of replication and promoter regions.

ataxia Loss of muscle control of voluntary movements.

ataxia-telangiectasia A rare fatal disease involving a damaged immune system, unsteady walk, premature aging, and a strong predisposition to some kinds of cancer. People who possess only one copy of the gene, called ATM, do not have the disease, but may be predisposed to cancer and unusually sensitive to radiation.

ataxia-telangiectasia mutated (ATM) A serine-threonine kinase in the nucleus that is a key signal transducer for recruitment of factors involved in double-strand break repair by homologous recombination; e.g. ATM kinase activity is induced by exposure of cells to ionizing radiation.

atomic force microscopy (AFM) A technique in which the deflection of a cantilever tip as it scans the surface of a molecule is used to produce a topographic image of the molecule's surface.

ATP synthase An enzyme that generates ATP using the free energy of an electrochemical gradient of protons across a membrane.

ATPase An enzyme that hydrolyzes ATP, releasing energy for other cellular activities.

att **sites** Sites on bacteriophage and host DNA where recombination occurs, allowing integration of the phage DNA into the host genome.

autonomous replicating sequence (ARS) A yeast origin of replication.

autoradiography A technique in which a radioactive sample is allowed to expose a photographic emulsion; e.g. an X-ray film.

autosomal dominant A pattern of Mendelian inheritance whereby an affected individual possesses one copy of a mutant allele and one normal allele. Individuals with autosomal dominant diseases have a 50–50 chance of passing the mutant allele and hence the disorder onto their children.

autosome Any chromosome other than a sex chromosome. Humans have 22 pairs of autosomes.

auxotroph An organism that requires specific supplementary compounds in order to grow (e.g. vitamins, amino acids, nucleotides, etc.) because it is unable to synthesize them.

B

backbone (DNA) The repeating sugar-phosphate groups that form the two strands of the DNA double helix.

backbone damage Types of DNA damage that include the formation of abasic sites and double-strand breaks.

bacteria A single-celled prokaryotic organism. Bacteria are found throughout nature and can be beneficial or pathogenic.

bacterial artificial chromosome (BAC) A vector based on the *E. coli* F plasmid capable of holding foreign DNA inserts up to 300,000 base pairs. Once the foreign DNA has been cloned and transformed into the host bacteria, many copies of it can be made.

bacteriophage A virus that infects bacteria (literally "bacterium eater"); often abbreviated as phage for short.

basal transcription Low levels of gene transcription determined by the frequency with which RNA polymerase binds a promoter and initiates transcription in the absence of transcription activators. The basal level is some 20–40-fold lower than activated levels of transcription.

base In general, a substance that accepts a H^+ ion or protein in solution. In molecular biology, the term typically refers to a cyclic, nitrogen-containing compound linked to deoxyribose in DNA and to ribose in RNA. See also nitrogenous base.

base analog A compound that substitutes for normal bases; e.g. 5-bromouracil.

base excision repair (BER) A repair pathway that removes a damaged base by a DNA glycosylase, then cleaves the 5′ side of the resulting abasic site by an AP endonuclease. The abasic sugar-phosphate and downstream bases are removed and the gap is filled by DNA polymerase and DNA ligase.

base pair (bp) A single pair of complementary nucleotides from opposite strands of the DNA double helix (A-T or G-C). The number of base pairs is used as a measure of the length of a double-stranded DNA molecule.

base pairing Weak bonding between purine and pyrimidine bases within nucleic acids. Normally adenine (A) pairs with thymine (T) by two hydrogen bonds in DNA, or with uracil (U) in RNA. Cytosine (C) pairs with guanine (G) by three hydrogen bonds in both DNA and RNA. Such base pairing is also referred to as "Watson–Crick" or "complementary" base pairing.

base stacking The tendency of the hydrophobic faces of the paired nitrogenous bases in DNA to stack on top of one another in such a way in solution as to minimize contact with the water molecules. A double-stranded DNA molecule thus has a hydrophobic core composed of stacked bases that exclude the maximum amount of water from the interior of the double helix.

basic helix-loop-helix (bHLH) motif An HLH motif coupled to a basic motif. When two bHLH proteins dimerize through their HLH motifs by forming a coiled coil, the basic motifs form a DNA-binding domain and are in position to interact with a specific region of DNA.

basic leucine zipper (bZIP) motif A leucine zipper motif coupled to a basic motif. When two bZIP proteins dimerize through their leucine zippers the basic motifs form a DNA-binding domain and are in the correct orientation to interact with a specific region of DNA.

B-DNA The standard Watson–Crick model of DNA. The predominant form of the DNA double helix within cells, favored at high relative humidity and in solution. A right-handed helix in which the bases are stacked almost exactly perpendicular to the main axis with 10.5 bases per turn.

beta (β)-galactosidase An enzyme that breaks the bond between the two constituent sugars of lactose.

beta (β)-lactamase An enzyme that breaks down ampicillin and renders a bacterium resistant to the antibiotic.

beta (β)-pleated sheet (or β-sheet) A fundamental unit of protein folding that involves extended amino acid chains in a protein that interact by hydrogen bonding. The chains are packed side by side to create an accordian-like appearance.

beta (β)-strand One of the polypeptide strands in a β-pleated sheet.

bidirectional DNA replication Replication that occurs in both directions at the same time from a common starting point, or origin of replication.

bioinformatics A discipline combining information technology with biotechnology; i.e. an area of computer science devoted to collecting, organizing, and analyzing DNA and protein sequences, and the data being generated by genomics and proteomics laboratories.

biolistics A method in which tiny metal pellets are coated with DNA and shot into cells.

BLAST A program that searches a database for DNA or protein sequence and displays how the query sequence lines up with the database sequences.

blue–white screening A method for distinguishing between bacteria transformed with recombinant and nonrecombinant vector DNA when the vector contains a multiple-cloning site that interrupts the *lacZ* gene. When grown on selective medium containing a colorless chromogenic compound X-gal, recombinant colonies are white. Nonrecombinant colonies are blue because β-galactosidase is produced and hydrolyzes X-gal to produced a blue-colored product.

bovine spongiform encephalopathy "Mad cow" disease; a prion infection in cattle.

branch migration Lateral motion of the branch of a Holliday junction during recombination.

BRCA1/BRCA2 The first breast cancer susceptibility genes to be identified. Mutated forms of these genes are believed to be responsible for about half the cases of inherited breast cancer, especially those that occur in younger women. Both are tumor suppressor genes involved in repair of DNA double-strand breaks by homologous recombination.

BRE TFIIB recognition element; a core promoter element in some RNA polymerase II gene promoters.

bromodomain A protein domain that binds specifically to acetylated lysine residues on other proteins, such as histones.

C

cancer Diseases in which abnormal cells divide and grow unchecked. Cancer can spread from its original site to other parts of the body and can also be fatal if not treated adequately.

candidate gene A gene, located in a chromosome region suspected of being involved in a disease, whose protein product suggests that it could be the disease gene in question.

canonical sequence See consensus sequence.

5′ cap A methylated guanosine bound through a 5′ → 5′ triphosphate linkage to the 5′ end of a eukaryotic mRNA, an hnRNA, or an snRNA.

CAP (catabolite activator protein) A protein which, together with cAMP, activates operons that are subject to catabolite repression. Also known as CRP.

cap-binding protein (CBP) A protein that associates with the 5′ cap on eukaryotic mRNA and allows the mRNA to bind to a ribosome. Also known as eIF4F.

carboxyl-terminal domain (CTD) The carboxyl-terminal region of the largest subunit of RNA polymerase II. Consists of dozens of repeats of a heptamer rich in serines and threonines and acts as a "landing pad" for RNA processing factors.

carboxyl terminus The end of a polypeptide with a free carboxyl group.

carcinogen Chemical compounds that are linked to sporadic human cancers, such as benzo(*a*)pyrene in cigarette smoke.

carcinoma Any of the various types of cancerous tumors that form in the epithelial tissue, the tissue forming the outer layer of the body surface and lining the digestive tract. Examples of this kind of cancer include breast, lung, and prostate cancer.

carrier An individual who possesses one copy of a mutant allele that causes disease only when two copies are present. Although carriers are not affected by the disease, two carriers can produce a child who has the disease.

cartilage hair hypoplasia A rare autosomal recessive form of dwarfism caused by mutations in the RNA component of RNase MRP; characterized by short limbs, short stature, fine sparse hair, impaired cellular immunity, anemia, and predisposition to several cancers.

catalytic center The active site of an enzyme, where catalysis takes place.

cathode Negative (black) electrode.

CCAAT-binding transcription factor (CTF) A transcription factor that binds to the CCAAT box.

CCAAT box An upstream DNA sequence motif, having the sequence CCAAT, found in many eukaryotic proximal promoters recognized by RNA polymerase II.

Cdc6 Cell division cycle 6. A component of the pre-replication complex that is essential for the initiation of DNA replication in eukaryotes.

cDNA (complementary DNA) A DNA copy of an RNA, made by reverse transcription.

cDNA library A collection of DNA sequences generated from mRNA expressed in a particular cell or tissue type at a given time. This type of library contains only protein-coding DNA sequences and does not include any noncoding DNA (e.g. regulatory elements or introns).

cell The basic structural unit of any living organism. It is a small, watery, compartment filled with chemicals and a complete copy of the organism's genome.

cell cycle The changes that take place in a cell in the period between its formation as one of the products of cell division and its own division and which in all cells includes DNA replication. In eukaryotic cells the cell cycle is divided into phases termed G1, S, G2, and M. G1 is the period immediately after mitosis and cell division when the newly formed cell is in the diploid state. S is the phase of DNA synthesis. This phase is followed by G2 when the cell is in a tetraploid state. Mitosis (M) follows to restore the diploid state, accompanied by cell division.

cell division splitting of a cell into two complete new cells, by binary fission in bacteria and other prokaryotes, and by division of both nucleus and cytoplasm in eukaryotic cells. The rapid cell divisions that occur during early embryogenesis are referred to as cleavages.

centimorgan A measure of genetic distance that tells how far apart two genes are and that yields a 1% recombination frequency between two markers. Generally 1 centimorgan equals about 1 million base pairs.

Central Dogma, the The hypothesis stated by Francis Crick in 1957 that "once information has passed into protein it cannot get out again." Describes the flow of information involving the genetic material.

centromere The constricted region near the center of a human chromosome. This is the region of the chromosome where the two sister chromatids are joined to one another. The centromere attaches chromosomes to the spindle fibers during mitosis and ensures that sister chromatids segregate correctly to daughter cells.

chaperones Proteins that bind to unfolded proteins or RNA molecules and help them fold properly.

Chargaff's rules The three regular relationships among the molar concentrations of the different bases in DNA: (1) The number of A residues in all DNA samples is equal to the number of T residues; i.e. [A] = [T], where the molar concentration of the base is denoted by the symbol for the base enclosed in square brackets. (2) The number of G residues equals that of C; i.e. [G] = [C]. (3) The amount of purine bases equals that of the pyrimidine bases: [A] + [G] = [T] + [C].

charging Coupling a tRNA with its cognate amino acid.

chemiluminescent The property of the product of an enzyme-catalyzed reaction that emits light and is therefore easily assayed.

chloramphenicol acetyl transferase (CAT) An enzyme encoded by a bacterial gene that adds acetyl groups to the antibiotic chloramphenicol. The CAT gene is frequently used as a reporter gene in eukaryotic transcription and translation experiments.

chloroplast DNA (cpDNA) The genetic material of the chloroplast, the organelle involved in photosynthesis in higher plants, some protozoans, and algae.

chromatids Copies of a chromosome produced during cell division.

chromatin Chromosomal DNA with its associated proteins.

chromatin assembly factor 1 (CAF-1) A protein that brings histones to the DNA replication fork via direction interaction with PCNA.

chromatin immunoprecipitation (ChIP) assay A method for purifying chromatin containing a protein of interest by immmunoprecipitating the chromatin with an antibody directed against that protein or against an epitope tag attached to the protein.

chromatin modification complexes Multiprotein complexes that modify histones post-translationally in ways that allow greater access or restrict access of other proteins to DNA.

chromatin remodeling ATP-dependent alterations of the structure of the nucleosomes that either move the nucleosome, remove them entirely, or modify their composition.

chromatin remodeling complexes Multiprotein complexes of the yeast SWI/SNF family (or their mammalian homologs BRG1 and BRM) and related families that contain ATP-dependent DNA unwinding activities.

chromatography A group of techniques for separating molecules based on their relative affinities for a mobile and a stationary phase; e.g. in ion-exchange chromatography, the charged resin is the stationary phase, and the buffer of increasing ionic strength is the mobile phase.

chromodomain A conserved region found in proteins involved in heterochromatin formation that binds to methylated histones.

chromogenic substrate A substrate that produces a colored product when acted on by an enzyme.

chromosome One of the threadlike "packages" of genes and other DNA in the nucleus of a cell. Different kinds of organisms have different numbers of chromosomes. Humans have 23 pairs of chromosomes, 46 in all. Each parent contributes one chromosome to each pair, so children get half of their chromosomes from their mothers and half from their fathers.

chronic wasting disease A type of transmissible spongiform encephalopathy that infects elk and deer.

cis-**acting** A term that describes a DNA sequence element, such as an enhancer, a promoter, or an operator, that must be on the same chromosome in order to influence a gene's activity.

cis-**splicing** Ordinary splicing in which the exons are on the same pre-mRNA.

cleavage factors I and II (CFI and CFII) RNA-binding proteins involved in cleavage of pre-mRNA at the polyadenylation site.

cleavage poly(A) specificity factor (CPSF) A protein that recognizes the AAUAAA part of the polyadenylation signal in a pre-mRNA and stimulates cleavage.

cleavage stimulation factor (CstF) A protein that recognizes the GU-rich part of the polyadenylation signal in a pre-mRNA and stimulates cleavage.

cloning The process of making copies of a specific piece of DNA, usually a gene. When molecular biologists speak of cloning, they generally do not mean the process of making genetically identical copies of an entire organism. Sometimes called molecular cloning to distinguish it from cloning by nuclear transfer.

cloning by nuclear transfer The production of genetically identical animals by nuclear transfer from adult somatic cells to unfertilized eggs.

coactivator A protein that increases transcriptional activity through protein–protein interactions without binding DNA directly. Operationally defined as a component required for activator-directed (activated) transcription, but dispensable for activator-independent (basal) transcription; includes chromatin remodeling and modification complexes.

Cockayne syndrome A disease characterized by photosensitivity, cataracts, deafness, and severe mental retardation; linked to defects in transcription-coupled repair.

codon A 3-base sequence in mRNA which specifies a single amino acid to be added into a polypeptide chain or causes termination of translation.

codon bias The frequency with which different codons are used varies significantly between different organisms and between proteins expressed at high or low levels within the same organism.

cofactor Any nonprotein substance required by a protein (or ribozyme) for biological activity such as prosthetic groups and, especially in enzyme-catalyzed reactions, other compounds which are not consumed in the process and are found unchanged at the end of the reaction.

cognate An antigen recognized by its corresponding antibody, a tRNA recognized by a particular aminoacyl-tRNA synthetase, etc.

coiled coil A protein motif in which two α-helices (coils) wind around each other. Coiled coils may form within one protein or between two separate proteins.

coimmunoprecipitation assay A technique used to analyze protein–protein interactions *in vivo*. Proteins from cell extracts are reacted with a specific antibody or antiserum against one of the proteins, then immunoprecipitated by centrifugation. The precipitated proteins are usually detected by Western blot analysis.

colony hybridization A procedure for selecting a baterial clone (colony) containing a gene of interest. DNAs from a large number of clones are tested with a labeled probe that hybridizes to the gene of interest.

competent cells Host bacterial cells chemically treated to make their membranes leaky; used for transformation with foreign DNA.

complementary base pairing See base pairing.

complementation group Various mutations which do not form a wild-type (normal) phenotype after crossing.

concatemers DNAs of multiple genome length.

congenital Any trait or condition that exists from birth.

consensus sequence The "ideal" form of a DNA sequence found in slightly different forms in different organisms, but which is believed to have the same function. The consensus sequence gives for each position the nucleotide most often found. Sometimes called a canonical sequence.

constant region (C) The region of an antibody (immunoglobulin) that is basically the same from one antibody to the next.

constitutive gene expression A gene is always turned on.

contig A chromosome map showing the locations of those regions of a chromosome where contiguous DNA segments overlap. Contig maps are important because they allow study of a complete, and often large, segment of the genome, by examining a series of overlapping clones which then provide an unbroken series of information about that region.

core histone One of four highly conserved histones (H2A, H2B, H3, H4) that are present in the nucleosome as an octamer composed of a dimer of histones H2A and H2B at each end and a tetramer of histones H3 and H4 in the center, around which 146 bp of genomic DNA are wound.

core histone octamer (or core octamer) A particle composed of a dimer of histones H2A and H2B at each end and a tetramer of histones H3 and H4 in the center, around which 146 bp of genomic DNA are wound. The core octamer lacks the linker histone and linker DNA that would be present in the complete nucleosome.

core promoter An approximately 60 bp DNA sequence overlapping the transcription start site (+1) that serves as the recognition site for RNA polymerase II and general transcription factors.

corepressors Proteins that decrease transcriptional activity through protein–protein interactions without binding DNA directly; include chromatin remodeling and modification complexes.

cos The cohesive ends of the linear lambda (λ) phage DNA.

cosmid A vector designed for cloning large DNA inserts. It contains the *cos* sites of lambda (λ) phage DNA, so it can be packaged into λ heads, and a plasmid origin of replication so it can replicate in bacteria as a plasmid.

cotranscriptonal cleavage (CoTC) The cleavage of a growing transcript downstream of the polyadenylation site that is part of the transcription termination process.

counts per minute (cpm) The average number of scintillations detected per minute by a liquid scintillation counter. Generally, this is disintegrations per minute (dpm) times the efficiency of the counter.

coupling The triggering of a large conformational change in both an enzyme and its substrate upon binding. Also known as induced fit.

CpG island A region of DNA containing many unmethylated CpG sequences that is usually associated with active genes.

Creutzfeldt–Jakob disease (CJD) A sporadic form of transmissible spongiform encephalopathy. In this disease PrP^C misfolds spontaneously and then by "autoinfection" generates more prions.

cross–over Physical exchange between DNAs that occurs during recombination.

cruciform A cross-shaped paired stem-loop formation that can form in regions of inverted repeats in DNA.

cryoelectron microscopy A protein or other macromolecule sample is snap-frozen at extremely low temperature in liquid helium so that it is fully hydrated without water crystals; it is then examined in a cryoelectron microscope.

cyclic AMP (cAMP) An adenine nucleotide with a cyclic phosphodiester linkage between the 3′ and 5′ carbons.

cyclin-dependent kinases (CDKs) A family of kinases whose activity is regulated by both phosphorylation and interaction with regulatory protein called cyclins. The CDKs are important in regulating progression through the cell cycle.

cyclins A family of proteins that accumulate gradually during interphase and are abruptly destroyed during mitosis; regulatory subunits of the cyclin-dependent kinases.

cyclobutane-pyrimidine dimers See pyrimidine dimers.

cystic fibrosis A hereditary disease whose symptoms usually appear shortly after birth, caused by mutations in the cystic fibrosis transmembrane regulator (CFTR) protein. They include faulty digestion, breathing difficulties and respiratory infections due to mucus accumulation, and excessive loss of salt in sweat. In the past, cystic fibrosis was almost always fatal in childhood, but treatment is now so improved that patients commonly live to their twenties and beyond.

cytidine A nucleoside containing the base cytosine.

cytogenetic map The visual appearance of a chromosome when stained and examined under a microscope. Particularly important are visually distinct regions, called light and dark bands, which give each of the chromosomes a unique appearance and allows characterization of chromosomal alterations. This feature allows a person's chromosomes to be studied in a clinical test known as a karyotype.

cytoplasmic polyadenylation element (CPE) A sequence in the 3′ UTR of an mRNA (consensus, UUUUUAU) that is involved in cytoplasmic polyadenylation.

cytosine One of the four bases in DNA that make up the letters ATGC, cytosine is the "C." The others are adenine, guanine, and thymine. Cytosine always pairs with guanine.

cytosine deamination The removal of an amino group (NH_2) from a cytosine in DNA, in which the amino group is replaced by a carbonyl group (C–O). This converts cytosine to uracil.

D

D loop (mtDNA) Displacement loop. A region of 500–600 nucleotides where replication begins in mitochondrial DNA.

Dalton The term used to describe the molecular weight of proteins. 1 Dalton is equivalent to 1 atomic mass unit.

DEAD box protein A member of a family of proteins containing the sequence Asp-Glu-Ala-Asp and having RNA helicase activity.

deadenylation The removal of AMP residues from poly(A) in the cytoplasm.

deamination The conversion of one base to another by replacement of the amino group with, for example, an oxygen; e.g. the conversion of cytosine in DNA to uracil, a base that should only be present in RNA, and the conversion of 5-methylcytosine to thymine.

decoding Interactions between codons and anticodons on the ribosome that lead to binding of the correct aminoacyl-tRNA.

degenerate code In the genetic code more than one codon can stand for a single amino acid.

deletion A mutation involving a loss of one or more base pairs of DNA from a chromosome. Deletion of a gene or part of a gene can lead to a disease or abnormality.

denaturation of DNA The unwinding and separation of the two strands of double-stranded DNA; also referred to as melting.

denaturation of protein Disruption of the three-dimensional structure of the protein without breaking any covalent bonds.

density gradient centrifugation A solution of cesium chloride (CsCl) containing a DNA sample is spun in an ultracentrifuge at high speed for several hours. An equilibrium between centrifugal force and diffusion occurs, such that a gradient forms with a high concentration of CsCl at the bottom of the tube and a low concentration at the top. DNA forms a band in the tube at the point where the density is the same as that of CsCl.

deoxynucleoside triphosphates (dNTPs) The building blocks of DNA: dATP, dCTP, dGTP, dTTP.

deoxyribonucleic acid (DNA) The chemical inside the nucleus of a cell that carries the genetic instructions for making living organisms.

deoxyribose The pentose (five-carbon) sugar present in a nucleotide subunit of DNA.

diabetes mellitus Two types of a highly variable disorder in which abnormalities in the ability to make and/or use the hormone insulin interfere with the process of turning dietary carbohydrates into glucose, the body's fuel. Type I is known as insulin-dependent diabetes mellitus, and type II is known as non-insulin-dependent diabetes mellitus.

Dicer The member of the RNAse III family that cleaves double-stranded RNA into pieces about 21 bp long (called siRNAs) during the RNAi process.

dideoxynucleotide A nucleotide, lacking the oxygens at both the 2′ and 3′ positions (hence "dideoxy"), used to terminate DNA chain elongation in DNA sequencing.

digoxygenin A plant steroid isolated from foxglove (*Digitalis*) used in nonradioactive labeling techniques.

dimer (protein) A complex of two polypeptides. These can be the same (homodimer) or different (heterodimer).

dimerization domain The region of a protein that interacts with another protein to form a dimer.

diploid The number of chromosomes in most cells except the gametes. In humans, the diploid number is 46.

directional cloning Insertion of a foreign DNA into two different restriction sites of a vector, such that the orientation of the insert can be predetermined.

disintegrations per minute (dpm) The average number of radioactive emissions produced each minute by a sample.

DNA-binding domain The part of a DNA-binding protein that makes specific contacts with a target site on the DNA.

DNA glycosylase An enzyme that breaks the glycosidic bond between a damaged base and its sugar to form an abasic site.

DNA ligase An enzyme that joins two double-stranded DNAs end to end.

DNA looping The process by which DNA-binding proteins can interact simultaneously with one another and with remote sites on DNA, by causing the DNA in between the sites to form a loop.

DNA melting The unwinding and separation of double-stranded DNA; also referred to as denaturation.

DNA methyltransferase 1 (DNMT1) An enzyme that adds methyl groups to hemimethylated DNA substrates during DNA replication.

DNA methyltransferase 3 (DNMT3) Together with its cofactor DNMT3L, this enzyme is required to establish genomic imprinting *de novo* in sperm and oocytes.

DNA microarray A way of studying how large numbers of genes interact with each other and how a cell's regulatory networks control vast batteries of genes simultaneously. The method uses a robot to precisely apply tiny droplets containing functional DNA to glass slides. Researchers then synthesize fluorescently labeled cDNA from mRNA isolated from the cells they are studying. The labeled probes are allowed to bind to cDNA strands on the slides. The slides are put into a scanning microscope that can measure the brightness of each fluorescent dot; brightness reveals how much of a specific DNA fragment is present, an indicator of how active the gene was in the original cell type.

DNA photolyase The enzyme that catalyzes photoreactivation by breaking pyrimidine dimers.

DNA polymerase An enzyme that synthesizes DNA by linking together deoxynucleoside monophosphates (dNMPs) in the order dictated by the complementary sequence of nucleotides in a template DNA strand. There are two major classes of DNA polymerases: those involved in genome replication (replicative polymerases) and those involved in DNA repair.

DNA polymerase I An *E. coli* DNA polymerase that plays a role in primer removal and gap filling between Okazaki fragments and in the nucleotide excision repair pathway.

DNA polymerase II An *E. coli* DNA polymerase that is involved in DNA repair mechanisms.

DNA polymerase III The *E. coli* replicative polymerase that catalyzes genome replication.

DNA polymerase IV The *E. coli* DNA polymerase that mediates translesion DNA synthesis. Also known as DinB (encoded by the *dinB* gene).

DNA polymerase V The *E. coli* DNA polymerase that mediates translesion DNA synthesis. Also known as the UmuD$'_2$C complex (encoded by the *umuDC* operon).

DNA polymerase α The eukaryotic DNA polymerase involved in priming DNA synthesis during replication and repair.

DNA polymerase β The eukaryotic DNA polymerase involved in high fidelity base excision and double-strand break repair.

DNA polymerase σ The eukaryotic DNA polymerase involved in high fidelity replication of the leading strand during replication and repair.

DNA polymerase ε The eukaryotic DNA polymerase involved in high fidelity replication of the lagging strand during replication and repair.

DNA polymerase γ The eukaryotic DNA polymerase involved in high fidelity replication and repair of mitochondrial DNA.

DNA polymerase η The eukaryotic DNA polymerase involved in high fidelity translesion DNA synthesis (relatively accurate replication past thymine–thymine dimers).

DNA polymerase ι A eukaryotic DNA polymerase involved in error-prone translesion DNA synthesis (required during meiosis).

DNA polymerase κ A eukaryotic DNA polymerase involved in error-prone translesion DNA synthesis and in double-strand break repair by nonhomologous end-joining.

DNA polymerase λ A eukaryotic DNA polymerase involved in error-prone translesion DNA synthesis.

DNA polymerase μ A eukaryotic DNA polymerase involved in error-prone double-stranded DNA break repair by nonhomologous end-joining.

DNA polymerase υ A eukaryotic DNA polymerase thought to be involved in error-prone DNA cross-link repair.

DNA polymerase θ A eukaryotic DNA polymerase involved in error-prone repair of DNA interstrand cross-links.

DNA polymerase ξ A eukaryotic DNA polymerase involved in error-prone translesion DNA synthesis (thymine dimer bypass).

DNA replication The process by which the DNA double helix unwinds and makes an exact copy of itself.

DNA sequencing Determining the exact order of the base pairs in a segment of DNA.

DNA transposons Transposable elements with a DNA intermediate during transposition.

DNA typing The use of highly variable regions of DNA to identify particular individuals (also called DNA fingerprinting).

DNA-PK (DNA protein kinase) The key enzyme in eukaryotic double-strand break repair by nonhomologous end-joining.

DNA-PK$_{CS}$ The catalytic subunit of DNA-PK.

DNase Deoxyribonuclease, an enzyme that degrades DNA.

DNase footprinting A method of detecting the binding site for a protein on DNA by observing the DNA region this protein protects from cleavage by DNase.

DNase hypersensitive sites Regions of chromatin that are about a hundred times more susceptible to cleavage by DNase than the rest of the chromatin. Active genes tend to be DNase-sensitive.

dominant An allele or trait that expresses its phenotype when heterozygous with a recessive allele; for example, a disease, even though the patient's genome possesses only one copy. With a dominant gene, the chance of passing on the gene (and therefore the disease) to children is 50–50 in each pregnancy.

dominant negative mutation A mutation that yields a protein that is not only inactive but disrupts the activity of wild-type protein made in the same cell, by forming a heterodimer, for example.

double helix The structural arrangement of DNA, which looks something like a very long ladder twisted into a helix, or coil. The sides of the "ladder" are formed by a backbone of sugar and phosphate molecules, and the "rungs" consist of nucleotide bases joined weakly in the middle by hydrogen bonds and stabilized by base stacking interactions.

double-strand break Damage to the DNA backbone in which nicks are made in both strands of the DNA double helix; induced by agents such as reactive oxygen species, ionizing radiation, and chemicals that generate reactive oxygen species (free radicals).

double-strand break repair Repair of double-strand breaks in DNA either by homologous recombination or nonhomologous end-joining.

downstream DNA sequence after the start of transcription (+1) in the direction of the 3′ end; numbered with positive numbers (e.g. +35).

DPE Downstream promoter element; an RNA polymerase II core promoter element in some gene promoters.

Drosha The endonuclease the cleaves a primary miRNA transcript in the nucleus to form a shorter hairpin pre-miRNA.

Ds *Disassociation*, a defective transposable element found in maize, which relies on an *Ac* element for transposition.

dsDNA Double-stranded DNA.

dsRNA Double-stranded RNA.

duplex DNA Double-stranded DNA.

duplication A particular kind of mutation: production of one or more copies of any piece of DNA, including a gene or even an entire chromosome.

duplicative transposition Transposition in which a DNA sequence replicates, so one copy remains in the original location as another copy moves to the new site. Also called replicative transposition.

dyskeratosis congenita A rare inherited, premature aging disease linked to loss of telomerase activity either through mutations in the dyskerin gene or the telomerase RNA gene.

dyskerin A pseudouridine synthase that binds to many snoRNAs and is proposed to play a role in ribosomal RNA base modification.

E

E site The exit site to which deacylated tRNAs bind on their way out of the ribosome.

EF-Tu The prokaryotic translation elongation factor that, along with GTP, carries aminoacyl-tRNAs (except fMet-tRNA$_f^{Met}$ to the ribosome A site).

eIFs The set of eukaryotic translation initiation factors.

electrophoretic mobility shift assay (EMSA) An assay for DNA–protein binding. A short labeled DNA is mixed with a protein and electrophoresed. If the DNA binds the protein, its electrophoretic mobility is greatly decreased.

electroporation The use of a strong electric current to introduce DNA into cells.

Elongator A protein complex that facilitates transcript elongation by RNA polymerase II.

empty vector A term sometimes used to refer to a cloning vector without an insert.

encode To contain the information for making an RNA or polypeptide. A gene can encode a functional RNA or a polypeptide.

end-filling Filling in the recessed 3′-end of a double-stranded DNA using deoxynucleoside triphosphates (dNTPs) and a DNA polymerase; often used to label the 3′-end of a DNA strand.

endonuclease An enzyme that cleaves the phosphodiester bond joining adjacent nucleotides at an internal site within a polynucleotide.

endoplasmic reticulum A network of membranes in the cell on which proteins destined for secretion from the cell are synthesized.

enhanceosome The complex formed by enhancers coupled to their activators, involving protein–protein interactions and DNA looping.

enhancer A DNA regulatory element that increases gene promoter activity. Enhancers are usually 700–1000 bp or more away from the start of transcription and can be downstream, upstream, or within an intron, and can function in either orientation relative to the promoter.

enzyme A molecule that catalyzes or increases the rate, or velocity, of a biochemical reaction without itself being changed in the overall process. Most enzymes are globular proteins, but some RNA molecules have the properties of enzymes (see ribozyme).

enzyme-linked immunosorbent assay (ELISA) An immunoassay used to quantify antigen–antibody reactions by combining the specificity of antibodies with the sensitivity of simple enzyme assays.

epigenetics The study of mitotically and/or meiotically heritable changes in gene expression without changes in gene sequence; e.g. through changes in the pattern of DNA methylation and histone modifications.

epitope A region of an antigen to which an antibody can bind.

epitope tagging Using genetic means to attach a small group of amino acids (an epitope tag) to a protein. This enables the protein to be purified readily by immunoprecipitation with the antibody that recognizes the epitope tag.

eRF1 The eukaryotic release factor that recognizes all three stop codons and releases the completed polypeptide from the ribosome.

eRF3 The eukaryotic release factor with ribosome-dependent GTPase activity that interacts with eRF1 in releasing polypeptides from the ribosome.

Escherichia coli (E. coli) An intestinal bacterium; the conventional host bacterium used in molecular biology research.

ethidium bromide A fluorescent dye that intercalates between the bases of nucleic acids.

euchromatin Chromatin that is open and accessible to RNA polymerase, and at least potentially active. These regions stain lightly and are thought to contain most of the genes.

eukaryote An organism whose cells have nuclei.

exons Sequences that are ligated together after excision during RNA splicing; typically the expressed sequences of an mRNA. The term is used to refer both to the corresponding sequence in the DNA and the RNA.

exonuclease An enzyme that removes dNMPs or NMPs from the end of a polynucleotide chain inward by breaking the terminal phosphodiester bond.

exosome A protein complex that degrades RNAs. Different exosomes are found in the nucleus and the cytoplasm.

exportin A soluble receptor that mediates nuclear export of a NES-bearing cargo.

expressed sequence tag (EST) Partial cDNA sequences generated by amplifying cellular mRNA by RT-PCR.

expression platform The domain of a riboswitch that converts metabolite binding into changes in gene expression via changes in RNA folding. The expression platform has the potential to form alternative antiterminator and terminator hairpins (see transcriptional attenuation).

expression site A locus on a chromosome where a gene can be moved to be expressed efficiently. For example, the expression site for VSG genes in trypanosomes is at the end (telomere) of a chromosome.

expression vector A cloning vector that allows expression of a cloned gene in bacterial or eukaryotic cells.

extrachromosomal A molecule of DNA, such as a plasmid, that remains separate from the host cell chromosome.

F

F plasmid An *E. coli* plasmid that allows conjugation between bacterial cells.

F₁ The progeny (i.e. the first filial generation) of a cross between two parental types that differ at one of more genes.

F$_2$ The progeny (i.e. the second filial generation) of a cross between two F$_1$ individuals or the progeny of a self-fertilized F$_1$.

FACT A protein that facilitates transcription elongation by eukaryotic RNA polymerases through nucleosomes arrays.

familial dysautonomia An autosomal recessive disorder of the sensory and autonomic nervous system caused by a splice site mutation in a subunit of the Elongator complex.

fatal familial insomnia An inherited autosomal dominant form of transmissible spongiform encephalopathy which involves a mutated PrP^C gene with a greater tendency to spontaneously misfold to the prion form.

fibroblasts A type of cell found just underneath the surface of the skin. Fibroblasts are part of the support structure for tissues and organs.

fibrous protein A protein with a long filamentous or "rod-like" structure; major structural components of cells.

five (5)-prime (5′) The ends of a DNA or RNA chain are chemically distinct and are designated by the symbols 3′ and 5′. The symbol 5′ refers to the carbon in the ribose or deoxyribose sugar ring to which a phosphate (PO$_4$) functional group is attached.

FlapEndoNuclease I (FEN-1) An nuclease that acts in association with PCNA to degrade the RNA primer during eukaryotic DNA replication.

fluorescence *in situ* hybridization (FISH) A process which hybridizes a fluorescent probe to whole chromosomes to determine the location of a gene or other DNA sequence within a chromosome. This technique is useful for identifying chromosomal abnormalities and gene mapping. Also used to determine the precise localization of RNA transcripts within a cell.

fluorescence resonance energy transfer (FRET) A method for detecting protein–protein interactions *in situ* by a nonradiative process whereby energy from an excited donor fluorophore is transferred to an acceptor fluorophore that is within approximately 10 nm of the excited fluorophore.

fluorophore A group of atoms in a molecule responsible for absorbing light energy and producing the color of the compound; i.e. the region of a molecule exhibiting fluorescence.

fragile X syndrome After Down syndrome, the second most frequent genetic cause of mental retardation. The disorder is one of a group of diseases that results from trinucleotide repeat expansion. In fragile X, the repeating triplet is CGG in a gene called FMR1 on the X chromosome. There are ordinarily fewer than 55 copies of the repeat. When the number of repeats exceeds 200, the expansion is called a "full mutation" and expression of the FMR1 gene is inhibited.

frameshift mutation An insertion or deletion of one or two bases in the coding reigon of a gene, which changes the reading frame of the corresponding mRNA.

free radicals Very reactive chemical substances with an unpaired electron that can cause DNA damage (e.g. reactive oxygen species such as O$_2^-$ and H$_2$O$_2$).

Friedreich's ataxia A rare inherited neurological disease characterized by the progressive loss of voluntary muscular coordination and heart enlargement. The disorder is caused by a GAA trinucleotide repeat expansion in the first intron of the frataxin gene.

fusion protein A protein resulting from the expression of a recombinant DNA containing two open reading frames (ORFs) fused together. One or both of the ORFs can be incomplete.

G

G + C content The base composition of DNA, defined as the "percent G + C," differs among species but is constant in all cells of an organism within a species. The G + C content can vary from 22 to 73%, depending on the organism.

G protein A protein that is activated by binding to GTP and inactivated by hydrolysis of the bound GTP to GDP by an inherent GTPase activity.

gamete A haploid sex cell.

gametogenesis The process by which gametes are produced.

gel electrophoresis The process in which molecules (such as proteins, DNA, or RNA fragments) can be separated according to size and electrical charge by applying an electric current to them. The current forces the molecules through pores in a thin layer agarose or polyacrylamide. The gel can be made so that its pores are just the right dimensions for separating molecules within a specific range of sizes and shapes. Smaller fragments usually travel further than large ones.

gel filtration A column chromatography method for separating molecules according to their sizes. Small molecules enter the beads of the gel and so take longer to move through the column than larger molecules, which cannot enter the beads.

gene The basic unit of heredity. A complete chromosomal segment responsible for making a functional product. Three essential features of a gene are: the expression of a product, the requirement that it be functional, and the inclusion of both coding and regulatory regions.

gene amplification An increase in the number of copies of any particular piece of DNA. A tumor cell often amplifies or copies DNA segments as a result of cell signals and sometimes environmental events.

gene cloning Generating many copies of a gene by inserting into an organism, such as a bacterium, where it can replicate along with the host.

gene conversion The conversion of one gene's sequence to that of another by homologous recombination.

gene expression The process by which gene products (RNA transcripts or protein) are made from the instructions encoded in DNA.

gene mapping Determining the relative positions of genes on a chromosome and the distance between them.

gene targeting The replacement or mutation of a particular gene that provides a means for creating strains of "knockout" organisms with mutations in virtually any genes.

gene therapy An evolving technique used to treat inherited diseases. The medical procedure involves either replacing, manipulating, or supplementing nonfunctional genes with healthy genes. Also referred to as somatic cell gene therapy.

gene transfer Insertion of unrelated DNA into the cells of an organism. Most techniques involve the use of a vector, such as a specially modified virus that can take the gene along when it enters the cell.

general transcription factors A set of five RNA polymerase II transcription factors, denoted TFIIB, TFIID, TFIIE, and TFIIH that are responsible for promoter recognition and for unwinding the promoter DNA. RNA polymerase II is absolutely dependent on these auxiliary transcription factors for the initiation of transcription.

general transcription machinery General, but diverse, components of large multiprotein RNA polymerase machines required for promoter recognition and the catalysis of RNA synthesis.

genetic code (ATGC) The set of 64 codons and the amino acids (or termination) for which they stand. Each gene's code combines the four bases in various ways to spell out three-letter "words" that specify which amino acid is needed at every step in making a protein.

genetic counseling A short-term educational counseling process for individuals and families who have a genetic disease or who are at risk for such a disease. Genetic counseling provides patients with information about their condition and helps them make informed decisions.

genetic linkage The physical association of genes on the same chromosome.

genetic map Also known as a linkage map; a chromosome map of a species that shows the position of its known genes and/or markers relative to each other, rather than as specific physical points on each chromosome.

genetic marker A segment of DNA with an identifiable physical location on a chromosome and whose inheritance can be followed. A marker can be a gene, or it can be some section of DNA with no known function. Because DNA segments that lie near each other on a chromosome tend to be inherited together, markers are often used as indirect ways of tracking the inheritance pattern of a gene that has not yet been identified, but whose approximate location is known.

genetic screening Testing a population group to identify a subset of individuals at high risk for having or transmitting a specific genetic disorder.

genome One complete set of genetic information contained within an organism or a cell; i.e. the single, circular chromosome of a bacterium is its genome. The term is often used interchangebly with the terms genomic DNA, chromosomal DNA, or nuclear DNA (to distinguish it from organelle or plasmid DNA).

genomic imprinting The nonrandom expression of a gene from only one of the two parental chromosomes, regulated by differential methylation.

genomic library A set of clones containing DNA fragments derived directly from a genome, rather than from mRNA.

genomics The comprehensive study of whole sets of genes and their interactions rather than single genes or proteins.

genotype The genetic identity or allelic constitution of an individual that does not show as outward characteristics. The genotypes at locus B in a diploid individual may be BB, Bb, or bb.

germline Inherited material that comes from the eggs or sperm and is passed on to offspring.

Gerstmann–Sträussler–Scheinker syndrome An inherited autosomal dominant form of transmissible spongiform encephalopathy which involves a mutated PrP^C gene with a greater tendency to spontaneously misfold to the prion form.

global genome repair Nucleotide excision repair pathway responsible for recognizing DNA damage in the whole genome.

globular protein A protein which folds into a roughly spherical shape. Most enzymes are globular proteins.

glucocorticoid receptor A nuclear receptor that mediates a highly abbreviated signal transduction pathway by activating target genes in response to ligand, leading to many diverse cellular responses, ranging from increases in blood sugar to anti-inflammatory action.

glucose A simple, six-carbon sugar used by many forms of life as an energy source.

glycoprotein A protein with a carbohydrate group attached post-translationally.

Golgi apparatus A membranous organelle that packages newly synthesized proteins for export from the cell.

green fluorescent protein A naturally fluorescent protein from the jelly fish *Aequorea victoria* commonly used as a reporter gene.

Greig cephalopolysyndactyly syndrome A very rare autosomal dominant disorder that is characterized by physical abnormalities affecting the fingers, toes, head, and face; caused by mutations in the GLI3 gene which is part of the Sonic hedgehog signal transduction pathway.

group I introns Self-splicing introns in which splicing is initiated by a free guanosine or guanosine nucleotide.

group II introns Self-splicing introns in which splicing is initiated by formation of a lariat-shaped intermediate.

guanine (G) One of the four bases in DNA. Guanine always pairs with cytosine.

guanosine A nucleoside containing the base guanine.

guide RNAs (editing) Small RNAs that bind to regions of an mRNA precursor in trypanosomes and serve as templates for editing a region upstream.

gyrase A topoisomerase that introduces negative superhelical turns into DNA and relaxes the positive superhelical strain created by unwinding the *E. coli* DNA during replication.

H

H strand Heavy strand of mitochondrial DNA, based on the relative buoyant density by density gradient centrifigution.

H5N1 virus An avian influenza virus strain that poses the threat of an influenza pandemic if it gains the ability to transmit readily from human to human.

hairpin A structure that resembles a bobby pin (hairpin), formed by intramolecular base pairing in an inverted repeat of a single-stranded DNA or RNA.

hairpin ribozyme A small ribozyme found in some virusoids that has a hairpin secondary structure.

half-life The time it takes for half of a population of molecules to disappear (turn-over).

hammerhead ribozyme A small ribozyme so called for its secondary structure of three helices in a T-shape; the most frequently found catalytic motif in plant pathogenic RNAs such as viroids.

haploid The number of chromosomes in a sperm or egg cell (gamete), half the diploid number.

haploinsufficiency A situation in which the protein produced by a single copy of an otherwise normal gene is not sufficient to assure normal function.

haplotype A cluster of alleles on a single chromosome.

Hayflick Limit The point at which cultured human and animal cells normally stop dividing because they have a limited capacity for replication.

heat shock proteins Proteins whose expression is significantly increased in response to environmental stress, including heat; promote protein folding and aid in the destruction of misfolded proteins.

helicase An enzyme that unwinds a polynucleotide double helix (either DNA or RNA).

helix-turn-helix (HTH) A structural motif in certain DNA-binding proteins, especially those from prokaryotes, that fits into the DNA major groove and gives the protein its DNA-binding capacity and specificity. A specialized type of HTH in eukaryotes is the homeodomain, present in DNA-binding proteins important for development.

helper virus A virus that supplies the functions lacking in a defective virus, allowing the latter to replicate.

hematopoietic stem cell An unspecialized precursor cell that will develop into a mature blood cell.

hemizygous When a new transgenic locus is present in only one member of a particular chromosome pair.

hemoglobin The oxygen-carrying protein in red blood cells.

hemophilia A sex-linked inherited bleeding disorder that generally only affects males. The disorder is characterized by a tendency to bleed spontaneously or at the slightest injury because of the lack of certain clotting factors in the blood.

hepatitis delta virus (HDV) RNA ribozyme A small ribozyme that is a viroid-like satellite virus of the human hepatitis B virus, which when present causes an exceptionally strong type of hepatitis in infected patients.

hereditary nonpolyposis colon cancer (HNPCC) A common form of human hereditary colon cancer, caused by defects in mismatch repair genes.

heredity The transmission of characteristics from parent to offspring by means of genetics.

heterochromatin Chromatin that is condensed and inactive.

heteroduplex A double-stranded polynucleotide whose two strands are not completely complementary.

heterogeneous nuclear RNA (hnRNA) A class of large, heterogeneous-size RNAs found in the nucleus, including unspliced mRNA precursors.

heterologous probe A probe that is similar to, but not exactly the same as, the nucleic acid sequence of interest.

heteroplasmy A condition in which both mutant and normal mitochondrial DNA coexist within the same cell.

heterozygous Possessing two different forms of a particular gene, one inherited from each parent.

high-throughput analysis Methods for whole genome and proteome analysis on a large scale.

highly conserved sequence A DNA sequence that is very similar in several different kinds of organisms. These cross-species similarities are regarded as evidence that a specific gene performs some basic function essential to many forms of life and that evolution has therefore conserved its structure by permitting few mutations to accumulate in it.

Hispanic thalessemia A rare disorder characterized by partial or complete deletions of the locus control region of the β-globin gene cluster.

histone Any one of a set of small, positively charged, basic proteins (H1, H2A, H2B, H3, H4), rich in arginine and lysine, bound to DNA in eukaryote chromosomes to form nucleosomes (see also linker histone and core histone).

histone acetyltransferase (HAT) An enzyme that transfers acetyl groups from acetyl CoA to core histones.

histone code A hypothesis proposing that covalent post-translational modifications of histones are read by the cell and lead to a complex, combinatorial transcriptional output.

histone deacetylase (HDAC) An enzyme that removes acetyl groups from core histones.

histone demethylase An enzyme that removes methyl groups from core histones (e.g. lysine specific demethylase 1, LSD1).

histone methyltransferase (HMT) A chromodomain-containing enzyme that transfers methyl groups to core histones.

hit and run model A model proposing that transcriptional activation reflects the probability that all components required for activation will meet at a certain chromatin site (the "hit"), i.e. a transcription complex is assembled in a stochastic fashion from freely diffusible proteins, and that the binding of these proteins is transient (the "run").

HMG protein An architectural nuclear protein with a high electrophoretic mobility (high-mobility group). Some of the HMG proteins are involved in transcriptional regulation.

Holliday junction The branched DNA structure formed by the first strand exchange during recombination.

holoenzyme A complete, fully functional enzyme molecule, consisting of a protein portion (apoenzyme), a nonprotein prosthetic group(s), or any other regulatory or accessory protein subunit if appropriate.

homeobox A sequence of about 180 bp that encodes a homeodomain; found in homeotic genes and other development-controlling genes in eukaryotes.

homeodomain A 60-amino acid domain of a DNA-binding protein with a type of helix-turn-helix domain that allows the protein to bind tightly to a specific DNA region.

homeotic gene A gene in which a mutation causes the transformation of one body part to another.

homologous chromosomes Chromosomes that are identical in size, shape, banding pattern and, except for allelic differences, genetic composition.

homologous probe A probe that is exactly complementary to the nucleic acid sequence of interest.

homologous recombination (DNA repair) Repair of double-strand breaks by retrieving genetic information from an undamaged homologous chromosome. In cases where the two chromosomes are not exactly homologous, gene conversion may take place.

homologous recombination (meiosis) The exchange of pieces of DNA during the formation of eggs and sperm. Recombination allows the chromosomes to shuffle their genetic material, increasing the potential of genetic diversity. Homologous recombination is also known as crossing-over.

homologs Genes that have evolved from a common ancestral gene. Includes orthologs and paralogs.

homoplasmy The normal condition in which all the mitochondrial DNA (mtDNA) within the cells of an individual are identical.

homozygous Possessing two identical forms of a particular gene, one inherited from each parent.

Hoogsteen base pairs A-T and G-C base pairs that have altered patterns of hydrogen bonding compared with Watson–Crick base pairs.

hopping (protein) A protein moves between binding sites on DNA through three-dimensional spaces, by dissociating from its initial site before reassociating elsewhere in the same DNA chain; the main mode of translocation over long distances.

hormone response elements Enhancers that respond to nuclear receptors bound to their ligands; e.g. the glucocorticoid response element (GRE).

hotspots Sites on chromosomes where mutations arise at a higher frequency than other regions of the DNA.

housekeeping genes Genes that code for proteins needed for basic processes in all kinds of cells.

human artificial chromosome (HAC) A vector used to transfer or express large fragments of human DNA. HACs behave and are constructed like human chromosomes.

human endogenous retrovirus (HERV) Transposition-defective LTR-containing retrotransposons in human cells.

Human Genome Project An international research project to map each human gene and to completely sequence human DNA.

human immunodeficiency virus (HIV)/acquired immunodeficiency syndrome (AIDS) AIDS was first reported in 1981 in the USA and has since become a major epidemic, killing nearly 12 million people and infecting more than 30 million others worldwide. The disease is caused by HIV, a retrovirus that destroys the body's ability to fight infections and certain cancers.

Huntington's disease A degenerative brain disorder that usually appears in mid-life, caused by a trinucleotide repeat expansion. Its symptoms, which include involuntary movement of the face and limbs, mood swings, and forgetfulness, get worse as the disease progresses. It is generally fatal within 20 years.

Hutchinson–Gilford progeria syndrome A premature aging syndrome caused by a splicing mutation in the lamin A gene.

hybrid dysgenesis A phenomenon observed in *Drosophila* in which the hybrid offspring of two certain parental strains suffer so much chromosomal damage that they are sterile, or dysgenic.

hybridization (or DNA or RNA) Complementary base pairing of two single strands of DNA or RNA from two different sources.

hydrogen bonds Very weak bonds that involve the sharing of a hydrogen between two electronegative atoms, such as an oxygen and nitrogen.

hydrophilic Water-attracting or attracted to water, as polar groups on compounds such as lipids and proteins.

hydrophobic Water-repelling or repelled by water (literally "water-hating"), as nonpolar groups on lipids, proteins, etc. which tend to aggregate, excluding water from between them.

hydroxyl radicals OH units with an unpaired election. They are highly reactive and can cause DNA damage.

hyperchromicity The phenomenon in which the absorption of UV light increases as double-stranded DNA denatures to become single-stranded. Native double-stranded DNA absorbs less light at 260 nm by about 40% than does the equivalent amount of single-stranded DNA.

I

ionizing radiation Radiation that can attack (ionize) the deoxyribose sugar in the DNA backbone directly or indirectly by generating reactive oxygen species; e.g. X-rays, radioactive materials.

immunocytochemistry *In situ* analysis of intracellular protein expression and localization using enzyme-conjugated secondary antibodies to the primary antibody against the target protein.

immunofluorescence assay *In situ* analysis of protein expression and localization using fluorescently labeled antibodies. When fluorescently tagged primary antibodies are used for detection, the technique is called direct immunofluorescence assay. When fluorescently tagged secondary antibodies are used for detection, the technique is called indirect immunofluorescence assay.

immunoglobulin (antibody) A protein that binds very specifically to an invading substance and alerts the body's immune defenses to destroy the invader.

immunohistochemistry *In situ* analysis of protein expression and localization in organs using enzyme-conjugated secondary antibodies to the primary antibody against the target protein.

immunoprecipitation A technique in which labeled proteins are reacted with a specific antibody or antiserum bound to resin beads, then precipitated by centrifugation. The precipitated proteins are usually detected by electrophoresis and autodiography.

importin A soluble receptor that mediates nuclear import of nuclear localization sequence (NLS)-bearing cargo.

imprinting control regions (ICRs) Specific intergenic regions responsible for establishing the differential imprint and for maintenance of genomic imprinting during development.

***in situ* hybridization** The base pairing of a labeled probe to metaphase chromosomes on a microscope slide, or to RNA to determine the precise localization within a cell.

in vitro Studies performed in cells or tissues grown in culture, or in cell extracts or synthetic mixtures of cell components.

in vivo Studies performed within a living organism.

incision Nicking a DNA strand with an endonuclease.

independent assortment A principle discovered by Mendel, which states that genes on different chromosomes are inherited independently.

induced mutation A mutation that occurs as a result of interaction of DNA with an outside agent or mutagen that causes DNA damage.

inducer A substance that releases negative control of on operon.

induction The synthesis of enzymes in response to the appearance of a specific substrate.

inherited Transmitted through genes from parents to offspring.

initiation factor A protein that helps catalyze the initiation of translation.

initiator (Inr) A core promoter element surrounding the transcription start site that is important in the efficiency of transcription from some RNA polymerase II gene promoters, especially those lacking TATA boxes.

inosine (I) A nucleoside containing the base hypoxanthine, which base pairs with cytosine; a common modified base found in RNA.

insert A foreign DNA molecule ligated into a vector.

insertion A type of chromosomal abnormality in which a DNA sequence is inserted into a gene, disrupting the normal structure and function of that gene.

insertion sequence (IS) A type of transposon found in bacteria, containing only inverted terminal repeats and the genes needed for transposition.

insulator A DNA regulatory element that acts as a chromatin boundary marker between regions of heterochromatin and euchromatin and can block enhancer or silencer activity of neighboring genes.

integrase An enzyme that integrates one nucleic acid into another; e.g. the provirus of a retrovirus into the host genome.

intellectual property rights Patents, copyrights, and trademarks.

intercalate To insert between two base pairs in DNA.

interferon A double-stranded RNA-activated antiviral protein with various effects on the cell.

intermediate A substrate–product in a biochemical pathway.

internal ribosome entry sequence (IRES) A sequence to which a ribosome can bind and begin translating in the middle of a transcript, without having to scan from the 5′-end.

interphase The stage of the cell cycle during which DNA is synthesized but the chromosomes are not visible.

introns The sequences that remain physically separated after excision during RNA splicing; may encode snoRNAs or miRNAs. The term is used to refer both to the corresponding sequence in the DNA and the RNA.

inverted repeat A symmetrical sequence of DNA, reading the same forward on one strand and backward on the opposite strand. Sometimes referred to as a palindrome.

IPTG Isopropylthiogalactoside. A sulfur-containing lactose analog that is not metabolized by β-galactosidase; used as an inducer of the *lac* operon in the laboratory.

isoelectric focusing Electrophoresing a mixture of proteins through a pH gradient until each protein stops at the pH that matches its isoelectric point. The proteins can no longer move toward the anode or cathode because they have no net charge at the isoelectric point.

isoelectric point The pH at which a protein has no net charge.

isoschizomers Two or more restriction endonucleases that recognize and cut the same restriction site.

ISWI A family of coactivators (imitation SWI/SNF) that help remodel chromatin by moving nucleosomes.

J

J A modified base in trypanosomes that replaces thymine.

joining region (J) The segment of an immunoglobulin gene encoding the last 13 amino acids of the variable region. One of the several joining regions is joined by a chromosomal rearrangement to the rest of the variable region, introducing extra variability into the immunoglobulin gene.

K

karyopherins Family of soluble receptors that mediate nuclear import and export.

karyotype The chromosomal complement of an individual, including the number of chromosomes and any abnormalities. The term is also used to refer to a photograph of an individual's chromosomes.

Kearns–Sayre syndrome A disease linked to mitochondrial DNA (mtDNA) mutations that is characterized by paralysis of eye muscles, progressive muscle degeneration, heart disease, hearing loss, diabetes, and kidney failure.

kilobase pair (kb or kbp) A unit of length use for DNA corresponding to 1000 base pairs.

kilodalton (kD) 1000 Daltons.

kinase An enzyme that catalyzes the addition of phosphate groups to a substrate.

kinetic experiment An experiment that measures the speed (kinetics) of a reaction.

kinetoplast DNA The mitochondrial DNA of trypanosomes; consists of minicircles and maxicircles.

kink A sharp bend in a double-stranded DNA made possible by unstacking of bases.

kink-turn motif An asymmetrical internal loop embedded in a RNA double helix that has a sharp bend in the phosphodiester backbone of the three-nucleotide bulge associated with this motif.

Klenow fragment A fragment of *E. coli* DNA polymerase I, created by cleaving with a protease, that lacks the $5' \rightarrow 3'$ exonuclease activity of the parent enzyme.

knockdown Repression of gene expression by RNAi.

knockout Inactivation of specific genes by removal of the sequence. Knockouts are often created in laboratory organisms such as yeast or mice by gene targeting so that scientists can study the knockout organism as a model for a particular disease.

known genes Genes from a genomic sequencing project whose sequences are identical to previously characterized genes.

Kozak consensus sequence The sequence context of a eukaryotic translation initiation signal.

Ku The ATPase regulatory subunit of DNA-PK. Binds to double-stranded DNA ends created by chromosome breaks and protects them until repair by nonhomologous end-joining can occur.

Kuru The first form of transmissible spongiform encephalopathy described in humans that was rampant at one time in New Guinea as a result of ritual cannabalism.

L

L strand Light strand of mitochondrial DNA; relative buoyant density determined by density gradient centrifigution.

L1 An abundant human LINE (long interspersed nuclear element) that comprises about 15% of the human genome.

***lac* operon** The operon that encodes enzymes that permit a bacterial cell to metabolize the milk sugar lactose.

Lac repressor A protein, the product of the *E. coli lacI* gene, that forms a tetramer that binds the *lac* operator and thereby represses the *lac* operon.

lacA The *E. coli* gene that encodes galactoside transacetylase.

lacI The *E. coli* gene that encodes the *lac* repressor.

lactose A dissaccharide composed of two simple sugars, glucose and galactose.

lacY The *E. coli* gene that encodes galactoside permease.

lacZ The *E. coli* gene that encodes β-galactosidase.

lagging strand The strand that is made discontinuously in semidiscontiuous DNA replication.

large T antigen The major product of the SV40 viral early region. A DNA helicase that binds to the viral origin of replication and unwinds DNA during replication. Causes malignant transformation of mammalian cells.

lariat The name given the lasso-shaped intermediate in certain splicing reactions.

leader A sequence of untranslated bases at the 5′ end of an mRNA, the 5′ untranslated region (5′-UTR).

leading strand The strand that is made continuously in semidiscontinuous DNA replication.

Leber's hereditary optic neuropathy (LHON) A form of young–adult blindness linked to a small inherited mutation in a mitochondrial gene.

leucine zipper A domain in a DNA-binding protein that includes several leucines spaced at regular intervals. Involved in dimerization with another leucine zipper protein to form a dimer that can then bind DNA.

leukemia Cancer of the developing blood cells in the bone marrow. Leukemia leads to rampant overproduction of white blood cells (leukocytes); symptoms usually include anemia, fever, enlarged liver, spleen, and/or lymph nodes.

library A collection of cloned DNA, usually from a specific organism.

licensing protein complex Mcm2-7 bound to an origin of replication; only licensed origins containing Mcm2-7 can initiate a pair of replication forks.

ligation The joining of linear DNA fragments together with covalent (phosphodiester) bonds.

linkage The association of genes and/or markers that lie near each other on a chromosome, and the likelihood of having one gene and/or marker transmitted with another through meiosis. Linked genes and markers tend to be inherited together.

linker histone One of the more variable histones (H1, H5, H1°) that occurs between core octamers, where the DNA enters and exits the nucleosome.

linker scanning mutagenesis Creation of clustered mutations by replacing small segments of DNA with synthetic oligonucleotides (linkers).

linking number (L) The number of times each strand (chain) crosses the other in a double-stranded DNA circle.

lipoprotein A protein with a lipid group attached post-translationally.

liposome A phospholipid-bound vesicle; often used to introduce DNA into cells.

liquid scintillation counting A technique for measuring the degree of radioactivity in a substance by surrounding it with scintillation fluid, a liquid containing a fluor that emits photons when excited by radioactive emissions.

liver cirrhosis Heavy scarring of the liver; common in chronic alcoholics.

locus (loci, pl.) The place on a chromosome where a specific gene is located; used synonymously with the term gene in many instances.

locus control region (LCR) A chromatin region, such as that associated with the globin genes, that ensures activity of the associated genes, regardless of chromatin location.

LOD score A statistical estimate of whether two loci are likely to lie near each other on a chromosome and are therefore likely to be inherited together. A LOD score of 3 or more is generally taken to indicate that the two loci are close.

long interspersed nuclear elements (LINEs) The most abundant non-LTR retrotransposons in mammals.

long-range regulatory elements Regulatory DNA sequences in multicellular eukaryotes that can work over distances of 100 kb or more from the gene promoter.

long terminal repeats (LTRs) Regions of several hundred base pairs of DNA found at both ends of the provirus of a retrovirus.

LTR-containing retrotransposons A retrotransposon with LTRs at both ends. Replicates its DNA like the provirus of a retrovirus except no transmissible viral particle is involved.

luciferase An enzyme that converts luciferin to a bioluminescent product that emits light and is therefore easily assayed. The firefly luciferase gene is often used as a reporter gene in eukaryotic transcription and translation experiments.

lymphocyte A small white blood cell that plays a major role in defending the body against disease. There are two main types of lymphocytes: B cells, which make antibodies that attack bacteria and toxins, and T cells, which attack body cells themselves when they have been taken over by viruses or become cancerous.

lysis Rupturing the membrane of a cell, as by a virulent bacteriophage.

lysogen A bacterium harboring a prophage.

M

Mad cow disease See bovine spongiform encephalopathy.

Mad–Max A mammalian transcriptional repressor; a heterodimer of Mad and Max proteins.

MALDI–TOF Matrix Assisted Laser Desorption/Ionization – Time Of Flight mass spectrometry. A technique for measuring the mass of peptides. The time of flight of ionized peptides down a tube toward a detector is inversely proportional to their mass and directly proportional to the charge on the protein.

mammalian artificial chromosome (MAC) A vector that contains a multiple cloning site for very large foreign DNA inserts, centromeric sequences, sequences that can initiate DNA replication and telomeric sequences. MACs segregate with the host cell's chromosomes during cell division.

major groove In B-form DNA, the larger of two continuous indentations running along the outside of the double helix.

maple syrup urine disease A metabolic disorder inherited as an autosomal recessive which affects the metabolism of the amino acids leucine, isoleucine, and valine leading to an accumulation of keto acids that gives the urine a sweet odor resembling maple syrup and interferes with brain function; caused by a defect in a component of the multienzyme branched-chain α-keto acid dehydrogenase complex.

mapping The process of deducing schematic representations of DNA. Three types of DNA maps can be constructed: physical maps, genetic maps, and cytogenetic maps, with the key distinguishing feature among these three types being the landmarks on which they are based.

mariner A defective, inactive human transposon that once transposed by direct DNA replication.

marker Also known as a genetic marker, a segment of DNA with an identifiable physical location on a chromosome whose inheritance can be followed. A marker can be a gene, or it can be some section of DNA with no known function. Because DNA segments that lie near each other on a chromosome tend to be inherited together, markers are often used as indirect ways of tracking the inheritance pattern of genes that have not yet been identified, but whose approximate locations are known.

maternal genes Genes that are expressed during oogenesis in the mother.

matrix attachment regions (MARs) AT-rich regions of DNA typically located near enhancers that organize the genome into chromatin loops with an average loop size of 70 kb; involved in regulating tissue specificity and developmental control of gene expression by recruiting transcription factors and providing a landing platform for chromatin-remodeling enzymes.

Mcm2-7 Minichromosome maintenance proteins. A hexameric (six-subunit) component of the pre-replication complex that is essential for the initiation of DNA replication in eukaryotes and has helicase activity.

Mediator A 20-subunit complex which transduces regulatory information from activator and repressor proteins to RNA polymerase II.

medium Nutritive material in which microorganisms, cells, and tissues are grown in the laboratory, plural "media."

megabase pair (Mb or Mbp) The unit length used for DNA corresponding to 1,000,000 base pairs.

meiosis Cell division that produces gametes (or spores) having half the number of chromosomes of the parental cell.

melanoma Cancer of the cells in the skin that produce melanin, a brown pigment. Melanoma often begins in a mole.

melting temperature (T_m) The temperature at which half the bases in a double-stranded DNA sample have denatured.

membrane protein A protein that folds into characteristic transmembrane helical structures and is embedded in the cell membrane.

Mendelian inheritance Manner in which genes and traits are passed from parents to children. Examples of Mendelian inheritance include autosomal dominant, autosomal recessive, and sex-linked genes.

messenger RNA (mRNA) The template for protein synthesis that binds to ribosomes. Each set of three bases, called codons, specifies a certain amino acid in the sequence of amino acids that comprise the protein. The sequence of a strand of mRNA is based on the sequence of a complementary strand of DNA.

metacentric chromosome A chromosome with the centromere in the middle.

metalloenzyme An enzyme in which binding of divalent cations (e.g. Mg^{2+}) in the active site is critical for its folding into an active conformation.

metaphase The phase of mitosis, or cell division, when the chromosomes align along the center of the cell. Because metaphase chromosomes are highly condensed, these chromosomes are used for gene mapping and identifying chromosomal aberrations.

metastasis The process by which cancer cells travel from the tissue of origin to other parts of the body where they begin to grow and replace normal tissue.

metazoan A multicellular animal.

7-methyl guanosine The capping nucleoside at the beginning of a eukaryotic mRNA.

micrococcal nuclease A nuclease that degrades the DNA between nucleosomes, leaving the nucleosomal DNA intact.

microRNA (miRNA) A short RNA molecule encoded by a cellular gene that folds into a hairpin to create a double-stranded RNA that then triggers the RNAi machinery. miRNAs play a central role in post-transcriptional gene regulation.

microsatellite See short tandem repeats.

microsatellite instability An increase in the accumulation of mutations in the microsatellite regions of DNA, leading to variability in the number or repeats; common in tumors.

mid-blastula transition The point in embryonic development at which the zygote's own genes become active.

minimal medium Culture medium containing a basic set of nutrients only, on which normal wild-type organisms can grow, but which cannot suport the growth of metabolic mutants.

minisatellite A short sequence of (usually) 12 bp or more repeated over and over in tandem; also known as variable number tandem repeats.

minor groove In B-form DNA, the smaller of the two continuous indentations running along the outside of the double helix.

minus 10 box (−10 box) An *E. coli* promoter element centered about 10 bp upstream of the start of transcription.

minus 35 box (−35 box) An *E. coli* promoter element centered about 35 bp upstream of the start of transcription.

mismatch repair The correction of a mismatched base incorporated by mistake during DNA replication; mediated by MutSα or MutSβ which recognize the mismatch, MutLα, exonuclease EXO1 which excises a large region of DNA around the machinery, and the DNA replication machinery which fills in the gap.

missense mutation A change in a codon that results in an amino acid change in the corresponding protein.

mitochondrial DNA (mtDNA) The genetic material of the mitochondria, the organelles that generate energy for the cell.

mitogen A substance, such as a growth factor that stimulates cell division.

mitogen-activated protein kinase (MAPK) A protein kinase that is activated by phosphorylation as a result of a signal transduction pathway initiated by a mitogen such as a growth factor.

mitosis Cell division that produces two daughter cells having nuclei identical to the parental cell.

model organism An organism chosen for study of another organism because of any or all of the following factors: small genome size, short generation time, and ease of manipulation in genetic experiments. A model organism is useful for medical research because it has specific characteristics that resemble a human disease or disorder. Scientists can create animal models, usually laboratory mice, by transferring new genes into them.

molecular biology The study of biological phenomena at the molecular level, in particular the study of the molecular structure of DNA and the information it encodes, and the biochemical basis of gene expression and its regulation.

molecular chaperones See chaperones.

molecular cloning See cloning.

molecular machines The large dynamic macromolecular assemblages that regulate the expression of genetic information in eukaryotes.

monoallelic gene expression Preferential transcription from a single allele in a cell; also known as allelic exclusion.

monoclonal antibodies Identical antibodies to a specific epitope of a protein produced by a clone originating from one cell.

monosomy Possessing only one copy of a particular chromosome instead of the normal two copies.

morpholino oligonucleotides Modified DNA analogs with an altered backbone linkage that lacks a negative charge; used in antisense-mediated inhibition of gene expression.

mouse model A laboratory mouse useful for medical research because it has specific characteristics that resemble a human disease or disorder. Mouse models can be created by transferring new genes into mice or by inactivating certain existing genes in them.

MRN complex A complex comprising exonuclease Mre11/Rad50/NBS1 that is recruited to double-strand break sites in DNA and initiates repair.

MTE Motif ten element; an RNA polymerase II core promoter element found in a few gene promoters.

multiple cloning site (MCS) A region in a cloning vector that contains several unique restriction sites in tandem. Any of these can be used for inserting foreign DNA.

multiplex PCR Simultaneous amplification of many targets of interest in one reaction by using more than one pair of primers.

mutagen Any chemical agent that causes an increase in the rate of mutation above the spontaneous background.

mutant An organism (or genetic system) that has suffered at least one mutation compared to the wild-type (normal).

mutation A permanent structural alteration in DNA. In most cases, DNA changes either have no effect or cause harm, but occasionally a mutation can improve an organism's chance of surviving and passing the beneficial change on to its descendants.

N

negative control A control system in which gene expression is turned off unless a controlling element (e.g. repressor) is removed.

neurofibromatosis An inherited progressive disorder in which tumors form on peripheral nerves. The tumors can result in loss of hearing and vision, cancer, epilepsy, bone deformities, and learning disabilities.

N-formyl methionine The initiating amino acid in prokaryotic translation.

nick A single-stranded break of a phosphodiester bond in DNA.

nitrogenous base Nitrogen-containing molecules having the chemical properties of a base that are components of nucleotides. Two of the bases, adenine (A) and guanine (G), have a double carbon-nitrogen ring structure. The other three bases, thymine (T), cytosine (C), and uracil (U), have a single-ring structure. Thymine is found in DNA only, while uracil is specific for RNA.

nonautonomous retrotransposon A non-LTR retrotranposon that encodes no proteins, so it depends on other retrotransposons for transposition activity.

noncanonical (non-Watson–Crick) base pairs Unconventional base pairs that form in RNA double helices such as the GU wobble, the sheared GA pair, the Reverse Hoogsteen pair, and the GA imino pair.

nonhomologous end-joining (NHEJ) A eukaryotic mechanism for repairing double-strand breaks in DNA. Double-strand breaks are rejoined via direct ligation of the DNA ends without any requirement for sequence homology.

non-LTR retrotransposons A retrotransposon that lacks LTRs and replicates by a mechanism different from that used by the LTR-containing retrotransposons.

nonpermissive conditions Those conditions under which a conditional mutant gene product cannot function (e.g. the inhibitory temperature for a temperature-sensitive mutant).

nonsense codons UAG, UAA, UGA These stop codons terminate protein synthesis by the ribosome.

nonsense-mediated mRNA decay (NMD) A eukaryotic system for degrading mRNAs with premature termination (stop) codons.

nonsense mutation A single DNA base substitution resulting in a stop codon.

nonspecific binding (protein) The first contact of a DNA-binding protein with DNA involves interactions with the the sugar-phosphate backbone and not the bases.

nontemplate DNA strand The strand complementary to the template strand. Also called the coding strand or sense strand.

nontranscribed spacer (NTS) A DNA region lying between two ribosomal RNA precursor genes in a tandemly repeated cluster of such genes.

Northern blot Transfer of RNA fragments to a support medium (see Southern blot). A technique used to identify and locate mRNA sequences that are complementary to a piece of DNA (or antisense RNA) called a probe.

novel variant of CJD (vCJD) A human form of transmissible spongiform encephalopathy linked to eating beef products from a cow with bovine spongiform encephalopathy (mad cow disease).

nuclear export sequence (NES) A specific amino acid sequence that interacts with an exportin and mediates nuclear export of the protein via the nuclear pore complex; typically hydrophobic leucine-rich sequences.

nuclear factor kappa B (NF-κB) A transcription factor that activates expression of the many genes involved in the immune system and stress response in cells.

nuclear lamina A protein meshwork underlying the nuclear membrane that is primarily composed of the intermediate filament proteins lamins A, B, and C.

nuclear localization sequence A specific amino acid sequence that interacts with an importin and targets the protein to the nucleus via the nuclear pore complex; typically rich in lysine and arginine.

nuclear magnetic resonance (NMR) spectroscopy A technique in which a concentrated protein solution is placed in a magnetic field and the effects of different radiofrequencies on the resonances of different atoms are measured.

nuclear matrix A structural organizer within the cell nucleus. Operationally defined as a branched meshwork of insoluble filamentous proteins within the nucleus that remains after digestion with high salt, nucleases, and detergent.

nuclear pore complexes Large multiprotein complexes embedded in the nuclear membrane that serve as a selective gateway for the exchange of material between the nucleus and cytoplasm.

nuclear receptor A transcription factor that interacts with ligand, such as glucocorticoids, thyroid hormone, vitamin D, or retinoic acid and binds to an enhancer or silencer to stimulate or repress transcription, respectively.

nucleic acids A long chain-like polymer (DNA or RNA) of repeating subunits called nucleotides.

nucleoid In prokaryotes, the condensed ovoid region of the cell containing the chromosomal DNA.

nucleolus A nonmembrane-bound cell organelle found in the nucleus that contains the ribosomal RNA genes and is the site of ribosome assembly.

nucleoside A nitrogenous base chemically linked to one molecule of a five-carbon sugar at the 1′ carbon of the sugar – either ribose (RNA) or deoxyribose (DNA).

nucleoside triphosphates (NTPs) The building blocks of RNA: ATP, CTP, GTP, UTP. Also referred to as ribonucleoside triphosphates.

nucleosome A repeating structural element in eukaryotic chromosomes, composed of the core octamer of histones (two each of histones H2A, H2B, H3, and H4) plus one molecule of the linker histone (H1) with approximately 180 bp DNA wrapped around.

nucleotide (nt) One of the structural components, or building blocks, of DNA and RNA. A nucleotide is composed of three parts: a five-carbon sugar, at least one phosphate group, and a nitrogenous base.

nucleotide excision repair A repair pathway in which DNA damage such as a pyrimidine dimer is recognized by the cooperative binding of RPA, XPA, XPC, and the TFIIH complex (including the helicases XPB and XPD). XPG endonuclease then cuts the DNA strand on either side of the damaged base, removing an oligonucleotide that contains the damage. The gap is filled in with DNA polymerase and DNA ligase.

nucleotide substitution A type of mutation in which a nucleotide pair in a DNA duplex is replaced with a different nucleotide pair. Mutations that alter a single nucleotide pair are called point mutations.

nucleus The central cell structure that houses the chromosomes.

O

O$_6$-methylguanine methyl transferase A suicide enzyme that accepts methyl or ethyl groups from alkylated DNA bases and thereby reverses the DNA damage.

Okazaki fragments Small DNA fragments, 1000–2000 bases long, created by discontinuous synthesis of the lagging strand.

oligo(dT) cellulose affinity chromatography A method for purifying poly (A+) RNA by binding it to oligo(dT) cellulose in buffer at relatively high ionic strength, and eluting it at low ionic strength.

oligonucleotide (oligo) A short sequence of single-stranded DNA or RNA. Oligos are often used as probes for detecting complementary DNA or RNA because they bind readily to their complements.

oncogene A gene whose product is capable of causing the transformation of normal cells into cancer cells through gain of function.

open reading frame (ORF) A reading frame that is uninterrupted by translation stop codons.

operator A DNA element found in prokaryotes that binds tightly to a specific repressor and thereby regulates the expression of adjacent genes.

operon A group of genes coordinately controlled by an operator.

oriC The *E. coli* origin of replication.

origin firing Initiation of a bidirectional pair of replication forks.

origin of replication The unique site on chromosomal DNA (replicon) where replication begins, and a bidirectional pair of replication forks initiate.

origin recognition complex A eukaryotic ATP-regulated DNA binding complex composed of six polypeptide subunits that binds to origins of replication and then recruits cdc6, Cdt1, and Mcm2-7 proteins, other components of the pre-replication complex that are essential for the initiation of DNA replication. The SV40 large T antigen functions as a viral ORC comparable to the cellular ORC.

orthologs Homologous genes in different species that have evolved from a common ancestral gene.

overexpression Production of a large quantity of recombinant proteins in a bacterial or eukaryotic host cell.

8-oxoguanine (oxo8) A damaged guanine base containing an extra oxygen atom; it can form a Hoogsteen base pair with adenine, thereby giving rise to a GC to TA transversion.

P

P element A transposable element of *Drosophila*, responsible for hybrid dysgenesis; used as a tool by molecular biologists for mutagenesis studies (transposon tagging).

P site The ribosomal site to which a peptidyl tRNA is bound at the time a new aminoacyl-tRNA enters the ribosome.

p53 A tumor suppressor gene which normally regulates the cell cycle and protects the cell from damage to its genome. Mutations in this gene cause cells to develop cancerous abnormalities.

palindrome See inverted repeat.

pandemic A worldwide epidemic of an infectious disease.

panediting Extensive editing of a pre-mRNA; occurs in trypanosome kinetoplast transcripts.

paralogs Homologous genes that have evolved by gene duplication within a species.

Parkinson's disease Common progressive neurological disorder that results from degeneration of nerve cells in a region of the brain that controls movement. The first symptom of the disease is usually tremor of a limb, especially when the body is at rest.

patent When applied to molecular biology, the government regulations or requirements conferring the right or title to an individual or organization to genes if there has been substantial human intervention.

pathogenic Causing disease; e.g. a parasite (especially a microorganism) in relation to a particular host.

pathway (biochemical) A series of biochemical reactions in which the product of one reaction (an intermediate) becomes the substrate for the next reaction.

pause sites DNA sites where an RNA polymerase pauses before continuing elongation.

pBR322 One of the original plasmid vectors for gene cloning.

pedigree A simplified diagram of a family's genealogy that shows family members' relationships to each other and how a particular trait or disease has been inherited.

peptide Two or more amino acids joined by a peptide bond.

peptide bond Covalent bond joining the α-amino group of one amino acid to the carboxyl group of another with the loss of a water molecule, and which is the bond linking amino acids together in a polypeptide chain.

peptidyl transferase An enzymatic activity that is an integral part of the large ribosomal RNA in the ribosome that catalyzes the formation of peptide bonds during protein synthesis.

permissive conditions Those conditions under which a conditional mutant gene product can function (i.e. the permissive temperature for a temperature-sensitive mutant).

Pfu polymerase A thermostable DNA polymerase from *Pyrococcus furiosus* used in PCR that has greater fidelity than Taq polymerase.

phage See bacteriophage.

phage mu A phage of *E. coli* that replicates by transposition.

phenotype The biochemical, behavioral, morphological, or other properties of an organism.

phosphatase An enzyme that catalyzes the removal of a phosphate group from a substrate.

phosphodiester bond The covalent sugar-phosphate bond ($-O-P-$) that forms the linkage between adjacent nucleotides in a nucleic acid (DNA or RNA). The hydroxyl group on the $3'$ carbon of a sugar of one nucleotide forms an ester bond to the phosphate group on the $5'$ carbon of another nucleotide, eliminating a molecule of water. The bond is also referred to as a $5' \rightarrow 3'$ phosphodiester bond, indicating the polarity of the strand.

phosphorimaging A technique for measuring the degree of radioactivity of a substance (e.g. on a blot) electronically without using X-ray film, using an instrument called a phosphorimager.

phosphorylation (protein) The post-translational modification of a protein by addition of a phosphate group at serine, threonine, or tyrosine residues.

photoreactivation Direct repair of a pyrimidine dimer by DNA photolyase.

physical map A chromosome map of a species that shows the specific physical locations of its genes and/or markers on each chromosome; e.g. based on physical characteristics such as restriction sites. Physical maps are particularly important when searching for disease genes by positional cloning strategies and for DNA sequencing.

plaque A hole that a virus makes on a layer of host cells by infecting and either killing the cells or slowing their growth.

plasmid A small, double-stranded circular (or linear) DNA molecule carried by bacteria, some fungi, and some higher plants. They are extrachromosomal, independent, and self-replicating.

pleiotropy Multiple phenotypic manifestations.

point mutation An alteration (substitution) of one, or a very small number, of contiguous bases.

polar molecule A molecule with an asymmetrical distribution of charge across the molecule.

poly(A)-binding protein (PABP) A protein that binds the poly(A) tails at the end of the pre-mRNA. There are nuclear (PABPN) and cytoplasmic (PABPC) forms.

poly(A)+ RNA RNA that contains a poly(A) tail at its $3'$ end.

poly(A)− RNA RNA that does not contain a poly(A) tail at its $3'$ end.

poly(A) tail Polyadenylic acid tail. The string of about 200 As added to the end of a typical eukaryotic mRNA.

polyacrylamide A cross-linked polymer of acrylamide; used in electrophoretic separation of nucleic acids and proteins.

polyadenylation Addition of poly(A) to the $3'$ end of an RNA.

polyadenylation signal The set of RNA sequences that mediate the cleavage and polyadenylation of a transcript. An AAUAAA sequence followed 20–30 nucleotides downstream by a GU-rich region, then a U-rich region is the canonical cleavage signal. After cleavage, the AAUAAA sequence is the polyadenylation signal.

polycistronic message An mRNA containing information from more than one gene.

polyclonal antibodies A mixture of antibodies with different specificities to different epitopes.

Polycomb group proteins A family of proteins that mediate silencing of homeobox genes by altering the higher order structure of chromatin.

polydactyly An abnormality in which a person is born with more than the normal number of fingers or toes.

polymerase chain reaction (PCR) Amplification of a specific region of DNA using primers that flank that region and repeated cycles of DNA polymerase activity. A fast, inexpensive technique for making an unlimited number of copies of any piece of DNA.

polymerase switching The hand-off ("trading places") of the DNA template from one DNA polymerase to another between primer synthesis and elongation, and during translesion DNA synthesis.

polymorphism A common variation in the sequence of DNA among individuals.

polynucleotide A polymer composed of nucleotide subunits; either DNA or RNA.

polypeptide A polymer composed of amino acid subunits; a single protein chain.

polysome A messenger RNA attached to several ribosomes.

position effect A phenomenon in which expression of a transgene is unpredictable; it varies with the chromosomal site of integration.

positional cloning A process which, through gene mapping techniques, is able to locate a gene responsible for a disease or other genetic traits when little or no information is known about the biochemical basis of the disease or trait.

positive control A control system in which gene expression depends on the presence of a positive effector such as CAP (and cAMP).

post-transcriptional control Control of gene expression that occurs when transcripts are processed by capping, splicing, editing, nuclear export, and degradation pathways.

post-transcriptional gene silencing (PTGS) See RNA interference.

post-translational modification The set of changes that occur in a protein after it is synthesized (e.g. modification by phosphorylation).

Prader–Willi syndrome A neurodevelopmental disorder caused by defects in genomic imprinting in a region of chromosome 15.

preinitiation complex (PIC) The combination of eukaryotic RNA polymerase and general transcription factors assembled at a promoter just before transcription begins.

pre-replication complex A complex of ORC, cdc6, Cdt1, and Mcm2-7 assembled at a eukaryotic origin of replication that is required for the initiation of a replication fork.

primary antibody The first set of antibodies made for a specific epitope.

primary structure The sequence of amino acids in a polypeptide, or of nucleotides in DNA or RNA.

primary transcript The initial, unprocessed RNA product of a gene.

primase The enzyme that catalyzes the formation of an RNA primer during DNA replication.

prime See five (5)-prime (5′) and three (3)-prime (3′).

primer (DNA replication) A small piece of RNA that provides the free 3′-OH end needed for DNA replication to begin.

primer (techniques) A short oligonucleotide sequence (DNA) used in laboratory techniques that involve DNA polymerase, e.g. the polymerase chain reaction, DNA sequencing, random primed DNA labeling, etc.

prion Proteinaceous infectious particle. The causative agent of transmissible spongiform encephalopathies. A prion is a misfolded form of a normal cellular protein that propagates through promoting misfolding of the normal protein into the infectious prion form. Prions can be sporadic, inherited, or infectious.

probe (nucleic acid) A piece of DNA or RNA, labeled with a tracer (typically radioactive), that allows a molecular biologist to track the hybridization of the probe to an unknown DNA or RNA. For example, a radioactive probe can be used to identify an unknown DNA band on a Southern blot.

processivity The tendency of an enzyme to remain bound to one or more of its substrates during repetitions of the catalytic process. For example, the longer an RNA or DNA polymerase continues making its product without dissociating from its template, the more processive it is.

programmed gene rearrangements Rearrangements of DNA that regulate the expression of some genes.

prokaryotes Microorganisms that lack nuclei; comprising eubacteria, cyanobacteria, and Archaea.

proliferating cell nuclear antigen (PCNA) A eukaryotic protein that acts as a "sliding clamp" to increase processivity of DNA polymerase during DNA replication, along with many other functions in the cell.

promoter The collection of DNA sequence elements, including the core promoter and promoter proximal elements, that are required for initiation of transcription or that increase the frequency of initiation only when positioned near the transcriptional start site.

promoter clearance The process by which an RNA polymerase moves away from a promoter after initiation of transcription.

promoter strength The relative frequency of transcription initiation; related to the affinity of RNA polymerase for the promoter region.

pronucleus The nucleus of a sperm or an egg prior to fertilization. Sperm and egg cells carry half the number of chromosomes of other nonreproductive cells. When the pronucleus of a sperm fuses with the pronucleus of an egg, their chromosomes combine and become part of a single nucleus in the resulting embryo, containing a full set of chromosomes.

proofreading (aminoacyl–tRNA synthetase) The process by which aminoacyl adenylates and, less commonly, aminoacyl–tRNAs are hydrolyzed if their amino acids are too small for the synthetase.

proofreading (DNA) The process used by DNA polymerase to check the accuracy of DNA replication as it occurs and to replace a mispaired nucleotide with the correct one. DNA polymerase has $3' \rightarrow 5'$ exonuclease activity which mediates this process.

proofreading (protein synthesis) The process by which aminoacyl–tRNAs are cross-checked on the ribosome for correctness before the amino acids are incorporated into the growing polypeptide chain.

proofreading (RNA) The backtracking of RNA polymerase along the DNA template followed by cleavage of the most recently added base(s).

prophage A phage genome integrated into the host's genome.

prophase Early stage of cell division in which chromosomes condense and become visible.

prosthetic group Non-protein chemical group (e.g. metal ion or heme) bound to a protein, as in many enzymes, usually forming part of the active site, and essential for biological activity (see also cofactor).

protease An enzyme that cleaves other proteins.

proteasome A collection of proteins (sedimentation coefficient 26S) that proteolytically degrades a ubiquitinylated protein.

Protection of Telomeres (POT1) A protein involved in regulation of telomerase activity; it binds the $3'$ single-stranded DNA tail of telomeres and participates in forming a folded chromatin structure that prevents access of telomerase to telomeres.

protein A large complex polymer or polypeptide made up of one or more chains of amino acid subunits. Proteins perform a wide variety of activities in the cell. Sometimes the term protein denotes a functional collection of more than one polypeptide (e.g. the hemoglobin protein as a quaternary structure that consists of four polypeptide chains).

protein arrays A high-throughput technique that allows rapid analysis of protein activity on a proteomic scale.

protein kinase A A serine-threonine-specific protein kinase whose activity is stimulated by cAMP.

protein sequencing Determining the sequence of amino acids in a protein.

proteome The complete set of proteins encoded by the genome.

proteomics The comprehensive study of the full set of proteins encoded by a genome.

proto-oncogene A cellular gene with the potential to give rise to an oncogene through inappropriate activation.

provirus A double-stranded DNA copy of a retroviral RNA genome, which inserts into the host cell genome.

proximal promoter element Regulatory element located just upstream of the core promoter and usually within 70–200 bp of the start of transcription (e.g. the CCAAT box and GC box); increases the frequency of initiation of transcription but only when positioned near the transcriptional start site.

PrP^C A normal cellular protein that when misfolded can become an infectious prion.

PrP^{Sc} Scrapie prion protein, the misfolded infectious form of the PrP^C protein.

pseudogene A sequence of DNA that is very similar to a normal gene but that has been altered slightly so it is not expressed. Such genes were probably once functional but over time acquired one or more mutations that rendered them incapable of producing a protein product.

pseudoknot A motif that forms in RNA when a single-stranded loop base pairs with a complementary sequence outside this loop and folds into a three-dimensional structure by coaxial stacking.

pseudouridine A modified nucleoside found in RNA, in which the ribose is joined to the 5-carbon instead of the 1-nitrogen of the uracil base.

pUC vectors Plasmid vectors based on pBR322, containing an ampicillin resistance gene, and a multiple cloning site that interrupts the *lacZ* gene, which enables blue/white screening for inserts.

pull-down assay A procedure for analysis of protein–protein interactions *in vitro*. For example, a glutathione S-transferase (GST) pull-down assay tests interaction between a GST-tagged protein (the bait) and another protein (the prey).

pulse-chase The process of giving a short period (the pulse) of radioactive precursor so that a substance such as protein becomes radioactive, then adding an excess of unlabeled precursor to "chase" the radioactivity out of the substance as protein turnover occurs.

pulse-field gel electrophoresis (PFGE) An electrophoresis technique in which the electric field is repeatedly reversed. Allows separation of very large pieces of DNA, up to several Mb in size.

pulse labeling Providing a radioactive precursor for only a short time. For example, DNA can be pulse labeled by incubating cells for a short time in radioactive thymidine.

purine The nitrogenous bases adenine (A) and thymine (T) which have a double carbon-nitrogen ring structure.

pyrimidine The nitrogenous bases thymine (T), cytosine (C), and uracil (U) which have a single carbon-nitrogen ring structure.

pyrimidine dimers Two adjacent pyrimidines in one DNA strand linked covalently, interrupting their base pairing with purines in the opposite strand. The main DNA damage caused by UV light is the formation of T-T dimers.

Q

quaternary structure The way two or more polypeptides interact to form a functional complex protein.

quenching Quickly chilling heat-denatured DNA to keep it denatured and prevent the single strands from annealing.

R

R-looping A classic technique for visualizing hybrids between DNA and RNA by electron microscopy. An R loop is formed when an RNA hybridizes to one strand of DNA and displaces the other strand as a loop.

Rad50, Rad52, Rad55, Rad57 Proteins involved in repair of DNA double-stranded breaks by homologous recombination.

Ran A member of the superfamily of GTP-binding proteins that act as molecular switches cycling between GDP- and GTP-bound states.

random primed labeling A method of incorporating radioactive nucleotides along the length of a fragment of DNA.

random walk (protein) The equal probabilities for forward and reverse steps as a protein slides by linear diffusion along DNA over short distances from a nonspecific binding site.

RAPD Randomly amplified polymorphic DNA analysis. A PCR reaction where the primers consist of random sequences. Pronounced "rapid."

reactive oxygen species See free radicals.

reading frame One of three possible ways the triplet codons in an mRNA can be translated, depending on where translation begins. A natural mRNA generally has only one correct reading frame.

recessive An allele or trait that does not express its phenotype when heterozygous with a dominant allele. Also, a genetic disorder that appears only in patients who have received two copies of a mutant gene, one from each parent.

recognition helix The α-helix in a DNA-binding motif of a DNA-binding protein that fits into the major groove of its DNA target; the sequence specific contacts define the specificity of the protein.

recombinant DNA The product of recombination between two (or more) fragments of DNA. Can occur naturally in a cell through genetic processes such as crossing-over, but the term is generally reserved for DNA molecules produced by joining segments derived from different biological sources (i.e. constructed by molecular biologists *in vitro*).

recombinant DNA technology A variety of techniques that molecular biologists use to manipulate DNA molecules to study the expression of genes.

recombination Reassortment of genes or alleles in new combinations. Occurs by crossing-over between or within DNAs.

recombination signal sequence (RSS) A specific sequence at a recombination junction, recognized by the recombination apparatus during immunoglobulin and T-cell receptor gene maturation.

recruitment Promoting the binding of a substance to a complex. Often refers to enhancing the binding of RNA polymerase or transcription factors to a promoter, or the binding of replication factors to an origin of replication.

relaxed DNA A DNA molecule that is not supercoiled.

release factor A protein that causes termination of translation at stop codons.

renaturation of DNA The process by which when heated solutions of denatured DNA are slowly cooled, single strands often meet complementary strands and form a new double helix. This is also referred to as annealing.

repetitive DNA DNA sequences that are repeated may times in a haploid genome.

replica plating Transferring colonies of bacteria or other cells from one culture plate to another, usually using a plating tool coated with a soft material that can pick up cells from one plate and place them in the same relative positions on the second plate (or membrane).

replication factor C (RFC) The ATP-dependent "clamp loader" that loads PCNA onto the DNA during replication.

replication factories Clusters of 40 to many hundreds of active replication forks in discrete subnuclear compartments or foci.

replication fork The point where the two parental DNA strands separate to allow replication.

replication initiation point (RIP) mapping A procedure that allows the detection of start sites for DNA synthesis at the nucleotide level.

replication licensing A system in eukaryotes that ensures that DNA replicates only once per cell cycle. Mediated through tight regulation of the formation and activation of pre-replication complexes by the levels of cyclin-dependent kinases.

replication protein A (RPA) A mammalian single-stranded DNA-binding protein that binds to single-stranded DNA and keeps it from base pairing with a complementary strand, used during DNA replication.

replicative transposition See duplicative transposition.

replicon All the DNA replicated from one origin of replication.

reporter gene A gene attached to a promoter or translation start site and used to measure the activity of the resulting transcription or translation. The reporter gene serves as an easily assayed counterpart for the gene it replaces.

repressor A protein that regulates a gene by turning it off.

resolution The final step in recombination, in which the second pair of strands is broken.

resolvase An endonuclease that nicks two DNA strands to resolve a Holliday junction after branch migration.

restriction endonuclease An enzyme that recognizes specific base sequences of double-stranded DNA and cuts the DNA at or near thoses sites. Restriction endonucleases are often called restriction enzymes.

restriction fragment A piece of DNA cut from a larger DNA by a restriction endonuclease.

restriction fragment length polymorphism (RFLP) A variation from one individual to the next in the number of cuttting sites for a given restriction endonuclease in a given genetic locus. Such variations affect the size of the resulting fragments. These sequences can be used as markers on physical maps and linkage maps. RFLP is pronounced "rif-lip."

restriction map A map that shows the locations of restriction sites in a region of DNA.

restriction-modification system The combination of a restriction endonuclease and the DNA methylase that recognizes the same DNA site in bacteria.

restriction site A sequence of nucleotides recognized and cut by a restriction endonuclease.

retinoblastoma (Rb) protein A tumor suppressor protein that acts as a cell cycle master switch.

retrohoming The process by which a mobile group I or group II intron in one gene can transpose to an intronless version of the same gene elsewhere in the genome.

retrotransposon A transposable element such as Ty in yeast that transposes via a retrovirus-like mechanism.

retrovirus A type of virus that contains RNA as its genetic material and whose replication depends on formation of a DNA provirus by reverse transcription. Retroviruses can cause many diseases, including some cancers and AIDS.

Rett syndrome A neurodevelopmental disorder caused by mutations in the gene coding for methyl-CpG-binding protein 2.

Rev1 A eukaryotic DNA polymerase involved in error-prone repair of abasic sites; its deoxycytidyl transferase activity inserts a C across from a nucleotide lacking a base.

reverse transcriptase An RNA-dependent DNA polymerase which is the enzyme commonly found in retroviruses that catalyzes reverse transcription.

reverse transcriptase PCR (RT-PCR) A PCR method that begins with the synthesis of cDNA from an mRNA template, using reverse transcriptase. The cDNA then serves as the template for conventional PCR.

reverse transcription Synthesis of a DNA using an RNA template.

rho (ρ) A protein that is needed for transcription termination at certain terminators in *E. coli* and its bacteriophages.

rho-dependent terminator A terminator in *E. coli* and its bacteriophages that requires rho for its activity.

rho-independent terminator A terminator in *E. coli* and its bacteriophages that does not require rho for its function. Also called an intrinsic terminator.

ribonuclease (RNase) An enzyme that degrades RNA.

ribonucleic acid (RNA) A polymer composed of ribonucleotides linked together by phosphodiester bonds. In RNA, the letter U, which stands for uracil, is substituted for T in the genetic code.

ribonucleoprotein particle (RNP) An RNA–protein complex.

ribonucleoside triphosphates (NTPs) See nucleoside triphosphates.

riboprobe A labeled RNA probe, commonly used in RNase protection assays and *in situ* hybridization.

ribose The pentose (5-carbon) sugar present in a nucleotide subunit of RNA.

ribose zipper motif An RNA motif that involves hydrogen bonding between the 2′-OH of a ribose in one helix and the 2-oxygen of a pyrimidine base (or the 3-nitrogen of a purine base) of the other helix between their respective minor groove surfaces.

ribosomal RNA (rRNA) The RNA molecules contained in ribosomes.

ribosomal subunit (30s) The small bacterial ribosomal subunit, involved in mRNA decoding.

ribosomal subunit (40s) The small eukaryotic ribosomal subunit, involved in mRNA decoding.

ribosomal subunit (50s) The large bacterial ribosomal subunit, involved in peptide bond formation.

ribosomal subunit (60s) The large eukaryotic ribosomal subunit, involved in peptide bond formation.

ribosome An RNA–protein complex that translates mRNAs to produce proteins (70S in bacteria and 80S in eukaryotes).

riboswitch A switchable "on–off" RNA element that selectively binds metabolites or functions as a sensor for signals as diverse as temperature or salt concentration and controls gene expression.

riboswitch ribozyme A riboswitch that is a metabolite-responsive ribozyme.

ribozyme A catalytic RNA molecule (RNA enzyme). Ribozymes can catalyze a number of chemical reactions that take place in living cells, ranging from cleavage of phosphodiester bonds to peptide bond formation.

RNA-directed RNA polymerase (RdRP) The enzyme that elongates the siRNA primers, using target mRNA as a template, thus providing more substrate for Dicer and amplying the siRNA.

RNA helicase An enzyme that can unwind a double-stranded RNA, or a double-stranded region within RNA.

RNA-induced silencing complex (RISC) The RNase complex that degrades target mRNA during RNAi; contains endonuclease "Slicer" activity.

RNA interference (RNAi) Control of gene expression by specific mRNA degradation or translational repression caused by insertion of a double-stranded RNA into a cell.

RNA ligase An enzyme that can join two pieces of RNA, such as the two pieces of pre-tRNA created by splicing out an intron.

RNA polymerase In general, the enzyme that directs transcription, or synthesis of RNA. Specifically, the bacterial RNA polymerase.

RNA polymerase I The eukaryotic RNA polymerase that directs transcription of ribosomal RNA.

RNA polymerase II The eukaryotic RNA polymerase that directs transcription of mRNA, snoRNAs, some snRNAs, and miRNAs.

RNA polymerase III The eukaryotic RNA polymerase that directs transcription of tRNA, 5S rRNA, and some snRNAs.

RNA polymerase IV The eukaryotic RNA polymerase (described in plants only) that directs transcription of siRNAs involved in heterochromatin formation.

RNA polymerase, chloroplast The multiple eukaryotic RNA polymerases (both chloroplast and nuclear-encoded) that direct transcription of chloroplast genes.

RNA polymerase, mitochondrial The eukaryotic RNA polymerase that directs transcription of mitochondrial genes.

RNA processing Modifying an initial transcript to its mature form by cleavage, splicing, capping, polyadenylation, editing, etc.

RNA splicing The process of removing introns from a primary transcript and attaching the exons to one another.

RNA world A hypothetical stage in the evolution of life some four billion years ago when RNA both carried the genetic information and catalyzed its own replication.

RNase MRP RNase mitochondrial RNA processing. The enzyme that cleaves the RNA primer during mtDNA replication and plays a role in processing 5.8S rRNA in the nucleolus.

RNase P The enzyme that cleaves the extra nucleotides from the 5′ end of a tRNA precursor. Most forms of bacterial RNase P have a catalytic RNA subunit.

RNase protection assay (RPA) A method for detecting and quantifying specific mRNA transcripts in a complex mixture of RNA and for mapping internal and external boundaries in mRNA, such as the start of transcription. Hybridization of a labeled RNA probe to a transcript protects a part of the probe from digestion with RNase. The amount of probe protected by the transcript is proportional to the concentration of the transcript.

rolling circle replication A mechanism of replication in which one strand of a double-stranded circular DNA remains intact and serves as the template for elongation of the other strand at a nick.

rRNA (5S) The smaller of the vertebrate rRNAs found in the large (60S) ribosomal subunit; transcribed from a separate gene. Sometimes referred to as 5S RNA.

rRNA (5.8S) The smallest of the eukaryotic rRNAs derived from the 45S or 40S precursor. Found in the large (60S) ribosomal subunit; base paired to the 28S rRNA.

rRNA (18S) The vertebrate rRNA found in the small (40S) ribosomal subunit.

rRNA (28S) The largest vertebrate rRNA found in the large (60S) ribosomal subunit, base paired to the 5.8S rRNA.

rRNA precursor (45S or 40S) The large rRNA precursor in vertebrates which contains the 28S, 18S, and 5.8S rRNA sequences.

Rubenstein–Taybi syndrome A rare autosomal dominant disease characterized by facial abnormalities, broad digits, stunted growth, and mental retardation, caused by mutations in the gene coding for CBP which is a coactivator with HAT activity.

run–off transcription A method for quantifying the extent of transcription of a particular gene *in vitro*. The RNA polymerase "runs off" the end of a truncated linear gene template to give a short RNA product of predictable length. The abundance of the run–off product is a measure of the extent of transcription of the gene *in vitro*.

S

S1 nuclease A nuclease specific for single-stranded RNA and DNA.

satellite RNA A subviral pathogen found in plants and animals that replicates only in the presence of a helper virus. Some of the larger satellite RNAs may encode a protein.

scanning A model of translation initiation in eukaryotes that invokes a 40S ribosomal subunit binding to the 5′ end of the mRNA and scanning, or sliding along, the mRNA until it finds the first start codon in a good context for initiation.

scrapie The prototype prion disease first described in sheep and goats.

screen A genetic sorting procedure that allows one to distinguish desired organisms from unwanted ones, but does not automatically remove the undesired ones.

SDS–PAGE SDS-polyacrylamide gel electrophoresis; the electrophoretic separation of proteins in the presence of a strong, negatively charged detergent (sodium dodecyl sulfate).

secondary antibody Antibodies directed against all primary antibodies of a given species (e.g. anti–rabbit); typically conjugated (covalently bonded) to detectable tags.

secondary structure The local folding of a polypeptide or nucleic acid.

sedimentation coefficient (S) A measure of the rate at which a molecule or particle travels toward the bottom of a centrifuge tube under the influence of a centrifugal force. Also known as a Svedberg unit.

selection A genetic sorting procedure that eliminates unwanted organisms of cell types, usually by preventing their growth or killing them.

semiconservative replication DNA replication in which the two strands of a parental duplex separate completely and pair with new progeny strands. One parental strand is therefore conserved in each progeny duplex.

semidiscontinuous replication A mechanism of DNA replication in which one strand is made continuously and the other is made discontinuously.

senescence An irreversible state of cellular aging, characterized by continued cell viability without further cell division.

sequencing Determining the amino acid sequence of a protein, or the base sequence of a DNA or RNA.

severe combined immunodeficiency (SCID) A disease affecting the immune system. SCID is fatal if affected individuals do not receive bone marrow transplants.

sex chromosome One of the two chromosomes that specify an organism's genetic sex. Humans have two kinds of sex chromosomes, one called X and the other Y. Normal females possess two X chromosomes and normal males one X and one Y.

sex-linked Located on the X chromosome. Sex-linked (or X-linked) diseases are generally seen only in males.

SH2 domain A phosphotyrosine-binding domain found in many signal transduction proteins.

SH3 domain A proline-rich-helix binding domain that mediates protein–protein interactions.

Shine–Dalgarno sequence A G-rich sequence (consensus = AGGAGGU) that is complementary to the 3′ end of *E. coli* 16S rRNA. Base pairing between these two sequences helps the ribosome bind an mRNA.

short interspersed nuclear elements (SINEs) Nonautonomous retrotransposons that do not encode the enzymatic activities necessary for their retrotransposition. The only active SINEs in the human genome are members of the abundant *Alu* element family.

short tandem repeats (STRs) Repetitive stretches of short sequences of DNA (usually 2–4 bp) used as genetic markers. A given STR is found in varying lengths scattered around a eukaryotic genome. STR analysis is currently the most widely used DNA typing procedure in forensic genetics.

shotgun sequencing An approach used to decode an organism's genome by cutting it into smaller fragments of DNA which can be sequenced individually. The sequences of these fragments are then ordered, based on overlaps in the genetic code, and finally reassembled into the complete sequence. The 'whole genome shotgun' method is applied to the entire genome all at once.

sickle cell disease A genetic disease in which abnormal β-globin is produced, seen most commonly in people of African ancestry. The disorder is caused by a single amino acid change. This mutation causes the protein to aggregate under low-oxygen conditions and the red blood cells take on a sickle shape, rather than their characteristic donut shape. Individuals who suffer from sickle cell disease are chronically anemic and experience significant damage to their heart, lungs, and kidneys.

sigma (σ) The prokaryotic RNA polymerase subunit that confers specificity of transcription; i.e. the ability to recognize specific gene promoters.

σ^{70} The principle sigma factor of *E. coli*.

signal peptide A stretch of about 20 amino acids, usually at the amino terminus of a polypeptide, that helps to anchor the nascent polypeptide and its ribosome in the endoplasmic reticulum. Polypeptides with a signal peptide are destined for packaging in the Golgi apparatus and are usually secreted from the cell.

signal recognition particle (SRP) Mediates protein targeting to the endoplasmic reticulum.

signal transduction pathway A biochemical pathway that connects a signal, such as a growth factor binding to the cell surface, with an intracellular effect, usually gene activation or repression.

silencer A DNA regulatory element that decreases gene promoter activity. Silencers are usually 700–1000 bp or more away from the start of transcription and can be downstream, upstream, or within an intron, and can function in either orientation relative to the promoter.

silent mutation Mutations that cause no detectable change in an organism, even in a haploid organism or in a homozygote.

single-copy DNA DNA sequences that are present once, or only a few times, in a haploid genome.

single nucleotide polymorphisms (SNPs) A single-nucleotide difference between two or more individuals at a particular genetic locus. Common, but minute, variations that occur in human DNA at a frequency of one every 1000 bases. SNP is pronounced "snip."

single-stranded DNA-binding protein (SSB) Binds to single-stranded DNA and keeps it from base pairing with a complementary strand, used during DNA replication.

SIR2, SIR3, SIR4 Proteins associated with, and required for, the formation of yeast heterochromatin, including at the telomeres.

siRNAs (short interfering RNAs) The short pieces (21–28 nt) of double-stranded RNA created by Dicer during the RNAi process.

site-directed mutagenesis A method for introducing specific, predetermined alterations into a cloned gene.

site-specific recombination Recombination that always occurs in the same place and depends on limited sequence similarity between the recombining DNAs.

7SL RNA A small RNA involved in recognizing the signal peptides on proteins destined for secretion.

sliding (protein) Helical movement of a DNA-binding protein along the DNA due to tracking along a groove of the DNA over short distances.

slipped structures An unusual secondary structure in DNA that can occur at tandem repeats in which misalignment of repeats leads to single-stranded loops of DNA.

small nuclear RNAs (snRNAs) A set of small uracil (U)-rich RNAs found in the nucleus, associated with proteins to form small nuclear ribonucleoproteins (snRNPs), which participate in splicing of pre-mRNAs.

small nucleolar RNAs (snoRNAs) A set of hundreds of small RNAs found in the nucleolus. A small subset of the snoRNAs, associated with proteins in small nucleolar ribonucleoproteins (snoRNPs), participate in processing and base modification of the large ribosomal RNA precursor. Pronounced "snow" RNA.

snoRNP An snoRNA with its associated proteins. Pronounced "snorp."

snRNP An snRNA with its associated proteins. Pronounced "snurp."

somatic cells All body cells, except the reproductive (sex) cells.

somatic mutation A mutation that affects only somatic cells, so it cannot be passed on to progeny.

SOS response The activation of a group of genes that helps *E. coli* cells respond to environmental stress such as chemical mutagens or radiation that can cause DNA damage.

Southern blot Transferring DNA fragments separated by gel electrophoresis to a support medium such as nitrocellulose or nylon membrane, in preparation for hybridization to a labeled probed. Used to identify and locate DNA sequences which are complementary to the probe.

spacer DNA DNA sequences found between, or sometimes within, repeated genes such as the ribosomal RNA genes.

spliced leader (SL) The independently synthesized 35 nt leader that is trans-spliced to surface antigen mRNA coding regions in trypanosomes.

spliceosome The large RNA–protein complex on which splicing of nuclear pre-mRNA precursors occurs; composed of five snRNPs and ~200 proteins.

splicing The process of linking together two RNA exons while removing the intron that lies between them.

splicing factors Proteins in addition to snRNPs that are essential for splicing nuclear pre-mRNAs.

spontaneous mutation A mutation that occurs as a result of natural processes in cells; e.g. DNA replication errors.

spore A specialized haploid cell formed sexually in plants or fungi, or asexually by fungi. The latter can either serve as a gamete, or germinate to produce a new haploid cell. In bacteria, the term refers to a specialized cell formed asexually in response to stressful conditions. Such a spore is resistant to environmental stress.

sporulation Formation of spores.

squelching Inhibition of one activator by increasing the concentration of a second one. This may occur through competition for a common factor that is at a limiting concentration.

SR proteins A group of RNA-binding proteins having an abundance of serine (S) and arginine (R).

stem cells Undifferentiated cells that can undergo unlimited division and can give rise to one or several different cell types.

sticky ends Single-stranded ends of double-stranded DNAs that are complementary and can therefore base pair and "stick" together by hydrogen bonding.

stop codon One of three codons (UAG, UAA, UGA) that code for termination of translation.

strand-coupled model An alternate model for mammalian mitochondrial DNA replication in which replication initiates at multiple sites and occurs by a semidiscontinuous, bidirectional mode of DNA replication.

strand-displacent model An alternate model for mammalian mitochondrial DNA replication in which replication is unidirectional around the circular DNA molecule and there is one replication fork for each strand. Also known as the strand-asynchronous model.

streptavidin A protein made by *Streptomyces* bacteria that binds to biotin.

stringency (of hybridization) The combination of factors (temperature, salt, and organic solvent) that influence the ability of two nucleic acid strands (DNA or RNA) to hybridize. At high stringency, only perfectly complementary strands will hybridize. At reduced stringency, some mismatches can be tolerated.

SWR1 complex A chromatin remodeling complex family named after its ATPase subunit Swr1 (for Swi2/Snf2 related) that transfers H2A.Z-H2B dimers to nucleosomes in exchange for H2A-H2B dimers.

substitution Replacement of one nucleotide in a DNA sequence by another nucleotide or replacement of one amino acid in a protein by another amino acid.

suicide gene A strategy for making cancer cells more vulnerable to chemotherapy. One approach has been to link parts of genes expressed in cancer cells to other genes for enzymes not found in mammals that can convert a harmless substance into one that is toxic to the tumor.

SUMO (small ubiquitin-related modifier) A small polypeptide that can be attached to other proteins and modify their function.

sumoylation The attachment of SUMO to other proteins.

supercoiled DNA A circular DNA molecule or loop doman within linear DNA that is under torsional stress, and the double helix coils around itself like a twisted rubber band. The DNA has become either overwound (positive supercoiling) or underwound (negative supercoiling) with respect to the number of complete turns of the DNA double helix.

supershift The extra electrophoretic gel mobility shift observed when a new protein (such as an antibody) binds a protein–DNA complex.

suppression Compensation by one mutation for the effects of another.

suppressor mutation A mutation that reverses the effects of mutation in the same or another gene.

SV40 Simian virus 40. A DNA tumor virus (papovavirus) with a small circular genome, capable of causing tumors in certain rodents. Used extensively as a model organism for eukaryotic molecular biology.

SWI/SNF A family of coactivators that help to remodel chromatin by disrupting nucleosome cores.

syndactyly Webbing and/or fusion of the fingers and toes.

syndrome The group or recognizable pattern of symptoms or abnormalities that indicate a particular trait or disease.

synteny Preservation of gene order in different organisms.

systemic lupus erythematosus (SLE) An autoimmune disorder in which individuals form antibodies against their own proteins. Symptoms range from mild to severe and include swollen glands, joint pain, fatigue, skin rashes, damage to internal organs, etc.

T

T antigen See large T antigen.

T cell receptor Antigen-binding proteins on the surface of T cells. Composed of two heavy (β) and two light (α) chains.

t loop A loop formed in the telomere at the end of a eukaryotic chromosome.

T loop The loop in a tRNA molecule that contains the nearly invariant sequence TΨC, where Ψ is pseudouridine.

tandem mass spectrometry (MS/MS) A process in which peptide ions are bombarded with rare gas atoms which provide energy required to cleave amide bonds of the peptide backbone. The process produces a collection of peptide ion fragments that differ in mass by a single amino acid. Measurement of the mass-to-charge ratios of the fragments allows the amino acid sequence to be read.

tandem repeat Two or more adjacent, approximate copies of a pattern of nucleotides, arranged in a head to tail fashion (sometimes called a direct repeat).

Taq polymerase A heat-resistant DNA polymerase obtained from the thermophilic bacterium *Thermus aquaticus*.

target mRNA The mRNA that is targeted and degraded during RNAi.

TATA box An element with the consensus sequence TATAAA that begins about 30 bp upstream of the start of transcription in many eukaryotic promoters recognized by RNA polymerase II.

TATA-box binding protein (TBP) A subunit of the TFIID complex in RNA polymerase II transcription. Binds to the TATA box.

TBP–associated factor (TAF) A protein associated with TBP in the TFIID complex.

T-DNA The tumor-inducing part of the Ti plasmid.

telocentric chromosome A chromosome with the centromere at one end.

telomerase An enzyme that can extend the ends of telomeres after DNA replication. The protein subunit (TERT) has reverse transcriptase activity, and the RNA component (TERC) serves as the template for synthesis of new telomere repeats.

telomerase reverse transcriptase (TERT) The protein component of the enzyme telomerase.

telomerase RNA component (TERC) The RNA component of the enzyme telomerase.

telomere Specialized structures that cap the end of chromosomes and prevent them from being joined to each other. Telomeres are composed of tandem repeats of a simple guanine (G)-rich sequence (e.g. in humans, several thousand repeats of the sequence TTAGGG).

telomere shortening The progressive shortening of the ends of chromosomes in cells without telomerase activity.

telophase The last stage of cell division, in which the nuclear membrane forms and encloses the chromosomes in the daughter cell.

temperature–sensitive mutation A mutation that causes a product to be made that is defective at high temperature (the nonpermissive temperature) but functional at low temperature (the permissive temperature).

template A polynucleotide (RNA or DNA) that serves as a guide for making a complementary polynucleotide.

template DNA strand The DNA strand of a gene that is complementary to the RNA product of the gene; that is, the strand that serves as the template for making the RNA. Also called the anticoding strand or antisense strand.

teratogen A substance that causes abnormal development of an organism.

terminal transferase An enzyme that adds deoxynucleoside triphosphophates, one at a time to the 3′ end of DNA.

terminal uridylyl transferase (TUTase) An enzyme that adds UMP residues to pre-mRNAs during RNA editing in trypanosomes.

tertiary structure The overall three-dimensional shape of a polypeptide or RNA.

tetraloop An RNA stem-loop structure that is stabilized due to special base-stacking interactions in the loop.

tetramer (protein) A complex of four polypeptides.

thermal cycler A machine that performs PCR reactions automatically by repeatedly cycling among the three temperatures required by DNA denaturation, primer annealing, and primer extension.

theta (θ) structure A replication intermediate formed during replication of circular double-stranded DNA in bacteria.

three (3)-prime (3′) The ends of a DNA or RNA chain are chemically distinct and are designated by the symbols 3′ and 5′. The symbol 3′ refers to the carbon in the ribose or deoxyribose sugar ring to which a hydroxyl (OH) functional group is attached.

thymidine A nucleoside containing the base thymine.

thymine (T) One of the four bases in DNA. Thymine always pairs with adenine.

thymine dimer Two adjacent thymines in one DNA strand linked covalently, whose base pairing with adenines in the opposite strand is interrupted; the most common form of DNA damage induced by UV irradiation.

thyroid hormone receptor A nuclear receptor that in the presence of thyroid hormone either activates or respresses the expression of thyroid hormone-responsive genes.

Ti plasmid The tumor-inducing plasmid from *Agrobacterium tumefaciens*. Used as a vector to carry foreign genes into plant cells.

toe print assay A primer extension assay that locates the edge of a protein (e.g. the ribosome) bound to DNA or RNA.

topoisomer A form of DNA that has the same sequence as another DNA molecule yet differs in its linkage number. Also known as a topological isomer.

topoisomerase An enzyme that converts (isomerizes) one topoisomer of DNA to another.

***trans*-acting** A term that describes a genetic element, such as a repressor gene or transcription factor gene, that can be on a separate chromosome and still influence another gene. These *trans*-acting genes function by producing a diffusible substance that can act at a distance.

transactivation domain The part of a transcription factor that stimulates transcription by protein–protein interactions.

transcribed spacer A region encoding a part of an rRNA precursor that is removed during processing to produce the mature rRNAs.

transcript An RNA copy of a gene.

transcription The process by which an RNA copy of a gene is made.

transcription bubble The region of locally melted DNA that follows the RNA polymerase as it synthesizes the RNA transcript.

transcription-coupled repair Nucleotide excision repair pathway responsible for recognizing DNA damage in the transcribed strand of active genes.

transcription factors Sequence-specific DNA binding proteins that bind to gene promoters and long-range regulatory elements and mediate gene-specific transcriptional activation or repression.

transcription terminator A specific DNA sequence that signals transcription to terminate.

transcription unit A region of DNA bounded by a promoter and a terminator that is transcribed as a single unit; it may be polycistronic and contain multiple coding regions.

transcriptional attenuation A regulatory process in some bacterial biosynthetic operons in which translation of a leader sequence in mRNA terminates transcription at a terminator hairpin instead of at the competing antiterminator hairpin.

transcriptome The sum of all the different transcripts an organism can make in its lifetime.

transcriptomics The global study of an organism's transcripts.

transesterification A reaction that simultaneously breaks one ester bond and creates another; for example, the reactions that take place during RNA splicing.

transfection Introduction of foreign DNA into eukaryotic cells, either for a short duration (transient) or for long-term analysis (stable integration).

transfer RNA (tRNA) A relatively small RNA molecule that is "charged" with an amino acid and delivers to the ribosome the appropriate amino acid via interaction of the tRNA anticodon with the mRNA codon.

transferrin An iron-carrier protein that imports iron into cells via the transferrin receptor.

transformation The introduction of foreign DNA into bacterial cells.

transgene A foreign gene transferred into an organism, making the recipient a transgenic organism.

transgenic organism An experimentally produced organism that carries transferred genetic material (the transgene) that has been inserted into its genome at a random site, usually by injecting the foreign DNA into the nucleus of a fertilized embryo.

transition A mutation in which a pyrimidine replaces a pyrimidine, or a purine replaces a purine.

translation The process by which ribosomes decode an RNA message (mRNA) to synthesize a protein.

translesion synthesis (TLS) A mechanism for bypassing DNA damage by replicating through it; mediated by specialized low-fidelity, "error-prone" DNA polymerases.

translocation (chromosome) Breakage and removal of a large segment of DNA from one chromosome, followed by the segment's attachment to a different chromosome.

translocation (protein synthesis) The translation elongation step, following peptide bond formation, that involves moving an mRNA one codon length through the ribosome and bringing a new codon into the ribosome A site.

transmissible spongiform encephalopathies A rare brain-wasting disease in mammals caused by infectious prions and characterized by sponge-like holes in the brain, dementia, and loss of muscle control of voluntary movements (ataxia).

transmission genetics The study of the transmission of genes from one generation to the next.

transposable element A DNA element that can move from one genomic location to another.

transposase The name for a collection of enzymes, encoded by a transposon, that catalyze transposition.

transposition The movement of a DNA element from one DNA location to another.

transposon See transposable element and DNA transposons.

trans-**splicing** Splicing together two RNA fragments transcribed from separate transcription units.

transversion A mutation in which a pyrimidine replaces a purine, or vice versa.

trichothiodystrophy An autosomal recessive disorder in which individuals have an exaggerated sensitivity to light, brittle hair and nails, and dry scaly skin; linked to defects in nucleotide excision repair genes.

trinucleotide repeats Repetitive regions of three bases of DNA with unusual gene instability. Dynamic expansion of trinucleotide repeats (e.g. GAAGAAGAA) leads to certain genetic neuorological disorders such as fragile X syndrome, Huntington's disease, Friedreich's ataxia, etc.

triple helix DNA A secondary structure that can form at purine-pyrimidine stretches in DNA with mirror repeat symmetry in which a third strand of DNA joins the first two to form a triplex DNA.

trisomy Possessing three copies of a particular chromosome instead of the normal two copies.

tritium A radioactive isotope of hydrogen (^{3}H).

tRNA charging The process of coupling tRNA with its cognate amino acid, catalyzed by aminoacyl-tRNA synthetase.

tRNA endonuclease The enzyme that cuts an intron out of a tRNA precursor.

tRNA$_f^{Met}$ The tRNA responsible for initiating protein synthesis in prokaryotes.

tRNA$_i^{Met}$ The tRNA responsible for initiating protein synthesis in eukaryotes.

tRNA$_m^{Met}$ The tRNA that inserts methionines that are coded for within a protein sequence.

trp **operon** The operon that encodes the enzymes needed to make the amino acid tryptophan.

trypanosomes Protozoa that parasitize both mammals and tsete flies; the latter spread the disease by biting mammals; infection leads to the disease called sleeping sickness.

TTAGG Repeat Binding Factor 1 and 2 (TRF1, TRF2) Proteins involved in regulation of telomerase activity; they count the number of G-rich repeats and when telomeres become overly long, inhibit further telomerase activity by transferring POT1 to the single-stranded overhang at the telomere tip.

tumor suppressor gene A protective gene that normally limits the growth of tumors. When a tumor suppressor is mutated and undergoes a loss of function, it may fail to keep a cancer from growing. BRCA1, p53, and the retinoblastoma protein are well-known tumor suppressor genes.

turns (protein) Relatively short loops of amino acids that do not exhibit a defined secondary structure themselves, but are essential for the overall folding of a protein.

two-dimensional gel electrophoresis A high-resolution method for separating proteins. The proteins are separated in the first dimension by isoelectric focusing. Then they are separated in the second dimension by SDS-PAGE.

U

U1 snRNP The first snRNP that recognizes the 5′-splice site in a nuclear pre-mRNA.

U2-associated factor (U2AF) A splicing factor that helps recognize the correct AG at the 3′-splice site by binding to both the polypyrimidine tract in the 3′-splice signal and the AG; composed of two subunits, U2AF35 and U2AF65.

U2 snRNP The snRNP that recognizes the branchpoint in a nuclear pre-mRNA.

U4 snRNP The first snRNP whose RNA base pairs with the RNA in the U6 snRNP until U6 snRNP is needed to splice a nuclear pre-mRNA.

U4atac A minor snRNA that participates in splicing variant introns and plays the same role as U4 snRNA.

U5 snRNP The snRNP that associates with both the 5′ and 3′ exon–intron junctions, thus helping to bring the two exons together during nuclear pre-mRNA splicing.

U6 snRNP The snRNP whose RNA base pairs with both the 5′ splice site and with the RNA in U2 snRNP in the spliceosome during nuclear pre-mRNA splicing.

U6atac A minor snRNA that participates in splicing variant introns and plays the same role as U6 snRNA.

U11 snRNA A minor snRNA that participates in splicing variant introns and plays the same role as U1 snRNA.

U12 snRNA A minor snRNA that participates in splicing variant introns and plays the same role as U2 snRNA.

ubiquitin A small polypeptide that can be attached to proteins and modifies their function. Polyubiquitinylation marks a protein for destruction by the proteasome.

ubiquitinylation The addition of ubiquitin to a protein. Sometimes called ubiquitination.

ultraviolet (UV) radiation Radiation found in sunlight. Can induce pyrimidine dimers in DNA.

undermethylated region A region of a gene or its flanking region that is relatively poor in, or devoid of, methyl groups. Also known as hypomethylation.

unidirectional DNA replication Replication that occurs in one direction, with only one active replication fork.

untranslated region (UTR) A region at the 5′ or 3′ end of an mRNA that lies outside the coding region, so it is not translated (i.e. before a start codon or after a stop codon).

upstream DNA sequence before the start of transcription (+1) in the direction of the 5′ end; numbered with negative numbers (e.g. −35).

upstream activating sequence (UAS) An enhancer for yeast genes.

uracil (U) One of the four bases in RNA. Uracil replaces thymine, which is the fourth base in DNA. Like thymine, uracil always pairs with adenine.

uridine A nucleoside containing the base uracil.

V

variable number tandem repeat (VNTR) A type of restriction fragment length polymorphism (RFLP) that includes tandem repeats of a minisatellite between the restriction sites.

variable region (V) The region of an antibody that binds specifically to a foreign substance or antigen. It varies considerably from one specific antibody to another.

variable surface glycoprotein (VSG) The antigen that coats a trypanosome. The parasite can vary the nature of this coat by switching which VSG gene is expressed in a telomere expression site.

V(D)J joining The assembly of active immunoglobulin or T-cell receptor genes by recombination involving separate V and J or V, D, and J segments in the embryonic genes.

vector DNA An agent, such as a virus or a plasmid, phage, or artificial chromosome, that carries a modified or foreign gene. Used as a carrier in gene cloning experiments. When used in gene therapy, a vector delivers the desired gene to a target cell.

vegetative cell A cell that is reproducing by division, rather than sporulating or reproducing asexually.

viroid A subviral RNA pathogen that causes infectious disease in higher plants.

virus A minute intracellular obligate parasite. A virus particle consists of a core of nucleic acid, which may be DNA or RNA, surrounded by a protein coat, and in some viruses a further lipid/glycoprotein envelope. It is unable to multiply or express its genes outside a host cell as it requires host cell enzymes to aid DNA replication, transcription, and translation.

virusoid A subviral pathogen that causes infectious disease in higher plants. It consists of an RNA molecule that does not encode any proteins and depends on a helper virus for replication.

VS ribozyme Varkud Satellite RNA. A small ribozyme transcribed from plasmid found in the mitochondria of some strains of *Neurospora crassa*, a filamentous fungus.

W

Watson–Crick base pairing See base pairing.

Western blot Electrophoresing proteins, then blotting them to a membrane and reacting them with a specific antibody or antiserum. The antibody is detected with a labeled secondary antibody. This technique is used to identify and locate proteins based on their ability to bind to specific antibodies.

wobble The ability of the third base of a codon to shift slightly to form a non–Watson–Crick base pair with the first base of an anticodon, thus allowing a tRNA to translate more than one codon.

wobble base pair A base pair formed by wobble (e.g. a G–U or A–I base pair).

wobble position The third base of a codon, where wobble base pairing is permitted.

wyosine A highly modified guanine nucleoside found in tRNA.

X

X chromosome inactivation The process by which one of the two X chromosomes in females is randomly inactivated in each cell.

X inactivation specific transcript (XIST) An RNA transcript that is expressed from and coats the inactive X chromosome.

xeroderma pigmentosum (XP) A disease characterized by extreme sensitivity to sunlight. Caused by a defect in nucleotide excision repair genes (XPA–XPG) or translesion synthesis (XP-V).

X-ray crystallography A method for determining the three-dimensional structure of molecules by measuring the diffraction of X-rays by crystals of a molecule or molecules.

X-ray diffraction See X-ray crystallography.

X-rays High-energy radiation. X-rays can ionize cellular components and can cause DNA double-strand breaks.

Y

yeast artificial chromosome (YAC) Extremely large segments of DNA from another species spliced into artificial chromosomes of yeast. YACs are used to clone up to one million bases of foreign DNA into a host cell, where the DNA is propagated along with the host cell chromosomes.

yeast mating type switching A simple form of cellular differentiation in which two mating types are defined by the expression of one of two gene cassettes; switching occurs through DNA rearrangement by homologous recombination.

yeast two-hybrid assay An assay for interaction between two proteins. One protein (the bait) is produced from a fusion protein with a DNA-binding domain from another protein. The other protein (the prey) is produced from a fusion protein with a transactivation domain. If the two fusion proteins interact within a yeast cell, they form a transcriptional activator that can activate one or more reporter genes.

Z

Z-DNA An alternate left-handed form of the DNA double helix with 12 base pairs per turn whose backbone has a zig-zag appearance. This form is stabilized by stretches of alternating purines and pyrimidines. Short sections of Z-DNA may function as regulatory elements within cells.

zinc finger A DNA-binding motif that contains a zinc ion complexed to four amino acid chains, usually the side chains of two cysteines and two histidines, or four cysteines. The motif is roughly finger-shaped and inserts into the DNA major groove, where it makes specific protein–DNA contacts.

zygote cell formed from the union of two gametes or reproduction cells.

Index

Page numbers in *italics* refer to figures; page numbers in **bold** refer to tables or boxes.